THE SOUNDING OF THE WHALE

D. GRAHAM BURNETT

THE SOUNDING OF

THE WHALE

SCIENCE & CETACEANS IN THE
TWENTIETH CENTURY

THE UNIVERSITY OF CHICAGO PRESS
CHICAGO & LONDON

D. GRAHAM BURNETT
is professor of history and history of science at Princeton University.

The University of Chicago Press, Chicago 60637
The University of Chicago Press, Ltd., London
© 2012 by D. Graham Burnett
All rights reserved. Published 2012.
Printed in the United States of America
21 20 19 18 17 16 15 14 13 12   1 2 3 4 5
ISBN-13: 978-0-226-08130-4 (cloth)
ISBN-10: 0-226-08130-3 (cloth)

Library of Congress Cataloging-in-Publication Data

Burnett, D. Graham.
The sounding of the whale: science & cetaceans in the twentieth century/
D. Graham Burnett.
p. cm.
Includes bibliographical references and index.
ISBN-13: 978-0-226-08130-4 (cloth: alkaline paper)
ISBN-10: 0-226-08130-3 (cloth: alkaline paper)
1. Whales—Research—History—20th century.
2. Whales—Research—Great Britain—History—20th century.
3. Whales—Research—United States—History—20th century.
4. Whaling—History—20th century.
5. Whaling—Law and legislation—History—20th century.
6. International Whaling Commission—History.   I. Title.
QL737.C4B86 2012
599.5072—dc23   2011030694

♾ This paper meets the requirements of ANSI/NISO Z39.48-1992
(Permanence of Paper)

For Consuelo Gaudes Burnett

And the whale himself—which whale was I thinking
of?...The whale of legend, or the whale of science, or the
whale of industry? For all are different, and from my height,
which covered sea and land, I could see the difference
between the sublime and the sordid.
Or so it seemed...

F. V. MORLEY & J. S. HODGSON
Whaling North and South, 1927

CONTENTS

ILLUSTRATIONS

FIGURES

PLATES

(*plates follow page 394*)

PREFACE

For in the mere act of penning my thoughts of this Leviathan, they weary
me, and make me faint with their out-reaching comprehensiveness
of sweep, as if to include the whole circle of the sciences . . .

HERMAN MELVILLE, *Moby-Dick*, 1851

FRIENDS, HOLD MY ARMS

I've eaten a bit of whale. Not a lot, but a bit. Both smoked (I thought it
quite like prosciutto of boar) and as a fresh steak, rare (indistinguishable
from elk, in my view). This was in Norway, in the company of a particularly
bloodthirsty spokesman for the industry who tried, as we chewed, to sell me
on the idea that every species that has any sense kills other species by way
of enlarging the ambit of its own vitality. He seemed, pressing this point, to
deem whale conservation a kind of race suicide, which was a disorienting
theory to be offered by a heavily accented German wielding a steak knife.
But he was such a companionable fellow, gregarious and enormously like-
able in other respects—not to mention abundantly knowledgeable about
whale matters. So, thinking of Ishmael ("Not ignoring what is good, I am
quick to perceive a horror, and could still be sociable with it—would they
let me—since it is but well to be on friendly terms with all the inmates of
the place one lodges in"), I raised my glass to human fellowship, and we
drank a long draught in the beery twinkle of an endless Scandinavian summer's eve.

By contrast, I have also wept in the presence of a living whale. This was in Baja California, in the Sea of Cortez, when our sputtering panga suddenly found itself in the middle of a boiling cauldron of crazed skipjack and terrified pilchards. Interesting enough. We cut the engine to watch. And then, some forty feet from the boat, without warning, up rose the towering bulk of a healthy young Bryde's whale, which launched itself skyward, mouth gulping in a colossal uprushing swallow. It seemed to continue rising for a count of ten before falling back into the cold blue, now pin-drop silent. In those suspended moments I had seen clearly the loose folds of striated skin that made up the expandable gape of that giant mouth, and their jowly openness had been touched with the faintest fresh pink. And seen from that side, suspended at the apex of its bolting leap, the creature reminded me of nothing so much as some fantastic and gargantuan frog, puffing its huge belly to the sky in a mad frolic of power and joy. In "Converse at Night in Copenhagen," Isak Dinesen writes of three kinds of perfect happiness, and the first is "to feel in oneself an excess of strength." There was some of that in this apparition, and that may be why I cried. I cannot say. The whole thing was simply too much to bear.

More whale moments? Most of them are in Mexico. I have the clearest memory of picking my way over a rocky stretch of island shore to investigate the extremely rotten carcass of a bull sperm whale, whose slow blasting under a tropical sun had left a slick of stench half a mile out to sea. The slightest shift in the wind meant strangled gags; sea lice in obscene hordes swarmed the strand, and the blowflies tormented the bold vultures that picked at strips of leathery yellow fat. Even broken by surf and decay, the animal's head was thicker than I am tall. Lying on its side, it towered over me.

And then, of course, there was that silent and moonless night in a small kayak, paddling about in terrified awe as, somewhere impossibly near, one of those giants—a fin, presumably—sucked up sudden, room-sized breaths and expelled them in deep and plosive gusts. I felt (alone in the inky dark, bobbing far from land) something of the basic, unmitigated, almost suicidal fear that one does well to recall while waxing eloquent about the beauties of untrammeled nature.

The field station in Bahia where I was staying had been used by generations of itinerant naturalists and students of the things that live in the sea. Some years back, on the occasion of the stranding of a small fin whale in the bay, a group of them had taken on the daunting task of recovering, preparing, and articulating the 35-foot skeleton. In the end, the project took years, but the fruit of the labor still stood when I visited, bleached to a crumbly lightness, strung out on a rusted armature of pipe, the beast's nose pointing due east out over the sea, to where the sun rose every morning. Awakening at dawn on a cot perched below this looming scaffold of bone, it was impossible not to think of the "Bower in the Arascides," the temple-skeleton of a whale Ishmael writes of having explored on the island of Tranque, and which affords him so rare an access to the measure of his prey: diligently he had its dimensions tattooed on the skin of his right arm in order to preserve these data for the world of learning, though he elected to omit the odd inches in order to save space on his flesh for a poem upon which he was then at work. A good idea, that. One must not let whale knowledge take over everything.

. . . .

I am moved to note, here at the outset, that most of the work on this book looked nothing like these scenes. Most of it involved sitting perfectly still in a chair, sometimes reading and sometimes writing. Sometimes I would lie down. In this respect, I tend to think that the making of this book has amounted to a kind of extended spiritual exercise: a project of self-denial and self-abnegation; a minor-key rendition of the ascetic ideal. No sun, no waves, no tattoos. Wanderings of the mind from the austerities of the task at hand were often fruitful, but the better for being brief, and stolen. The best part of the process, I think, was the extremely strange way that everything could look at the end of a workday as one went outside, say, or saw another person, or wondered what life looked like to those who had spent the day in full career with the actual world, as opposed to bookish resignation from its affairs. I associate the most memorable of such moments with feelings approaching hysterical glee, and thus it is probably all to the good that these shivers of addled euphoria were generally fleeting. One mustn't have too much fun writing whale books. Or reading them either. But that is probably easier.

TIME, STRENGTH, CASH, AND PATIENCE

Solitary as the task of making whale books can be, they do not happen absent various emoluments and sociabilities. Princeton University afforded me both the time (in the form of a pair of generous leaves) and the cash (including research funds from the dean of the faculty, the History Department, and other internal sources) to realize this project. I am very grateful. In addition, portions of this work were supported by a grant from the National Endowment for the Humanities (FA-37754–03) and a Howard Foundation Fellowship. Over the years I have presented sections of this material to a number of helpfully critical audiences, including the 2005 HMAP "Oceans Past" conference at the Syddansk Universitet, Kolding; the Rutgers Center for Historical Analysis "The Sea in Global History" conference in 2006; "The Decimation of Whales," an international symposium at the Hvalfangstmuseet, Sandefjord, in 2007; the "Knowing Global Environments" celebration conference at the University of Pennsylvania in the same year; the American Cetacean Society's 2008 conference; and a variety of history of science and environmental history workshops, including gatherings at the annual meetings of the History of Science Society, York University (Toronto), University of California (Berkeley), Harvard, Yale, and Princeton. Teaching, too, has been important to the development of this project as a whole: several classes of students who participated in the Stanford Summer Session at the Vermilion Sea Field Station, Baja California, heard and commented on chapters 2 and 6; graduate students at Princeton helped me familiarize myself with relevant literatures in a pair of seminars ("Science Across the Seas," in 2002, and "Humans and Animals," in 2005); and the freshmen in my "Beast in the Sea" seminar in 2008 soldiered through chapters 4 and 5 and gave me valuable feedback. My colleagues in the Program in History of Science—Angela Creager, Michael Gordin, Helen Tilley, Keith Wailoo, and, of course, the late Michael Mahoney (who gave me my first training in the field)—offered collegial advice and generous readings. Along the way, other colleagues took the time to read and respond to portions of the material that appears in this book. The following deserve special mention for making such time: Dan Rogers, John Krige, James Schulz, and Lorraine Daston. In the endgame, I received the benefit of two close analyses of the whole manuscript by expert readers for the University of Chicago Press, Kurk Dorsey and Gary Kroll, both of whom delivered me generous comments, specific corrections, and helpful amplifications. At about the same time, Henry Cowles went through the text line by line, pressing me on secondary literature and catching a number of errors. Finally, Bill Perrin afforded the grace of a technical reading, pen in hand, by one of the most distinguished marine mammal biologists living. To all of them, my sincere thanks.

A study like this one requires a great deal of assistance from archivists, librarians, practicing scientists, research assistants, friends, and others who pitch in with references, recollections, leads, or sources. I am sure I am omitting many of those who afforded me such aid, but here is at least a partial list: John Bannister, Jeff Breiwick, the late Sidney G. Brown, Robert Bruesewitz, Anne Datta, Deborah Day, Jeff Dolven, Greg Donovan, Michael Dyer, Richard Ellis, Stuart Frank, Ray Gambell, Anthony Grafton, Catherine Hansen, Judy Hanson, Robert Headland, Aaron E. Hirsh, Paula Jenkins, Henrik Stissing Jensen, Christine Kim, Sonja Kromann, Richard Laws, Steven Mandeville-Gamble, Debbie Macy, Rosalind Marsden, Scott McVay, James Mead, Ed Mitchell, Domingo Monet, the late Lara Moore, Jac Mullen, Joe Nardello, Naomi Oreskes, Dmitri Petrov, Joanna Rae, Randy Reeves, Norman Reid, Sam Ridgway, Pauline Simpson, Tim D. Smith, Janani Sreenivasan, William Tavolga, Roberto Trujillo, Polly Tucker, the late David Van Keuren, Veronica Volny, the late William Watkins, and Emma Woodason, along with the whole staff of Article Express, Interlibrary Services, and Printing and Mailing at Princeton University, without whom this work would have been impossible.

Finally, whale books do not happen without at least an even measure of succor. Yes, it is true that my mother told me whale stories as a small boy, and yes, she and my father took me out on campus at Indiana University in the mid-1970s to meet an earnest, bearded grad student who gave me a copy of the 1975 *Audubon* issue on whales. These things, I presume, stuck. I have a sister who was for several years my scuba-diving partner (we went our separate ways underwater when she started cage diving with great whites), and much of my sense of the sea was shaped in her company. By the end of all this, I myself had a few children, including a daughter who could ask me, clear as a bell, at the dinner table, "Y dada, ¿qué hiciste en tu oficina hoy?" To which the ritual answer was, "Hoy escribí mi libro de ballenas," a predictable reply always greeted with a patronizingly theatrical "¡Oh, qué bien!" I took great courage from these reliable little parleys.

As I did from the remarkable support of a beloved wife, Christina, who laid her hands on this manuscript and on its maker—making each, in its season, whole in its way. *Gracias.*

ABBREVIATIONS

AIBS American Institute of Biological Science

ASA / NCTM American Statistical Association / National Council of Teachers of Mathematics

ASM American Society of Mammalogists

ASW antisubmarine warfare

BBS Bureau of Biological Survey

BIWS Bureau of International Whaling Statistics

BMNH British Museum (Natural History)

BRI Brain Research Institute

BWU blue whale units

CCW Council for the Conservation of Whales

CDW catcher's day's work

CPUE catch per unit effort

CRI Communication Research Institute

EMBTEL Embassy telegram

FAO Food and Agriculture Organization

HMAP History of Marine Animal Populations

IATTC Inter-American Tropical Tuna Commission

IBP International Biological Program

ICES International Council for the Exploration of the Sea

ICNAF International Convention of the Northwest Atlantic Fisheries

ICRW International Convention for the Regulation of Whaling

IICRW	Informal Inter-agency Committee on the Regulation of Whaling
IUCN	International Union for Conservation of Nature
IWC	International Whaling Commission
LINC	Laboratory Instrument Computer
MAF	Ministry of Agriculture and Fisheries
MAFF	Ministry of Agriculture, Fisheries, and Food
MSY	maximum sustainable yield
MSYL	maximum sustainable yield level
NAM	Neil Alison Mackintosh
NARA	National Archives and Records Administration
NHM	Natural History Museum
NIMH	National Institute of Mental Health
NIO	National Institute of Oceanography
NMC	Naval Missile Center
NOL	National Oceanographic Library
NOTS	Naval Ordnance Test Station
NSF	National Science Foundation
NUC	Naval Undersea Center
ONI	Office of Naval Intelligence
ONR	Office of Naval Research
RU	Record Unit
SCAR	Scientific Committee on Antarctic Research
SETI	search for extraterrestrial intelligence
SGB	Sidney G. Brown
SOFAR	Sound Fixing and Ranging
SOSUS	Sound Surveillance System
SMRU	Sea Mammal Research Unit
SPRI	Scott Polar Research Institute
UNCLOS	United Nations Conference on the Law of the Sea
WHOI	Woods Hole Oceanographic Institution
WRU	Whale Research Unit

ONE

INTRODUCTION

Like the boy on the burning deck the little Herr Professor (as he came to be
called) stood on the flensing stage. . . . Between his boots and the planking there
existed a layer of viscous yellowish grease: whence, doubtless, the apprehension
betrayed at his bearded lips, the awkward stiffness of his bodily attitude. But
his eyes, under beaded brows, were brightly alert, for the spirit was gaining
mastery over the flesh, as it so often does when Science is goddess.

J. J. BELL, *The Whalers*, 1914

During these months at sea, I have watched the sperm whales, looking for
keys to an understanding. I have found it impossible to function simply as an
impassive machine, turning the actions of the whales into scientific truths. . . .
I lower the hydrophone, and hear the whales: "Click . . . click . . . click . . ."

HAL WHITEHEAD, *Voyage to the Whales*, 1990

SCIENCE AND THE WHALES

This is a book about whales, but there are relatively few whales in it. Indeed, let's start with a basic truth: there is not a single cetacean of any sort
in these pages. You knew that, of course, since even the smallest dolphin
needs much more room than the largest trim size of the most voluminous
scholarly tome. And though they breathe air, cetaceans basically like being in the water, while books are mostly written on paper, a substance that
fares poorly when submerged. In this sense books and whales are, in an

important way, immiscible. I tried to keep this in mind as I wrote, and it will be good to keep it in mind as you read.

So let me start again: this is a book about knowledge of whales. And to be still more precise, it is a book about the knowledge of whales garnered and mobilized by experts over the course of the twentieth century. Experts like the two men who appear in the epigraphs for this introduction, two whale scientists (a tribe sometimes known by the Melvillean moniker "cetologist," sometimes by the more sedate professional designation "marine mammal biologist") whose labors—one slogging through the gruesome residue of a whaling station with knife and notebook, the other bronzing himself on the bow of a hydrophone-equipped sailboat in the Indian Ocean—mark out the chronological (and perhaps also the spiritual) endpoints of this book as a whole. Two whale scientists pursuing knowledge of whales in different ways, at different times, for different purposes. Their work and its effects—this is my subject.

Knowledge is a funny thing. It is hard to explain what it is, hard to explain how we get it, hard to explain how it works in the world. It is characteristic of knowledge that it takes different forms than the thing known, and this means that the known thing is consistently absent from knowledge of it. One feels this, sometimes, even painfully. This book is interested in all these problems, and it frets about them, even as it recapitulates and reenacts them. In this sense, at least, the writing of whale books and the doing of whale science are more alike than different. Both go into the world absent their whales. If it is the whale you want, you will have to go to sea, where, because of the events I recount in this book, you are likely to have a considerable wait. Bring a book. You might bring this book, since it is long.

Like knowledge, whales are also funny, and a little hard to pin down. It would be difficult to pick a set of creatures that have been subjected to a more dramatic reimagining over the course of the last century: once seen as monstrous dwellers in the abysmal depths, shelled with explosives, melted for industrial commodities, and gunned as target practice by gleeful flyboys, these peculiar beasts eventually came to be understood by many as soulful, musical friends of humanity, symbols of ecological holism, bellwethers of environmental welfare, and even totems of a movement to transform the world and our attitude toward it. How did this happen? This book offers an answer to that question, and in sifting out that answer, it traces almost a hundred years of human efforts to understand these fugitive and mysterious animals. At the beginning of the chronology of this book, the most significant scientific publication in the world, *Nature*, could prominently and grossly misidentify the species of a whale depicted in its pages—and go uncorrected. Such was the extent of general scientific ignorance of these animals.[1] By the end of the period surveyed below, there was hardly a schoolchild in North America who had not been obliged to write up a whale report for science class. Because these superlatives of organic organization have taken up a great deal of space in the collective imagination, and because of the remarkable trajectory of their reconception since 1900 (a process in which the sciences played a significant role), I contend that a history of whale science can shed considerable light on the changing understanding of nature in the twentieth century. That is my claim, and the pages that follow represent my best effort to deliver thereupon.

I have various (imagined) readers in mind for this work, which is situated at the intersection of several different disciplinary literatures. For starters, my primary approach is that of the history of science. It is—after all, and for better or worse—the scientists' techniques for producing knowledge of nature that have proved most robust and authoritative in the modern world. How do those techniques work? How do they develop? And how do the findings of the scientists help make the world in which we live? These are, I think, the central questions that concern any historian of science, and they are questions that motivate and organize this study. I am, therefore, preoccupied throughout with showing what it meant to have scientific knowledge of cetaceans at different moments in the twentieth century, and I work to demonstrate who succeeded in making such claims, how they did so, and what larger consequences followed on their efforts. The range of different kinds of "cetology"—from sloppy slaughterhouse anatomy conducted under macabre and trying conditions to fiddly bioacoustics work performed by tidy military scientists wearing headphones (or stoned hippies playing synthesizers)—proves surprising, and the conflicts between these different sorts of whale science ended up playing a significant role

1. See Roy Chapman Andrews, "What Shore Whaling Is Doing for Science," *Nature* 88, no. 2200 (28 December 1911): 280–82, at p. 281, where the caption on figure 3 reads "'Cutting in': a Right Whale," but the animal in the photograph is clearly a rorqual. A year later, Theodore Salvesen, lecturing on the explosive growth of the modern whaling industry at the Royal Society of Arts, in London, could be met by a pressing question from the chairman, Lord Sanderson, "whether a whale was really a fish?" Salvesen, "The Whaling Industry of Today," *Journal of the Royal Society of Arts* 60, no. 3097 (29 March 1912): 515–23, at p. 523.

both in the history of whaling, and in the history of whale conservation, which was in turn an important component of the rise of the modern environmental movement.

It is the fraught history of modern whaling (of which more later in this introduction) that gives the story of whale science much of its significance, not to mention its poignancy. The bulk of chapter 2, for instance, deals with the emergence of an extensive and well-funded program of biological research on the large whales of the Southern Hemisphere in the early part of the century—work that aimed to lay the foundations for the "rational regulation" of the whaling industry, which was then rapidly expanding into new waters in the Antarctic. The failure of this initial scientific-cum-regulatory undertaking—and it was a complicated sort of failure, as I show in some detail—had lasting repercussions, I argue, for the later history of efforts to control the commercial exploitation of the world's whales. And for the scientists who were charged to do "biology"—the science of *life*—in the stygian swamps where their subject organism underwent Brobdingnagian dismemberment and rendering, field research came to mean a demanding acculturation to industrial-scale killing, grinding, and cooking. It is my hope that this aspect of my investigation—a portrait of a life science at work in the maw of death, a set of scientific investigations inextricably entangled with a highly remunerative and destructive activity—will hold the attention of traditional historians of biology as well as historians who work on the field sciences, natural history, agricultural research, and science in commercial settings. The changing relationship between science and industry is a significant theme in this study.

Because much of the early research into the life histories, migration patterns, and basic biology of the large whales was conducted by Great Britain as part of a major multi-vessel scientific initiative (known as the "Discovery Investigations"), chapter 2 also engages the larger history of oceanography in the first half of the twentieth century. And because Britain had designs on the ice, islands, and waters of the Southern Hemisphere (where the vast majority of the whaling in this period was conducted), I have also gestured, if passingly, at issues of science and imperialism in writing about Discovery and the ways that whale research served to advance various geopolitical strategies in the period before World War II.

The question of what it would mean to be "rational" about the fantastically lucrative circum-Antarctic killing fields dogged the work of whale scientists and the policy makers who hoped to make use of their findings. This issue is central to chapters 3, 4, and 5, where I am concerned to unfold the changing relationship between science and regulation from 1930 to 1965. It is my hope that this material will be of interest not only to historians of science but also to political scientists, environmental activists, and others concerned to understand how expert knowledge functions in the complex arena of collective decision making.[2] Because whales were a unique, open-ocean commercial quarry, they raised from early on unprecedented problems for regulators, diplomats, and international lawyers, and these challenges eventually led to the formation of the first formal international body dedicated to the management of a biological resource, the International Whaling Commission (IWC), founded shortly after World War II. This organization was explicitly committed to building a mechanism whereby "scientific findings" about whales would serve as the basis for sound regulatory policies that could be implemented on a global scale. A Panglossian techno-scientific optimism spangled the early years of the IWC, a touch of which can be found in this paean to whaling "factories" (the big blue-water whale-processing vessels that roamed the oceans digesting large cetaceans into commercial fats, waxes, and fertilizers) offered by a leading member of the IWC's "Scientific Committee" in 1952:

> In the course of time, the floating factory has become more and more of a technical marvel. It is an oil-plant and a meat-meal factory. It is also a canning factory. It is a very well-equipped chemical works, with a most ingenious and varied routine. It is in fact a scientific institute of the first rank.[3]

Chapters 4 and 5 take up the fate of this dream in some detail. In doing so they not only lay out a revised history of one of the great debacles of

---

2. For helpful introductions to this literature, consider the following studies (which intersect with this one at several points): Stephen Bocking, *Nature's Experts: Science, Politics, and the Environment* (New Brunswick, NJ: Rutgers University Press, 2004); Edward Christie, *Finding Solutions for Environmental Conflicts: Power and Negotiation* (Cheltenham, UK: Edward Elgar, 2008); Radoslav S. Dimitrov, *Science and International Policy: Regimes and Nonregimes in Global Governance* (Lanham, MD: Rowman & Littlefield, 2006); and Robert F. Durant, Daniel J. Fiorino, and Rosemary O'Leary, eds., *Environmental Governance Reconsidered: Challenges, Choices, and Opportunities* (Cambridge, MA: MIT Press, 2004).

3. This was Birger Bergersen, in a speech preserved in his private papers: Birger Bergersen Papers, Hvalfangstmuseet, Sandefjord, box 3, folder "taler." Note that he subsequently revised this to read, "It can in fact be used as a scientific laboratory of the first rank."

twentieth-century natural resource management, but also suggest a way of approaching the larger problem of telling suitably nuanced stories about the intersection of science and politics in a regulatory setting. In chapter 4, for instance, I trace out the evolution of the scientific advising system in the IWC, paying particular attention to the ways that scientists themselves functioned as savvy political actors sensitive to the need for careful "boundary work" between the questions that would be defined as "scientific" and those that would be defined as "political." An analytic focus on the elaboration of these boundaries leads to some larger conclusions about what it meant to "do science" in a new and challenging environment: the committee rooms of the post–World War II international organizations for global governance, geopolitical diplomacy, and international regulation.[4]

There were new sciences in play as well. Chapter 5 examines the mobilization of mathematical models of population dynamics in the regulatory arena in an effort to show how these models were made into powerful tools for forcing consensus among conflicting actors. This section of the book may be of interest to those historians concerned to understand the ways in which numbers, calculations, and computational systems have come to affect public life. And if there is a chapter of this book that I think could be profitably read by a student of politics, I think this would be it. Though, to be fair, it would have to be a more than ordinarily patient student of the discipline, since my treatment of this episode cannot easily be reduced to the sort of "finding" that one could readily mobilize in a think tank working group: there is a narrative here, there are characters, and there are some mathematical models too. It is the (tacit) contention of the chapter that one cannot really understand what happened without rolling up one's sleeves and working to make sense of the math, the people, and the specific sequence of historical events. What is the take-home point once one has subjected oneself to this exercise? Well, the most important lesson may simply be that one must do this actual work; that without this work one cannot really understand what happened. In that sense, while I would like this material to be read by political scientists (particularly those with an interest in science, society, and environmental problems), I am aware that

---

4. See Joseph E. Taylor III, "Boundary Terminology," *Environmental History* 13, no. 3 (July 2008): 454–81. My interest in boundaries is different from Taylor's, but his concern with terminological specificity in the move to "global" and "international" historical framings is right on point. It is an open question whether I have been careful enough, but I have certainly been preoccupied with this problem.

some of them may find its historical (and scientific?) detail tedious, even rebarbative.

And that points to a larger fact about this book and its approach: this is an archival history of a somewhat demanding variety. It has been written out of reams of published and manuscript material—personal letters, scientific notebooks, technical reports, diplomatic correspondence—from dozens of archival collections in half a dozen countries. It is not unreasonable to ask some hard questions about the ultimate value of such studies, which are difficult to research and compose and often by no means especially pleasant to read. I am, as I give this volume to a world increasingly concerned with Twitter-scale texts, acutely conscious of these sorts of questions and feel them with great force—particularly when, say, I glance from the walls of my office (crammed with unwieldy binders and an unholy proliferation of old books) to the screen of my iPhone (which quietly insists that the relevant world can stream bright and clean through a glassy lozenge of responsive obsidian). This, however, is not the place to mount a full-scale defense of the culture of the book, or, for that matter, a plea for the future of the bricklike academic monograph. Suffice it to say that the satisfactions of the latter are an acquired taste, and I, having tasted, would happily share my morsel with any comer.

Including environmental historians. Chapter 6, which attempts to explain—by reference to changing scientific ideas and practices—much of the extraordinary shift in attitudes toward whales and dolphins that occurred across the 1960s and 1970s, is at least a contribution to the history of environmentalism in Europe and North America in the period associated with the Vietnam conflict and the rise of the counterculture. If I am right, this story is a remarkable instance of crossing lines of biology, linguistics, information theory, and acoustics, all of which get tangled up in an unlikely hot tub churning Cold War bioscience, ocean theme park entertainment, sexual liberation, and mind-altering drugs. The story of learning to love the whales is an adult swim, as it turns out, and I very much hope that this chapter makes the case for pushing the links between the history of science and the history of environmental ideas and movements.[5] Is it, or are any of the other parts of this book, really engaged with environmental history?

5. The story I tell in chapter 6 can be usefully read in conjunction with Adam Rome's work on the relationship between the environmental movement and the culture of the 1960s: Adam Rome, "'Give Earth a Chance': The Environmental Movement and the Sixties," *Journal of American History* 90, no. 2 (September 2003): 525–54.

I would like to think so. In important ways, for instance, I have accepted the arguments of a set of pioneering scholars over the last two decades who have insisted that animals and our relations with them constitute a crucial subject for historical investigation. This study seeks to contribute to a robust literature on human-animal relations and the historical construction of the human-animal boundary. By rearranging a history of several quite disparate modes of scientific research in the twentieth century (reproductive physiology, psychology, biological oceanography, population dynamic modeling, acoustics) around a specific taxon, I aim to show the value of thinking with animals. Some would argue, I think, that this historiographic move (which I am by no means the first to make) does not really bring us into the heartland of environmental history. But there is more to my story than that: the tapping of the ocean resources of the Antarctic Convergence in the first half of the twentieth century, for example, certainly represents an instance of human-driven environmental change that can vie with the most salient and historically significant episodes of such phenomena, and here we would seem to be very squarely on the environmental historian's terrain. Though of course we are not on terra firma at all, but out upon the oceans, which have to date proved somewhat recalcitrant historical subjects. There is reason to think this is changing: American historians recently heard a clarion call for new work in the environmental history of the oceans, and I would be delighted if this book found readers intending to make new contributions in this area—not least because I have benefited from my exposure to this scholarship and have presented much of this work to colleagues in this field over the last several years.[6]

But here too I am aware of the challenges. There are, for instance, some fundamental differences in approach that militate against easy synthesis of history of science and environmental history, despite their shared terrain. A slightly caricatured account of the problem would run something like this: environmental historians are inclined to deploy as historical *explanans* some of the very findings that historians of science consider the *explanandum*. This tends to frustrate the historian of science. At the same time, the arguably exaggerated preoccupation with treating "nature" as endlessly and ineluctably constituted by human discourse or practices can

---

6. I am thinking of W. Jeffrey Bolster, "Opportunities in Marine Environmental History," *Environmental History* 11, no. 3 (July 2006): 567–97, and Bolster, "Putting the Ocean in Atlantic History: Maritime Communities and Marine Ecology in the Northwest Atlantic, 1500–1800," *American Historical Review* 113 (February 2008): 19–47.

(not wholly unreasonably) strike the practicing environmental historian as either sophomoric, paranoiac, quixotic, or downright nuts—or, I suppose, as some combination of all of the above.[7] I must say that I am inclined to think this problem basically insurmountable. At any rate, I have not surmounted it. But I have reconnoitered the escarpment, thrown a grapple or two, hollered over the ridge. I hope the environmental historians who make their way through this book will discern evidence of my attempts, and that they can find things of use herein. There remain, though, a number of environmental-historical approaches I have not even attempted. Just one example: Do I give the whales "agency" in this book? Not really. There aren't any whales in this book, remember? Only words about whales. Though many of those words, particularly in chapter 6, are exactly about the agency of the cetaceans—about their inner lives, their minds, their efforts to "tell us what they are thinking." But it is the emergence of much talk on this subject that I am trying to *explain* in this chapter. What were the whales saying? I have no idea. Do I give too much agency to (human) words? Maybe. It is ever thus with bookish folk. If it is whales you want, you have to go to sea.

And with that, let me turn to a brief history of those who did just that: the whalers. The scientists would follow in their wake.

LEVIATHAN AND PUMPED AIR: THE ORIGINS OF MODERN WHALING

Five distinct (if overlapping) episodes of intensive commercial whaling, distinguished by the pursuit of different stocks using different technologies, can be readily identified.[8] The first of these, the pursuit of the "right" whales (several species of the family Balaenidae) in the temperate and northern waters near western Europe, ran from the Middle Ages through to the early

7. There is a considerable literature that goes after these problems patiently and seriously. For a useful recent position paper, consider Sverker Sörlin and Paul Warde, "The Problem of the Problem of Environmental History: A Re-reading of the Field," *Environmental History* 12, no. 1 (January 2007): 107–30.

8. This sketch omits prehistoric whaling, whaling by "aboriginal peoples" in various regions (including the Inuit and several Indonesian communities), and the distinctive history of whaling in East Asia. A history of world whaling might be periodized differently. There are several books that aspire to survey whaling at a global scale, but they are, of necessity, uneven in their depth and coverage. Richard Ellis, *Men and Whales* (New York: Knopf, 1991) is the best; see also Daniel Francis, *A History of World Whaling* (New York: Viking, 1990).

twentieth century, though the heart of the enterprise lay in the seventeenth, eighteenth, and early nineteenth centuries, when French, Dutch, and British ships pursued bowheads (*Balaena mysticetus*) from Spitsbergen to the Greenland Sea and the Davis Strait in what came to be known collectively as the "northern fishery." By the late eighteenth and early nineteenth centuries, a "southern fishery" had emerged to rival the waning productivity of this icy enterprise. Characteristically defined by the American ascendancy in the pursuit of sperm whales (*Physeter macrocephalus*) in the Pacific Ocean in the 1820s through the 1860s, this second phase of whaling properly includes the pursuit of "great" right whales (several species of *Eubalaena*) as well, primarily in the lower latitudes, in addition to the chase for the cosmopolitan sperms in various other waters, including the Atlantic and the Indian Oceans. Both the northern and the southern fisheries were conducted from sailing vessels and involved attacking whales in light, fast skiffs powered by oarmen who brought the wielder of a hand harpoon as close as possible to a whale in the open sea. If successful, the harpooner affixed the skiff to the animal by means of a strong line bent to a toggle on the shaft of his harpoon, and the boat then served as a drogue, retarding the flight of the injured whale, which would eventually be dispatched by means of repeated thrusts from a long-handled lance. Processing techniques involved "cutting in" and removing blubber and (in the case of the baleen whales) "whalebone," or baleen plates, while the carcass lay in the water, lashed to the ship. Storage and transport varied. In the northern fishery, where ships stayed relatively close to ports and seldom made voyages longer than a single season, blubber was simply trimmed and barreled raw; the cold conditions adequately preserved it for shore processing into an oil suitable for household illumination and lubrication. The practice in the southern fishery, where long voyages in tropical conditions were the norm, evolved into an elaborate system for "trying out" the oil in rendering stoves placed amidships, permitting casks of relatively stable, liquid oil to be laid in the hold.

Some (debatable) combination of resource depletion and resource substitution—namely, the development of the techniques of petroleum processing in the second half of the nineteenth century—brought the southern fishery to a protracted demise after the 1860s, by which time there had been little left of the northern fishery for decades.[9] The toll of

9. For an econometric analysis that leans toward the resource substitution account, see Lance E. Davis, Robert E. Gallman, and Karin Gleiter, *In Pursuit of Leviathan:*

these first two phases of sustained commercial whaling on the right whales of the world's oceans was very high indeed—essentially commercial extinction in the northern Atlantic and severe depletion everywhere. The effect of nineteenth-century whaling on sperm whale stocks is a subject of active dispute among conservation biologists using genetic techniques to reconstruct population sizes, economic historians analyzing price data, and whaling historians tabulating the surviving catch records of the industry.

It is necessary to recall that these first two phases of commercial whaling—along with several smaller enterprises, not dealt with here, that focused, for instance, on gray and humpback whales (*Eschrichtius robustus* and *Megaptera novaeangliae*), species that could be consistently pursued from shore in a number of coastal regions—were basically unable to touch the larger rorquals (genus *Balaenoptera*), meaning chiefly the whales now called fins (or finners; *B. physalus*) and blues (*B. musculus*).[10] These powerful, fast-swimming animals, distinguished by their slim profiles and grooved throats, were, by and large, too difficult to approach and too hard to kill to repay the efforts of open-boat whalers using hand-harpoon techniques. Even when, by the late eighteenth century, experiments were made with explosive ordnance in harpoons and lances, these instruments did not prove adequate to add the large rorquals to the list of species seriously pursued in the first seventy years of the nineteenth century; while such

Technology, Institutions, Productivity, and Profits in American Whaling, 1816–1906 (Chicago: University of Chicago Press, 1997). This revisionist account has met with some resistance among whaling historians who are not economists. For a more traditional view, see Phillip J. Clapham and C. Scott Baker, "Whaling, Modern," in *The Encyclopedia of Marine Mammals*, ed. W. F. Perrin, Bernd G. Würsig, and J. G. M. Thewissen (San Diego: Academic Press, 2002), 1328–32. An older article nicely opens the issues: James A. Ruffner, "Two Problems in Fuel Technology," *History of Technology* 3 (1978): 123–61, at p. 131.

10. For an introduction to the use of historical sources to reconstruct whale populations, see Tim D. Smith, "Examining Cetacean Ecology Using Historical Fishery Data," *Research in Maritime History* 21 (2001): 207–14. See also Poul Holm, Tim D. Smith, and David J. Starkey, eds., *The Exploited Seas: New Directions for Marine Environmental History* (Saint John's, Newfoundland: International Maritime Economic History Association / Census of Marine Life, 2001). There has been some controversy surrounding the work of J. Roman and S. R. Palumbi, who have used genetic diversity data to reconstruct pre-exploitation whale populations (see, for instance, "Whales before Whaling in the North Atlantic," *Science* 301, no. 5632 [25 July 2003]: 508–10; but compare the critical letters by Sidney Holt and Edward Mitchell, "Counting Whales in the North Atlantic," *Science* 303, no. 5654 [2 January 2004]: 39–40).

devices did see limited service in the period, they were generally deemed too unwieldy, unpredictable, and dangerous (to their users) to come into wide circulation. While a number of sea captains, fed up filling logbooks with disgusted comments on the profusion of tantalizing rorquals that blew around their ships, actually made efforts to kill them, and a few succeeded, the slim odds of reward for the effort were further reduced by the propensity of rorqual carcasses to sink, a problem seldom encountered with sperm and right whales, and one that—in the absence of massive tackle and larger ships—robbed hunter of prey at best, and in the worst case, could actually threaten his vessel.

The solution to these problems in the late 1860s—largely as a result of the dogged efforts of a single individual, the pious Norwegian sealer Svend Foyn—gave rise to the "modern" whaling industry. Foyn developed a mounted cannon that fired explosive "grenade harpoons" from the bow of a screw-driven steamer (figures 1.1 and 1.2). This combination—with certain improvements, including the use of massive shock-absorbing accumulator winches and, by the early 1880s, compressed air to "inflate" carcasses to keep them from sinking (hence my puckish section heading above)— brought rorquals within reach. For about forty years—a period that can be understood as the third episode in whaling history—the center of this enterprise was the northern Atlantic (and especially the waters of Finnmark). With the discovery of the unprecedented fecundity of the Antarctic waters in the early twentieth century—and the development of practical techniques for hydrogenating whale oil for use in margarine, creating a large new market beyond soap and lubricant manufacturers—the bonanza fourth phase of whaling reinvigorated an increasingly anemic northern rorqual industry. Shore bases in South Georgia (in the subantarctic Atlantic) and elsewhere in the deep south docked catcher boats taking record numbers of whales almost immediately; within five years these boats were catching more whales and generating more oil than the entire Atlantic north of the equator (figure 1.3)[11].

11. For the standard account of the birth and development of Foyn's system, see J. N. Tønnessen and Arne Odd Johnsen, *The History of Modern Whaling* (Berkeley: University of California Press, 1982); this book is an edited revision of the four-volume Norwegian work *Den Moderne Hvalfangsts Historie*, published between 1959 and 1970. There are several single-company histories that are of considerable value as well, including the most recent work by Ian B. Hart, *Pesca: The History of Compañia Argentina de Pesca Sociedad Anónima of Buenos Aires: An Account of the Pioneer Modern Whaling and Sealing Company in the Antarctic* (Salcombe: Aidan Ellis, 2001). See also Wray Vamplew,

FIGURE 1.1
(*above*) The gun: Svend Foyn's harpoon cannon. (From Morley and Hodgson, *Whaling North and South*, opp. p. 162.)

....

FIGURE 1.2
(*right*) The hunt: A grenade harpoon strikes its target. (From Morley and Hodgson, *Whaling North and South*, opp. p. 178.)

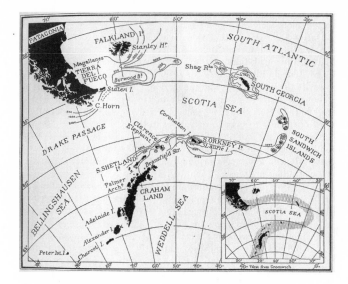

FIGURE 1.3 The grounds: A map of South Georgia and the surrounding Antarctic waters. (From *Report on the Progress of the
Discovery Committee's Investigations,* p. 11.)

Expansion was hampered by the appalling conditions of wind, water,
and weather and by the difficulty of building and maintaining processing
factories in the rugged and frozen archipelago of the peri-Antarctic islands.
"Factory ships" came into increasing use. These large vessels, frequently
refitted merchant or even passenger ships, housed the hardware of a shore
station: steam pressure boilers at the least and, with increasing sophistication in the second decade of the twentieth century, an array of other
refining equipment that could process flesh and bone, producing both additional oil and dry meals used as feed additives and fertilizers. Docked in
the lee of an islet or iceberg and serviced by smaller catcher boats, these
factory vessels helped extend the reach of the industry in forbidding waters.
Whales were again cut up in the water—as in the old northern and southern
fisheries—and the pieces hoisted into the deckside mouths of the boilers
and kilns. Because of the need for calm conditions for "cutting in" (effected
from small floating platforms moored to the ship's hull) as well as adequate
supplies of fresh water for the boilers, it was not until the development of

Salvesen of Leith (Edinburgh: Scottish Academic Press, 1975), and Gerald Elliot, *A Whaling Enterprise: Salvesen in the Antarctic* (Norwich: Michael Russell, 1998).

full-scale "pelagic factories" after 1925 that the whaling companies were entirely liberated from a dependence on harbors (figure 1.4). These new-style factory vessels—most following the design of the *Lancing*, the first factory ship equipped with a stern slipway—made it possible to draw the whole carcass out of the water for processing on deck. As such vessels grew larger and more stable, the processing factory was liberated from the need

FIGURE 1.4 (and see PLATE 1) Into the belly of the beast: A stern-slipway factory vessel engulfs its prey. (From the collection of the Natural History Museum, London.)

for sheltered waters and set free to roam the oceans, attended by fleets of catcher boats, and sometimes even serviced (particularly after World War II) by spotter aircraft. Without any dependence on land, the industry moved out from under a variety of regulatory systems that had been imposed by national governments—particularly Great Britain's—on the basis of their territorial jurisdiction over much-needed harbors; international accords became the only means of controlling whaling activities. There were considerable repercussions. The abandonment of shore stations and the global pursuit of whales by pelagic factory fleets mark the last major episode in the history of whaling, a period that came to a close only with the implementation of the "moratorium" on commercial whaling by the member nations of the IWC, an agreement passed in 1982 and broadly effective by 1987. Some whaling still goes on under various exceptional provisions (chiefly aboriginal and subsistence whaling and what is known as "scientific whaling") and by nations operating in formal objection to the moratorium (notably Norway and Iceland). While activists make much of these ongoing fisheries, the age of intensive commercial whaling appears to have come to an end in 1987.

This book is concerned with the science of whales from the opening of the Antarctic in 1904 through to the demise of large-scale commercial whaling in the early 1980s, a period that encompasses much of period four and all of period five as I have laid them out above; that is, with the rise of the "modern" whaling industry in the Southern Ocean, at first in the South Atlantic via shore stations at South Georgia and elsewhere, and then increasingly (via pelagic factories) throughout the teeming waters of Antarctica. Not until 1963 did world production of whale oil outside the Antarctic exceed that from Antarctic waters, reversing the pattern established in 1910. By that time the majority of the earth's cetacean biomass had been immolated, though it took some time before this could be said with certainty.

THE ANTARCTIC INDUSTRY AND
BRITISH ADMINISTRATION

A closer look at the opening of the modern southern fishery will afford helpful context for much of what follows in this book. The transfer of Foyn's whaling techniques and technologies to the Antarctic in the first decade of the twentieth century was largely the result of efforts by another Norwegian whaler and sealer, Carl Anton Larsen. By the time Captain Larsen undertook his first whaling expedition in northern waters—an 1884 trip to eastern Iceland in pursuit of bottlenose whales (he was 24)—there was broad consensus among whalemen that both the Finnmark and Greenland whale fisheries were in decline and that a shortage of whales was a significant part of the problem (they were generally thought to have moved farther offshore). At the same time, the reports that trickled back with Antarctic sealers and explorers led to speculation about potentially vast numbers of whales in southern waters.[12] Several exploratory ventures were mounted, and Larsen himself captained the *Jason* on two voyages into the southern Atlantic between 1892 and 1894. Seals offset expenses, but both expeditions were financial failures. While Larsen saw blue, fin, and humpback whales aplenty, he had come equipped for right whales, and they proved scarce. The possibility of mounting a capital-intensive voyage for large rorquals lingered, but the challenges of distance, cost, and conditions gave pause to potential investors. After several years back in Finnmark whaling, where he experienced firsthand the dwindling numbers of profitable large whales (and witnessed the drift of the industry toward ever more remote stations in pursuit of fresh grounds), Larsen—whose reputation as a pioneering navigator in the southern ice had earned him considerable regard in an era increasingly infatuated with polar exploration—received an invitation to serve as captain of the *Antarctic*, the flagship of the Swedish South Polar Expedition, under the leadership of the geologist Baron N. Otto G. Nordenskjöld.[13] Larsen accepted, not least because he hoped again to assess the possibilities of taking whales profitably in the south.[14] The historian of science Aant Elzinga has examined what he calls the "mutually reinforcing interests: whaling, exploration and science" that shaped this expedition, and he has reviewed a number of Scandinavian sources that show how blubber hunting was made to fit into a research program already negotiating between the contesting ideals of ecumenical scientific internationalism (on the one hand) and heroic exploration under a national flag (on the other).[15]

12. Ian B. Hart, *Pesca: The History of Compañia Argentina de Pesca Sociedad Anónima of Buenos Aires: An Account of the Pioneer Modern Whaling and Sealing Company in the Antarctic* (Salcombe: Aidan Ellis, 2001), 12, 13.

13. Ibid., p. 19. Larsen managed a station at Mafjord that saw its annual take slide from over a hundred to fewer than fifty whales in the last six years of his tenure.

14. Ibid., p. 22.

15. Aant Elzinga, "Beyond the Ends of the World: Nationalism and Internationalism in Antarctic Exploration and Imagination, 1895–1914" (paper presented at the History of Science Society Conference, Milwaukee, WI, 7–9 November 2002).

Larsen was, in a manner of speaking, extricated from this tangle when the *Antarctic* sank, crushed by ice near Snow Hill Island in February and March of 1903.[16] He and the other members of the expedition were rescued by an Argentine ship and landed at Buenos Aires in December of that year. It was there, in the heady exuberance of a hero's welcome, that Larsen secured the interest of a group of Argentine financiers who were prepared to back a modern whaling venture in the deep south. Thus the pioneer southern whaling company, the Compañia Argentina de Pesca, Sociedad Anónima (or CAP or, often, "Pesca"), was born. Returning to Norway in pursuit of other backers and equipment, Larsen rapidly assembled the necessary ships, men, and material for the undertaking, but was unsuccessful in attracting Scandinavian capital. By November 1904 Larsen's first two ships had dropped anchor in the harbor of Grytviken, on the icy island of South Georgia, a site Larsen had scouted on earlier passages. The first whale was taken before the month was out. A rudimentary factory was erected ashore almost immediately, and nearly two hundred whales (primarily humpbacks) were taken in the first season.[17] The era of modern whaling in the Antarctic had begun.

It had done so outside of any formal governmental structure. British claims to South Georgia—an uninhabited, mountainous ridge slightly smaller than Long Island, which pricks the South Atlantic some 800 miles east of the Falklands—derive from a visit in 1775 by Captain Cook, who landed and took possession for the crown, but in 1904 there still remained no trace of effective British authority on the island. Three-quarters of South Georgia's landmass is frozen throughout the year, but its good ice-free harbors (particularly on the eastern shore) had attracted numerous visits from sealers and elephant seal hunters throughout the nineteenth century; a handful of landings by scientific expeditions and surveys from several nations completed the roster of pre-twentieth-century visitors. The Letters Patent of 23 June 1843, which made provisions for British colonial governance in the Falkland Islands "and their Dependencies," while making no explicit mention of South Georgia, came to be understood (by the Colonial Office) to have placed the island under the emerging authority of the Falk-

16. Robert Headland, *The Island of South Georgia* (Cambridge: Cambridge University Press, 1984), 64.

17. Ian B. Hart, *Pesca: The History of Compañia Argentina de Pesca Sociedad Anónima of Buenos Aires: An Account of the Pioneer Modern Whaling and Sealing Company in the Antarctic* (Salcombe: Aidan Ellis, 2001), 44, 53, 55.

land Islands government. But it was not until 1908 that an amendment to those Letters Patent actually listed South Georgia formally as a component of the Falkland Islands' administrative entity—along with an archipelago of still smaller and more forbidding islands limning the Scotia Sea and reaching down to the Antarctic continent: the South Sandwich, South Shetland, and South Orkney Islands.[18]

The late date is telling, as it reveals the degree to which the "gold rush" of Antarctic whaling (to use a notion invoked at the time) shaped British colonial policy in the region in the early twentieth century. An exhaustive account of the process and means by which British control was asserted and ultimately ratified in these regions is beyond the scope of this book, but by 1906 delicate diplomatic negotiations (of both the gunboat and non-gunboat variety) had resulted in Larsen's CAP station at Grytviken receiving legal sanction in the form of a twenty-one-year lease conferred by the Crown.[19] This precedent established (and not immediately challenged), the governor of the Falkland Islands from 1904 to 1915, William Lamond Allardyce, soon found himself reviewing a flood of applications for other whaling concessions on the island and its neighbors. After the expansive gesture of the clarificatory Letters Patent of 21 July 1908, he and his successors were in a position to control access and set conditions for the industry throughout a vast area of the Antarctic by means of similar leases.[20] As Allardyce, who would eventually be knighted for his colonial service, boasted in 1911 to a well-placed friend in London (while nudging him to help make sure he was not passed over for a Coronation Medal this time around):

18. The same amendments also laid claim to territory on the continents of Antarctica and South America. For a (not unbiased) review of this contentious matter (entangled as it is with the not-so-distant Falklands War), see Robert Headland, *The Island of South Georgia* (Cambridge: Cambridge University Press, 1984), chap. 9.

19. See D. W. H. Walton, "The First South Georgia Leases: Compañia Argentina de Pesca and The South Georgia Exploring Company Limited," *Falkland Islands Journal* (1983): 14–25. See also Hart's account of the negotiations in *Pesca: The History of Compañia Argentina de Pesca Sociedad Anónima of Buenos Aires: An Account of the Pioneer Modern Whaling and Sealing Company in the Antarctic* (Salcombe: Aidan Ellis, 2001), 64–95.

20. Robert Headland lists all the early leases in *The Island of South Georgia* (Cambridge: Cambridge University Press, 1984), app. 1. There is also a strong chapter on South Georgia whaling in Stephen Palmer, "Far from Moderate" (PhD diss., University of Portsmouth, 2004).

Today we get a revenue of some £2400 from S. Georgia, and a still larger sum for the whaling licenses issued for the territorial waters of the South Shetlands, South Orkneys, Graham's Land, and the Sandwich Group, all of which if I did not personally annex are now ours by Letters Patent.[21]

It had been his practice since 1905, he gloated, to maintain that "all the whales in these seas were ours . . . declaring that those captured on the high seas were the misguided ones which had been foolish enough to leave territorial waters."[22] The result, in rents and fees, had been a massive influx into the coffers of a colony almost entirely dependent on the ranching of sheep.[23] By the 1910–1911 season, he reported, some 5,000 whales had been taken in the waters of the Dependencies "of a value of close on half a million, the above as against nothing when I came here."[24] And this "extraordinary catch"—already exceeding the whaling returns of the rest of the world combined—"has . . . created a furor and excitement not unlike that of a mining boom."[25]

If Allardyce's account of the explosive growth of the industry perhaps overcredits his own hand in the development (he was by this time quite desperate for a governorship in the Bahamas or a position in an Australian state—indeed, just about anywhere but his outpost at the end of the world), his pencil sketch of a brazen and highly remunerative British colonial gambit captures the essential contours of the early boom years, during which an almost entirely Norwegian industry, underwritten by capital from South

21. Allardyce to Sir John Anderson, 5 August 1911, Scott Polar Research Institute, University of Cambridge (SPRI) MS 1212/1/1–11, file "Allardyce, William L., Correspondence with Sir John Anderson, 1911–13."

22. Ibid. Territorial waters were understood to extend three miles beyond shore in this period, though this practice was in flux at the time. For more on these issues, see chapter 5 below.

23. An income of £2,000 per annum (as in 1908) represented an increase of 12 percent of the whole revenue of the Falklands in 1904; see Allardyce to Crewe, confidential dispatch, 28 August 1908, SPRI MS 240/1–2, file "Dispatches to Secretary of State for the colonies from Governor W. L. Allardyce, 1908–1915." Through 1910 the total revenue of the colony seldom exceeded £17,000; see Allardyce dispatch of 24 December 1910, in ibid.

24. Allardyce to Sir John Anderson, 5 August 1911, SPRI MS 1212/1/1–11, file "Allardyce, William L., Correspondence with Sir John Anderson, 1911–13."

25. Allardyce to Crewe, 6 July 1911, SPRI MS 240/1–2, file "Dispatches to Secretary of State for the colonies from Governor W. L. Allardyce, 1908–1915."

America, Scandinavia, and elsewhere, effected an enormous expansion of British territorial claims in the Antarctic while also, conveniently, paying for the privilege.[26] By 1909 revenue from the industry was being used to support a stipendiary magistrate on South Georgia, who reported to HM Government in Port Stanley, giving Allardyce (who never visited the island) and the British Empire a judicial and administrative presence in the region for the very first time.

There was, by 1909, plenty of administrative work to be done: the whalers needed to be managed, and so, it was increasingly suspected, did the whales.

---

26. Though profits for those early companies were immense, and dividends of 50 percent per annum and more for initial investors were not uncommon. See J. N. Tønnessen and Arne Odd Johnsen, *The History of Modern Whaling* (Berkeley: University of California Press, 1982), 185ff.

INTO THE BELLY OF THE BEAST

MODERN WHALING AND
THE BIOLOGY OF THE BIG WHALES

1910 – 1942

A plot was hatched some time before
(By whom and when I'm not quite sure)
To send an expedition forth
To turn its back upon the North,
And seeking oceans on whose foam
The grizzly southern icebergs roam,
To try, as for the Holy Grail,
To probe the secrets of the whale . . .

From "The Discovery Expedition—Prologue,"
N. A. MACKINTOSH, 1925[1]

INTRODUCTION

On 20 September 1928, the popular British *Daily Express* ran a lead cartoon by the distinguished political satirist "Poy" (Percy Hutton Fearon)
to accompany a story and photo spread on the ongoing work of the Discovery Investigations—then very much in the public eye (figure 2.1). In

---

1. This quotation is drawn from the prologue of a poem that appears never to have
been completed. See "The Discovery Expedition," the second of four bits of doggerel by
Mackintosh, in SPRI MS 1090 / 2;D. In a letter to Stanley Kemp, Mackintosh promises that
he will "work up" the poem. See Mackintosh to Kemp, 18 June 1925, National Oceanographic Library (NOL) Discovery Manuscripts, Southampton Oceanography Centre,
Southampton, England, green filing cabinet, file "Marine Station—Correspondence,
etc. 1924–1927."

A LARGE SUBJECT.

Sighted—But Not Yet Caught.

FIGURE 2.1 The pursuit of knowledge: The *Discovery* sails for the whale. (From the collection of the National Oceanographic Library, Southampton.)

it, a bespectacled and bearded boffin clinging to the topmast of a tall ship (he would seem to be a cross between Galileo and Ahab, but he is labeled "science") sings out "There she blows!" on sighting the looming bulk of a cetacean on the horizon. That this creature is labeled "Life's Problem" says a good deal about early twentieth-century popular fascination with the idea of unraveling the mystery of these fugitive beasts.[2] That the whale scientist is lodged in the crow's nest nicely speaks to the whalerish dimensions of early whale knowledge.[3] The accompanying article, laid out under the title "To the Pole for Whales: Great Adventure of the Year Begins," played up the vigor and excitement of this unusual scientific research program and emphasized the industrial scale of the expanding whaling fleet in the Antarctic. Here lay, as the cartoon caption would have it, "A Large Subject" for the enterprising scientists, who would tail the whalemen "into the heart of the ice" where the catcher vessels and factory ships, isolated by forbidding geography, would labor in the season "as a self-contained colony." It is a choice testimony to the perceived convergence of the whalers and the whale researchers in these years that one of the photographs accompanying the article depicts a catcher boat hauling several bloated dead rorquals to a shore factory—and misidentifies the vessel as the *Discovery* herself.[4] Oops. With the realities of this convergence we will be much concerned in the pages that follow.

This chapter takes up the early development of scientific research on the large whales of the Southern Hemisphere, paying particular attention to the origin of the institution under whose auspices the most important whale study in the first half of the twentieth century was done—the British Discovery Investigations, a series of shipboard and land-based expeditions to the Southern Ocean undertaken between 1925 and the 1940s, which were supported by a durable eponymous administrative body and research center in London, a body that both preceded and outlived the fieldwork phase of the undertaking (the "Discovery Committee" met from 1923 to 1949; the *Discovery Reports* were published into the 1970s), and which eventually became part of the National Institute of Oceanography formed after World War II.[5]

2. That its spout is labeled "oxygen," moreover, would seem to indicate no small confusion not only about whales, but about some fairly basic science—as if cetaceans were giant motile plants!

3. I have written about the nineteenth-century roots of this subject in *Trying Leviathan* (Princeton, NJ: Princeton University Press, 2007).

4. I have this article and the cartoon from a scrapbook of press cuttings in the NOL Discovery MS, p. 36.

5. A note on conventions: Because the "Discovery" in "Discovery Investigations" and "Discovery Committee" was originally the ship *Discovery*, it is not unusual to see this specific reference signaled by means of italics or inverted commas; hence, for instance, "the *Discovery* Investigations." This usage was adopted (if not uniformly) by

In reviewing the emergence of the Discovery Investigations and assessing the achievements of its research program on whales (the central explicit purpose of the whole enterprise), I make use here of previously untapped archival sources to correct and elaborate existing literature on the program's origin and development. It was a major undertaking, among Britain's most heavily funded scientific research projects before 1940 (absorbing as it did about a million pounds sterling in less than twenty years).[6] In particular, I attend in detail to the formative role played by Sir Sidney Harmer, keeper of zoology at, and eventually director of, the British Museum (Natural History), showing how his concerns shaped the scientific program for a whale research expedition and detailing the network of correspondents and committees through which he and others built governmental support for the undertaking (figure 2.2). In the pages that follow, I offer a fine-grained account of the institutional politics, international negotiations, and scientific exigencies that gave rise to this important (if not wholly successful) initiative, and I turn, in conclusion, to an assessment of its legacy for the history of cetacean science and industry regulation.

What was that legacy? Briefly stated, Harmer and the Discovery Investigations scientists, drawing on and elaborating Scandinavian precedents, codified a "system of research" on the large whales that became the dominant mode of whale science for most of the first half of the twentieth century—a system that was widely emulated by other nations with an interest in whale stocks. At the core of this work lay two kinds of field biology: first, flensing-platform investigations into gross anatomy and reproductive physiology; second, open-ocean marking expeditions that plugged living

---

active participants in the enterprise in the early years (including Harmer) and can still be found in the most recent discussions of the work (see, e.g., Keith Rodney Benson and Philip F. Rehbock, *Oceanographic History: The Pacific and Beyond* [Seattle: University of Washington Press, 2002], 550). However, on balance, I believe it is less correct, in view of the fact that the "Discovery Committee" itself did not generally italicize or otherwise set off the word "Discovery" in referring to the work of the Discovery Committee or the Discovery Investigations from its second annual report (of 1928) forward. I have thus opted to italicize the term "Discovery" only when referring to the actual vessel *Discovery*, although I have not changed alternate usage in quoted sources.

6. The official governmental overview of these matters (Sir Charles Joseph Jeffries, ed., *A Review of Colonial Research, 1940–1960* [London: Her Majesty's Stationery Office, 1964]) sums the whole contribution of the British government to nonmedical scientific research from 1920 to 1940 at a mere £500,000.

FIGURE 2.2
Sidney F. Harmer: The prime mover behind the Discovery Investigations fretted over his decision to trade whales for whale knowledge. (Courtesy of the Royal Society, London.)

whales with a numbered steel "dart" that could be recovered in the course of commercial whale processing should the animal subsequently be captured. The aim of these research activities was mastery over the life cycle of these animals: their growth and reproductive biology, their geographic distribution and movement. Direct stock assessment (as in establishing total population numbers) was never really the formal project of this sort of work, though it was certainly hoped that physiological and marking data could provide insights into the extent of the hunting pressure and its sustainability. From the outset it was hoped that the information gathered could be used to help frame suitable regulations for the industry, regulations intended to guarantee in perpetuity a profitable supply of commercially suitable whales.

Sidney Harmer did much initially to conceptualize and promote this system, and the young man who became the chief scientific officer of the Discovery Investigations (and subsequently the director of research), Neil Alison Mackintosh, refined and practiced the system for more than thirty years, becoming the most prominent whale scientist in the world in the critical years after World War II entirely on the basis of his leadership of Discovery and the publications he generated from its data (figure 2.3). While Discovery as a whole gradually took within its mission an increasingly broad set of physical and biological oceanographic investigations, these more expansive undertakings were always (at least nominally) tied

FIGURE 2.3
Neil Alison Mackintosh: Custodian of
"a system of research." (From his obituary
by Swithinbank in the *Journal of Glaciology,* p. 335.)

to the biology of whales. So, for instance, hydrologic work was defended as essential to a proper understanding of the environment and habitat of the large, commercially exploited rorquals of the Southern Hemisphere. The reason for this insistent "whaliness" lay, as I show below, in the administrative and fiscal origins of the investigations, which were funded out of taxes levied on whale oil landed by companies operating out of the Dependencies of the Falkland Islands. These unwelcome fees were justified to the whalemen on the grounds that the funds were being used to study the stocks—in order, in the end, to protect the industry. As the money piled up in London, various ways to spend it came to mind, but some link to the conservation of the valuable whale stocks of the Antarctic was always sought, if only to spring the lid of a considerable coffer. The rise of fully pelagic whaling (which meant oil was no longer landed on British soil, since factory vessels working freely in the open ocean could simply steam home when the season ended) rapidly decreased the revenues of the industry tax and progressively choked off Discovery's initially generous budget. But that was not the only challenge faced by Discovery's whale biologists: already by the mid-1930s, significant chunks of the Discovery budget were being siphoned away from "direct work on whales" and toward the kinds of survey work, flag showing, and Admiralty-oriented ocean science that felt increasingly urgent in the years leading up to World War II.

Harmer himself inclined to the sort of preservationist sentiments that had some period currency among British metropolitan elites, particularly those with an interest in natural history and responsible custodianship over the "fauna of the empire." Though he was careful to frame his public calls for work on the whales in terms of the need for economically rational "conservation" of the exploited stock, he could, privately, express a genuine indignation concerning the depravity of a commercial rapaciousness that courted actual extermination of creatures he did not hesitate to call (despite very limited personal experience) "noble beasts . . . which are among the most marvelous productions of nature."[7] Those who followed him in leadership of the enterprise he built—men like Mackintosh, among other Discovery staff who would go on to prominence in the scientific study of cetaceans—did not share much, if any, of this sensibility. Indeed, one of the things this chapter sets out to show, particularly in its conclusion, is the extent to which the practical elaboration of Harmer's "system of research" brought an unintended consequence of some perversity: this type of science drew its practitioners deep into complicity with the expanding industry, and trained up a generation of what I will call "hip-booted cetologists," men very much acculturated to the ways of the modern whalers. Several hip-booters would become important figures in the emerging international institutions of whaling regulation (especially the International Whaling Commission, created after World War II), but these men brought to their roles as scientific experts years of fieldwork in the forbidding environments of the Antarctic whale fishery, fieldwork that afforded them real knowledge of whale biology to be sure, but that at the same time gave them an intimacy with, and a sympathy for, the world of the whalemen (and their paymasters). I argue that this fieldwork culture was part of what made Mackintosh a decidedly ambivalent champion of conservation measures when he assumed the chairmanship of the Scientific Committee (and several of its subcommittees) of the IWC in the critical years of the 1950s. That will be a story for chapters 4 and 5. For now, suffice it to say that a kind of tragedy haunts the story of whale biology in the first half of the twentieth century: the science that developed between 1910 and 1940 for the purpose of protecting whale populations from excessive exploitation by whalers became, along the way, a science so deeply entangled with the whaling industry—dependent upon it, bound to it, acculturated to its physical labor, and finally, constituted on its operations—as to become, finally, nothing less than an *obstacle* to many conservation policies. It was a reasonably complete

7. Harmer to Lord Crawford, 3 March 1931, Natural History Museum, London (NHM) DF 1004 / CP / 747 / 2, file "Whaling 1925–1939."

science of whales but was ultimately incapable of realizing the aims of its founders: checking the progressive destruction of the world's large cetaceans. The making of a whale science eventually made whalemen of scientists, with deeply unsatisfactory results for the world's populations of large whales.

EARLY WARNINGS FROM THE SOUTH

By the end of the 1907–1908 season, it had become widely apparent that the scramble for what Governor Allardyce had taken to calling (gleefully) "the Whaler's El Dorado" was producing a reckless and wasteful urgency.[8] Calls for control began to be heard. In August 1908 a Norwegian whaling engineer and inventor who had spent several boom seasons in the south, Jens Andreas Mørch, published a brief notice in *Scientific American* pointing out that the vast majority of whaling in the Antarctic was being conducted from primitive factory ships equipped exclusively to process the easily accessible oil in the animals' blubber "blanket." As a result, he explained, "huge masses of meat and bones are left to drive before the wind and tide." He estimated that some 1,600 stripped carcasses had been set adrift in the last season alone, and he lamented the extraordinary profligacy of such processing, which abandoned some 50 percent of the oil that could be extracted from the catch (if chopped meat and bones received treatment in pressure cookers), not to mention the discarded value that lay in fertilizer and other potential derivatives. He added that "mankind of to-day does not take kindly to the wholesale waste of such proportions, and it must be because the British Government is uninformed about the matter that this state of things is allowed to pass on." He advocated a license provision requiring that any company whaling from stations in the Dependencies work up the full carcass. Mørch was not alone in his concern: mainstays of the northern industry, which was by no means glutted with whales in those years, looked on with some dismay at the improvident proceedings in the south. Mørch quoted a comment in the March 1908 issue of the *Norwegian Whaling Gazette* that echoed his objections to the careless haste of floating factory whaling:

8. The "El Dorado" quote is from Allardyce to Crewe, confidential dispatch, 3 October 1908, SPRI MS 240/1–2, file "Dispatches to Secretary of State for the colonies from Governor W. L. Allardyce, 1908–1915."

The unfortunate condition of the whole affair is that the carcasses are so poorly utilized and that large values are wasted. The whole business must therefore be characterized only as a depredation which must mark only a transitory stage. When only half the value of an animal is utilized there must be something deficient and this must be remedied if the business shall be able to satisfy our economical sense.[9]

Much has been made of Mørch's role in stimulating movement toward conservation-oriented legislation and research in the Antarctic whale fishery, particularly in view of the important letter he wrote on 7 June 1910 to the new keeper of zoology at the British Museum (Natural History), Sidney F. Harmer.[10] This letter, advocating scientific research on whale stocks and suggesting earmarking some portion of license fees for that purpose, certainly did stimulate Harmer's interest and immediate action (which, as I will show, were essential to the development of the Discovery Investigations), but it is important to emphasize that colonial officials in both the Falklands and the Colonial Office had intelligence from multiple sources on whaling conditions in the Dependencies and the need for restraints. In fact, the earliest regulation of whaling in the region had been passed by the Legislative Council of the Falklands in October 1906 (establishing licensing and affirming the broad regulatory powers of the Governor in Council), and the Whaling Ordinance of 1908 (promulgated on 8 August) reaffirmed and enhanced these provisions.[11] Moreover, Mørch's interest in

9. This and the quotes above are from Jens Andreas Mörch, "Improvements in Whaling Methods," *Scientific American* 99 (1 August 1908): 75. It stands to reason that some of these voices of concern proceeded from anxiety that a voracious and unfettered southern industry might well drive down prices.

10. See, e.g., J. N. Tønnessen and Arne Odd Johnsen, *The History of Modern Whaling* (Berkeley: University of California Press, 1982), 344, and Rosalind Marsden, "Expedition to Investigation: The Work of the Discovery Committee," in *Understanding the Oceans: A Century of Ocean Exploration*, ed. Margaret Deacon, A. L. Rice, and C. P. Summerhayes (London: University College of London Press, 2001), 69–86, at p. 70.

11. The Whaling Ordinance of 1908 shifted the revenue system away from royalty payments to fixed fees and removed a requirement that vessels shipping oil call at Port Stanley. See Legislative Council of the Falkland Islands, "Ordinance No. 3 of 1906: An Ordinance to Regulate the Whale Fishery of the Colony of the Falkland Islands," *Falkland Islands Gazette* 16, no. 11 (5 October 1906): 118–20, and Legislative Council of the Falkland Islands, "Ordinance No. 5 of 1908: An Ordinance to Regulate the Whale Fishery of the Colony of the Falkland Islands," *Falkland Islands Gazette* 18, no. 9 (1 September 1908):

seeing better processing of cetacean carcasses in this fast-growing fishery cannot be wholly separated from his work as an engineer and inventor of processing equipment: between 1904 and 1906 alone, he secured fourteen patents on whaling technologies, a number of them on devices for more profitable extraction of raw materials from the catch. Indeed, in 1909 he published another essay in *Scientific American*, "Manufacture of Whale Products," a technical review of the mechanical and chemical processes in use at best-practice facilities.[12]

As a result, then, of pressure from several directions, regulations were in effect by the end of 1908 limiting the number of catcher boats permitted under new licenses (in order to restrict overexploitation and to create an incentive for the better processing of carcasses). In December of that year, Allardyce wrote to the Earl of Crewe, secretary of state for the colonies, that "it is highly desirable in my opinion that all new leases granted should be conditional on the leasees providing themselves with the necessary plant for utilising the whole of the whales, and thus avoid the enormous waste which occurs at present owing to the blubber only being used."[13] It was a waste that a Falkland Islands customs inspector, visiting the South Orkneys a few years later, would liken to "taking the cream and throwing away the milk."[14]

Such expressions of concern should not be taken to indicate that Allardyce himself had any particularly developed concern for the conservation of whales, though he did have a whole-hearted preoccupation with the conservation of the whaling *industry* (in which he would eventually become financially involved) and a desire to see capital investments on shore in his jurisdiction (thus he had no love for the factory ships that floated in the bays

---

115–17, for the full text of both laws. See also Allardyce to Crewe, 8 August 1908, SPRI MS 240/1–2, file "Dispatches to Secretary of State for the colonies from Governor W. L. Allardyce, 1908–1915," explaining the differences.

12. John Mørch (Jens Andreas Mörch), "Manufacture of Whale Products," *Scientific American*, supplement, no. 1772 (2 January 1909): 15–16.

13. On catcher restrictions, see Allardyce to Crewe, confidential dispatch, 3 October 1908, SPRI MS 240/1–2, file "Dispatches to Secretary of State for the colonies from Governor W. L. Allardyce, 1908–1915." On lease requirements, see Allardyce to Crewe, 22 December 1908, in ibid.

14. Allardyce to Harcourt, 11 April 1913, SPRI MS 240/1–2, file "Dispatches to Secretary of State for the colonies from Governor W. L. Allardyce, 1908–1915," enclosing a report by W. Moyes on the South Orkney season of 1912–1913.

of the islands, boiling up oil and churning out glue-water).[15] These interests led him to countenance, and even encourage, regulations consistent with what he considered the orderly and durable development of the industry. He was, after all, acutely aware that careless hunting could rob the colony of potential revenues: the commercial extermination of the southern fur seals had drawn his attention by 1906, and he had supported the passage of a number of ordinances outlawing poaching and restricting access to firearms (later regulations would protect various breeding birds in the colony).[16] It is, however, by no means clear that Allardyce believed that overhunting of the stocks of Antarctic whales was even possible. Pressing the Colonial Office to free his hand in the granting of additional leases on South Georgia in 1908, Allardyce reported the assurances of C. A. Larsen's brother, Lauritz E. Larsen (also a whaling captain and a pioneer in the south), that "the waters around the Dependency were literally teeming with whales."[17] And by 1911, when pressure had begun to mount for more substantive limitations, lest the industry "kill the goose that laid the golden egg," Allardyce was dismissive, a stance he maintained throughout his tenure. In July of that

15. See J. N. Tønnessen and Arne Odd Johnsen, *The History of Modern Whaling* (Berkeley: University of California Press, 1982), 340, and Ian B. Hart, *Pesca: The History of Compañia Argentina de Pesca Sociedad Anónima of Buenos Aires: An Account of the Pioneer Modern Whaling and Sealing Company in the Antarctic* (Salcombe: Aidan Ellis, 2001), 87; but see communications from Allardyce to Lewis Harcourt in the summer of 1914, SPRI MS 240/1–2, file "Dispatches to Secretary of State for the colonies from Governor W. L. Allardyce, 1908–1915" (in defense of charges against the stipendiary magistrate), in which Allardyce asserts that no one in a position of responsibility in the government has a conflict of interest where the whaling industry is concerned. For Allardyce as a "strong proponent" of conservation policies, see Robert Headland, *The Island of South Georgia* (Cambridge: Cambridge University Press, 1984), 114.

16. Allardyce to Elgin, 6 February 1908, SPRI MS 240/1–2, file "Dispatches to Secretary of State for the colonies from Governor W. L. Allardyce, 1908–1915," referencing dispatch of 29 December 1906. On birds, see Ordinance No. 1 of 1913 in the *Falkland Islands Gazette*. See also Headland's discussion of the "Ordinance to provide for the preservation of certain Wild Animals and Birds," passed in September 1912, in *The Island of South Georgia*, 235. Allardyce was at the same time interested in stimulating, for instance, a trade in penguin eggs (for food) to Africa, as shown in Allardyce to Elgin, 29 June 1908, SPRI MS 240/1–2, file "Dispatches to Secretary of State for the colonies from Governor W. L. Allardyce, 1908–1915."

17. Allardyce to Crewe, 22 December 1908, SPRI MS 240/1–2, file "Dispatches to Secretary of State for the colonies from Governor W. L. Allardyce, 1908–1915."

year, transmitting the report of the magistrate on South Georgia, Allardyce pointed out that gross profits of some half a million pounds had already been amassed and the flow showed no sign of abating. "Therefore," he assured the secretary of state for the colonies, "the statements which interested parties have from time to time put forward as to the diminution of whales owing to the number of companies operating around South Georgia have no foundation in fact."[18] When pressed to ensure, at the very least, that the stipendiary magistrate enforce the existing restrictive lease provisions, Allardyce gave the Colonial Office a brisk lesson in South Georgia realpolitik: the stipendiary magistrate, he informed the secretary of state for the colonies, "occupies two rooms in a cottage owned by the Compania Argentina de Pesca and boards at the managers mess," placing him "in a most delicate and difficult position" when it came time to deliver sanctions; his nearest ally was some 800 miles of rough sea away—and he had no boat.[19] After two years on the island, the first magistrate had a nervous breakdown.[20]

Moreover, as far as Allardyce was concerned, there was no issue. In personal and official correspondence alike, he repeatedly insisted that the supply of whales appeared inexhaustible. In 1912, for instance, he reassured a friend in London: "With regard to the whales, I think you need have no apprehension of their numbers becoming unduly diminished. As a matter of fact the latest reports show that they are quite as numerous round South Georgia as in the South Shetlands and Graham's Land waters."[21] And parrying further concerned queries from the Colonial Office, Allardyce returned several times to the insinuation that any expressions of anxiety about the pace of the whaling forwarded to their attention could only be the work of "interested parties."[22]

HARMER AND THE BRITISH MUSEUM

In fact, it appears that the major reason for the mounting pressure was the dogged labor of the gifted and administratively uncompromising zoologist

18. Allardyce to Crewe, 6 July 1911, in ibid.

19. Allardyce to Harcourt, confidential dispatch, 29 September 1911, in ibid.

20. Allardyce to Crewe, 9 July 1911, in ibid., reporting a four-and-a-half-month leave for James Innes Wilson and his temporary replacement by Edward B. Binnie.

21. Allardyce to Sir John Anderson, 15 April 1912, SPRI MS 1212/1/1–11, file "Allardyce, William L., Correspondence with Sir John Anderson, 1911–1913."

22. See, e.g., Allardyce to Crewe, 2 August 1911, SPRI MS 240/1–2, file "Dispatches to Secretary of State for the colonies from Governor W. L. Allardyce, 1908–1915."

Sidney F. Harmer, who had recently left the Cambridge University Museum of Zoology, where he had served as superintendent since 1892, to assume the newly re-created post of keeper of zoology at the British Museum (Natural History).[23] Harmer had come up to Cambridge in 1881, where he came under the influence of Francis M. Balfour and threw himself into taxonomy and systematics, receiving first class honors on both parts of the Natural Sciences Tripos. After research at the Marine Biological Station at Roscoff, he advanced through postgraduate degrees in zoology and was elected a fellow at King's in 1886, becoming a university lecturer shortly thereafter.[24] While his area of specialty was invertebrate morphology, his expansive interests, Norfolk origins, and taste for the sea led him to whale work, and in 1883 he coauthored a paper on an unusual species of beaked whale that had washed ashore not far from his family home in Norwich.[25] On arriving at the British Museum, Harmer found himself the inheritor not only of a superior collection of anatomical specimens from Cetacea, but also of a strong tradition of interest in the taxonomy, morphology, and evolution of whales and dolphins.[26] In this context, his response upon receiving, in

23. Under the directorship of the zoologist Sir Ray Lankester (with whom Harmer had studied at University College of London in the late 1870s), the position of keeper of zoology had been combined with the directorship, but they were again separated after Lankester's departure. The keepership of zoology remained a stepping stone to the director's office, as evidenced by Harmer's own succession in 1919. For details on these offices, their purviews, and their holders, see William Thomas Stearn, *The Natural History at South Kensington: A History of the British Museum, 1753–1980* (London: Natural History Museum, 1998).

24. W. T. Calman's obituary of Harmer, "Sidney Frederic Harmer, 1862–1950," *Obituary Notices of the Royal Society* 7 (1951): 359–71, provides additional detail on Harmer's youth and early research.

25. See Thomas Southwell and Sidney F. Harmer, "Notes on a Specimen of Sowerby's Whale (*Mesoplodon bidens*) Stranded on the Norfolk Coast," *Annals and Magazine of Natural History*, ser. 6, vol. 11 (1893): 275–84.

26. John Edward Gray, who had served as keeper of zoology from 1840 to 1874, had a strong interest in whales and published a 400-page catalogue of the cetological collections he helped amass; see his *Catalogue of Seals and Whales in the British Museum*, 2nd ed. (London: Taylor and Francis, 1866). The zoologist William Henry Flower, who took over the directorship of the whole museum in 1884, succeeding Richard Owen, also published extensively on whales and dolphins (see, for instance, his edition for the Ray Society of a series of Scandinavian monographs on cetacean anatomy, physiology, and taxonomy: Daniel Frederik Eschricht, Johannes Theodor Reinhardt, and Wilhelm Kukkjeborg, *Recent Memoirs on the Cetacea*, ed. William Henry Flower [London: Robert Hardwicke, 1866]), and he presided over the creation

June 1910, less than a year after taking up his new post, a lengthy letter from J. A. Mørch about the excesses of the southern whale fishery can perhaps be better understood. But even granting that Harmer was a ready audience for the warning, there is little hint or precursor of the extraordinary intensity of Harmer's ongoing subsequent attention to the subject; his unyielding, morally inflected insistence that extermination was under way—and unforgivable—appears sui generis, deeply felt, and largely unique.

The original correspondence between Harmer and Mørch concerned specimens of marine invertebrates, apparently whale parasites, that Mørch forwarded to the museum in the spring of 1910, taking advantage of the enclosure letter to expand on the pressing need for scientific study of the large whales increasingly targeted by the modern whaling industry around the world: "Those of the genus *Balaenoptera* and the humpback whale have had too little attention from the scientific world from a biological point of view considering their present commercial importance and also other interest attaching to these large animals." Raising the specter of depletion, Mørch alleged that the originally prolific grounds near the Bransfield Strait had already seen reduced catches after a mere five years of work by a nascent industry. Where were the whales? What was needed, he asserted, was research into the migration patterns and movements of the animals, work that might involve the collation of reports by gunners and captains; such an information-gathering enterprise might be supplemented by improved record keeping at stations. While some such work had been done in the north by a Norwegian scientist (G. A. Guldberg), Mørch saw little hope for future research in Finnmark and the northern waters, where, he asserted cryptically, "more general investigations would have been attended with certain obstacles"—apparently a suggestion that the national industry of Norway was not likely to come under overly close scrutiny at home. British moves in the Antarctic therefore created, as Mørch saw it, a unique opportunity: "But the case is altogether different now the whaling grounds of the southern hemisphere have become the centers of attraction and subject to the license and under the jurisdiction of the British

---

of the much-admired "Whale Gallery," hung with articulated skeletons of several of the great whales. He considered the study of the origin of whales essential to the study of evolution. See Flower, "On Whales, Past and Present, and Their Probable Origin," chap. 15 in *Essays on Museums and Other Subjects Connected with Natural History* (London: Macmillan, 1898). This essay was originally a lecture at the Royal Institution in 1883.

government." Perhaps now something could be done, and the costs offset by licensing fees.[27]

Harmer's bureaucratic efficiency remains enshrined in his surviving files, which are meticulously ordered and indexed. It was this thorough conception of administrative order that steered the Mørch memo through the interlocking institutions of the Edwardian civil service establishment.[28] On 24 June 1910 Harmer included a discussion of the letter in his quarterly report to the Trustees of the British Museum, who in turn, via the secretary of the establishment, forwarded a letter on 3 May 1911, "with reference to the slaughter of whales and sea elephants in the southern seas," to the Colonial Office, addressed to Lewis Harcourt, secretary of state for the colonies.[29] The Colonial Office replied promptly, forwarding a collection of

27. Mørch to Harmer, 7 June 1910, NHM DF 1004 / CP / 749 / 2, file "Whales—1913–1916 (also 1910–11)."

28. Calman comments that admiring junior colleagues were not above ribbing Harmer for his "precise attention to details" and his penchant for "elaborately indexed" records of everything (see W. T. Calman, "Sidney Frederic Harmer 1862–1950," *Obituary Notices of the Royal Society* 7 [1951]: 366). Sadly, however, Harmer's files are now broken up and sprinkled (more or less uncatalogued) among a number of archives. A brief note on sources follows: Harmer's minutes of the Discovery Committee meetings, annotated and arranged with tables of contents, are currently in the possession of the Scott Polar Research Institute (MS 1284 / 4), where they were deposited by N. A. Mackintosh. The legitimacy of his possession (and conveyance) remains open to dispute, since the National Oceanographic Library at Southampton is supposed to be the repository of the Discovery archives. Also at SPRI are Harmer's minutes and correspondence related to the "Interdepartmental Committee on Research and Development in the Dependencies of the Falkland Islands" (MS 1284 / 1 and 2), which ran from 1918 to 1919 and yielded the 1920 interdepartmental report, and his minutes of the "Interdepartmental Committee regarding Research Ships for Employment in the Dependencies of the Falkland Islands" (MS 1284 / 3), which ran from 1922 to 1923. However, Harmer systematically removed papers from these documents to file them in other series, most importantly the "Scientific Work" files, which were considered lost until I located volumes 1–3 and 17 in the then uncatalogued archive of the Sea Mammal Research Unit (SMRU) at the University of St. Andrews, Scotland (MS 38600, box 644); that institution also holds several files of Harmer's correspondence (boxes 637 and 645). However, by far the largest volume of Harmer material remains at the Natural History Museum, where it is, unfortunately, spread out: in the actual archives there are a large number of his administrative files in the DF series (DF 1004 / CP / 749 / 1ff.), but an enormous additional amount of material, particularly the "official files" on whales and a series of boxed files of cetological work and notes, resides in the autonomous archive of the Mammal Division.

29. See NHM DF 1004 / CP / 749 / 1.

documents—notably, the first few reports by the stipendiary magistrate on South Georgia, along with cover letters by Governor Allardyce—together with a note reassuring the trustees that "the question of the preservation of the whale has already engaged the attention of the Secretary of State," but noting that any real action had proved difficult, owing to "the lack of scientific information and of any international agreement." An invitation followed: "Mr. Harcourt will be glad to cooperate with the trustees as far as possible in the collection of such particulars as might be of scientific and practical value."[30] Forwarded back to Harmer, this letter set in motion more than a decade of his scientific labors on the Antarctic whales—labors that would lead, in the early 1920s, to the Discovery Investigations.

A general summary of the bureaucratic process initiated in this exchange would go as follows: Harmer immediately took on the task of gathering information on the whaling industry in the south, assimilating the catch data into a series of a dozen substantial reports (some of them more than 100 pages) and many shorter ones, which became the primary impetus for an overlapping sequence of governmental committees before and after World War I. To these committees Harmer gave evidence and testimony, and he eventually served on the one that brought forward the recommendation for a large-scale oceanographic research program to investigate the whales of the Antarctic waters. Looking back nostalgically in 1937, about a million pounds later and in the thick of disputes over money and the leadership of what had become the massive Discovery enterprise, Harmer mused to a fellow old hand about the origin of the whole program:

> I look back with much satisfaction to days, before the "Discovery" had been thought of, when I had already begun to take an interest in the protection of Whales. I served on two Committees (I think I am right in saying that there were two) which prepared the way for setting up the 'Discovery' Committee. I think I am right in saying that these came into existence as a result of numerous Reports, on the necessity of taking some action for the protection of whales, which I wrote for the Trustees of the British Museum, who sent them to the Colonial Office.[31]

30. NHM DF 1004/CP/749/1.

31. Harmer to Wordie, 5 June 1937, SMRU MS 38600, box 637, file "Whaling Research Expedition—Discovery Expedition—Harmer Correspondence."

Harmer's own uncertainty about the actual number of committees expresses eloquently the complex sequence of events, but his outline of the process and its origin is entirely fair.

Returning to this origin, and to his earliest exchange with the Colonial Office: Harmer reviewed the documentation forwarded by the office of the secretary of state for the colonies to the trustees in the summer of 1911. It neither persuaded him that the "preservation of the whale" was in safe hands nor allayed his more general concern. He drafted a reply to the trustees of the museum, announcing that "it is impossible to read the correspondence [from the Colonial Office] without very serious misgivings." Seizing on the language of "El Dorado" percolating through the reports of the stipendiary magistrate and the governor, Harmer warned in no uncertain terms of the prospect of a collapse in the industry, advising the trustees that "strong representations should be made to the Colonial Office calling attention to the grave danger of a repetition in the southern waters of the process of extermination which has been so effective in other localities."[32]

Harmer had clearly already developed, by 1911, his own account of the history of the whaling industry in the nineteenth century. This history—a litany of boom and bust cycles in which the discovery of new grounds and techniques brought profitable windfalls, followed by commercial extermination at best and quite possibly extermination in an absolute sense—strongly informed his push for research and regulation at South Georgia. Since this view of whaling's past was by no means universally held (a popular alternative explanation for the same facts posited that whales were particularly spooky and that when persecuted, they rapidly learned new techniques of evasion and sought out new waters), it is reasonable to inquire the source of Harmer's "historical argument." Two likely textual sources can be traced: first, the work of his distinguished cetological predecessor at the museum, Sir William Henry Flower, whose "Whales, and British and Colonial Whale Fisheries" of 1898 amassed evidence that whaling had been a lengthy sequence of exercises in "cupidity" and "ruthless extermination"; and second, the writings of the leading figure in Scandinavian fisheries sciences at the time, the Norwegian Johan Hjort, who by 1902 had composed an unflinching account of the decline of the northern whaling industry since the rise of Foyn's techniques.[33] Hjort, who would focus with increas-

32. Harmer's report to trustees, 14 July 1911, NHM DF 1004/CP/749/1.

33. See William Henry Flower, "Whales, and British and Colonial Whale Fisheries," chap. 14 in *Essays on Museums and Other Subjects Connected with Natural His-*

ing energy on whale research in the 1910s and 1920s, would become a kind of bête noire for Harmer, who did not trust the independence of the Norwegian's judgment when it came to the regulation of the industry; in later years Harmer would regularly cite Hjort's own early works on the northern industry against the Norwegian scientist's later optimism about the durability of the southern stocks.[34] Since Flower had expressed in the 1890s the opinion that "the probability of any large whale being again met with in the Antarctic regions is very remote," Harmer may well have approached the early news of the rise in the southern industry with a heightened sense of the fragility and scale of the stocks.[35]

Harmer also had other, more direct sources of information about the course of the industry: the museum served as a clearinghouse of information on animals around the world, and particularly around the empire. Thus telling queries and correspondence filtered through the hands of the

---

tory (London: Macmillan, 1898), a lecture delivered to the Royal Colonial Institute in 1895. See also Johan Hjort, "Fiskeri og hvalfangst i det nordilige Norge," *Aarsberetning Vedkommende Norges Fiskerier* 4 (1902): 1–251, which was not translated. Harmer, however, was certainly able to make his way though Norwegian texts. I am not certain when he acquired this language, but his Russian was serviceable by the late 1880s (see W. T. Calman, "Sidney Frederic Harmer 1862–1950," *Obituary Notices of the Royal Society* 7 [1951]: 359–71), and he offered a detailed synopsis of Hjort's work by 1917 (SMRU MS 38600, box 637, Harmer's memoranda on whales, binder 1), though it is possible he did so with the help of M. A. C. Hinton, who was at the museum at that time and who did translations of Norwegian whaling material by Risting and Olsen at about this time (see Misc. 298, pp. 180–93; for full citation information on this document, see n. 64 below). The best evidence I have found for his own abilities is in a loose letter preserved in the Sidney G. Brown (SGB) Papers, Harmer to Kemp, 30 October 1930, in which Harmer offers his own translations of several lines from an article on plankton in the *Norsk Hvalfangst-Tidende*.

34. In many ways Hjort's sense of the magnitude of the southern population was closer to the truth, and Harmer's early sifting of the catch reports for evidence of declines in blue and fin catches was quixotic. Where the humpbacks were concerned, however, Harmer's anxiety was well placed. By the late twenties (with the rise of pelagic factory ships) the toll on all the whales of the Southern Hemisphere had risen precipitously, and (nearly) everyone recognized that it could not be maintained indefinitely.

35. William Henry Flower, *Essays on Museums and Other Subjects Connected with Natural History* (London: Macmillan, 1898), 207. A reader familiar with the history of modern whaling may be surprised to find Flower writing of depletion of *southern* stocks at the end of the nineteenth century (by which time *northern* stocks of rorquals had come under very serious pressure from the modern industry), but it is a reminder that there was a widespread sense that the old "southern" whale fishery (the open-boat

zoology keeper. For instance, at the same time that Harmer was assessing the Colonial Office correspondence about whaling in South Georgia, he received from the office of the director of the museum an inquiry from the merchant firm of Brangwin, Clark, Wise & Co., who were seeking information for "a business friend in Durban" on the expanding whaling industry there; could the officials at the museum provide any advice as to the likelihood of "the course that is being taken bringing the business to an early and abrupt conclusion, from the destruction of the whole supply"?[36]

Queries like this expressed concern—increasingly widespread in the period—about potential links between the coastal whaling enterprises of southern Africa and the explosive fishery in South Georgia. A number of commentators broached the possibility that the same whales were being hunted in both places; if this was the case, assessment and regulation needed to proceed accordingly. Harmer placed this issue prominently in his letter to the trustees, and he encouraged them to press the Colonial Office to implement a system for recording data from all the whaling establishments in the colonies.[37] Information on sex, species, and date taken "might do much to settle the important question of whether the schools of humpbacks found off South Georgia have anything to do with those which are hunted in South Africa." If there was evidence of migrations or schooling,

---

pursuit of right and sperm whales—i.e., everything that was *not* the bowhead fishery of Spitsbergen, etc.) had significantly depleted the populations of the great whales in the Southern Hemisphere. Whether that was true (at least as far as rights and sperms are concerned) remains a subject of some dispute; the enormous *rorqual* populations of the Antarctic were clearly unknown to Flower.

36. Brangwin, Clark, Wise & Co. to Lydekker, 11 July 1911, NHM DF 1004/CP/749/1.

37. Reading the sequence of Harmer's reports through 1927 provides a striking view of the way an imperial reach facilitated a global view of a problem in marine science. There is more to be said about the relationship between the British colonies and the sciences of the seas generally (and of whales in particular). Flower's opening remarks to the Royal Colonial Institute on 8 January 1895 are suggestive in this regard: "When asked by your Council to lecture on some subject connected with natural history, it occurred to me that the great link between Britain and her Colonies was the ocean and that, therefore, something concerning its animal inhabitants might be interesting to those whose avocations and situation in life call them to traverse its pathless ways. Even in an ordinary voyage, such as is necessitated by any intercourse between a Colony and the Mother Country, some familiarity may be acquired with the gigantic denizens of the deep." William Henry Flower, *Essays on Museums and Other Subjects Connected with Natural History* (London: Macmillan, 1898), 185.

then perhaps there was segregation in these "herds." If classes could be identified by sex or age, then certain prohibitions could be implemented of the sort "essential to regulate in some way the number killed if the Antarctic whales are not to be decimated in number in the next few years."[38]

The resulting letter to the Colonial Office from the trustees of the museum established the pattern of exchange and collaboration for the decade to come: the whaling reports from the Falkland Islands and Dependencies (and other colonies) would be forwarded from the Colonial Office to the trustees, and from the trustees to Harmer; Harmer would tabulate the returns, graph them, and write up his analysis in the form of a memo to the trustees; the secretary of the museum would then compose a letter to the Colonial Office from the trustees, drawing heavily on Harmer's document.[39] In later years Harmer would refer to the Mørch letter as "the first of my series of reports"—a telling comment on the importance of the

38. Harmer's report to trustees, 14 July 1911, NHM DF 1004 / CP / 749 / 1. The clear analogy here is to the work that had been done on the North Sea herring fishery by Friedrich Heincke, Einar Lea, and Hjort himself. For an overview, see Helen M. Rozwadowski, *The Sea Knows No Boundaries: A Century of Marine Science under ICES* (Copenhagen: International Council for the Exploration of the Sea in association with University of Washington Press, 2002), 81–90; for detailed discussions of the development of year-class accounts of the fluctuation of the fishery, see Tim D. Smith, *Scaling Fisheries: The Science of Measuring the Effects of Fishing, 1855–1955* (Cambridge: Cambridge University Press, 1994).

39. Though Harmer's reports were not generally passed on verbatim until later (the first handled this way was the report of 22 January 1913) and not over his signature until 1916. By 1919, when Harmer replaced Sir Lazarus Fletcher as the director of the whole British Museum (Natural History), his reports were very much his own and went across to Whitehall as such. The manuscript documents at NHM from the early years, however, reveal pencil markings that appear to be the excisions of the trustees or secretary as they modified Harmer's memos into letters of their own to the Colonial Office. In general, these modifications reflect a softening of what might be taken to be stridency or overly specific recommendations for policy action. See Harmer's report to trustees, 19 October 1912, NHM DF 1004 / CP / 749 / 1. The major Harmer reports were (by date): 22 January 1913 (4 pages, no figures); 16 July 1915 (8 pages, 2 figures); 2 July 1917 (61 pages, 31 pages appendix, bibliography, and figures); 17 December 1918 (87 pages, 51 pages appendix and figures); 17 July 1920 (52 pages); 19 May 1921 (29 pages); 28 September 1921 (22 pages, 5 pages appendix and figures [postdated 18 October]); 19 March 1923 (28 pages); 19 July 1924 (47 pages, 2 figures); and the synthetic "final report" of 22 February 1927 (69 pages, 9 pages appendix and figures). I finally turned up the full run of these documents in three sequential binders in SMRU MS 38600, box 637.

letter, to be sure, but also a reflection of its circumscribed place in the larger sequence of activities that gave rise to the Discovery Investigations.[40]

A most significant, but until now apparently overlooked, episode in this sequence was the unsuccessful Swedish effort, beginning in 1912, to undertake a research expedition to study the whales of the southern waters. The importance of this push (which collapsed for good only at the end of World War I) in any history of the Discovery whale studies is twofold. First, it directly stimulated British action—specifically, the research expedition of Major G. E. H. Barrett-Hamilton to South Georgia in 1913–1914, which provided the fieldwork template as well as the empirical basis for the early whale biology of the Discovery Investigations. Second, the Swedish proposals laid the groundwork for the principle of a science of whales riding on the largess and the infrastructure of the whaling industry. This was an essential feature of what would become cetology in the first half of the twentieth century, and it was an arrangement that gave pause to early practitioners like Harmer. The correspondence around the failed Swedish proposal thus affords a unique opportunity to watch the earliest misgivings, considerations, and justifications that percolated around this all-important decision for twentieth-century whale science.

A FATAL COMPACT?

The initial proposal took shape at the hands of J. G. Andersson, a geologist who directed the Geological Survey of Sweden and served as the president of his country's geographic society. Andersson had been second in command of the Swedish South Polar Expedition of 1901–1903, the voyage of the *Antarctic* that had given rise to C. A. Larsen's Compañia Argentina de Pesca. Teaming up again with the commander of that expedition, O. Nordenskjöld, Andersson had begun to pursue private funding for renewed Swedish research in the Antarctic. The separation of Norway from Sweden

---

40. Harmer to Kemp, 23 February 1925, NHM Zoology Collections, box files (Harmer), "Whales and Whaling—Official Papers." This reference nicely balances what has been an overemphasis on the Mørch letter: it was, yes, the start of things, but it was only the first (very small) step in what became an enormous enterprise over the next decade. That enterprise is entirely overlooked by Tønnessen and Johnsen, for instance, who make only a single brief allusion to Harmer (see J. N. Tønnessen and Arne Odd Johnsen, *The History of Modern Whaling* [Berkeley: University of California Press, 1982], 342), and this about a speech he made in 1922!

in 1905, and the waxing Norwegian domination of the increasingly lucrative southern whaling industry, created a nexus of commercial and nationalist interests on which Andersson could draw. By late 1911, he had assembled a set of Swedish backers keen to break into the South Georgia industry, but whose commercial fate lay in the hands of Allardyce and the officials of the Colonial Office, who controlled leases and licenses. These plums were by this time going to a narrow circle of applicants (friends of the governor received particular attention). In the face of these unfavorable odds, the would-be Swedish venture wrapped itself, with the help of Andersson and Nordenskjöld, in the mantle of science. By the end of January 1912, Governor Allardyce found himself considering a novel license application on the letterhead of the Sveriges Geologiska Undersökning, over the signature of the director:

> We have now succeeded in interesting some capitalists in our country, and they would be willing to float a company for carrying on whaling in the Southern seas, having at the same time for their object to give an opportunity for further geographical and scientific researches in those areas where whaling will be carried on. The new company intends especially to devote particular attention to the investigation into hydrographic conditions for the occurrence of whales in different parts of the seas and at different seasons of the year, a problem whose solution may clearly be of the very greatest importance for *a rational exploitation of whaling*.[41]

The language here closely mirrors that of the early planning conference in Stockholm in 1899 for what became the International Council for the Exploration of the Sea (ICES), which similarly espoused "rational exploitation of the sea" and whose members were committed to a research program linking hydrography and fisheries biology.[42] At the same time, the letter expressed a tinge of indignation concerning the explosive exploitation of

---

41. J. G. Andersson to Allardyce, 19 January 1912, SPRI MS 240 / 1–2, file "Dispatches to Secretary of State for the colonies from Governor W. L. Allardyce, 1908–1915." Emphasis added.

42. See the introduction and chap. 1 in Helen M. Rozwadowski, *The Sea Knows No Boundaries: A Century of Marine Science under ICES* (Copenhagen: International Council for the Exploration of the Sea in association with University of Washington Press, 2002), in particular the draft documents for the 1899 meeting, from which the quote is taken (p. 34).

Antarctic resources by citizens of what many Swedes considered a break-away republic—a massively profitable enterprise from which the Swedes themselves, who could claim precedence, had been excluded:

> We venture to point out that the entire whaling industry now going on and financially so successful, traces its origin back to the experiences gained in the course of our expedition; and that the new Swedish Whaling Enterprise intends methodically to advance future scientific explorations that may be of importance for a rational utilization of the natural resources within your colony.[43]

Further letters of support and clarification followed, detailing a plan for an elaborately "systematic research into the life of the whales and their distribution in various parts of the Antarctic Ocean at different seasons, their dependence upon the occurrence of different kinds of food, and the hydrographic factors, which determine the occurrence of whale food in the sea." With all this in the offering, which would, it was often repeated, be of "the very greatest importance for a rational exploitation of whaling," Andersson swelled to his theme: "In the interests of *science*, it would therefore be most desirable if we could get our two applications granted."[44]

Allardyce, who had just had a difficult time getting Colonial Office approval for an additional South Georgia lease that he hoped to confer on a local favorite (and who already had the British Museum sniffing inquisitorially at the administration of the industry), was in no mood to be wooed by Swedish whalemen bearing the gift of rationality. The request was denied. But the formal and informal networks of communication and administration began to spark. Further representations of the plan, forwarded by Andersson to the Colonial Office in London in March 1912, were in turn sent on to the British Museum and directed to Harmer's attention. In this way Harmer learned of a sweetened proposal, by which "the proposed whaling company" funded by a "group of Swedish businessmen," promised, in return for licenses, not only to afford transportation and communications facilities to an Antarctic scientific expedition (complete with land stations), but even to "set aside annually a considerable portion [10%] of the expected

---

43. Rozwadowski, *The sea Knows No Boundaries*, p. 34.

44. J. G. Andersson to Allardyce, 7 February 1912, SPRI MS 240/1–2, file "Dispatches to Secretary of State for the colonies from Governor W. L. Allardyce, 1908–1915." Emphasis added.

profits on the whaling operations, which reserve would form a fund for continuous and systematic exploration of the parts of the Antarctic region in question."[45] The fund would be managed by the Geographical Society of Sweden and would be used in large part for whale research: "their migration, their food, their dependence on hydrological factors, their rapidity of propagation, etc."[46]

Soon Andersson and Nordenskjöld had letters out directly to the secretary of the British Museum, explaining their revised suit pending at the Colonial Office and suggesting the possibility of collaboration with British naturalists should the plan go forward (perhaps with the support of the museum's well-connected directors). In this letter, the issues of depletion, whale biology, and stock analysis received much more detailed attention:

> It may of course be questioned whether the present large development of the whaling industry will not rather be of detriment to the stock of whales in the southern seas. But this is a question which can only be answered by careful investigations, and should whaling in any special detail prove detrimental, this might be obviated through practical legislation. An any rate, a careful research into the occurrence of the whales, their biology and migrations, will be of exceedingly great and practical significance, and it has always been our intention to carry out such studies.[47]

The irony of a whaling expedition setting out with the stated aim of reducing the "risk of decimating the shoals of whales" was not lost on Harmer, whose response to these overtures was to sound his acquaintances

45. Nordenskjöld and Andersson to Harcourt, 5 March 1912, NHM DF 1004/CP/749/1. The tone of this letter shifted firmly to the plaintive: "we venture . . . to remind your Excellency of the fact that the expedition led by us in 1901–1903 perhaps more than any other expedition to the said regions contributed toward their scientific exploration, and that this expedition has given rise to the extensive whaling operations now being carried on in the Antarctic seas, for which reason it might perhaps not be too bold to respectfully bid for a place also for Swedish subjects to participate in this whaling industry."

46. Ibid.

47. Nordenskjöld and Andersson to Fagan, circa March 1912, NHM DF 1004/CP/749/1.

in Scandinavian science for their assessment of the value, virtue, and seriousness of the plan.[48] Expressing his misgivings about buying whale science with whale oil, he solicited in particular the opinion of the distinguished Swedish zoologist and paleontologist Einar Lönnberg, the director of the natural history museum of the Swedish Academy of Sciences. Lönnberg (who had written and lectured on cetology) replied promptly, spelling out the urgency of the situation in no uncertain terms: "The result of the present state of affairs will be that the whales sooner or later will be exterminated, or at least so decimated that whaling no longer pays the expenses, which will probably be identical with the complete extermination of the southern right whale (*Balaena australis*) and the southern humpback (*Megaptera lalandii*)."[49] From such an arrangement, he pointed out, "nothing is gained," and he cited the well-worn adage of the premature death of the hen that laid the golden eggs. Reviewing the Swedish proposal, Lönnberg admitted that he saw the grounds for Harmer's concern, but he inclined toward support nevertheless:

> If on the other hand a new license is issued on the condition that a certain sum is paid for scientific researches, among which ought to be stipulated a thorough study of the life history of the whales, their habits, food, propagation, time of gestation, wanderings, etc.—the whales are pursued and killed by still another company, this is true. The extermination of the whales would seem to be hastened still more by this . . . but knowledge about the whales and scientific material has been won.

48. In fairness, Andersson and Nordenskjöld themselves tried in their letter to allay concern about putting more pressure on the stock, noting that adding two more catchers when there were already more than fifty working British waters in the Antarctic could add only a small percentage to the annual take.

49. The species designations of these "southern whales" have been superseded. The southern right whale is now known as *Eubalaena australis*, and the humpback (*Megaptera novaeangliae*) is treated as a cosmopolitan species. It is perhaps worth noting that the resolution of the question of specific identity hinged on the southern researches that were here being debated, though by the time they began, humpbacks and rights were immensely rare on the South Georgia grounds. For a sense of Lönnberg's whale research, see his "Om Hvalarnes Harstamning," *Svenska Veteskaps-Akademiens Årsbok* (1910): 219–59, translated by Mark D. Uhen as "On the Origin of Whales," accessed 11 March 2011, www.paleoglot.org/files/Lönnberg 10.pdf.

And the great appeal of this knowledge was that it provided the foundations of a regulatory regime:

> I have no jurisprudence, but I think it is clear that, since a certain knowledge of the whales has been won, the British government could say to the holders of the licenses: "This wholesale butchering cannot go on any longer, for the maintenance of the stock of whales certain regulations and restrictions must be made."[50]

It is clear from Harmer's notes and correspondence files that he struggled mightily in assessing the costs and potential benefits of the proposed arrangement. In drafting his recommendations to the secretary of the museum, he weighed his words, acknowledging that "I feel very reluctant to lend support to any scheme which might have the effect of hastening the process of extermination," but going on to say that "after very careful consideration, I think the Museum might support the Swedish scheme."

It was a case, in his judgment, of the ends justifying the means. Or, at least, there was a reasonable chance they might: "Though we might be unwilling to lend our influence to the establishment of yet another center of destruction, the information which would be gained thereby is so important that it would probably more than compensate for the number of whales killed." At the heart of the matter, for Harmer, was the issue of the leverage of his own position and that of the museum as a scientific establishment. As he put it to the secretary, the repeated communications from the trustees of the museum to the Colonial Office concerning the need for protection were obviously well-intentioned, but if the officials there actually *took* the recommendation and asked the museum to advise formally on a regulatory policy, the situation would be awkward, because "we should find ourselves in the position of having nothing very definite to say." Since it was unclear to Harmer how any "sound advice on the subject can be given, except as the result of a study of the problem on the spot," the question was how to underwrite such a study, and the Swedish proposal had the virtue of "offering a way of having the scientific work done on a self-supporting basis. . . . I doubt whether it could be done equally well any other way."

This was not wholly true. Harmer *could* imagine a better arrangement: the same setup, conducting the same investigations, but run as a "completely British organization." He had to admit, ruefully, that "it would, how-

---

50. Lönnberg to Harmer, 20 April 1912, NHM DF 1004/CP/749/1.

ever, be very difficult to do this now, without laying oneself open to the charge (which could not be denied) of having taken the Swedish ideas and of having used them for our own purposes."[51]

In fact, this constitutes a superb précis of how the Discovery Investigations actually began. Not that the Swedish scheme was purloined all at once, or even explicitly; rather, the elements of the program were gradually enacted by the Colonial Office, with Harmer's support, even while talk of future collaboration (and even an "Anglo-Swedish expedition") kept a hopeful Andersson and Nordenskjöld from complaining. As World War I descended and diplomatic relations between England and Sweden soured, the grounds were laid for a British abandonment of the joint scheme.[52] By 1919 a slightly testy Harmer—whose eldest son sustained serious injuries in France in 1918—could write to Nordenskjöld that any prospect for collaborative work was stone dead. It was too bad about the war, yes, and the missed opportunities for cooperation on whale research (which was needed more urgently than ever), but "a scheme has accordingly been in preparation for some time, by which it is hoped that the British Government will take up the question by itself, without external assistance."[53] Harmer himself, by this time, was serving on the committee drawing up the plans for a major expedition, and plans were in the works to tap the whaling industry as a whole for the program by means of export duties on whale oil shipped from the Falkland Islands and Dependencies. The kind of arrangement proposed by the Swedes in 1912 was, by 1920, being enacted on the grandest scale.

The war made Harmer much less deferential to his Scandinavian colleagues; back in July 1913 he had been considerably more delicate when he found himself in the awkward position of needing to explain to Lönnberg and Andersson that the immediate plan for some sort of joint Anglo-Swedish study of the whales of the Antarctic had to be put on hold because the "Colonial Office . . . have decided to send out a Naturalist to South Georgia, etc. to make enquiries and to report to them." Though Harmer

51. All of the above quotes are from Harmer to Fagan, 24 June 1912, NHM DF 1004 / CP / 749 / 1.

52. On the relationship between Britain and the Scandinavian countries in this period, I have found the following work helpful: Patrick Salmon, *Scandinavia and the Great Powers, 1890–1940* (Cambridge, Cambridge University Press, 1997), particularly chap. 4, "Economic Warfare and the Northern Neutrals 1914–1918."

53. Harmer to Nordenskjöld, 4 March 1919, SMRU MS 38600, box 645, file "Anglo-Swedish Antarctic Station."

tried to separate this sudden move from the pending Swedish proposal—even suggesting that this new development "would be quite in the direction of helping forward the larger scheme in which you and your colleagues in Stockholm are interested"—he also found himself obliged to apologize for letting them know so late that independent British plans so much like those proposed by Andersson had developed so far, even as schemes for preliminary joint Anglo-Swedish expeditions had advanced to detailed logistics (including how to divide up the study of the region and how to share data on migrations and breeding biology). The Colonial Office naturalist, Major G. E. H. Barrett-Hamilton, would be leaving for the Antarctic in a mere matter of weeks. "I should have liked to have been able to tell you what was proposed at an earlier stage," Harmer wrote to Lönnberg, "but the matter was confidential, and I could say nothing."[54]

Since the Barrett-Hamilton expedition clearly aimed to resolve the very questions on which Lönnberg and Harmer had by this time been corresponding for more than a year, Lönnberg's reply was cool. "The cooperation concerning the biology of the Antarctic Whales has thus to a degree already been checked," he pointed out, and added with some restraint that "it is perhaps the best thing that you take the whole thing in your care."[55] Harmer made a number of efforts to repair the damage, including trying to arrange for Barrett-Hamilton to be escorted by a Swedish naturalist, but the Swedes, piqued, turned down the tentative offer before it could be firmed up. And while Harmer continued over the next several years to promote some form of collaborative Anglo-Swedish Antarctic research station—even raising some funds from the trustees of the museum, the Royal Society, and the Zoological Society of London for that purpose—this ongoing effort cannot but look, in hindsight, a bit like Harmer's attempt to conjure up a consolation prize for Lönnberg and the other Swedes, a gift that might help him assuage a conscience made slightly uneasy by his hand in the British expropriation of the whale research plans of 1912.[56]

And he did have a hand in that expropriation. Although he never wholly

54. Harmer to Lönnberg, 16 July 1913, in ibid.
55. Lönnberg to Harmer, 18 July 1913, in ibid.
56. The bulk of this fund-raising work happened between March 1914 and August, when war brought plans to a stop (see below). The sequence of letters and proposals related to this research station can be found in the continuation of SMRU MS 38600, box 645, in the file entitled "Anglo-Swedish Antarctic Station."

put aside his misgivings about commandeering the Swedish plan and placing it under British aegis—writing to the trustees in early 1913 that the Swedes had a "sort of prescriptive right to participation" in the research, given that they had been "the first to suggest a means of carrying out the work required"—his concerns about the private whaler funding never dissipated: "By consenting to it," he pointed out in a lukewarm internal endorsement at the museum, "the Trustees may be held hereafter to have *assisted in the extermination of the whales*."[57] When the whaling capitalists in Sweden revised their proposal, limiting their funding to a fixed amount (£2,000), rather than a percentage of profits, and seeking a five-year embargo on the publication of scientific results of the whale research (presumably to allow their company the time to profit from whatever might be learned about migration patterns while also forestalling regulatory action based on the findings of the research), it became much easier to balk at the whole arrangement.[58] Hence, when in April 1913 Lewis Harcourt, the secretary of state for the colonies, wrote to the secretary of the museum offering £600 to support "a marine biologist to visit South Georgia and possibly the South Shetlands during the ensuing Antarctic summer" in order to undertake an initial study of whales, Harmer jumped at the opportunity, noting privately in a museum memorandum that "although closely related with the Swedish scheme," he did not feel this study was directly in conflict with it, and that it would be possible to avoid the appearance of a usurpation.[59] Government funding was strongly to be preferred, in Harmer's view, and he had himself played an active role over the intervening months in pressing the Colonial Office to take on the responsibility of the research. In January, in his report on whaling in the British Empire (which had been forwarded to Harcourt), he had written that nearly 10,000 whales appeared to have been killed in the last Antarctic season alone, and that companies working at such a pace were "probably capable of exterminating whales by their own unaided efforts." Without better information it was impossible to say for certain if these rates were excessive, and therefore "[a] scientific examination on the

57. Harmer's report to trustees, 20 January 1913, NHM DF 1004/CP/749/1. Emphasis added.

58. For these details of changes to the plan, see Andersson to Harmer, 19 December 1912, in ibid.

59. Harcourt quote: Harcourt to Fagan, 15 April 1913, NHM DF 1004/CP/749/2. Harmer quote: Harmer's report to trustees, 22 April 1913, in ibid.

spot, carried out by capable naturalists might do much" to resolve the uncertainty.[60] In his report to the trustees of the museum at about the same time, Harmer went even further, pointing out that "the British Government is . . . under the strongest *moral obligation* to do all that can be done for the protection of whales."[61] To buttress this extraordinary claim, Harmer adduced language in the 1911 edition of the *Colonial Office List*, which noted that in 1832, the Falkland Islands "were taken possession of by the British Government for the protection of the whale fishery."[62] Since South Georgia and the other islands were dependencies of the Falklands, he continued, the government ought to consider itself bound by this original charter in the region. It is not clear from this memo whether Harmer understood that "protection of the whale fishery" here meant protection of the British *industry* (to wit, access to the Pacific for the prosecution of the booming southern sperm fishery of the 1830s, when competition with the United States was intense), not protection of the *whales*, but it is likely that he did, and that he elided the issue: protection of the whales was, after all, a kind of protection of the industry.

EARLY SCIENTIFIC INVESTIGATIONS

The Colonial Office's support for the Barrett-Hamilton whale biology expedition of 1913–1914 was directly related to an emerging consensus among several branches of the British government that preparations needed to be made for the convening of a diplomatic conference that would seek international consensus on whaling regulations. The problem of a "lack . . . of international agreement" preoccupied the Colonial Office officials from the start of their correspondence with the British Museum (and Harmer) in 1911. Without such agreement, unilateral regulatory actions were not only unlikely to be effective; they risked driving a mobile industry elsewhere.[63] Over the next two years, dozens of letters on whales and whaling circulated among the executive offices of the British Empire and its diplomatic outposts. This correspondence traversed a wide web of infor-

---

60. "Protection of Whales—memorandum from B.M.(N.H.) to C.O.," 22 January 1913, SMRU MS 38600, box 637, Harmer's memoranda on whales, binder 1.

61. Harmer's report to trustees, 20 January 1913, NHM DF 1004/CP/749/1. Emphasis added.

62. Ibid.

63. Colonial Office to "B.M.(N.H.)," circa July 1911, NHM DF 1004/CP/749/1.

mants and interested parties, including colonial secretaries in the Falk-lands, Africa, Australia, New Zealand, Newfoundland, Canada, India, and elsewhere; the Foreign Office; the Board of Trade; the British Museum; and consuls and attachés in some two dozen foreign nations. All of this material passed through the hands of E. Rowland Darnley in the Colonial Office, who became the administrative center of the effort to get control of the scope of the industry in the empire and beyond.[64] Out of these exchanges, Darnley consolidated the legal provisions bearing on whales in the United Kingdom, the Dominions, the Colonies, and "uninhabited Possessions,"

64. The most important elements of this correspondence, with ancillary documen-tation, were reprinted by the Colonial Office in a pair of "Confidential" briefs: *Cor-respondence [October, 1911–December, 1913] respecting the Protection of Whales and the Whaling Industry*, Miscellaneous no. 278, March 1914 (hereafter Misc. 278) and *Further Correspondence [January 1914–March 1915] relating to Whaling and the Protection of Whales*, Miscellaneous no. 300, October 1915 (hereafter Misc. 300). The change in title is interesting, suggesting as it does that the "protection of the whaling industry" fell out of favor as a way of describing the enterprise, at least within the Colonial Office, immediately before the war. I have been surprised at the absence of these documents, and the accompanying—and in some ways even more valuable—Miscellaneous No. 298 (*Inter-departmental Committee on Whaling and the Protection of Whales, Minutes of Evidence* [the hyphen fell out of use in the designation "Inter-departmental" over the next five years], October 1915, also "Confidential"; hereafter Misc. 298), from the literature on the history of modern whaling. They do not appear in J. N. Tønnessen and Arne Odd Johnsen, *The History of Modern Whaling* (Berkeley: University of California Press, 1982); Ian B. Hart, *Pesca: The History of Compañia Argentina de Pesca Sociedad Anónima of Buenos Aires: An Account of the Pioneer Modern Whaling and Sealing Company in the Antarctic* (Salcombe: Aidan Ellis, 2001); Rosalind Marsden, "Expedition to Investigation: The Work of the Discovery Committee," in *Understanding the Oceans: A Century of Ocean Exploration*, ed. Margaret Deacon, A. L. Rice, and C. P. Summerhayes (London: University College of London Press, 2001), 69–86; Robert Headland, *The Island of South Georgia* (Cambridge: Cambridge University Press, 1984); or Gordon Jackson, *The British Whaling Trade* (London: A. and C. Black, 1978). In fact, I cannot find them cited any-where. I discovered two of the volumes, Misc. 300 and Misc. 298, in the private collection of SGB (almost certainly Mackintosh's personal volumes), and there are also copies of the set (labeled as Borley's) at NOL GERD.B5/3. At the National History Museum (which does not hold copies), there is excellent evidence for how hard these documents are to secure: taped in the front of NHM DF 1004/CP/749/1 is an interlibrary loan form filed from the British Library (which also appears not to hold copies) requesting help from the NHM librarians to fulfill a patron request for Misc. 278; the form dates from 1985. Even so, Misc. 298 appears to have been declassified in October 1932, when Kemp and Borley agreed to share it with an American researcher, F. G. Benedict, who wrote from Washington, DC, to request a copy. See SMRU MS 38600, box 636, Disc. 18736.

assembled an overview of the capital invested in the industry around the globe, and reviewed up-to-date lists of known cetacean species provided by the British Museum. By 19 March 1913 Darnley, using these materials, had drafted a four-page "Memorandum" outlining the formation of an "Inter-Departmental Committee to consider measures to be laid before an International Conference with a view to the protection of whales and the regulation of the whaling industry."[65] The idea was to set in motion the internal British process of preparing for a whale summit modeled on the recent (successful) International Fur Seal Conference of 1911.[66] On 12 August 1913 this first of a series of whale committees met for the first time at the Colonial Office, equipped with this mandate.[67]

By that time, Major Barrett-Hamilton was in the thick of preparations for his departure for South Georgia, having his commission in hand to undertake a program of researches on the whales of the Southern Hemisphere at Colonial Office expense. Darnley reported to the assembled committee members at their inaugural meeting that this "biological mission" was already under way, and that Barrett-Hamilton "would probably be available to give evidence before the Committee about May next year."[68] As it turned out, in February 1914, Barrett-Hamilton's body, steeped in formaldehyde, encased in a leaden shell, and wrapped in a large British ensign, would be shipped back to Liverpool in the ship *Orwell*.[69] His death (from cardiac arrest) at South Georgia after a stay of only two months represented a considerable blow to the plan for British whaling research and gave a preliminary indication of the rigors of the fieldwork that would be involved in Antarctic cetology.[70] In an effort to salvage the investment (both his and

65. For the minutes of this meeting, see Misc. 278, pp. 67–68.

66. There was also a precedent in the Scottish inquest into the whaling industry in 1908.

67. The first meeting was attended by James Maurice, Lamb, Sperling, Baker, Darnley, and Williams, secretary. Missing: Holt and Vernon.

68. Misc. 278, pp. 67–68.

69. For details on the repatriation, see James Innes Wilson to Secretary of State for the Colonies, 21 January 1914, SPRI MS 1228 / 14 / 4, file "Biological Mission." This file and the others in this SPRI register are the Magistrate's Archive for South Georgia, which provides a valuable view from the Dependencies of the activities of the London committees and the British Museum.

70. The cause of death was given as heart failure, but his two months of intense physical labor on the flensing platform at the Leith Harbour whaling station clearly took a toll.

theirs), arrangements were made through Harmer, Oldfield G. Thomas, FRS (a personally wealthy old hand in the Zoology Department), and officials at the Colonial Office to convey Barrett-Hamilton's surviving notebooks and records to one of his colleagues, M. A. C. Hinton (who would go on to coordinate a great deal of whale research at the British Museum in the 1920s), who found that they were in good enough shape to be worked into a 150-page confidential report. The final version of this report, drafted in less than a year, came into the hands of the Inter-departmental Committee in April 1915, becoming a major portion of the committee's evidentiary brief.[71] Because the Barrett-Hamilton/Hinton report constituted the point of departure for the postwar whale work of the Discovery Investigations, it will be worth taking a moment to look at Barrett-Hamilton's fateful expedition, and the document it generated, in more detail.

Major G. E. H. Barrett-Hamilton had little expertise in whale biology

71. The publishing history of this report is complicated enough to merit a note: Hinton, who had been working with Barrett-Hamilton on his incomplete "History of British Mammals," and who was a familiar in the mammal section of the British Museum, reviewed Barrett-Hamilton's scientific effects (rough field notes, a summary of whale measurements, his reading notes on Cetacea, notes on specific whales, his journal, and several copies of his dispatched letters to Fagan and Harmer) in March 1914, and by April had taken on the assignment to work up a report based on these materials. He did so, initially, on a volunteer basis, but Harmer secured him an honorarium of £25 in the first instance (I am not sure if he received additional payment, but it seems likely, given the amount of work he ended up doing). By 10 July 1914 Hinton had put together a ten-page "Preliminary Memorandum on the Papers Left by the Late Major Barrett-Hamilton, Related to the Whales of South Georgia," and by March 1915 it had been expanded into a 150-page final report. Both documents were printed in "Confidential" Misc. 298 in October of that year. Misc. 298 was of considerable importance to the work of the Interdepartmental Committee on Research and Development in the Dependencies of the Falkland Islands, which began meeting in April 1918 and issued, late in 1919, a major public report calling for a large-scale oceanographic expedition to study the whales of the Southern Hemisphere. (This document, spelling out the structure of what would become the Discovery Investigations, was printed in April 1920 as Cmd. 657, Colonial Office [United Kingdom], *Report of the Interdepartmental Committee on Research and Development in the Dependencies of the Falkland Islands* [London: His Majesty's Stationery Office, 1920], henceforth Cmd. 657.) Harmer, who served on this committee (and its successors), cited Hinton's report in his "Memorandum on the Present Position of the Southern Whaling Industry," app. VIII of Cmd. 657; the citation gave rise to calls for access to Hinton's work, and in 1925, when the Discovery Investigations were under way, the Colonial Office obliged, printing *Reports on Papers Left by the Late Major G. E. H. Barrett-Hamilton relating to the Whales of South Georgia* for public use.

per se (there were very few people in the United Kingdom with such hands-on expertise at this point), but he was a hardy and avid field naturalist and a recognized authority on mammalian zoology. He had been at work for some time on a history of British mammals, and he had considerable experience with Arctic research on seals, having served as assistant to the prominent Scottish marine scientist D'Arcy W. Thompson during his appointment as delegate to the international negotiations over the North Pacific fur seal trade several years earlier.[72] Moreover, Barrett-Hamilton was personally acquainted with a number of individuals at the British Museum, particularly the then-assistant secretary, C. E. Fagan, and it was this connection that resulted in his commission. Barrett-Hamilton's manuscript journal of the South Georgia expedition begins with a brief summary of the serendipitous encounter that led to his appointment:

> This trip arose out of a conversation which I had with C. E. Fagan, Assistant Secretary BMNH, whom I met on the Waterloo station platform when proceeding to Aldershot to join the 2nd R.I. Rifles for the maneuvers of 1912. In May Fagan wrote me offering an appointment under the Colonial Office to proceed to South Georgia on a mission as marine biologist, to investigate certain questions connected with the whale fishery there.[73]

Barrett-Hamilton accepted and began reviewing Colonial Office documentation on the southern industry while preparing a work plan in consultation with Harmer.

Although there seems to be no surviving manuscript that explicitly lays out the research program of the expedition, it is possible to reconstruct

---

The text (and even the pagination) is identical to that in the confidential Misc. 298. For an explicit discussion of Harmer's decision to cite the Hinton report in Cmd. 657, see his letter to H. T. Allen of 20 May 1919 in SPRI MS 1284/2, in which Harmer expresses surprise that the 1915 "Minutes of Evidence" (Misc. 298) would not be published as part of Cmd. 657, but rather would be "suppressed," as he put it.

72. On his unfinished work, see Harmer's report to trustees, 4 April 1914, NHM DF 1004/CP/749/2; on work with Thompson and the fur seal negotiations, see Harmer's report to trustees, 27 May 1913, in ibid.

73. This manuscript is held at the NHM Zoology Library as 89qH: "Journal of a Trip to South Georgia by G. E. H. Barrett-Hamilton." As far as I can tell, it has not hitherto been the subject of scholarly attention. A copy of Barrett-Hamilton's acceptance letter to Fagan from 25 May 1913 can be found at NHM DF 1004/CP/749/2.

what Barrett-Hamilton set out to do at South Georgia from several sources: first, there are his accounts of his fieldwork and Hinton's summaries of the data he collected; second, there are Harmer's contemporaneous writings and testimony about the biological problems that needed to be solved in order to advance a regulatory scheme for the southern whales; and finally, there is an inventory of the books and papers found in Barrett-Hamilton's possession upon his death, which provides insight into his reading and thus gives a picture of the literature to which he saw himself working to contribute.

Most significant in this last category is the reference to a copy of "Cock's [*sic*] Fin Whaling" that rested in Barrett-Hamilton's trunk (beside a copy of Wilde's *De Profundis* and a *Life of Frank Buckland*, suggesting the diversity of the major's tastes). The magistrate on South Georgia, James Innes Wilson, who prepared the inventory, did not specify which of A. Heneage Cocks's articles on the fin whale fishery of the northern waters Barrett-Hamilton had to hand, and there are several possibilities.[74] It is, however, not essential to know exactly which, since all Cocks's pieces were similar: they were zoological investigations of the large rorquals being taken by the catcher boats of the modern Finnmark whaling industry of the late nineteenth century. Taken together, his papers were concerned with the size and reproductive condition of 284 fin whales captured in the northern waters. Tallied body measurements offered a point of departure for work on the specific identity of the whales (Were they identical to those taken in the western Atlantic? If anatomical indices were disparate, this was strong evidence that they were not); data on fetuses found in females promised to reveal important information about the life cycle of the animals (Were all the fetuses about the same size at any given time? If so, this pointed to a definite breeding season. Did they grow throughout the season in a manner that could be plotted? If so, the growth curve could be used to estimate the timing of that season). This latter effort to collate fetuses by size and date in order to reconstruct the breeding cycle was the richer and more novel aspect of this research, and here Cocks was following a technique pioneered by his contemporary, G. A. Guldberg (deceased by 1913), who had first at-

74. For instance: "The Finwhale Fishery on the Coast of Finmark," *Zoologist*, ser. 3, vol. 8 (1884): 366–70; "Additional Notes on the Finwhale Fishery on the North European Coast," *Zoologist*, ser. 3, vol. 9 (1885): 134–43; "The Finwhale Fishery of 1885 of the North European Coast," *Zoologist*, ser. 3, vol. 10 (1886): 121–36; "The Finwhale Fishery of 1886 on the Lapland Coast," *Zoologist*, ser. 3, vol. 11 (1887): 1–16.

tempted to use whalers' catch data to ascertain the gestation period of the North Atlantic great whales.[75]

Barrett-Hamilton's investigation represented an extension of this kind of work to the Southern Hemisphere. Measurements were the essential daily undertaking, along with the search for fetuses and the examination of the internal reproductive organs of each carcass.[76] For instance, in one of his first communiqués from South Georgia, Barrett-Hamilton wrote back to Fagan on 2 December, after examining a total of nine fins, two blues, and two humpbacks over two slow weeks at the station. Remarking that the blues and fins appeared to be, on average, considerably larger than the figures given for the northern species (though more specimens were needed to say this with confidence), he went on to offer a preliminary discussion of the investigations into reproductive biology:

> The two female [finners] which I have examined each contain fetuses of about 5 feet long and no indication of suckling calves. The breeding season thus corresponds to that in the north, but the fact suggests that a female does not pair until after she has reared her calf. This means slow propagation, perhaps a calf every second year. As in the north, males are smaller than females.

He then went on to give similar information on the humpbacks:

> Of *Humpbacks* I have examined two females, and they recently weaned calves and contained embryos respectively of 6 and 9 inches. The

75. See G. A. Guldberg, "Bidrag til Cetaceernes Biologi. Om fortplantningen og draegtigheden hos de Nordatlantiske Bardehvaler," *Christiania-Videnskabs-Selskabet-Forhandlinger*, no. 9 (1886): 56. It is Hinton who acknowledges Guldberg's priority in this area. See also the discussion in Jens Andreas Mörch, "On the Natural History of Whalebone Whales," *Proceedings of the General Meetings for Scientific Business of the Zoological Society of London* 2 (1911): 661–70, particularly p. 663. I have not consulted Guldberg, but am basing my discussion of the work on Harmer, Mörch, and Hinton. I will discuss Harmer's development of the techniques for statistical work with fetuses below, but it is worth noting that it appears neither Cocks nor Guldberg actually collected their data themselves, relying instead on whalers at the shore stations. In this sense, Barrett-Hamilton's field project was distinctive.

76. Stomach contents, internal and external parasites, and other collections and observations were of lesser importance, but also received attention.

Humpback therefore breeds later than the Finner (as in the north) &, again, like the Finner, seems to have a calf only every second season.[77]

From Harmer's internal memos and his correspondence with the Scandinavian scientists, discussed above, we have already seen the larger importance of this kind of information: breeding details were essential to establishing the fecundity of the stock, and specific identity was the point of departure for any effort to establish how large that stock was. Along these lines, the very clearest statements of what Barrett-Hamilton was supposed to discover and why can be found in the invaluable transcript of Harmer's scientific testimony before Darnley's Inter-departmental Committee on Whaling and the Protection of Whales.[78] Called as the first witness to give evidence, Harmer appeared before the eight-member committee on 10 December 1913 to answer questions about the state of scientific knowledge of whales and how such knowledge might be deployed in regulation of the industry. Again and again Harmer was obliged to acknowledge how little was actually known, and several times he explicitly mentioned how much depended on the work of the biological mission currently conducting its investigation. For instance, the very first question—posed by the committee's chairman, H. G. Maurice, from the Board of Agriculture and Fisheries—spanned a range of life history questions: "Have we got any certain knowledge about the rate of reproduction of whales, the period of gestation, and things like that?" To which Harmer replied, "My impression is that we have no certain knowledge, and that opinions have been very widely different on that particular subject." The same could be said, he added, for age determination, age at sexual maturity, and migratory habits—all these basic issues remained to be worked out.

Reviewing the species lists prepared for the committee by the British Museum, Maurice pressed on the issue of specific identity and its connection to movements and stock analysis:

MAURICE: I fancy we have no very complete knowledge about the area of migration of whales, which again bears on the question of protection, and it seems to me that if there was some uncertainty about the

---

77. Barrett-Hamilton to Fagan, 2 December 1913, NHM DF 1004 / CP / 749 / 2.

78. Presented in Misc. 298, pp. 3–9, and based in part on Harmer's memorandum to the committee, dated 7 November 1913 and reproduced as #41 in Misc. 278.

precise identity of different species (I mean their particular identity as distinguished from others very like them), and if it might be the case that a number of whales were called by different names which were in fact the same species, we might discover that there was more inter-migration than one had previously suspected?[79]

To which Harmer replied that he agreed, adding, "I do not think that we know quite definitely at present whether the Antarctic species are the same as our own," and that this question ought to be possible to resolve "by col-lecting material which could be compared, and partly by making use of the services of some person like Major Barrett-Hamilton, as is being done at the present time, to examine the question on the spot."

When the direct question of scientific advice for regulation came up at the end of Harmer's testimony, he again gestured to Barrett-Hamilton's investigation:

MAURICE: I think the last point I want to ask you about is a point blank question to begin with: Can you suggest a satisfactory method of pro-tection for whales or alternative methods? I mean from the scientific point of view: we know the administrative difficulties would have to be combated if possible?

HARMER: I feel I would rather not commit myself very definitely on that subject, because I really have not got the necessary information. I hope that when this mission of Major Barrett-Hamilton's returns he will be about to tell us a little more about it. I think it would be very desirable to know what he had to say on the subject.

So, asked another committee member, seeking clarification, "until more is known about the habits of the whales it is almost impossible to say what form of protection is best?" And Harmer conceded that this was his view, prompting the chairman to throw a conundrum back to the witness:

MAURICE: If I might put the point to you, the British Museum authori-ties began by pressing for protection for whales, and we look rather as if we were to move in a vicious circle. We do not know enough about the nature of the methods of protection to apply, and on the other hand we are told, and there is a good deal of evidence for it, that the

79. Misc. 298, p. 4.

rate of destruction is such that the matter is extremely urgent. If you are to start on an investigation, and to get the information we have not got, you will find that there are no whales to investigate on your own hypothesis really?

HARMER: Yes, we have been saying, at any rate to ourselves, for some time past, that we completely realise the urgency of the question, and we have foreseen that we might be asked by the Colonial Office what remedies we should suggest. That has made us extremely anxious to get observations made of the kind that Major Barrett-Hamilton is carrying out now, in order to be able to be in a position to give some satisfactory answer which at present I do not think we are able to do.[80]

Maurice had here touched a point on which Harmer was acutely sensitive. Indeed, so great was his concern about the need for preservation (despite the inadequate basis of scientific knowledge) that initially, he had even warned the trustees of the museum against getting involved in the governmental committees proposing to investigate the whaling issue. As early as the autumn of 1912, he was already asserting that "It is in fact difficult to avoid fearing that the mischief [large-scale depletion of the southern whales] has already been done," and he therefore suggested that it might be better for the museum to stay out of any too-little, too-late Colonial Office efforts. The danger of participation in any such committees, he warned, was that they "may be regarded hereafter as bearing part of the blame, if the anticipated result [extermination] should result." His prophecy—misguidedly premature, to be sure—is nevertheless remarkable in light of the fate of the International Whaling Commission after World War II, whose reputation suffered in just this way.[81]

Given that his mission was under the indirect guidance of Harmer and the Zoology Department of the British Museum, it is not surprising to find Barrett-Hamilton expressing a good deal of concern about the scale of the industry he was witnessing, along with skepticism about the views of the station managers and whalemen with whom he discussed the problem. As

---

80. Misc. 298, p. 8.

81. Harmer's report to trustees, 19 October 1912, NHM DF 1004/CP/749/1; it is interesting that this passage is crossed out in the manuscript, suggesting that it was to be deleted (unsurprisingly) from the version of the report sent on to the Colonial Office by the trustees.

the season progressed into the peak months of December and January—
and the waters around the island offered blues and fins in abundance, but
almost no humpbacks (which in previous years had served as the mainstay
of the industry)—Barrett-Hamilton wrote again to Fagan:

> A lot of whales are being killed, and the season looks like being a
> good one. But the catch is of finners and blue whales chiefly, not, as
> formerly, of humpbacks. Unless these latter arrive in greater numbers
> later on, it will be difficult to avoid the conclusion that the whalers,
> having killed off the humpbacks, are now doing the same thing to the
> other whales.[82]

He offered Harmer the same bleak news directly in a similar letter written the following day. The opinion circulated at the Leith Harbor whaling
station—that the whalemen had taken no more of the whales than herring
fishermen took of the herring in the North Sea, and that the whales had
simply gone away in 1911 "and have not come back since"—could inspire
little confidence, Barrett-Hamilton decided.[83]

Before undertaking his trip, Barrett-Hamilton was well aware that
Harmer and others (including Mørch) had raised the question of whether
the humpbacks taken at the southern African coastal stations were the
same humpbacks being hunted in the Antarctic, and whether these two
whaling grounds focused on different points in a single annual migration.
The timing of the two seasons fit together in a suspicious way: the richest
months at the South African stations fell during the southern winter—May
to November—and the South Georgia season picked up in November and
ran through February—the southern summer. While he was still aboard
the floating factory ship *Thor*, which conveyed him to the south in the autumn of 1913, Barrett-Hamilton was already making inquiries among the
whaling personnel on board about possible sites on the South African coast
where further research could be conducted—research that, paired with his
investigations at South Georgia, promised to establish whether, in fact, the
whales were the same.[84] Such a program of geographically distributed researches had been contemplated in the preparations for the joint Anglo-

82. Barrett-Hamilton to Fagan, 11 January 1914, NHM Zoology Collections, "Whales
and Whaling—Official Papers"; quoted in Misc. 298, p. 65.

83. Ibid., p. 65.

84. Ibid., pp. 66–67.

Swedish expedition (indeed, the Swedes had hoped to work the Antarctic stations, leaving the British naturalists to work at Durban or Cape Town), but it appears that, had he lived, Barrett-Hamilton would likely have taken a lead in the first stages of this work in the temperate African waters. In the end, this investigation, like other elements of Barrett-Hamilton's abortive expedition, would be taken up after the war by the zoologists of the Discovery Investigations.[85]

LIFE SCIENCE IN A DEATH ZONE

Having reconstructed the questions Barrett-Hamilton set out to answer, as well as why they mattered, it remains to consider how his work was actually done. This, I believe, is the aspect of the expedition that was most novel and that had the greatest importance for establishing the kind of fieldwork that would dominate the Discovery whale research, particularly in its early years. Zoologists like Cocks and Guldberg, after all, had relied almost entirely on measurements and samples conveyed to them by personnel at whaling stations. With very few exceptions (notably the Smithsonian's Frederick W. True, who had spent several weeks at a station in Newfoundland in 1899 and 1901), lengthy stays at whaling outposts were not yet a standard part of research on whales.[86] In this sense, even as he pioneered research in a new geographic region, Barrett-Hamilton was also pioneering a research approach and a distinctive kind of fieldwork: hip-booted whale science, a kind of biological investigation that

85. For instance, Barrett-Hamilton and Harmer discussed the possibility of whale marking as well as the need for scientists to "go out on the whale-boats, and study the whales at sea." Ibid., p. 177. They had discussed the possibility of marking whales in 1913. See ibid., p. 3.

86. True's work resulted in a very considerable monograph: Frederick W. True, *The Whalebone Whales of the Western North Atlantic: Compared with Those Occurring in European Waters, with Some Observations on the Species of the North Pacific*, Smithsonian Contributions to Knowledge 33 (Washington, DC: Smithsonian Institution, 1904). For a discussion of this investigation, see Gary Kroll, *America's Ocean Wilderness: A Cultural History of Twentieth Century Exploration* (Lawrence: University Press of Kansas, 2008). Cocks, Collett, Lillie, Guldberg, Stanley T. Burfield, and others had made visits to stations in Finnmark and elsewhere (and Guldberg, at least, had also made sorties on whaling vessels and seen the animals taken at sea), but extended stays at the stations to collect data remained a rarity, and such visits generally served to collect museum specimens and oversee their transfer back to metropolitan institutions.

demanded sloppy anatomical intimacy with tons and tons of large whale carcasses, and, as such, a mode of inquiry absolutely impossible to imagine before the rise of industrial-scale whaling. Nineteenth-century naturalists with an interest in whales were limited to work with a few unwieldy bones; a stranding or two afforded a rare and career-defining glimpse of the object of inquiry, and as a rule, such glimpses were caught through a miasma of stink and decay. Modern whaling changed all that and gave Barrett-Hamilton and the men who followed him a swamping flood of specimens, a gargantuan conveyor belt of meat over which to pick. A statistical approach to biological features of these organisms became possible—indeed, given the amount of data and the speed at which it passed, perhaps necessary—for the first time. One could begin to think about a given whale as a sample of a population, rather than as a carnival singularity washed onto the beach. The enterprise took a considerable toll on him, as it would on others.

To understand the character of this work demands a review of the labor at a shore whaling station in the south. Figure 2.4, a general plan for such a factory, will give a sense of the layout and organization of the physical plant at these outposts. In the early years at South Georgia, factories like this one, tucked on the narrow strand against the base of the looming mountains of the island, typically sheltered some hundred and fifty men, almost all of them Norwegian, and most of them on tours of duty ranging from a single season to several years.[87] Their dormitories were within a stone's throw of the central "plan," or planked wooden cutting deck, situated at the center of the processing buildings, a sloped deck that dropped down into the water on the low side. In the shallows beside this ramp, the hunting boats deposited their catches, mooring them to buoys before heading back out for more—in the flush of the season, with several boats bringing in multiple whales a day, some forty carcasses could await attention at the factory, where production easily bottlenecked. On the high side of the plan, steam winches provided the power to draw each carcass up onto the deck, where skilled "flensers" would immediately go to work, using scythelike blades to score the blubber and cut it from the body, assisted by winch lines hooked into the blubber itself. Scaling the carcass like mountaineers, cutting steps up the flesh as footholds for their boots (which were armed with long metal nails—the only way to stay upright on a platform slicked with grease and icy fluids), the flensers arranged for further hoisting to roll the

---

87. There were a few "permanent residents."

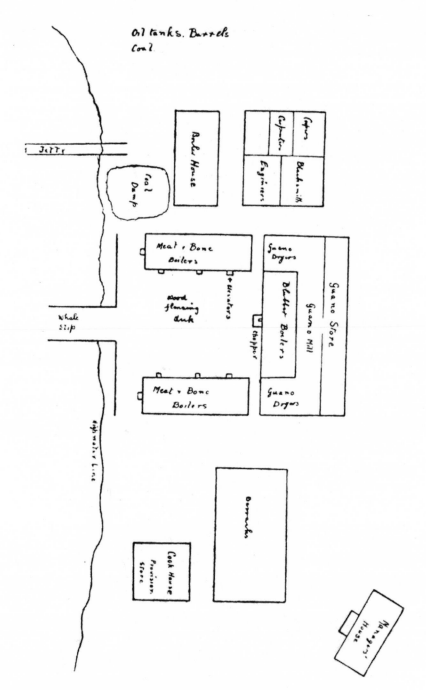

FIGURE 2.4 The catafalque and its plan: Layout of a shore whaling station. (From Bennett, *Whaling in the Antarctic*, p. 104.)

animal over, and the process continued, the blubber strips being hoisted off, trimmed, and dragged to the boiling facilities that rimmed the plan. Blubberless, the remaining carcass generally came under the ministrations of the "lemmers" (or butchers), who used a combination of knives and saws to break down the body itself, clipping the giant bones and slabs of meat into chunks that could be processed in the pressure cookers and kilns. The whole enterprise, from drawing an 80-foot carcass out of the water to its disappearance into the maws of the "digesters," might take two hours. Work generally ran from 6:00 a.m. to 6:00 p.m. but when necessary could continue for as long as the Antarctic summer light held.[88] There is much more that could be said about the social and technical minutiae of a shore station (the processes were intricate, and they changed in numerous small ways over the period considered in this book), but this general outline will suffice to sketch the setting into which Barrett-Hamilton and other hip-booted whale biologists came.[89]

It is not merely from the vantage point of contemporary mores that such a place and such an activity may seem grotesque. Even for hardened Nordic sailors and men not squeamish about the inside of an abattoir, whaling factories were reputed to be a testing environment, and the southern factories were known to offer the most extreme exemplars of an enterprise already hedged with unflattering superlatives. The smell of such stations, wrote one commentator in the period, was "loathsome" and might only be compared to "a charnel house boiling wholesale in vaseline," or, hazarded another, "a mixture of the smell of a tanning factory and that of a fish meal and manure works together with a sickly and almost overpowering odour of meat extract."[90] And this miasma—which cured the bright white paint

88. See, for instance, Barrett-Hamilton's journal entry for 29 December 1913, when there were forty-two whales waiting to be flensed but the station could manage only about ten a day, necessitating a full night shift running from 6:00 p.m. to 10:00 p.m. and then from 2:00 a.m. to 6:00 a.m. See "Journal of a Trip to South Georgia by G. E. H. Barrett-Hamilton," NHM Zoology Library, item 89qH.

89. For more details see J. N. Tønnessen and Arne Odd Johnsen, *The History of Modern Whaling* (Berkeley: University of California Press, 1982), 169ff, and A. G. Bennett, *Whaling in the Antarctic* (New York: Henry Holt, 1932), 100ff.

90. F. V. Morley and J. S. Hodgson, *Whaling North and South* (London: Methuen, 1927), 10; Alister Clavering Hardy, *Great Waters: A Voyage of Natural History to Study Whales, Plankton, and the Waters of the Southern Ocean* (New York: Harper and Row, 1967), 161.

FIGURE 2.5 A day's work: Autolysis set in rapidly, even under icy conditions. (From the collection of the Natural History Museum, London.)

on ships to a haggard parchment—was more than the smell of flesh and fat upon "a kind of smoking high altar."[91] In less than half a day, a dead whale, insulated by its thick blubber, began to undergo significant decomposition. Rising internal temperatures accelerated this process of autolysis, and the mephitic gases generated within the body cavity swelled the animals into stinking balloons, heaved them up out of the water, and not infrequently everted their internal organs (figure 2.5). If it burst, such a "burned" or "blown" carcass could easily kill a man, a happening by no means unknown at stations choked by plenty in the crazed intensity of the early years; still more macabre, the forcible ejection of an unborn fetus (which could be larger than an automobile) represented a real danger on a flensing plan that had fallen behind on its workload. Many more mundane dangers— from the kick of massive bone saws to the snapping of high-tension wires—

91. F. D. Ommanney, *Lost Leviathan: Whales and Whaling* (New York: Dodd, Mead, 1971), 102.

exacted annual casualties; drink and disease made regular contributions to the list of invalids.

The biological oceanographer Alister Clavering Hardy (later Sir), who served as one of the leaders of the Discovery Investigations and carried a belletristic sensibility into these desolate regions, left this description of his first encounter with the Grytviken station from the deck of the research ship in February 1926:

> It is a fantastic scene. The water in which the whales float, and on which we too are riding, is blood red. On the platform itself are whales in all stages of dismemberment. Little figures, busy with long-handled knives like hockey sticks, look like flies as they work upon the huge carcasses; and from time to time these massive remains are pulled about the platform by steel cables. . . . Then there are the factory buildings themselves, ejecting clouds of steam, and, sheer above it all, is a mountain peak of dark rock splashed with brilliant sunlit snow towering against a clear blue sky. This blue of the sky reflects on the blood-red water below us to give a curious lilac tint . . .[92]

His effort to capture the scene in a watercolor sketch is shown in figure 2.6.

Framed by a seascape distance and word-painted with an evocative palette, such scenes might even have about them elements of a Dantean sublimity. Another early Discovery scientist, recollecting biological work on the plan, would reach for language that quivered between the biblical and the pagan archaic: "What penalty, I used to wonder, would the gods in due time inflict for such a sacrilege?"[93] And a whaling company medical doctor visiting the Leith Harbor station on South Georgia in later years similarly resorted to a mythic solemnity to convey the scenes of his tour: "This was the meat-meal store, the catafalque where reposed in deep silence and eternal stillness the ashes of a thousand monstrous sea beasts.

92. Alister Clavering Hardy, *Great Waters: A Voyage of Natural History to Study Whales, Plankton, and the Waters of the Southern Ocean* (New York: Harper and Row, 1967), 161.

93. F. D. Ommanney, *Lost Leviathan: Whales and Whaling* (New York: Dodd, Mead, 1971), 130.

FIGURE 2.6 (and see PLATE 2) A sea of blood: Hardy's watercolor of a South Georgia whaling station. (From Hardy, *Great Waters: A Voyage of Natural History to Study Whales, Plankton, and the Waters of the Southern Ocean*, p. 161.)

This huge building was shaped like an ancient cathedral and exuded the same atmosphere."[94] The American naturalist Robert Cushman Murphy, who landed at Grytviken in 1912, having taken passage as a collector for the American Museum of Natural History aboard one of the last operational sail-powered New England whalers, described the whole cove as "a great cauldron so filled with the rotting flesh and macerated bones of whales that they not only bestrew its bottom, but also thickly encrust its rim to the highest watermark," and in grim despair declared the whole district "an enormous sepulcher."[95]

To one in such funerary surrounds, booted and slipping in the sticky-slick organic residue coating the plan, the whole craggy island could seem, as another company doctor visiting in 1911 put it grimly, "like a decorated coffin."[96] His simile gains force when read in conjunction with a letter written in the same year by the first magistrate on South Georgia, James Innes Wilson, as he tried to describe the conditions on the island to Governor Allardyce, back in the Falklands. Wilson explained that the companies' haste to get the easy pickings meant "an utter disregard for cleanliness" and a willingness to abandon carcasses stripped of only the most accessible blubber; these giant skinned bodies (known as *skrot* in Norwegian) crowded the harbor and washed ashore in various states of decomposition, producing "sights of the most gruesome nature."[97] At least at South Georgia the water and air temperatures checked the fetor of this charnel-house scene; those working the floating factories anchored in the protected harbor of Deception Island in the South Shetlands (more than 400 nautical miles farther south than South Georgia) had no such luck. There, in the circular bay formed by a volcanic crater, hydrothermal activity elevated water temperatures, keeping thousands of tons of offal in a putrefying state, while relatively constant southwest winds piled all the residue on one side of the beach. During an inspection visit in the late 1910s, an envoy from the Falkland Islands wrote to explain that it was nearly impossible to land on this part of the island, since a visitor "risks sinking knee deep in a de-

94. Robert Blackwood Robertson, *Of Whales and Men* (New York: Knopf, 1954), 61.

95. Robert Cushman Murphy, *Logbook for Grace: Whaling Brig Daisy, 1912–1913* (New York: Macmillan, 1947), 139.

96. James Innes Wilson to Allardyce, 18 July 1911, quoting Dr. Norgren, SPRI MS 240 / 1–2, file "Dispatches to Secretary of State for the colonies from Governor W. L. Allardyce, 1908–1915."

97. James Innes Wilson to Allardyce, 6 July 1911, in ibid.

posit of rank fat," the accumulation of twelve years' waste.[98] Later, when Discovery whale scientists moved onto pelagic factory vessels to pursue their researches, they found environments that were, if anything, still more confining and obscene. The first of them wrote a letter back to his superiors in London apologizing for his failure to make any oceanographic observations during the voyage—not because he was so exhausted (though he was), but because "the ship was constantly surrounded with hot glue-water, entrails, and various discharges incident to the working up of whales."[99] The ocean itself was transformed wherever they went. In all, the southern whaling grounds were, as the same commentator put it later, "a land of death," and he asked his readers to consider the human cost of the whole enterprise: "Is it surprising that powerful men break down under the strain of such unnatural conditions?"[100]

It was into this "unnatural" world that Barrett-Hamilton and (later) the Discovery whale biologists stepped, assigned the unenviable task of trying there to study, of all things, nature. Harmer summed up the scenario with considerable understatement when he noted in one of his early reports that "the investigator on the flensing platform is certainly entitled to sympathy, in view of the unpleasant nature of his work."[101] Barrett-Hamilton, however, displaying the fortitude and pluck expected of Englishmen in ice (and of science militant at the colonial periphery), showed little inclination to complain of the hardships or risks, despite the fact that just two years earlier, a major avalanche had crushed half of the station where he now found himself at work.[102] To Fagan he wrote of his daily routine: "When

98. Quoted in Harmer, "Report on Whales and Whaling," 17 July 1920, SMRU MS 38600, box 637, Harmer's memoranda on whales, binder 3. The inspector was A. G. Bennett, who spent a total of eight whaling seasons at Deception Island and had several different titles. He did his own research on whale movements, discussed below. For details on his life, see Stephen Canon Palmer, "A. G. Bennett—Naturalist, Whaling Officer, Customs Officer and Postmaster," *Falkland Islands Journal* 6, no. 4 (1995): 91–105.

99. SMRU MS 38600, box 636, file "Mr. Laurie's Visit on a Pelagic Whaler" (Disc. 15021). This was Alec H. Laurie, reporting on his work on the *Southern Princess* in 1932–1933.

100. A. G. Bennett, *Whaling in the Antarctic* (New York: Henry Holt and Company, 1932), 84.

101. Harmer's report of 19 March 1923, SMRU MS 38600, box 637, Harmer's memoranda on whales, binder 3.

102. J. N. Tønnessen and Arne Odd Johnsen, *The History of Modern Whaling* (Berkeley: University of California Press, 1982), 189. On the place of ice in conceptions of

there are whales to be cut up I am out and on the platform at 6 a.m., and as far as blizzards and snow showers will allow I believe I am doing useful work." His private journal expressed a mixture of exhaustion at the strenuous physical conditions and exhilaration at the opportunity to study such little-known animals at close hand. On 1 December, for instance, he noted, "A cold unpleasant day, but splendid for whales, as we had a 91-foot Blue whale; a 73.6-foot Finner and a 69-foot Finner on the slips, and others. The blue whale was a monstrous mountain of flesh." In the end, he spent nearly twelve hours tromping around, over, and inside this veritable organic landscape, and he closed his journal entry, "out all day . . . and had to write up notes in the evening; to bed very tired."[103]

Even in the face of such labors, Barrett-Hamilton was proud to report that he felt he could keep up the pace without help. He wrote to Fagan that he had, as yet, spent none of the museum's money, and moreover, that he had managed to figure out how to do all the scientific labor on the plan alone, freeing his assistant—a skilled preparator and taxidermist named Percy Stammwitz, seconded on the expedition by the British Museum—to work on the collections of anatomical specimens:

> I have tried to systematize the measuring of whales and I have produced a form for the purpose, of which I enclose a copy. This, and a special measuring machine, makes the work easier and more rapid, so that I am able to let Stammwitz attend to other things. Some such form might meet with the approval for distribution to stations all over the world.

The proposed form is here included as figure 2.7.[104]

British manhood in this period, see Frances Spufford, *I May Be Some Time: Ice and the English Imagination* (London: Faber and Faber, 1996).

103. Barrett-Hamilton to Fagan, 2 December 1913, NHM DF 1004/CP/749/2. "Journal of a Trip to South Georgia by G. E. H. Barrett-Hamilton," entry for 1 December 1913, NHM Zoology Library, item 89qH.

104. While forms like this never received worldwide distribution, a similar figure was incorporated by Harmer into the later editions of his BMNH form 136, "Stranded Whales, Porpoises, and Dolphins" (second edition, 1920, SGB Papers), a document he developed to systematize the collection of information on cetaceans that washed ashore in the British Isles. This program, the "Stranding Reports," began in 1912; reports were regularized the following year, and Harmer put them out until 1927, after which they were continued into the 1970s by the former Discovery whale biologist Francis C. Fraser,

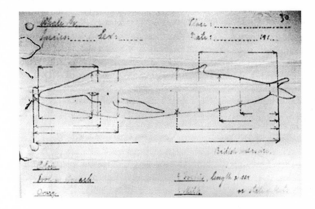

FIGURE 2.7
How to span a
whale: Barrett-
Hamilton's measure-
ment form. (From
the collection of
the Natural History
Museum, London.)

This concern with forms reveals the ways in which Barrett-Hamilton's expedition combined bloody hands-on fieldwork with administrative and diplomatic activities. In fact, before the expedition's departure, Harmer had hastily prepared BMNH form 132, "Report on Whales Captured" (figure 2.8), and asked Barrett-Hamilton to convey copies of this document to South Georgia, where they were distributed by James Innes Wilson and became a standard tool for the collection of data on the southern fishery. Filled out by the companies and sent back to Harmer each season via the colonial secretary in the Falklands, these forms—together with the magistrate's an-nual whaling reports—were the basis of Harmer's sequence of scientific reports on whaling over the following decade.[105] So closely engaged did the systems of bureaucratic oversight and scientific research become that in later years no one could remember if the forms were the property and responsibility of the Colonial Office or the British Museum, or whence they had originally come.[106]

---

who himself rose in 1957 to become keeper of zoology. Note that when Mackintosh and the other Discovery whale biologists began their work at shore stations, they took up a very similar sequence of measurements.

105. SPRI holds, in the South Georgia Archives, a file of correspondence from the office of the South Georgia magistrate relating to form 132; this file includes requests for forms from various companies, cover letters that went out with the forms, and a list of the companies to which they were sent. See SPRI MS 1228 / 6 / 20, South Georgia Archives, file "British Museum of Natural History."

106. On the development of form 132, see Harmer's memorandum, circa April–October 1918, SMRU MS 38600, box 637, file "Whaling Research Expedition—Discovery Expedition—Harmer Correspondence." Many of the completed forms survive in the NHM Zoology Collections.

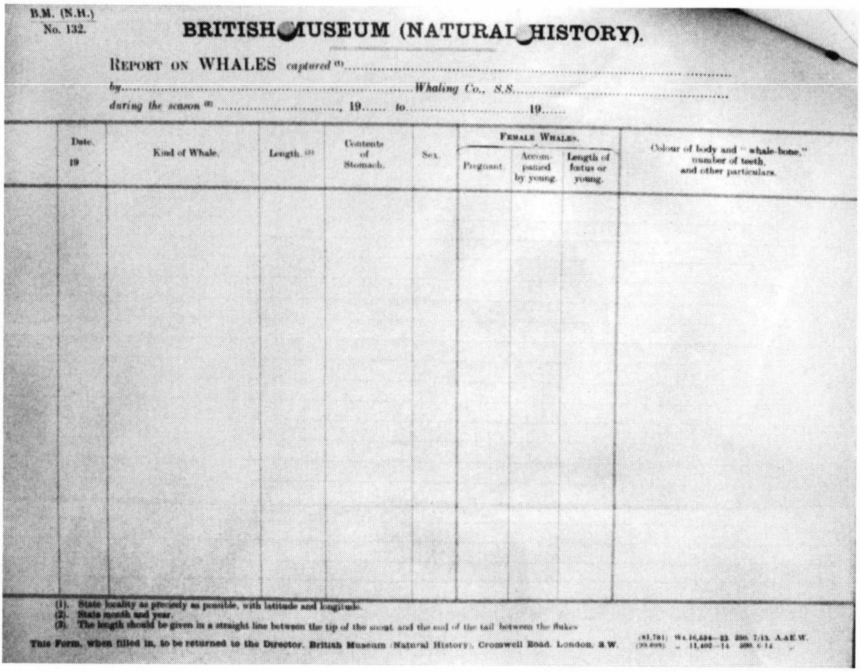

B.M. (N.H.)
No. 132.

BRITISH MUSEUM (NATURAL HISTORY).

REPORT ON WHALES *captured* (1)..
by...*Whaling Co., S.S.*..
during the season (2) ...., 19...... *to*......................... 19.....

Date. 19	Kind of Whale.	Length. (3)	Contents of Stomach.	Sex.	Female Whales. Pregnant.	Accompanied by young.	Length of fœtus or young.	Colour of body and "whale-bone," number of teeth, and other particulars.

(1). State locality as precisely as possible, with latitude and longitude.
(2). State month and year.
(3). The length should be given in a straight line between the tip of the snout and the end of the tail between the flukes.
This Form, when filled in, to be returned to the Director, British Museum (Natural History), Cromwell Road, London, S.W.

FIGURE 2.8 Report on Whales Captured: BMNH form 132, prepared by Sidney Harmer. (From the collection of the Scott Polar Research Institute, University of Cambridge.)

Barrett-Hamilton's outbound trip had been coordinated through the largest British whaling interest, the Salvesen Company, which ran the Leith Harbor station at South Georgia. The manager of the local operation ("one of the most intelligent people on the island") eventually put Barrett-Hamilton up in the company's lodgings at the station and ensured that he and Stammwitz had what they needed for their work, including the whaling implements they used for their dissections (which had to be conducted like a careful ballet dance threaded in and out of the scrum of flensers and lemmers, who worked at breakneck pace). As for the Falkland Islands administration, it was rapidly apparent to Barrett-Hamilton that they had less to offer than the company:

> The government is in a curious position here. Its representatives have no means of obtaining supplies or transport except from the whaling companies, and the latter are not backward in pointing out that a

trip from one station to another for one of their whalers conveying a government official costs a lot of money. The government is responsible for sending me about, but its representatives have no means of doing so.[107]

It had been Barrett-Hamilton who had negotiated his own arrangements with the whaling personnel he met on the *Thor* voyage, and he pointed out that it was lucky he had done so, since it was unclear how things would have worked out otherwise. The magistrate had advanced him some Colonial Office money and "placed part of the customs shed at the disposal of the Mission for the preparation of specimens" (as Wilson wrote to Allardyce, informing him about Barrett-Hamilton's progress), but there was little else he could offer, being himself, in effect, a guest of the Compañia Argentina de Pesca at Grytviken.

Nor is it clear that Allardyce would have encouraged his representative to provide greater support. The Falkland Islands' governor was not well disposed to the whole program of the investigations, having written an irritated letter to the Colonial Office the previous year, declining to ask the magistrate to collect data for the British Museum and pointing out,

> with all due respect to the representations of the British Museum, that there are many authorities with a varied, extensive, and practical experience of whaling who maintain that there is not risk at present either of exterminating the whale or of unduly reducing its numbers. . . . These gentlemen, scientists and others, support their personal knowledge of the whale fishery obtained in different parts of the world, by comparing the smallness of the world's catch and the restricted areas operated upon, with the vastness of the oceans in which these animals are found.[108]

It was to counter arguments like this one—vastness of the oceans equals vast stocks of whales—that Harmer was particularly interested in demonstrating that there were specific migration patterns and that the industry

107. Barrett-Hamilton to Fagan, 2 December 1913, NHM DF 1004/CP/749/2.

108. On the customs shed: Wilson to Allardyce, 9 December 1913, SPRI MS 1228/14/4, file "Biological Mission." Allardyce quote: Allardyce to Harcourt, confidential dispatch, 11 April 1912, SPRI MS 240/1–2, file "Dispatches to Secretary of State for the colonies from Governor W. L. Allardyce, 1908–1915."

had positioned itself along those routes. If this were the case, the vastness of the oceans might be irrelevant: the rest of the oceans might contain very few whales.

Allardyce had no illusions about the hostility to the industry that lurked under the ceramic vaults of the South Kensington museum. When the Colonial Office informed him that the Barrett-Hamilton expedition was going to take place, Allardyce suggested that the visitor might usefully be employed, while at South Georgia, by conducting an audit of some customs accounts that had troubled Whitehall—a proposal, perhaps disingenuous, that displayed the governor's studied disinterest in the specialized services of a "marine biologist."[109]

Picking his way diplomatically between an unsympathetic (if weak) government and a powerful industry not at all well disposed to regulation, Barrett-Hamilton managed to land on his feet at Leith Harbor.[110] His accommodations, he wrote back to London, were perfectly fine—though, he noted, it was by no means easy staying properly groomed at the station, given the filth. The observation is particularly poignant coming from a man who had in his luggage not only two combs, a nail buffer, and a clothes brush, but also an elegant pair of gauntlet gloves. Having packed such civilized essentials, it is perhaps little wonder that, after a day on the plan dissecting the genital organs out of hulking, ripe carcasses, he felt that he was "not so well *washed* as at home!" despite his best efforts.[111] At the same time, the major did retain a sense of decorum robust enough to prompt a detailed critique (in his private journal) of the Christmas dinner served at the home

109. Allardyce to Harcourt, 6 May 1914, in ibid.

110. Though his position was tested by an incident on 17 December, when he served on the jury of inquest examining the death of a "poor delicate boy" in the employ of the South Georgia Company (the name for the Salvesen operation at Leith Harbour). Barrett-Hamilton noted in his journal that there was a rivalry among several of the doctors called to give evidence in the inquiry and that he (Barrett-Hamilton) "crossed words with the magistrate, who was acting as coroner." Always the naturalist, Barrett-Hamilton noted that at the funeral the next day he had been able to see, in the fresh cut of the grave, that the matted tussock grass of South Georgia extended to a depth of five feet underground. The primary medical officer at this inquest, H. C. Titterton, would, a month later, perform an autopsy on Barrett-Hamilton himself. See "Journal of a Trip to South Georgia by G. E. H. Barrett-Hamilton," entry for 17 December 1913, NHM Zoology Library, item 89qH.

111. Barrett-Hamilton to Fagan, 2 December 1913, NHM DF 1004/CP/749/2.

of C. A. Larsen, manager of the Grytviken station: from Barrett-Hamilton's perspective, the whole affair betrayed a "lack of attention to detail"—the fish was poor, the seating arrangement flawed, the Norwegian punch "not very appropriate," and "also they forgot cheese."

That was still in December. By January, when the company had several "record" production days and clogged the harbor with bloated animals, the tone of the notebook turned grim:

> Some of the carcasses are absolutely rotten, the stench, especially on New Year's Day, was disgusting—the whole slip slippery and filthy beyond words. Somehow or other I managed to keep my legs, and my slip clothes having been washed, I feel better. Before that the stink of my clothes nearly drove me out of my bedroom.[112]

And a marginal note, further on, gives a sense of the increasingly foul surroundings: "The shore is getting greasy from the escaping blubber from the slip and the moored carcasses."

As Stammwitz worked on cleaning and packing complete specimens of baleen and flipper bones for a humpback, blue, and fin, along with several vertebrae and scapulae, Barrett-Hamilton continued to hack his way over and into a total of 294 great whales, mostly finners, but also including 3 sperms, 50 blues, and 57 humpbacks. Fetuses he bottled and passed to Stammwitz, who prepared them for dispatching. Whenever possible Barrett-Hamilton cut out the two large ovaries in each animal (whose position deep in the posterior body cavity necessitated the clearing of a passage through the intestines) so they could be inspected for evidence bearing on the breeding cycle. His journal entries grew more sparse as work on the plan intensified, and by 16 January he had a week's worth of runic rough notes backed up in the pencil book he carried during the day, not having had time to transfer and expand them into his clean journal.

On that afternoon he worked up four blue whales, including an 85.5-foot female (no. 293) that appeared to be in the earliest stages of pregnancy, on which he wrote the following specimen notes: "Milk, nil; mammae absolutely retracted. Ovaries not seen. Uterus small-bored, hence could not have held large foetus; courna red and folded as if impregnated; ? very small

112. "Journal of a Trip to South Georgia by G. E. H. Barrett-Hamilton," entry for 2 January 1914, NHM Zoology Library, item 89qH.

foetus found and bottled."[113] The question mark was well placed, since the bit of tissue he collected was apparently not a fetus.[114]

Then, that evening, the major "retired to bed . . . at his usual time, with the intention of commencing work at 6 o'clock the following morning—as indeed had been his custom throughout when whales were on the plan." But when the steward called with coffee the next morning, he found the naturalist quite dead. The company doctor, conducting the postmortem, found evidence of a heart attack, and the magistrate filed a report addressed to the Colonial Office: "I much regret to report that Major G. E. H. Barrett-Hamilton, Zoologist, died suddenly of heart failure early on Saturday morning the 17th instant, at the whaling station of the South Georgia Co. Ltd., Leith Harbour, where he had been conducting whaling investigations."[115] In life he had been a thorn in Allardyce's side, and in death he continued to irritate: the governor wrote back to Wilson, reprimanding him for drafting the death report directly to London instead of to Port Stanley, and then Allardyce saw to it that the company's surgeon received one-fifth of the fee he requested for his autopsy.[116]

NEW PLANS

Barrett-Hamilton's death constituted a significant setback to Harmer's plans for southern whale research as well as to Darnley's regulatory inquest at the Colonial Office. But as M. A. C. Hinton began to sift Barrett-Hamilton's preliminary results and Harmer and others debriefed Stammwitz, the urgency of the problem came into clearer focus, prompting renewed efforts by the British Museum to gather information. Meanwhile, the Inter-departmental Committee on Whaling and the Protection of Whales pressed on, soliciting testimony from a range of individuals associated with the industry, where concern was mounting about the government's intentions.

113. Misc. 298, p. 138.

114. Hinton does not list it in his final tabulation of blue whale fetuses. See ibid., p. 140.

115. Wilson to Secretary of State for the Colonies, 21 January 1914, SPRI MS 1228/14/4, file "Biological mission."

116. Allardyce was also irritated not to receive specimens from the expedition (though it is not clear for what purpose he wanted them). Harmer argued that the Barrett-Hamilton mission was not really a "collecting expedition" and that there were too few specimens to share. See Harmer's report to trustees, 23 November 1914, NHM DF 1004/CP/749/2.

In fact, even as Barrett-Hamilton's body was being prepared for repatriation back in South Georgia, Governor Allardyce, visiting London in February 1914, stood before the committee to offer his version of the situation with Antarctic whaling. By this point it had become clear that the 1913–1914 humpback season was shaping up to be a failure, and Allardyce, responding to questioning, offered the whalemen's explanation: "In the case of the Humpback whale, I think that is a very clever mammal indeed, and he scents danger, knows danger, and if we press him unduly he will withdraw," hence any regulation should be understood not as preserving the stock from extermination, but merely as "not discouraging" the creatures from making their ordinary peregrinations. On the question of scientific research, Allardyce downplayed shore station anatomy and physiology, calling instead for oceanic reconnaissance with an eye toward determining where all the whales were going:

> I suggested some time ago that the best way in which we could ascertain information about the whale would be to charter a vessel and let her follow the whales, and then we shall know something about the habits of the whale. It is not sufficient to send a man to South Georgia or to the South Shetlands to study dead whales. They are much more interesting to us while alive, because we want to know their habits, where they come from, and where they go to. So far as the Colonial Government is concerned it is immaterial to us how many different kinds there are; what the fishermen want are whales, and we want so to protect the whales as to spin out the period of whale fishing for as long as ever we can . . .[117]

Here—besides getting in digs at Harmeresque preoccupations with taxonomy and at meddlesome naturalists tromping about on the flensing plans of South Georgia—the governor was sounding a note heard several times by the committee over the first half of 1914: if any regulation was contemplated, it should be aimed at the optimization of the commerce in whale products.[118]

For instance, on 24 April, Holman Kingdon, managing director of Joseph

117. Misc. 298, p. 22.

118. It seems likely that he was also keen to get access to a decent ship, the want of which was (unsurprisingly) a perpetual irritation to the governor of an island archipelago.

Crosfield & Sons, Ltd., a soap and chemical company in Warrington, laid a traveling case of samples before the committee. After reviewing the different grades of oil and the processes of "hardening" (hydrogenation) that made it possible to produce bar soaps from whale oil (a new development that he called "a revolution" in his lifetime), he made the interests of his industry clear: "We do not want the supply of whale oil stopped; we want it regularly, and we are entirely in sympathy with any proposals you may make, or any steps you may take, if the result of these steps is to secure more regular production."[119] Both gluts and shortages were a source of instability for companies like his that made use of whale oil. And Kingdon noted that there was a real possibility that the market for this oil would be expanding shortly: "There is no doubt," he informed the committee, "that at the present time a tremendous amount of scientific work is being done on these processes, and on the improvement of products generally with the object of getting some edible result." He demurred on the question of when a whale-oil margarine might appear in grocery stores, but he added that the day would likely come: "I can imagine that in a severe state of hunger, I would not mind spreading *that* (*indicating the sample of hardened oil*) on a piece of biscuit and eating it myself."[120] In fact, shortly after the war, a reconstituted Inter-departmental Committee would get a chance to taste some successful whale-oil margarine, and by the time the Discovery Investigations were well under way, margarine derived from high-grade whale oil was a major consumer product.[121]

119. Misc. 298, p. 32.

120. Ibid., p. 30. Emphasis in original. In fact, it appears that whale oil was already being incorporated into margarines and artificial lards at this time. The history of this development is difficult to trace, however, because of the reluctance of firms to acknowledge the practice, which, it was thought, would alienate consumers. Vegetable oils remained the preferred edible fats for hardening, but the low cost of whale oils in those years encouraged experimentation and mixing. For a review of these matters in the period 1905–1920, see J. N. Tønnessen and Arne Odd Johnsen, *The History of Modern Whaling* (Berkeley: University of California Press, 1982), 230–40; see also Charles Henry Wilson, *History of Unilever: A Study in Economic Growth and Social Change*, 2 vols. (London: Cassell, 1954), and J. H. van Stuyvenberg, *Margarine: An Economic, Social and Scientific History, 1869–1969* (Liverpool: Liverpool University Press, 1969).

121. Consolidation in the soap and margarine industries was a direct result of the "fat wars" immediately following World War I. By 1929 a series of buyouts and collapses had left two main competitors, Lever Brothers and Margarine Unie; the next year these companies merged into the monopolistic giant Unilever, which bought the majority of

   Other whaling interests expressed similar enthusiasm for a regularized, if not overly "regulated" industry. C. O. Johnson, the director of the Southern Whaling and Sealing Co., a company with stations in South Georgia, Durban, and Port Alexander (on the western African coast), was willing to countenance some restrictions on whaling, but he strongly advocated limits calculated by end product (barrels of oil, bags of guano) rather than by number of whales, which he deemed both impracticable and undesirable. In his testimony on the relative yields of the different species ("you will get some Blue whales which will amount to four Humpbacks, and a Fin whale may be less") and the principle of catch limits set in production units, it is possible to see the rudiments of what would become, by the early 1930s, the first "barrel quotas," the purpose of which was, above all, to smooth out production spikes and control prices. Cartels dislike gluts. These barrel quotas would become "whale quotas" only derivatively, by means of a calculation based on the relative oil productivity of the different species. Thus a given barrel quota divided by 110—the agreed-upon average number of barrels to be had from a blue whale—yielded the same quota in "Blue Whale Units"; hunting boats could then fill this "BWU" quota by killing any commercial species of whale, each of which was assigned, for the purpose of tabulation, a fractional value reflecting its average yield as a proportion of an average blue whale; hence "one BWU" could be made up by taking a single blue whale, or two fin whales, or two and a half humpbacks, etc. The conversion of the BWU from an *industry metric designed to regulate oil production* to a *conservation metric used for regulating whaling* proved disastrous for the southern stocks, as I discuss in chapters 4 and 5 below.[122] Testifying before the committee in the 1910s, several of the operational whaling companies took self-serving proposals for regulation to their natural conclusion, suggesting that a worthy first step would be to prevent the entrance of new ventures into the Antarctic grounds, "so that the existing companies might have a fair chance of escaping an amount of competition that makes operations unprofitable to all."[123]

---

the whale oil that came to market. See chaps. 12 and 13 of Gordon Jackson, *The British Whaling Trade* (London: A. and C. Black, 1978).

   122. See Misc. 298, p. 35, for Johnson's testimony. On the emergence of the earliest quota schemes, see J. N. Tønnessen and Arne Odd Johnsen, *The History of Modern Whaling* (Berkeley: University of California Press, 1982), 402–3.

   123. Misc. 298, p. 48, quoting a letter from the major whale-oil merchants in the United Kingdom, David Geddes & Sons.

As for the notion of extermination, representatives from numerous whaling concerns collectively expressed deep skepticism. In support of their position, they tried to convey, among other things, how hard it was to shoot a whale. F. Cook, the managing director of the Southern Cross Whaling Company, based in Cape Town, pointed out (on the basis of much personal experience) that "you may chase for hours and never be able to get within striking distance, while the numbers are so great in the small area in which we work, about 50 miles radius, that the quantity in the ocean outside our area must be enormous and quite impossible to exterminate." Buttressing this observation with a cartographic argument, Cook invited the members of the committee to mark off on a world map six feet square all the coastal whaling stations on earth, and then to notice that "the area worked round such land stations would not be more than the size of the point of a pencil," leaving, he insisted "the enormous expanse of ocean in which the whales are free to come and go"; by these lights "it seems ridiculous to talk about protection." Moreover, returning to Allardyce's claim, Cook insisted that whaling was a great asset to colonial economies, where a whale provided jobs, money, and fertilizer, "while in the water it is of value to no one," he declared, "and you cannot make pets of them."[124]

As rumors spread about the growing possibility of restrictions, a number of leading figures in the whaling industry met to present a united front to the committee. Significant in this respect was a London meeting in early May of 1914 among the English whaling magnate T. E. Salvesen (then director of the largest whaling firm in the world), H. Korgh Hansen (a leading Norwegian director, and head of the Norwegian Whaling Association), and Dr. Johan Hjort (the director of fisheries at Bergen and probably the best-known marine scientist of the day, who had made his name by unraveling the fluctuations in the North Sea herring fishery and had just coauthored a semipopular book in English on the biology and physics of the seas, *The Depths of the Ocean*, with *Challenger* hero Sir John Murray). Hjort and Hansen were in England on the occasion of Hjort's imminent testimony before Darnley's committee, with which Hansen would file a letter on behalf of the Norwegian Whaling Association, warning that "any new laws affecting the regulation of this industry might easily prove fatal to the trade." Upon discussing Hjort's opinions on the question, Salvesen found himself so wholly in accord with the views this distinguished scientist intended to

124. Misc. 298, p. 47.

put forward to the committee (views that squared seamlessly with those of the Norwegian whalemen) that he elected to decline an invitation to appear before the committee himself, offering instead a letter deputizing Hjort to represent the views of Mssrs. Chr. Salvesen & Co. As he put it in that missive, "it would be sufficient to state that we concur with the evidence led by Dr. Hjort," adding that "we are willing to assist in gaining further information regarding whales, which we consider absolutely necessary before any steps could be taken."[125]

Hjort's stature in oceanography and fisheries research, and his familiarity with (and previous publications on) Norwegian whaling, made him a very significant voice in the emerging debates over whale science and the whale industry in the southern waters. Where northern whaling was concerned, he was considerably better informed than any comparable British expert, having intensively reviewed several dozen logbooks of Finnmark whaling voyages as part of a study undertaken on behalf of the Norwegian government at the turn of the century.[126] Booms and busts in economically vital fisheries had since become his specialty. His pioneering work on age classes, and his hypothesis that most fisheries problems would ultimately be found to turn on a better understanding of plankton ("the blood of the sea"), had put him at the nexus of an extraordinarily productive new area of research that involved linking fisheries science to the emerging "Kiel School" of biological oceanography.[127] Along the way he had risen to prominence as an adviser and investigator of fishery problems in several nations, won the Agassiz Medal for excellence in ocean science, and become a leading figure in the International Council for the Exploration of the Sea.

On 7 May 1914 Hjort gave extensive testimony to the Inter-departmental Committee at the Colonial Office, evidence he subsequently supplemented

125. Misc. 298, p. 49.

126. Ibid., p. 38; this study of distributions and life histories resulted in Hjort's "Fiskeri og hvalfangst i det nordilige Norge," *Aarsberetning Vedkommende Norges Fiskerier* 4 (1902): 1–251.

127. Susan Schlee, *The Edge of an Unfamiliar World: A History of Oceanography* (New York: Dutton, 1973), 222–32; Helen M. Rozwadowski, *The Sea Knows No Boundaries: A Century of Marine Science under ICES* (Copenhagen: International Council for the Exploration of the Sea in association with University of Washington Press, 2002), 75–90; Tim D. Smith, *Scaling Fisheries: The Science of Measuring the Effects of Fishing, 1855–1955* (Cambridge: Cambridge University Press, 1994), chap. 4; and Eric L. Mills, *Biological Oceanography: An Early History, 1870–1960* (Ithaca, NY: Cornell University Press, 1989).

with a formal "Memorandum on the Distribution of the Whales in the Waters about the Antarctic Continent." On the issue of migrations, Hjort both expressed his conviction that the "plankton whales" would be found to make considerable journeys in pursuit of their food and adduced concrete proofs that such migrations had indeed occurred—evidence in the form of old harpoons found in the bodies of whales taken by Norwegian whalers (old harpoons whose provenance could be roughly established). If a program of marking could be developed, a project he claimed was under consideration in Norway, more data could be collected about these routes, as well as about aging and development—areas in which, he acknowledged, much remained unknown.

On the basis of these and other uncertainties, Hjort, like Harmer, expressed reluctance to advise the committee concerning protective regulations. But, unlike Harmer, Hjort made it clear that he was by no means convinced that world's whales were in pressing need of protection—the southern whales least of all, given their vast domain, so remote from Europe and so expensive to work. Again and again his presentation emphasized the multicausal nature of the fluctuations in fisheries, a theme he extended throughout his brief historical review of whaling: once-productive grounds might fail in a particular season as a result of changing physical conditions (such as water temperature), which might mean whale food was elsewhere; furthermore, the price of oil had a direct effect on effort (and thus on results). In his view, no serious effort had ever been made to study the rise and fall of the various whale fisheries in a way that attended to such ramifying and recursive factors. That particular whaling industries in the past had seen decreased catches (for instance, the southern fishery for sperm whales and the Arctic right whale fishery) was, he conceded, beyond question, but he insisted that, as best he could determine, "the decrease in whaling was always local," and pointed out that "a local decrease may occur although the whole stock may still be abundant."[128] Moreover, on each aspect of whale biology relevant to assessing the fecundity of the stock, Hjort inclined to optimistic views: reproduction appeared to be annual, growth rapid, sexual maturity achieved very young. While more research was needed, and distinctions between the various species were surely important, Hjort's tabulations of sightings of large whales in southern waters by expeditions over the previous century gave him confidence that there were enormous numbers of cetaceans in the circum-Antarctic regions. He

128. Quotes from Misc. 298, pp. 40, 44.

could therefore "see no possibility of extermination of the whales" unless, somehow, "the technics of the whaling industry could be so developed that whaling could be extended to the whole Antarctic."[129] Here, of course, Hjort was prescient, since it was in fact only after the development of a full-scale pelagic southern fishery that the stocks of southern blues and fins came under overwhelming pressure.[130]

Confident that there were plenty of whales, and that the scale of southern hunting was comparatively small, Hjort expressed, when pressed by the committee, only very modest support for a handful of regulatory initiatives, such as the protection of the Greenland whale in the Arctic (bowheads were commercially irrelevant in the north by this point, and never found in the south) and a prohibition on shooting a cow with calf (already prohibited in the Falkland Islands and Dependencies). As for the notion of a closed season, or some requirement forcing companies to work up the full carcass of a captured whale, Hjort demurred, citing in response to the first proposal that there was not yet adequate evidence of a discrete breeding season for most species, and gesturing at the costs of mounting an expedition when confronted with the second. The most radical suggestion he was prepared to offer was the idea that should the southern stocks actually come to be threatened, they might be maintained if international consensus could be achieved on the designation of a no-hunting zone in some portion of the Antarctic. A "Preservation?" asked Darnley. "Yes," Hjort replied, "like Yellowstone Park for the bison, or something like that." Though it would be necessary to ensure that such a zone did not interfere with regions where the industry was actually going on.[131]

---

129. Misc. 298, p. 44. These tabulations were based in large part on the remarkable ten-page collation by Émile G. Racovitza in his volume of the zoology of the *Belgica* expedition: *Expedition Antarctique Belge: Résultats du Voyage du S.Y. Belgica en 1897–1898–1899, Rapports Scientifiques, Zoologie, Cétacés* (Anvers: J.-E. Buschmann, 1903), 110–20. He listed by geographic coordinates more than two hundred fifty sightings of whales in the peri-Antarctic waters over more than a century.

130. Though pelagic factory ships remained more than a decade away, the possibility of this innovation was a live topic before the committee in 1914. Captain L. Brun, manager of the Belmullet station in Ireland (who had extensive experience in the south), announced to the committee, "I believe the day will come when you will catch whales anywhere. . . . If the whale oil goes up to £30 I believe they will go round the whole world after the whales, wherever they can find them, and do the whole thing on board." Misc. 298, p. 28.

131. Misc. 298, p. 44.

In light of these sanguine stock assessments and sympathetic perspectives on Norwegian (and British) whaling concerns, it is little surprise to learn that Harmer harbored grave reservations about Hjort's views. Aware that the keeper of zoology had become (indirectly) the primary scientific adviser to Darnley's committee, Hjort arranged to visit Harmer at the British Museum during his London stay, shortly after giving his testimony. By this time Harmer, having reviewed the pessimistic correspondence from Barrett-Hamilton and received intelligence from other sources on the situation in the south, had grown increasingly persuaded both of the threat to the southern whales and of the unacceptably unregulated character of the free-for-all industry in the Dependencies of the Falkland Islands.

Significant new information had come via correspondence from the United States. On his visit to South Georgia on the American whaling brig *Daisy* in 1912–1913, Robert Cushman Murphy witnessed the carnage of whaling and sealing in the south firsthand and came away with a bleak view of the future for living creatures in the Dependencies. Particularly troubling to him was the role of his own vessel in prosecuting this destruction, in that the *Daisy*'s Captain Benjamin Cleveland, accustomed to old-style unregulated Antarctic depredations, unhesitatingly misrepresented his intentions to the magistrate at South Georgia and then proceeded to lead his men on a campaign of indiscriminate slaughter, killing elephant seals and penguins by the hundreds, without regard to the various restrictions ostensibly in effect throughout the island. Upon his return to the Brooklyn Museum (where he had become the curator of mammals and birds), Murphy embarked on a letter-writing campaign, seeking to draw the attention of British authorities to the misdeeds of his own captain and, more generally, to the poor prospects for Antarctic fauna in the region. Several of these letters passed to Harmer at the British Museum, where they became part of his brief for the urgency of the whaling situation. According to Murphy, "single steamers sometimes come into port with ten carcases [*sic*] in one day," and thus "the abundance of the various species cannot long continue"; moreover, he alleged that one of the secretaries of the Compañia Argentina de Pesca had even calculated that "if the present carnage continues," whaling would die at South Georgia within six years.[132]

---

132. Murphy to Fagan, 12 August 1913, cited in Misc. 278, p. 86, where it was appended to Harmer's memorandum. It is interesting to note that Murphy's protestations were eventually laid before Allardyce's government, obliging Wilson and Allardyce

Murphy's mentor, Frederic A. Lucas, the distinguished American zoologist at the American Museum of Natural History, also wrote the British Museum to echo Murphy's expressions of concern about the South Georgia whaling industry.[133]

Armed with this additional evidence, Harmer had continued to push for research on the situation. After learning of Barrett-Hamilton's death, Harmer returned to the sidelined Swedish proposal with renewed vigor, writing privately to the council of the Royal Society to call attention to "a matter of National importance" and seeking immediate support, given the "urgency" of the situation: "The destruction of the whales is undoubtedly proceeding at a very alarming rate; and . . . there appear to be indications

---

to defend themselves to the Colonial Office against charges of neglect. For these exchanges, see Allardyce to Harcourt, 8 August 1914, with enclosures, SPRI MS 240 / 1–2, file "Dispatches to Secretary of State for the colonies from Governor W. L. Allardyce, 1908–1915." Harmer included a critique of the response of the colonial government to these charges in his report to the trustees, 23 November 1914, NHM DF 1004/ CP / 749 / 2. Murphy's account and analysis of the events appear in several places, most lyrically (but also with greatest hindsight) in Robert Cushman Murphy, *Logbook for Grace: Whaling Brig* Daisy, *1912–1913* (New York: Macmillan, 1947), but see also his article "A Desolate Island of the Antarctic," *American Museum Journal* 13, no. 6 (October 1913): 243–60, and "South Georgia, an Outpost of the Antarctic," *National Geographic Magazine* 41, no. 4 (April 1922): 409–44.

133. Lucas had published a very early warning about the overkill of great whales: Frederic A. Lucas, "The Passing of the Whale," originally published as a "supplement" to *Bulletin of the New York Zoological Society* 30 (July 1908): 445–48, subsequently reprinted in *Forest and Stream* 71, no. 8 (22 August 1908): 291–93. Lucas's interest, like that of a considerable number of American naturalists in this period, seems to have originated in the rapid growth of the Newfoundland whaling industry. There seems to have been a burst of conservation-oriented agitation concerning marine mammals among American zoologists in the first years of the twentieth century (seals, of course, were an important part of the story): circa 1907 the American Society of Vertebrate Paleontologists passed a resolution (subsequently ratified by the AAAS) that read, "*Resolved*, the American Society of Vertebrate Paleontologists will aid in any way practicable those measures legislative, international and local which will prevent the now imminent extermination of the great marine vertebrates, especially the cetaceans and manatees, seals, green and other turtles on the coasts of the United States, or on the high seas." For a flavor of this enthusiasm, see G. R. Wieland, "The Conservation of the Great Marine Vertebrates: Imminent Destruction of the Wealth of the Seas," *Popular Science Monthly* 72, no. 5 (May 1908): 425–30, from which this quote was taken.

that the persecutions which these magnificent animals are experiencing is already producing the results which were feared."[134]

Given his desire rapidly to secure better intelligence on the situation in the south, Harmer was very much of two minds on meeting with Hjort in May 1914. On the one hand, as he put it in a letter to Nordenskjöld written just after Hjort left the museum, "Dr. Hjort is at present inclined to take the view that there is not much danger of the extermination of the whales. In this I am not prepared to agree with him." On the other hand, any real action on the problem demanded more knowledge about the biology of the whales, and "in trying to arrive at the facts, it would be essential for us to get into communication with some of the Norwegian companies; and this is a thing Dr. Hjort can do far better than we can."[135]

Which, of course, Hjort knew, and this was his calling card not only with Harmer, but also with Darnley's committee, where he offered the chairman his services as a broker with the industry, declaring, "I shall be very glad to organise the collection of all the material which you may want during the next season."[136] To Harmer personally, Hjort renewed this offer, expressing a willingness to "take advantage of his exceptional opportunities for obtaining information from the Norwegian Whaling Companies operating in South Georgia and the South Shetlands (or elsewhere)" and promising to "put the results of his inquiries at our [i.e., Harmer's] disposal."[137] All this could be had, moreover, for free. Harmer could see no reason to turn away such an offer, which if acted on immediately would secure material from the 1914–1915 season, and he even suggested to Nordenskjöld that the program "might be regarded as a continuation of the work begun by Major Barrett-Hamilton last season."

What Harmer overlooked was a warning Barrett-Hamilton himself had forwarded to the museum before he passed away: tensions between Swedes and Norwegians might cause difficulties for an Anglo-Swedish research program.[138] On reviewing Harmer's description of the possible collaboration, Nordenskjöld fired off a hasty and heated defense of Swedish priority

134. Harmer to "Sir John" (no further information given), 9 March 1914, SMRU MS 38600, box 645, file "Anglo-Swedish Antarctic Station."

135. Harmer to Nordenskjöld, 11 May 1914, in ibid.

136. Misc. 298, p. 44.

137. Harmer to Nordenskjöld, 11 May 1914, SMRU MS 38600, box 645, file "Anglo-Swedish Antarctic Station."

138. Barrett-Hamilton to Fagan, 11 January 1914, NHM Zoology Collections, file "Whales and Whaling—Official Papers."

against Hjort's meddling: "I hope he will not take such steps at the *South Shetlands*, our chosen region, and when I understand it has always been the intention that our expedition should, by the aid of *British* and *Swedish* workers, study also the biology of the whales, that he does not *there!* without agreement also with the Swedish Committee, take up any work by sending out Norwegian scientists . . ." A spat ensued, with Nordenskjöld immediately shooting off another letter to Hjort himself, all of which was immensely awkward for Harmer, who had told Hjort nothing about the Swedish scheme (treating those developments as private). The dust soon settled, however, with no permanent damage done: the Swedes secured, in principle, priority in the South Shetlands, and Hjort undertook to have Harmer's form 132 translated into Norwegian and distributed to the dozens of whaling companies around the world that were members of the Norwegian Association.[139] Shortly thereafter he began preparing to send crates of glass vials with instructions to Norwegian whaling captains "for collection of foetus'es [*sic*] stomach-contents and plancton [*sic*]," and he pursued plans for marking experiments, musing to Harmer that "the more I consider the problem before us, the more I think that marking is the only way of getting exact knowledge about the migrations and number of whales present (cfr. f.i. my investigations of the cod)."[140]

In the same letter, however, Hjort went further, nudging Harmer for any insights into the British government's plans for whaling restrictions and going so far as to suggest that the whole program of research might be *contingent* on the continuation of a largely unregulated environment for the Norwegian firms. It is a remarkable letter, and it largely accounts, I believe, for the care with which Hjort was kept at bay by Harmer and others engaged in the intensive preparations for the Discovery Investigations after the war. After noting "that it is our duty, while whaling is going on, to get as much information as possible" about the life histories of whales, Hjort proceeded to point out that guarantees of future access to the whales were urgently needed to help him make his case to the companies participating in the collection of scientific data:

> All these works will necessitate considerable expense, which will have to be paid by the whalers. It is therefore only possible to get the inves-

139. Nordenskjöld to Harmer, 18 May 1914, SMRU MS 38600, box 645, file "Anglo-Swedish Antarctic Station." Hjort to Harmer, 12 June 1914, SMRU MS 38600, box 645, file "Anglo-Swedish Antarctic Station."

140. Hjort to Harmer, 29 June 1914, in ibid.

tigation done, if whaling can go on undisturbed for a certain time[,] at least two seasons more, three would be better, because then we might be able to consider the problems on the enlarged information. This was my reason for proposing to you, that we should cooperate in the collection of information and in the studying of the information obtained.

And he went on to seek specifics on the work of the Inter-departmental Committee:

> I have written to Mr. Maurice [the chairman], that it would greatly help me, if I could tell them [the whalers] something about the position of the British Government to whaling of the first 2 seasons. It would also be valuable to me, if you could send me a few words about jour [sic] position and we would be very thankfull [sic] to you for any suggestions in the matter.

Harmer's reply focused on the prospects of research and carefully disavowed any knowledge of the government's intentions or the prospects of industry regulation. While Harmer clearly understood that collaboration with the industry was the sine qua non of whale biology under the current conditions, it was equally clear to him that Hjort's close ties to the whalers raised the specter of conflicted interests, and Harmer was absolutely unwilling to serve—as Hjort clearly hoped—as Hjort's inside line or ally during the Colonial Office's unfolding inquiry.

For his part, Hjort made no apologies for his links to the industry and instead positioned himself on the offensive where such questions might come up. In his testimony to Darnley's committee, Hjort explained that the paucity of facts on the emerging Antarctic fishery reflected the absence of exactly such links: in the Antarctic, he explained, "science has not yet come into that collaboration with the whalers which is absolutely necessary for the advancement of knowledge," and in the supporting documentation he filed with the committee, he included an entire section entitled "The Finwhale industry largely responsible for increased knowledge of whales."[141]

But Harmer's skepticism only grew several weeks later, when M. A. C. Hinton submitted his preliminary report on Barrett-Hamilton's surviving notes. Making use of that data, and of additional documentation—including a confidential report by Commander F. E. K. Strong, a senior naval officer with the West African Coast Fleet, who described "wholesale

141. Misc. 298, pp. 40, 50.

butchery" of humpbacks in Angola and the French Congo and lamented the practice, reputed to be widespread, of killing calves—Hinton assembled a pessimistic portrait of the situation, going so far as to state that the whole southern enterprise was headed toward "what appears at present to be the inevitable and speedy extinction of all the larger *Cetacea*."[142] In particular, Hinton found evidence in Barrett-Hamilton's notebooks very much at odds with Hjort's optimistic pronouncements on the reproductive biology of the fins, blues, and humpbacks.[143] While about half the females of these species were pregnant, according to Barrett-Hamilton's reconnaissance, he found very small numbers of lactating specimens: less than 2 percent of mature fins, no blues, and less than 10 percent of humpbacks. These observations pointed to a two-year reproductive cycle at best (as opposed to annual), and suggested that young whales were not surviving: "The rate of propagation is thus a slow one, and though nothing is definitely known of the subject it is not improbable that the natural rate of mortality among infant and adolescent whales is a high one." On top of this, clear statistical evidence from the amassed catch data of the past five seasons indicated that humpbacks were being taken in smaller and smaller numbers. While there had been suggestions that this reflected whalers' increasing preference for the larger blues and finners, evidence from Barrett-Hamilton's notes (as well as a debriefing of his assistant, Stammwitz) pointed the opposite way: gunners *sought out* humpbacks over the larger species, which tested the capacities of the station's equipment and were in several respects more difficult to secure and process. Persuaded that the humpbacks were indeed being hunted at both ends of an annual migration, and that the peak season corresponded with (and thus likely disrupted) their breeding, Hinton concluded, "The humpback thus seems to stand in urgent need of protection if its early extinction is to be avoided." He went on to advocate a total ban at South Georgia, or

142. A draft of this document, "Preliminary Memorandum on the Papers Left by the Late Major Barrett-Hamilton, Relating to the Whales of South Georgia," dated 10 July 1914, can be found in NHM DF 1004/CP/749/2; the preliminary report also appears in Misc. 298 and in the freestanding Colonial Office publication of Hinton's work in 1925.

143. It is interesting to see Harmer explicitly discussing the use of Hinton's report to counter Hjort; this came up in Harmer's memorandum to the trustees on 18 May 1915, NHM DF 1004/CP/749/2, in which Harmer pressed for help securing additional copies of the (by then confidential) report. The Colonial Office had sent six. Harmer explained that he might need more, "particularly if we come to discussing the position with Dr. Hjort."

an extended closed season at least. While the blues and fins appeared to be less immediately in danger, Hinton ended his report with a general expression of concern: once whalers had wiped out the humpbacks, they would turn to these larger species, and the Finnmark fishery gave little reason for confidence that they would weather such an attack. In all, Hinton left his readers with "a grave warning against the unscientific and indiscriminate slaughter to which these animals are at present subject."

Filed in July 1914, Hinton's grave preliminary memo was rapidly overshadowed by world events. The rumblings of a European war in August dramatically shifted the prospects for whale research, as well as the strategic significance of all knowledge of the whaling industry: the glycerin derived from the saponification of whale (and other) oils was a significant ingredient in military explosives. A hasty German expedition against the Falkland Islands and Dependencies was defeated, and arrangements were hastily made to relax what restrictions were in place on the industry in an effort to maintain wartime production.[144] Harmer broke off his

144. For details on these "temporary" arrangements (which proved difficult to undo), see the confidential memorandum "The Whaling Industry and the War," presented at meeting 8, 27 August 1918, SPRI MS 1284/1, explaining in part that "the Governor, with the sanction of the Secretary of State, may during the war authorise the departure from any or all of the provisions of the existing regulation as to him may seem fit and expedient." This policy meant that additional catcher boats were added, so that by the 1916–1917 season, Salvesen had an additional thirteen catchers operating at South Georgia, bringing their total to twenty. The company also secured an additional lease on the island, which they retained after the war. There is also a valuable account of the government's perspective on whales and World War I in G. Grindle (for the Colonial Office) to Trustees of the British Museum, 22 February 1917, SPRI MS 1284/1, which reads in part (concerning South Georgia), "In that dependency, as also in the South Shetlands, the urgency of the demand for the better qualities of whale oil for munitions purposes has unfortunately made it necessary to relax the regulations for the prevention of waste . . . and to permit the temporary employment of extra whale catchers . . . and to incur certain conditional obligations as regards the grant of additional whaling facilities after the war." Since other oils (palm kernel, linseed, cottonseed, coconut, etc.) and tallows also yielded glycerin, the central issue during the war was to keep the production of explosives at its maximum while not creating shortages in edible fats. Thus, in the words of a leading figure in wartime fat processing, "without whale oil the Government would have been unable to carry out both its food and munition campaigns." See Cmd. 657, p. 67. According to a statement by Sir Starr Jameson of the Empire Resources Development Committee (cited in "Empire Resources," *Times* [London], 30 January 1917), British control in the Falklands meant that 600,000 barrels of whale oil came into England during the war at less than £40 a ton; by contrast, German glycerin

correspondence with Nordenskjöld in some impatience, invoking "a crisis without parallel in our history," and wrote to Lucas at the American Museum of Natural History, where plans were afoot to send Roy Chapman Andrews to South Georgia ("I know of no better man to continue Major Barrett-Hamilton's work than Andrews," wrote Robert Cushman Murphy), to say that "all such matters will have to be postponed" in light of the recent German attacks.[145] Darnley's committee was effectively suspended, though not before the documentation it had collected and generated—including Hinton's reports on Barrett-Hamilton's work—was codified into a set of highly confidential government briefs, tucked away for future reference.[146]

SPECIFIC IDENTITY, GEOGRAPHIC
SEGREGATION, AND MIGRATIONS

While the official interdepartmental work on the whale problem ground to a halt in 1915 and 1916, as governmental attentions turned to the war

---

producers were able to secure only small lots and were known to have paid as much as £300 a ton. See Harmer's comments on the Colonial Office letter of 22 February 1917, which are found in the minutes of meeting 4, circa 10 June 1918, SPRI MS 1284/1. There were about 6 barrels (40 imperial gallons each) to the ton. Harmer says "approximately six" in his review in *Geographical Journal* ("The Scientific Development of the Falkland Island Dependencies," *Geographical Journal* 56, no. 1 [July 1920]: 61–65); L. Harrison Matthews says six 40-gallon barrels to the ton in his *South Georgia, the British Empire's Subantarctic Outpost: A Synopsis of the History of the Island* (Bristol: John Write & Sons, 1931), 147; Tønnessen and Johnsen say 40 imperial gallons in 1920; see J. N. Tønnessen and Arne Odd Johnsen, *The History of Modern Whaling* (Berkeley: University of California Press, 1982), 301; see also p. 233 for a review of different barrel sizes used in this period.

145. Quotes from Murphy to Fagan, 10 June 1914, NHM DF 1004/CP/749/1, and Harmer to Lucas, 20 January 1915, NHM DF 1004/CP/749/2. The quote "crisis without parallel" comes from a letter from Harmer to Nordenskjöld after August 1914, the exact date of which I seem to have failed to record. Mea culpa! It can be found in SMRU MS 38600, box 645, file "Anglo-Swedish Antarctic Station." There is a very detailed treatment of the German side of this question in Lars U. Scholl, "Whale Oil and Fat Supply: The Issue of German Whaling in the Twentieth Century," *International Journal of Maritime History* 3, no. 2 (December 1991): 39–62. For an early treatment of the interwar development of these problems, see Karl Brandt, *The German Fat Plan and Its Economic Setting* (Stanford, CA: Food Research Institute, 1938).

146. On the "confidential" treatment of the reports, see Colonial Office to British Museum, July 1914, NHM DF 1004/CP/749/2.

and the relevant agencies focused on meeting national needs for edible fats and military glycerin, Harmer himself continued to collect data and draft his reports, never ceasing to call attention to what he saw as the imminent threat of extermination of the southern whales. In May 1915, for instance, he submitted to the trustees of the museum a summary of all the available catch data from South Georgia, rendering them in graphic form (figures 2.9 and 2.10) to dramatize the precipitous collapse of the humpback fishery. Rejecting as "an assumption of the most arbitrary nature" the opinion hazarded in the reports coming out of the Falklands government (namely, that the humpbacks must be finding so much food to the north that they were being delayed in their passage to South Georgia), Harmer instead insisted that a proper interpretation of the figures "gives rise to serious anxiety," in that they appeared to confirm Barrett-Hamilton's suspicion that the whalers were moving on to blues and fins, having done away with the humpbacks. The trustees forwarded a version of Harmer's report to the Colonial Office and included the plunging graphs.[147]

Moreover, despite wartime interruptions of shipping and communication, Harmer succeeded in maintaining transmission of form 132. Using the fetal data recorded by southern whalers on this form, he embarked on an ambitious study of the reproductive cycles of the three main commercial species of the southern industry.[148] The resulting report, circulated to the Colonial Office and elsewhere in October 1916, gave significant impetus to the formation (again, largely by Darnley), as the war came to a close, of a reconstituted interdepartmental committee dedicated to developing a major program of whale research in the southern waters.[149] Because this 1916

147. The manuscript version of this report to the trustees, dated 10 May 1915, can be found in NHM DF 1004/ CP / 749 / 2; the version forwarded to the Colonial Office by Fagan, labeled "B.M.(N.H.) 1016 / 15" and dated 16 July 1915, can be found in SMRU MS 38600, box 637.

148. On Harmer's efforts to keep the forms coming, see Harmer's letter to the director of the museum, 29 July 1915, NHM DF 1004/ CP / 749 / 2.

149. I found this document in SPRI MS 1228 / 3 / 1, South Georgia Archives, in a packet of materials that had clearly made the circuit from Harmer to the trustees to the Colonial Office to Port Stanley in the Falkland Islands to the magistrate at South Georgia. This document was complete, including all thirty-three appendices, and through the kind permission of Robert Headland, I was able to secure a copy. I subsequently found others (not all complete) at NHM and SMRU, indicating the wide distribution of the document (which has the B.M. register number 5042a / 16).

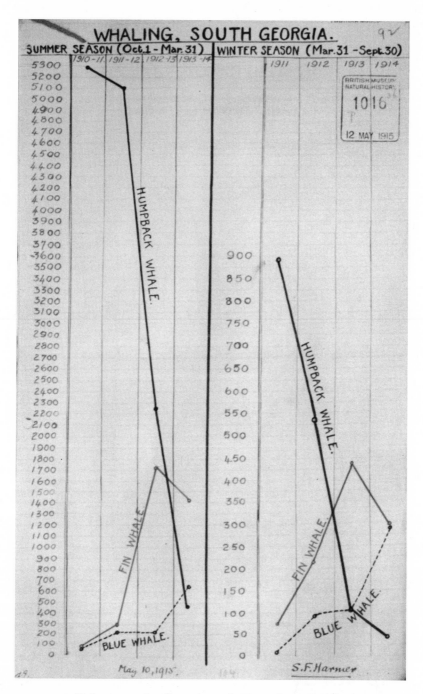

FIGURE 2.9 Trying to get a handle on the catch: Harmer's graph of South Georgia catch data. (Courtesy of the Natural History Museum, London.)

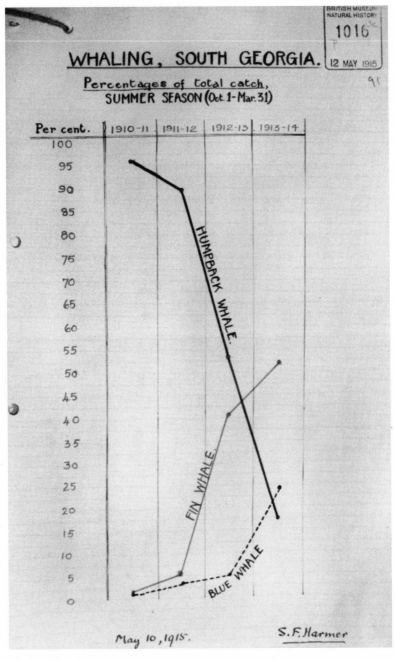

FIGURE 2.10 The catch, continued: Harmer worked to document what he took to be a dangerous collapse of the humpback stocks. (Courtesy of the Natural History Museum, London.)

report deployed the techniques of fetal growth analysis that would become the standard method in the postwar period, I will take a moment here to show how Harmer went about this significant investigation.

Tabulating the returns from the four South Georgia companies that filed form 132 for the 1913–1914 season and the five that did so for 1914–1915, Harmer recorded a total of 23 humpback fetuses, 93 from fin whales, and 105 from blues (two of which were twins). Fetuses of all sizes were found throughout the year. The question was, could these data be made to yield a clearer picture of the breeding cycle of these animals—and, in particular, of the timing of the mating season? Harmer believed they could. Taking the fin data to exemplify his approach, the first step was to plot the fetuses by date of capture and size, as in figure 2.11, where length (in feet and inches) runs up the y axis and the season, divided into half-month segments, runs across the x axis. Then, settling on a gestation period of a little over a year (an estimate based on the work of Guldberg and the guesses of whalers in the Finnmark waters) and assuming a linear growth curve for the fetus, Harmer correlated different fetal sizes with stages of the gestation period. Superimposing this linear curve onto the plotted data for the South Georgia fetuses (figure 2.12) permitted Harmer to group these specimens by (calculated) conception month. Acknowledging that the results pointed to no absolutely fixed breeding period, Harmer was nevertheless encouraged by the way this analysis correlated the majority of his specimens with pairings in July, and configured the whole data set into a plausible-looking distribution (the curve at the right in figure 2.13). Recognizing that there were plenty of sources of error in his analysis, he decided to do the same thing with the fetal data available in the earlier Norwegian literature on the northern fishery. To his delight, this yielded a local peak of impregnation shifted by almost exactly six months from the southern maximum. He plotted these curves (number of conceptions by month) for both the northern and southern stocks to dramatize this shift, which seemed to reflect that the breeding cycles of the whales in the two regions were strongly correlated with the seasons in the two hemispheres (the curve at the left in figure 2.13 represents the northern fins).[150] As he put it in the report to the trustees, "If the conclusion is correct it follows that in each Hemisphere the period

150. Harmer considerably elaborated this analysis in his report of 17 December 1918, working up a larger data set using several additional techniques. That report can be found in SMRU MS 38600, box 637, Harmer's memoranda on whales, binder 2.

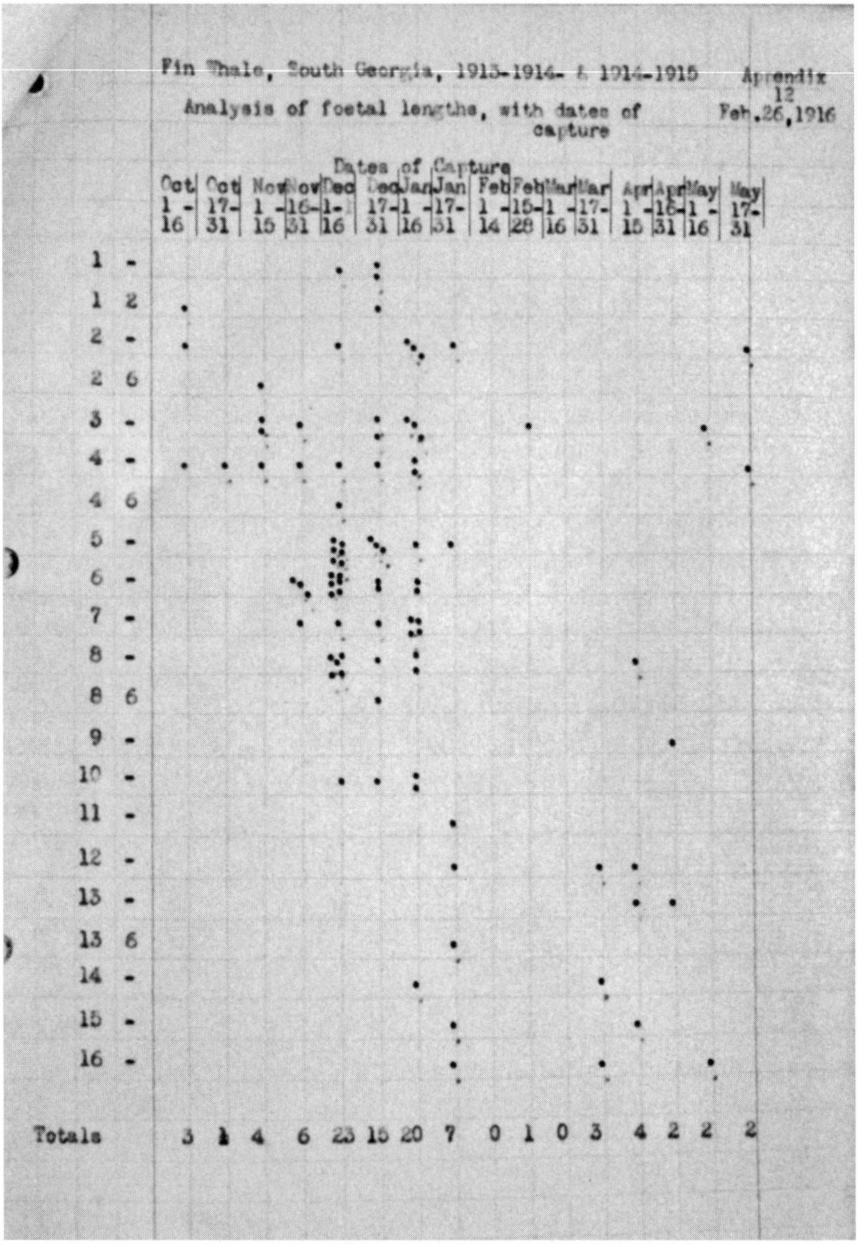

FIGURE 2.11 Statistical analysis of the reproductive cycle: Harmer's scatter plot of fetus length and capture date. (From the collection of the Scott Polar Research Institute, University of Cambridge.)

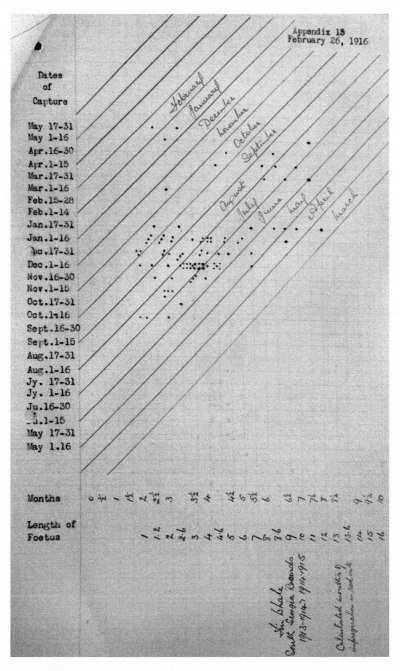

FIGURE 2.12 A mating season? Harmer's scatter plot of estimated conception months. (From the collection of the Scott Polar Research Institute, University of Cambridge.)

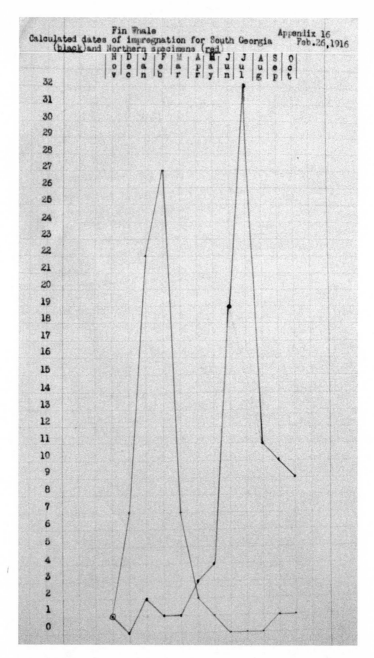

FIGURE 2.13 Stock segregation? Harmer's analysis of reproductive cycles led him to postulate separate northern and southern stocks of fin whales. (From the collection of the Scott Polar Research Institute, University of Cambridge.)

of maximum pairing activity falls in one of the winter months, when the whales have migrated away from the neighbourhood of the poles and are in warmer water near the equator. In each case the curve is at its lowest at a period coinciding with the highest point of the curve for the other hemisphere."[151]

A similar analysis for the blue whales of the north and south yielded a similar seasonal shift, but also a less focused "high season" for mating. In an effort to disaggregate this seemingly long mating period, Harmer tried breaking the data up by maternal size classes, producing some (not terribly strong) evidence that larger, older females were impregnated earlier than smaller, younger ones, a tentative correlation that suggested some stock segregation potentially useful to future regulatory efforts. With a much smaller set of humpback specimens, Harmer still succeeded in drawing a tidy curve for the peak mating period of the southern humpbacks, which he placed in September, later than both the fins and the bulk of the blues.

Summarizing his report, Harmer drew attention to the need for better data, and more of it: "In framing legislation for the control of whaling operations, with the view of preventing the extermination of the species, it is of the first importance to arrive at reliable result[s] with regard to the reproductive processes of the whales."[152] Having undertaken the first stages of this work, Harmer pointed out that his initial assessments "enable some prediction to be made as to the months in which foetuses of a given age may be expected to occur in the greatest number."[153] Only more data, preferably from regions to the north and south of South Georgia, would make it possible to "test these conclusions more fully." Perhaps the companies working in the South Shetlands and on the southern continental coasts could be persuaded to collect additional fetal observations? Given, however, that the aim of such work was progress toward closed seasons during the breeding periods, it was by no means obvious that this program would be successful. Regulations were never far from Harmer's mind; indeed, one of the concrete recommendations that he was able to offer in the 1916 report—after correlating the recorded fetuses with the sizes of the females carrying them—was that minimum size limits were needed and that they needed to be set higher than had previously been considered. There was, as he concluded in an internal memorandum to the trustees sent shortly

151. Ibid., p. 14.
152. Ibid., p. 7.
153. Ibid., p. 33.

after he filed the 1916 report, a "grave danger which threatens the whales," and more research was needed urgently.[154]

In view of the many fearsome matters pressing on the Colonial Office by the winter of 1916–1917, the response to Harmer's report was both prompt and manifold: not one letter, but two, adopting dramatically different stances. One was a "formal" letter that stayed firmly within the channels of semiofficial advisory communications (by this time well established) between the Colonial Office and the museum. Which is to say that, just as Harmer had submitted his report to the trustees, who forwarded it via the secretary of the museum to the secretary of state for the colonies (at this time Bonar Long) as a communication from the museum as an institution, the official Colonial Office response consisted of a letter from Long's assistant addressed to the director of the British Museum reviewing the Colonial Office position on the matter of whales and whaling. In sum, this letter, dated 22 February 1917, was dismissive: it reviewed the wartime rationales for reduced protection and increased hunting concessions; it pointed out that even were they desirable, restrictions were nearly impossible to enforce; and it rejected the insinuation that humpback stocks had collapsed (noting archly that "considerable caution would appear to be required in drawing inferences as to the frequency of local occurrence from the figures of whales caught"), backing up this assertion by generously forwarding a copy of Hjort's "Memorandum on the Distribution of the Whales in the Waters about the Antarctic Continent" from 1914. Indeed, the secretary offered several citations of Hjort's opinions, including his suggestion that the blues and fins might not even see an overall "reduction in numbers" (much less "extermination") under the current hunting regime, and concluded by assuring the museum that the Colonial Office would be sure to stay in close contact with Dr. Hjort concerning all future whaling questions.[155] Which was not to say that Long had not read Harmer's long and detailed report: he perused it closely enough to note that the conclusions about breeding (which he summarized as Harmer's finding that "the principal pairing season of all these whales falls within the period from June to October") ought

154. Harmer's report to trustees, 14 July 1916, NHM DF 1004/CP/749/4.

155. Hjort had spent much of World War I in England under a kind of self-imposed exile that emerged out of disputes over wartime Norwegian fishery regulations. See Tim D. Smith, *Scaling Fisheries: The Science of Measuring the Effects of Fishing, 1855–1955* (Cambridge: Cambridge University Press, 1994), 215.

to have reassured the concerned doctor a good deal; after all, "during this period no whaling is carried on at the South Shetlands and little at South Georgia, so that the whales would appear already to receive a high degree of protection in this respect."[156]

It is, of course, quite clear that this was not the conclusion Harmer intended his readers to draw, and the repeated references to Hjort were not calculated to set the keeper of zoology at ease. However, by the time Harmer held this perfunctory official letter in hand, he already knew that other wheels were turning at Whitehall.

THE ORIGIN OF THE DISCOVERY INVESTIGATIONS

On 19 February 1917, Darnley forwarded Harmer a shadow reply to his report:

I enclose a copy of a memorandum relating to a scheme which as yet has no official status, and consequently is not referred to in our official letter. I think it can hardly fail to interest you and we should be grateful for your criticisms and advice, not only as a biologist, but also *in view of your intimate acquaintance with a somewhat similar Swedish scheme.*[157]

Folded with the letter was a five-page "Memorandum on Whaling Research Expedition," drafted by the Scottish botanist and geographer R. N. Rudmose Brown, whose reputation as a polar explorer had been built on his service as a naturalist on the 1902–1904 Antarctic voyage of William Bruce's *Scotia*, which had done extensive surveying in the region that became the Dependencies of the Falkland Islands.[158] Since then, Rudmose Brown had

156. Grindle to British Museum, 22 February 1917, reproduced in the packet of materials filed in SPRI MS 1228/3/1, South Georgia Archives (see n. 149 above).

157. Darnley to Harmer, 19 February 1917, SMRU MS 38600, box 637, file "Whaling Research Expedition—Discovery Expedition—Harmer Correspondence." Emphasis added.

158. Rudmose Brown (with J. H. Pirie, R. C. Mossman, and W. S. Bruce) had published the best-known account of that expedition: *The Voyage of the "Scotia": Being the Record of a Voyage of Exploration in Antarctic Seas* (Edinburgh: W. Blackwood and Sons, 1906). The book, which helped make Rudmose Brown's reputation, has been republished (Edinburgh: Mercat Press, 2002).

done oceanographic research at Bruce's Edinburgh laboratories, served as vice president of the International Polar Congress of 1907, and eventually become a professor of geography at Sheffield University. During the war he had been seconded to Admiralty Intelligence, and it appears that it was in this capacity that he drafted a memo for Darnley outlining a postwar scientific expedition into the region of the Dependencies. A brief "Note on the Economic Resources of the Dependencies of the Falkland Islands" prepared by the Colonial Office and sent to the Admiralty at about this time makes it clear that considerations strategic, political, and fiscal all thrust the region to the attention of the British government at this moment: "This area," the note points out, "contains about 3,100,000 square miles or 1½% of the entire surface of the globe"; furthermore, "the area contains the only antarctic or sub-antarctic region containing good harbours open all the summer, and in the case of South Georgia throughout the year, and it is the only such region where any industry exists or is likely to exist for a long time to come."[159] While the "known wealth of this area is almost entirely marine," it was possible that mineral deposits on the islands—perhaps even coal—would come to light with future exploration. At present, however, the whaling industry eclipsed all other sources of revenue from this vast British territory: the document estimated gross returns of some £14,000,000 over the preceding ten years.

It was in this context that Darnley had begun to envision an all-British reincarnation of the suspended plans for an Anglo-Swedish whaling re-search expedition. Following up on testimony from the 1913–1914 hearings of the Inter-departmental Committee on Whaling and the Protection of Whales, Darnley conceived a major oceanographic undertaking, with both a ship component and a whaling station component, tasked to resolve the scientific issues Harmer had been pressing for more than half a decade. At Darnley's suggestion, Rudmose Brown's memo laid out, in a preliminary way, the logistics of such an enterprise, calling for the acquisition of a rein-forced wooden whaler of a type "now obsolete," but which would be able to go deep into the Antarctic ice (on account of a reinforced hull) and would have extended range (on account of being rigged to sail, using auxiliary steam only when necessary). In his heart, Rudmose Brown was calling for a ship not unlike his beloved *Scotia*, and he named as possible vessels such

---

159. The note can be found in NHM DF 1004/CP/749/4; it is unsigned and undated, but it appears to have been prepared by Darnley, and was forwarded by Long to the Admiralty in August 1917.

Antarctic-exploring workhorses as the *Terra Nova*, the *Aurora*, and the *Pourquoi Pas?* For the modern iron whalers he had no affection, declaring them not truly iceworthy and in any case "too small to accommodate the scientific staff, their laboratory, and stores" (though he could imagine such a vessel, built for the purpose at twice the normal size and strongly reinforced, as another option). While the use of the larger, slower, sailing vessel would mean that the expedition would be "unsuited for the chase of the quick moving finners and some of the smaller whales such as the dolphins" (thus striking one of the potential programs of research, tracking whale migrations directly by following the animals for long distances), Rudmose Brown suggested that the vessel might still be "fitted with whaling gear" for catching whatever whales came within range, in addition to more standard oceanographic equipment like trawls, dredges, and sounding devices.[160] Detailed whale biology, he noted, would need to happen on shore:

> The occasional use of a whaling station would be required for cutting up a captured whale. It would also be well to have the entrails of all whales examined at as many stations as possible with a view to obtaining information about their food. This might be done either by arrangement with the medical officer at the station, if there was one, or some other competent person, or by landing a scientist or trained laboratory assistant.

It is clear throughout this memorandum that Rudmose Brown had in mind a sort of *Challenger* expedition (lots of trawling, dredging, sounding, and collecting) that would pay particular attention to cetaceans. He also made sure to put in a plug for a Scottish base for the work (doubtless thinking that an influx of additional research resources might help shore up the publication of the *Scotia's* scientific work, which was lingering back in Edinburgh):

> Arrangements should be made with some university zoological laboratory in Britain to receive, unpack, and report on the material or distribute it when necessary to specialists. It would be preferable to choose a laboratory of one of the Scottish universities because of their

160. There was no mention, however, of plankton nets, demonstrating that Rudmose Brown's grasp of contemporary oceanography was as shaky as his sense of how one might pursue whale biology.

specialization in marine biology and Scotland's interest in whaling past and present in all east coast towns.

He suggested a scientific staff of four—two biologists and two oceanographers—and offered a back-of-the-envelope calculation: capitalization costs might run to £15,000, and annual expenses (he presupposed a multiyear program) would be in the neighborhood of £8,000.[161]

Harmer marked up his copy of this memorandum, and his marginalia, read together with his reply to Darnley, offer a clear picture of his role in shaping the emerging plans. "Cordially I welcome your proposal," Harmer wrote to Darnley on 5 April 1917, for a "post-war investigation of the questions connected with whaling in the Falkland Islands and their Dependencies." Harmer added that he particularly liked the idea of a "purely British undertaking"; if this went ahead, the lingering talk of a joint Anglo-Swedish plan could finally be put to rest. On specifics, Harmer had a number of suggestions: in the margins of Rudmose Brown's memo, he wrote in "Foetuses"—an aspect of whale biology that Rudmose Brown did not discuss—beside the paragraph dealing with whaling station dissections, and he annotated Rudmose Brown's paragraph on the use of Scottish institutions in some detail, writing in a large question mark and the letters "B.M. (N.H.)"—noting that his own institution was not only better equipped to deal with large osteological specimens but could also serve as the hub for dissemination and publication of results. He leaned on this latter point in his letter to Darnley, and also added that it was "essential" that such an expedition focus on the "obtaining of information about the reproductive processes." This meant, in his view, sending "an observer who has some knowledge of the reproductive processes in mammals," and who would thus be able to carry out detailed examinations of "the condition of ovaries and mammary glands." In turn this pointed to the desirability of establishing "a laboratory on shore where work of this kind can be done," presumably at a whaling station. As far as the ship was concerned, Harmer was prepared to defer to naval experts, but he suggested that both the *Discovery* (Scott's old ship) and the *Nimrod* might be available after the war, even as he also expressed enthusiasm for a vessel that might "follow and even capture the largest Blue Whales"—a vessel that could be used not just for plankton work (which he called "obviously essential") but also for

161. Rudmose Brown, "Memorandum on Whaling Research Expedition," 3 February 1917, NHM DF 1004 / CP / 749 / 4.

marking whales in order to secure "incontrovertible evidence with regard to the migration."[162]

This "unofficial" letter thus contains all the essential elements of the Discovery Investigations, right down to the naming of the eponymous vessel. In sum, Harmer was calling for a whaling station laboratory (for biological and anatomical work focused on the reproductive cycle of the commercial species) and a research ship for biological oceanography and whale marking (with a view to studies of distribution and migration). By the end of 1926, all the elements of this plan had been realized. A young zoology graduate from University College London named Neil A. Mackintosh was by then running a marine biological station at South Georgia and, in the company of two other young researchers, was working up whale carcasses on the plan at Grytviken every morning. And not one but two research vessels were engaged in the work at sea: the refurbished RRS *Discovery* was pursuing plankton surveys and general oceanography around South Georgia under sail, and the newly built RRS *William Scoresby*, a purpose-built research-scale whale catcher, was readying for its first efforts to shoot marking darts into the southern rorquals.

. . . .

To understand how Harmer's scientific wish list, sketched in this brief letter, translated into the largest British scientific expedition of the interwar period, it is necessary to turn to the formation and activity of the second major governmental committee to take up the issue of southern whaling: the Interdepartmental Committee on Research and Development in the Dependencies of the Falkland Islands. This committee, on which Harmer served prominently, met for the first time at the Colonial Office in April 1918 and issued, two years later, a detailed proposal for the whole project, including plans for financing and logistics.[163] Of this important document—the 1920 "Report of the Interdepartmental Committee on Research and De-

---

162. There are two copies of this document with what appear to be Harmer's annotations: in NHM DF 1004 / CP / 749 / 4, and in SMRU MS 38600, box 637, file "Whaling Research Expedition—Discovery Expedition—Harmer Correspondence." Both locations also house copies of Harmer's reply to Darnley, 5 April 1917.

163. The committee was composed of C. V. Smith (Admiralty), P. C. Lyon, Harmer, J. O. Borley, Darnley, and H. T. Allen. It issued the "Report of the Interdepartmental Committee on Research and Development in the Dependencies of the Falkland Islands," which was circulated in late 1919 and published in April 1920 as Cmd. 657 (see n. 71 above).

velopment in the Dependencies of the Falkland Islands," or Cmd. 657—
Darnley would later write with affection, declaring that "this Blue Book
had a soul."[164] As indeed it did: that "soul" was a vast renewal of British
Antarctic vigor, full of the "splendour of the opportunity for spiritual adventure which will be afforded to fresh minds eager to penetrate into the
unknown." To its defenders, Cmd. 657 represented a scientific undertaking
that would "develop men not unworthy to be mentioned in such company
as that of Darwin and Hooker, Ross and Nares, Murray and Bruce."[165] The
costs were high, but so were the stakes. The proposal won the attention of a
young secretary of state for the colonies, Winston Churchill, who kept attention focused on it during his tenure.[166] By early 1923 the proposal had been
authorized, and there was money to be spent. The clique of individuals who
had steered the program through the government for a number of years—
notably Darnley and Harmer—became the powerful and deep-pocketed
"Discovery Committee," which would oversee thirty-five years of whale research in the southern oceans.[167] Reviewing the push and pull that gave rise
to the 1920 report—with particular attention to Harmer's role in shaping
the scientific mission of the proposed expedition—will demonstrate how
emerging ideas about whale science gave form to the program that became
Discovery and how conflicting agendas were negotiated as a collaborative
research expedition came into focus.

The substance of Harmer's "unofficial" letter to Darnley of 5 April 1917
was essentially repeated, and ratified, in a formal letter sent by the trustees

164. See Rowland Darnley, "A New Antarctic Expedition," *Nineteenth Century and
After* 93 (May 1923): 718–28, at p. 718.

165. Ibid., p. 728. On the conflation of oceanographic science and the grand tradition of geographic exploration, see Helen Rozwadowski, *Fathoming the Ocean: The
Discovery and Exploration of the Deep Sea* (Cambridge, MA: Harvard University Press,
2005).

166. In fact, Churchill formed yet another committee, the "Interdepartmental Committee regarding Research Ships for Employment in the Dependencies of the Falkland
Islands." This committee, which met six times between 18 May 1922 and 7 February
1923, was composed of Harmer, Borley, Fortescue Flannery, J. M. Wordie, A. W. Johns,
I. Steed, B. Harvey, E. R. Darnley, and H. T. Allen.

167. This group, originally known as the "Whaling Research Executive Committee,"
met from 19 April 1923 until 1949, when the Discovery Investigations were folded into
the new National Institute of Oceanography. It was originally composed of Darnley,
Harmer, Borley, Wordie, R. W. Glennie, Allen, and Flannery; note that all but one of
these men (Glennie) had served on at least one of the preceding committees, and several
had served on two or more.

of the museum to the Colonial Office in December. In that correspondence the trustees gave their seal of approval to the plan for a research expedition to "conduct investigations into whales" in the waters of the Dependencies.[168] By that time Darnley had also secured the initial formal support of the Admiralty for the project, which had been approached in the late summer with a draft of the plan and invited to consider the opportunity to conduct "a hydrographic survey of the waters in which whaling is principally carried on." Since these waters were little known, and "abound in reefs and sunken rocks," a research vessel working on whales might plausibly serve as the platform for a variety of secondary investigations that would be of interest to the Admiralty and the Royal Navy, including "meteorological, geological, magnetic, and tidal work," not to mention studies of ice formation and marine currents.[169] In response to this invitation, the hydrographer of the navy filed a memo with the Colonial Office affirming that there was a need for this work: though the purpose of the expedition, he recognized, was "primarily to afford data for whaling policy," it would, he confirmed, "undoubtedly give an opportunity for scientific investigations of considerable value" to the Admiralty.[170] Hydrographic information, particularly on bathymetry and seasonal shifts in the pack ice, was seen to have potential value to the whale work as well. The idea that whale migrations might follow ocean floor contours retained some currency in the interwar period; moreover, since most whales were understood to be dependent on plankton, and it was supposed that plankton would prove sensitive to water temperature (among other things), the location of the ice edge was deemed a significant factor in whale movements.[171]

From the outset of the planning process, the Colonial Office, the Brit-

168. Fagan to Colonial Office, 31 December 1917, NHM DF 1004/CP/749/4.

169. Address to the Secretary of the Admiralty, 1 August 1917, in ibid.

170. O. Murray to Undersecretary of State for the Colonies, 9 November 1917, in ibid.

171. Perhaps the most fanciful statement of the contour hypothesis was that of A. G. Bennett, who wrote in the early 1930s, "Remembering the great antiquity of whales, it seems possible that modern generations are now merely following by inherited instinct the exceedingly ancient tracks of their forefathers; and these tracks may have led through comparatively shallow waters on the fringe of ancient land masses that have disappeared." See Bennett, *Whaling in the Antarctic* (New York: Henry Holt, 1932), 47. Harmer noted that the manager of the Scottish station at Harris, Carl F. Herlofson, was of the opinion that the whalebone whales in the northern waters tended to follow the 100-fathom line in their movements. See Harmer's report of 19 May 1921, SMRU MS

ish Museum, and the Admiralty alike saw the appeal for the undertaking from "an Imperial point of view" (as the secretary of state put it in a letter of 1917). In the supporting documentation there were gestures at meteorological theories that considered the Antarctic as the engine of weather in the Southern Hemisphere, and suggestions that the resources of the Union of South Africa and the government of India, as well as other southern colonies—all of which had an interest in weather prediction—ought to be marshaled to the emerging plan for the expedition.[172]

With these three powerful metropolitan institutions rallied to the cause, Darnley secured, in late January of 1818, "terms" for the formation of a new interdepartmental committee, which was charged to

> consider what can now be done to facilitate prompt action at the conclusion of the war in regards to the preservation of the whaling industry and the development of the other industries in the Dependencies of the Falkland Islands; and to consider not only the economic questions above referred to, and the scheme for the employment of a research vessel, but also what purely scientific investigations are most required in these regions and whether any preliminary enquiries by experts in this country should be instituted.[173]

And before the first meeting Darnley circulated to the prospective members of the new committee copies of Miscellaneous 298, the confidential evidentiary brief prepared by the prewar committee in 1915. Because it was a closely guarded document, he instructed each of the recipients to read it over and then return it to the Colonial Office without sharing any of the contents with anyone.[174]

At the first meeting of the new committee, Harmer moved to amend its charge in a way that met with universal approval. The change sheds helpful light on the links between a broad proto-ecological approach to the whal-

---

38600, box 637, Harmer's memoranda on whales, binder 3. For a brief discussion of ice, temperature, and whaling, see Cmd. 657, pp. 21–23, 132, etc.

172. The quote is from the address to the Secretary of the Admiralty, 1 August 1917, NHM DF 1004/CP/749/4. On meteorology and empire, see Katherine Anderson, *Predicting the Weather: Victorians and the Science of Meteorology* (Chicago: University of Chicago Press, 2005).

173. See SPRI MS 1284/1, citing Colonial Office letter of 22 January 1918, no. 228/1918.

174. See Allen to Harmer, 4 May 1918, SPRI MS 1284/1.

ing problem and the perennial issue of British reach in the deep south. Harmer informed the assembled representatives of the Admiralty, Colonial Office, and Fisheries Board and the other members that "the reference to investigations 'in these regions' [see the quote above] was too narrow in its application, since it would doubtless be necessary to conduct enquiries elsewhere" in carrying out researches on these migratory animals that continuously moved through the oceans of the world.[175] His motion carried, and the "terms" were redlined to include the phrase "in connection with" immediately before "these regions." The first move of the Interdepartmental Committee on Research and Development in the Dependencies of the Falkland Islands was thus, on a scientific rationale, to extend their mandate to encompass essentially the whole of the Southern Hemisphere. Wherever whales went, wherever their food went, or wherever factors affecting their food might be operating—these were the regions within Harmer's purview; upon his recommendation, they came within the purview of the committee as a whole. This expansive gesture served Harmer's research mission, to be sure, but it was also entirely consistent with the larger geopolitical interests that informed the membership of the committee. As Darnley put it when he convened the group, "the Dependencies of the Falkland Islands formed one of the larger undeveloped estates of the Empire, and that, as the Dependencies included open ports, it was possible to control marine industries over a large area."[176] Reach and control were originary preoccupations of the committee, and they provided the impetus for many aspects of its work (if not for its very existence). The question of the "political" rationales for the Discovery Investigations remains contested, but there can be little doubt that such issues were strongly at play in the emergence of the work and in its subsequent justifications.[177]

175. A reading of these moves in the broader context of British imperial ambitions is very much at odds with Marsden's (in my view inadequately historical) effort to understand the early Discovery work as "an early attempt at ecologically sustainable development." See her detailed and otherwise useful "*Discovery* Investigations," *Archives of Natural History* 32, no. 2 (2005): 161–76. NB: This is a special issue on the topic of the Discovery Investigations.

176. Minutes of meeting 1, 30 April 1918, SPRI MS 1284/1.

177. On the question of the political dimensions of the work, see Peter J. Beck, *The International Politics of Antarctica* (London: Croom Helm, 1986). But compare Rosalind Marsden, "Expedition to Investigation: The Work of the Discovery Committee," in *Understanding the Oceans: A Century of Ocean Exploration*, ed. Margaret Deacon, A. L. Rice, and C. P. Summerhayes (London: University College of London Press, 2001),

How else to explain, for instance, why the members of the Interdepartmental Committee on Research and Development in the Dependencies of the Falkland Islands devoted considerable attention to considering how to increase the representation of British personnel in the whaling industry in the deep south? Indeed, schemes were considered (though ultimately abandoned) for having the proposed "whaling research vessel" double as a kind of training boat for British harpoon gunners, who might, it was hoped, spearhead a British displacement of the overwhelming representation of Norwegians in this extensive and remote British colony. The docking of forty-one foreign vessels, against seventeen British ones, "upon territory under the British flag" was "a matter of grave national concern," as one preparatory document put it, particularly in light of Britain's ambivalent relations with the Scandinavian neutrals during the war.[178] Old military hands forwarded advice to the committee concerning the transition from a wartime footing to a whale-focused research and development program in the south: "I have remarked many times," noted the RAF wing commander H. M. Stanley Turner, MD, "that a seaman gunner from a destroyer would make an ideal whaling gunner."[179] And as late as 1923, when the shipbuilding firm of Flannery, Baggally & Johnson Ltd. were drafting their reply to the emerging specifications for a research vessel to undertake the expedition, they made it clear that they understood that the vessel would also be used "for educational purposes . . . to train British workers for the industry." They therefore proposed to construct a large catcher boat with a harpoon gun, accumulator apparatus, winches, and compressed air device for inflating carcasses—all this to allow green hands to practice whaling.[180] Only when the shipbuilding was under way would this plan be dropped. For a while the plan was to use the mounting developed for the harpoon gun to anchor a 6-foot-long whale-marking crossbow developed by Harmer and the Cambridge physicist C. V. Boys, but in the end this too fell away (a different marking device was eventually adopted), and no full-sized whale gun

69–86. For a helpful overview of the place of science in the political negotiations over the southern continent, see Aant Elzinga, "Antarctica: The Construction of a Continent by and for Science," in *Denationalizing Science*, ed. Elisabeth Crawford, Terry Shinn, and Sverker Sörlin (Dordrecht: Kluwer Academic Press, 1993), 73–106.

178. Flannery, Baggally & Johnson Ltd. to members of the committee, circa 1923, SPRI MS 1284/3.

179. Turner to Allen, 10 July 1918, SPRI MS 1284/1.

180. Flannery, Baggally & Johnson Ltd. to members of the committee, circa 1923, SPRI MS 1284/3.

sat on the bow of the early Discovery vessels.[181] A fallback plan to grant a Norwegian firm an extra license for a whaling vessel on the condition that it be used as a training ship for British personnel under the tutelage of a Norwegian "instructor" (with the company to reap the returns, but only if the instructor was prohibited from firing the gun) also came to nothing.[182]

That one committee could work simultaneously to stimulate a home-grown whaling industry and to stem overexploitation of whale stocks suggests, to be sure, that contradictory agendas were intersecting in the Whitehall conference room where the committee held its fortnightly meetings. As might therefore be expected, a review of even a small portion of the committee's correspondence and meeting minutes confirms that the 1920 report was the result of significant compromises among members holding divergent views. Particularly telling in this regard was the process by which a memorandum by Harmer, entitled "Danger of Extermination of the Whales," was broken up and transformed as the final document came together.[183] From the outset the committee's task called for different members to take responsibility for drafting memos on discrete subjects within the committee's general mandate: the Admiralty representative, for instance, was assigned the memo on hydrography; J. O. Borley, from the Board of Agriculture and Fisheries, wrote on the potential for developing a fishing industry in the region. These memos became the draft text for the various sections of the thirty-page, jointly authored final report. However, in a number of cases, more detailed single-author memos were also included in Cmd. 657, where they were printed in the extensive appendix. Harmer accepted the task of writing a memo encompassing the current state of scientific knowledge of whales (particularly their breeding and migration) and outlining research plans. While much of this material

181. A small harpoon cannon was eventually adopted for *Discovery* and *Discovery II* and, I believe, for the *William Scoresby* as well. For a discussion of the issue in relation to *Discovery*, see "Equipment Sub-Committee Meeting," 28 January 1925, item 219, subhead 7, SPRI MS 1284/4. There is a picture of the small cannon in use on *Discovery II* in SMRU MS 38600, box 643 (which contains several smaller boxes of negatives and prints), and in the *Star* of 3 October 1931 (reproduced in the NOL Discovery MS scrapbook of press cuttings, p. 95).

182. See Cmd. 657, p. 14, for an outline of this proposal.

183. I am here addressing in detail a point noted by Rosalind Marsden in "Expedition to Investigation: The Work of the Discovery Committee," in *Understanding the Oceans: A Century of Ocean Exploration*, ed. Margaret Deacon, A. L. Rice, and C. P. Summerhayes (London: University College of London Press, 2001), 69–86, at p. 82 n. 9.

was incorporated directly into the main text of the collective final report, his most charged section, which dealt sharply with the rapid fall in humpback catches at South Georgia and placed this decline in the larger context of his "historical argument" about the sequential self-destruction of all whaling industries from the sixteenth century forward, was transformed beyond recognition as the final document took shape. As it evolved from a 1,200-word essay, packed with statistics, into two paragraphs (a total of three hundred words), Harmer's contribution lost not only its edge of urgency but also the harsh language of "extermination" in its monitory title, becoming the tame subsection g of Cmd. 657, "Danger of Depletion of the Stocks of Whales," which noted irenically that "careful watching" of industry returns (and perhaps some nonspecific "measures" to keep up stocks) were "in the interests of the whaling companies as much as of the whales themselves."[184]

Harmer accepted this rewrite, but exported the stiffer material into his extensive individual essay for the appendix, a "Memorandum on the Present Position of the Southern Whaling Industry," where, not beholden to co-signatories, he pulled no punches.[185] There, in addition to reviewing his findings on breeding habits and fetal growth in detail, he explicitly addressed (critically) Hjort's optimistic work on distribution and intoned that there were "already grave signs of depletion" in some subantarctic stocks.[186] And Harmer went even further, warning presciently that "these are matters in which present prosperity is apt to induce those who are financially inter-

184. Cmd. 657, p. 12. Harmer's more pointed memo appears in the minutes of the sixteenth meeting of the committee, 23 May 1919, SPRI MS 1284/1. Particularly telling in any effort to parse the composed committee text are Harmer's marginalia in SPRI MS 1284/1; for instance, in the records associated with meeting 17, 12 June 1919, Harmer annotated his memorandum "Danger of Extermination of Whales" in pencil, striking out the word "Extermination" and noting, "Much modified by the committee June 12/19."

185. See Harmer to Allen, 25 May 1919, SPRI MS 1284/2: "The relegation of my section on the Danger of Extermination to the appendix makes it desirable to have something shorter on this subject for the report." In this way Harmer pushed to have the two paragraphs on stock depletion inserted into the main body of the collective report.

186. Cmd. 657, p. 81. Harmer also noted that the near-disappearance of the Greenland whale in the eastern Arctic seas had not been reversed after nearly a century without commercial pressure: "This is indeed, a legitimate answer to Dr. Hjort's argument based on the existence of a widely distributed Antarctic stock of whales."

ested to take an unduly optimistic view," on which account he encouraged the committee (and the government) to "proceed with caution" in assessing the evidence presented to them.

HEDGING AGAINST HJORT

The ad hominem quality of these remarks suggests the degree to which Harmer composed his appendix as a kind of counter-appendix, structured to riposte Hjort's 1914 "Memorandum on the Distribution of the Whales in the Waters about the Antarctic Continent," the essay Hjort prepared for Darnley's prewar committee. It is not clear precisely how the postwar committee decided to reprint this memo in its own report of 1920, but it is safe to assume that Harmer did not support doing so, and that it was included by those within the committee who did not wish to alienate the Norwegian and English whaling companies. A typescript copy of Hjort's memo is preserved in Harmer's notes from the meetings, and his irritated annotations ("No!"; "very doubtful"; etc.) make it clear that he viewed the memo's confident assertions about a vast circumpolar stock with as much skepticism as ever. Aware, as he put it to the trustees of the museum in late 1917, that "the Colonial Office seems inclined to adopt . . . [the] optimistic view which has been maintained by Dr. Hjort," Harmer went out of his way in the late war years to build a brief against Hjort's analysis.[187] In one of his whaling reports forwarded to the trustees of the museum in this period (another of the reports passed by them to the secretary of state for the colonies), Harmer devoted several pages to a direct criticism of Hjort's 1914 memo, calling it little better than a brief for whalemen: "Dr. Harmer thinks he is not misrepresenting the general purport of the memorandum by describing it as an attempt to prove that whales have been and are so numerous in Antarctic waters that the extent of the whaling operations now in progress need give no cause for immediate anxiety"—a position Harmer asserted was easily refuted by Hjort's own work on the northern industry.[188] As the postwar committee expanded the scope of its investigations, Harmer continued to bulldog not only Hjort but an array of other Scandinavian scientists and whaler-witnesses.

187. Harmer's report to trustees, 18 October 1917, NHM DF 1004 / CP / 749 / 4.

188. See Harmer, "Report on Whaling Operations, Principally in South Georgia and South Shetlands, 1915–1916," 2 July 1917, SMRU MS 38600, box 637, Harmer's memoranda on whales, binder 1.

In 1918 word of the renewed Colonial Office investigations spread rapidly throughout the whaling communities of England, Scotland, Norway, and even the United States, particularly after the committee began soliciting new testimony from individuals in the industry. Scientific attention, too, was forthcoming: in the late summer of 1918, *Nature* abstracted an article by R. C. Murphy on the scale of the Antarctic whaling industry and inserted an editorial comment calling for "the advice of scientific experts, who . . . have not been consulted in the matter." This charge brought a prompt response from Harmer, whose correction letter—reviewing Colonial Office exchanges with the British Museum since 1910 and mentioning both of the relevant committees on whales and whaling—was published in a subsequent issue, further publicizing the government's commitment to the principle of scientifically informed regulations.[189] By autumn the *Times* of London was reporting, under a Christiana dateline, that "whaling interests in Norway . . . have been much interested in an announcement recently made here of the appointment of a committee by the British Colonial Office to investigate the subject of whales and the whaling industry." It was the intention of the Norwegian association of whalers, the piece went on, to provide their own position paper on the subject: "A complete scientific report of the whaling industry undertaken with a view to the rational utilization of the present stock of whales is in course of preparation here, and will shortly be published and dispatched to the British Committee."[190] When the report—"Studies on the Modern Whale Fishery in the Southern Hemisphere," authored by Hjort's junior scientific collaborator, the fisheries biologist Einar Lea—appeared, Harmer wrote letters to members of the committee that helped to ensure it was buried, having been deemed (in a damning phrase) "to emanate more from the bar than the bench."[191]

Harmer orchestrated similar assessments of the testimony offered by the leading "practical whalemen," whose opinions were solicited by the committee in view of their extensive direct experience. A questionnaire on the operations of the industry (finances, pay scales, technologies, etc.) was

189. See "Notes," *Nature* 101, no. 2546 (15 August 1918): 468–72, at p. 470 (for Murphy's article), and "The South Georgia Whale Fishery," *Nature* 102, no. 2552 (26 September 1918): 65 (for Harmer's answer).

190. "Late War News," *Times* (London), 9 September 1918, clipping in SPRI MS 1284/1.

191. The quote actually comes from Borley, who wrote to Harmer that "I have to agree that his utterances seem to emanate . . ." See Borley to Harmer, 14 June 1919, SPRI MS 1284/2.

prepared by the committee in October 1918, to which Harmer appended an additional series of tailored questions about whale biology: Were the northern and southern species the same in the view of whalemen who had worked both regions? Why were the humpback catches dwindling to zero? What were the breeding habits of the commercial species? When and where did they pair? How long was the gestation period? How big were the young at birth? How long did they nurse? Harmer also wanted to know if the whalers believed that the data on form 132 were recorded accurately by the companies. This question suggests that he had begun to suspect the quality of this information (a suspicion he would confirm mathematically several years later).[192]

The founder and dean of the southern industry, Captain C. A. Larsen himself, provided written replies to all these questions and also came to London to give personal testimony to the committee. During the course of that visit, he met independently with Harmer, who questioned him closely on the breeding issues. Harmer's notes of that meeting survive, and they reveal his particular preoccupations at this period, as well as his misgivings about industry expertise. For instance, when Larsen expressed "no doubt" that the commercial species in the north and south were identical, Harmer put before him his recently assembled evidence of a six-month shift in the breeding cycles in the two hemispheres, noting to himself that "Captain Larsen admits" this difference. Larsen did not think it mattered, however, since the reindeer that the Norwegians had introduced to South Georgia (to provide fresh meat) rapidly shifted their breeding cycle by six months. Pressing the issue of pairing, Harmer was apparently gratified to find that Larsen concurred with some of his other findings as well: "Captain Larsen agrees completely with the conclusions I have come to, from the study of statistics; namely that the fin whale and the humpback [in the south] have well-marked periods of maximum frequency of pairing, which fall in or about July for the fin whale and a month or two later for the humpback; that the blue whale also has a maximum but that the pairing period is more spread out over the year than in the other two species."[193]

It is of course possible that Harmer was all too ready to hear such a wholesale ratification of his views (painstakingly reached by means of the

192. The full list of questions is enclosed in Allen's letter to Knudsen, 2 October 1918, SPRI MS 1284/1.

193. Minutes of meeting 11, 26 October 1918 (including discussion of "Memorandum of an Interview with Captain Larsen"), in ibid.

proxy of form 132), but the fact remains that Larsen did not give this response when he testified before the whole committee, where he was cited as
announcing that "Fin whales and Blue whales apparently have no definite
pairing season" and (in a statement more directly damaging to Harmer's
efforts) that "it is not possible to judge [when pairing takes place] from the
size of the embryo, since at the same period a very small embryo may be
found in one whale, and a very large one in another." Harmer lost no time
in writing to the secretary of the committee to complain:

> There are certain important respects, particularly with regards to
> the length of period of gestation, the occurrence of a definite pairing
> period, and the length of the young at birth, in which Captain Larsen
> told the committee one thing while he told me something different,
> in a personal interview that I had with him at the museum within
> a day or two of the date when he gave evidence before the commit-
> tee. I feel that his credibility is seriously affected by these discrep-
> ancies; and I am very doubtful whether we ought to publish all his
> statements.[194]

Despite this protest, much of Larsen's testimony did make it into Cmd.
657, including his assertion that he was "absolutely of the opinion that no
danger threatens" even the humpbacks and, further, that there was no need
for a closed season because economic exigencies would afford such protections automatically should they ever prove necessary. In the end, however,
most of his views, and those of the other whalemen, were shunted into a
pair of appendices.

INPUT FROM THE AMERICANS

Harmer served the interdepartmental committee as the clearinghouse
for advice and information from a different set of sources, more remote
from the industry and more sympathetic to his own position. Prominent
among these correspondents were the American naturalists with whale
interests, particularly Murphy, Andrews, and Lucas. In 1916 Harmer had
been a prompt reader of the newly published *Whale Hunting with Gun
and Camera* (subtitled *A Naturalist's Account of the Modern Shore-Whaling*

194. Harmer to Allen, 16 January 1919, SPRI MS 1284/2.

Industry, of Whales and Their Habits, and of Hunting Experiences in Various Parts of the World), a narrative account of manly whale research by the cowboy spirit of the American Museum of Natural History, Roy Chapman Andrews.[195] Part Teddy Rooseveltesque sea romp, part field journal of a collecting expedition, this volume recounts Andrews's exploits in the company of whalers pursuing humpbacks and grays (along with several other species) on both coasts of the Pacific. Harmer's reading notes on the book indicate that he was particularly taken with the final chapter, "The Passing of the Whale," and that he copied out several of the bleaker passages in which Andrews predicted the fate of what he called a "marvelously specialized group of animals." For instance, Harmer recorded Andrews's opinion that "with the introduction of the modern harpoon gun the commercial extinction of whales is inevitable if the slaughter continues unchecked."[196] Not only would Harmer quote from this chapter in his appendix to Cmd. 657 (warning, in Andrews's words, that "the world hunt for whales 'bids fair to end ere the close of the twentieth century'"), but he even brought his personal copy of the book to one of the committee's early meetings, where it circulated; it was agreed that copies would be purchased for all the members.[197]

In the summer of 1918, Harmer wrote separate letters to Lucas, Murphy, and Andrews (praising the last for his "fascinating book"), seeking their advice and recommendations. He mentioned to all of them that plans were afoot for a government-sponsored research expedition, and he expressed his fears, writing, "I regard the present rate of destruction of whales with very serious anxiety" and adding that "unless something can be done to improve the position, I am convinced that the whales are likely to be either exterminated or reduced in numbers to an extent from which they may never recover." He encouraged each of the Americans to pass along sug-

195. Roy Chapman Andrews, *Whale Hunting with Gun and Camera: A Naturalist's Account of the Modern Shore-Whaling Industry, of Whales and Their Habits, and of Hunting Experiences in Various Parts of the World* (New York: D. Appleton, 1916). On Andrews and the "gaming of the whale," see Gary Kroll, *America's Ocean Wilderness: A Cultural History of Twentieth Century Exploration* (Lawrence: University Press of Kansas, 2008).

196. I found a file of Harmer's reading notes on Cetacea in the NHM Zoology Collections, file "Whales—miscel. notes on."

197. The quote is from Cmd. 657, p. 81. On the circulation of the book at the meetings, see the minutes of meeting 3, 30 May 1918, SPRI MS 1284/1.

gestions for protection measures, admitting that it was unclear whether the best scenarios were within reach, since the "question is complicated by the existence of vested interests which cannot be altogether ignored."[198]

Andrews was out of the country and unable to reply, but both Lucas and Murphy composed long letters outlining their ideas about the future of marine conservation generally and protection of whales in particular. Murphy's response was the more detailed, and it is clear that Harmer was much taken with it, since he immediately forwarded it to the committee chairman with a note explaining that it was "an important letter" from "an authority on the subject" who had "himself visited South Georgia."[199] While Murphy concurred with Harmer that the rise of modern whaling had created a "crisis" that "calls for heroic measures" (and went on to affirm that the history of sequential collapse of earlier whaling enterprises "permits us to draw no other conclusion"), he also spelled out explicitly the need for commercial realpolitik in confronting the problem:

> I take it that all of us who are interested in this important and even critical subject, wish to put conservation on a business basis. In other words, it is not our intention to stop whaling but rather to regulate the industry that it will give promise of a reasonable, *perpetual* yield.

Moving from this premise, Murphy saw the gravest problem not in "vested interests" but, as he put it, in "our more or less complete ignorance of such subjects as longevity, age at sexual maturity, frequence of breeding, the breeding 'season' (if there be one), period of gestation, rate of growth, seasonal segregation of the sexes, whether the animals are monogamous or polygamous, routes and distances of migrations, etc." For this reason, Murphy expressed the opinion that "the most profitable undertaking for permanently bettering the condition of the whales would undoubtedly be a scientific expedition to the grounds of the far south to carry on and extend the work of the lamented Major Barrett-Hamilton." With the reports of such an expedition in hand, he continued, "we might then be able to tell just where we stand, and to construct graphs of the prospects. If we could talk to legislators in terms of figures and certainties, action might fol-

---

198. The letters from Harmer to Lucas and Andrews were both dated 5 June 1918; the two sent to Murphy came later, dated 12 July and 30 August. All four can be found in SPRI MS 1284/2. Emphasis in the original.

199. Harmer to Allen, 20 March 1919, in ibid.

low more easily than we think." Necessary "bag limits" and closed seasons might be established, along with total protection for various species. The North Pacific Fur Seal Convention of 1911, which limited the exploitation of seals in the Pribilof Islands, suggested success was possible, and the ending of the war ("and the enormous demand for the glycerin content of the whale oils") meant the time was right: "It would be splendid if the conservation of marine mammals, including the seals as well as the whales, could be made one of the features of the reconstruction." International action, based on better scientific information, was the long-term solution.[200]

For Lucas, a Wilsonian internationalism loomed even larger, and he outlined an intergovernmental scheme not entirely unlike what would eventually take shape as the International Whaling Commission (though not until after World War II):

It is suggested that a small international commission or bureau be established with headquarters in London, Washington or New York to gather information relative to whaling and issue a report thereon, such bureau to be under the control or form part of some institution already established, as one of the fishery boards of Great Britain or the British Museum, or the Bureau of Fisheries, the U.S. National Museum or the American Museum of Natural History in the United States. Such a bureau to be supported by grants from the various governments interested and so far as possible to use the means for gathering information as are already available in the way of consuls, consular agents or custom house officials. Such a report as suggested would show the number and locality of whaling stations, the date of their establishment, the number of steamers employed and number of whales taken annually for as many years past as convenient. Such information would give some idea of the number of whales killed and show whether the number taken has increased or decreased in the last decade. Information as to former stations and the reasons for their abandonment is most important as indicating commercial extermina-

200. Though Murphy held to the notion of individual nations controlling local activities, and indeed, even suggested that whale resources ought to be, in the end, "made a national monopoly" controlled by a "civil government agency" that could either exploit them directly or lease the right to do so to bidders "exactly as Peruvian guano contracts or other privileges are leased." It is unclear to me how he thought this idea fit with the "business basis" of conservation. All quotes from Murphy to Harmer, 5 March 1919, in ibid.

tion of whales in that immediate vicinity. Increased catch is usually the first evidence of overfishing and increase of number of stations and no increase in catch is proof conclusive of the decline of the fishery.

Interestingly, Lucas's focus on collation of general information about the industry on a worldwide scale, using the diplomatic infrastructure already in place, appears to have proceeded from his concern that more sophisticated biological investigations—for all their interest and scientific value— might prove too labor-intensive and intricate to marshal to the cause: to "conserve" the whales. As he put it, "information is needed as soon as possible, as an effort to obtain too much or too detailed information would probably defeat the end in view." It was a position not clearly articulated by other scientists involved in the issue; wholly disregarded, it proved remarkably prescient.[201]

In contrast to Lucas (but like Murphy), Harmer manifestly believed that scientific investigation of whales—the more of it the better—was the necessary point of departure for any political action. From the earliest meetings of the committee, he worked to keep whaling research at the forefront of its attention.[202] In 1919, as the committee worked with increasing attention on the final composition and revision of their soon-to-be-issued report, Harmer labored to ensure that the scientific questions at stake in the whale research came sharply into focus. At the very top of his list was the need to "test the theory which has been advanced by Dr. Hjort, Captain C. A. Larsen, and other Norwegian experts, that the Antarctic circumpolar waters constitute a vast reservoir of whales."[203] Harmer, as I have shown, was inclined to believe that all the species moved north and south annually in regular migrations, a pattern best exemplified by the humpbacks. If this was the case, there was no reason to think that the circumpolar waters were a

201. Lucas to Harmer, 10 March 1919, with enclosed two-page document "Suggestions for Control of the Whale Fishery," in ibid. There were also two other letters from Lucas to Harmer (28 June 1918 and 13 August 1918), but while I found Harmer's references to them, he appears to have passed them on and failed to keep copies (or they have been lost).

202. At the second meeting, for instance, Harmer was described in the minutes as having "expressed the view that it was desirable to proceed first with the investigation of the whaling question" and pressing for the collection of additional data for the British Museum. See minutes of meeting 2, 14 May 1918, SPRI MS 1284/1.

203. Cmd. 657, p. 12.

vast "reservoir" of the commercial species. Instead, the zones frequented by the animals might well be circumscribed, their passages temporary and local, and their numbers, in the end, much smaller than might be supposed by someone looking at a polar projection map of the southern seas and assuming that those expansive regions were all inhabited by great numbers of great whales.[204] Regarding the location of the Dependencies of the Falkland Islands on such a map with migrations in mind led to the observation that the outlying islands of South Georgia and the South Shetlands were strategically placed to intercept a southward stream of whales funneling from the whole of the Atlantic, or even a mass migration circulating through the southern waters (if indeed this was what they did); these were worrying visions to anyone concerned with the vulnerability of the stock.[205] It is likely that Harmer's preoccupation with such discrete migrations proceeded at least in part from his reading of Andrews's work, since Andrews had done the majority of his Pacific research on the gray whale, whose coastal migrations were both highly focused and extremely regular.[206]

204. Darnley summed up the issue this way: "Two theories have been propounded; one is that the same stock of whales is attacked year after year at South Georgia and the South Shetlands. The other is that the whales are drawn from a stock living all around the borders of the Antarctic continent. If the first theory is true the industry would be very unlikely to survive on its present scale. If the second is true probably the catch could be increased without danger to the stock. I would suggest that the information so far obtainable is wholly insufficient to enable us to decide whether either of these theories is correct, and that further information is particularly needed as to the geographical distribution of the stocks, their migrations, whale food and reproduction." Minutes of meeting 1, 30 April 1918, "Memorandum of statement by Mr. Darnley on the whaling industry," circa 1917–1918, SPRI MS 1284/1.

205. Harmer said as much in a note in *Nature* in 1923, published on the occasion of the mobilization of the *Discovery*: "It will be seen, however, by consulting a map, that South Georgia and the South Shetlands lie in the region where the Antarctic Ocean is narrowest, and that they are admirably situated to intercept the stream of whales. . . . It would not be surprising if operations at these stations alone were found capable of depleting very seriously the entire stock of Antarctic whales." "Scientific Investigation of the Whaling Problem," *Nature* 111, no. 2790 (21 April 1923): 540–41.

206. Roy Chapman Andrews, "Monographs of the Pacific Cetacea I—The California Gray Whale," *Memoirs of the American Museum of Natural History*, n.s., vol. 1, part 5 (March 1914): 231–87. It seems likely that Harmer's ideas about migration were also informed by the work of Thomas Southwell, whose article on the migration of (what he called) the right whale included two maps illustrating seasonal movements in the

FINALIZING AND IMPLEMENTING THE PROGRAM

To resolve the question of whether the Antarctic species behaved similarly, Harmer pressed the committee to include within the research plan for the proposed expedition not only shore station whale biology and whale-food-oriented biological oceanography, but also a dedicated "whale-marking scheme." He drafted the language for the proposed scheme, which was incorporated into the final committee report almost unaltered:

> The adoption of the method of marking whales, probably by the use of a small projectile which could afterwards be recovered when the whale is finally captured, appears to us the most practical means of obtaining direct and conclusive evidence with regard to many important facts that need determination. Some of the migrations undertaken by individual whales could thus be definitely established.

With time and luck, the same program might also yield insights into maturation and longevity. So promising did this plan of research seem to Harmer that, as he went on to state, "it is in the highest degree desirable that future experiments should be undertaken at an early date in order to ascertain whether this method can be successfully employed." He imagined "a small barbed spike" in the form of a "cartridge or harpoon," which would be shot into the animals with force sufficient to penetrate the blubber. Records of the location of the marking would be kept and a reward offered for every mark recovered.[207] It was a scheme that had been discussed by Harmer and Barrett-Hamilton as early as 1913 and which Hjort and the Norwegians were already reportedly testing, though as yet without success.

Harmer's insistence on the importance of this aspect of whale research had a large and direct influence on the final form of the research expedition, since it became clear that such a marking regime would be difficult (if not impossible) to conduct from the kind of large oceanographic vessel—complete with mobile laboratories—most suitable for studying plankton, ocean chemistry, and general hydrography. By August 1918 Darnley had begun to muse about the need for a second vessel, a

---

seas east and west of Greenland: see Southwell, "The Migration of the Right Whale (*Balaena mysticetus*)," *Natural Science* 12 (June 1898): 397–414.

207. See minutes of meeting 17, 12 June 1919, Harmer draft document "Marking of Whales," SPRI MS 1284/1. For the published version, see Cmd. 657, p. 9.

steam launch of some sort, that might be used for "stalking and marking whales."[208] While the report of 1920 suggested that motor launches carried by the main research vessel might be adequate for the marking work, this idea was later dropped as infeasible. By 1923, when actual preparations for the investigations were under way, a tailored two-ship plan had been adopted: one would be an oceanographic vessel (the *Discovery*), and one would be a "whale catcher," built largely for the purpose of marking (the *William Scoresby*).

In the endgame leading up to the committee's 1920 report, Harmer remained vigilant for last-minute alterations, particularly those that touched on the presentation of the "whale problem." For instance, given that Hjort's South Polar distribution map, speckled thickly with reported whale sightings all around the Antarctic continent, was to be printed in the report, Harmer pressed (though unsuccessfully) for permission to include graphs of catch data that he believed would buttress his warnings about humpback stocks. He had better luck, however, in a telling exchange with J. O. Borley, the committee member from the Board of Agriculture and Fisheries, who had continued to massage the South Georgia catch data and had found a different way to present the numbers. Just a few weeks before the final draft of the report went off for proofing, Borley wrote Harmer to suggest the inclusion of a chart showing the number of whales caught *per catcher vessel* between 1909 and 1917. This presentation changed the overall pattern of the data, since in place of a plunging total humpback catch, Borley could show that "the decline in the catch of Humpbacks per boat has not been continuous."[209] Harmer's reply was diplomatic: Borley should proceed as he wished. But Harmer went on to stake out the terrain for his own view in the body of the report: "I confess, however, that I should be sorry to have your paper given in the immediate neighborhood of what I have myself said on the subject of the reduction of the number of the Humpbacks. From the way you put it there seems to me a danger that the average reader might infer that after all the reduction is not serious, whereas I myself take the view that it is hardly possible to exaggerate its gravity." Even presenting the data as Borley sought to do, Harmer noted, left a 90 percent slide in the figures over a mere five seasons. The framing of the data, Harmer insisted,

208. Darnley memorandum, 27 August 1918, SPRI MS 1284/1.

209. Borley to Harmer, 4 July 1919, SPRI MS 1284/2. The trend was still downward, dropping from 200 humpbacks per boat in 1909–1910 to 12 in 1916–1917, but there were several brief upward trends in between.

would affect its reception, and this mattered a great deal to the long-term success of the undertaking.[210]

By the late summer of 1919, as the final document went to press, Harmer slipped away from London with his family, making off to the Atlantic Hotel in Antrim for vacation and to await the reception of the report.[211] He had good reason to take a rest. In March 1919, after a messy and disputed process, Harmer had been appointed to the directorship of the British Museum (Natural History), and he had decided to take on those new administrative responsibilities while maintaining his post as keeper of zoology.[212] Knighthood, customarily bestowed on the director, followed, and Sir Sidney took his place firmly on the plinth of South Kensington as a highly public scientific figure, arguably the nation's most commanding presence in matters of natural history.

This elevation gave Harmer a powerful new position from which to draw attention to the whaling question and to push for support for the recommendations of Darnley's committee. These initiatives percolated through the government between 1919 and 1923, stalling several times (largely due to fiscal anxieties—the conservative budget estimate for the whale research expedition exceeded £200,000). During this period Harmer missed no opportunity to speak on behalf of the expedition and to rally support within scientific and learned communities in Britain and beyond, and he went so far as to withhold his approval from other oceanographic research programs proposed during the period for fear that they would detract from the momentum gathering behind the southern whale expedition.[213] Even

210. Harmer to Borley, 6 July 1919, in ibid.

211. Harmer to Allen, 29 July 1919, in ibid.

212. For an account of the controversy after the announcement of Sir Lazarus Fletcher's retirement, see William Thomas Stearn, *The Natural History at South Kensington: A History of the British Museum, 1753–1980* (London: Natural History Museum, 1998), 110–18. The dispute involved the proposed elevation of Charles E. Fagan, the "assistant" secretary of the museum (there was no "secretary"). The prospect of the appointment of a director from the administrative side rather than the scientific sections met with strong opposition from university-based natural scientists around Britain, and the trustees were obliged to withdraw the plan. Stearn concludes that "Harmer behaved most honourably throughout," and notes that he maintained his good relationship with Fagan, who, elevated to "secretary" as a consolation prize, died shortly thereafter.

213. There was talk of a "*Challenger*-style" expedition at the Cardiff meeting of the British Association for the Advancement of Science in 1920. Harmer (who was

as the committee's report was still in the final stages of production, Harmer took advantage of an invitation to address the Royal Institution to promote the emerging proposals, choosing as the topic of his lecture "Subantarctic Whales and Whaling." In this presentation (to the institution's regular Friday evening assembly) he outlined his monitory history of successive whaling industry collapses before turning the audience's attention to the recent boom in the modern Antarctic fishery. Reviewing the statistics he had amassed at the British Museum, Harmer proceeded to detail the conflicting arguments around the diminution in the humpback catches. "The question of the disappearance of the whales is not merely a sentimental one," he asserted. Not only would "zoologists . . . naturally view their extermination with deep concern on scientific grounds," but there was also the issue of their economic importance; during the war the whales had been of "vital importance," since the glycerin they afforded had proved "essential for our National security." Scientific researches promised to provide the information necessary for "protective measures," Harmer assured his listeners, and to illustrate how he took them through his recent analysis of fetal data, showing how such measurements could be used to calculate breeding seasons. He concluded by explaining the role of the British government and the Colonial Office in initiating the necessary study, and he revealed that plans were afoot for a "scheme for an Expedition which is to investigate the whaling problem on the spot, with the view of obtaining information on which legislation may be based."[214] He offered similar lectures at the British Association for the Advancement of Science and elsewhere, including more specialized scientific settings like the annual meeting of the Association of Economic Biologists and a gathering of the *Challenger* Society. He also published notes on the program of the expedition and its importance in *Nature* and elsewhere, going so far as to write a review of the committee's report for the *Geographical Journal* praising its recommendations (while

---

on the council) wrote Allen to say that he felt he could not support it if it conflicted with the emerging whale work. Harmer to Allen, 30 October 1920, SMRU MS 38600, box 637, file "Whaling Research Expedition—Discovery Expedition—Harmer Correspondence."

214. This presentation, to which members of the interdepartmental committee were invited, was printed in the *Notices of the Proceedings of the Royal Institution* 22 (1922): 537–41. It is interesting to note that Harmer had the text of the presentation vetted by the secretary of the committee; see Harmer, draft of "Discourse to the Royal Institution," 16 May 1919, SPRI MS 1284/2.

acknowledging that, as an author, he perhaps had to be "pardoned" for his enthusiasm).[215]

This continuous agitation, coupled with sustained and diplomatic advocacy by Darnley and others within the Colonial Office (who had to thread the plans through delicate negotiations with business leaders and Foreign Office hands concerned with Scandinavian relations), succeeded in winning approval for the initial, indispensable element of the plan: a significant hike in the export duty charged on whale and seal oil landed in the Falkland Islands and Dependencies. Cmd. 657 had noted that the tax in effect in 1919 (3½d. per barrel) represented a real rate of less than one-half of one percent at prewar prices, which had more than doubled by the war's end; the report thus asserted that "a substantially increased revenue can be obtained from the industry without imposing any undue burden upon it, and it may be anticipated that it will prove sufficient to cover the salaries of a staff of scientists, a considerable portion, if not the whole, of the cost of the initial scientific equipment, and also the expenses of maintaining this equipment."[216] The committee declined to suggest a specific rate. The general principle of their recommendation, however, gained force when the 1919–1920 season proved the most lucrative ever and Norwegian companies, flush with cash, paid dividends of up to 240 percent.[217] Pressure from the report, coupled with this bonanza, led to a series of increases in the tax, with a considerable portion of the revenue to be assigned to a fund earmarked for research

215. A partial list: "History of Whaling in the South and Statistical Evidence of Breeding Periods Derived from Foetal Records," *Challenger Society* 1, no. 16 (1919): 6–7; "Modern Whaling," *Reports of the British Association* 89, sec. D (Edinburgh, 1921): 422–23; "Scientific Investigation of the Whaling Problem," *Nature* 111, no. 2790 (21 April 1923): 540–41. Harmer's review of the committee's report was published as "The Scientific Development of the Falkland Island Dependencies," *Geographical Journal* 56, no. 1 (July 1920): 61–65.

216. Cmd. 657, p. 28.

217. See J. N. Tønnessen and Arne Odd Johnsen, *The History of Modern Whaling* (Berkeley: University of California Press, 1982), 301. Note that the contracting system for payments was hugely complex, and made more so by wartime contingencies, fluctuating currencies, and so forth. Tønnessen and Johnsen write stoically that "it would be possible to write a large book on the basis of the tremendous number of documents between the Norwegian Foreign Ministry and the [British] Foreign Office, between the two countries' legations, between oil brokers, whaling companies, the Whaling Association, De-No-Fa [De Nordiske Fabriker, a major buyer-producer], Lever Bros., the inter-departmental correspondence in Britain and Norway, the minutes of numerous conferences, etc." (p. 296).

and development in the Dependencies.[218] By 1921 H. T. Allen could write to Harmer that "we are building up a nice little nest-egg from the export duties on whale oil," and therefore it might not matter whether the Admiralty was prepared to kick in a substantial financial contribution to the whale research expedition.[219]

With money in the bank, detailed planning got under way, and Harmer was appointed to yet another Colonial Office committee—this one charged with making specific recommendations about the acquisition of research vessels.[220] As a result of the report issued by this group, by 1923 the Colonial Office owned Scott's storied vessel, and a costly refitting was in the works. Shortly thereafter the newly named "Discovery Committee"—the executive group assembled to oversee the expedition—had arranged for a contract of tender to buy a custom-manufactured "Steel Screw Whale-Marking Vessel," the ship that would be christened the *William Scoresby*.[221]

218. The expedition was not the only item earmarked for support out of these monies: there were plans for a variety of improvements to South Georgia, including a wireless station. The tax was raised as high as 5s. a barrel, which would have represented something like 8 percent of 1920 prices, but a slump in 1921 made this rate prohibitive, and a series of refunds and corrections were applied, yielding a sliding tax scale pegged to the price of oil itself. J. N. Tønnessen and Arne Odd Johnsen review the system (which was replaced with a fixed rate of 2s. per barrel in 1928–1929) in *The History of Modern Whaling* (Berkeley: University of California Press, 1982), 301–2. For the actual text of the ordinance, see Legislative Council of the Falkland Islands, "Ordinance No. 6, 1924: An Ordinance to Provide for the Establishment of a Research and Development Fund for the Dependencies," *Falkland Islands Gazette* 34, no. 1 (1925): 15–16, as well as "Ordinance No. 7, 1924: An Ordinance to Provide for the Refund of a Portion of the Duties of Customs Levied and Collected in Respect of Whale Oil and Seal Oil, under section 7 of the Tariff Ordinance, 1900," 17–18, which deals with the refund provisions.

219. Allen to Harmer, 11 January 1921, SMRU MS 38600, box 637, file "Whaling Research Expedition—Discovery Expedition—Harmer Correspondence."

220. See n. 166 above for details on this committee.

221. Given that the *Discovery* was purchased for £5,000, and that the refurbishment alone (keel to topsail) cost no less than £45,000, it is tempting to suggest that what was actually purchased was little more than an escutcheon, and thereby a genealogical link to the paragon of British polar heroism. For these numbers, see Darnley's memorandum of 10 March 1923, SMRU MS 38600, box 637, file "Whaling Research Expedition—Discovery Expedition—Harmer Correspondence." The *Scoresby* contract was drafted in 1924 but apparently not signed until 1925 (according to the Discovery Committee's first annual report of 1926: Great Britain, *"Discovery" Expedition: First Annual Report, 1926* [London: His Majesty's Stationery Office, 1927]). For a copy of the contract itself, see SMRU MS 38600, box 642 (see figure 2.31 below).

Darnley, with £155,000 suddenly at his disposal, was imbued with a sense of history in the making, and he wrote to the inner circle of individuals who had helped see the plan realized, intoning the magnitude of the undertaking that now lay before them:

> We are now embarked on an enterprise of great economic and scientific importance, and one which is likely to attract considerable public interest. Its scale is not inferior to that of the voyage of the *Challenger* or of the previous voyage of the *Discovery*, while the work is likely to last longer, at least five years, and perhaps ten.[222]

Darnley was correct that the plan for a whale research expedition, and the prospect of a new Antarctic voyage for the legendary *Discovery*, would capture the attention of the British press. Shortly after the announcement of the purchase of the vessel, the *Times* ran a story under the banner "The Progress of Science" detailing the proposed research program and the economic significance of the questions at issue. Harmer's role in bringing the issue to light, as well as his cautionary views, also received considerable space.[223] From 1923 to 1925, as the vessel received its overhaul and personnel were selected for the mission, dozens of papers followed the developments closely, and many of the reports were infused with the romantic glamour of a new Antarctic expedition under the Union Jack. The *Daily Chronicle*, for instance, remarked that the "highbrows" headed for the roaring forties to study whaling problems would need "training both mentally and physically" to conduct their researches through three years of "hardship"; the language of "exploration" and "adventure" found its way into many articles, and references to Scott's legacy were ubiquitous.[224] At the same time, however, a good deal of the coverage did note that this was a "New Type of Expedition" (as the *Liverpool Daily Courier* put it)—not another polar dash, but an unprecedented "Effort to Save the Whales."[225] Several papers

222. Darnley to members of the committee, 10 March 1923, SMRU MS 38600, box 637, file "Whaling Research Expedition—Discovery Expedition—Harmer Correspondence."

223. "Whales and Whale Fisheries: Need of Investigation," *Times* (London), 10 July 1923, in NOL Discovery MS, scrapbook of press cuttings.

224. "Three Years in Antarctic: Highbrows Under Forty for New Expedition," *Daily Chronicle*, 1 Feb 1924, in ibid. For other examples of polar glamour, see "*Discovery* on a New Antarctic Trip: Rebuilding Scott's Famous Old Vessel," *Evening News*, February 1924, and "Preparing for Next Year's Adventure," *Times* (London), 15 August 1924, both also filed in ibid.

225. "Efforts to Save Whales," *Liverpool Daily Courier*, 5 December 1924, in ibid.

took up the compelling idea of the whale-marking scheme, reporting that "Labels for Whales" promised to "save the whales of the southern seas from extermination."[226] If "whales' identity discs" were successful, then, "like salmon," their breeding migrations might be traced.[227] Other dailies described the aim of the program in still more familiar terms, explaining to readers that it was hoped that some day whales might roam the oceans, "living the artificial life of a sort of marine pheasant, with a close season, a whaling license, and a specially protected maternity home for mother whales in some suitable Antarctic sound." Promising as this scenario might sound, a considerable number of articles expressed concern about the real prospects of conservation, and a few, invoking the long-lost British east-coast whaling industries of Dundee and Hull, even broached the subject of extermination: "All the more useful sorts of whales," wrote one editorialist, "are like to go the way of American bison and the carrier pigeon long before their life histories have been fully investigated."[228]

Early reportage on the Discovery Investigations seized on these specific elements—the danger of extermination and the need for scientific research—in large part because of Harmer's own sustained efforts to communicate the significance of the program as widely as possible in these inaugural years. In addition to keeping up a regular series of short reports in scientific journals, he also made appearances in a variety of municipal halls and at sociable gatherings to promote the undertaking: for instance, he turned up on the dais of the Art Workers' Guildhall, Queen Square, with the visiting governor of the Falklands (J. Middleton) to talk about the whale research expedition; and he lectured on the need for "intensive study of the whole whaling problem" at a hospital benefit at Woolwich Polytechnic.[229] Indeed, the extensive publicity eventually brought Harmer and the Discovery Committee under critical scrutiny from the most unlikely corners:

226. "Labels for Whales," *Daily Mail*, 20 November 1923, in ibid.

227. "Whales' Identity Discs," *Yorkshire Telegraph and Star*, 21 May 1925, in ibid.

228. This and the previous quote are from "The Doom of the Whale," *Bridlington Free Press*, 20 June 1925, in ibid.

229. See, for instance, "The Blue Whale," *Nature* 115, no. 2886 (21 February 1925): 283ff, in ibid. Harmer's presentation at the Art Workers' Guildhall (he was their annual Gilbert White lecturer) evidenced a comity with the Falkland Islands administration that would not last; it is reported in "Whaling Research: Plans of the 'Discovery' Expedition," *Morning Post*, 3 March 1924. For the presentation at Woolwich, see "Sir Sidney Harmer on Need for Regulation," *Times* (London), 9 February 1925. Both articles are also filed in NOL Discovery MS, scrapbook of press cuttings.

word of the whale-marking program reached English antivivisectionists, some of whom protested the whole undertaking, and denied that the animals would remain "unconscious" of the steel darts fired into their bodies, since "the whale is sensitive enough to undergo 'more than sufferable anguish.'"[230]

If much of England thus looked on in the autumn of 1925, as the rechristened *Discovery* set sail (after a considerable delay) on its southern expedition, the Scandinavian whaling interests had been watching nervously for some time. As soon as word got out in early 1923 that the expedition had received a green light, Hjort sent a personal note to Harmer with another direct plea for inside information on the prospects of new regulatory efforts:

> As the whalers often have consulted me it would interest me very much—in spite of my position as a private individual—to receive any information in the matter, which you might be kind enough to send me. You will understand that I at any time should be more than glad to offer such services in a question like this, which might be considered useful.

In the same note he complained that though his "little laboratory" at the University of Christiania was running smoothly (and he had no shortage of students in fisheries science and oceanography), he remained, sadly, stranded: "No boat I am sorry to say," in palpable contrast to the emerging British undertaking.[231]

230. Letter from E. H. Visiak to *Daily News*, 27 January 1925, citing an article in *London Calling*. Visiak mentioned *Moby-Dick* as suggestive evidence for the sensitivity of the animals. A copy of this letter found its way into Harmer's set of Discovery Committee minutes: minutes of meeting 25, 1 January 1925, item 216, SPRI MS 1284/4.

231. Hjort to Harmer, 31 January 1923, SMRU MS 38600, box 637, file "Whaling Research Expedition—Discovery Expedition—Harmer Correspondence." Hjort had read London dispatches in the Norwegian papers that morning, which were telegram summaries of a report in the *Morning Post*; the *Morning Post* was in turn reporting on an abstract published in *Nature* of Harmer's lecture on 10 November 1922 to the Association of Economic Biologists, in which Harmer was reported to have said, "The efforts of all lovers of Nature should be directed to the restriction of whaling to an amount which is not inconsistent with the permanent preservation of these magnificent marine mammals and of the industry which they are so unfortunate as to support." *Nature* 110, no. 2772 (16 December 1922): 827.

Harmer's reply was prompt, though not inviting. "The information given by the Norwegian papers, as referred to in your letter of January 31," he wrote, "appears to be perfectly correct, so far as you report it to me." He went on to confirm the substance of the rumors:

> The Colonial Office are making definite arrangements for sending an Expedition to South Georgia and other localities, in order to make a scientific examination of the whole Whaling problem. This is a matter in which I have taken a special interest for some years, and I have been very anxious not to see the whales unduly reduced in numbers. The history of Whaling throughout the world shows that this result has been almost invariably attained as the direct consequence of whaling operations.

Harmer minced no words concerning his feelings about Hjort's place in this problem: "I confess to being specially afraid of your own view that the Oceans are large enough to make the activities of Whaling Companies, in restricted localities such as South Georgia, a matter of small importance." Pressing the issue of migration, Harmer reiterated his argument that the islands of the Dependencies were uniquely positioned to cut the whales off at a geographically constricted passage point in their hypothesized movements, and he went on to chide his Norwegian colleague for his sanguine (if not sanguinary) views: "It seems to me our duty as Zoologists to make every effort to prevent extermination."

Reviewing the intentions of the British expedition, Harmer then outlined the research program he was in the process of finalizing. It will be worth giving his description in full, since the next sections of this chapter will examine in greater detail the direct work on whales described here:

> There is thus every reason for obtaining the most accurate and reliable evidence with regard to the natural history of whales, with the view of ascertaining to what extent whaling ought to be restricted, and in short whether any practical measures can and ought to be taken, in order to avoid extermination or anything like it. Information seems to be specially required with regards to the migrations of whales and their causes. It seems obvious that these movements are partly influenced by the supply of food, and are partly connected with their breeding. It will accordingly be a special object of the Expedition to obtain information on these subjects, by an examination of plankton and hydrographic

conditions, by the recording of foetal specimens and if possible by the marking of individual Whales.[232]

As for Hjort's offer of his "services," Harmer was noncommittal: perhaps he might contact the executive committee when it was assembled. It appears that Hjort did not write back.

Norwegian anxiety was not allayed by the continuing developments at the Colonial Office, particularly as Harmer's own views became increasingly public and his administrative role in the forthcoming expedition was finalized.[233] Indeed, Harmer was obliged to address the issue of his personal views explicitly in a statement made in *Nature* in April 1923, where he sought to quell the tremors of industry resistance, writing, "Although I do not conceal my personal conviction, as at present informed, that whaling is being conducted on too large a scale, I do not deny that a study of the subject by competent investigators on the spot may lead to a different conclusion." Acknowledging that "the willing cooperation of the whaling companies will be of the greatest value to the expedition," he further sought to reassure them that the Colonial Office had not been hijacked by a cabal of anti-whaling naturalists:

> The Trustees of the British Museum have acted in an advisory capacity to the Colonial Office since they first became interested in Antarctic whaling, not long after its inception. I am authorised to state that they do not desire to take up an extreme position in the matter, but that their efforts are directed to the restriction of whaling to an extent which is not inconsistent with the permanent preservation of whales. This is a moderate view, with which it may be hoped that the representatives of the whaling industry will agree in principle.[234]

232. This and the quotes above are from Harmer to Hjort, 14 February 1923, SMRU MS 38600, box 637, file "Whaling Research Expedition—Discovery Expedition—Harmer Correspondence."

233. See Sidney Harmer, "Is the Whale Being Exterminated?" *Anglo-Norwegian Trade Journal* 9, no. 98 (February 1923). For the finalization of Harmer's position on the Discovery Committee (he refused to accept an honorarium, believing that it was necessary for him to be unpaid in order to discharge his full responsibilities to the trustees and the museum), see Harmer to Darnley, 18 February 1923, SMRU MS 38600, box 637, file "Whaling Research Expedition—Discovery Expedition—Harmer Correspondence."

234. Harmer, "Scientific Investigation of the Whaling Problem," *Nature* 111, no. 2790 (21 April 1923): 540–41.

It appears that Hjort used the mounting pressure around the whaling question to lobby for new resources in order to undertake a Norwegian expedition to parallel the British one, since by the autumn of 1923 there was talk of bringing the distinguished research vessel *Michael Sars* back into commission for whale work under Hjort. With that prospect in view, Hjort made another approach to the Discovery group, writing this time not to the inflexible Harmer, but to J. O. Borley at the Ministry of Agriculture and Fisheries, who also served on the Discovery Committee and held the UK equivalent of Hjort's old fisheries office. "I have begun to draw up the plans for an eventual expedition," he explained, and then outlined a northern program nearly identical to that sketched by Harmer for the *Discovery* in the Antarctic:

> My ideas do mainly centre about the following groups of problems:—
> The feeding grounds and migrations of whales in the waters between Iceland and Norway in connection with marking experiments.
> The quantitative distribution of those animal plankton organisms, on which the herring and whales feed.

Hjort also explained that he had spent six weeks over the summer at a Norwegian whaling station, where he had already begun some preliminary investigations. He closed with yet another request for information about the plans for the *Discovery*, coupled with a suggestion that collaboration might be fruitful: "I consider a co-operation very desirable in all these groups of problems and should very much like to go over to London for a few days provided that I may have a chance of discussing these problems with you, your staff, and the Committee for the Antarctic expedition."[235]

Borley passed the query to Harmer, asking how he thought they ought to respond. Harmer favored caution, reading Hjort's enthusiasm as strategic while at the same time acknowledging his skill as an oceanographer and the potential value of his expertise. "Judging from the line he formerly took I suspect that he is feeling anxious lest our Committee should make recommendations adverse to the recent interests of Norwegian Whaling Companies," Harmer suggested, and thus "he would obviously be in a stronger position if he could point to investigations of his own else-

---

235. Excerpts of Hjort's letter as forwarded to Harmer by Borley on 27 November 1923, in SMRU MS 38600, box 637, file "Whaling Research Expedition—Discovery Expedition—Harmer Correspondence."

where." His thoughts on the plankton work and marking might certainly be solicited, Harmer thought, since "the two methods of investigation he specially indicates are practically identical with what we had in our own minds." He added, damning with faint praise, that "Hjort is a first-rate observer, and except in cases where he may be prejudiced in a particular direction, I should have a high opinion of the value of his assistance."[236] Harmer could afford to be somewhat smug, since by this time he had in hand a full-scale prototype of an English whale-marking device and plans for a private display of the weapon in the courtyard of the British Museum. The Discovery Committee's plans were clearly considerably ahead of Hjort's.

Borley (who was looking forward to the demonstration of Harmer's whale-marking crossbow) fell right into line, writing to Harmer that he had "admirably" summed up the "situation in regard to Hjort": "It did not occur to me that there might be any second motive for instituting his investigations quite in the way you framed it. . . . I confess it did not occur to me that he wanted to be on the inside, so to speak, of English whaling work, so that he might know how to value it, if not perhaps influence it."[237] In the end, efforts were made to ensure that the plankton nets and hydrographic instruments carried by the *Discovery* were compatible with those used by Hjort in the north in order to ensure comparable results, and the chief zoologist for the British expedition, Alister C. Hardy, would make a short voyage with Hjort in the *Michael Sars* in July 1924, both to field-test three different whale-marking guns and to gain familiarity with the techniques of biological oceanography in use on the ship.[238] But despite these exchanges, and despite Hjort's later work through ICES and the Bureau of International Whaling Statistics in Norway, Harmer never forgave Hjort for his early views on whales. There is an irony here, in that the Norwegian scientist's work on mathematical models of population dynamics would profoundly influence cetacean stock assessment across the twentieth century, leaving a legacy to later whale biology in many ways more lasting and significant than Harmer's dogged efforts (as I will show in chapter 5).[239] In

236. Harmer to Borley, 29 November 1923, in ibid.

237. Borley to Harmer, 30 November 1923, in ibid.

238. For a report of this expedition, see A. C. Hardy, "Report on a Voyage with Professor Hjort on the 'Michael Sars,' July 1924," NOL Discovery MS, Allen Minutes.

239. Hjort was the lead author on one of the most significant papers in twentieth-century fisheries biology, "The Optimum Catch." This paper, the fruit of Hjort's work at Cambridge in physiology and zoology in the late 1910s (and reflecting his exposure to

a private letter drafted in 1930 to Stanley Kemp, the director of research for the Discovery Investigations, Harmer (by this time retired from the museum, though still involved in the management of the ongoing Discovery Investigations) lamented what he saw as Hjort's betrayal of what should have been their common cause:

> I have throughout been very disappointed with Hjort. As a Zoologist he must know as well as you or I do that disaster is likely to come. As a Politician he has to support the operations of the whaling Companies. But I cannot help feeling that if he had resisted the temptation to conceal the views I feel sure he must hold as a Zoologist he could have done much to educate public opinion in Norway.[240]

By 1930, however, it had become abundantly clear that Hjort had been right, and Harmer wrong, about the scope and scale of the Antarctic stocks: there were vast numbers of large rorquals all the way around the polar landmass, not tight migrations of small populations that streamed by the headlands of the Scotia Archipelago, as Harmer had feared. Or, rather, it was a little more complicated than that: Harmer's vision of concentrated and vulnerable migrations *had* been correct as far as the humpbacks of the southern Atlantic were concerned, but those stocks had been more or less wiped out by the early 1920s, and the whalers had shifted their focus to fins and blues. Moreover, by 1930, shore whaling from South Georgia was in terminal decline, and with it the fortunes of the Discovery Investigations. Pelagic whaling from massive free-wheeling stern-slipway factory vessels had gradually taken commercial whaling beyond the jurisdictional frameworks of the Colonial Office and beyond the revenue-generating reach of the whale-oil tax levied on landings in the Dependencies. Since the inception of the research program a decade earlier, the situation had rapidly grown much more grave and considerably less tractable.

---

novel statistical techniques for population analysis), "completely reoriented the thinking of a generation of fisheries biologists" by showing the potential for predictive models of population change. For a detailed account of this paper and its argument, see chap. 6 of Tim D. Smith, *Scaling Fisheries: The Science of Measuring the Effects of Fishing, 1855–1955* (Cambridge: Cambridge University Press, 1994). The original piece appeared in a special "Essays on Population" issue of the *Hvalrådets Skrifter*: see Johan Hjort, Gunnar Jahn, and Per Ottestad, "The Optimum Catch," *Hvalrådets Skrifter* 7 (1933): 92–127. See also the general piece on the history of whales and whaling in the same issue: Johan Hjort, "Whales and Whaling," *Hvalrådets Skrifter* 7 (1933): 7–29.

240. Harmer to Kemp, 30 October 1930, SGB Papers.

THE SCIENTIFIC WORK OF THE
DISCOVERY INVESTIGATIONS

The Discovery Investigations were, as I have indicated, a massive and complex series of interlocking research initiatives, one of the very largest investments of public funds in science made by the United Kingdom before World War II. Across the interwar period this program deployed three major research vessels: Scott's old wooden *Discovery* (figure 2.14), the purpose-built whale catcher–like *William Scoresby* (figure 2.15), and the 1,036-ton *Discovery II* (figure 2.16), a floating oceanographic laboratory capable of traveling 10,000 miles without refueling. Together, these ships made more than a dozen lengthy expeditions into the oceans of the Southern Hemisphere (figures 2.17 and 2.18), employing in the neighborhood of a thousand sailors, scientists, and administrators, who collectively burned through more than a million pounds sterling (at a time when the sum total of British support for other nonmedical scientific activity in the colonies was less than £500,000). Discovery served as the training ground for essentially every British oceanographer and marine biologist to achieve international standing before 1965, and its published output—in the form of the booklike series of *Discovery Reports* volumes—totaled more than 15,000 large-format pages.[241] There is obviously a great deal to be said about such an enterprise, and it is surprising how little attention it has received in the secondary literature. At this point I have shown in some detail how this undertaking arose and described the bureaucratic mechanisms, research questions, and preliminary investigations that shaped its mission and de-

241. There is an enormously valuable volume-by-volume summary of the sales and distribution of the *Reports* through 1937 in SMRU MS 38600, box 644, Harmer's "Scientific Work," file 17. It should be noted that the scope of the *Reports* can be thought of as, to some extent, an indictment of the Discovery Investigations as a whole. By keeping so much of the published output of its work "in-house" and in something like book form, the Discovery Investigations hearkened back to the era of the great nineteenth-century ocean expeditions (above all, the *Challenger* expedition, with which Discovery personnel were inclined to compare their own work). But by the 1920s and 1930s, peer-reviewed journal publication had almost entirely superseded such monographic publishing arrangements, and the scientific results of the Discovery Investigations ended up coming into the world in a peculiarly incestuous and closeted mode that doubtless hurt their reception. Marsden makes a version of this point in her article "Expedition to Investigation: The Work of the Discovery Committee," in *Understanding the Oceans: A Century of Ocean Exploration*, ed. Margaret Deacon, A. L. Rice, and C. P. Summerhayes (London: University College of London Press, 2001), 69–86, at p. 78.

FIGURE 2.14
(*left,* and see PLATE 3) A
good wooden ship: The
Discovery. (From Great
Britain, Discovery Committee, *Report on the Progress
of the Discovery Committee's
Investigations,* opp. p. 8.)

FIGURE 2.15
(*below,* and see PLATE 4)
A research ship to run
down whales: The *William
Scoresby.* (From Great Britain, Discovery Committee,
*Report on the Progress of the
Discovery Committee's Investigations,* opp. p. 12.)

FIGURE 2.16 (and see PLATE 5) The floating laboratory: The *Discovery II*. (From Great Britain, Discovery Committee, *Report on the Progress of the Discovery Committee's Investigations*, frontispiece.)

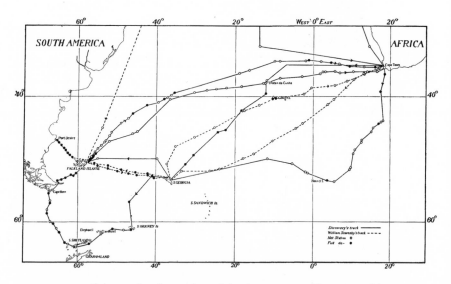

FIGURE 2.17 Crisscrossing the southern Atlantic, 1925–1927: The routes of the *Discovery* and the *William Scoresby*. (From Great Britain, *Discovery Investigations: Second Annual Report*, p. 12.)

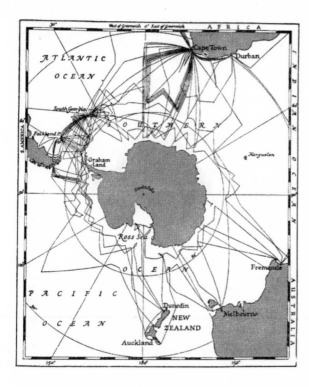

FIGURE 2.18
Circumnavigat-
ing the South Pole:
Routes of the *Dis-
covery II*, 1930–1939.
(From Mackin-
tosh, "The Work
of the Discovery
Committee," p. 7.)

fined its objectives. It will be the aim of this final section of this chapter to summarize and assess the scientific work of the undertaking as a whole. That undertaking, as I have shown, began as a "whale research expedition" and went on to shape the study of commercially exploited cetaceans for much of the twentieth century.

So how did it all go? Well, in some sense Harmer's vision for a science of the great whales of the Southern Hemisphere was fully realized (though whether he would have been entirely happy with the result is a question I shall take up in more detail below). In 1942 Neil A. Mackintosh— the man Harmer hired back in 1924 to follow in Barrett-Hamilton's hip-booted footsteps as Discovery's flensing-platform cetologist in South Georgia—published an authoritative summary of the research initiative he had come to lead. His monograph, "The Southern Stocks of Whalebone Whales," which appeared in volume 22 of the *Discovery Reports*, admirably summed up almost two decades of biological investigations and exempli-fied the achievements of Discovery whale research. By this time Harmer's health was failing, and in December of the same year he withdrew formally from the Discovery Committee. The truth was that he had not been active for some time and his interests had passed increasingly to the care of his garden (the great man's mind softened considerably in his later years). But had he been wholly compos mentis, he would surely have recognized in Mackintosh's pages the answers to many of the questions that had been posed in the course of the British Museum's dealings with the Colonial Office back before World War I. Indeed, the table of contents of Mackintosh's summa could pass for a direct point-by-point response to Harmer's wish list from the days of his service on the Interdepartmental Committee on Research and Development in the Dependencies of the Falkland Islands: specific identities of the southern species of large whalebone whales; their food, parasites, and pathologies; their breeding cycles, growth curves, and life spans; their distributions and migrations, and the seasonal changes in the composition of the stocks seen on the hunting grounds. To be sure, definitive answers on a number of these points had not been reached (there were, for instance, several techniques for establishing the age of a dead whale, and they did not all consistently agree, leading to some ambiguity concerning longevity; Mackintosh was a hesitant person and tended to emphasize that there remained many unanswered questions, particularly where migrations and stock segregation were concerned), but the twofold attack on the problems of whale biology originally laid out by Harmer—gross anatomical investigations oriented to an understanding of reproductive physiology on the one hand, marking expeditions that relied on an expanding industry for mark recovery on the other—had succeeded in resolving a good deal of the basic natural history of the large whales of the Southern Hemisphere. So, for instance, life cycle graphs like those shown in figures 2.19–2.22 could be drawn for blues and fins, representing a visual epitome of the work Mackintosh and his Discovery partner F. G. Wheeler carried out over several years in the mid-1920s on the flensing platform at Grytviken, South Georgia, during which time they measured, sampled, and probed some 1,683 large whales as they were broken down for the boilers. It had been grueling work, work that to a considerable degree defined Discovery whale biology, and work that others would emulate.

Young Mackintosh had been working as a curator at the Huxley Museum of the Royal College of Science when Harmer and the Discovery Committee plucked him up to lead the first anatomical/physiological work at South Georgia.[242] And while he was a plausibly sturdy character (having won his

---

242. Minutes of meeting 15, 19 March 1924, SPRI MS 1284/4. Mackintosh, like L. Harrison Matthews, was to some extent an inside candidate; both had been in contact with

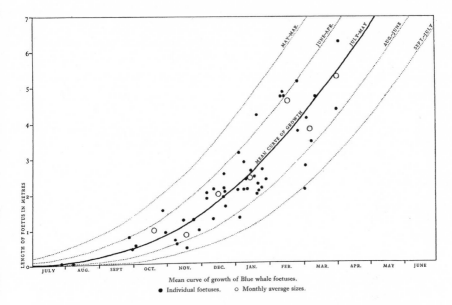

FIGURE 2.19 Life cycle table: Blue whale gestation. (From Mackintosh and Wheeler, "Southern Blue and Fin Whales," p. 423.)

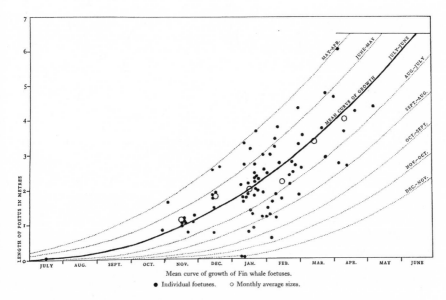

FIGURE 2.20 Life cycle table: Fin whale gestation. (From Mackintosh and Wheeler, "Southern Blue and Fin Whales," p. 425.)

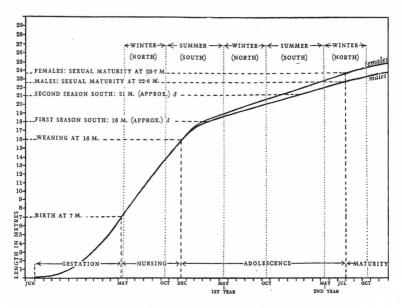

FIGURE 2.21 Blue whales: Mean curve from conception to sexual maturity. (From Mackintosh and Wheeler, "Southern Blue and Fin Whales," p. 442.)

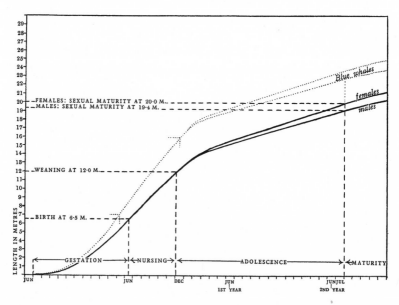

FIGURE 2.22 Fin whales in comparison: Mean curve from conception to sexual maturity. (From Mackintosh and Wheeler, "Southern Blue and Fin Whales," p. 443.)

colors for football at the Westminster School and become the captain of the athletic club at the Royal College of Science), there was nothing that could prepare him for the exigencies of an Antarctic whaling station, for which he would ship just a few months after being hired in the spring of 1924.[243] He spent the summer furiously trying to get on top of the science (he had in fact taken only a high second class degree in zoology, indicating ability, but not standout excellence).[244] To this end, Harmer sent Mackintosh and his partner-to-be, F. G. Wheeler, off for a two-week crash course in the reproductive physiology of mammals at Cambridge (taught by F. H. A. Marshall, author of a massive treatise on the topic), where "Mack" crammed, learning how to dissect, section, and stain an ovary, how to identify the different physical marks—like scars—that indicate ovulations, and how to interpret the form of those scars for clues to the reproductive history of a given animal.[245] Shortly thereafter Mackintosh headed off to the Outer Hebrides for a brief stay at the Harris whaling station in order to get at least a little hands-on experience measuring and sampling large whales on a flensing platform.[246] Then, by October 1924, he was en route to South Georgia, in

Harmer before the Discovery positions were advertised. See Harmer to Mackintosh, 27 September 1923, and Harmer to Matthews, 20 February and 6 June 1923, all letters housed in SGB Papers, file "Institute of Civil Servants."

243. Mackintosh actually wrote out a timeline of this early work, which survives: NOL GERD.M2/1, "Record of events in formation of Disc. Collections." Harmer's notes of his original interviews with Mackintosh and the other candidates also survive: SGB Papers, file "Institute of Civil Servants," (loose sheet in Harmer's hand). Harmer noted that Mackintosh had done some mountaineering in Switzerland and claimed not to have a problem with seasickness. He added that he "photographs and draws," and that his drawings were "very good." The text of the original advertisement for the positions can be found in the minutes of meeting 12, 17 December 1923, document 70, SPRI MS 1284/4.

244. He subsequently did take a first in biology.

245. On the need for Mackintosh to receive this training, part of which took place at the university's agricultural station, see "Extract from Letter, November 25, 1924, S. F. Harmer to W. Heape, FRS," SMRU MS 38600, box 679, file "NAM—Breeding, Growth, Age, etc.," together with Mackintosh's own notes on "Problems connected with Breeding" at the same location (presumably sent to Harmer after the course).

246. As luck would have it, only a single whale was taken during his visit. See Mackintosh's report to Harmer on his efforts to work out a protocol for body metrics on the plan in SMRU MS 38600, box 639, file "NAM visit to the Harris Whaling Station." Interestingly, Mackintosh had already briefly stayed at such a whaling station in Olna Firth, in the Shetlands, in 1923, where he was looking at coastal plankton. See SGB Papers, file "Institute of Civil Servants" (loose sheet in Harmer's hand). Harmer himself had paid a weeklong visit to the Harris station back in 1920, where he got a feel for

advance of the *Discovery* vessel itself, Harmer having wanted to get the flensing-platform biological work under way as soon as possible.

Arrival in the smoking bays of South Georgia was a rude shock, even for someone who had gotten up close to a Scottish whaling station. Mackintosh kept a private notebook of his early work in South Georgia, and in it he recorded listening to the unearthly booming of the spring avalanches on the first night of his stay. He spent that night aboard the vessel that delivered him, since there were as yet no shore accommodations for the Discovery party. The Marine Biological Station, a small Discovery-funded laboratory-cum-residential facility some 800 feet from the flensing platform of the whaling station (figures 2.23 and 2.24), was not yet complete, and Mackintosh would have to shack up initially in company quarters at the whaling station itself. One of his early diary entries describes a fatal accident at a neighboring factory in which a whale jaw slipped from the hoist and fell on a flenser, killing him instantly. Another recounts his negotiations with several lemmers keen to sell him a foot-long fin whale fetus for a ration of whisky—he initially demurred, but a few days later, more acclimatized to the culture of the station, he traded a quarter bottle for another.[247] Once Wheeler arrived and the season picked up, work on the landed whales began in earnest, with Wheeler, who had more experience in histology, taking the lead in the laboratory analysis of the ovaries removed from the animals on the plan.[248] Out at five o'clock in the morning in the cold air of the southern summer, Mackintosh and Wheeler spent their days

---

the thrill of the hunt. He wrote to H. T. Allen at the Colonial Office that he got to see three fin whales taken and that one of them "charged and hit our boat, with its head, immediately under where I was standing. It was an unforgettable experience." Harmer to Allen, 20 October 1920, SMRU MS 38600, box 637, "Whaling Research Expedition—Discovery Expedition—Harmer Correspondence."

247. The journal, entitled "Diary kept whilst based at MBL, Grytviken, 15 October 1924 to 26 June 1927," can be found at the Scott Polar Research Institute: SPRI MS 1268/1;BJ. My references to it hail from the entries for 22 November and 23 December 1924, and 31 January and 7 February 1925.

248. For a detailed outline of this program of work (which involved counting the corpora and investigating their size in the hopes of determining the structure and timing of the estrous cycle), see Mackintosh's "Investigations at the Shore Station on the Breeding of Whales," minutes of meeting 19, 13 August 1924, item 134, SPRI MS 1284/4. The Marine Biological Station was outfitted with a kit that perfectly represents the macro/micro nature of the flensing-platform work: in addition to "a half-dozen felling axes" and "leather thigh boots, 10 pairs," the laboratory also boasted both a microtome for paraffin-embedded samples and another for frozen specimens, along with a

FIGURE 2.23
(*top*) An Antarctic laboratory: The South Georgia Marine Biological Station. (From Great Britain, Discovery Committee, *Report on the Progress of the Discovery Committee's Investigations,* opp. p. 20.)

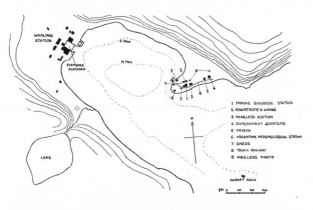

1 MARINE BIOLOGICAL STATION
2 MAGISTRATE'S HOUSE
3 WIRELESS STATION
4 GOVERNMENT QUARTERS
5 PRISON
6 ARGENTINE METEOROLOGICAL STATION
7 SHEDS
8 TRUCK RAILWAY
9 WIRELESS MASTS

FIGURE 2.24
(*middle*) The lab and the cutting platform: Map of the Station grounds. (From Kemp, Hardy, and Mackintosh, "Discovery Investigations," p. 223.)

FIGURE 2.25
(*bottom*) The reckoning: Discovery scientists had to excavate biological specimens on the plan. Note the elevators on the building behind—they were used to raise chunks of the carcass up to the top of the boilers within the shed. (Courtesy of the Sea Mammal Research Unit, University of St. Andrews.)

making dozens of measurements of the carcasses and excavating gonads from steaming sides of gore (figure 2.25). Once the day's catch had been broken down for the cookers, work turned to laboratory investigation of the tissue samples and clean copying of the field notebooks. A separate sheet was used for each whale (figure 2.26), and each recovered fetus (figure 2.27), and summary logs of all the biometrics were compiled as well (figure 2.28).[249] While Mackintosh was initially frustrated by the meager scientific results to be wrung from bone-wearying hours on the plan (he complained in his journal that his first report back to London looked "very thin" and that it "required all my ingenuity to make it look as though" they had been able to do some useful work), by July 1926 he and his collaborators had worked up no fewer than 738 whales and could already offer some striking and concrete findings concerning ovarian anatomy and physiology. For example, it appeared that in both blues and fins, the "corpus luteum b," as it was called (the nodule marking the residue of the follicle of an ovulation), persisted indefinitely and gradually shrank in size through reabsorption into the ovarian tissue. The durability of these nodules was unusual (at least in comparison to what was known about the process of ovulation in rabbits and cows) and suggested a possible means of aging female whales, if not in absolute terms, at least in terms of the relative age composition of exploited stocks—information that could, over time, give an indication of an overtaxed population.[250]

---

half-dozen microscopes of various types. See minutes of meeting 18, 2 June 1924, item 110, in ibid.

249. The full run of these field notebooks (from which the images in figures 2.26–2.28 come) can be found at SMRU, where they are catalogued as WR/DC/1–4. The roughest materials appear in a subseries, WR/DC/4/1–5.

250. For an outline of this work in its early stages, see "Fifth Scientific Report of the Marine Station," 1 July 1926, NOL Discovery MS, green binder, "Scientific Reports Marine Station." But compare Harmer's extensive notes on this document (and the other such reports submitted by the whale biologists working at Grytviken in the late 1920s) in SMRU MS 38600, box 644, file "Discovery Expedition: Reports on the Scientific Work." Interestingly, Harmer's marginalia suggest that he was initially skeptical of the value of this technique, though his own files show that he collected a great deal of material on the development and attrition of the corpus luteum in mammals (see, e.g., his clippings from Nature in SMRU MS 38600, box 637, file "Whaling Research Expedition—Discovery Expedition—Harmer Correspondence"). For a later elaboration of the method and its limits, see Mackintosh and Wheeler, "Southern Blue and Fin Whales," Discovery Reports 1 (1929): 257–540, particularly pp. 382–96. The use of corpus counts would come to be sharply disputed as an aging technique, and several

No. 184 | Date 1.4.25 | Species Blue | Sex ♀

Internal Genitalia.

Ovaries deeply convoluted. Some 7mm follicles below the surface.

Ovary 1) 34 x 11 x 7.5. Much convoluted. Convolutions deep. Numerous small (10 mm) follicles, not showing on surface.

Corpus luteum B. 8.5 x 6 x 2.5. Scar. Very old. No corona. No cavity. Normal.

Corpus luteum B. 3.5 x 3.5 x 2.5 White scar. No corona. No cavity. Old. Normal.

Ovary 2) 61.5 x 13 x 6. Very convoluted. Some 7mm follicles below surface.

Corpus luteum B? 22 x 21.5 x 8. Scar & 4cms corona, apical.

Corpus luteum B. 7 x 6 x 3. Scar. No corona. No cavity. Rather of "racemose gland" type.

Weight of Ovary or Testis. 1) 8 lbs 8 oz 2) 11 lbs. 8 oz.

External Genitalia.

Genital region considerably swollen. Genital groove wide open.

Vulva not much congested.

Lactation.

Mammary grooves open but teats not quite showing. Glandular tissue extensive but not secreting milk and not swollen. No fluid in milk bag.

Fœtus.

Foetus present. Length 6.3 m. ♀.

Further Remarks.

An exceptionally bulky whale. Girth 11.5 m. Probably the biggest up to date.

Crescents in the form of well marked depressions open at the posterior end.

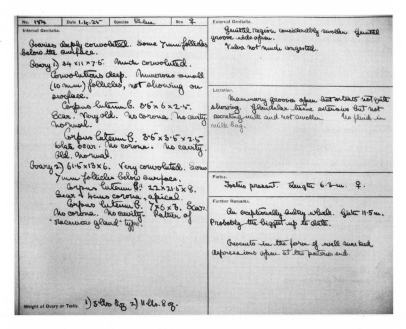

FIGURE 2.26 Keeping track: Profile sheet for adult whale. (Courtesy of the Sea Mammal Research Unit, University of St. Andrews.)

FŒTUS

Whale No. 276 | Date 17.11.25. | Species Fin | Length of Parent 22.45 | Sex of fœtus male

1	1.02	2	.01 (8.98)	3	.14 (13.73)	4	.165 (16.18)
5	.17 (16.67)	6	.4 (39.22)	7		8	.28 (27.45)
9	.075 (7.35)	10	.317 (31.08)	11	.49 (48.04)	12	~~.073~~
13	.073 (7.16)	14	.025 (2.45)	15	.05 (4.90)	16	.095 (9.31)
17	.12 (11.76)	18		19	.03 (2.94)	20	
21	.15 (14.71)	22		23		24	.09 (8.82)
25	.105 (10.29)	26	.21 (20.59)	27		28	

Preserved? Yes. Formalin.

Pigmentation: Pigment developing on head & back, inside lower jaw and on flipper.

Baleen: None

Hair: Well developed.

Ventral Grooves: Just beginning

Palate and Tongue

Blubber: 1) 3 mms 2) 3 mms

Genitalia: Penis external.

Further Remarks: Twin found in same cornu of the uterus.

FIGURE 2.27 Collecting notes: Profile sheet for whale fetus. (Courtesy of the Sea Mammal Research Unit, University of St. Andrews.)

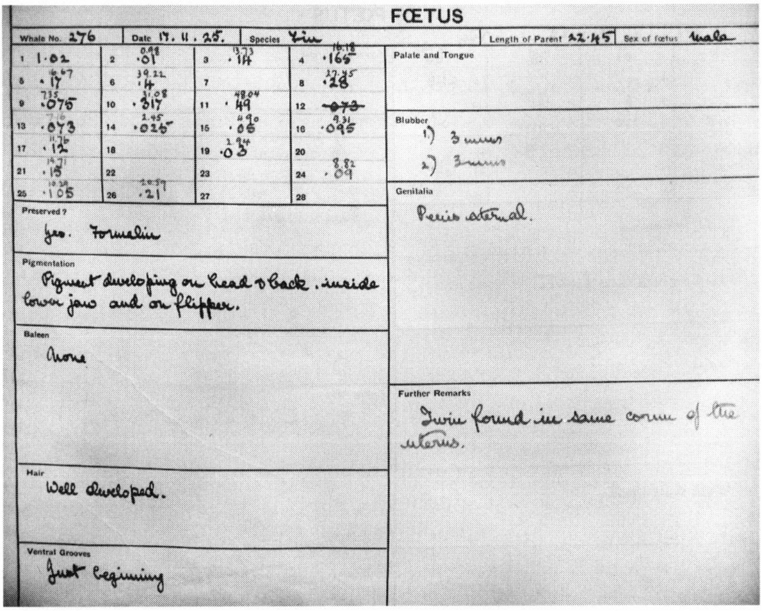

In an effort to address the problems of growth and development in a different way, the Discovery biologists experimented with a variety of techniques for assessing physical (and sexual) maturity: average length for females showing at least one corpus luteum was an important general index, but eventually a way was worked out for checking the fusion (ankylosis) of the epiphyses with the centra in the vertebral column, an indication of the cessation of growth in mammals.[251] Initially this had seemed impossible, and Mackintosh wrote in one of his early reports that he and his colleagues had been unable to figure out how to get at the cartilaginous junctures of the vertebrae "owing to the tough fibrous tissue which covers the bone." Some sort of special "cleaning" seemed necessary.[252] But this fussiness was soon overcome by greater intimacy with flensing-platform butchery. By 1930 every Discovery biologist knew the procedure, in which one waited for the steam winch to tear the meat from the back (in effect, the massive tenderloin) and then swooped in, axe in hand: "When the vertebral column is being stripped of muscle, two or three vertebrae of the dorsal and lumbar series can be cut at the edge with a hatchet" (figure 2.29).[253] The absence of a cartilage layer under the epiphyses indicated maturity (figure 2.30).[254] By the mid-1930s it was already clear that at many stations a very small percentage of the sexually mature blue whale females being processed had yet

---

Discovery researchers would disagree about the proper way of converting numbers of corpora to estimated ages (which was in part about determining the number of ovulations to occur, on average, between pregnancies). For takes on the question through the 1930s, see J. F. G. Wheeler, "The Age of Fin Whales at Physical Maturity with a Note on Multiple Ovulations," *Discovery Reports* 2 (1931): 403–34, and Alec H. Laurie, "The Age of Female Blue Whales and the Effect of Whaling on the Stock," *Discovery Reports* 15 (1937): 223–84. For an inside glimpse of the argument, consult the letter from Laurie to Kemp, 12 April 1933, SMRU MS 38600, box 680, file "Biologists in Factories, 1932–33."

251. There had in fact been some debate in the nineteenth century over whether such fusion ever occurred in whales, given their need for great bodily flexibility. In the 1850s Richard Owen, on the basis of a set of immature skeletons, argued for their exceptional status, but his conclusions were subsequently disproved by Flower in 1864.

252. "Fifth Scientific Report of the Marine Station," 1 July 1926, NOL Discovery MS, green binder, "Scientific Reports Marine Station."

253. "Thirteenth Scientific Report," 2 January 1930, in ibid.

254. Figure 2.29 is from SMRU MS 38600, box 643 (which contains several smaller boxes of negatives and prints). Figure 2.30 is from SMRU MS 38600, box 680, file "Biologists in Factories, 1939, 1940, 1941."

Date	Whale No.	Sex	1	2	3	4	5	6	7	8	9	10	11	12	13	14
27.10.25	252	male	19.6		3.5		4.08	8.0	1.05	4.87	0.95	5.44	8.87	8.6	1.34	0.48
27.10.25	255	male	21.77		3.8	4.13	4.3	8.9	1.0	5.57	1.1	6.17	10.0	9.1	1.57	0.47
29.10.25	257	female	22.57		4.67	4.96	5.12	9.6	1.08	5.33	1.1	6.25	9.7		0.59	0.42
2.11.25	260	female	21.3		2.15		4.6	8.75	1.0	5.33	1.1	6.0	9.6	9.3	0.7	0.43
3.11.25	262	female	22.4		4.8	4.97	5.2	9.9	1.08	5.4	1.1	6.2		9.4	0.15	0.6
4.11.25	263	female	24.0		4.8	5.28	5.4	10.37	1.17	5.48	1.08	6.8	10.9	10.3	0.7	0.47
5.11.25	264	female	21.05		3.9		4.5	8.7	1.1	5.1	1.05	5.85	9.6	9.2	0.6	0.52
6.11.25	266	male	20.55		4.0	4.3	4.4	8.8	1.05	4.65	1.07	5.85	8.15	8.8	1.61	0.52
7.11.25	265	male	20.0		3.7	3.95	4.2	8.6	0.95	5.05	1.1	5.6	9.2	8.7	1.4	0.65
7.11.25	269	male	20.3		4.0	4.1	4.4		1.02	4.55	1.0	5.45	8.8		1.35	0.55
11.11.25	271	female	22.1		4.3	4.65	4.9	8.65		5.2	1.1	6.0	9.65		0.5	0.05
17.11.25	276	female	21.45		4.3	4.7	4.8	9.25	1.05	5.55	1.05	6.3	10.35	9.9	0.75	1.6
18.11.25	277	male	20.2		4.08	4.36	4.5	8.7	0.95	4.72	0.96	5.35	8.55		1.65	3.15
18.11.25	278	female	21.85		3.85		4.5	8.8	1.0		1.13	6.15	9.6		0.6	
24.11.25	279	male	20.6		4.0	4.3	4.48	8.9	1.02	4.8	1.0	5.7	9.2	9.05	1.6	0.5
26.11.25	280	female	22.15								1.3					0.55
27.11.25	281	female	21.75		4.05	4.3	4.4		4.15	1.05	6.2	9.3				0.45
30.11.25	283	male	20.9		3.9	4.33	4.45	8.9	0.96		1.1	6.0	9.5			
16.11.25	285	female	21.9		4.6	4.85	5.15	10.17	1.18	5.4	1.08	6.1	10.0	9.6	0.65	0.5

FIGURE 2.28 The summary: Compilation logs of all whale biometrics. (Courtesy of the Sea Mammal Research Unit, University of St. Andrews.)

achieved physical maturity—an observation suggesting (to some, at least) that the stock was already under considerable pressure.[255]

By 1942 Mackintosh had spent many years poring over the data sets generated by this sort of work (his own and that of other Discovery scientists who followed in his footsteps), plotting frequency distributions of corpora against total length of animal, total length of animal against extent of fusion in the vertebrae, relative size of corpora against fusion, and examining a host of other combinations in the hopes of firming up the details of the life histories of both the blues and the fins.[256] His conclusions on these matters, as presented in "The Southern Stocks of Whalebone Whales," could nevertheless feel a little meager. Sexual maturity for female fins came in

255. Though optimists could claim that there was a selection bias in the sample, since younger animals—it was alleged—were inexperienced and more likely to fall prey to the harpooner.

256. For a flavor of this work, see Mackintosh's holograph graph sheets and associated material in the file "NAM—Regression of Corpora" in SMRU MS 38600, box 679.

FIGURE 2.29
(*right* and see PLATE 6) The postmortem: A hip-booted cetologist
checking fusion in the vertebral
column. (Courtesy of the Sea Mammal Research Unit, University of
St. Andrews.)

....

FIGURE 2.30
(*below*) Cetological pedagogy:
Worksheet on how to properly
perform research cuts. (Courtesy
of the Sea Mammal Research Unit,
University of St. Andrews.)

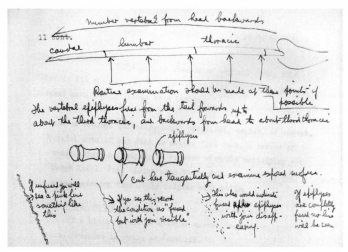

at about 65 feet on average; for female blues, at 77 feet. With males it was
tougher to say, since all one had to go on was the size of the testes, and
there seemed to be a good number of intermediate instances where it was
a difficult call whether a given testicle hailed from a sexually mature animal
or not (Mackintosh eventually settled on 63 and 74 feet for fins and blues,
respectively). In general, it seemed that both males and females achieved
sexual maturity at about two years of age (Harmer had guessed as much)
and that the breeding cycle was, as a rule, a two-year affair (again, this had
been surmised before the Discovery Investigations were launched). About
sizes at physical maturity, Mackintosh felt he could say nothing too definite, other than that there seemed to be a surprising correlation between a

specific number of corpora in the ovaries and physical maturity (for fins, more than fifteen corpora almost always meant that the epiphyses would be found to be fused; for blues, the number was eleven). And this observation pointed directly to the vexatious matter of the link between corpora and age. Such reliable correlations certainly buttressed the idea that the corpora accumulated regularly over time and therefore ought to be possible to use as an index of actual age. But even after almost fifteen years of work, a conversion factor for turning numbers of corpora into years of life remained elusive. Mackintosh had to admit as much in his monograph, where he reviewed the state of the art circa 1942. Wheeler, on the basis of the number of corpora that seemed to accumulate in the period between sexual and physical maturity in fins (and believing he had detected three frequency peaks in a distribution of total numbers of corpora observed in these animals), settled on the idea that there were two ovulations per year, on average, with animals reaching physical maturity three or four years after sexual maturity. Alec H. Laurie, working later and on blues, adopted a technique more focused on the relative sizes of the corpora themselves (which shrank as they got older) and came up with an estimate closer to a single ovulation per year, a number buttressed by sophisticated German histological work conducted in the late 1930s.[257] In short, then, Mackintosh could say only that more work on ovaries was sorely needed and that no other definite technique for aging had yet been found.

So much for flensing-platform anatomy and reproductive physiology. What about the second major aspect of the "direct work on whales" outlined by Harmer: the marking campaign aimed at the resolution of questions of distribution and migration? Recall the centrality of this program to Harmer's original vision: back in the 1910s he was already pressing the Colonial Office on the importance of marking to any consideration of "the size of the whale stock," not least because, in his own view, such work was likely to reveal discrete migration patterns—patterns he thought would give the lie to Hjort's notion that the southern rorquals were ubiquitous in the circum-Antarctic

257. This work proceeded from a new technique that purportedly made it possible to distinguish a corpus luteum of pregnancy from one that resulted from the release of an ovum that did not see fertilization and placental development. Given that a cycle of pregnancy and nursing took two years, this enabled the researchers to sift out one set of corpora according to a timed cycle, clarifying the total rate of accumulation. On the basis of such research, N. Peters claimed to see corpora building up at a rate slightly below one per year.

waters.[258] The inspiration for this kind of study hailed from several different precedents. The oceanic movements of large whales had for several centuries been deduced by whalemen themselves (and naturalists with an interest in their work) from a handful of serendipitous recoveries of harpoons buried in the blubber or flesh of animals wounded in one locale and subsequently taken in another.[259] By the early twentieth century, more formal marking programs were being used in research on several species of fish and birds in Europe and the United States.[260] Hjort himself made considerable contributions in this area where fish were concerned and had a strong interest in initiating whale-marking work as well (he had in fact developed an experimental dart for whales before World War I). Harmer was also aware of other sorts of data on whale movements, material that informed his thinking about a marking program: the amateur naturalist A. G. Bennett, who served as a seasonal whaling officer in the Falkland Islands administration (and eventually as an "acting government naturalist" as well) and spent season after season on the whaling grounds beginning in the early 1920s, implemented an original system for documenting the movements of the whales

258. See, for instance, Harmer, "Report on Whaling Operations, Principally in South Georgia and South Shetlands, 1915–1916," 2 July 1917, SMRU MS 38600, box 637, Harmer's memoranda on whales, binder 1.

259. The early nineteenth-century whaler-FRS William Scoresby discusses some of these instances in his *Account of the Arctic Regions, with a History and Description of the Northern Whale-Fishery*, vol. 1 (Edinburgh: Archibald Constable, 1820), 10–12. The American hydrographer and oceanic biogeographer Matthew Fontaine Maury developed a sweeping theory of right whale distribution on the basis of later harpoon recoveries. For a discussion, see my chapter "Sea of Fire" in *Tropical Visions in an Age of Empire*, ed. Felix Driver and Luciana Martins (Chicago: University of Chicago Press, 2005), 113–134.

260. For instance, a Dane, Johannes Petersen, did tagging work on plaice in the North Sea in the 1880s and 1890s, and by the early twentieth century (1904–1905) Walter Garstang (working for the Marine Biological Association of the United Kingdom) was studying fish migrations using tagging methods. See Helen M. Rozwadowski, *The Sea Knows No Boundaries: A Century of Marine Science under ICES* (Copenhagen: International Council for the Exploration of the Sea in association with University of Washington Press, 2002), 17, 51–52. Where birds are concerned, the earliest examples date back to the early 1800s (Audubon, for instance), but systematic banding experiments can be traced to Christian Mortensen, who banded 160 starlings in Denmark in 1899. Further work was done by the Smithsonian's Paul Bartsch (on night herons in the United States) circa 1902. By 1909 French, German, and British scientists were all involved in bird banding programs. Hjort worked extensively on fish tagging, and his pioneering "harpoon-marks" for whales were used in 1914 by Shackleton on the *Endurance* voyage.

that passed through the local waters. Distributing preprinted notecards to Norwegian whaling captains, Bennett requested that they create a record of every encounter with a large rorqual, noting the species together with its compass heading (and various other bits of environmental information). On the basis of more than 20,000 card returns, Bennett distilled average "vectors" for the movements of the major commercial species and plotted that information in polar coordinates; geographic segmentation of the data was also attempted in an effort to reconstruct migration patterns.[261]

In reviewing the early documentary history of the Discovery Investigations, it is striking to note the extent to which Harmer's focus on marking gave shape to the evolution of the undertaking as a whole. As early as the summer of 1918, recall, E. R. Darnley at the Colonial Office was already taken with the idea that the expedition would need a vessel to chase and tag whales, and it was clear that this was in addition to a larger laboratory ship for oceanographic work related to cetacean food.[262] As one committee member put it, "for marking purposes whales would have to be approached as if they were going to be killed," and since "the modern 'whale catcher' had developed during a period of about forty years for the purpose of permitting close approach to whales and shooting them," there was a clear argument for buying such a vessel for the marking work. This would eventually be the purpose-built *William Scoresby*, which was, as I mentioned above, originally contracted as a "Steel Screw Whale-Marking Vessel" (figure 2.31).[263] While the committee corresponded with whaling companies and the Admiralty and contracted with shipwrights to realize a

261. For an early summary of this work, see SGB Papers, file "Whale Movements, Bennett, etc.," which contains Bennett's "Memorandum on the Movements of Whales at South Shetlands" together with confidential memos between the secretary of state for the colonies and Government House, Port Stanley. There was obviously some administrative concern that "the whaling gunners should gather no hint of the results (directions of movements) expected," though it is not clear whether this is primarily out of anxiety concerning the spoiling of future data collection or the revelation of whale movements to interested parties. For later work by Bennett along the same lines, see SMRU MS 38600, box 678, which contains a large amount of Bennett material, including a number of his original card returns and his notes on the geographic synthesis of his results. I have found nothing on this work, which is peculiar and interesting in a number of ways. For more general information on Bennett, see n. 98 above.

262. See Darnley's memorandum of 27 August 1918, SPRI MS 1284/1.

263. The quotes are from Middleton's address to the ninth meeting on 29 October 1923; see SPRI MS 1284/4, document 52.

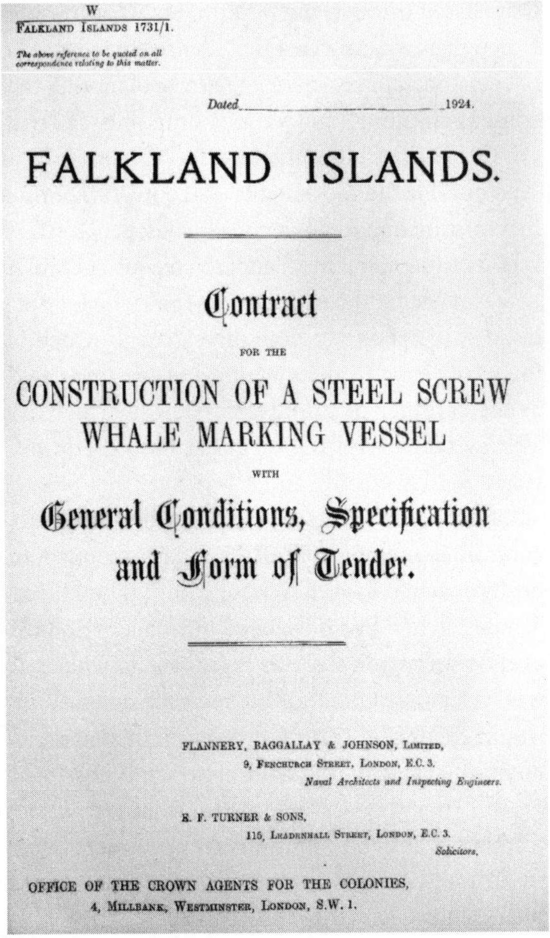

W
FALKLAND ISLANDS 1731/1.

The above reference to be quoted on all correspondence relating to this matter.

Dated_____1924.

FALKLAND ISLANDS.

Contract

FOR THE

CONSTRUCTION OF A STEEL SCREW WHALE MARKING VESSEL

WITH

General Conditions, Specification and Form of Tender.

FLANNERY, BAGGALLAY & JOHNSON, Limited,
9, Fenchurch Street, London, E.C. 3.
Naval Architects and Inspecting Engineers.

R. F. TURNER & SONS,
115, Leadenhall Street, London, E.C. 3.
Solicitors.

OFFICE OF THE CROWN AGENTS FOR THE COLONIES,
4, Millbank, Westminster, London, S.W. 1.

FIGURE 2.31
A whale-marking vessel: Contract for the construction of the *William Scoresby*. (Courtesy of the Sea Mammal Research Unit, University of St. Andrews.)

larger-than-usual catcher vessel (one with expanded fuel capacity for long hauls and some forward bracing for work in the ice), Harmer threw himself into the problem of the mark itself and the best means of delivering it.

Of his seriousness of purpose in this enterprise there can be no doubt, but several of his critical early decisions on these matters had more or less disastrous consequences for the research program over the decade that followed. In the first place, on the basis of what can only be described as a museum zoologist's preoccupation with bookish knowledge of his research organism, Harmer committed early to the idea that the marking device had to be *silent*. Presumably this reflected his belief that the animals could hear and his respect for the mammalian savvy of what he thought of as majestic creatures. That the large whales were not consistently, as whale-men knew, especially jumpy about noises in the air in their vicinity did little to push Harmer off of his vision of the whale-marking "crossbow" to which I have already alluded. The realization of this device—steel, nearly seven feet long, and requiring a special winch to be cocked—absorbed a great deal of Harmer's attention in the early 1920s. As I noted above, he drew into the project not only the Cambridge physicist C. V. Boys but also a cohort of sporting archers, colonial old hands, and creepy amateur medievalists who together made up the "Royal Toxophilite Society" (figures 2.32 and 2.33). Giving this group of die-hard bowmen an essentially unlimited budget and access to the leading materials scientists of the age in a quest to realize the ultimate "modern" crossbow (the prototype device was mounted on a counterbalanced swivel like those used for antiaircraft guns) proved the recipe for some exceptionally entertaining afternoons but did not lead to a workable mechanism for shooting whales at sea.[264] Accuracy was certainly one issue: the unfletched "arrows" were balky, and though the Admiralty tried for a time to get Harmer and Boys to figure out a way to *rifle* the "barrel" of the bow (in the hopes that the spin would stabilize the projectile), this proved impossible. But the more grave issue was the unwieldiness of the mechanism itself—that, and its inherent dangers. The recoil of the spring arms was terrifying, and an early experimental user was nearly cut in two by the whipping action of the bladelike spring steel. A large "shield" was briefly contemplated as the only way to protect the gunner, particularly

264. Boys threw himself into the project, and by April 1922 he had declared himself the "Secretary of the St. Mary Bourne Cow Shooting Club" and was sending out invitations to the prototype tests in Andover (the "Great Event" where the "Cow Shooting Begins"). Harmer's file on this project makes Monty Python look like simple journalism (for all this, see NHM Zoology Collections, file "Whale-marking, 1919–1924"). For instance, Harmer tapped "the Tox" for the insights of Ingo Simon, then holder of the world distance record on the Turkish bow (459 yards, 8 inches), and collected featherless Tibetan arrows gathered by a connoisseur in Peshawar. The springs on Boys's device were provided by a company that made sprung arm-cuffs for English bartenders (to hold up their sleeves). Before the dream came unraveled, the inner courtyard of the British Museum was being used as a shooting range by a gathering of grandees, using a dummy whale fashioned from oilcloth. Eventually slabs of fetid whale blubber were delivered to the Bond Street shop of Holland and Holland (gunmakers to the HRH &tc.) so that they might experiment with the ballistics. For some of this, see chap. 2 of Alister Clavering Hardy, *Great Waters: A Voyage of Natural History to Study Whales, Plankton, and the Waters of the Southern Ocean* (New York: Harper and Row, 1967).

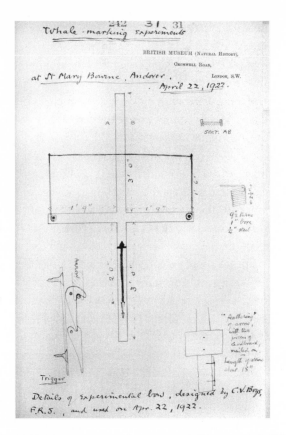

FIGURE 2.32
(*right*) Edwardian atavism
meets Cambridge physics:
Harmer's whale-marking
crossbow. (From the
collection of the Natural
History Museum, Zoology
Collections Archive.)

FIGURE 2.33
(*below*) A Norse fantasy:
A second diagram of
Harmer's crossbow. (From
the collection of the Natural History Museum, Zool-
ogy Collections Archive.)

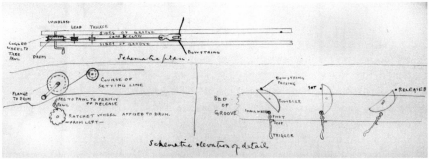

in rough southern waters, where heavy seas could pitch him forward into
harm's way.

All this might have been dismissible as a short-lived Edwardian folly
because, by the time the *William Scoresby* set out on its first commission to
mark some whales, it was not carrying a crossbow of any kind, but instead

a more or less standard-issue twelve-gauge shotgun, fitted out to shoot a dart into the whale.[265] But the crossbow episode had left its mark on the Discovery marking gun: rather than shooting some sort of cylinder-like slug (the original idea for a shotgun-delivered tag, based on Hjort's early work, and advanced by the ballistics experts consulted by Harmer), the Discovery shotgun had been fitted out to shoot a *modified version of the crossbow arrow* (figure 2.34). Assembled in two "stages" (a light, detachable wooden shaft loosely coupled to the actual mark, which looked a bit like an oversize thumbtack with barbs), this arrow had never behaved well in flight, not least because of the peculiar disklike configuration of its head, which was as nonaerodynamic as could possibly be imagined. There had been much gnashing of teeth over how to improve the profile of this whale dart (a conelike batting of stiff cloth, and various other fixes, were tried), but the basic problem was that the awkward "arrowhead" in question was itself the actual whale mark that Harmer had settled on; how to deliver it was very much treated as a secondary problem. So how had Harmer decided on this particular configuration for the mark? In the first place, he believed it was necessary to have the mark be visible to a flenser working on the animal on the plan—hence the buttonlike disk that would remain on the surface of the skin. (As it turns out, this was mostly misguided, since the processing of whale blubber required mincing it fairly finely, in the course of which a metal slug buried in the "blanket layer" would generally be discovered.) It also seems likely that Harmer was inspired in this unhappy configuration by the traditional "Petersen disks" that had become by this time the primary form of fish mark used in the North Sea and elsewhere.[266] Finally, and fatally, Harmer had been impressed by an exchange with A. G. Bennett, who had told him, in May 1920 (on the basis of much time spent looking at

265. This gun was, ironically, a return to the very first proposal for a marking device and followed Hjort's lead. The Admiralty director of naval ordnance had proposed an eight-bore shotgun with a nine-inch dart back in October 1919. See NHM Zoology Collections, file "Whale-marking, 1919–1924." It should be noted, however, that by the mid-1920s the committee members, feeling a certain patriotic hauteur, were not inclined to follow Hjort too closely; they went out of their way to indicate in their records that when Hjort lent them his gun "it did not prove satisfactory on the range." See "Equipment Sub-Committee Meeting," 28 January 1925, item 218, subhead 5, SPRI MS 1284/4.

266. On the Petersen disk, see Tim D. Smith, *Scaling Fisheries: The Science of Measuring the Effects of Fishing, 1855–1955* (Cambridge: Cambridge University Press, 1994), 78. Also of interest, though curious, is a published letter by Sanelma Nicht on the tagging of *Pleuronectes*: "Plaice and Place," *Cabinet* 32 (Winter 2008/2009): 45–46.

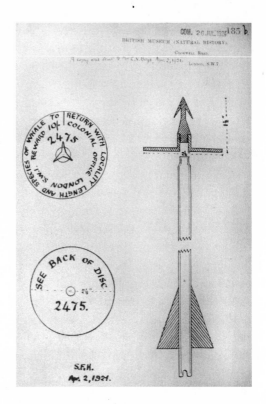

FIGURE 2.34
The bolt: Drawing of the prototype for Harmer's marking disk and its shaft. (From the collection of the Natural History Museum, Zoology Collections Archive.)

whale cadavers in the deep south), that wounds in the skin of cetaceans had a tendency to become septic, and that on no account should the proposed mark penetrate deeper than a scant *one and a half inches* into the animal's body, at the risk of introducing serious infection.[267] Not for more than a decade would it become clear that a mark of this configuration—which looks more than a little like the parasitic copepods of the genus *Pennella*, which whales have evolved to slough off with some efficiency—would be ejected from the skin very rapidly.[268]

267. Harmer's notes on this conversation are recorded in the NHM Zoology Collections, file "Whale-marking, 1919–1924."

268. To his credit, Harmer did worry about whether the marks would stick, but his concern focused on the mechanical action of the passing water "lifting" the leading edge of the disk. Having experimented with the hold of different barb arrangements, he decided this was unlikely. That the animals might have a physiological process for rejecting such an irritant is an idea he seems not to have considered seriously, though it was suggested by a medical doctor named A. H. Cheatle in the summer of 1924. See the

That was the long-term bad news. It would be some time coming, since there was no way to know for sure, simply on the basis of *not yet* having seen a returned mark, that the marks were not working. Indeed, there was at least tacitly a notion that the absence of returns could point to the total hunting pressure being much lower than Harmer himself had feared. In this sense, ironically, the technical errors of the whale-marking scheme must be understood to have contributed to delaying a sense of urgency concerning the expansion of pelagic whaling in the later 1920s. In September 1930, with four years of no mark returns, the committee recorded a rather defensive memo on the idea of switching to the proven Norwegian "Sorlle-pattern" mark: "It has, however, never been really established that the darts are ineffective, and any adoption of a new pattern would mean both expense and delay in making experimental samples, testing them on the range, and finally manufacturing throughout."[269] As late as 1931 the Discovery Committee was still placing orders for the thumbtack-style mark and somewhat stubbornly resisting a switch to the Norwegian pattern (which was configured as a tubelike slug).[270]

The shorter-term bad news was that equipping the crossbow arrows with cuffs of "wadding" and firing them from a shotgun made for a projectile only marginally less awkward and inaccurate than the original fletch (figure 2.35). The slightest breeze played fearfully with the dart in flight, and changes in humidity and temperature caused the shafts to warp—in some cases so badly that they could not even be loaded, much less delivered to a

---

postscript to A. C. Hardy's "Report on a Voyage with Professor Hjort on the 'Michael Sars,' July 1924," NOL Discovery MS, Allen Minutes. Some whalemen who saw the early mark noted that the barbed shaft should have been made much longer: D. Dilwyn John to Secretary of the Discovery Committee, 9 February 1926, SGB Papers, file "Work at Whaling Stations—Hebrides, 1924 and 1928; Algeciras, 1926."

269. NOL Discovery MS, file "Indent *William Scoresby* 224."

270. The committee noted placing such an order on 8 June 1931. See ibid. There was clearly concern, however; by July, the committee had instructed Stanley Kemp, Discovery's director of research, to write to Hjort "in view of the difficulties which have hitherto attended whale marking." See "Minutes of the Scientific Subcommittee, 1927–1949," meeting of 2 July 1931, SPRI MS 1284/6. I turned up the letter—written the very next day—in Oslo, in Hjort's papers: Kemp to Hjort, 3 July 1931, Johan Hjort Papers, Nasjonalbiblioteket (National Library), Oslo, Norway, Manuscripts Collection MS 4° 2911: XIX A. Kemp was candid about the difficulties, and he noted his concern that the marks were being ejected by a physiological process like that used in clearing *Pennella*.

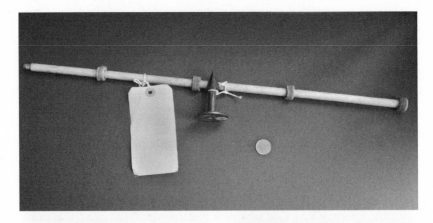

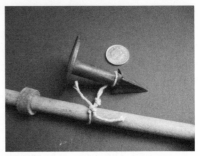

FIGURE 2.35 (and see PLATE 7) Rethinking delivery: Photographs of a modified fletch (to be fired from a shotgun). (From the collection of the Scott Polar Research Institute, University of Cambridge.)

moving target under the adverse conditions of the roaring forties.[271] Loose heads came off in midair, causing the shaft to spin harmlessly into sea, a twig in the tempest.

And the situation was even worse than that. Since by early in 1927 it had become clear that the just-launched *William Scoresby* was, at least under an English captain, effectively incapable of putting a "gunner" within striking distance of a whale: the vessel was too big, rode too awkwardly in the water (partly as a result of its ice armoring), and was basically underpowered to chase big whales. Moreover, it turned out to be a matter of some pilot-

271. For examples of these problems, see report 1, J. W. S. Marr aboard the *A. W. Sorlle*, November 1928, NOL Discovery MS, green binder, "Scientific Reports: Whale Marking, South Georgia, Hired Catchers." See also SMRU MS 38600, box 635, "Whale Observation Notebooks," particularly that from the work of the *William Scoresby*, 1926–1929.

ing skill to run down a fleeing rorqual. The report on her maiden voyage elaborated the pursuit of several dozen whales in the South Georgia waters, with a grand total of zero hits, despite chases that lasted for hours under grueling conditions as the vessel took water over the bow in heavy swell.[272] Disheartened, the Discovery researchers invited the manager of the Grytviken whaling station out with them to "consult" on the hunt (he was a distinguished Norwegian whaling captain himself), but even he could not help them and concluded that the boat was basically not equal to the task.[273] Various adjustments were made in the hopes of reducing engine noise (which was perhaps spooking the animals) and improving the handling of the vessel (which was slow to respond to the helm), but the gains were marginal. An entry by George Rayner in his "Whale Observation Notebook" from a marking voyage aboard the *William Scoresby* in September 1929 is characteristic:

16:10 Two large blue whales were observed to starboard and pursuit given. At 16:45 the vessel was getting close to them. After sounding they swam just beneath the surface, so that it was possible to see them and so pursue them very readily. Two darts were fired at one, but missed, falling alongside the whale due to being deflected by a strong cross wind. The second whale at this time was much nearer the ship than the one fired at, being almost under the bows of the ship and just beneath the surface, but it failed to offer a target, suddenly becoming frightened and swimming away at a great pace.[274]

272. See J. E. Hamilton, "First Report on the Scientific Work of the R.S.S. 'William Scoresby,'" SMRU MS 38600, box 644, Harmer's "Scientific Work," file 2. To be fair, three successful markings were recorded in the calmer waters off the West African coast, but the success rate was abysmally low, and one (unsuccessful) chase took seven hours.

273. For the opinions of several Norwegian captains on the suitability of the *William Scoresby* for this sort of work, see report 1, J. W. S. Marr aboard the *A. W. Sorlle*, November 1928, NOL Discovery MS, green binder, "Scientific Reports: Whale Marking, South Georgia, Hired Catchers."

274. See entry for 5 September 1929, SMRU MS 38600, box 634, file "Whale Observation Notebook—W.S. 1928–1929." Box 635, which continues this series, includes several quite lyrical passages on whale-marking expeditions by Eugene Rolfe Gunther, in many respects the prose stylist of the Discovery scientists. See his interleaved page in the notebook of observations from the marking voyage of the *Skua*, 1936–1937. Gunther includes some similar passages from an earlier phase of the work in *Notes and Sketches*

And so it went. In that same year the king of the British whaling moguls, Captain Harold Keith Salvesen himself (recently back from a tour of the Salvesen stations in the deep south), wrote a rather cruel note to Sidney Harmer, asking him to come clean about the fiasco: "I am very interested," he began, cordially enough, " in the experiments marking whales." But then the other shoe fell:

> It is, as you must know, agreed by the Officers of the Scoresby (upon whom I called at Grytviken) that these have hitherto proved an absolute failure, but nevertheless I gather at home that the public have been misled by the lectures of and articles inspired by the members of the Discovery Expedition, and believe that the reverse is the case. The failure of the Scoresby was I believe responsible for the sending out of two whale-catchers last season from South Georgia, with whale gunners under instruction to mark whales.[275]

Which was exactly right. Probably the most important development to come out of the difficulties with the *William Scoresby* as a whale chaser was the shifting of the marking program from an in-house operation, run by Discovery scientists out of Discovery vessels, to an increasingly outsourced endeavor, relying on hired commercial whale catchers and their skilled Norwegian captains (and even gunners). This shift had considerable implications for the long-term entanglements of whale research with the whaling industry. And it is also true that the Discovery Committee was a bit disingenuous about this revision of their plans, in effect putting a good face on the problem by insisting that their own vessel was so busy doing important plankton survey work that there was hardly any time left for whale marking, necessitating supplementary work for hire by company catcher boats.[276] Even with the faster vessels and experienced captains, hit

---

Made During Two Years on the "Discovery" Expedition, 1925–1927 (Oxford: Holywell Press, 1928); see, e.g., p. 39.

275. Salvesen to Harmer, 19 May 1929, SGB Papers.

276. For an example of this line on the issue, see "Draft Memorandum to Mr. John (as revised by Subcommittee 28 / 9 / 29)," NOL Discovery MS, file "Scoresby, Second Commission," which is larded with comments to the effect that the results of marking from the *William Scoresby* had been "very satisfactory" when there had been time to attempt such work. Or consider the way the matter is fudged in a more public document: Great Britain, *Discovery Investigations: Second Annual Report, January, 1927–May, 1928* (London: His Majesty's Stationery Office, 1929), 7. Salvesen himself is almost surely responding to the lecture given by A. C. Hardy to the Royal Geographical Society in

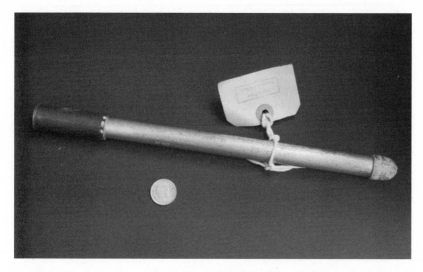

FIGURE 2.36 (and see PLATE 8) The "slug": Photo of an early "Discovery Mark." (From the collection of the Scott Polar Research Institute, University of Cambridge.)

rates stayed very low until 1932, when the committee capitulated and began experimenting with a slug-style dart designed to penetrate the skin entirely and rest in the blubber. These new darts came to be known as "Discovery marks," and their use was standard from the mid-1930s on (figure 2.36).[277] Harmer's old arrow dart—"the laughing stock of the Southern Ocean," as Salvesen uncharitably put it—was retired.[278]

No sooner had the switch to the new dart been made than the committee received word of a trio of returns—though these could not be said to

early April of 1928, which received a great deal of favorable, if also misleading, press; see, e.g., "Marked Whales, Shot with Steel Discs, Tracing Migration," *Daily Telegraph*, 3 April 1928, in NOL Discovery MS, scrapbook of press cuttings.

277. The best résumé of the work from hired catchers can be found in the synthetic volume of voyage reports in the NOL collection: NOL Discovery MS, green binder, "Scientific Reports: Whale Marking South Georgia, Hired Catchers, 1928–1937." Hit rates using the arrow-style dart hovered around 30 percent, and they rose to closer to 75 percent with the new slug-style mark. Several different designs of the latter (made of different metals, and with different head configurations) were tested before the final form was settled on. For a published review of some of this work, see A. C. Hardy, "Whale-Marking in the Southern Ocean," *Geographical Journal* 96, no. 5 (November 1940): 345–50.

278. Salvesen to Harmer, 19 May 1929, SGB Papers.

convey any information about migration patterns, in that all three were recovered just a few days after they had been shot into whales, before the creatures in question could make any significant transit.[279] The committee restrained its enthusiasm (noting somewhat grudgingly that "until more marks have been returned the committee cannot be assured that the new type of mark is successful"), but did authorize more hired catchers and more of the new darts. By 1934, with a healthy trickle of darts coming back from the factories, the committee was prepared to declare that they now had conclusive evidence "one, whales can be marked without injury; two, that an effective and permanent method of marking a whale has been discovered; and three, that the marks can be recovered during the process of flensing."[280] This third aspect of the program never proved wholly satisfactory, it must be said, in that a considerable percentage of the marks came to light only after the whale in question had been entirely worked up—as in, when the boilers were being cleaned out or when the "grax" (a hideous organic residue of the factory process) was being raked over; this timing generally made definitive information about the specific marked animal (and its final location) impossible to recover.[281]

Nevertheless, the darts kept coming in (sped on their way by a hand-

279. These were marks 41, 206, and 462/66, all recovered in the 1932–1933 season, from eight to eleven days after being delivered. For details on these returns, and on the full range of prewar marking returns, see George W. Rayner, "Whale Marking: Progress and Results to December 1939," *Discovery Reports* 19 (1940): 245–84. Also of interest is Rayner, "Preliminary Results of the Marking of Whales by the Discovery Committee," *Nature* 144, no. 3659 (December 16, 1939): 999–1002; and Rayner, "Whale Marking II: Distribution of Blue, Fin and Humpback Whales Marked from 1932 to 1938," *Discovery Reports* 25 (1948): 31–38.

280. Both of these quotes are drawn from "Minutes of the Scientific Subcommittee, 1927–1949," SPRI MS 1284/6, specifically from the minutes of the meetings of 30 November 1933 and of 18 April 1934.

281. Rayner discusses some of these problems in his "Whale Marking: Progress and Results to December 1939," *Discovery Reports* 19 (1940): 245–84, at pp. 254–56, but it is also interesting to consult some of the returned mark forms, a number of which are held at SPRI, and which include tidbits like this one, for mark no. 6376: "mark was recovered from the grax having worked up 12 finwhales." Early work on the slug-style mark involved efforts to prevent the cylinder from passing through the blubber into the muscle tissue, both because of the risk of injury to the animal and because of the reduced chances of good recovery. See Buckle's "Report from the Southern Empress," circa 1936, SMRU 36800, box 636, file "Whaling Industry Regulation Act, 1934."

some finder's fee that went to the flenser or boiler-room worker who put his hands on the thing and sent it along to the Discovery Committee in London), and by 1942, when Mackintosh was summing up the scientific labors of the Discovery Investigations for his synthetic report, "The Southern Stocks of Whalebone Whales," he could mobilize the data from more than 200 returned Discovery marks.[282] A fair percentage of these were not especially interesting returns (with marking catcher vessels and catching catcher vessels working out of the same ports, piloted in many cases by the same men, and cruising the same grounds, more than a few whales ended up on the flensing platform when the wounds from their marks were still fresh), but a small number had proved quite dramatic, affording important information about the life cycles of the large rorquals of the Southern Ocean.[283] There was, for instance, the young female fin whale tagged as no. 3482, marked as a calf at 65° S in February 1935—almost directly below Cape Town as the crow flies (though it would have to be a hardy crow, as no. 3482 was plugged not far from the Antarctic ice edge). She came to an end a little more than two years later, in July 1937, 1,900 miles due north, falling to a catcher vessel working the waters of Saldana Bay off the African coast. By this time she was a little short of 69 feet, and though her ovaries were not recovered, her recorded length was above the accepted index for sexual maturity established by Mackintosh and Wheeler on the plan at Grytviken in the 1920s. By these lights, then, whale no. 3482 appeared to offer concrete confirmation that sexual maturity could be achieved in something less than

282. More than 5,000 marks had been registered as delivered to a cetacean target in the ocean.

283. For an explicit discussion of the problems of marking on the whaling grounds, see Mackintosh's note to J. John Hart, encouraging him to figure out how to mark some animals "off the beaten track of the whalers, as they [the whales] are less likely to be killed before they have moved or survived until next season." Mackintosh to Hart, 11 November 1936, SMRU MS 38600, box 679, file "J. John Hart, Notes—Whale Marking Cruise, 1936–1937, W.S." Also interesting in this regard is Herdman's "Report on Whale-Marking at South Georgia, Nov. 20th 1935–Jan. 19th 1936," NOL Discovery MS, green binder, "Scientific Reports: Whale Marking South Georgia, Hired Catchers, 1928–1937," where he writes of the effort to stay away from the catcher boats while marking. The other side of this story would be the irritation of the whalers at a bunch of English scientists forever spooking the whales. For that perspective, see D. Dilwyn John to Secretary of the Discovery Committee, 9 February 1926, SGB Papers, file "Work at Whaling Stations—Hebrides, 1924 and 1928; Algeciras, 1926." John notes that the whalemen prohibited them from trying to dart a whale until the gunner had fired the harpoon and missed!

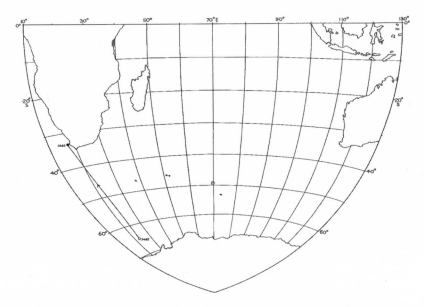

FIGURE 2.37 The journey: Map of a north–south fin whale migration. (From Rayner, "Whale Marking: Progress and Results to December 1939," pl. LIX.)

three years, and quite possibly two. Moreover, her long-haul north–south trajectory was the firmest proof yet of a general seasonal movement of the fins from the warmer northern waters (in the southern winter) to the ice edge in the deep south (in the southern summer) (figure 2.37). Such migrations among the blues and fins had been suspected for a long time on the basis of indirect evidence—parasite patterns, the actual successes of the industry in different locations at different times, and analogy to humpbacks (in which such movements were established relatively quickly using both marking and visual data)—but here was a plottable instance of the path.

Mackintosh was still more excited by one of the very last marks to come in before his monograph went to press: no. 696, fired into a female fin off South Georgia by Alec H. Laurie aboard the Pesca catcher vessel *Skua* in December 1934 and recovered in February 1941 by a catcher feeding the pelagic factory vessel *Svend Foyn* in the same general area. By this time Mackintosh was actively pursuing a policy of shipping Discovery-linked biologists aboard pelagic factories whenever possible, and it was one of his plants (who was in fact working as a whaling inspector at the time—still further evidence of the deepening entanglements of whale science with the whaling industry) who succeeded in recovering the ovaries of this particular animal. A corpus count yielded a total of eight, which—given that the whale was not, as Mackintosh put it, "conspicuously immature" when marked—pointed to a rate of accumulation of corpora not exceeding one and one-third per annum.[284] This finding seemed to suggest that Laurie was right and Wheeler wrong about the proper way of converting corpus counts into years. It was an example of how the two prongs of Discovery-style whale research—flensing-platform anatomy and reproductive physiology, and marking surveys—could in fact intersect, and potentially even reinforce each other.

Summarizing more generally what had been figured out about the animals from marking, Mackintosh emphasized the humpback data, which was the most dramatic, since it was indeed possible to build a compelling case for well-segmented stocks making tight, Harmeresque north–south migrations on a consistent seasonal basis. Mackintosh could even map their patterned movements, generalizing on the basis of more than 30 recovered marks (figure 2.38). Satisfying as that was, it was, in a way, a pyrrhic victory as far as regulatory matters were concerned, given that the humpbacks by this time represented less than 15 percent of the southern catch, and much of that was the result of relatively small-scale coastal whaling.[285] The core species of the industry by the latter 1930s, of course, were the blues and the fins—and increasingly the fins, as blues had become a noticeably smaller percentage of the annual take.[286] The boom years of the humpback industry at South Georgia, and the 6,000-animal annual yields that had so concerned Sidney Harmer back before World War I, were ancient history—the stocks no longer existed. To trace the distribution and

284. N. A. Mackintosh, "The Southern Stocks of Whalebone Whales," *Discovery Reports* 22 (1942): 197–300, at p. 227. On the subject of sending Discovery-trained men aboard factory vessels as "second inspectors," see SMRU MS 38600, box 680, file "Biologists in factories, 1939, 1940, 1941," and specifically the letter from A. T. A. Dobson to Mackintosh, 8 August 1939. This box contains a number of files related to the integration of Discovery scientific work with commercial factory-ship exigencies.

285. An international agreement in 1938 banned the taking of humpback whales by factory vessels in the Antarctic.

286. While many in and outside the industry believed that this change pointed to a real decline in the blue whale population, there were those who felt that the reduction could be accounted for by changes in hunting techniques and regions. For a discussion, see George L. Small, *The Blue Whale* (New York: Columbia University Press, 1971), chap. 3.

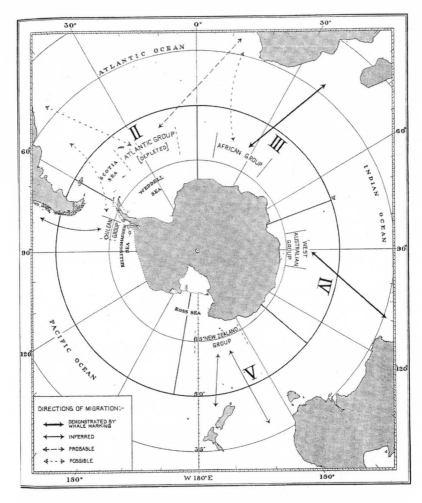

FIGURE 2.38 Movements of humpbacks: Mackintosh's map showing the segregation of humpback stocks. (From Mackintosh, "The Southern Stocks of Whalebone
Whales," p. 247.)

migration patterns of the blues and the fins, Mackintosh tried to reason
analogically, stretching the available marking data as far as it would go
(and maybe a little farther) to suggest somewhat discrete stocks of those
species as well.[287] On the basis of a number of marks that indicated animals

287. In this context, it is interesting to review his continued efforts to establish relatively "tight" migrations for fins in the immediate postwar period. Some of his notes
from the mid-1950s show him actually drawing "corridors" on maps of the Southern

could be found in the same general regions even after intervals exceeding a year, Mackintosh argued that, although both species were clearly "distributed more or less continuously around the Antarctic continent," circum-Antarctic movements were nevertheless "to some extent restricted." And he went on to call this point "perhaps the most important fact which has emerged from whale marking," since, in his view, it implied that "there is no indiscriminate shuffling of the stocks, either in the Antarctic or in warmer waters."

Why did he think this was a significant finding? One that could really make an impact on the gathering initiatives to frame international regulation of the industry? Because, as Mackintosh put it, this Discovery discovery meant that "a reduction of the stock, for instance as a result of whaling, is to be regarded, at least temporarily, as *a local reduction rather than a subtraction from the general pool of whales in the Southern Ocean*."[288]

. . . .

It is a curious sentence, and one on which we would do well to pause for a moment as we think about how to sum up the work of the hip-booted whale biologists before World War II. How are we to understand this proposition? On the one hand, reading charitably, it would be possible to argue that Mackintosh is here gesturing toward the need for genuinely geographically tailored, stock-specific management regimes—a notion that would not, as it happens, gain real traction in the regulatory arena for the better part of fifty years.[289] Such an interpretation seems more than a little extravagant, but it is not absolutely impossible, just as it is not impossible (and indeed, possibly even more credible) to detect in Mackintosh's some-

---

Hemisphere and tallying lists of facts and factors under two headings, "in favour of a broad front" vs. "in favour of channeled migrations." This is a clear indication of the durability (perhaps even the *quixotic* durability) of what is best thought of as "the Harmer question." For these manuscript notes, see SMRU MS 38600, box 675, file "Whales: Observations on Distribution."

288. N. A. Mackintosh, "The Southern Stocks of Whalebone Whales," *Discovery Reports* 22 (1942): 197–300, at p. 251. Emphasis added.

289. It is important to remember that even species-specific regulatory arrangements were, with few exceptions, a long time coming: the use of the "Blue Whale Unit" for catch quotas throughout the 1950s and 1960s would effectively obscure the shifting species composition of the annual takes and enormously blunt the effectiveness of various protective measures. For more on this, see chapters 4 and 5 below.

what tortured phrasing a gesture back toward the Harmer-Hjort dispute of the early 1920s. Perhaps Mackintosh's insistence on the "localness" of depletions represents an effort, however awkward, to stump one last time for a Harmeresque notion of tight, discrete groupings of large whales in the Southern Ocean. Perhaps.

But one can be quite certain that Harmer, with his doomsday concerns and his tireless activism on behalf of whale conservation, would never have written the sentence that way. After all, Mackintosh's nearly oxymoronic formulation encourages the reader to think of all stock reductions as some-how "local" *rather than* as "subtractions from the general pool" of southern rorquals. But this makes no sense. If a reduction is local, it cannot but be, at the same time, some sort of reduction from the total "pool," given that the pool is itself composed, by Mackintosh's own (implicit) definition, of a set of geographically "local" subpopulations. One can certainly see, if some-what dimly, what Mackintosh is getting at, but one would be hard pressed to say that he has said it very clearly.

Or maybe we need to put this a little more strongly: he has really given the proposition in a form that is tailored to facilitate a distressing mis-understanding, since he has summed up an important point about the *vulnerability* of local stocks in such a way that it reads like a comforting reassurance concerning the *invulnerability* of the total population. How very odd. And it is safe to say that it is odd in a way that we should probably call symptomatic, rather than exceptional—which points to the larger issue at stake in the work of the Discovery Investigations through the 1930s and, more generally, to the larger issue at stake in the work of the hip-booted whale biologists of the first half of the twentieth century.

A modern reader of Mackintosh's 1942 study finds it surprisingly mealy-mouthed about the extent of the problem it is ostensibly structured to ad-dress. What can be said about those "stocks" of "whalebone whales"? Well, the animals get about so long; they weigh, on average, about such and such; and their life cycles work something like this. Yes, but what about the actual *stocks*? Well, the relative abundance of the commercially pursued species was roughly 15 percent blues, 75 percent fins, and 10 percent humpbacks—estimates based largely on the catch statistics themselves.[290] And yes, the

---

290. Weighted for gunner preference. To be fair, Mackintosh also cleverly wrung the marking data to get to these numbers, and he did so in a way that helped salvage a good deal of wasted marking effort: he treated the marking voyages as, in effect, sighting surveys, making use of the marking logbooks to tally the total whales sighted

blues would seem, on the basis of changing catch composition, average length, and number of immature animals taken, to be overtaxed. But as for general recommendations concerning how the biological work could help shape regulation, Mackintosh offered only the very mildest and most self-evident propositions (better to shoot males than females, large whales rather than small ones, etc.), together with the hopeful valediction that "it is thought that when more precise information on the habits and distribution of whales is available" some "useful approximation" of the absolute magnitude of the stock might be achieved. More work on ovaries was also needed. And more marking too.[291]

A modern reader with some historical sensitivity will naturally wish to be cautious about retrojecting post-1972 cetacean conservation concerns into an assessment of Mackintosh's 1942 article and will therefore probably try to understand the relatively bloodless character of his treatment of these matters as, in some sense, reflective of the moment. And all that is fine, as far as it goes. But such dutiful historicism can be rattled by a quick look back at Sidney Harmer's forceful blasts from the late 1920s.[292] They are a

---

by species, regardless of whether or not they were actually marked. He also, of course, had by this time a good deal of "straight" sighting data from the last commission of the *Discovery II*. These data, as I will show in chapter 5, proved important in later stock assessment work.

291. Mackintosh also mentioned, gingerly, the possibility of an Antarctic "sanctuary," a notion already a commonplace in international negotiations. And he noted that the outbreak of war was likely, at least temporarily, to reduce hunting pressure.

292. See, for instance, Harmer's final report to the Trustees of the British Museum before retiring in 1927, a thick sheaf of typescript that surveyed whaling in the whole of the British Empire, culling his reams of statistics from form 132 (by this time distributed from Angola to Grenada, Natal to the South Orkneys) and abstracting more than a hundred industry and Colonial Office documents. Harmer left his post with a grim note to all concerned. "It is not possible to feel satisfied with the present position," he noted in conclusion, since "the rate of killing is too great." Harmer, "Report on Whales and Whaling," February 1927, part 2, NHM DF 1004 / CP / 747; the quotes are from p. 68. Particular thanks to Polly Tucker for helping me find this document, to which I had found references, but which was not filed with Harmer's other material. Harmer reprised much of the contents of this report in a more public venue, namely, his presidential addresses to the Linnean Society, which he led for several years after stepping down from the museum. In 1928 he reviewed for the membership the bleak history of whaling from its earliest days before declaring that "to the naturalist the extermination of a species must appear a crime." Harmer, "Presidential Address," 24 May 1928, *Proceedings of the Linnean Society of London* 140, no. 1 (November 1928): 51–95. Two years later he delivered a massively detailed analysis of southern whaling to buttress his allega-

powerful reminder that there was nothing inevitable about the industry irenicism and hangdog caution of Mackintosh's stance on the same problems some fifteen years later. What had happened?

The short answer is that over the course of those fifteen years, the whale biology that Harmer had planned and projected (explicitly for the purpose of reining in a runaway whaling industry) had been gradually drawn into something like a commensal relationship with the whaling industry itself. On the one hand, a set of institutional challenges—hostility from competing administrative units within the British government, decreasing revenues from the whale-oil tax, personnel problems—weakened the "all-powerful" Discovery Committee and forced realignments in the larger mission of the organization. These realignments pulled the Discovery Investigations away from a central focus on regulation-relevant biological investigations of the southern whales (and did so while compromising the independence of the investigations themselves). On the other hand, the actual fieldwork—the "direct work on whales" outlined by Harmer—tended over these same years to acclimatize the Discovery biologists to the ways and means of the whalers. While Sidney Harmer originally intended to train a community of "whale scientists," a generation of demanding field research in the Southern Hemisphere had created, instead, a community of "whalemen-scientists," men who had spent years living as whalers, among whalers, and doing the things that whalers did. After all, the two fundamental "prongs" of the Discovery system of whale research—marking surveys and anatomically oriented studies of reproductive physiology—precisely recapitulated, in their actual practices, the two basic activities of the whaling industry: shooting whales in the open ocean, and disarticulating their massive bodies on a blood-slicked plan. Moreover, the institutional story (a weakened Discovery that needed to reach out to the industry more and more) and the fieldwork story (Discovery researchers gradually became honorary whalemen, proud of their pidgin Norwegian and anxious to show off their calluses) were mutually reinforcing, as we have already seen: from the late 1920s on, a chastened Discovery Committee was forced to draw the industry more directly into whale marking, using hired catchers and professional personnel, with real implications for the daily experience of

---

tion that, in view of the rapid expansion of pelagic whaling, things were getting even worse, presenting the world with an "alarming" situation of "extreme gravity." Harmer, "Presidential Address," 24 May 1930, *Proceedings of the Linnean Society of London* 142, no. 1 (January 1930): 85–163 (substantially revised from the presented form).

whale biologists involved in marking. Similarly, when, later in the 1930s, international agreements necessitated shipping "inspectors" on pelagic whaling vessels, Discovery biologists were seen by the industry as promising candidates: they were already acculturated to the demands of the flensing platform, and anyway, Mackintosh was hard-pressed to find the money to employ them otherwise.[293] Along the way, whale biology was being sucked into the belly of the beast.

I would like, in the pages that follow, to buttress these claims. First, I will review a few of the institutional challenges faced by the Discovery Investigations over the years in question, and then turn, in closing, to a brief discussion of the evolution of a kind of complicity between the Discovery whale biology community (led, of course, by Mackintosh) and the industry it had in some sense been conjured up to call to account.

The Discovery Committee found itself, from relatively early on, mounting a defensive operation within the overlapping structures of British administrative authority. Perhaps the punchiest indication of this fact can be found in the "confidential" (but soon widely discussed) dispatch sent in 1928 from the then-governor of the Falkland Islands, Arnold Hodson, to the secretary of state for the colonies.[294] This blistering send-up of the first three active years of the Discovery Investigations makes for hilarious reading and offers, among other things, an elegant case for the forms of rhetorical training native to the British administrative elites before World War II. Hodson, who had traveled to South Georgia and had direct dealings with both the Discovery personnel and the scientists based at the Marine Biological Station, gave no quarter in his 29-page denunciation of the incompetence and pathos of the whole Discovery undertaking, which he declared a "grave public scandal" that had wasted upwards of £300,000 (an amount, he was quick to note, that represented more than six times the

293. This worked both ways: inspectors hired from the outside were made correspondents of the Discovery whale research program and given instruction in various techniques of specimen collection and measurement making (for details on these arrangements, see SMRU MS 38600, box 636, file "Employment of Whaling Inspectors on Scientific Work" [Disc. 22039]), and actual Discovery scientists were seconded to factory vessels, both to keep these men employed when the financial situation became dire and, later, to afford them suitable training grounds (for a feel for this practice, see F. D. Ommanney's account of a season on the *Salvestria* in 1939–1940 in *Lost Leviathan: Whales and Whaling* [New York: Dodd, Mead, 1971], chap. 9).

294. I discovered the full "binder" of the dispatch and responses at NOL Discovery MS, misc. unsorted files 4, "Governor's Despatch 8 / 'v' / 1928."

annual revenue in toto of the colony for which he was responsible). Review-ing in painful detail a sequence of awkward mishaps (e.g., two temporary groundings of the *William Scoresby*, requiring the help of an Argentine whaling vessel, "to the eternal shame of British Seamanship," plus a separate incident involving the striking of a well-known rock in a South Georgia harbor) and humiliating derelictions of duty by Discovery personnel (e.g., public drunkenness, insubordination, and harassment of women in Port Stanley), Hodson went on to mock the fiasco of the (totally unsuccessful) marking efforts and to accuse the Discovery Committee of general errors of judgment that were "little short of criminal." They understood nothing of what they were getting themselves into, he said, and had outfitted their ves-sels with sleds and tents as if they were off on a "dash to the pole" while sat-ing their natural historical curiosities and manias at obscene public cost.[295] The whole notion of an investigation into the whales of the Antarctic by a committee of intellectuals and administrative personnel sitting in London struck Hodson as equally logical as that a similar "Committee might be appointed and sit in the Falkland Islands to enquire into the causes of the decline of the ostrich industry in South Africa or to report on the methods of elephant hunting followed in the several parts of the world."

Hodson proposed, perhaps predictably, that the whole of the whale-oil tax revenue be conveyed immediately to the use of his own administra-tion in the Falklands, that Norwegian whalemen be hired for pennies on the pound to conduct whatever investigations were required, and that the remainder of the available monies be used "in the manner best calculated to secure the moral and material prosperity of this section of His Majesty's subjects, and not to acquire the doubtful advantage of a more intimate insight into the generative organs of the whale." After all, the whales them-selves would surely move on at some point, once they came to sense the

295. In his jibe about a "dash to the pole," Hodson put his finger on that besetting sin of the whole Discovery undertaking (discussed above): its romantic preoccupation with the high heroic era of British polar exploration. One sees this perhaps most clearly in the decision to refurbish Scott's old vessel (which was a dubious undertaking financially and difficult to defend—as would soon be shown—as a practical matter for oceanographic exploration), but also in the whole rhetoric around the Discovery Investigations, which were endlessly touted as the proper successor to the voyages of Cook and Shackleton. Even Harmer's preoccupation with the notorious whale crossbow can be thought of as lightly tainted by a kind of Anglo-Norse ice-atavism that quietly perfused the whole program, generally to its detriment.

dangers of the region (here, of course, Hodson was borrowing a line from his predecessor, Allardyce). "May I be pardoned, therefore," he wrote in conclusion, "if I express the hope that under the new order of things the leviathan will be left to disport himself in peace, at any rate, from the inquisitive eye of science?"[296]

In a flurry of activity in the summer of 1928, the Discovery Committee closed ranks, drafted a suite of careful rebuttals and rejoinders, and solicited its allies in the various ministries with a stake in the program.[297] In the end, the governor's (palpably self-interested) proposals for a rearrangement of fiscal control were relatively easy to quash, but his ringing disparagements of Discovery—that it was, on the whole, "the common laughing-stock of the whaling fleet"—hung in the air for some time, especially when, thwarted by his superiors at the Colonial Office, Hodson began to mobilize allies in Parliament and in the press.[298]

A little negative publicity was irritating, but the Committee packed with powerful figures from the critical ministries and offices could hold its own in backroom backbiting around Westminster. But only as long as it had lots of money in the bank, and by the early 1930s the cash cow of the whale-oil tax was looking very anemic indeed. The reason, of course, was the rapid expansion of fully pelagic factory-ship whaling in the open oceans. Between 1928 and 1931, Norway and the United Kingdom added eighteen large stern-slipway factory vessels to the Antarctic fleet, along with more

296. NOL Discovery MS, misc. unsorted files 4, "Governor's Despatch 8 / 'v' / 1928."

297. It is significant, as far as public relations were concerned, that the British Museum rapidly pulled together an elaborate "Discovery Exhibition," which opened on 1 December 1928. Planning for this exhibition began in the spring of that year, exactly as the controversy was heating up. See Kemp to Regan, 24 April 1928, NHM DF 1004 / CP / 748, file "Whale Room, 1923–1955."

298. For a sense of this episode, consult the broadside launched by Sir Robert Donald in the *Daily Mail* on 3 September 1928, "The 'Discovery' Muddle, £130,000 spent on £5,000 Ship, Waste of Public Money," and the flurry of follow-up pieces in all the major papers in the United Kingdom in the month that followed, in NOL Discovery MS, scrapbook of press cuttings. The issue grew to full scandal proportions, and Kemp spent a good deal of that autumn putting out fires. A visit to London by a member of the Legislative Council of the Falklands, W. A. Harding, in the same month helped feed the news cycle. Some of Hodson's frustration, of course, was bound up with more general questions of colonial governance, and it is worth remembering that the entire center-periphery showdown unfolded in the shadow of the 1926 Imperial Conference and the recently promulgated Balfour Declaration.

than a hundred new catcher boats. Joining the already existing flotilla of vessels from five countries, these new floating factories gave the southern industry of 1931 the capacity to lay down at sea a total of almost two and a half million barrels of whale oil. The 1930–1931 season went on to produce the largest amount of whale oil ever collected from the world's oceans in a single year—the liquefied remains of more than 28,000 whales.[299] This volume of production—which would by 1932 so overwhelm the market as to produce a price crash and a year-long industry lay-up—dwarfed, and essentially displaced, the declining yields from the aging shore stations of the Scotia Archipelago. Who would bother producing whale oil under the geographic restrictions and high-tax conditions of a place like South Georgia if a little capital infusion could produce a pelagic factory affording open access to the whole Antarctic and the freedom to land oil tax-free? The results of that reasoning were inscribed in the revenue figures forwarded to the Discovery Committee in April 1933 by the undersecretary of state for the colonies:

1928	£198,153
1929	£122,814
1930	£84,546
1931	£48,153
1932	£36,695 (*estimated*)
1933	£30,245 (*estimated*)

Little wonder that the same letter concluded on a cool note: "In the circumstances, it is evident that the whole question of the future utilization of the Fund calls for review."[300] It was back to the scramble for the Discovery Committee, which still had some £385,000 in the bank, but was in effect being told that there was no more where that came from, and that future commissions of the expensive Discovery vessels would need to be reconsidered.[301] The committee was asked to draft a plan for wrapping up the

299. For these statistics, see J. N. Tønnessen and Arne Odd Johnsen, *The History of Modern Whaling* (Berkeley: University of California Press, 1982), chap. 22 and table 54.

300. "Copy of letter from the Under Secretary of State, Colonial Office, dated 27th April 1933," SMRU MS 38600, box 645, file "Future of the Discovery."

301. A two-season commission for both the *Discovery II* and the *William Scoresby* cost at this time in the neighborhood of £70,000, but the committee had a variety of other expenses, including salaries for its own staff, costs of upkeep of specimens, laboratory work, and administration.

investigations as a whole in short order. Piqued, the group fell back on the claim that since some £600,000 had already been spent, "to abandon the work now," when their main objects remained unrealized (but visible on the horizon), "would, the Committee feel, amount to a deplorable waste." But that was, of course, a dangerous tack, as it seemed both to concede that the half million pounds spent had produced scant returns, and to rely on the logic of sending good money after what could be easily described as bad.[302]

To strengthen their case, the committee also assembled a mass of supporting documentation aimed at putting the best face on the work of the investigations to date. But the financial challenge came at a difficult time in that it was still effectively impossible to point to any important progress on the marking and migration studies, and the anatomical work on whales had yielded results that were difficult to sell to a member of parliament.[303] Not that Harmer didn't try. He worked back channels furiously, penning, for instance, a long letter to Stanley Baldwin (both an MP and a trustee of the British Museum) that warned of what seemed to be a Colonial Office coup in the offing. It appeared to Harmer (and others) that someone had an eye on the remaining Discovery funds, and Harmer wanted Baldwin to stick up for the principle that this money had been raised from the whalers for work on the whales and that there was "a moral obligation to respect that promise."[304] But Harmer was savvy enough to press another line of

302. The committee did touch lightly on this argument, however, in their cover response of 18 May 1933. See SMRU MS 38600, box 645, file "Future of the Discovery."

303. These documents include "Position of the 'Discovery' Investigations and Future Policy," and "Note on the Present Position in Regards to the Regulation of Whaling," together with "Note of Representation to the Secretary of State," all of which can be found in ibid.

304. Harmer to Baldwin, 28 June 1933, ibid. These anxieties concerning a takeover would be sharpened at the end of the year, when Darnley would be hastily deposed from his decade-long chairmanship of the committee and replaced by an outsider with close ties to the secretary of state for the colonies. This change led to something close to panic among the members of the committee, who fretted over how to respond. Interestingly, in the wake of several newspaper columns on the matter (*Morning Post*, 22 December 1933, 11, and 29 December 1933, 8–9; NOL Discovery MS, scrapbook of press cuttings), Harmer was approached by the editor of *Nature*, which had received solicitation from Darnley directly in the hopes that the journal would do an exposé on his ouster; the editor sought Harmer's advice about how to proceed. Harmer held the editor off long enough to meet the new chairman (the Earl of Plymouth) and determine that he seemed "sympathetic" to the aims of the Discovery Investigations. Deciding, in consultation

argument as well, one that would rise to increasing prominence as the Discovery Investigations struggled for protection within the shifting political landscape of the 1930s. As Harmer put it,

> The British Government lays claim to the major part of Antarctica and to most of the Subantarctic Islands, which are of special importance. This country is the one which should examine their great potentialities and it should accept responsibility for the protection of their resources. Apart from our Research Ship "Discovery II" there is no instrument of British policy in these waters; and it is only by such activities as these that we can maintain our territorial claims.[305]

It is as clear an indication as could be imagined of the extent to which Discovery began to wrap itself in the flag with greater fanfare in the run-up to World War II.[306] Pragmatically speaking, this strategy met with considerable success, and *Discovery II* and *William Scoresby* both ended up recommissioned several more times before the war, but in the process also found themselves drawn into several flag-showing exercises of limited scientific value.[307] The whales—whose oil no longer played a role in the making of

---

with other committee members, that there was nothing to be gained by using *Nature* to humiliate the government, Harmer gently recommended that the editor turn down Darnley's plea, calling it "prejudicial to the interests of the Discovery Committee, and therefore to scientific research." *Nature*'s reply was deliciously supine: "The editor certainly does not desire to have this matter discussed in the columns of *Nature* unless naturalists in the position of Sir Sidney Harmer considered that a useful purpose would be served in doing so." For all of this, see the small bundle of "confidential" papers at the end of SMRU MS 38600, box 637, file "Whaling Research Expedition—Discovery Expedition—Harmer Correspondence."

305. Harmer to Baldwin, 28 June 1933, SMRU MS 38600, box 645, file "Future of the Discovery."

306. This strategy was used with particular blatancy during the controversy around Darnley's removal. Harmer proposed that if the new chairman made any trouble, "it might be desirable to take up his challenge at once," and to do so by giving him a collection of recent statements in the Norwegian press that tended to call into question British sovereignty in the Antarctic; these statements were, in Harmer's view, "surely a strong reason for continuing the work at sea of the 'Discovery II.'" See Harmer to Maurice, 3 January 1934, SMRU MS 38600, box 637, file "Whaling Research Expedition—Discovery Expedition—Harmer Correspondence."

307. Including the notorious non-rescue of Lincoln Ellsworth and H. Hollick Kenyon. For more on this phase of the life of the Discovery Investigations, see John Coleman-Cooke, *Discovery II in the Antarctic: The Story of British Research in the Southern Seas*

military explosives—were ceasing to be the priority. Physical oceanography was increasingly upstaging the "direct work on whales," in a pattern that would culminate, after the war, in something very much like a coup (whereby the Discovery Investigations as a whole would be folded into a new research structure, the National Institute of Oceanography, where Mackintosh would play second fiddle to physical oceanographers who had once been his junior Discovery colleagues).[308] In a general way, something like this had been on the minds of the oceanographers from the outset: there was a telling exchange to this effect in April 1928 in the lecture hall of the Royal Geographical Society, when the distinguished senior ocean scientist H. R. Mill responded to a presentation by Discovery's biological oceanographer Alistair Hardy. Mill was enthusiastic about the Discovery Investigations

> not because in my restricted outlook I consider the abundance of whales in the southern ocean is of any importance in itself, but the whales have induced people to go out and make observations on the character of the ocean in which they live. In other words, I would use the whale simply as bait to induce clever young men to spend their time in acquiring information as to the nature of the oceans and the movements that go on there.[309]

(London: Odhams Press, 1963), and the discussion by Rosalind Marsden in "Expedition to Investigation: The Work of the Discovery Committee," in *Understanding the Oceans: A Century of Ocean Exploration*, ed. Margaret Deacon, A. L. Rice, and C. P. Summerhayes (London: University College of London Press, 2001), 69–86.

308. I take up this transition in more detail in chapter 4 below, but it is perhaps worth noting here that the rapid rise to dominance of the physical oceanographers in the interwar work of the Discovery Investigations feels, from a historical perspective, rather overdetermined: with increasing awareness in the 1930s of the threats presented by German submarine development, the Admiralty (an important stakeholder in Discovery) had every reason to step up its commitment to the profiling of the deep-sea environment; at the same time, in the wake of the work of the German *Meteor* expedition of 1925–1926, there were major new basic discoveries to be made concerning the structure and dynamics of what was in these years coming to be called the "Antarctic Convergence" (discoveries that had implications for the understanding of upwelling, krill blooms, etc.). Physics and chemistry trumped morphology and biogeography across these years in a way that will feel familiar to any student of the history of science in the first half of the twentieth century. For more on this, see Gordon Elliott Fogg, *A History of Antarctic Science* (Cambridge: Cambridge University Press, 1992), esp. chaps. 6, 7.

309. The quote comes from the front matter in the George Deacon Papers at NOL.

Whalebait. It is a notion that goes some way toward explaining what actually happened.

For an interesting window onto the strategic rethinking undertaken by the Discovery Committee in the context of the financial rumblings of the early 1930s, it is worth consulting an internal document generated by the Scientific Subcommittee in 1933, which outlines budget cuts and possible shifts in research emphasis. This document is telling in its shift away from an emphasis on "direct work on whales" and toward a robust concern with general oceanography and hydrographic surveying. And indeed, as in Harmer's letter, the whole analysis ends with a rousing invocation of Britain's historical primacy in polar exploration and drums a patriotic tattoo for the future of Discovery:

> Since the days of Captain Cook Great Britain has taken the leading part in discovery and exploration in the Antarctic, and to-day the Empire lays claim to more than half the sub-antarctic islands and to about three-fifths of the Antarctic continent, a vast territory of almost unknown potentiality. . . . To occupy heavily glaciated lands in a permanent manner is probably impossible, but our claims could be greatly strengthened, if not secured, by a patrolling vessel kept in commission in Antarctic waters.[310]

In addition to this sort of rhetorical posturing, the document looked in some detail at money-saving measures, and in doing so proposed several revisions to the way the whale work was being conducted. For instance, the memo noted clearly that the work at the Marine Biological Station would be of limited value going forward, given the change in whaling methods seen over the late 1920s, and that, given this new situation, it would be worth getting Discovery personnel working on anatomical and physiological issues stationed aboard whaling vessels. The best part about such "hospitality" arrangements was that they "can be undertaken at very low cost." Similarly, the best way to continue promoting the whale-marking venture would be "to enlist the assistance of whaling companies," once again because in this way "it is possible that the work can be done at comparatively low cost." Both of these moves reflect very clearly the ways in which shifting financial

---

310. "Note on the Future Course of the Discovery Investigations," with manuscript annotations in Harmer's hand (marked "not to be filed"), dated 5 April 1933 and appearing over Kemp's signature. It can be found in SMRU MS 38600, box 637, file "Whaling Research Expedition—Discovery Expedition—Harmer Correspondence."

conditions forced the committee to shift its direct work on whales into an increasingly cooperative / parasitic position with respect to the industry. It is worth recalling that Mackintosh was at this time the chief scientific officer of the Discovery Investigations and that his outlook on the future of whale research was shaped in these years of diminished resources and increased dependence on the industry. The care with which he would navigate around industry prerogatives for much of the rest of his career must be understood to have its origin in the challenges of this moment.

One might well ask, however, whether Mackintosh would genuinely have thought of himself as navigating *around*, rather than simply navigating *with*. After all, in the period after World War II Mackintosh would be very interested in the idea of actually chartering a *whole factory vessel* (with its attendant catcher boats) for the purpose of a globe-encompassing cetological expedition—a whaling voyage that would permit the whale scientists to hunt and collect whales all over the world without regard to regulation or commercial necessity.[311] It is a fantasy that throws into high relief just how much of a whaleman Mackintosh (and his colleagues) had become through the interwar work. As one former Discovery whale biologist put it with considerable pride in the early 1960s, the training of a true "cetologist" created men who were a "race apart from other zoologists," since they had to be "undismayed by the task of dissecting a hundred tons of highly flavored carrion."[312] Over the course of two decades chasing down and cutting up large whales in the Antarctic, the hip-booters had become a fraternity, sealed nostalgically in the rituals of the Southern Ocean. Their

311. Mackintosh's own file on this topic is listed in the NOL finding aid as Discovery MS file 4679, "Proposal to Charter Floating Factory and Whale Catchers," but, like a large number of other (potentially controversial) files, it is missing from the actual collection; a note in the register reads, "File held by Dr. Mackintosh." I was able to locate some of this missing material in the SGB Papers, but not this file. The desire for such an arrangement (the forerunner, it should be noted, of the current Japanese "scientific whaling" fleets) was nevertheless no secret: the former Discovery whale biologist L. Harrison Matthews made an explicit plea to a "philanthropist" to bankroll a scientific factory vessel at the "First International Symposium on Cetacean Research" in Washington, DC, in 1963. See L. Harrison Matthews, "Chairman's Introduction to First Session of International Symposium on Cetacean Research," in *Whales, Dolphins, and Porpoises*, ed. Kenneth Stafford Norris (Berkeley: University of California Press, 1966), 4. For a further discussion of this gathering, see chapter 6 below.

312. This is L. Harrison Matthews again: "Chairman's Introduction to First Session of International Symposium on Cetacean Research," in *Whales, Dolphins, and Porpoises*, ed. Kenneth Stafford Norris (Berkeley: University of California Press, 1966).

work was dependent on and inextricable from that of the whalemen, to be sure, and the Discovery whale scientists had come to emulate the labors of their whaler-kin with assiduous pride and cultivated fortitude.

It is not difficult to trace the evolution of this peculiar persona in the archival remains of the Discovery Investigations, and nowhere is it more obvious than in the ways that the Discovery biologists "gamed" the pursuit of whales in the marking campaigns of the 1920s and 1930s. This activity was explicitly referred to as "hunting" in the correspondence and diaries of these scientists, who maintained cheerful rivalries as marksmen, scoring each gunner on a given crew and ribbing each other for poor shooting.[313] They even practiced their skills on towed targets for sport when no whales were about, declaring such exercises "particularly valuable for teaching us to shoot quickly with the ship in a seaway."[314] And by the mid-1930s, following the model of the Norwegian catcher vessels (from which they were often working), actual *bonuses* were being paid to the marking gunners for every hit (figure 2.39).[315] Along the way the Discovery staff developed a connoisseur's taste for the quirky minutiae of the whalemen's craft: they learned, for instance, the finer points of the different chase strategies that could be used under different circumstances (the brute-force *prøsser jag*, or "Prussian attack," which was all straight-ahead and full throttle, versus the sneaky *luse jag*, or "stalk of the louse," where finesse and silence and strategic positioning replaced horsepower).[316]

And with these forms of intimacy and appreciation came a certain amount of face-to-face quid pro quo dealing as well. Recall that Harmer had, from the outset, fretted about the dangers of trading actual whales for

313. This sort of thing is pervasive; for a taste, see the early correspondence from the Marine Biological Station in NOL Discovery MS, green filing cabinet, files "Marine Station—Correspondence, etc. 1924–1927" and "Marine Station—Correspondence, etc. 1928." Of particular value are Francis C. Fraser's letters to Kemp through the 1927–1928 season, especially that of 15 January 1929 (which can be found in the latter file, despite the later date). See also Gunther's youthful ejaculation, "Royal sport is whale hunting," in Eugene Rolfe Gunther, *Notes and Sketches Made During Two Years on the "Discovery" Expedition, 1925–1927* (Oxford: The Holywell Press, 1928), 29.

314. J. John Hart's holograph notebook (no date, but circa 1936), SMRU MS 38600, box 679, file "J. John Hart: Notes—Whale Marking Cruise 1936–1937, W.S."

315. See "Minutes of the Scientific Subcommittee, 1927–1949," particularly documentation from the meeting of 23 March 1936, in SPRI MS 1284/6.

316. For some of this, see "Note on approaching whales" in SMRU MS 38600, box 635, notebook "Whale Observation, W. Scoresby 1926–1929."

FIGURE 2.39
The whaleman-scientist:
Marking whales from a
bow-lashed barrel. (From
the collection of the
National Oceanographic
Library, Southampton.)

whale knowledge, but such arrangements proved impossible to resist. In the late 1920s, for example, the Discovery whale biologists pulled strings to get the whaling companies an extra five hundred humpbacks at South Georgia ("in the interests of the scientific enquiry"). More generally, there were efforts to work out a standing arrangement that would allow the Discovery scientists to issue "occasional permits" to various companies, "subject to the condition that the staff at the Marine Station are given facilities to measure the animals captured."[317] When it was time to hire catcher vessels

317. See the letter from the Colonial Office to the governor's office in the Falkland Islands, 30 May 1927, NOL Discovery MS, green filing cabinet, file "Marine Station— Correspondence, etc. 1924–1927." It is not clear if the governor approved the request, since in 1928 Fraser would express dismay that the magistrate at South Georgia denied permission to take an extremely rare southern right whale when Captain Esbensen's hunting boat was hot on its trail. "I was very sorry it was refused, because I would have liked to examine a right whale," wrote Fraser to Kemp, and closed playfully, "I've no doubt Esbensen would have been pleased too, altho' probably not for the same reason."

to extend the marking campaign, it was not unheard of to bargain "an extra whale or two" as part of the payment.[318] The integration of Discovery personnel with whaling inspection staff in the later 1930s was perhaps even more corrupting. It was not merely that the *William Scoresby* began to do service ferrying inspectors from one vessel to another and otherwise running errands for the factory vessels—this was perhaps only fair, in that the *William Scoresby* relied on the kindness of the pelagic whalers to stay refueled during her long passages (she bought fuel on friendly terms from the mother ships when needed).[319] There were the deeper entanglements of the monitoring itself. The Discovery biologist F. D. Ommanney, who had put in his seasons on the plan at South Georgia, shipped as an inspector on the *Salvestria* in 1939, where he split his time between collecting reproductive tissues and monitoring that the gunners were adhering to the catching regulations in effect (various restrictions had by then been promulgated by international agreement). He found it impossible to be strict where the latter duty was concerned, given that he was spending all afternoon on the cutting deck "mountaineering . . . over mounds of red flesh and piles of bones" while "skillfully dodging over taut and criss-crossing steel hawsers to the accompaniment of the usual warning shouts." Under such death-defying circumstances, where the trap doors to the boilers opened directly down through the deck, Ommanney made sure to make "an ally of the plan foreman," who now and again even "delayed the work for a minute or two" to let the hip-booted biologist in crampons grab a fetus. One did not then raise a hue and cry when a lactating female came up through the slip-way (shooting whales accompanied by calves was at this time forbidden). Who wanted to make enemies of shipmates on whom one was so perfectly dependent—particularly when every single sailor was paid on a "lay" system, so that declaring a violation picked everyone's pocket? Ommanney described his routine as petty theater:

> Usually, in these cases, I had to listen to a long explanation by the gunner when he next came on board the factory. Sometimes he came alongside and visited me specially to tell me about it. It had been dark

Fraser to Kemp, 27 April 1928, NOL Discovery MS, green filing cabinet, file "Marine Station—Correspondence, etc. 1928."

318. See Mackintosh's "Diary kept during the Voyage of the R.R.S. William Scoresby (3rd commission), 1930–1932," entry for 16 February 1931, SPRI MS 1268/11;BJ.

319. For mentions of some of these arrangements, see SMRU 38600, box 636, file "Employment of Whaling Inspectors on Scientific Work" (Disc. 22039).

or foggy, he did not see the calf. I smiled and felt like an indulgent schoolmaster and reached for the Red Label Johnnie Walker.[320]

So crucial, in fact, was whisky to the lubricated dealings on board these vessels that when the first Discovery whale biologist to ship on a pelagic whaler assembled his kit, the Discovery Committee itself footed the bill for three cases of Johnnie Walker Black, to be used as the coin of the realm (together with a case of gin).[321] It was used, one supposes, to buy some specimens and to toast a few mistakes. One could also use it to wash down a little whale sausage, a staple aboard the vessels and an acquired taste—though a number of the Discovery men learned to think it superior to beef. Just like the whalemen.

CONCLUSION

It has been the aim of this chapter to trace the development of a particular kind of whale science across the first half of the twentieth century and to offer an account of the institution that did the most to codify its methods and generate its results. The Discovery Investigations, which arose out of conservation-oriented concern about the rapid expansion of modern whaling in the Southern Hemisphere, codified what Mackintosh himself would call a "system of research" on the large whales, and, using it, generated the preponderance of the biological knowledge of these animals available before World War II. As I have shown, the culture of Discovery fieldwork and the fiscal-cum-administrative challenges faced by the Discovery Committee across the long 1930s led this "hip-booted" whale science into entanglements with the industry that would ultimately make it very difficult for such scientists to do the sort of research that Sidney Harmer had intended—research that would facilitate regulation and help bring the whalemen to heel.

In the course of doing the research for this chapter, I had the occa-

---

320. When the captain filed the final report of infractions, he stopped by to tell Ommanney the results: "He winked, and so did I." This and the quotes above are from F. D. Ommanney, *Lost Leviathan: Whales and Whaling* (New York: Dodd, Mead, 1971), 189–91.

321. This was Laurie, who took passage on a Lever Brothers vessel, the *Southern Princess*, for the 1932–1933 season, visiting briefly on the *Southern Empress* and the *Skytteren* as well. For the anecdote about the liquor, together with much other interesting material, see SMRU MS 38600, box 636, file "Mr. Laurie's Visit on a Pelagic Whaler" (Disc. 15021).

sion to talk with several of the surviving men who were part of the work of Discovery. In conversing with one of them—an impossibly gentle and kind old man who had shipped on a factory vessel as a Discovery-attached inspector shortly after World War II and who had worked with Mackintosh for more than a decade—I chanced to inquire about his recollection of the heady days in the 1970s, when an increasingly vociferous international call for a moratorium on commercial whaling threatened the future of the industry. The older gentleman, balancing his teacup on his saucer, looked at me with moist eyes: "We were against them," he declared solidly, invoking the memory of his old Discovery boss (Mackintosh himself, as it happens). "They took the position that there should be no whaling whatsoever, and we were working on whales—we would have no material!"[322]

Indeed. Spoken like a true hip-booter. And by the early 1970s, the environmental activists pressing for the moratorium had their number. As one young rabble-rouser put it in 1974, "Cetology has for a long time been a 'dead' science," in the sense that "the bulk of the scientific reports are based on data taken from dead whales." For this reason, he continued, "the whale scientist may often be in a parasitic relationship to the whaling industry."[323] It was an accusation that infuriated the great majority of the whale biologists then living, many of whom even then took pride in their ability to wield a flensing knife. The legacy of the Discovery Investigations died very hard.

But by 1974, of course, with the rise of a new cohort of ecologically sensitive and behavior-oriented marine mammal biologists (the subject of chapter 6 below), the Discovery system was indeed a dying science. Or so it seemed. In fact, interestingly, as momentum gathered for an actual global moratorium on commercial whaling in the late 1970s, Mackintosh's system would get a new lease on life as a legalistically unassailable excuse (offered by Japan) to continue to hunt and kill whales: international agreements featured an exception clause that allowed for the taking of whales "for scientific purposes" (see chapter 5 below). There is something close to perfect

322. Sidney Brown, interview by the author, 23 March 2004, Cambridge, UK.

323. This was Scott McVay, in the published version of his address to the International Conference on the Biology of Whales in 1971 (the "Shenandoah Conference"), which resulted in William Edward Schevill, G. Carleton Ray, and Kenneth S. Norris, eds., *The Whale Problem: A Status Report* (Cambridge, MA: Harvard University Press, 1974). The quotes are from p. 381 of what became chap. 17 of that volume: McVay, "Reflections on the Management of Whaling," 369–82.

irony in the fact that Harmer's initial vision of a research program on large whales that would lead to the protection of these animals eventually gave rise to a unique mechanism for prosecuting their destruction.

Harmer would have been very disappointed. Even at his most pessimistic, he could hardly have imagined that the compact between biologists and whalers over which he fretted in the 1910s would eventually prove the ancestor of "scientific whaling" in its twenty-first-century form.

THREE

« THE PRINCE OF WHALES »

A. REMINGTON KELLOGG
AND AMERICAN CONSERVATION

1919 – 1940

> If one had to pick out just one sort of beast as the most interesting from the
> purely scientific standpoint, what would the verdict be? Undoubtedly the
> animal of the greatest potentialities in this respect is the whale.
>
> "The Value of Whales to Science," press release by the
> Council for the Conservation of Whales, 1929[1]

INTRODUCTION

On 8 February 1928, the greasy and sluggish mail boat out of Manteo, North
Carolina, received six unusual passengers bound for the rugged Outer
Banks fishing village at Cape Hatteras. Bespectacled, accoutred in old suits
and overcoats, and heavily laden with fragile luggage, this party of young
scientists represented a diverse array of medical and biological expertise.
Dr. George B. Wislocki had already made a name for himself as a talented
anatomist with a gift for fine dissections of the sensory organs of vertebrates; he would shortly leave Johns Hopkins for Harvard. His Hopkins
colleague Orthello Langworthy, a neuroanatomist with a strong interest in
histological work on the brain, was also along for the expedition. So too
were the German émigrés Ernst Huber and Adolph Hans Schulz, compara-

---

1. "The Value of Whales to Science," Smithsonian Archives, Record Unit (RU) 7170,
Remington Kellogg Papers, 1871–1965 and undated, box 10, file "Information—Whale
Press Releases." The chapter title comes from Kellogg to Witmer Stone, 5 May 1929,
Smithsonian Archives, RU 7170, box 8, file "Stockt–Stre."

tive anatomists and embryologists who would both go on to make significant contributions to the study of primate evolution. A. Brazier Howell, the
independently wealthy and free-spirited secretary of the American Society
of Mammalogists, had left Washington, DC, the previous morning to join
up for the journey. Originally little more than an amateur ornithologist
with a smattering of training at the Sheffield Scientific School at Yale, Howell had matured into a talented field naturalist through years of collecting
expeditions on the West Coast, eventually joining the Bureau of Biological Survey as a "dollar-a-year man"—kept on payroll so he could have a
title and office but mostly left to his own devices. The post left him ample
time to pen technical briefs (with a neo-Lamarckian flavor) on mammalian morphology and myology and to exercise his prodigious talents as a
general facilitator, lobbyist, and gadfly reformer. At forty-one, he was the
oldest of the six.

Rounding out the group was a thirty-three-year-old, long-boned biologist
from Iowa who had organized the whole show: Arthur Remington Kellogg.
Like his friend and confidant Howell, Kellogg had an official institutional
affiliation with the Bureau of Biological Survey; unlike Howell, however,
"Remmy" did not enjoy the enviable freedom of a sinecure. Rather, Kellogg's post at the bureau involved tedious work in the "food habits" division
of the agency, cataloging the stomach contents of tailless amphibians from
around the country. This research was intended to advance the bureau's core
mission of economic biology: identifying pest species and figuring out how
to kill them. It was, however, an open secret among the staff at the bureau
that Kellogg aspired to a higher destiny than shuffling 10,000 index cards
on toad spoor for the rest of his life.[2] Thanks to the zealous patronage of
John C. Merriam, the paleontologist and powerful president of the Carnegie
Institution (and, significantly, a fellow Iowan), Kellogg benefited from a line
of supplementary support: Carnegie money and connections allowed him
to work nights and weekends in the vast paleontological collections of the
Smithsonian, where he had for the last seven years tirelessly pursued research originally begun in the 1910s as a graduate student in zoology under
Merriam back at the University of California, Berkeley.[3]

2. Kellogg to Franklin Metcalf, 31 August 1925, Smithsonian Archives, RU 7170, box 6,
file "Merriam, M. L.–Met." Several of his colleagues also, apparently, aspired to higher
callings: Kellogg strongly suspected one of them of dabbling in bootlegged liquor.

3. On Merriam, see Chester Stock, "John Campbell Merriam 1869–1945," *National
Academy Biographical Memoirs* 26 (1948): 209–17.

FIGURE 3.1 At Hatteras: The bottlenose fishery. (From Townsend, "The Porpoise in Captivity," following p. 299.)

Kellogg must have been in a good mood as the vessel pulled away from the dock. He had just sent off the final proofs of a major two-part monographic article, the product of those seven years of research. The piece, which would appear in the March and June issues of the *Quarterly Review of Biology* under the title "The History of Whales—Their Adaptation to Life in the Water," would earn Kellogg his PhD later that summer, and it would serve as his ticket out of his desk job at the Bureau of Biological Survey and into a full-time research position at the United States National Museum (as an assistant curator in the Division of Mammals). In February 1928 there was no one in the country who knew more about the evolutionary development and fossil history of whales than A. Remington Kellogg.

And whale research was the mission of the Hatteras expedition. Since the middle of the nineteenth century, the beaches of Hatteras Island had been home to a small but resilient fishery for bottlenose dolphins, whose hide and head oil remained paying commodities in the twentieth century (figure 3.1).[4] In their luggage the debarking scientific interlopers carried all

4. The terms "dolphin" and "porpoise" will, in keeping with the usage of the period, be considered interchangeable. In the 1960s it became increasingly standard in the United States (particularly in the scientific community) to reserve the term "porpoise" for members of the family Phocoenidae, while all small toothed whales could be called "dolphins" (i.e., not only the "true dolphins," the Delphinidae).

the equipment necessary to erect a mobile physiological and general biological laboratory: galvanic cells, probes, voltage regulators, surgical instruments, a microscope and the necessary means to prepare and mount slides,
a fragile manometer, lumbar puncture needles, photographic equipment,
anesthetic, stains, specimen vials in a host of sizes, drums of preservatives
(rock salt, formaldehyde), crates of gauze, tablets of paraffin, and sundry
other biomedical accoutrements. They planned to set themselves up in the
fishing village of Hatteras, tucked behind the dunes, for two weeks, board-
ing with kindly older ladies (there was no inn on Hatteras Island, not even
a drugstore) and spending their days chasing, catching, and experimenting
on the Atlantic bottlenose, *Tursiops truncatus*.

. . . .

This chapter, which aims to show how cetaceans became objects of scientific
inquiry in the United States in the interwar period (and with what effects),
departs from the Hatteras expedition of 1928, an undertaking that marked
a new level of American scientific interest in these powerful, elusive, and
little-understood organisms. By first looking back from this episode (to
reconstruct how Kellogg, Howell, and the others found their way to the
sands of Hatteras Island and what they hoped to discover there) and then
forward beyond it (to trace the broad repercussions of their investigations
and the resultant alliances), I will sketch the distinctive development of
whale science in North America before World War II. In view of the foundational role that Remington Kellogg played during the postwar period in
the creation and administration of the International Whaling Commission (he was the "father" of the Commission, according to Tønnessen and
Johnsen, and effectively the draftsman of the foundational "International
Convention for the Regulation of Whaling" in 1946; he remained the pre-
eminent scientific-cum-diplomatic figure in cetological circles through the
1950s), a better understanding of this period is badly needed.[5] This chapter
will also usefully complement my treatment of the roughly contemporary
British Discovery Investigations: in that case, a huge and rapidly expanding
whaling industry created very particular conditions for the development of
whale science, as we have seen in chapter 2. In the United States, different
institutions and different preoccupations (from silent film to the Scopes

---

5. J. N. Tønnessen and Arne Odd Johnsen, *The History of Modern Whaling* (Berkeley:
University of California Press, 1982), 672.

trial to controversies over the Interior Department's predatory mammal control program) shaped the study of cetaceans.[6]

As I will show in the pages that follow, the Hatteras expedition reflected the confluence of a number of divergent research programs, and the different participants hoped that beached dolphins would help them solve very different problems. In the end, however, shared interest in the taxon—and a growing awareness of threats to the long-term survival of the world's whales, large and small—annealed a community of American scientists in an ambitious campaign to draw broad popular attention to the significance of these animals. The Hatteras group formed the nucleus of what would eventually become a remarkable (and now totally forgotten) organization, the Council for the Conservation of Whales (ccw), which appears to have been the first conservation association anywhere in the world entirely dedicated to "saving the whales"; that it was the first such undertaking in the United States is certain.[7] Mobilizing scientists, journalists, and philanthropists in an elaborate strategy that specifically used science to garner public interest, the ccw—numbering among its members many of the dignitaries of American Progressive Era wildlife preservation, men like Gifford Pinchot, William T. Hornaday, and Henry Fairfield Osborn—succeeded in playing an important role in national legislation and international negotiations regulating the whaling industry before World War ii.[8] Given that one very able commentator on the history of twentieth-century whaling and ecology has written that "the American whale conservation movement began in the late 1960's," closer attention to the Council for the Conservation of Whales is clearly in order.[9]

6. Though, as this chapter will show, there were similarities as well, including symbiotic relationships with commercial enterprises and conservation concerns linked to more general anxieties about game depletion. There were also connections, since Kellogg came to be well acquainted with both the personnel and the work of the Discovery Investigations.

7. Klaus Barthelmess has written an interesting essay on the early work of Paul Sarasin and the Schweizerische Naturschutzkommission, or Swiss Commission for the Protection of Nature, which drew attention to the need for whale protection before World War I. See his essay "An International Campaign against Whaling and Sealing Prior to World War One," in *Whaling and History ii: New Perspectives*, ed. Jan Erik Ringstad (Sandefjord, Norway: Sandefjordmuseene, 2006), 147–67.

8. For the members of the first advisory board, see n. 202 below.

9. R. Michael M'Gonigle, "The 'Economizing' of Ecology: Why Big Rare Whales Still Die," *Ecology Law Quarterly* 9, no. 1 (1980): 119–237, at p. 189. J. N. Tønnessen and

Among other things, the topic presents an exceptional opportunity to work at the intersection of the history of science and environmental history—an aim of this book as a whole.

In surveying this period, then, the present chapter proceeds as follows. First, I trace the origins of the Hatteras expedition, examining the emergence of the scientific interests and connections that eventually made the trip both desirable and possible. Telling this story provides an opportunity to sketch Remington Kellogg's intellectual development in his formative years, since, as prime mover of the enterprise, his expanding research agenda and professional alliances largely account for the sudden flurry of whale interest in the United States in (and after) 1928. I then briefly review what actually happened when the scientists got their hands on some dolphins in February of that year, before turning to the rapid sequence of events that attended the Hatteras expedition's return: a symposium on whale research in Washington, DC, resolutions expressing conservation concern, and shortly thereafter a widening public campaign that used scientific findings to galvanize public support. Kellogg's own career trajectory remains important here: his promotion from the Bureau of Biological Survey to a full-time position within the Smithsonian both reflected the success of his whale work and facilitated his movement into service as a government adviser on regulation. That service, in turn, sent him to Europe three times before World War II, each time as a US representative in international negotiations. Those trips permitted him long stays in the cetological collections of museums in London, Amsterdam, Berlin, and the other cathedrals of continental natural history, where Kellogg did the research that would thrust him definitively into international prominence in the scientific study of whales. By the time World War II shattered the emerging structures of multilateral whaling regulation, sundered scholarly alliances, and destroyed many of the museums in which he had worked, A. Remington Kellogg had come to be uniquely positioned—diplomatically, scientifically—to mobilize the renewed conservation efforts that would follow the conflict. Those efforts, and the fate of the institution they shaped, the International Whaling Commission, will be the subject of chapters 4 and 5.

Arne Odd Johnsen, in *The History of Modern Whaling* (Berkeley: University of California Press, 1982), do a slightly better job, noting that there was an American push toward conservation of whales in the late 1920s and early 1930s (see p. 401). They do not, however, discuss the CCW itself.

"FILLED WITH ENTHUSIASM FOR THE WHALE"

By the time Kellogg and his party arrived at Hatteras in 1928, the dolphin fishery was no longer a thriving concern: the animals were increasingly scarce, the catches paltry, the lookouts and boatmen notably desultory as they waited through each day, scanning the surf without great vigilance. But in 1928 there were not many options for an East Coast naturalist who wanted to get his hands on a live cetacean.

Bones, by contrast, were easier, and because Kellogg had, since the late 1910s, gradually built for himself a reputation as an up-and-coming young vertebrate paleontologist-cum-zoologist with a specialty in marine mammals, whale bones had been his stock in trade for nearly a decade. They trickled into the museum in an unpredictable stream of railroad crates dispatched from a variety of sources: dedicated collectors, energetic Coast Guard officers, California petroleum engineers, and an expanding coterie of museum curators who exchanged duplicates and swapped finds. After arriving in Washington, DC, from Berkeley in 1920, Kellogg had gradually (with Merriam's help) gained access to the massive collections of the two great late nineteenth-century American naturalist-morphologists to work extensively on whales: Edward D. Cope and Frederick W. True. By the late 1920s, Kellogg had won the connections and the backing to reach out and secure hard-to-find specimens when he found himself in need of something exotic. His diligence and early track record of no-nonsense publications had put him in friendly contact with several luminary figures in American science in the period, men with lifelong interests in marine mammals. David Starr Jordan, the increasingly venerable president of Leland Stanford Junior University, could put a little pressure on a California mining company that had turned up something interesting and seemed to be sitting on it (figure 3.2). Leonard Stejneger, the head curator of biology at the National Museum, had, like Jordan, given a large part of his professional life to the study and conservation of the fur seals of the northern Pacific before, again like Jordan, settling down to life as a kind of ambassador of erudition.[10] He too was will-

10. For Jordan's assistance to Kellogg, see Kellogg to McKinley Stockton, 22 February 1928, Smithsonian Archives, RU 7170, box 8, file "Stockt–Stre"; for Stejneger's interest in the National Museum's cetacean collections, and for evidence that these materials were filtered to Kellogg, see the sheaf of letters in the Smithsonian Archives, RU 7170, box 8, file "Wa–Wh."

[Palo Alto, Calif., June 6, 1929]

FIGURE 3.2 Scientist-statesmen: David Starr Jordan (*left*) and Remington Kellogg (*right*). (Courtesy of the Smithsonian Institution.)

ing to assist a young protégé looking for material, and he drew Kellogg's attention to the vast Calvert Cliffs deposits in Maryland, where many of the best whale fossils on the East Coast had been found. Moreover, Kellogg's main patron, Merriam, possessed enviable connections for carrying out the occasional special mission: when, in 1926, Kellogg realized that there appeared to be no complete skull of a bowhead whale (*Balaena mysticetus*) in any collection in the United States—a lacuna that undermined his efforts to construct a sequence of morphological changes in the small bones associated with hearing—Merriam promptly wrote a letter to none other than Andrew W. Mellon, the treasurer of the United States, suggesting that a revenue cutter might be engaged to secure such a skull from the beach at Point Barrow, Alaska. Mellon was happy to oblige.[11]

It had not always been quite so simple. Kellogg, who hailed from Davenport, Iowa, was the son of a printer. He found his way to a professional life in natural history via the University of Kansas, where he distinguished himself as an assistant to Charles Bunker, curator of mammals and birds in the university's natural history museum. He won the title of "systematic assistant" even before he had been awarded his master's degree, and the enthusiastic support of his instructors in zoology secured him summer work as a field collector for the Bureau of Biological Survey, in whose service he rambled widely through Wyoming, Montana, and Northern California, shooting birds and trapping shrews, moles, wood rats, and larger backwoods creatures.[12] By 1916, when Kellogg enrolled as a PhD student in zoology at Berkeley, the natural sciences there—shaped and subsidized by the pioneering philanthropic activities of Annie Montague Alexander—yielded pride of place to no institution in the country.[13] Joseph Grinnell had shaped the Museum of Vertebrate Zoology into a tight-knit (though somewhat old-fashioned) research community, and John C. Merriam's stewardship of the Department of Paleontology, if autocratic, had nevertheless succeeded in raising him to national prominence as a scientific figure; the reputation of his program rose accordingly.[14] By the time Kellogg returned in 1919,

11. Though the skull proved harder to find than expected, in view of the fact that the species was very nearly extinct. See the exchange of letters in the Smithsonian Archives, RU 7170, box 6, file "Merriam, John C. (2/3)."

12. Professor John Sundwall called Kellogg "one of the most brilliant students we have ever had at the University of Kansas." Smithsonian Archives, RU 7170, box 9, file "Department of Agriculture." For details on Kellogg's Bureau of Biological Survey work, it is useful to review this file, which includes his specific and general instructions for collecting. His notebooks from this work also survive (Smithsonian Archives, RU 7434, Remington Kellogg Papers, 1903–1969, and related to 1982, box 2) and include his recipes for trap baits and coyote and wolf poisons.

13. See, among other things, Richard M. Eaken, "History of Zoology at the University of California, Berkeley," *Bios* 27 (1956): 67–90.

14. In 1915 its fossil collection was the largest west of the Mississippi and held several hundred type specimens. See Jere H. Lipps, "Success Story: The History and Development

however, after a two-year leave necessitated by military service in France, the simmering tensions between Berkeley's scientific benefactress and her suite of colleague-dependents had come to a boil. Despite Alexander's projecting and generosity, Merriam's focus had wandered from the West Coast in the war years. Promoted to chairman of the National Research Council at the close of the conflict, he shortly thereafter accepted the prestigious directorship of the Carnegie Institution and left for Washington.[15] The simultaneous retirement of Benjamin Wheeler, Berkeley's president, contributed to the sense of upheaval. The future of paleontology, geology, and zoology at the University of California (which had never been easy office mates, and which were all struggling in different ways for attention and support as new kinds of biology increasingly colonized American research universities) grew distressingly uncertain.[16]

---

of the Museum of Paleontology at the University of California, Berkeley," *Proceedings of the California Academy of Sciences* 55, supplement 1, no. 9 (October 2004): 209–43, at p. 228. See also Sally Gregory Kohlstedt, "Museums on Campus: A Tradition of Inquiry and Teaching," in *The American Development of Biology,* ed. Ronald Rainger and Jane Maienschein (Philadelphia: University of Pennsylvania Press, 1988), 15–47, particularly app. 3.

15. For an overview of the rising importance of these philanthropic institutions in American science in this period, see Robert E. Kohler, "Science, Foundations, and American Universities in the 1920s," *Osiris* 3 (1987): 135–64.

16. For an excellent survey of the institutional history of this period, see Barbara R. Stein, *On Her Own Terms: Annie Montague and the Rise of Science in the American West* (Berkeley: University of California Press, 2001). Also of use is Jere H. Lipps, "Success Story: The History and Development of the Museum of Paleontology at the University of California, Berkeley," *Proceedings of the California Academy of Sciences* 55, supplement 1, no. 9 (October 2004): 209–43. On the struggles of vertebrate paleontology in these years, see Ronald Rainger, *An Agenda for Antiquity: Henry Fairfield Osborn and Vertebrate Paleontology at the American Museum of Natural History, 1890–1935* (Tuscaloosa: University of Alabama Press, 1991). For a more general sense of the changing institutional and disciplinary landscape of the life sciences in this period, see Garland E. Allen, *Life Science in the Twentieth Century* (New York: Wiley, 1975), and the sequence of interesting challenges to his thesis that appear in the *Journal of the History of Biology* 14, no. 1 (Spring 1981), as well as in the volume edited by Ronald Rainger and Jane Maienschein, *The American Development of Biology* (Philadelphia: University of Pennsylvania Press, 1988), and a companion volume, edited by Keith Rodney Benson, Jane Maienschein, and Ronald Rainger, *The Expansion of American Biology* (New Brunswick, NJ: Rutgers University Press, 1991). For making sense of Kellogg's development, Ronald Rainger's pieces are of particular value. See, for instance, "Vertebrate Paleontology as Biology:

Kellogg was one of those who, in the breakup, marched with Merriam, a move that burned a number of the student's bridges back to his alma mater and simultaneously reshaped his research plans. What role his new adviser played in securing Kellogg's first Washington position at the Bureau of Biological Survey is not precisely clear, but it is worth recalling that John Merriam's cousin, C. Hart Merriam, had not long before directed the Survey, and that John Merriam's new position at Carnegie ensured Remmy a gracious reception in the scientific circles of the nation's capital. Kellogg was by no means naïve about the importance of such things. Writing to a graduate school pal seeking a university post in 1926, Kellogg laid out the importance of powerful allies in the pursuit of scientific opportunity: "If you have good backing you should land an associate [position]. It all depends upon those who recommend you for the place."[17]

While at Berkeley, Kellogg had already begun a research program on marine mammals, though his focus had not been on whales, but rather on the pinnipeds (seals and sea lions). This interest is not difficult to situate historically: for several decades seals had been a major scientific preoccupation in California. The four-nation North Pacific Fur Seal Convention in 1911—whereby the United States, Canada, Russia, and Japan agreed to limit their exploitation of the valuable Pribilof Islands herds—represented a landmark in international wildlife conservation, not least because it emerged out of sustained research on breeding habits, regional distribu-

---

Henry Fairfield Osborn and the American Museum of Natural History," in *The American Development of Biology*, ed. Rainger and Maienschein, 219–56, and "The Continuation of the Morphological Tradition: American Paleontology, 1880–1910," *Journal of the History of Biology* 14, no. 1 (Spring 1981): 129–58. Also useful is Benson's chapter "From Museum Research to Laboratory Research: The Transformation of Natural History into Academic Biology," in *The American Development of Biology*, ed. Rainger and Maienschein, 49–83; Jane Maienschein, "Shifting Assumptions in American Biology: Embryology, 1890–1910," *Journal of the History of Biology* 14, no. 1 (Spring 1981): 89–113, and (especially where the movement of life sciences to the seashore is concerned) Philip J. Pauly's "Summer Resort and Scientific Discipline: Woods Hole and the Structure of American Biology, 1882–1925," also in *The American Development of Biology*, ed. Rainger and Maienschein, 121–50. Finally, Léo Laporte's essay on George G. Simpson, "George G. Simpson, Paleontology, and the Expansion of Biology," in *The Expansion of American Biology*, ed. Benson, Maienschein, and Rainger, 80–106, also helps sketch the scientific world into which Kellogg matured.

17. Kellogg to Franklin Metcalf, 31 May 1926, Smithsonian Archives, RU 7170, box 6, file "Merriam, M. L.–Met."

tion, and stock analysis.[18] David Starr Jordan, Joseph Grinnell's mentor, had served on the Fur Seal Advisory Board and had conducted extensive investigations into how the population might be exploited in such a way as to avoid eventual extinction, going so far as to joust publicly with scientists who made opposing recommendations.[19] For Jordan, Grinnell, Stejneger, and others, human-induced marine mammal extinction in the northern Pacific loomed as one of the largest scientific problems of the day. Money and violence had marked the fur seal controversy, which unfolded against ongoing scientific disputes over the notorious case of Steller's sea cow, a creature akin to a manatee that reached a length of twenty-five feet—it had been found (and eaten) in great numbers by Russian traders and explorers in the Bering Sea in the mid-eighteenth century. By the twentieth century its extinction was an accepted fact, but the questions of how fast it vanished, and above all, why, remained contentious. Positions in these debates were very much colored by participants' views on the dangers to the modern fur seal fishery.[20] Shortly before Kellogg arrived in Berkeley for the first time as a young graduate student, Annie Alexander had returned from an expedition to Alaska with a (very rare) sea cow skeleton, which she presented to Grinnell, and the Museum of Vertebrate Zoology on campus—not otherwise given to displays—featured several dioramic installations of Pacific pinnipeds.[21]

In short, marine mammals were the focus of considerable interest at Berkeley and Stanford in this period, and both institutions were repositories of expertise on the subject. Merriam's suggestion that Kellogg apply himself to the fossil history of this group of organisms surely owes something to the broad attention that they had recently attracted in these scientific communities. At the same time, Merriam's success in enticing Kellogg—who had at this time all the makings of an American field naturalist (a love of birding, camp life, and critters)—away from the zoology of vital creatures (he had toyed with the idea of entomology) and toward

18. For a valuable discussion of this episode, see Kurkpatrick Dorsey, *The Dawn of Conservation Diplomacy: U.S.-Canadian Wildlife Protection Treaties in the Progressive Era* (Seattle: University of Washington Press, 1998).

19. Henry Elliott's views on pelagic sealing came under particularly withering scrutiny.

20. Stejneger and Nordenskjöld engaged in a heated polemic on this issue in the 1880s.

21. Barbara R. Stein, *On Her Own Terms: Annie Montague and the Rise of Science in the American West* (Berkeley: University of California Press, 2001), 87.

paleontological topics must be understood in the context of the extensive California and Oregon Tertiary fossil deposits discovered and tapped in the first decades of the twentieth century. Petroleum exploration and the pursuit of other mineral resources helped bring half a dozen major sites to the attention of Merriam and his colleagues in this period: the La Brea tar pits were perhaps the most famous of these, but the marine deposits of the Temblor Formation and particularly the region around Sharktooth Mountain yielded the richest hauls of marine mammal remains. These were a unique opportunity for an enterprising graduate student, and before he took his leave for Washington, Kellogg made a number of field expeditions to these sites to gather materials; after he left he stayed in close contact with those still working these hauls.[22]

On arriving in Washington in 1920—and gaining access, via Merriam, to the partially indexed fossil whale materials gathered by Cope decades earlier—Kellogg shifted firmly away from pinnipeds, and codified a program of research on the paleontology of the Archaeoceti (the prehistoric whales). This was the project that he wrote up for his chief at the bureau in order to secure permission to continue his thesis work (after hours only) in the National Museum. As he put it in his first memo to his boss, "an investigation of the fossil Cetacea should throw considerable light upon various considerations involved in the distribution of the living forms, and confirm our ideas as to the evolution of this group."[23]

Initially, Kellogg conceived the project as essentially comparative. He would work up the East Coast specimens held at the Smithsonian and others he would collect from the nearby Calvert Cliffs (he bought a used car to make it easier to get out to the site on the weekends), and he would write a

22. "We are still gathering in whale material for you. Perhaps Furlong has already told you of the prospects of whale material at King City. We will shortly go after this." Chester Stock to Kellogg, 23 January 1922, Smithsonian Archives, RU 7170, box 8, file "Stock, Chester."

23. Kellogg to E. W. Nelson, 24 November 1920, Smithsonian Archives, RU 7170, box 6, file "Merriam, John C. (3/3)." It is interesting to note that there were other examples of Carnegie research associates working out successful arrangements in the National Museum in this period. For instance, Oliver P. Hay wrote almost thirty papers in the 1920s based on specimens in the National Museum, but he was never officially employed there, having a formal appointment with the Carnegie Institution. For a discussion, see Charles W. Gilmore, "A History of the Division of Vertebrate Paleontology in the United States National Museum," *Proceedings of the United States National Museum* 90, no. 3109 (1942): 305–77, particularly p. 317.

dissertation analyzing the zonal occurrence of the different types of Cetacea in the Calvert Cliffs deposit, comparing his findings with the West Coast materials he had gathered, along with whatever else might be forwarded to him by his remaining Berkeley collaborators.[24] Merriam, increasingly preoccupied with administrative obligations, and never given to hand-holding with students, signed off on the project and made Kellogg a "research associate" of the Carnegie Institution. Though it carried no stipend, the position folded Kellogg into the research group that Merriam sought to maintain, despite his own disengagement from active research.[25]

For Remington Kellogg, Merriam's actual mentorship in Washington was haphazard at best, but connections and money were easily secured. Above all, the association with Carnegie and Merriam meant a reliable line of subsidy for an essential aspect of this kind of research: high-quality line drawings. Most of Kellogg's early publications were brief descriptive notices on marine mammal fossils, contributions that had value to a considerable degree because of the fine line drawings that illustrated the novel specimens.[26] These drawings, most of them executed by an accomplished draftsman named Sidney Prentice, were expensive and highly prized. Merriam ensured that Kellogg had ready access to the necessary funds, helping him build an impressive roster of morphologically oriented technical publications on the paleontology of the prehistoric whales.[27]

Even as that roster grew in the early 1920s, Kellogg himself floundered

24. Draft of thesis outline sent to Stock, Smithsonian Archives, RU 7170, box 8, file "Stock, Chester." Eustace Furlong and Stock continued to help Kellogg get specimens from California in this period.

25. It was this kind of empire building that, in conjunction with Merriam's ongoing efforts to continue running the Berkeley program after his departure, infuriated Alexander (and others) and significantly destabilized the development of paleontology at the University of California in those years.

26. For a discussion of the importance of these kinds of illustrations in this sort of work in the life sciences in these years, see Philip J. Pauly, *Biologists and the Promise of American Life: From Meriwether Lewis to Alfred Kinsey* (Princeton, NJ: Princeton University Press, 2000), 119.

27. For example, see Remington Kellogg, "Description of the Skull of *Megaptera miocaena*, a Fossil Humpback Whale from the Miocene Diatomaceous Earth of Lompoc, California," *Proceedings of the United States National Museum* 61, no. 2435, art. 14 (1922): 1–18, pls. 1–4; and Kellogg, "Description of Two Squalodonts Recently Discovered in the Calvert Cliffs, Maryland; and Notes on the Shark-toothed Cetaceans," *Proceedings of the United States National Museum* 61, no. 2462, art. 16 (1923): 1–14.

somewhat, and he periodically despaired, privately, about how to configure his thesis in a way that would enable him to finish it expeditiously while positioning himself for an escape from the bureau.[28] Yet despite the wandering focus of his whale project, Kellogg was never less than relentless about putting hours and energy into whatever aspect of it currently held his attention. It is in part this casting about for a workable project, under an unusual absence of constraints (a more or less absentee adviser always willing to send money or open a door), that drove Kellogg to reach out to a wide variety of scientists interested in—or potentially interested in—whales, weaving the network that would subsequently play an important role in the push for conservation and legislation in the late 1920s and 1930s.

To make sense of Kellogg's meandering investigations, it is helpful to recall that in throwing his lot in with Merriam, Kellogg made a relatively precipitous transfer from zoology, the discipline in which he had enrolled at Berkeley, to vertebrate paleontology, an area in which he had considerably less formal training and which was by no means a particularly promising career direction in those years.[29] It was an adventitious switch, and one that left Kellogg repeatedly crossing the borderlands between the two areas, always looking for a way to keep a hand in what seemed to be active developments in adjacent fields: physiology, invertebrate paleontology, even biogeography. His attention was prone to wander.[30]

28. At the bureau, Kellogg experienced what he took to be the hostile machinations of co-workers jealous of his special arrangements. As early as June 1923 he could write, "I have access to the most complete collection of skeletons of living cetaceans in the United States.... I would have left Washington long ago had it not been for this study." Kellogg to S. J. Holmes, 2 June 1923, Smithsonian Archives, RU 7170, box 4, file "Hea–Hon." He did receive several queries about alternative jobs in this period, including academic posts at Minnesota and back at Kansas.

29. "I am glad to note that you have considered it best to change your major from zoology to paleontology. I think this is quite a good move." Stock to Kellogg, 23 January 1923, Smithsonian Archives, RU 7170, box 8, file "Stock, Chester." The change required a reconstitution of Kellogg's dissertation committee and weighted it heavily toward those still close to Merriam. It is also important to note in this regard that though Kellogg was working with fossils in the National Museum, he did not actually have much contact with the Division of Paleontology there, but rather eventually found his institutional home in the Division of Mammals. This captures, I think, his efforts to stay out of a wholesale commitment to paleontology in these years.

30. Keith R. Benson offers a useful characterization of a general shift among American morphologists at the turn of the century from more descriptive to more functional

Practically speaking, Kellogg embarked in 1921 and 1922 on a broad review of the available literature on the fossil whales, with an eye toward satisfying Merriam's expressed curiosity about "aquatic adaptation as it relates
to the marine mammals."[31] Relatively rapidly, Kellogg thought he spotted a
way that he might contribute to a revision of the available ancestral series:
most of the work he had found concentrated on the changing forms of the
teeth of the archaic cetaceans; Kellogg, by contrast, was persuaded that a
more general examination of skull structure would better tether the sequence to a coherent account of progressive adaptations necessitated by life
in the water.[32] Here Kellogg was surely influenced by the work of one of the
permanent staff at the Smithsonian, Gerrit S. Miller Jr., assistant curator of
mammals, who in these same years developed the core insights described
in his monograph on the unusual "telescoping" of the bones in cetacean
skulls (and the use of this feature in systematics), a paper published in 1925
that remains a classic.[33] While Miller focused on the taxonomy of living
whales, his account of the progressive modification of the major bones in
the skull and jaw drew heavily on fossil specimens. For this reason, even as

---

preoccupations, and he suggests that this shift drew them increasingly to physiology.
Ronald Rainger, in discussing the work of Henry Fairfield Osborn and William King
Gregory, demonstrates a move between 1910 and 1920 among some paleontologists
to make their discipline more "biological." It seems safe to say that Kellogg's own ex-
panding interests can be linked to these broader shifts. See Keith R. Benson, "Ameri-
can Morphology in the Late Nineteenth Century: The Biology Department at Johns
Hopkins University," *Journal of the History of Biology* 18, no. 2 (1985): 163–205, and
Ronald Rainger, "Vertebrate Paleontology as Biology: Henry Fairfield Osborn and the
American Museum of Natural History," in *The American Development of Biology*, ed.
Ronald Rainger and Jane Maienschein (Philadelphia: University of Pennsylvania Press,
1988), 219–56, particularly pp. 221, 237–45.

31. Kellogg to E. W. Nelson, 24 November 1920, Smithsonian Archives, RU 7170,
box 6, file "Merriam, John C. (3 / 3)."

32. Kellogg's interest in adaptation in particular is significant, since it reflects a distinctive focus of American morphologists and comparative anatomists in this period. See
William Coleman, "Morphology between Type Concept and Descent Theory," *Journal
of the History of Medicine and Allied Sciences* 31, no. 2 (April 1976): 149–75.

33. The skulls of modern cetaceans display an unusual set of morphological features
that involve the "sliding" of the anterior and posterior structures over (and under) each
other; the exact pattern differs between whales with teeth (odontocetes) and those with
baleen, or "whalebone" (mysticetes). Miller was not the first to notice this phenomenon,
nor did he coin the term "telescoping" to describe it.

Kellogg began to imagine the shape his contribution might take, he increasingly despaired of working up his comparative study of the Archaeoceti before Miller had completed his magnum opus. To do so would be to tread on adjacent (and better-placed) toes.[34]

There were other, even more immediate, problems as well. No one could dispute the advantages of a connection to the National Museum when it came to securing specimens, particularly new things that might come to light and find their way into the national holdings. But a condition of Kellogg's access to the collection as an outside researcher was a gentleman's agreement not to publish on new findings until preliminary papers were published by the National Museum itself. He complained several times to friends about the delays this caused.[35] And even that was a modest irritation in comparison with the frustrations that could arise in the pursuit of choice finds. When, in early 1924, one of Kellogg's old California friends turned up a superb squalodont skull in the Temblor Formation, the monster bones even attracted the attention of the newspapers. Soon thereafter, though, a set of geologists at the neighboring California Academy of Sciences got wind that much of the nearby good stuff was being shipped off to Washington. Working local connections, they contacted the Pacific Oil Company, which owned the property where Kellogg's collaborators had been digging. Along in the mail came a dreaded letter from G. Dallas Hanna, an officer of the academy, informing Kellogg that in the future a Hanna group would be working closely with Pacific Oil to assemble "as extensive a collection as possible" from the site. "Do not think we are trying to steal your thunder," Hanna noted in the friendliest way possible; it was just that "we are contemplating some intensive work on the marine mammals here and I expect to turn the bulk of the resources of the department of Paleontology toward collecting."[36] Wrote Kellogg dejectedly to Merriam, "It appears that our

34. Kellogg to Merriam, 14 December 1921, and Kellogg to Merriam, 26 May 1922, both in Smithsonian Archives, RU 7170, box 6, file "Merriam, John C. (3/3)." In the latter, Kellogg wrote: "The present harmonius [*sic*] relations can be better maintained I believe if I hold up further studies along that line until he completes his studies."

35. Kellogg to Franklin Metcalf, 31 August 1925, Smithsonian Archives, RU 7170, box 6, file "Merriam, M. L.–Met"; Kellogg to Merriam, 12 November 1924, Smithsonian Archives, RU 7170, box 6, file "Merriam, John C. (3/3)"; Kellogg to Stock, 14 April 1924, Smithsonian Archives, RU 7170, box 8, file "Stock, Chester."

36. Frederick H. Meisnest to Kellogg, 1 July 1928, Smithsonian Archives, RU 7170, box 5, file "Mc–Merriam, C. W."

expectations for obtaining an adequate fauna from the Temblor formation have just about vanished."[37]

Gentle politicking eased that situation somewhat, but the basic issues remained: working on fossils made one a hostage not only to fortune (and the postal services) but also to just about anyone who could get on site before you. Admittedly, the Carnegie Institution graciously helped cover freight charges. Nevertheless, as Kellogg wrote to a classmate, "It is rather disheartening to have to plod along hoping that important specimens will turn up."[38] Later, still harping on the same problem, Kellogg spelled it out more explicitly: "This business of hunting for fossils to fill in the necessary gaps is just one series of disappointments. The next time I tackle a problem of this sort, I will have the material in hand before I commit myself."[39] Worst of all, if you were a cetacean person, when something did turn up, it required a pneumatic hammer and a crane, not a brush and a pick. As he complained to a friend in 1922 concerning museum work on whales, "The skeletons of the living species are suspended in the air and the skulls themselves are usually about 18 feet from the floor. Skulls of the large fossil whalebone whales require a derrick to handle them."[40] This meant trying to get photographs or arranging to have a team of men on hand every time you wanted to look at the other side of a specimen.[41] "My recommendation to anyone after a PhD. [sic]," Kellogg wrote to a former University of Kansas buddy thinking about a career in research biology, "would be to take a problem that can be viewed under a microscope."[42] It was probably a doubly wry joke: on the one hand, it captured the frustration of a junior vertebrate morphologist watching the triumph of other, more laboratory-based forms of biology; on the other hand, it was the declaration of a scientist exasperated with the gigantism of his chosen organism.

37. Kellogg to Merriam, 16 July 1924, Smithsonian Archives, RU 7170, box 6, file "Merriam, John C. (3/3)."

38. Kellogg to Franklin Metcalf, 31 August 1925, Smithsonian Archives, RU 7170, box 6, file "Merriam, M. L.–Met." NB: Merriam taught Kellogg to list fossils in their matrix as "rough slab rock" in order to get the lowest shipping rates. Kellogg to Merriam, 19 March 1923, Smithsonian Archives, RU 7170, box 6, file "Merriam, John C. (3/3)."

39. Kellogg to Franklin Metcalf, 25 January 1928, Smithsonian Archives, RU 7170, box 6, file "Merriam, M. L.–Met."

40. Kellogg to Stock, 12 April 1922, Smithsonian Archives, RU 7170, box 8, file "Stock, Chester."

41. On the use of photos, see Kellogg to Stock, 12 April 1922, in ibid.

42. Ibid.

Confronting his most immediate difficulties—problems of access to the best of the large and unwieldy specimens both in the halls of the Smithsonian and in the hills of Bakersfield (on the one hand) and the danger of what amounted to imminent preemption by Miller (on the other)—Kellogg began to cast about for alternative approaches to the material, anything that would keep the ever-distant Merriam happy and increase his chances of securing a terminal degree. By 1923 a promising avenue had come into view. Among the most plentiful fossils associated with marine mammal deposits were the small, loose bones (of several types and enormously variable shape) associated with the region of the ear. Despite their sculptural forms, these bones were not particularly exciting finds: they were unspectacular (particularly in comparison to, say, the giant skull of a squalodont), there were many of them (often unaccompanied by larger bones), and their precise function was not well understood in prehistoric whales or their living descendants. For Kellogg, however, these unglamorous lumps began to look like very promising nuggets, like small bones that might pay a handsome reward (figures 3.3 and 3.4).

For starters, their size recommended them: they were small, and therefore they were easy to collect and transport, freeing him from reliance on the infrastructure necessary to secure and manipulate gargantuan skulls. Moreover, they were plentiful and largely overlooked: Miller, absorbed in the complex and plastic architecture of the maxillary and the rostrum in the cetacean skull, seemed indifferent to them as he elaborated his account of the taxonomic importance of telescoping. And yet, as Kellogg wrote to Merriam, "at present I am of the opinion that the earbones, and especially the periotic, will prove exceedingly important for making determinations and for confirming phyllogenetic [*sic*] relationships." The rationale for this hunch lay in the very preoccupation Kellogg knew to be uppermost in Merriam's mind: the history of mammalian aquatic adaptation. Kellogg continued in the same letter, "Since the combined periotic and tympanic is an organ which has been perfected for deep sea diving and one which last of all would be subjected to external strains and stresses, it should be of much value in any study of aquatic modifications."[43]

There were other features that drew these bones to the attention of an ambitious young student working in the suburbs of vertebrate paleontology.

43. Kellogg to Merriam, 26 May 1922, Smithsonian Archives, RU 7170, box 6, file "Merriam, John C. (3/3)."

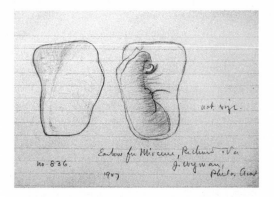

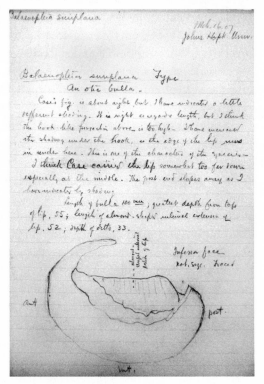

FIGURE 3.3
(*top*) Ear bones: Sketch by Remington Kellogg. (Courtesy of the Smithsonian Institution.)

....

FIGURE 3.4
(*right*) Ear bones, continued: Second sketch by Kellogg. (Courtesy of the Smithsonian Institution.)

Periotics were plentiful in loose marine deposits because they were remarkably dense. As Kellogg put it in a letter to his old friend Roy Moodie in 1923, "Since the bone is very dense and usually survives the destruction of the skull by natural causes, we expect it will be of much use." The "use" in question had to do with geology: "The problem is very closely connected with interconti-

nental geological correlation," Kellogg explained, "and from present indications marine mammals are going to be extremely useful in such studies."[44]

This idea may have come originally from Kellogg's after-hours reading
in the library of the National Museum. As soon as he got interested in ear
bones, Kellogg set to work gathering a voluminous bibliography of relevant
sources, and he deployed his characteristic energy in compiling, translating (where necessary), and abstracting these materials onto notecards.[45]
In one of these sources, he found a suggestive comment by the great nineteenth-century Belgian paleontologist Pierre Joseph Van Benden, who had
noted early that, because cetacean skulls (fossil and recent) were so large
and complex, periotic bones might serve as a superb anatomical proxy for
identifying strandings or new fossil finds. Van Benden went on to praise
their use in the determination of type specimens and to gesture toward the
potential use of these bones in more general geological and paleontological work.[46] Soon Kellogg had, by cordial correspondence, trawled half a
dozen museum collections across the United States and secured loans of
dozens and dozens of marine mammal ear bones (he preferred those from
the right side of the skull whenever possible, since having all the bones
from the same side made comparative work much easier). No one was the
least bit possessive about them: Princeton, Kansas, the American Museum
of Natural History, several California institutions, and even Gerrit Miller
himself, all offered up their periotics without a peep.[47] Word of his interest
carried to Olympian heights: by 1924 David Starr Jordan himself posted
marine mammal ear bones to Kellogg with his cordial regards.[48]

Merriam's enthusiasm for this program of research—the use of cetacean ear bones as index fossils in geological correlation work—must be

44. Kellogg to Moodie, 16 May 1923, Smithsonian Archives, RU 7170, box 6, file
"Mi–Moo." Kellogg wrote an almost exact copy of this paragraph to S. J. Holmes back at
the University of California, Berkeley, in June of the same year. Kellogg to S. J. Holmes,
2 June 1923, Smithsonian Archives, RU 7170, box 4, file "Hoa–Hon."

45. For examples of his notes, see Smithsonian Archives, RU 7170, box 9, file
"Information—Cetacea"; an example of his bibliography by 1925 can be found in the
same file.

46. Cited in Kellogg to Merriam, 12 January 1923, Smithsonian Archives, RU 7170,
box 6, file "Merriam, John C. (3 / 3)."

47. Kellogg to W. W. Nesbit, 15 March 1929, Smithsonian Archives, RU 7170, box 6,
file "Ne–Nu."

48. Kellogg to Stock, 26 September 1924, Smithsonian Archives, RU 7170, box 8,
file "Stock, Chester."

understood in the context of long-standing rivalries in the disciplines of paleontology and zoology.[49] Within paleontology there remained in this period a significant rift between those inclined to tie the field to geology (generally those oriented toward the study of invertebrates, whose fossil forms were invaluable to stratigraphy and correlation) and those who saw their task as continuous with the study of the anatomy and morphology of living creatures (generally those interested in vertebrates). Merriam's vigorous advocacy of the latter view had actually played a role in spawning a new department at Berkeley, and Kellogg was acutely aware of the attendant divisions. As he would later reflect sourly: "The meetings of the Paleontological Society that I have attended have been a sorry lot. The invertebrate men leave in a body as soon as a vertebrate paper is announced."[50] In view of the dominance of geologically oriented invertebrate paleontologists in the (ever important—particularly to petroleum companies) study of correlations and stratigraphic analysis, Kellogg probably suspected that Merriam would welcome with open arms a viable vertebrate index fossil for intercontinental geological correlation.[51]

He did. Soon Kellogg found himself invited by the chief of the Coastal Plain Investigation at the United States Geological Survey, T. Wayland Vaughan (soon to become the powerfully placed head of Scripps), to par-

---

49. See Barbara R. Stein, *On Her Own Terms: Annie Montague and the Rise of Science in the American West* (Berkeley: University of California Press, 2001), 111.

50. Kellogg to Alfred S. Romer, 9 January 1935, Smithsonian Archives, RU 7170, box 7, file "Ro–Ry."

51. See Kellogg's detailed two-page letter to Merriam on this topic, 10 November 1923, Smithsonian Archives, RU 7170, box 6, file "Merriam, John C. (3 / 3)." I have here given a somewhat contentious interpretation of the value of this work in disciplinary rivalries, which I think is defensible; however, it is worth noting that Merriam had himself done a considerable amount of early work in invertebrate zoology and understood its value in historical geology (particularly the use of the Tertiary echinoids as horizon markers; Chester Stock, "John Campbell Merriam 1869–1945," *National Academy Biographical Memoirs* 26 (1948): 209–17, at p. 210). In addition, much of Merriam's work at the Carnegie Institution can be seen to have been oriented toward encouraging interdisciplinary collaboration (e.g., his committee on the surface features of the moon). It would thus be possible to argue that Kellogg's desire to press for coordination between vertebrate paleontologists and invertebrate paleontologists by means of Archaeoceti fossils fit with respect to these other (and less fractious) aspects of Merriam's training and vision, even as Kellogg's plan would clearly advance Merriam's more militant commitments to vertebrate paleontology.

ticipate in a conference, "The Correlation of the Tertiary Formations of Southeastern North America, and the West Indies with the Tertiary Formations of Europe." Invertebrate specialists dominated the program.[52] The connection to Vaughan (an introduction made by Merriam) proved immensely valuable to Kellogg as far as securing further specimens was concerned. Vaughan knew numerous geologists and members of the Society of Economic Paleontologists who were stationed in various remote locales investigating Tertiary deposits on behalf of oil interests. He requested that Kellogg prepare instructions on securing marine mammal material and agreed to forward this document to prospective collectors. Kellogg went out of his way to emphasize the importance of the ear bones, giving detailed suggestions on what to look for and how to gather them.[53]

At the conference, Kellogg proposed a version of what had become his primary research agenda:

Adequate collections of fossil pelagic mammals, particularly those types which are wide ranging, from selected Tertiary formations will afford a basis for formulating a more precise and accurate method of correlating widely separated marine formations. In view of their wide distribution it may be predicted that fossil cetaceans will be *the deciding factor in determining contemporaneous marine formations in various parts of the world.*[54]

He gradually won some supporters, he thought, who, as he put it, "do not appear to have many doubts as to the value of cetaceans for intercontinental correlation." To deliver on his promises, Kellogg set to work in his notes on drafting large correlation charts that emphasized marine mammals and that were rooted in his growing mastery of the periotic bone and its myriad forms (figure 3.5).[55] But the success of the project, he increasingly came to recognize, hinged to no small degree on the assertion that cetaceans were, and had always been, cosmopolitan creatures, widely and more or less evenly distributed in the oceans of the world. Only then could nega-

52. See program at Smithsonian Archives, RU 7170, box 8, file "U–V."

53. For all this, see Kellogg to T. Wayland Vaughan, 21 February 1922, in ibid.

54. Smithsonian Archives, RU 7170, box 6, file "Merriam, John C. (3/3)." Emphasis added.

55. Kellogg to Stock, Smithsonian Archives, RU 7170, box 8, file "Stock, Chester."

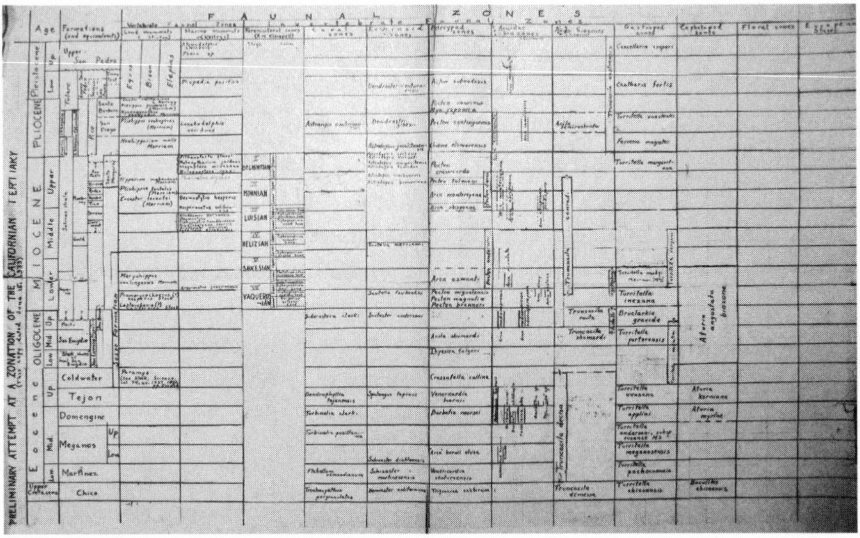

FIGURE 3.5 Discerning patterns in periotics: Kellogg's correlation charts. (Courtesy of the Smithsonian Institution.)

tive evidence be made meaningful and the geochronological value of the Archaeoceti be guaranteed.[56] But was this claim about cetacean biogeography correct?

It was with this question in mind that Kellogg began to collect notes on the distribution of living cetaceans.[57] He soon found out that he had

56. For Kellogg's sense of this matter, see Kellogg to Vaughan, 9 February 1922, Smithsonian Archives, RU 7170, box 8, file "U–V."

57. To see the connections between the live distribution work and the geological correlation program drawn clearly, see John C. Merriam et. al., "Continuation of Paleontological Researches," *Carnegie Institution of Washington Year Book* 27 (13 December 1928): 384–89, at p. 387. For a sense of the durability of this research, see Kellogg to Harmer, 1 March 1930, Smithsonian Archives, RU 7170, box 3, file " Harmer, Sidney F.," in which Kellogg wrote of the original efforts, "My main purpose was to show geologists interested in geological correlation that the larger whales are accustomed to make long journies [*sic*] during their migrations from breeding grounds to feeding grounds, and that some of the herds or individuals at least may visit the shores of one or more continents in the course of a year. . . . The chances of dispersal on opposite sides of any ocean of whales as contrasted to mollusks might be likened to the relative odds of a transatlantic steamer and a row boat." See also Kellogg to Hubert Schenck (in the

stumbled on an issue of heated interest, though not for what it might suggest about the periodization of the Cenozoic. Rather, it was a problem of the deepest interest to whalers, whose industry had, in the previous ten years, reinvented itself in the midst of an unprecedented Antarctic bonanza. By 1923 Norwegian whalers were on the cusp of technological developments (specifically the stern-slipway factory vessel) that would permit whales to be pursued, captured, and exploited on the high seas anywhere, anytime. Where whales were and where they went had become questions worth millions of pounds to English and Scandinavian whaling concerns. The same questions were also of interest to a small group of pioneering conservation-oriented scientists in Europe—men like Sidney Harmer at the British Museum (at this time actively engaged in the final planning for the Discovery Investigations)—who strongly suspected that industrial-scale exploitation would rapidly place whale populations at serious risk.[58]

Of these developments Kellogg could not have been wholly ignorant, even before his interests pulled him in the direction of the biogeography of living cetaceans: being bibliographically monomaniacal, tied to the Carnegie Institution, and increasingly (if, as I hope to suggest, somewhat reluctantly) a vertebrate paleontologist, there is no way Kellogg could have been unaware of G. R. Wieland's efforts, before the war, to draw attention to the destructive character of modern whaling. Wieland, associated with the Carnegie Institution and Yale's Peabody Museum, had written a piece in 1908 in *Popular Science Monthly* entitled "The Conservation of the Great Marine Vertebrates: Imminent Destruction of the Wealth of the Seas," which linked the ongoing North American fur seal crisis to the new profiteering spasms of Antarctic whaling. Wieland reported the collective concerns of the American Society of Vertebrate Paleontologists, who had gone

---

Geology Department at Stanford), 14 January 1933, Smithsonian Archives, RU 7170, box 7, file "Sap–Sc," in which Kellogg wrote, "All of our larger whales perform annually migrations from breeding grounds in tropics to feeding grounds in arctics and then return. Thus if a seaway existed between the Atlantic and Pacific in Helvetian time, we should expect to find identical types in both Pacific Coast horizons and European horizons." Kellogg went on to discuss the relative merits of foraminifera, mollusks, and cetaceans in this sort of correlation work.

58. Paul Sarasin is also significant in this regard. See Klaus Barthelmess, "An International Campaign against Whaling and Sealing Prior to World War One," in *Whaling and History II: New Perspectives*, ed. Jan Erik Ringstad (Sandefjord, Norway: Sandefjordmuseene, 2006), 147–67.

so far as to draft a resolution calling for legislative protections for whales and their kin. A similar resolution came out of the New York Zoological Society at this time, and Frederic A. Lucas (then at the Brooklyn Institute, but soon to join the Department of Mammalogy at the American Museum of Natural History) published an essay entitled "The Passing of the Whale" that reached a wide audience that year.[59] Lucas drew particular attention to the fact that "it required but a few years to wipe out the California gray whale," and (though it was not in fact extinct) Kellogg, as a student of the coastal Pacific marine mammals, would have been well aware of the story of its rapid demise. Other scientific authors—like Roy Chapman Andrews, whose 1916 *Whale Hunting with Gun and Camera* had been a commercial success—had also written recently and eloquently about the growth of the modern industry and the destruction of gray whales.

Still, up to 1924 or so, Kellogg had betrayed no particular interest in the issue of modern whaling or in the fate of living whales. It was only as he began to collect information on the distribution and migrations of the animals in connection with his research into geological correlation that he became aware of the global scope, historic depth, and modern intensity of the whaling industry. He sought out biogeographic data from three different sources—sightings, strandings, and catch records—but rapidly discovered that the vast majority of the information he could locate about where whales lived and when and how they moved around the seas hailed from the available histories of a sequence of more or less short-lived episodes of intensive exploitation. The numbers that had been tallied in the Antarctic in the last decade were nothing less than shocking.[60]

Kellogg again found his way to and through this information by means of his extensive reading (the article he eventually published on the topic, "What Is Known of the Migrations of Some of the Whalebone Whales," collated sightings and catch records from more than fifty sources in five languages), but he supplemented these researches by means of a widening correspondence. For instance, when he learned of the sequence of "stranding reports" gathered and published by the British Museum, Kellogg wrote

---

59. Frederic A. Lucas, "The Passing of the Whale," *Forest and Stream* 71, no. 8 (22 August 1908): 291–93.

60. In the mid-1920s the world catch of large whales exceeded 20,000 per year. See J. N. Tønnessen and Arne Odd Johnsen, *The History of Modern Whaling* (Berkeley: University of California Press, 1982), 293, 330.

a note to Sidney Harmer himself, including a set of offprints and requesting in exchange copies of Harmer's stranding data over the last several years. Not only was this query acknowledged, but Harmer and Kellogg were soon considering ways that the British system of standardized stranding forms might be extended to the much vaster coastal peripheries of the United States. Over the years to come, the young Kellogg and the aging Harmer would maintain an active and collegial epistolary relationship, one that was sealed into something very like a warm friendship on the occasion of Kellogg's visit to London in 1930.[61] Significantly, it is in an early letter to Harmer, written in 1925, when Kellogg was about a year into his efforts to make sense of the modern biogeography of whales—and had thus received a rapid tutorial in the history of human exploitation—that the first reference to the immediate threat to the animals appears in Kellogg's work. In September of that year, he wrote to Harmer, "If we ever expect to solve the distribution of some of our cetaceans we will have to commence as soon as possible. Within a few years the information may no longer be available."[62]

It was a concern that reflected a shift in Kellogg's interests. In the same letter, for instance, Kellogg noted that "my own interests in cetaceans lean more to the biological," something he would have been unlikely to suggest two years earlier, when he was working (to please Merriam) to reinvent himself as a nuts-and-bolts paleontologist. While in 1922 Kellogg complained regularly about not having enough of the right sort of fossils to work on, by September 1925 he could note in passing to Harmer that "we have considerable collections of fossil cetaceans from several localities, but I do not have much time to study them." What had happened?

Briefly stated, between 1923 and 1925 Kellogg's research interests turned increasingly to issues of cetacean physiology, development, and ecology. Several factors contributed to this drift. Above all, the general difficulties Kellogg initially faced making his way in the paleontology of the Archaeoceti pressed him, as I have suggested, to seek out adjacent, alternative projects. To this can be added Kellogg's long-standing engagement with general zoology as well as his day job, which required that he give the vast majority

61. The file of their correspondence is in the Smithsonian Archives, RU 7170, box 3, and runs through to Harmer's death.

62. Kellogg to Harmer, 7 September 1925, Smithsonian Archives, RU 7170, box 3, file "Harmer, Sidney F."

of every day to problems of habit, digestion, and reproduction.[63] This work surely kept him thinking about vital processes and the perennial problem of the "balance of nature" (particularly at the Bureau of Biological Survey, which was forever trying to engineer the biota of American fields and forests). Finally, there was the overarching problem that Merriam had set before him: aquatic adaptation. While Kellogg considered himself at heart a bone man and happily expressed a general preference for hard structures over soft anatomy, his ongoing efforts to make sense of the evolutionary history of the cetaceans led him, by steps that can be retraced, to the conclusion that real progress on this problem would be made only by means of sustained work on the sensory systems, embryology, and life habits of marine mammals.[64]

To understand this development (which was a prerequisite for the Hatteras expedition and everything that followed), a brief review of the debate around cetacean evolution in this period is in order. When Merriam

63. His publications in connection with this work are instructive: Remington Kellogg, "Are Moles Held in Check by Blacksnakes?" *Bulletin of the Green Section of the U.S. Golf Association* 2, no. 5 (17 May 1922): 157–59; "The Toad," U.S. Department of Agriculture, Bureau of Biological Survey Circular Bi-664 (August 1922): 1–7. He also contributed to Herbert L. Stoddard, "Report on Cooperative Quail Investigation: 1925–1926 (With Preliminary Recommendations for the Development of Quail Preserves)" (Washington, DC: Published by the Committee Representing the Quail Study Fund for Southern Georgia and Northern Florida, 1926). For all these materials, the best source is the series of Kellogg offprint volumes in the Remington Kellogg Library at the National Museum of Natural History, Smithsonian Institution.

64. As late as May 1924, Kellogg could write, "Comparative anatomy has many points of interest, but to be honest soft anatomy never struck a responsive chord in my makeup." Kellogg to L. A. Adams (at the University of Illinois), 10 May 1924, Smithsonian Archives, RU 7170, box 1, file "Ad–Al." For a discussion of others working at this time to integrate physiology and embryology with paleontological investigations, see Ronald Rainger's "Vertebrate Paleontology as Biology: Henry Fairfield Osborn and the American Museum of Natural History," in *The American Development of Biology*, ed. Ronald Rainger and Jane Maienschein (Philadelphia: University of Pennsylvania Press, 1988), 219–56, and Léo Laporte's "George G. Simpson, Paleontology, and the Expansion of Biology," in *The Expansion of American Biology*, ed. Keith Rodney Benson, Jane Maienschein, and Ronald Rainger (New Brunswick, NJ: Rutgers University Press, 1991), 80–106, particularly pp. 84–87. My sense is that Kellogg's emphasis on sensory physiology makes his program distinctive, though it would be interesting to understand to what degree other biologically oriented paleontologists were investigating sensory physiology at this time.

turned Kellogg to this problem in earnest in 1920, the elder scientist was launching his acolyte onto roiling waters.[65] The basic phylogeny of the cetaceans had been a topic of heated controversy for more than fifty years, and opposing views had, if anything, hardened in the early twentieth century (driven in part by the discovery of several tantalizing new fossils).[66] Early notions, popular in the 1870s, that posited primitive whales as the earliest mammals—"missing links" tying the emergence of the mammals to their reptilian forebears—fell from favor relatively rapidly, only to be replaced by a melee of competing theories.[67] William Henry Flower at the British Museum—who used the cetaceans as the very best example of the importance of comparative anatomy in the refinement of taxonomy—put into circulation by the early 1880s the suggestion that whales might have

65. Interestingly, Kellogg had been introduced to aspects of this problem while still at Berkeley. In his notes from E. B. Babcock's lectures in experimental zoology in 1919, Kellogg recorded, in connection with the general problem of whether ontogeny recapitulates phylogeny, the issue of "the persistence of teeth in the foetal whale." See Smithsonian Archives, RU 7434, box 5, folder 4, "Translation, W. Kükenthal."

66. Particularly Eberhardt Fraas's primitive archaeocete in Egypt circa 1904 (what he called Protocetus atavus), and Andrews's Pappocetus lugardi, found in Nigeria. For Kellogg's own view of Protocetus atavus, see his correspondence with Herbert J. Pack, where Kellogg wrote that P. atavus "has many characters which are morphologically intermediate between later whales and typical Eocene land Carnivore-insectivore stock." Kellogg to Herbert J. Pack, 12 December 1927, Smithsonian Archives, RU 7170, box 6, file "Pa–Pap." It is worth noting that a good deal of this work falls firmly into Garland Allen's category of "speculative" evolutionary morphology. See Garland E. Allen, Life Science in the Twentieth Century (New York: Wiley, 1975), 10; see also his clarification, "Morphology and Twentieth-Century Biology: A Response," Journal of the History of Biology 14, no. 1 (Spring 1981): 159–76.

67. The early position was espoused by J. F. Brandt ("Untersuchungen über die Fossilen und Subfossilen Cetaceen Europa's," Mémoires de l'Académie Impériale des Sciences de St.-Pétersbourg, ser. 7, vol. 20, no. 1 [1873]: 1–372 and plates); the remarkable similarity in form between the zeuglodonts and the ichthyosaurs lent support to this view. Indeed, the first Zeuglodon remains had been misidentified as reptilian. A key specimen was famously discovered by Harlan in the 1830s in Caldwell Parish in southeastern Louisiana. Harlan, thinking he had found a new dinosaur, gave the creature a reptilian name, Basilosaurus, an identification subsequently challenged and overturned by Owen. In 1929 Kellogg returned to this region of Louisiana to trace the location of Harlan's finds and to secure for the Smithsonian its own Zeuglodon, which is on display to this day in the main dinosaur hall.

evolved from a primitive pinniped ancestor, but the idea never won wide support among paleontologists.[68] Final refutation of this suggestion fell to a Danish anatomist, Adolph Herluf Winge, who by the early twentieth century had become the most outspoken advocate of the view (narrowly dominant) that the two main series of modern cetaceans—the *odontocetes*, or toothed whales, on the one hand and the *mysticetes*, or "baleen" whales, on the other—had both evolved from the same land-dwelling carnivorous mammal—in Winge's view, one of the doglike *hyaenodonts*.[69] As evidence, Winge could adduce the strikingly similar teeth and jaws of the earliest fossils of whale-like creatures (what he called the *zeuglodonts*) and several Hyaenodontidae (particularly *Pterodon* and *Hyaenodon*). In his estimation these similarities left "no doubt about the relationship."[70] Whether modern whales were direct descendants of any of the known zeuglodonts (as Winge thought they almost certainly were) or merely shared with this group a more primitive common marine mammal ancestor as yet unknown (as the aristocratic German paleontologist Karl Stromer von Reichenbach argued) was a topic on which much ink was spilled in the 1910s.[71]

68. See William Henry Flower, "Cetacea," in *Encyclopedia Britannica* 15, 9th ed., ed. T. S. Baynes (Edinburgh: A. and C. Black, 1875–1889): 391–400. See also Kellogg to Dix Teachenor, 12 November 1937, Smithsonian Archives, RU 7170, box 8, "Ta–Th." Flower's later suggestion that the cetaceans might be derived from ungulates fared little better. See Herluf Winge and Gerrit S. Miller, "A Review of the Interrelationships of the Cetacea," *Smithsonian Miscellaneous Collections* 72, no. 8 (Washington, DC: Smithsonian Institution, 1921), 47. It is this theory that has recently been revived by genetic approaches. For the most recent contribution to the subject, consider Jonathan H. Geisler and Jessica M. Theodor, "Hippopotamus and Whale Phylogeny," *Nature* 458 (19 March 2009): E1–E5.

69. Herluf Winge, "Udsigt over Hvalernes indbyrdes Slaegtskab," *Vidensk. Medd. fra Dansk naturh. Foren.* 70 (1918): 59–142; this paper was translated by Gerrit Miller with the help of Stejneger and was thus available to Kellogg as Herluf Winge and Gerrit S. Miller, "A Review of the Interrelationships of the Cetacea," *Smithsonian Miscellaneous Collections* 72, no. 8 (Washington, DC: Smithsonian Institution, 1921). Max Weber was an early and outspoken advocate of a single land mammal ancestor, though not, he thought, a carnivore. See Max Weber, *Die Säugetiere: Einführung in die Anatomie und Systematik der recenten und fossilen Mammalia* (Jena: Fischer, 1904).

70. Herluf Winge and Gerrit S. Miller, "A Review of the Interrelationships of the Cetacea," *Smithsonian Miscellaneous Collections* 72, no. 8 (Washington, DC: Smithsonian Institution, 1921), 1.

71. Othenio Abel and Eberhardt Fraas also espoused versions of the latter position.

But all these disputants shared a common adversary in the supporters of the durable contention that modern whales were in fact the product of several *different* episodes of aquatic adaptation, proceeding more or less in parallel in two or more different land-dwelling mammals. The most visible advocate of this position—generally known at the time as the *polyphyletic* or *multiserial* hypothesis—was Willy Georg Kükenthal, curator of the natural history museum at Jena, Germany.[72] For nearly thirty years this prominent student of Ernst Haeckel had relentlessly attacked the monophyletic hypothesis, and by the late 1910s he and Winge were locked in polemical opposition.[73]

In Kükenthal's view, those paleontologists who rested easy with their whale trees (which showed nicely branching divergence of the main types, both prehistoric and modern, from a single trunk rooted in a land mammal of some sort [figure 3.6]) had all fallen victim to the same delusion that prompted earlier investigators to tie the archaic whales and the marine dinosaurs together in the same genealogical sequence: they had been seduced by mere similarity of shape, of appearance. Where aquatic adaptation was concerned, however, this was very dangerous indeed, since there was ample evidence that the particular demands of locomotion in a liquid environment led to remarkable convergences of form. In addition, it was hardly a stroke in favor of the monophyletic hypothesis that you could not find (as Kükenthal was quick to point out) two advocates of the position who would draw you exactly the same tree. But put that aside. More grievous, in Kükenthal's view, was the fact that they all applied, reflexively, the worst version of the method of comparative anatomy, assuming that "the degree of similarity corresponds to the degree of relationship."[74] This, Kükenthal argued, was bound to lead to errors, particularly in the investigation of any set of organisms so subject to what he called "the phenomenon of conver-

---

72. On the status of polyphyletic hypotheses in these years, see Ronald Rainger, "The Continuation of the Morphological Tradition: American Paleontology, 1880–1910," *Journal of the History of Biology* 14, no. 1 (Spring 1981): 129–58, at pp. 150–52.

73. For background on the intellectual and institutional framework of these debates in German-speaking Europe, see Lynn K. Nyhart, *Biology Takes Form: Animal Morphology and the German Universities, 1800–1900* (Chicago: University of Chicago Press, 1995). Particularly useful on the conflicts within morphology in this period in Germany is Nyhart's earlier article, "The Disciplinary Breakdown of German Morphology, 1870–1900," *Isis* 78, no. 3 (1987): 365–89.

74. Smithsonian Archives, RU 7434, box 5, folder 4, "Translation, W. Kükenthal."

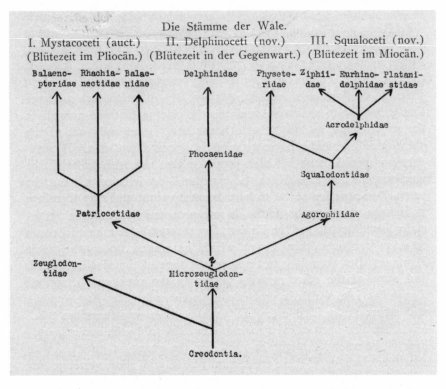

FIGURE 3.6 Family ties: Evolution of the Cetacea. (From Winge, "Udsigt over Hvalernes indbyrdes Slaegtskab," p. 61.)

gent breeding."[75] Distinguishing those traits that were diagnostic of genealogical relationships from those that were merely the product of adaptation to similar conditions of existence demanded more than just museum work on bones: as Kükenthal put it, "it is necessary to bring not only the *form*, but also the *functions* of the organs under investigation."[76] In his notebook Kellogg carefully underlined those two words.

Kellogg's discovery of Kükenthal's heated refutation of the monophyletic hypothesis pointed the young researcher in new directions. He was so taken with the text of Kükenthal's "Zur Stammesgeschichte der Wale," which he seems to have first encountered in 1923, that he sent it to his father-in-law in California (a German emigrant and thus a native speaker)

75. Ibid. (For "breeding" here, a modern translator would probably prefer "evolution.")
76. Ibid.

for a full translation. Kellogg could make his way through German texts (his *Doktorvater*, Merriam, had trained under Karl von Zittel in Munich, and Merriam expected his students to be diligent in the language), but Kellogg wanted to pore over Kükenthal's vigorous 1922 position paper.[77]

And pore he did, carefully annotating the document as it expanded on the need for further investigation of the embryology and sensory development of the different kinds of cetaceans. For Kükenthal, the final demise of the monophyletic hypothesis lay in the profound differences between the systems for smell, hearing, and (to a lesser degree) sight between the odontocetes and mysticetes. The innervation, musculature, and structure of the blowhole gave additional support to the polyphyletic interpretation. Further embryological research would confirm his position, he argued, in that it would reveal that the earlier stages of embryological development showed progressively greater divergence between toothed and baleen whales, as predicted by the polyphyletic hypothesis; a monophyletic origin for the modern whales would demand the opposite—ever greater similarity in ever smaller embryos.

Kükenthal's paper presented Kellogg with nothing less than a new and suggestive program of research, one that promised to shed light on a classic and hotly disputed problem in vertebrate evolution and to do so in a way that would not put him into direct conflict with Miller, since Miller himself was increasingly committed to the polyphyletic view and was configuring his work on the skull as comparative anatomical groundwork for a reinvigorated and novel polyphyletic hypothesis (one that would eventually posit three separate lines of land-mammal-to-cetacean evolution).[78] By

77. The identification of the translator is based on the hand of the manuscript in ibid., in conjunction with Kellogg's correspondence with "Dad" Heinrich. See Kellogg to "Dad" Heinrich, 1 March 1930, Smithsonian Archives, RU 7170, box 4, file "Hea–Hes." Heinrich would also later translate Boenninghaus. Kükenthal's original paper was published in *Sitzungsberichte der Preussischen Akademie der Wissenschaften (Physikalisch-mathematischen Klasse)*, Berlin, IX (4 May 1922): 72–87. Kellogg regularly made his after-hours researches a family affair. His wife Marguerite, whom he met at Berkeley and who was a student of art, hand-colored the lantern slides for his lectures. See Kellogg to Miles D. Pirnie, 7 September 1923, Smithsonian Archives, RU 7170, box 7, file "Pe–Pi."

78. Miller argued that the zeuglodonts were a wholly separate instance of aquatic adaptation. See Gerrit S. Miller, "The Telescoping of the Cetacean Skull (with Eight Plates)," *Smithsonian Miscellaneous Collections* 76, no. 5 (1925): 1–55 (also called "Publication 2822").

investigating the physiological and embryological aspects of the problem, Kellogg would thus fit into the research program of his superiors at the US National Museum, address the central problem of interest to his supervisor, avoid dependency on the catch-as-catch-can world of fossil finds, and maintain a connection to zoology, his first scientific love. The comparative paleontology of East Coast and West Coast Archaeoceti fossils fell by the wayside, and Kellogg sketched out a new plan emphasizing the sensory physiology of the modern whales and dolphins.[79]

By March 1923 he was writing to Merriam that, in connection with the study of the cetacean auditory system, he had "met with a number of puzzling details which can only be settled by dissection of the head of some cetacean."[80] Intrigued with the possibilities, Kellogg began canvassing a variety of possible sources for a (more or less) fresh whale, eventually zeroing in on his first fleshed specimen later that year: the head of an embryonic rorqual in storage at the Hopkins Marine Laboratory at Pacific Grove. He also began to collect other information about cetacean embryos, learning, for instance, that a three-inch-long fetal humpback looked exactly like a little full-grown whale. Wrote one of his Bureau of Biological Survey colleagues on learning this fact, "Whales must come from an ancient stem!"[81] Even as his physiological interests expanded, however, Kellogg continued to emphasize the links between these researches and more properly paleontological problems. He summed up his view for a colleague at Florida State (they were corresponding about a beached sperm whale): "I am more interested in the fossil forms than in the living. Nevertheless, one has to pay considerable attention to the living cetaceans in order to inter-

79. This plan is most clearly laid out in the four-page research statement "Aquatic Adaptation in the Cetacea and Its Effects on Bodily Activities," Smithsonian Archives, RU 7434, box 2, file "News Clippings"; but see also John C. Merriam et. al., "Continuation of Paleontological Researches," *Carnegie Institution Yearbook* 25 (December 1926): 403–7, at p. 405. Kellogg was notably deferential as he moved into physiological areas; see, for instance, his letter to Yandell Henderson at the Yale Medical School, 22 November 1926: "Physiology is a rather far cry from paleontology and I know very little about it." Smithsonian Archives, RU 7170, box 4, file "Hea–Hes."

80. Kellogg to Merriam, 19 March 1929, Smithsonian Archives, RU 7170, box 6, file "Merriam, John C. (3/3)."

81. This was Alexander Wetmore, with whom Kellogg was friendly from the University of Kansas and who would go on to become assistant secretary, and eventually secretary, of the Smithsonian. Wetmore to Kellogg, 19 March 1923, Smithsonian Archives, RU 7170, box 8, file "Wa–Wh."

pret the evidence afforded by the fossil forms."[82] It was a statement of his new creed.

The minor god of serendipity attended on Kellogg: as it happened, the cetacean bones that had become his minor area of expertise—the ear bones, particularly the periotic and the bulla—presented thorny physiological problems. The question of whether whales (which have no external "ears" at all, and many of which have no ear opening of any kind) could hear was an old one on which nineteenth-century naturalists (and whalemen) had expressed divergent opinions.[83] In the early twentieth century there remained zoologists who believed that whales could not really hear under water.[84] By that time, however, a number of efforts had been made to explain how the cetacean ear might detect sound waves propagated through a liquid medium, and in 1904 one such account had been published as part of a German thesis that reviewed the anatomy of whale ears in numbing detail: Friedrich Wilhelm Georg Boenninghaus's *Das Ohr des Zahnwales*, published with dissection plates so fine that they retain something like cult status among marine mammal biologists.[85] When Kellogg turned to the problem in the 1920s, new and interesting evidence on the topic had come to light: during World War I an experimental high-frequency submarine signaling device was observed to have very definite effects on dolphins, which appeared to be more or less indifferent to low-frequency sounds like the thud of artillery.[86]

82. Kellogg to T. Van Hyning, 12 May 1923, Smithsonian Archives, RU 7180, box 8, file "U–V."

83. For a useful review, see the introduction to W. H. Dudok van Heel, "Sound and Cetacea," *Netherlands Journal of Sea Research* 1, no. 4 (1962): 407–507.

84. See, for instance, D. G. Lillie, "Observations on the Anatomy and General Biology of Some Members of the Larger *Cetacea*," *Proceedings of the Zoological Society of London* 80, no. 3 (1910): 769–92.

85. Georg Boenninghaus, *Das Ohr des Zahnwales* (Jena: Gustav Fischer, 1903).

86. Kellogg discussed these findings in a summary of his "Studies on the Relation of Sense Organs to the General Problem of Aquatic Adaptation," *Carnegie Institution of Washington Yearbook* 25 (December 1926): 405. He was in correspondence with R. F. Blake on the use of the Fessenden oscillator. See Kellogg to Blake, 6 May 1926, Smithsonian Archives, RU 7170, box 1, file "Bla–Blo." See also R. F. Blake, "Submarine Signaling: The Protection of Shipping by a Wall of Sound and Other Uses of the Submarine Telegraph," *Transactions of the American Institute of Electrical Engineers* 33, no. 2 (1914): 1549–61. Also interesting on this subject is Gary L. Frost, "Inventing Schemes and Strategies: The Making and Selling of the Fessenden Oscillator," *Technology and Culture* 42, no. 3 (2001): 462–88.

Kellogg secured a copy of Boenninghaus, packed it off (all two hundred pages) to his father-in-law for translation, and began to think about how a better understanding of cetacean hearing might contribute to a study of aquatic adaptation. The periotics of the zeuglodonts seemed particularly suggestive, since the first one he got his hands on looked strikingly similar to those of modern whalebone whales.[87] This was unexpected, since the zeuglodonts were often thought to be closer to the toothed whales (they all had plenty of teeth, for starters). In their general skull architecture, zeuglodonts were so different from both whalebone and toothed whales (the zeuglodont skull showed no telescoping, and the braincase was elongated rather than foreshortened) that Miller was intent on arguing that they represented a wholly distinct mammalian adaptation to the sea and had no genealogical relationship to modern cetaceans of any kind. But what was most promising about Kellogg's observation lay elsewhere: if it could be shown that this particular morphology of the periotic was a functional adaptation to hearing deep under water, this might point toward a revision of the received wisdom on the habits and history of the earliest whale-like creatures. The zeuglodonts, large Eocene marine predators, were clearly aquatic mammals, but they were thought by most at the time to be shallow-water hunters, still closely related to their land-bound progenitors. If, however, they possessed ear bones very like those of modern whales, and that auditory apparatus could be shown to equip them for deep-water hearing, then it would be safe to infer both that they were very probably open-water creatures (less wallowing, carnivorous hippos than cruising mammalian sharks) and that they had been in the water much longer (geologically speaking) than anyone had surmised. It would be a significant finding. There were similarly enticing questions to be raised about smell and sight.

From 1925 to 1927, then, Kellogg embarked on a suite of investigations into these aspects of sensory physiology and their relationship to cetacean evolution. A new confidence possessed him in this period: after several years of diligent reading and letter writing, he was, as he pointed out to Merriam, "now corresponding with all who are now working on pelagic

87. Kellogg to Merriam, 13 March 1923, Smithsonian Archives, RU 7170, box 6, file "Merriam, John C. (3 / 3)." He would later discover other zeuglodont periotics closer to the odontocete type. See Kellogg to Merriam, 26 March 1925, Smithsonian Archives, RU 7170, box 6, file "Merriam, John C. (2 / 3)."

mammals."[88] And he meant this in the global sense: he was engaging with prominent paleontologists and zoologists from Yugoslavia to Sweden to New Zealand on a regular basis, and he was doing so with confidence and even a tone of authority.[89] After outlining the many prongs of his research plan to a colleague in this period, he made his new unifying vision explicit: "You may think that some of these studies are rather varied, but one of my chief aims has been to interpret the effect of aquatic adaptation on the body of a whale, particularly the sense organs."[90]

The first down payment on his promises for this undertaking came at the 1925 meeting of the American Society of Mammalogists, where Kellogg presented a paper on the sense of smell in the marine mammals.[91] Whale-

88. Kellogg to Merriam, 26 September 1925, Smithsonian Archives, RU 7170, box 6, file "Merriam, John C. (2/3)."

89. See, for instance, his correspondence with Ernst Stromer, Smithsonian Archives, RU 7170, box 8, file "Stromer, Ernst von," and particularly Kellogg's damning judgment of the work of the prominent Austrian paleontologist Othenio Abel, of whom Kellogg wrote, "Abel should be given credit for popularizing paleontology, but many of his theories regarding the Cetacea will not stand critical analysis." Kellogg to Stromer, 4 December 1922, Smithsonian Archives, RU 7170, box 8, file "Stromer, Ernst von."

90. Kellogg to Ira E. Cornwall, 29 June 1926, Smithsonian Archives, RU 7170, box 2, file "Coop–Cow." See also Kellogg to Yandell Henderson, 22 November 1926, Smithsonian Archives, RU 7170, box 4, file "Hea–Hes," in which Kellogg requests additional help from physiologists that "will enable those of us who are interested in paleontology to explain puzzling questions that arise in connection with the geological history of these mammals."

91. Kellogg to Merriam, 24 March 1925, Smithsonian Archives, RU 7170, box 6, file "Merriam, John C. (2/3)." The history of the American Society of Mammalogists (and particularly its relationship with other learned societies) provides some helpful insights into the structure of the American life sciences in this period. For instance, in 1925, when the Executive Committee of the Union of American Biological Societies and representatives from the Division of Biology and Agriculture of the National Research Council came together to try to reorganize how the different life sciences were represented in intersociety gatherings, they recommended eight subdivisions of the biological sciences: botany, zoology, applied zoology (dairy and poultry science, etc.), applied botany (forestry, horticulture), "specialized" botany (including bacteriology), botany-zoology (including, interestingly, both genetics and ecology), physiology, and, finally, "specialized" zoology. The American Society of Mammalogists was to be placed in that last category, together with, among other things, the entomology societies. More research on this restructuring might be interesting. See "Report of Committee on Group Representation," 9 May 1925, Smithsonian Archives, RU 7170, box 6, file "N–Nq."

bone whales, he showed, retained the organs of olfaction; modern toothed whales did not. While the skulls of zeuglodonts were constructed in such a way as to leave little doubt that they could smell (they had clear passageways—olfactory foramina—through the skull for the olfactory nerves, and a careful survey of the braincase revealed a well-marked bony sinus that would have contained an olfactory lobe), the skulls of early toothed whales could be arranged in a sequence that seemed to show the progressive occlusion of the olfactory foramen. This sequence, while not cleanly an evolutionary lineage, nevertheless appeared to illustrate the progressive loss of the olfactory sense in the odontocetes and even to suggest the cause: the remodeling of the anterior part of the skull as the narial passages moved back from the snout to the top of the head—a process associated with Miller's telescoping of the cetacean skull. It looked as if this had resulted in the choking off of the route by which the olfactory nerves passed through the skull wall.[92]

The next year, in 1926, Kellogg, gathering up his work on the ear bones, presented a similar paper on cetacean hearing at the American Society of Mammalogists' meeting in New York to extremely warm reviews. As he reported to Merriam afterward, a number of distinguished scholars (William King Gregory, Frederic A. Lucas, even C. Hart Merriam) took him aside to discuss his ongoing work on the "pelagic mammal problem," and he received particular praise for "the utilization of evidence derived from both living and fossil forms." Best of all, "several of the members who attended the meeting told my wife that last years [sic] paper on the sense of smell and the one of this year on hearing had opened up a surprisingly interesting field of investigation to them."[93] Finally, the following year, in 1927, he followed this pair of presentations with a paper on vision in the cetaceans, the product of a crash course in optics that left a trail of geometrical diagrams and algebra through his notebooks of the period.[94] This study too yielded immensely tempting fruit. On the basis of these investigations, Kellogg began to nurture an original hypothesis for the extinction of the zeuglodonts. He wrote to Merriam "that one of the causes for the disappearance of the

92. Kellogg reviews this fusing of the mesethmoid and ectethmoid bones into a continuous sheet spreading over the troughlike vomer in part 2 of his "History of Whales—Their Adaptation to Life in the Water" (PhD diss., University of California, Berkeley, 1928), 200–202.

93. Kellogg to Merriam, 19 May 1926, Smithsonian Archives, RU 7170, box 6, file "Merriam, John C. (2/3)."

94. See Smithsonian Archives, Accession 99–012, box 2, file "Eye notes, undated."

zeuglodonts may have been defective eyesight, and this is substantiated by the dwindling and constriction of the optic chiasmic region from early to late types."[95]

By the time he delivered that argument at the Philadelphia meeting of the ASM, Kellogg had become a young star in vertebrate zoology and paleontology. His distinctive program of physiological-cum-morphological research on a particularly difficult problem in the history of evolution and adaptation had attracted wide attention. He was juggling a number of job offers and fielding enthusiastic solicitations from such kingmakers as Henry Fairfield Osborn, who was anxious to attract him to the American Museum of Natural History in New York City. Wheels were turning in Washington as well, and by early 1928 a full-time curatorial position at the US National Museum had come on line. Kellogg was finally in a position to pack up his desk at the Bureau of Biological Survey. Merriam had deemed Kellogg's work sufficiently important to suggest to Raymond Pearl, editor of the *Quarterly Review of Biology*, that he ought to solicit from Kellogg a sweeping synthetic piece on the evolution of whales, a piece modeled explicitly on William Diller Matthew's landmark paper "The Evolution of the Horse," which had appeared in that journal in 1926.[96]

It is necessary to recall that Merriam's strong interest in the general health and standing of the sciences in the United States, and his position at the Carnegie Institution (which counted public education among its ambitions), gave him a particular appetite for Kellogg's innovative work in the mid-1920s. In 1925 the Scopes trial thrust evolution to the forefront of national consciousness and revealed to scientific grandees on both coasts a disheartening recalcitrance and even hostility simmering in the heartland.[97] It has largely been forgotten, in view of the obvious significance of

95. Kellogg to Merriam, 12 April 1927, Smithsonian Archives, RU 7170, box 6, file "Merriam, John C. (2/3)."

96. See William Diller Matthew, "The Evolution of the Horse: A Record and Its Interpretation," *Quarterly Review of Biology* 1, no. 2 (April 1926): 139–85. For a discussion of Matthew's horse work, see Ronald Rainger, *An Agenda for Antiquity: Henry Fairfield Osborn and Vertebrate Paleontology at the American Museum of Natural History, 1890–1935* (Tuscaloosa: University of Alabama Press, 1991), 90–94, 205–10. For Merriam's suggestion to Kellogg and Pearl, see Kellogg to Merriam, 4 November 1926, and Merriam to Kellogg, 12 November 1926, both in Smithsonian Archives, RU 7170, box 6, file "Merriam, John C. (2/3)."

97. Philip J. Pauly discusses the impact of the trial on several American scientists at the time in *Biologists and the Promise of American Life: From Meriwether Lewis to*

monkeys in evolutionary debates, that whales also featured prominently in the Scopes testimony: cetacean evolution was thrust onto the front pages of American newspapers in July 1925, when Professor Maynard M. Metcalf, a zoologist who would soon take up a position at Johns Hopkins, took the stand as Clarence Darrow's first scientific witness in the trial and entered with him into a parley on the evolutionary history of the emergence of land animals. The conversation soon turned to whales and seals—which had evolved back into the water, as Metcalf explained—a topic that gave rise to a more general consideration of the broad category of "mammals."[98]

Merriam's vision for Kellogg's projected piece in the *Quarterly Review* must be understood in light of this larger national debate around evolution. Merriam encouraged Kellogg to "take up a general statement of the present state of adaptation of the whales to their environment" along with a "brief outline of the evolution of the whales." Above all, it was necessary to "avoid technical language wherever possible" since a clear, accessible exposition "widens the audience enormously." From Merriam's perspective, whales "furnish some of the most interesting material for study of evolution in the whole palaeontological record," and thus Kellogg was positioned to offer a valuable—indeed, potentially, a *textbook*—summary of an evolutionary process.[99] It is a testimony to Merriam's sense of the importance and timeli-

---

Alfred Kinsey (Princeton, NJ: Princeton University Press, 2000), 213. For a full treatment see Edward J. Larson, *Summer for the Gods: The Scopes Trial and America's Continuing Debate over Science and Religion* (Cambridge, MA: Harvard University Press, 1998), and Jeffrey P. Moran, *The Scopes Trial: A Brief History with Documents* (New York: Bedford St. Martin's, 2002). For an interesting discussion of the cultural significance of evolution in the United States after World War I, see Gregg Mitman, "Evolution as Gospel: William Patton, the Language of Democracy, and the Great War," *Isis* 81, no. 3 (1990): 446–63.

98. See "Indictment is Sustained," special to the *New York Times*, 16 July 1925, 1. For complete testimony on whales, see Clarence Darrow and William J. Bryan, *The World's Most Famous Court Trial* (Clark, NJ: The Lawbook Exchange, 2008), 142. The subject of Jonah and the whale turned out to be one of the most dramatic episodes of the proceedings; see 284–85.

99. Merriam to Kellogg, 26 November 1926, Smithsonian Archives, RU 7170, box 6, file "Merriam, John C. (2/3)." Compare Osborn's interest in this period in encouraging research into similarly dramatic and visual evolutionary narratives using paleontological evidence: "Fossil vertebrates also had sheer entertainment value and could contribute to public education. As the documentary evidence for evolution, fossil vertebrates could convey to the public the importance of nature and nature's laws." Ronald Rainger, *An Agenda for Antiquity: Henry Fairfield Osborn and Vertebrate Paleontology at the*

ness of this line of research that when, in late 1928, the Carnegie Institution mounted its annual public exhibition, he chose Kellogg's whale research as the centerpiece of the show, working with him to create compelling exhibits of, for instance, the evolution of the cetaceans' distinctively handlike flipper and preparing and printing an elaborate illustrated press release summarizing Kellogg's findings with newspaper energy and brevity.[100] Some papers did pick up the story, and Merriam was so pleased that he wrote Kellogg a rare letter of uninhibited praise: "I know that I am expressing also the feeling of the Trustees in saying that this work is of great importance, both to the Institution and for the general movement to further development and appreciation of research in the country."[101] By 1928, Merriam's prediction had come true: Kellogg and a collection of new colleagues and friends were on the cusp of making whales a national affair.

To show exactly how, it will be necessary to stay with the period 1926–1928 for another moment, in order to examine how Kellogg's foray into cetacean physiology put him in contact with several new communities: physiologists above all, but also physicists and oceanographers, whalers and diplomats. I have already shown that Kellogg's early efforts to buttress his claims about the geological importance of whale fossils led him into correspondence with the British scientists of the Discovery Investigations and heightened his awareness of the rapid expansion of modern whaling. What remains to be examined is the way in which Kellogg's expanding interests of the mid-1920s made it increasingly urgent for him to get his hands on some living (or very recently dead) cetaceans. It was this undertaking that brought Kellogg into increasingly intimate contact with the world of whalers and the emerging international problems of whale conservation.

American Museum of Natural History, 1890–1935 (Tuscaloosa: University of Alabama Press, 1991), 88.

100. "Whales—Their Adaptation to Life in the Sea," released by the Carnegie Institution of Washington, DC, Smithsonian Archives, RU 7434, box 2, folder 26, "Whales—Adaptation to the Sea, 1928." This appears to be only the second such broadsheet press release prepared by the Carnegie Institution. By this point Kellogg's work in the period 1925–1927 had already garnered considerable popular attention. See "Ear Bones Indicate Whales' Evolution," *Baltimore Evening Sun*, 6 January 1927; "Evolution of Whales Traced," *Bridgeport (CT) Post*, 7 January 1927; and "Whale find helps inland sea theory" (no newspaper given), all in Smithsonian Archives, RU 7170, box 9, file "Information—Newspaper Clippings".

101. Merriam to Kellogg, 2 January 1929, Smithsonian Archives, RU 7170, box 6, file "Merriam, John C. (2/3)."

I mentioned above that Kellogg's first investigations into the periotic bone prompted him to seek out, as early as 1923, the head of a fetal rorqual for dissection. He did not stop there. By 1925 Kellogg had formed a jocular friendship with his colleague at the Bureau of Biological Survey, A. Brazier Howell, who, like Kellogg, can perhaps best be described as having been *in* the bureau but not really *of* it. The two men shared a taste for western rambling and birding, in addition to membership in the American Society of Mammalogists and a penchant for stamp collecting. Howell's interests were various, even eccentric (everything from oölogy to commercial aquariums to bat eradication to the restoration of old automobiles), but his independent means and excellent connections made him a valuable ally: he was actively engaged in the Cooper Ornithological Club and helped endow their journal, the *Condor*. He was regularly recruited to serve as treasurer or secretary of worthy scientific associations, to which he slipped much-needed funds.[102]

It was Kellogg who enlisted Howell in cetacean research in the 1920s. While Howell had lived in Washington, DC, since 1922, he retained close ties to the West Coast, where he spent almost a decade before the war and where he owned an orange grove nestled in the hills near Pasadena. Familiar with the rough camp-and-trawler ways of the California fishing industry in this period (he had made use of connections to the International Fisheries Company of Baja California to gain access to prime birding grounds in the Los Coronados Islands and elsewhere), Howell had invaluable experience insinuating the attentions of a naturalist into the dockside world of fishermen and canners. California was home to the only established shore whaling industry in the United States in this period, which consisted of two small seasonal stations that processed a few hundred whales a year—mostly fin whales—into oil, chicken feed, and fertilizer; the accumulated bones were exported to Hawaii, where they were burned into a fine charcoal used in the processing of sugar.[103] In 1926 Howell made a detour during a West Coast visit in order to spend two weeks in the stink of the ramshackle whaling station belonging to the California Sea Products Com-

102. On Howell, see Luther Little, "Alfred Brazier Howell, 1886–1961," *Journal of Mammalogy* 49, no. 4 (1968): 732–42, and Jack C. von Bloeker, "Who Were Harry R. Painton, A. Brazier Howell and Francis F. Roberts?" *Condor* 95 (1993): 1061–63.

103. For statistics on the West Coast industry in this period, see Charles H. Townsend, "Twentieth-Century Whaling," *Bulletin of the New York Zoological Society* 33, no. 1 (January–February 1930): 3–31.

pany, located in Trinidad, just south of the redwood forests on the Oregon border.[104] While the main purpose of the stay was to secure ear bones for Kellogg, it is likely that Howell had his own reasons for wanting to spend some time with the bones of cetaceans: he had recently published a small, strange article in the *Proceedings of the United States National Museum* entitled "Asymmetry in the Skulls of Mammals"—a close morphological and developmental examination of a handful of pathologically malformed mammalian skulls, including a pair of Pacific pinnipeds. In view of the fact that odontocetes—particularly sperm whales—display probably the most extreme cranial asymmetry of any mammal (and sperm whales were taken at Trinidad, if infrequently), it seems reasonable to suppose that Howell hoped to have a chance to examine the relationship between the musculature and the bone structure in one of these animals.[105]

Kellogg's haul from the journey amounted to several five-gallon soldered tins of salt-packed specimens and a pair of colorful letters about life among the carcasses. "Everything we eat tastes like dead whale," wrote Howell gleefully after five days at the station. He went on to describe his efforts to secure the bits that Kellogg had requested:

In separate pieces of cheese cloth are the bullae, incuses & stapes of the big finback. I had one beautiful time trying to get the whole works, and after hacking through a ton of meat, had to take to an axe, and I never did find the rest of the ear there were so many pieces of broken bone cluttering up the neighborhood. . . . I am very much afraid you wont get what you are after from this stuff, but will get better if I can.[106]

Later, he renewed his efforts:

I am going to try to saw out the whole side of a head with the ear in place—just big enough to fit in a 5 gal. can. Then I will inject preserva-

104. He made the trip with the ornithologist Laurence M. Huey, curator of the Department of Birds and Mammals at the San Diego Natural History Museum; for details, see A. Brazier Howell, "Visit to a California Whaling Station," *Smithsonian Miscellaneous Collections* 78, no. 7 (1926): 71–79.

105. How muscular development affected bone growth was at the center of the asymmetry article: see A. Brazier Howell, "Asymmetry in the Skulls of Mammals," *Proceedings of the United States National Museum* 67, no. 2599, art. 27 (1925): 1–18.

106. A. Brazier Howell to Kellogg, 22 [?] July 1926, Smithsonian Archives, RU 7170, box 4, file "Howell, A. Brazier, 1920–1929 and undated (3 / 3)."

tive everywhere that a [illegible] needle will penetrate and cover with 5% formalin and salt. Preservation will not be perfect throughout the whole mass, so get at it just as soon as you can.

In the end, it was a generally successful romp through the carrion. Unfortunately, the eye of the (enormously rare) gray whale arrived quite badly decayed, but the finback eyes, and the ear bones (a total of four variously complete sets), were intact, and it was possible to make out a bit of how they were configured in situ, with the help of Boenninghaus.[107]

By the time Kellogg had those specimens in hand, he had already brokered a set of collaborations that would help him examine the organs with a new set of instruments, taking advantage of a new range of expertise. In February 1926 he had written to Merriam, reporting his progress and asking for help: "I have been wondering whether or not any one connected with the Carnegie Institution could offer any suggestions in regard to a method for determining the efficiency of the tympanic bulla of cetaceans as a 'sounding box.'" Having established that many whales did not have an ordinary, open ear canal (or anything like the flexible ear "drum" common to most mammals), Kellogg had begun to toy with several different possible explanations for how whales might hear: "Sound vibrations are either transmitted to the tympanic cavity through the eustachian [sic] tube and associated complex of air cells or through the solid tissues of the head." The question was which of these paths was more likely and what, exactly, served as the receiver for the vibrations. Kellogg thought it might be possible to run a set of experiments to figure out the answer: "Since physicists

107. For a listing of Howell's specimens, see "Whaling station of California Sea Products Co; A. B. Howell collector," Smithsonian Archives, Accession 99–012, box 2, file "Eye notes, undated." Alexander Wetmore, secretary of the National Museum, also helped Kellogg secure specimens from Pacific Grove. See Wetmore to Kellogg from the Wiltshire Hotel, 19 March 1923, Smithsonian Archives, RU 7170, box 8, file "Wa–Wh." For a summary of Kellogg's views on Boenninghaus's theory of cetacean hearing (which is, admittedly, odd—see W. H. Dudok van Heel, "Sound and Cetacea," *Netherlands Journal of Sea Research* 1, no. 4 [1962]: 407–507), see Kellogg to "Dad" Heinrich, 1 March 1920, Smithsonian Archives, RU 7170, box 4, file "Hea–Hes," where he writes that it contains "much of value buried among a lot of vague theorising that confused one not accustomed to such methods of reasoning. He could have boiled his remarks down to 25 pages and then had room to spare. But when they write a thesis, they seem to think that bulk alone counts and consequently pad the paper as much as the editor will permit."

have made tests to determine the efficiency of various phonograph horns and loud speakers, it occurred to me that apparatus might be available to determine the range of the vibrations which affect the tympanic bulla."[108] Merriam put Kellogg in touch with R. W. G. Wyckoff, at the Carnegie Institution's Geophysical Laboratory, and Wyckoff tapped the expertise of a colleague at General Electric who had worked on the acoustics of the "Orthophonic Victrola" and who had access to the proper equipment.[109] Meanwhile, Kellogg had gained an entree to the Army Medical School and was using their X-ray machine to examine the convoluted inner structure of periotic bones, which contained within them the delicate passageways of the cochlea and the semicircular canals.[110] Even Miller was increasingly taken with Kellogg's ideas about resonance and cetacean hearing, and he pointed out that there were a pair of bony pedicles that linked the bulla to the periotic: "If they are set in vibration, the sound waves will reach the cochlea through the ossicular chain." Everything depended on proving that the bulla could function as a sounding box.[111]

As Kellogg thus reached out to a set of engineers and physicists for help tracing the path of sound vibrations through the middle and inner ear of the cetaceans, he simultaneously sought out the assistance of physiologists and embryologists who had studied human hearing. Here again Merriam was invaluable, connecting Kellogg with the talented embryologist George L. Streeter at the Johns Hopkins University School of Medicine (then president of the American Association of Anatomists), who had conducted an elaborate neurological and anatomical study of human hearing, a project supported not only by the Carnegie Institution but also by the Radio Corporation of America and General Electric.[112] Kellogg paid

---

108. Kellogg to Merriam, 8 February 1926, Smithsonian Archives, RU 7170, box 6, file "Merriam, John C. (2/3)."

109. Kellogg to Merriam, 15 February 1926, in ibid.

110. Kellogg to Merriam, 8 February 1926, in ibid.

111. Kellogg to Merriam, 15 February 1926, in ibid. Kellogg was basically correct in his hypothesis that the tympanic bone vibrates in response to sound waves and that these vibrations are transmitted via the ossicular chain (see Sirpa Nummela, Tom Reuter, Simo Hemilä, Peter Holmberg, and Pertti Paukku, "The Anatomy of the Killer Whale Middle Ear," *Hearing Research* 133, no. 1–2 [July 1999]: 61–70).

112. On Streeter, see George Washington Corner, "George Linius Streeter," in *Dictionary of Scientific Biography* 13, ed. Charles Coulston Gillespie (New York: Charles Scribner's Sons, 1970): 96–97, and Harold Speert, "Memorable Medical Mentors," pt. 9, "George L. Streeter (1873–1948)," *Obstetrical and Gynecological Survey* 60, no. 1 (2005):

a visit to Hopkins and showed Streeter what he had. Streeter was deeply impressed:

> After Dr. Streeter had examined my diagrams and the specimens, he stated that he thought it would be very important to investigate the organ of hearing in cetaceans because it would shed light on how animals with neither external ears, nor external auditory canals, nor movable auditory ossicles were able to recognize sounds. He said that I should obtain some porpoises and bring their brains to his laboratory where they could be sectioned and studied. It will be necessary to obtain fresh material.[113]

Streeter was ready to assign several researchers in his lab group to the problem, and he expressed a willingness to work on the problem himself as well. So taken was he with what he saw that he and Kellogg soon had several other sections of the Johns Hopkins medical school puzzling over porpoises.[114] Lewis Hill Weed, the forbidding young dean of the school (and one of the leading physiologists in the United States), also met with Kellogg and took an interest in the eyes and general neurological system of the animals: having recently completed a major work on the production and reabsorption of the spinal fluid, Weed was curious about how such deep-diving animals could handle rapidly changing pressures on these sensitive body cavities.[115] He, too, was prepared to throw researchers and equipment at cetaceans. At it happened, Weed summered at a fashionable coastal resort in New England, and he returned from his summer vacation after meeting

---

3–6. On Kellogg's visit to Streeter, see Kellogg to Howell, 4 August 1925, Smithsonian Archives, RU 7170, box 4, file "Howell, A. Brazier, 1920–1929 and undated (3/3)," and Kellogg to George L. Streeter, 17 July 1926, Smithsonian Archives, RU 7170, box 8, file "Stockt–Stre."

113. Kellogg to Merriam, 25 July 1926, Smithsonian Archives, RU 7170, box 6, file "Merriam, John C. (2/3)."

114. For useful background, see Gerald L. Geison, *Physiology in the American Context, 1850–1940* (Bethesda, MD: American Physiological Society, 1987).

115. On Weed, see Harold Speert, "Memorable Medical Mentors," pt. 1, "Lewis Hill Weed (1886–1952)," *Obstetrical and Gynecological Survey* 59, no. 2 (2004): 61–64. Note that in the early 1920s, Kükenthal suggested a link between the structure of the meninges in cetaceans and adaptations to deep diving (see Smithsonian Archives, RU 7434, box 5, folder 4, "Translation, W. Kükenthal"), so this interest is again consistent with Kellogg's research plan as I have laid it out above.

with Kellogg "filled with enthusiasm for the whale" and anxious to get his hands on some material to distribute to researchers in his laboratory.[116]

All of which left a pressing question: Where to get some cetaceans?

Small odontocetes like dolphins and porpoises were clearly the best bet. In theory these animals were available on the East Coast. But where? And who could catch them? How? The Carnegie Institution again offered possibilities. The institution, after all, had a tall ship under sail: the brigantine *Carnegie*, recently refitted, which would soon be plying the Atlantic making geomagnetic and oceanographic observations. Merriam could surely arrange for it to bring in small cetaceans from time to time.[117] But that still left open the question of just exactly how one went about catching the creatures.

Ideas came from some unexpected quarters. The silent film *Down to the Sea in Ships*—a modest success in the early 1920s—featured a live-action sequence in which a dolphin was harpooned from the bowsprit of a whaling vessel; the scene made a vivid impression on Kellogg when he went to see it. When the movie ended, Kellogg took note of the credits and wrote for advice to the producer (the film had been made in New Bedford, Massachusetts, once the metropolis of American whaling) for advice on where he might find a working harpoon and someone to wield it. He penned a similar letter the very same day to the dean of American whaler-naturalists, Robert Cushman Murphy (who had, recall, in 1913 made an Antarctic voyage on one of the last New England sperm whalers under sail, the *Daisy*, and had visited the burgeoning South Georgia whaling stations in their infancy, writing about them for American readers in *National Geographic*). Murphy, whose scientific expertise lay in ornithology, continued to coordinate US scientific expeditions in the Southern Hemisphere through his curatorial post at the American Museum of Natural History, and he relished his reputation as a hearty seafarer.[118] "Can you recommend a harpoon suitable for capturing the smaller porpoise?" wrote Kellogg. "There is a possibility that

116. Lewis Weed to Kellogg, 17 September 1927, Smithsonian Archives, RU 7170, box 8, file "Wa–Wh."

117. On the refitting of the *Carnegie*, see Susan Schlee, *The Edge of an Unfamiliar World: A History of Oceanography* (New York: Dutton, 1973), 265–72. For Kellogg's efforts to get the *Carnegie* to bring him specimens, see his correspondence with John A. Fleming in the Smithsonian Archives, RU 7170, box 2, file "F–Fo."

118. Robert Cushman Murphy to Kellogg, 23 March 1926, Smithsonian Archives, RU 7170, box 6, file "Mor–My." Murphy was on the board of the Whitney South Sea Expedition.

I may be able to obtain porpoises for dissection and study if the collector can be supplied with a suitable harpoon."[119] Did Murphy happen to know any actual harpooners?

Responses were prompt and detailed. Not only was the *Carnegie* outfitted with light tackle for collecting porpoises (toggle harpoons, floats, and line: figures 3.7 and 3.8), but soon Kellogg was in contact with one of the last remaining whale-oil companies working out of New Bedford: William F. Nye, Inc., "manufactures of sperm sewing machine oil, watch and clock oil, and especially fine lubricating oils for all kinds of machinery" (figure 3.9).[120] From correspondence with the Nye corporation, Kellogg learned that much of the commercial lubricating oil for sale in the United States did not come from sperm whales, but rather from smaller odontocetes—mostly dolphins and pilot whales. His best opportunity to get his hands on some of these creatures, he discovered, was the company's last East Coast "whaling" station: the Hatteras fishery. Enthusiastic about not having to wait for the *Carnegie*'s return, and intrigued by the prospect of fresh and possibly even living specimens, Kellogg succeeded in arranging an exploratory trip down to North Carolina. By May 1927 he had arrived, solo, on the sandy spit of Hatteras Island with four gallons of formalin and a dissecting kit.

To his dismay, the fishermen landed nothing the whole week he was there.[121] Small pods of bottlenose cruised the rising and falling tides, and Kellogg spent two weeks looking on from the beach, and from the skiffs, as the net boats positioned themselves in the surf and arrayed their funnel-like trap, following flag signals from the watchmen ashore.[122] But the animals were skittish, the weather rough, and ultimately Kellogg was obliged to content himself with glimpses of the quarry as they approached, paused, and, seemingly spooked, veered away from the nets. If anything, this intrigued the young naturalist all the more. As far as bones were concerned, he could have his fill at Hatteras, and he spent some time picking over the

119. Kellogg to Robert Cushman Murphy, 24 February 1926, in ibid. (note that Kellogg mentions the film as evidence of the feasibility of this mode of capture).

120. See also Smithsonian Archives, RU 7170, box 4, file "Hi–Hj."

121. For accounts of the first Hatteras trip, see Kellogg to Gilbert Hinsdale, 11 May 1927, and Kellogg to Gilbert Hinsdale, 29 May 1927, both in Smithsonian Archives, RU 7170, box 4, file "Hi–Hj." See also Kellogg to Stock, 29 May 1927, Smithsonian Archives, RU 7170, box 8, file "Stock, Chester."

122. For detailed discussion of the operation of the fishery, see Smithsonian Archives, RU 7170, box 10, file "Information—Whale Press Releases," the end of the lecture "The Migration of Whales."

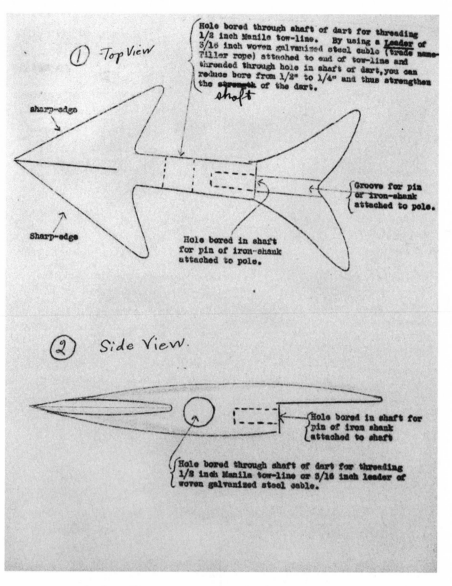

① Top View

sharp-edge

Hole bored through shaft of dart for threading 1/2 inch Manila tow-line. By using a Leader of 3/16 inch woven galvanized steel cable (trade name Tiller rope) attached to end of tow-line and threaded through hole in shaft of dart, you can reduce bore from 1/2" to 1/4" and thus strengthen the strength of the dart.

shaft

Groove for pin or iron-shank attached to pole.

Sharp-edge

Hole bored in shaft for pin of iron-shank attached to pole.

② Side View.

Hole bored in shaft for pin of iron shank attached to shaft

Hole bored through shaft of dart for threading 1/2 inch Manila tow-line or 3/16 inch leader of woven galvanized steel cable.

FIGURE 3.7 Outfitting the *Carnegie*: Diagram of porpoise tackle. (Courtesy of the Smithsonian Institution.)

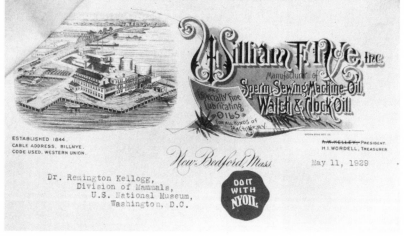

FIGURE 3.8
(*right*) *Carnegie* gear: Second diagram of porpoise tackle. (Courtesy of the Smithsonian Institution.)

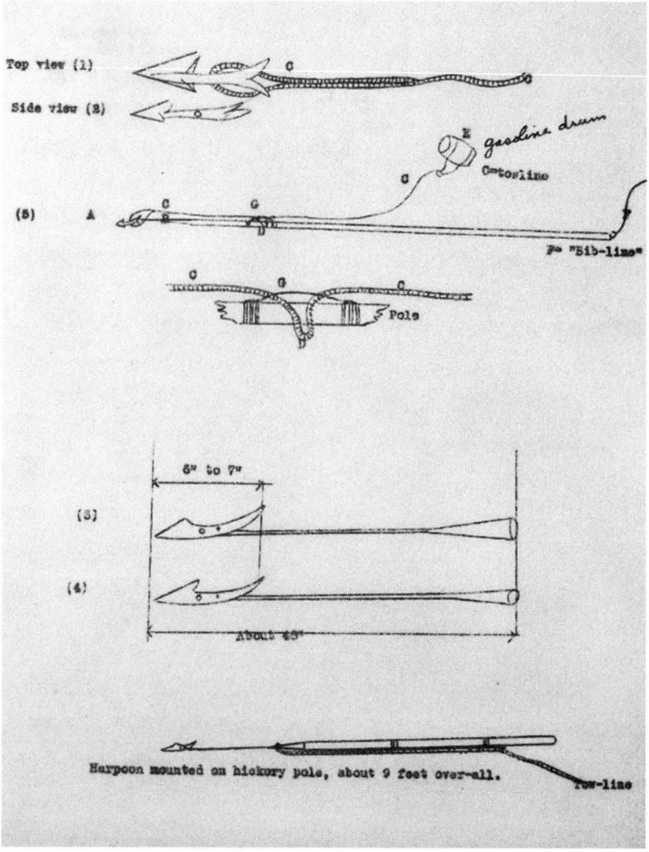

FIGURE 3.9
(*below*) Commodification of the whale: Advertisement for William F. Nye, Inc., manufacturer of lubricating oils. (Courtesy of the Smithsonian Institution.)

William F. Nye, Inc.
Manufacturer of
Sperm, Sewing Machine Oil,
Watch & Clock Oil
and Especially Fine Lubricating Oils for all kinds of Machinery

ESTABLISHED 1844.
CABLE ADDRESS, BILLNYE.
CODE USED, WESTERN UNION.

President.
H. I. WORDELL, Treasurer

New Bedford, Mass. May 11, 1929

Dr. Remington Kellogg,
Division of Mammals,
U.S. National Museum,
Washington, D.C.

DO IT WITH NYOIL

bone pile before he headed home left, having left his forwarding information and instructions about the preparation of specimens with the foreman
at the station. As fate would have it, no sooner had Kellogg departed than
the fishermen dragged a superb haul up onto the beach. A few severed
heads were en route to him almost immediately (tapped, naturally, of their
valuable jaw oil), adequately preserved for preliminary analysis.[123]

Streeter looked at several of the specimens delivered by the young researcher and was struck by the size and complexity of the brain of a bottlenose. The acoustic centers, Streeter wrote, "are relatively the largest of any
brain I am acquainted with," and it was difficult to imagine that the animals
could not hear very well indeed.[124] In fact, it seemed as if they put a great
deal more effort into hearing than into seeing, at least as far as neuroanatomy was concerned. Kellogg thought back to the spooky gray-black forms
he had seen arching through the water at Hatteras and darting away from
the skiffs, and he began to wonder. With momentum gathering among his
new group of Hopkins medical colleagues for more research on cetaceans,
Kellogg banged out a long letter to one of the principals at the Nye corporation, the treasurer Howard I. Wordell, suggesting the potential commercial
benefits of further dolphin investigations:

> I am still interested in the possibility of making some tests on the
> keenness of sight and hearing in the porpoise, for it is not only of
> great interest to us, but has a direct bearing on your methods of taking
> porpoises. Laboratory examination of the brains of these porpoises
> indicate that they have an extraordinary development of the center of
> hearing.[125]

Wordell wrote back, expressing guarded enthusiasm: while he claimed that
his company was "naturally interested in co-operating with you," he could
make no promises about just how long the Hatteras fishery would remain

123. Kellogg also secured a digestive tract, which he sent to Harry C. Raven at the
American Museum of Natural History in June 1927; this is a good example of how Kellogg used the distribution of cetacean specimens to entice a variety of zoologists and
paleontologists to consider problems of cetacean biology. See H. C. Raven to Kellogg,
22 June 1927, Smithsonian Archives, RU 7170, box 7, file "Pri–Ra."

124. George L. Streeter to Kellogg, 22 June 1927, Smithsonian Archives, RU 7170,
box 8, file "Stockt–Stre."

125. Kellogg to H. I. Wordell, 5 November 1927, Smithsonian Archives, RU 7170,
box 8, file "Wi–Wy."

in operation. With only seven animals captured in the previous month, at a cost of nearly $100 apiece, the Nye corporation was "making every effort to locate Porpoise Jaw or Black Fish Jaw Oil elsewhere" in the hopes of closing up "a losing venture."[126]

But Kellogg's ambitions for a renewed expedition continued to expand. In collaboration with Weed and others at Hopkins, he had begun to explore the possibility of bringing several live dolphins to a research facility somewhere—perhaps the federal laboratories of the Bureau of Fisheries in Beaufort, North Carolina.[127] A much broader battery of physiological experiments could be run on an animal in captivity, particularly those dealing with respiration.[128] With this possibility in mind, Kellogg wrote to Charles Haskins Townsend, the director of the New York Aquarium, which had briefly, before the war, held a number of bottlenose in a large open basin beside the docks at Battery Park in Manhattan. "Plans are now being formulated for some detailed physiological studies on living porpoises," Kellogg explained. "Several members of the Johns Hopkins University staff are cooperating in this undertaking and it will be necessary to transport one or two specimens to some point where electricity will be available."[129] How had Townsend managed to transport the animals he had brought to New York City?

Townsend sent along a copy of his detailed 1914 article, "The Porpoise in Captivity," which not only described how to move, feed, and entertain captive dolphins, but also gave a flavor for the remarkable public response to these playful and energetic additions to the aquarium. Not in recent memory had such animals been maintained in captivity before, and their leaping antics in their 7-foot-deep, 37-foot-wide pool brought droves of spectators. So strangely sociable did the creatures seem that onlookers and caretakers

126. H. I. Wordell to Kellogg, 18 November 1927, in ibid.

127. For a discussion of the importance of Beaufort and its links to Johns Hopkins, see Keith R. Benson, "American Morphology in the Late Nineteenth Century: The Biology Department at Johns Hopkins University," *Journal of the History of Biology* 18, no. 2 (1985): 163–205, at pp. 192–94.

128. Kellogg's growing interest in respiration is reflected in his correspondence with Yandell Henderson at Yale: Kellogg to Yandell Henderson, 4 November 1926, Smithsonian Archives, RU 7170, box 4, file "Hea–Hes." See also Kellogg to Francis G. Benedict, 12 November 1926, Smithsonian Archives, RU 7170, box 1, file "Bea–Bes."

129. Kellogg to Charles H. Townsend, 12 November 1927, Smithsonian Archives, RU 7170, box 8, file "To–Ty."

soon took to "regarding them as almost domesticated animals."[130] Unfortunately, dirty harbor water and uneven temperatures had cut short the experiment, and all the animals eventually died. Kellogg's inquiry, however, piqued Townsend's interest, and he asked to be kept abreast of the plans for the Hatteras expedition: if it looked like it was going to go ahead, perhaps he and some other aquarium personnel would tag along and try to secure a new stock of animals.

Spurred by this additional support, Kellogg nudged Wordell again, pointing out that the scientists would be happy to reimburse the Nye corporation at fair market value for any oil lost through dissections or the collection of specimens. Moreover, Kellogg insisted that further scientific research might be the key to reinvigorating a flagging enterprise:

> I believe that fishermen in general credit porpoises with too little intelligence. They are water-living mammals and not fish. From time to time as the opportunity afforded itself I have questioned whalers as to their experiences. They all credit cats, dogs, coyotes, and the like with some sense, but are reluctant to believe that porpoises are able to profit by past experiences. Porpoises have a highly developed brain, the center of hearing is highly developed, but sight is of secondary importance. Mammals learn by experience, and it is reasonable to suppose that porpoises do likewise.

Since some dolphins always slipped the netting operations, particular individuals would almost surely grow increasingly wary and might contribute to the skittishness of the pods in the future. Doubtless drawing on his old expertise as a trapper and field biologist with the Bureau of Biological Survey, Kellogg recommended to Wordell that the station be moved every so often, that the oars be muffled, and that the fishermen be encouraged to think more like hunters.[131] Reviewing the catch statistics at the station over the previous fifty years, Kellogg also noted population trends that were consistent with depletion:

130. Charles Haskins Townsend, "The Porpoise in Captivity," *Zoologica: N.Y. Zoological Society* 1, no. 16 (1914): 289–99, at p. 292.

131. These suggestions come from Kellogg to H. I. Wordell, 19 November 1927; H. I. Wordell to Kellogg, 4 February 1928; and Kellogg to H. I. Wordell, 22 February 1928, all in Smithsonian Archives, RU 7170, box 8, file "Wi–Wy."

We do not as yet know how long a porpoise may live, but we do know that they give birth to a single young and not oftener than once a year. If they are like the larger whales they have one young every two years. Hence it may be necessary to allow the breeding stock in that area several years to recuperate from the steady drain.[132]

Kellogg also went so far as to suggest that his research indicated that there might be better prospects for the fishery farther to the south; perhaps an oceanographic cruise of some sort along the coast would convey a clearer sense of the distribution of the dolphins in the area.

Wordell's interest was clearly heightened by the suggestion that scientific study might actually benefit the fishery, and he found time to make a visit to Washington in December 1927, where he met with Kellogg. The two men evidently hit it off—Wordell impressed with Kellogg's industriousness (he had dug out some details on brines for preserving porpoise hide, which he thought might help Wordell reduce losses on that side of his business); Kellogg pleased to make the acquaintance of someone very well connected with suppliers of whale products from around the world. Wordell threw his station open to Kellogg and the Hopkins group, which had expanded to half a dozen doctors and medical researchers. For his part, Brazier Howell, struck by a prepublication copy of Kellogg's *Quarterly Review* article, had begun to plan a book-length popular study on the evolution of whales, and he was keen to join the party. He was in the process of leaving the Bureau of Biological Survey for a faculty position in the Anatomy Department of the Johns Hopkins University School of Medicine, so he would be traveling with his new colleagues. Less than two months later, the Hatteras expedition stepped ashore in North Carolina.[133]

"SCIENCE SEEKING THE SECRETS OF THE FAST VANISHING WHALES"

Having now reviewed in some detail the shifting scientific problems and professional alliances that gave rise to the Hatteras expedition, I turn in this

132. Kellogg to H. I. Wordell, 19 November 1927, in ibid.
133. Though plans were made to seek $2,000 from the Carnegie Institution to fund the trip (Lewis Weed to Kellogg, 17 September 1927, Smithsonian Archives, RU 7170, box 8, file "Wa–Wh"), ultimately it was Hopkins that underwrote the cost of the expedition (Lewis Weed to Kellogg, 15 November 1927, in ibid.).

section to the actual visit and its aftermath. If part one of this chapter offered a survey of the scientific study of cetaceans in the United States from
World War I to 1928 (largely by means of a detailed analysis of the maturation of Remington Kellogg as a specialist in marine mammals), then this
second part will continue that survey through to the start of World War II
(again by staying close to Kellogg and his circle, a circle that began to make
widening waves in these years).

At Hatteras, Kellogg, Howell, and the others were well received, largely
on account of the enthusiastic preparations made by Wordell, who had
given his station foreman elaborate instructions on how to arrange things
for the unusual guests.[134] Moreover, evidently persuaded by Kellogg's ad-
monitions about the sophistication of cetacean hearing, Wordell had gone
so far as to insist that all the oars on the skiffs be muffled and that the
thole pins and gunwales be upholstered with sound-deadening pads so
as to increase the stealthiness of his catch boats.[135] The fishermen's view of
these unprecedented proceedings is nowhere recorded, though at the very
least they seem to have done no harm. Dolphins were spotted just outside
the surf on the visiting party's first morning on the beach, and the fifteen
crewmen had soon strung half a mile of net through the waves. A dexter-
ous closing maneuver sealed the corral successfully, and, a few hours later,
careful handling of the smaller sweep nets dragged a total of ten dolphins
up onto the sand. Looped at the tail with hemp lanyards, these animals
were dragged to the edge of the high-water mark. At this point the scientists
waved off the fishermen, who would normally have cut the throats of the
beached creatures (to avoid the risk of injury in handling them).[136]

The more ambitious plan of trying to transport the catch to the Bureau
of Fisheries laboratory at Beaufort for extended live testing had been jet-
tisoned: planning that operation looked as if it would take too long, and
with all the uncertainty about whether the Hatteras station would remain
operational, Kellogg and Weed had decided it was better to simply go immediately and see what could be done on the beach. But there seemed to
be no reason to think that the animals could not be kept alive on the sand

134. See H. I. Wordell to Kellogg, 4 February 1928, Smithsonian Archives, RU 7170,
box 8, file "Wi–Wy": among other things, this called for a meticulous cleaning of the butchering devices, presumably to ensure that specimens would not be contaminated.

135. Ibid.

136. The details in this paragraph come from "The Migration of Whales," p. 13,
Smithsonian Archives, RU 7170, box 10, file "Information—Whale Press Releases."

for some time, provided they were regularly doused with water. Townsend (who had ultimately pulled out of the whole project when provisions for live transport fell through) had warned Kellogg that they would overheat easily, and therefore every effort was made to keep them as cool as possible. Still, everyone realized that the stressed animals would not last long, so the whole group swung hastily into action—there was no guarantee that there would be any additional luck over the weeks to come, so this might be their only chance.

The first problem was how to do anything at all with approximately six hundred pounds of very angry muscle packed into a 10-foot torpedo. As Kellogg wrote, "A live porpoise can be handled about as readily as a satchel of dynamite," and this meant doing a lumbar puncture suddenly did not look like a very good idea.[137] Kellogg had given this problem some thought, however, as had the Hopkins doctors, who were considerably better acquainted with surgical interventions and anesthetic techniques. Since everyone agreed that the only way useful data on vital signs could be obtained (not even the heart rate of the animals was known) was "if the whale is helpless enough to allow one to take the necessary readings," they had arranged to try to force a wad of cotton saturated with ether down the blowhole in the hopes of putting the dolphin under.[138]

That perilous and messy endeavor failed, killing the first experimental animal much faster than any of the doctors expected. This removed the immediate danger to the scientists' life and limb, but sank their efforts to measure pulse and blood pressure.[139] Other investigations, however, remained possible: the animal was bled, to assess blood volume, and blood samples were taken in order to measure osmotic pressure, hemoglobin content, and other biochemical indices. One of the problems of greatest interest to the physiologists was how, exactly, cetaceans managed the buildup of carbon

137. Kellogg to Lewis Weed, 10 November 1927, Smithsonian Archives, RU 7170, box 8, file "Wa–Wh."

138. The quote is from Kellogg to Lewis Weed, 16 February 1927, in ibid. On the need for temperature data, see Kellogg to Yandell Henderson, 22 November 1926, Smithsonian Archives, RU 7170, box 4, file "Hea–Hes"; on the need for pulse data, A. J. Clark to Yandell Henderson, 1 December 1926, Smithsonian Archives, RU 7170, box 4, file "Hea–Hes."

139. This would prove a problem well into the 1950s; basically, dolphins do not breathe when anesthetized. The construction of an artificial respirator for an animal that takes in and expels air in sharp blasts tested medical and veterinary practitioners into the 1960s.

dioxide that would necessarily occur during long dives; control of nitrogen (which could cause the "bends" in ordinary terrestrial animals) was a related problem.[140] One hypothesis suggested that the chemical properties of odontocetes' head oil might play a role in one or both of these processes, so fresh samples of this substance were also taken.[141] All these respiratory problems were closely bound up with the larger issue of the animals' general metabolic rate, and it was hoped that more physiological data might make it possible to explain the mechanisms of "the whale's continued existence in an aquatic environment."[142]

With the second animal the group was more careful, since Langworthy, the neuroanatomist, was particularly keen to attempt some gross cortical mapping. This meant exposing the live brain in order to probe its surface. By watching the animal's responses, it would be possible to get a general idea about the functional regions of the cerebellum. It was, as Streeter had already noted from Kellogg's early specimens, a strangely large brain, one with elaborate patterns of convolution such as were generally thought to be more or less unique to human beings. Langworthy had brought a portable galvanic probe, and he readied his equipment while anesthetic was delivered—sparingly this time—to the blowhole. Then, joining forces, the men fell to the unlovely task of restraining the furiously squealing animal in order first to expose the skull and then to saw into it to expose the brain. The dolphin, Kellogg later acknowledged, never really went under, but—

140. At this point there remained a great deal of uncertainty about just how deep different species of whales could dive and how long they could remain submerged.

141. Kellogg to Yandell Henderson, 22 November 1926, Smithsonian Archives, RU 7170, box 4, file "Hea–Hes."

142. Ibid. At bottom, the effort at this stage involved collecting enough data to determine if the respiratory and circulatory metabolism of cetaceans was consistent with the general mammalian pattern: to put it another way, did whales and dolphins fit on the curves already established for other mammals of similar size? For instance, the rule of thumb was that metabolic rate varied inversely as the cube root of the body weight: see A. J. Clark to Yandell Henderson, 1 December 1926, in ibid. Calculation of oxygen consumption demanded knowledge of the surface area of the animal and the heat production per square meter: Yandell Henderson to Kellogg, 15 November 1926, in ibid. Circulation could be estimated if there was a known ratio of heart weight to body weight and if measurements had been made of the aortic cross section at its largest point: A. J. Clark to Yandell Henderson, 1 December 1926, in ibid. These are examples of the kinds of anatomical information that were necessary for understanding basic metabolic physiology, and the Hopkins group was interested in collecting this information on the dolphins.

overheated and traumatized—its resistance waned, and Langworthy managed to deliver a set of low-voltage shocks to the motor areas, generating various spasms in the animal, which was soon dead. The experiment was a modest success, as Kellogg later reported, though it seems to have blunted the group's enthusiasm for further vivisectional brain interventions—that was the only specimen submitted to this operation.[143]

In the end, the remaining animals stayed alive on the beach for almost two days before the weight of their own bodies dispatched them through asphyxiation, no doubt hastened by the ulcerating effects of the sun and general shock myopathy. Langworthy worked speedily to remove six brains before any decay could set in and to prepare them for transport back to Hopkins, where he and several colleagues would embark on a two-year project of sectioning, staining, and microscopic examination, with the aim of producing an atlas of the central nervous system of *Tursiops truncatus*.[144] The others wielded their dissection kits and cameras. Ernst Huber worked on the facial innervation, and particularly on the apparatus controlling the blowhole and its curious musculature.[145] George Wislocki

143. Kellogg to H. I. Wordell, 22 February 1928, Smithsonian Archives, RU 7170, box 8, file "Wi–Wy."

144. The atlas would eventually be published as Orthello Langworthy, "A Description of the Central Nervous System of the Porpoise (*Tursiops truncatus*)," *Journal of Comparative Neurology* 54, no. 2 (1932): 437–38, where Langworthy would provide evidence relevant to paleontologists, arguing that the frontal cerebral cortex of the porpoise looked a good deal like that of the undifferentiated cortex of the Insectivora and that the area frontalis was not unlike that of a cat. For a discussion, see Remington Kellogg, *A Review of the Archaeoceti* (Washington, DC: Carnegie Institution of Washington, 1936), 329.

145. Huber would die young (at the age of 40) in 1932, and Howell would eventually publish an expanded version of his study posthumously as Ernst Huber, "Anatomical Notes on Pinnipedia and Cetacea," *Carnegie Institution of Washington* 447 (1934): 105–36. The larger significance of this aspect of cetacean anatomy, from a paleontological point of view, lay in the large difference between the whalebone and toothed whales where this mechanism was concerned. Toothed whales have a single blowhole and a complex series of plugs and sinus cavities lining the upper airway; closure of the passage occurs by means of these plugs. Whalebone whales have a pair of tubular narial passages that are configured with spiral folds that can be twisted closed. Kellogg discusses these different arrangements and their evolutionary significance in his PhD dissertation, "The History of Whales—Their Adaptation to Life in the Water" (PhD diss., University of California, Berkeley, 1928), 193ff. Kellogg and Raven discussed the possibility of examining the anatomy of this complex region using X-rays: packing the blowhole with a suspension of bismuth, it was thought, might make it possible to secure images that

collected samples of a host of body tissues for histological examination, particularly the lungs, which—it was thought—must have unique powers to tap atmospheric air (hugely compressed by deep dives) of every last bit of oxygen. He also spent time on the reproductive system and examined the configuration of the uterus and ovaries.[146] Adolph Schulz and Kellogg, along with Howell, spent a great deal of time on skull anatomy, and Schulz would spend the slower last week of the trip—no further catches were, in fact, made—picking over the bone pile to collect no fewer than one hundred sixty *Tursiops* skulls, which he shipped back to Baltimore in order to conduct a statistical study of variation within the species.[147] Among other things, the group succeeded in getting a measurement of the spinal fluid pressure, something Weed had been very curious about, since he and Kellogg had discussed the possibility that the need to maintain compensating pressures in the cerebrospinal cavity might have played a role in reshaping

would clarify the three-dimensional form of the airways. See Kellogg to H. C. Raven, 16 July 1927, Smithsonian Archives, RU 7170, box 7, file "Pri–Ra."

146. For Wislocki's interest in lungs and the reproductive system, see Wislocki to Kellogg, 16 June 1927, Smithsonian Archives, RU 7170, box 8, file "Wi–Wy." He would publish on both of these topics: George Wislocki, "On the Structure of the Lungs of the Porpoise (*Tursiops truncatus*)," *American Journal of Anatomy* 44, no. 1 (1929): 47–77, and George Wislocki, "The Placentation of the Harbor Porpoise (*Phocoena phocoena*)," *Biology Bulletin* 65, no. 1 (1933): 80–89. Wislocki also prepared slides of a host of body tissues, including the skin and fat layers. See A. Brazier Howell, *Aquatic Mammals: Their Adaptations to Life in the Water* (Springfield, IL: C. C. Thomas, 1930), 57. On the issue of gas utilization at high pressure, see Yandell Henderson to Kellogg, 15 November 1926, Smithsonian Archives, RU 7170, box 4, file "Hea–Hes." It was understood that the chest cavity would not be able to withstand a dozen atmospheres, so the air in the lungs would be compressed during descent; this was confirmed photographically in the 1960s by Ridgway and Wood, working with a dolphin named Tuffy for the navy. See F. G. Wood and Sam H. Ridgway, "Utilization of Porpoises in the Man-in-the-Sea Program," in *An Experimental 45-Day Undersea Saturation Dive at 205 Feet*, ONR Report ACR-124 (Washington, DC: Office of Naval Research, 1967), 407–11.

147. Kellogg to Ernst Stromer, 20 January 1933, Smithsonian Archives, RU 7170, box 8, file "Stromer, Ernst von," and Kellogg to Merriam, 22 February 1926, Smithsonian Archives, RU 7170, box 6, file "Merriam, John C. (2/3)." Kellogg was interested in this problem as well, which had direct implications for paleontology: bone variation within a species (and structural changes associated with maturation) could trouble the identification of new paleontological specimens. Stromer, for instance, had argued that the fossil Andrews identified in 1906 as *Prozeuglodon atrox* was very likely an immature *Zeuglodon isis*. See Remington Kellogg, "The History of Whales—Their Adaptation to Life in the Water" (PhD diss., University of California, Berkeley, 1928), 39.

the skull structure of the cetaceans.[148] Kellogg and Howell also spent a good deal of time probing the finer points of skull anatomy, and, as Kellogg wrote later to Merriam, this yielded a number of surprises, particularly regarding the positions of the different passages that led nerves through the skull wall.[149] (Recall that Kellogg believed he could trace the gradual occlusion of some of the olfactory foramina in sequences of fossil odontocetes.) He was now able to see for himself the progressive occlusion of these same apertures in the development of *Tursiops truncatus* from youth to adulthood. It was a gratifying instance of ontogeny apparently recapitulating phylogeny right there on the beach.[150]

But it was certainly not all anatomy and physiology at Cape Hatteras: there were the pleasures of camaraderie, the excitement of the hunt, and the gusty invigorations of surf and dunes. Shortly after the party returned to their several offices and laboratories in Baltimore and Washington, Kellogg, already waxing nostalgic, jotted a note to George Wislocki: "Now that the expedition has returned safely after a venturesome sea voyage, I find it hard to get back into the accustomed daily routine."[151] To keep the spirit alive, Kellogg passed along a handful of additional references on cetacean anatomy to assist Wislocki in working up the material he brought back. For his part, Wislocki had been sufficiently inspired as to begin almost immediately to plan a trip to one of the West Coast whaling stations to pursue the same lines of research with some of the great whales.[152]

A still more immediate means of cultivating the fruits of the Hatteras

148. "I hope that you will be able to do a puncture through the occipito-atlantoid membrane so that you can measure directly in terms of millimeters of fluid the pressure of the cerebrospinal fluid. The data which you obtain from such an observation will be of the greatest service in determining the relation of pressure to the form of the dolphin's calvarium." Lewis Weed to Kellogg, 10 February 1927, Smithsonian Archives, RU 7170, box 8, file "Wa–Wh."

149. "By dissection I was able to discover and identify correctly the basicranial foramina of the skull, and found to my surprise that even these foramina have shifted from their normal position." Kellogg to Merriam, 22 February 1926, Smithsonian Archives, RU 7170, box 6, file "Merriam, John C. (2/3)."

150. He had noticed this in the bony anatomy of the *Tursiops* skull after his first Hatteras trip, and he discusses it in "The History of Whales—their Adaptation to Life in the Water" (PhD diss., University of California, Berkeley, 1928), 200–201.

151. Kellogg to Wislocki, 22 February 1928, Smithsonian Archives, RU 7170, box 6, file "Wi–Wy."

152. Kellogg to Merriam, 22 February 1926, Smithsonian Archives, RU 7170, box 6, file "Merriam, John C. (2/3)." Wislocki was also intent on arranging a return to North

expedition lay at hand: even before the group set out on the ferry bound for North Carolina, Kellogg had begun to plan a "whale symposium" for the 1928 meeting of the American Society of Mammalogists, scheduled to take place in Washington over the Easter holiday in early April, less than two months after their return.[153] Kellogg imagined this gathering as an opportunity to draw the ASM's attention to the range of scientific work now being done on these animals and to report on what he had learned over the last two years about the rapid expansion of the world's whaling industry. Building on the momentum of his *Quarterly Review* article and tapping his new circle of collaborators and colleagues, Kellogg soon had almost a dozen papers slated for the event. While the core Hopkins group was well represented on this list, not all of the presenters were scientists: Kellogg was keen to have talks by those most familiar with the situation in the whale-oil markets as well as by those who had worked alongside whalers. Wordell himself was featured prominently, and he ultimately gave a lecture to the mammalogists entitled "The Utilization of Whale Products," illustrated with slides taken by Kellogg at Hatteras.[154] Wordell had also put Kellogg in touch with several of his acquaintances in the business, and in the end Kellogg pulled in these men as well, slating a lecture by Lewis Radcliffe from the Bureau of Fisheries, entitled "The Economics of the Whaling Industry," and another paper on the Canadian fishery for beluga in the Matamek River.[155] Even as he thus reached out to those with commercial expertise, Kellogg also stirred up his community of museum-based zoologists, talking the respected ichthyologist John Treadwell Nichols at the American Museum of Natural History into proposing a paper on what he had seen of whales in Alaska and securing a commitment from Nichols's colleague Harry C. Raven, an assistant in William K. Gregory's Department of Comparative Anatomy.[156]

Carolina, conditions permitting. See Kellogg to H. I. Wordell, 14 August 1928, Smithsonian Archives, RU 7434, box 6, file "Correspondence, 1928–1940."

153. Kellogg to H. I. Wordell, 31 December 1927, Smithsonian Archives, RU 7170, box 8, file "Wi–Wy."

154. Kellogg to H. I Wordell, 19 March 1928, in ibid.

155. Kellogg to Copley Amory, 30 February 1928, Smithsonian Archives, RU 7170, box 1, file "Al–Am," and Kellogg to H. I. Wordell, 19 March 1928, Smithsonian Archives, RU 7170, box 8, file "Wi–Wy." NB: It is not absolutely clear that Radcliffe actually gave his paper.

156. On Nichols, see Kellogg to E. M. K. Geiling, 7 October 1933, Smithsonian Archives, RU 7170, box 3, file "Ga–Ge." Nichols was an important figure in the

Kellogg's timing with the invitation to Raven was particularly good, since Raven happened to be, at that moment—most unexpectedly—up to his neck in a whale. In March 1928 a baby sperm whale turned up in the Gowanus Canal, a stone's throw from the docks and warehouses at the foot of Court Street in Brooklyn. A gaggle of unemployed ironworkers and bargemen soon set upon the animal, which was found to have been wounded in the back (presumably from a collision with a vessel in the harbor). Using two lengths of cable, sharp boathooks, and assorted hardware at hand, the men succeeded in noosing the animal and dragging it into the muck, where they did it in. The *New York Times* reveled in the affair, running the news item on the front page under the sensational title "Two-Ton Whale Seized in Gowanus Canal; Puts Up Terrific Fight."[157] No sooner had word of these proceedings crossed the East River than curators from the American Museum of Natural History turned up to offer the captors (who had the carcass tied to a piling in the Clinton Basin) fifty dollars for their catch. The deal done, the "monster" (as the papers had it) was soon winched onto the back of a flatbed truck and delivered to New York's cathedral of natural history on Central Park West, where Raven dropped everything, emptied his stores of embalming fluid into the body (in the hopes of buying a few extra days), and set to work "trying to do as much dissecting as possible while it keeps."[158]

In view of this sudden bounty, Raven had no hesitation in accepting a place on the program for the whale symposium. In fact, only a few of Kellogg's invitations were declined. Glover Allen, at Harvard's Museum of Comparative Zoology, bowed out (others, he thought, could better handle the history of New England whaling), and Robert Cushman Murphy was sorry, but he was headed to the West Coast in early April and would not

American Society of Ichthyologists and Herpetologists (I assume Kellogg knew him through his Bureau of Biological Survey work on amphibians and reptiles). See H. C. Raven to Kellogg, 20 March 1928, Smithsonian Archives, RU 7170, box 7, file "Pri–Ra."

157. "Two-Ton Whale Seized in Gowanus Canal; Puts up Terrific Fight; Museum Will Get It," *New York Times*, 14 March 1928, 1.

158. H. C. Raven to Kellogg, 30 March 1928, Smithsonian Archives, RU 7170, box 7, file "Pri–Ra." The language of monstrosity in connection with whales was a commonplace in the newspapers of the period. See, for instance, the article in the *New York Herald Tribune* of 29 May 1927: "Modern Whaler Here with Oil Worth Million . . . Killed 530 Sea Monsters."

be able to attend the meeting.[159] Kellogg also spent a good deal of time trying—apparently without success—to get his hands on some film footage depicting live cetaceans in the sea. Murphy had some still images, but no moving pictures, and Townsend (who, Kellogg thought, had to have something on hand) was not overly forthcoming.[160]

A late addition to the program proved to be, in many ways, the most dramatic speaker. Waldo L. Schmitt, curator of the Division of Marine Invertebrates at the National Museum (and a fellow product of the University of California, Berkeley, who, like Kellogg, had experience working with the Carnegie Institution), had just returned to Washington after almost two years of travel and collecting on the east and west coasts of South America, an expedition sponsored by the Walter Rathbone Bacon Traveling Scholarship.[161] In the course of his research into the decapod crustaceans (crabs, shrimp, lobsters), Schmitt had taken advantage of connections in Argentina (home of the grandfather of all Antarctic whaling companies, the Compañía Argentina de Pesca) to tag along on a voyage through the Dependencies of the Falkland Islands, then, of course, the epicenter of the global whaling industry. For a student of decapods, the chance to see the staggering phenomenon of Antarctic krill blooms had to have been a draw. But what Schmitt found—besides the ice, wind, and barren geography of places like Deception Island—was a cyclopean slaughterhouse. It was in this period that the putrefying offal discarded from the factory vessels into the volcanic crater of Deception Harbor had piled up to depths that made the rocky shores impassable; thermal vents kept this stew well above freezing, creating a cauldron of death and decay. It was an encounter no naturalist could readily forget, and the scope of the industry—which was processing thousands of whales each season and culling enormous profits—made a

159. Kellogg to Glover M. Allen, 13 January 1928, Smithsonian Archives, RU 7170, box 1, file "Allen, Glover M."; Robert Cushman Murphy to Kellogg, 20 March 1928, Smithsonian Archives, RU 7170, box 6, file "Mor–My."

160. Robert Cushman Murphy to Kellogg, 20 March 1928, Smithsonian Archives, RU 7170, box 6, file "Mor–My." On Kellogg's approach to Townsend, see Kellogg to Glover M. Allen, 13 January 1928, Smithsonian Archives, RU 7170, box 1, file "Allen, Glover M."

161. Schmitt's MA was from Berkeley; he had completed his PhD at George Washington University in 1922. In the early 1920s he spent several summers at the Carnegie Marine Laboratory at the Dry Tortugas, Florida. See the finding aid to Smithsonian RU 7231 for more details on Schmitt's biography.

deep impression on Schmitt, who agreed to talk about modern Antarctic whaling at Kellogg's symposium.[162]

When the American Society of Mammalogists finally convened in early April of 1928, Kellogg had succeeded in putting together a diverse roster of individuals, each with some kind of expertise in cetaceans. The tenth annual gathering of the ASM became, in effect, a whale meeting: not only were the whale talks picked up by a daily news service and distributed on the wire (the idea that whales could be worth $10,000 a piece to the modern industry made a striking headline[163]), but Kellogg and Howell succeeded in galvanizing the ASM itself into taking action on the issue of whale conservation. The prime mover behind this initiative appears again to have been Kellogg himself, who, with the support of William B. Bell (a charter member of the ASM), proposed a resolution to the members that read as follows:

Whereas; it has been disclosed that the operations for the capture of whales and utilization of whale products as now conducted result in much economic waste and such slaughter of the animals as seriously endangers the existing supply of these mammals and the conservation of an important natural resource. Now, therefore, be it Resolved: That the American Society of Mammalogists at its tenth annual meeting recommends that steps be taken looking to the appointment of an international commission to investigate conditions and facts regarding the life history, habits, and commercial utilization of whales, and to make recommendations to the governments concerned as to procedure or regulations which will safeguard and perpetuate this resource for the benefit of all parties interested.[164]

162. It is worth noting that Schmitt was no stranger to commercial fisheries: in 1914 he was employed by the Bureau of Fisheries in the Halibut Surveys of Washington and Oregon, during which he spent time on the old ocean research vessel *Albatross*.

163. Daily Science News Bulletin, "Whales in Antarctic bring $10,000 a piece," Science Service, Washington, DC, no. 367, F Sheet 3, 7 April 1928; Smithsonian Archives, RU 7170, box 10, file "Information—Whale Press Releases."

164. This is the text of the resolution as it appeared in the "Comment and News" section of the *Journal of Mammalogy* (vol. 11, no. 1 [February 1930]: 100, Smithsonian Archives, RU 7170, box 10); but compare the manuscript version in RU 7434, box 8, which reads (here giving main variants): "Whereas: Information presented at the 1928 annual meeting of the A.S.M. disclosed the fact that . . . result in [economic omitted] waste for valuable products and . . . [mammals omitted; for 'conservation' read 'perpetuation' . . .]"

It was this resolution, passed by the ASM, that would become the chartering document for the Council for the Conservation of Whales, organized under the auspices of the ASM by Kellogg and Howell before the end of the year. It was an organization that, as I will show, played a distinctive and significant role agitating for the protection of whales in the interwar period, even as it served to leverage the professional career of Remington Kellogg, positioning him to become a scientific diplomat in the expanding arena of international regulation and conservation politics after World War II.

Since the remainder of this chapter will deal largely with the CCW, its membership, and its campaign to protect cetaceans, I would like to pause here to situate this unique body with respect to a number of broader trends in American conservation in this period. While the CCW was exceptional in many respects (it focused on a taxon hitherto of little general interest among Americans, and without any obvious "constituency" of defenders; it aimed from the outset at international action; and it succeeded in unifying diverse threads, drawing together a community of rod-and-gun sportsmen, rational-exploitation-oriented conservationists, sentimental preservationists, white-coated scientists, and industry executives), it also emerged out of, and drew on, earlier efforts to protect the natural environment and its denizens (it made use of veterans of Progressive Era campaigns to protect forests and wildlife, and it gathered momentum from judicious moves on the wider topography of nature politics in the United States in the late 1920s). Some consideration of these contextual matters will help frame the work of the CCW in the critical years between 1929 and 1936.[165]

To begin with, then, it will be worth zeroing in on a small but telling

---

At the end, the phrase "for the benefit of all parties interested" has been added in pencil on the draft text. It is worth pausing on these changes—particularly the insertion of the language of economics and conservation—for what they suggest about the preoccupations of those putting forward this recommendation (see further discussion below). It is tempting to suggest that Howell drafted this resolution, but the best evidence against his playing a significant role in it comes in a letter from Howell to Kellogg, 20 November 1928, Smithsonian Archives, RU 7170, box 4, file "Howell, A. Brazier, 1920–1929 and undated (3/3)," in which Howell reveals that "I have never received a copy, have never seen a copy, and don't believe the damned thing was ever passed."

165. For a brief overview of Progressive Era conservation, particularly as far as policy matters are concerned, see Richard N. L. Andrews, *Managing the Environment, Managing Ourselves: A History of American Environmental Policy* (New Haven, CT: Yale University Press, 1999), chap. 8. The classic interpretation remains Samuel P. Hays,

archival clue: Kellogg's personal working draft of the ASM's whale resolution was typed out on the back of a blank sheet of stationery bearing
the crest and charter of the Cooper Ornithological Club, an organization
dedicated (as the letterhead announced) to "the conservation of Birds and
Wildlife in general, for the sake of the future."[166] This juxtaposition—of
whale-concern and bird-concern—was not merely fortuitous. For starters,
it surely indicates that Howell and Kellogg hatched the whale conservation
plan together (Howell was an executive officer of the club). But reaching
beyond that merely biographical observation, the recto and verso of this
sheet capture a more general juxtaposition of interests. The protection of
birds had proved to be the most successful aspect of the early twentieth-
century Anglo-American movement for stronger legal measures pertain-
ing to the protection of wildlife.[167] It is possible to debate the relative contributions made to these efforts by the anticruelty-oriented contingent of
humane associations (one the one hand) and the shotgun-toting sportsmen
anxious to guarantee good shooting (on the other).[168] Putting that interesting question aside, however, the larger significance of bird conservation ef-

---

*Conservation and the Gospel of Efficiency: The Progressive Conservation Movement,
1890–1920* (Cambridge, MA: Harvard University Press, 1959).

166. Smithsonian Institution, James Mead Files, Marine Mammal Department,
National Museum of Natural History, file "Conservation—Whales."

167. For an overview of this story, see Robin W. Doughty, *Feather Fashions and Bird
Preservation: A Study in Nature Protection* (Berkeley: University of California Press,
1975), and Mark V. Barrow, *A Passion for Birds: American Ornithology after Audubon*
(Princeton, NJ: Princeton University Press, 1998). Also very useful are chaps. 11 and 12
of James B. Trefethen and Peter Corbin, *An American Crusade for Wildlife* (New York:
Winchester Press, 1975). It is interesting to note how much overlap there is between
the group Trefethen and Corbin discuss on pp. 151ff. (advocates for the McLean Bill)
and the group that becomes the core of the CCW. See also Philip J. Pauly, *Biologists and
the Promise of American Life: From Meriwether Lewis to Alfred Kinsey* (Princeton, NJ:
Princeton University Press, 2000).

168. For an emphasis on the latter, see James B. Trefethen and Peter Corbin, *An
American Crusade for Wildlife* (New York: Winchester Press, 1975), and John F. Reiger,
American Sportsmen and the Origins of Conservation (New York: Winchester Press,
1975). For a more recent critical appraisal of the question, see Thomas R. Dunlap,
Saving America's Wildlife: Ecology and the American Mind, 1850–1990 (Princeton, NJ:
Princeton University Press, 1988); chap. 6 is particularly useful on the importance of
the humane societies. Also relevant on this topic is Anita Guerrini, *Experimenting
with Humans and Animals: From Galen to Animal Rights* (Baltimore: Johns Hopkins
University Press, 2003).

forts for the historian of science lies in the fact that very particular features of the lives of migratory birds—and greater knowledge of these habits— facilitated novel legal measures for bird protection in the 1910s and 1920s.[169] Important in this respect is the 1918 passage by Congress of the Migratory Bird Treaty Act, which buttressed the newly erected legal architecture that sheltered wide-ranging wildlife under federal, rather than state, law.[170] In addition, the fact that migratory birds had well-marked breeding grounds gave rise to distinctive conservation strategies, most importantly the use of "sanctuaries" to protect the reproductive cycle and ensure annual replenishment of the stock.[171]

Significantly, whales would raise these same issues: long migrations through multiple jurisdictions (necessitating international regulation) and identifiable breeding grounds (which called for sanctuaries). The prominent individuals that Kellogg and Howell would draw into the activities of the Council for the Conservation of Whales—men like George Bird Grinnell (founder, with Theodore Roosevelt, of the Boone and Crockett Club), T. Gilbert Pearson (longtime leader of the National Association of Audubon Societies), and Theodore Sherman Palmer (who had drafted the original US-Canadian bird treaty)—were the decorated heroes of a quarter-century of highly public campaigning on behalf of North American birds, and in connection with naturalists like Howell, they had helped parlay the

169. For an interesting bit of evidence for the perceived importance of these legal developments, see the dedication in William Temple Hornaday, *Thirty Years War for Wild Life* (New York: Arno, 1970 [1931]), which reads, "This volume is dedicated to the Congress of the United States as a small token of appreciation of its generous services to wild life during the decade from 1920 to 1930, in new legislation to provide game sanctuaries, and to reduce excessive killing privileges."

170. The act gave force to the US-Canada treaty on these birds and in doing so strengthened the Weeks-McLean Act of 1913, which (on the basis of the commerce clause) gave the secretary of agriculture the power to regulate the hunting of highly migratory species. See Thomas R. Dunlap, *Saving America's Wildlife: Ecology and the American Mind, 1850–1990* (Princeton, NJ: Princeton University Press, 1988), 37. See also Kurkpatrick Dorsey, *The Dawn of Conservation Diplomacy: U.S.-Canadian Wildlife Protection Treaties in the Progressive Era* (Seattle: University of Washington Press, 1998).

171. There were, of course, precedents for this strategy, such as hunting preserves. For a discussion, see William Temple Hornaday, *Thirty Years War for Wild Life* (New York: Arno, 1970 [1931]), chap. 30. For an introduction to the history of reserves, see W. M. Adams, *Against Extinction: The Story of Conservation* (London: Earthscan, 2004), chap. 4.

details of avian natural history into particular legislative actions. In a number of important ways, the emergent character of the whale problem in the United States would be assimilated, in its basic structure, to the migratory bird problem of the previous decade, and alliances formed in that earlier struggle would be called on again.

It is important to emphasize how close Kellogg himself was to these bird-related activities and the degree to which they shaped his conception of what he took to be an increasingly imminent crisis in world whale stocks. Not only had Kellogg made amateur contributions to such ornithophilic publications as *Bird Lore* in his youthful days as a rambling collector-naturalist, but more importantly, later, while at the Food Habits Research Division of the Bureau of Biological Survey, he contributed substantially to Herbert L. Stoddard's important work on the natural history of the bobwhite quail. Kellogg's contribution involved the discovery that marsh hawks were not significant quail predators, but rather feasted on cotton rats, whose depredations on quail nests were well known—all of which led to his (heretical) declaration that "the marsh hawk is probably the best benefactor the quail has in the region."[172]

All that may seem a long way from whale conservation, but work like this placed Kellogg in the thick of the debates that would be central to the emerging American program of conservation-oriented ecology. Stoddard's quail study—which originated as a series of recommendations about private game preserves and the maintenance of sanctuaries—can be understood as nothing less than a watershed in this area: as Dunlap argues in his extended essay "Ecology and the American Mind," the book set a new benchmark for a "rigorous science of game management." Perhaps even more importantly (as far as its long-term historical significance is concerned), Stoddard's *The Bobwhite Quail: Its Habits, Preservation, and Increase* came early into the hands of one Aldo Leopold, then an adjunct instructor in game management at the University of Wisconsin, who set out to perform a Stoddardesque analysis on the quail population in the farmlands north of Madison—work that would inform his increasingly moralized ideas about the interconnectedness of the biotic pyramid and the

172. Herbert L. Stoddard, "Report on Cooperative Quail Investigation: 1925–1926 (With Preliminary Recommendations for the Development of Quail Preserves)" (Washington, DC: Published by the Committee Representing the Quail Study Fund for Southern Georgia and Northern Florida, 1926), 39.

need for a comprehensive and cautious "land ethic."[173] The unfolding influence of this sort of intensive, quantitative analysis of natural communities forms a significant portion of the history of ecology in the United States. That story is well beyond the scope of this book, but it would be impossible to understand the kinds of concerns that motivated Kellogg's push for whale conservation without recognizing that his work at the Bureau of Biological Survey in the 1920s exposed him to novel ways in which biology could service wildlife management. In keeping with this vision, the Council for the Conservation of Whales aimed to place a science of cetaceans at the heart of a movement to protect marine mammals.

At the same time, their Bureau of Biological Survey work put both Kellogg and Howell quite close to the center of the sharp conflicts engendered by these new ecological ideas and management approaches. Perhaps the most important of these—the controversy surrounding the bureau's campaign against "predatory and noxious" animals—came to a head in the late 1920s and largely accounts for both Kellogg's and Howell's departure from the bureau in 1928. Because this issue had repercussions for the emerging whale conservation program, it will be worth reviewing briefly the outlines of this notable episode.[174]

The federal government had been active in the control of predatory species since before World War I, expanding a well-established tradition of private and state bounty hunters who pursued "pest" species (like coyotes, wolves, and mountain lions) in the service of cattlemen, sheep ranchers, and other western land users. By the mid-1920s these earlier ad hoc efforts had expanded into full-scale "warfare" against a wide range of "vermin," including nearly all the large carnivores. "Shooting, trapping,

173. See Thomas R. Dunlap, *Saving America's Wildlife: Ecology and the American Mind, 1850–1990* (Princeton, NJ: Princeton University Press, 1988), 71–74; Leopold's work would of course take shape in *A Sand County Almanac, and Sketches Here and There* (New York: Oxford University Press, 1987 [1949]), with its powerful ideas of a land ethic. See Susan Flader, *Thinking Like a Mountain: Aldo Leopold and the Evolution of an Ecological Attitude toward Deer, Wolves, and Forests* (Columbia: University of Missouri Press, 1974).

174. The episode is dealt with in Thomas R. Dunlap, *Saving America's Wildlife: Ecology and the American Mind, 1850–1990* (Princeton, NJ: Princeton University Press, 1988), chap. 4, "Worthless Wildlife"; for the quoted phrase, see Jenks Cameron, *The Bureau of Biological Survey: Its History, Activities and Organization* (Baltimore: Johns Hopkins University Press, 1929), 173.

killing in the den in puphood, and poisoning" were avidly pursued by an extensive network of Bureau of Biological Survey personnel, and a considerable bureaucratic structure had arisen to provide logistical support for these operations, to oversee the record keeping, and to maintain relations with the various interested parties—politicians, stockmen's unions, farmers—who had a stake in the activities.[175] These stakes were often more than notional: the well-organized sheepmen of Idaho were not alone in assessing themselves a small fee per head in 1925, which permitted them to contribute no less than $30,000 to the bureau in support of its efforts (the Idaho legislature, divided about the merits of such federal interventions, had been less forthcoming).[176] The increasingly cozy relationships between ranchers and bureau personnel engendered by these kinds of fee-for-service arrangements began to raise flags among the bureau's scientific staff in this period, who started to protest that conflicts of interest were inevitable and that unsound management practices were too common.

The American Society of Mammalogists—and their house organ, the *Journal of Mammalogy*—became an important forum for these debates, which were particularly contentious within this community, nearly all of whose members had some experience in the bureau.[177] As Howell put it in a letter to Kellogg, the bureau could not "combine destruction and conservation without piling into the ditch," and now it was "bleating for help." While it was admittedly sad that the many biologists employed by the survey might be "out of jobs" if the organization came under the axe, something had to be done to get a grip on the "racketeers now running control operations."[178] Howell went on to suggest maliciously that the bureau might consider creating some new divisions that better spelled out its mission,

175. Jenks Cameron, *The Bureau of Biological Survey: Its History, Activities and Organization* (Baltimore: Johns Hopkins University Press, 1929), 173.

176. Ibid., p. 174.

177. Dunlap points out that every president of the ASM between 1919 and 1954 (except one) either had been in the bureau or had protested publicly against some aspect of the "wildlife discouragement" program (the phrase is Cameron's). See Thomas R. Dunlap, *Saving America's Wildlife: Ecology and the American Mind, 1850–1990* (Princeton, NJ: Princeton University Press, 1988), 50. There were, for instance, angry resignations during the most contentious period.

178. Howell to Kellogg, 25 January 1934, Smithsonian Archives, RU 7170, box 4, file "Howell, A. Brazier, 1932–1943 (1/3)."

such as the "Conservation and Meat Department."[179] For his part Kellogg shared Howell's dismay, grimly referring to a gathering of bureau biologists as the "meeting of the . . . herpeticides."[180]

As these private comments suggest, many of the country's life scientists were increasingly discontented with the direction of federal wildlife policy in the period and with the role of biologists in wildlife management. Polemics were joined. Perhaps the most heated of these exchanges came in 1925, in a pair of position papers published in the *Journal*: Lee R. Dice's "The Scientific Value of Predatory Mammals" and a reply by E. A. Goldman, "The Predatory Mammal Problem and the Balance of Nature." These essays characterized the opposing positions staked out in the mid-twenties on the issue. Goldman clearly articulated the view of backers of the control programs, arguing that predatory mammals had caused upward of twenty million dollars' worth of damage to livestock in the western states and that rodents were responsible for devastating depredations on farmers. While one might fantasize about an unspoiled "balance of nature," Goldman suggested, "since civilized man has hopelessly overturned the balance by creating artificial conditions and contacts throughout his sphere of influence, practical considerations demand that he assume effective control of wild life everywhere." This meant that "rodents and other animals, as well as the carnivores, must be checked wherever they become too numerous or too injurious to human interests." In conclusion, he acknowledged that "as nature lovers we are loath to contemplate the destruction of any species, but as practical conservationists we are forced by the records to decide against such predatory animals as mountain lions, wolves, and coyotes" for the simple reason that game and livestock had to be protected.[181]

The contrary view, as outlined by Dice, emphasized that the bureau's program appeared to aim at nothing less than the "extermination" of dozens of North American mammals, a project that could not be looked on with equanimity by a professional mammalogist. What is perhaps most

---

179. Howell to Kellogg, 27 November (circa 1926), Smithsonian Archives, RU 7170, box 4, file "Howell, A. Brazier, 1920–1929 and undated (3 / 3)."

180. Kellogg to Arthur H. Henn, 30 April 1929, Smithsonian Archives, RU 7434, box 6, file "Correspndence, 1928–1940."

181. He stated explicitly that it was important not to "alienate the livestock industry." Quotes from E. A. Goldman, "The Predatory Mammal Problem and the Balance of Nature," *Journal of Mammalogy* 6, no. 1 (February 1925): 28–33, at pp. 31, 33.

striking in this editorial is the degree to which Dice (and others in his camp, like Howell and Kellogg) did not merely emphasize scientific uncertainty about the long-term consequences of this kind of meddling (Would wiping out coyotes lead to an explosion of jackrabbits? There was circumstantial evidence that it could), but placed an even stronger emphasis on the grave danger of losing "to science" any creature not yet fully understood. As Dice put it, "There is no substitute in scientific work for the living animal and the fresh specimen, and every species exterminated marks a decided loss for the scientific world." Because comparative anatomy and embryology "have an ultimate bearing on our elucidation of the facts of evolution . . . the loss of any subspecies will make more difficult, or perhaps impossible, a correct interpretation of the evidence presented by that series of forms." It was therefore urgent to remember that "our knowledge of neither anatomy, embryology, histology, nor cytology of any North American species of mammal can be considered complete at the present time," and thus "the extermination of any species, predatory or not, in any faunal district, is a serious loss to science."[182]

In building the case for whale conservation, Kellogg and Howell would borrow a page from this argument, insisting again and again on "the value of whales to science" in their press releases and gesturing at the ways that the study of these organisms would shed light on evolution, dive physiology, and a host of other medical and biological problems.[183] The use of science as itself a rationale for (and not merely a means to) proper conservation appears to mark the work of the Council for the Conservation of Whales in an exceptional way.

Dice's emphasis on the need for further knowledge of mammals in particular drew tacitly on a more general contemporary anxiety about the fate of warm-blooded creatures in the modern age. It has been largely forgotten that the early 1920s saw a burst of acute pessimism about the fate of mammals around the world. Henry Fairfield Osborn and Harold Elmer Anthony (both of whom would be tapped by the CCW) precipitated much of the interest in this issue with a pair of doomsday warnings, both of which appeared in the autumn of 1922: "Can We Save the Mammals?" published in *Natural History*, and "Close of the Age of Mammals," published in the

182. Quotes from Lee R. Dice, "The Scientific Value of Predatory Mammals," *Journal of Mammalogy* 6, no. 1 (February 1925): 25–27.

183. See n. 1 above.

Journal of Mammalogy.[184] Osborn and Anthony were explicit in seeking to mobilize the same forces that had been so successful in the conservation of birds, but they advocated redirecting those forces toward a new object of concern. As they put it, "The saving of the birds renders us hopeful that certain of the finest kinds of mammals, including those which are nearing extinction from natural causes, as well as many of the fur bearers that have been reduced in numbers through persistent persecution, can still be saved."[185]

The notion that there were considerable numbers of mammals nearing extinction in the early twentieth century due to "natural causes" will probably strike the modern reader as odd, but this idea was central to the mammal clamor of the day. For Osborn, Anthony, and others, deepening familiarity with the paleontology of the Pliocene—with its gargantuan woolly mammoths, massively racked Irish elk, and giant sloths—had led to the unsettling realization that "mammalian perfection reached its climax at the close of Pliocene time, about 400,000 years ago."[186] It had, in some sense, been downhill ever since.[187] Just as the great Age of Reptiles had passed, so too, the mammals would eventually become sorry remnants of their most glorious period, and they were, it could be argued, already in decline. Debates around this issue intersected with concerns about human activities in general (agriculture, expansion, civilization) and the work of the Bureau

184. See Henry Fairfield Osborn and Harold Elmer Anthony, "Can We Save the Mammals?" *Natural History* 22, no. 5 (September–October 1922): 388–405, and Henry Fairfield Osborn and Harold Elmer Anthony, "Close of the Age of Mammals," *Journal of Mammalogy* 3, no. 4 (November 1922): 219–37.

185. Henry Fairfield Osborn and Harold Elmer Anthony, "Can We Save the Mammals?" *Natural History* 22, no. 5 (September–October 1922): 388–405, at p. 389. Compare with Henry Fairfield Osborn and Harold Elmer Anthony, "Close of the Age of Mammals," *Journal of Mammalogy* 3, no. 4 (November 1922): 219–37, at p. 225: "The final cause of the Close of the Age of Mammals can be arrested only through the creation of sound sentiment and education of the children and of women, in the same manner in which the National Association of Audubon Societies has arrested the destruction of birds."

186. Henry Fairfield Osborn and Harold Elmer Anthony, "Close of the Age of Mammals," *Journal of Mammalogy* 3, no. 4 (November 1922): 219–37, at p. 219.

187. The rapid depletion of the bison of the Great Plains featured prominently in these discussions. This story has been told several times: see, for instance, James B. Trefethen and Peter Corbin, *An American Crusade for Wildlife* (New York: Winchester Press, 1975), chap. 11, and Andrew C. Isenberg, *The Destruction of the Bison: An Environmental History, 1750–1920* (Cambridge: Cambridge University Press, 2000).

of Biological Survey in particular. In fact, E. W. Nelson, the bureau's chief through 1927, served as a discussant at one conference dedicated to the discussion of the "Close of the Age of Mammals," at which he emphasized the grave difficulties confronting any effort to preserve wild animals in the American West, given the competing demands for access to grazing and farming lands.[188] For W. T. Hornaday, the challenges facing Africa and Asia, where there was more left to save, were perhaps even greater.[189]

While this mammal pessimism of the twenties generally focused on terrestrial quadrupeds (which could be, and with increasing frequency were, taxidermically posed in heroic attitudes surmounting dioramic promontories), its prophets seldom failed to make at least a mention of the marine mammals. The speedy destruction of fur seals was, as discussed above, well known, but whales too made a cameo appearance in these briefs. Osborn, familiar with the difficulties his museum had faced trying to find the skeleton of a bowhead, went so far as to spell out that "the California gray whale is nearly extinct. The Right whale is in danger of extermination, and the disappearance of the bowhead whale is also threatened."[190] Moreover, since the American Museum had, in Roy Chapman Andrews, one of the few naturalists to familiarize himself with the modern whaling industry before World War I, Osborn even understood the ignominious cause of this

188. See Nelson's discussion of Osborn and Anthony in Henry Fairfield Osborn and Harold Elmer Anthony, "Close of the Age of Mammals," *Journal of Mammalogy* 3, no. 4 (November 1922): 219–37, at pp. 235–36.

189. See Hornaday's discussion of Osborn and Anthony in Henry Fairfield Osborn and Harold Elmer Anthony, "Close of the Age of Mammals," *Journal of Mammalogy* 3, no. 4 (November 1922): 219–37, at pp. 231ff. Kellogg's view of conservation in these regions was marked by his own casual (if private) racism, as evidenced in this aside to Howell concerning a proposal for international wildlife protection (forwarded to them by Aldo Leopold): "It seems to me that this . . . outfit could more profitably direct their attentions to these United States of ours and let a few zulus, hindoos, and fuzzies fry their own fat." Kellogg to Howell, 13 August 1930, Smithsonian Archives, RU 7170, box 4, file "Howell, A. Brazier, 1930–1931 (2 / 3)." It is impossible to pass over the fact that there are several very dismaying folders in the Kellogg papers containing anti–civil rights doggerel and brutal low humor.

190. See Henry Fairfield Osborn and Harold Elmer Anthony, "Can We Save the Mammals?" *Natural History* 22, no. 5 (September–October 1922): 388–405, at p. 405. Rumor had it that Frederic A. Lucas had offered ship captains involved in arctic trade no less than $5,000 for a complete specimen, without success. See Kellogg to Merriam, 19 May 1926, Smithsonian Archives, RU 7170, box 6, file "Merriam, John C. (2 / 3)."

decline: "The *coup de grace* to marine life has been given," he explained, "by the fertilizer industry" which was "rapidly eliminating the Cetacea."[191]

Wide-ranging and shared interests—birds, rodents, marine mammals—put Kellogg and Howell at the crossroads of these different conservation controversies of the late 1920s. As the great Kern County mouse outbreak of 1927 spread from California throughout the country (to chiming protests that the plague should be laid at the feet of government coyote killers), as the Kaibab National Forest was picked clean by its starving, predatorless deer population (leading the chief of the bureau privately to declare the region a disaster area), and as Aldo Leopold shot his last she-wolf somewhere in the Gila headwaters (committing the "sin" whose redemption would enter the founding mythology of the modern environmental movement in the United States), Brazier Howell was working over his wealthy friends in Pasadena, "trying to get a million dollars or so . . . for whale research," and collecting dirt on the failures of the predatory mammal program.[192]

The deepening crisis at the Bureau of Biological Survey in this period (which involved a disruptive changeover in leadership and ever more vocal public concern about wildlife control programs) drove a growing number of biologists to turn against the organization and to denounce the relentless killing—in the view of many, the language of "conservation" had too easily come to service slaughter for the sake of commercial interests.[193] Kellogg

191. Henry Fairfield Osborn and Harold Elmer Anthony, "Close of the Age of Mammals," *Journal of Mammalogy* 3, no. 4 (November 1922): 219–37, at p. 220.

192. On the Kern County mouse outbreak, see Thomas R. Dunlap, *Saving America's Wildlife: Ecology and the American Mind, 1850–1990* (Princeton, NJ: Princeton University Press, 1988), 52–53, 66; on Kaibab there is a large literature, but the reference to Nelson here comes from Dunlap, p. 69; on Leopold's "sin against the wolves" (his phrase), see Susan Flader, *Thinking Like a Mountain: Aldo Leopold and the Evolution of an Ecological Attitude toward Deer, Wolves, and Forests* (Columbia: University of Missouri Press, 1974), 102 (and see also her introduction), and Aldo Leopold, *A Sand County Almanac, and Sketches Here and There* (New York: Oxford University Press, 1987 [1949]), 129–30. On Howell's pursuit of money for whale work, see Kellogg to Lewis Weed, 20 September 1927, Smithsonian Archives, RU 7170, box 8, file "Wa–Wh"; and for Howell on the predatory mammal program, see Howell to Kellogg, 15 July (year unknown, circa 1929), Smithsonian Archives, RU 7170, box 10, file "Information—Whale Press Releases": "Have been getting some good dope on the pred. mam. situation and should be able to induce quite a bit of insomnia in Wash. and vicinity next fall."

193. For a sense of this view, see the resolution passed by the ASM and others in 1930: Howell to Paul G. Redington, 14 April 1930, Smithsonian Archives, RU 7170, box 4, file "Howell, A. Brazier, 1930–1931 (2/3)." On the general crisis at the Bureau of Biological

and Howell, important agitators of this discontent, thus drew, as the Council for the Conservation of Whales took shape, on a widening community of naturalists and life scientists willing to express their frustration with the direction of conservation policies in the United States—a community increasingly dismayed with the scope of anthropogenic change in natural environments, worried about the mammals of the world, and particularly ill at ease with the extent to which biologists had permitted themselves to become complicit in the killing. For this group, the issue of federal mismanagement of terrestrial wildlife and the need for better regulatory protections for whales frequently came up together and reflected a more general concern for the use of good science in preserving besieged mammals from extermination. Wrote one supportive midwestern naturalist to Kellogg in a ranging letter that complained bitterly about the bureau, "Have they no respect for the Creator of the Universe? . . . It's good-bye whales, good-bye orangs, good-bye coyotes, and good-bye everything except damned old Homo insapiens."[194] It would be an uphill climb, he warned, "making the world safe for mammals."[195] Others agreed. Complaining of the emerging statistics on the Antarctic whaling industry, another supporter, Charles C. Adams, the director of the New York State Museum at Albany, asked why it was that no one seemed to be able to "face the problem of the *future!*" He went on to say, "I have no sympathy with this blind destructive policy. I have been giving some attention to the preservation of our land predators," since "the encroachments of agriculture, threaten the extermination of many of our fine species." He too volunteered his help on the whale work.[196]

It is telling that no sooner had Kellogg and Howell steered the whale resolution through the American Society of Mammalogists than they sent a copy along to the Bureau of Biological Survey with a letter pressing for its "cooperation and approval" in their "efforts to promote a more satisfactory method for conserving and protecting the future supply of whales and other pelagic mammals." The letter went on to suggest that, "in view

Survey in this period, see Keir Brooks Sterling, "Builders of the U.S. Biological Survey, 1885–1930," *Journal of Forest History* 33 (October 1989): 180–87, esp. p. 187.

194. This was Marcus W. Lyon Jr., treasurer of the Indiana Academy of Sciences. Marcus W. Lyon to Kellogg, 3 August 1931, Smithsonian Archives, RU 7170, box 5, file "Lyn–Lyc."

195. Marcus W. Lyon to Kellogg, 19 February 1932, in ibid.

196. Charles C. Adams to L. G. Romell, 18 July 1929, Smithsonian Archives, RU 7170, box 1, file "Charles C. Adams, 1926–1944."

of the extraordinary slaughter of the larger whales in southern waters at the present time and the rapid depletion of the industry along the North American coasts," it would be appropriate for the United States to "take the initiative in promoting an international conference with a view to arranging an agreement of some sort."[197] The embattled new bureau chief answered by proxy, suggesting that more information would help advance the cause.[198] In the end, Kellogg and Howell would find a more sympathetic ear in Lewis Radcliffe, the acting commissioner of the Bureau of Fisheries in the Department of Commerce, and as momentum gathered rapidly toward regulatory action in the early 1930s, Howell would work to ensure that the whale conservation issue was kept away from the Bureau of Biological Survey, in which he and his conservation-oriented colleagues had come to have little faith.[199]

·   ·   ·   ·   ·   ·   ·   ·   ·   ·   ·

I have sketched the struggles over the Bureau of Biological Survey's conservation policies in the 1920s because they provided the most immediate context for the emergence of the Council for the Conservation of Whales. As I have argued above, disputes over wildlife control programs, anxieties about the fate of mammals in North America (and the world), the emergence of new legal means (national, international) for the protection of migratory animals, and the recent successes in campaigns for bird preservation all informed Kellogg and Howell as they thought up, planned out, and recruited for a campaign—led by scientists, driven by scientific concerns, and

197. Text of the letter from A. Brazier Howell to Commissioner, Bureau of Fisheries, 17 April 1928, Smithsonian Institution, James Mead Files, file "Conservation—Whales"; but note that versions of this letter were also sent to the Bureau of Biological Survey and the US Tariff Commission as well as the National Museum of Canada.

198. H. P Sheldon to Howell, 28 April 1928, in ibid.

199. For Radcliffe's response, see Lewis Radcliffe to Howell, 20 April 1928, in ibid.; he had been involved (peripherally) in the planning for the Hatteras expedition, since it was to him that requests for use of the Beaufort laboratory had to be directed. It was perhaps this early contact with the question of scientific research on whales that led him to suggest that the proposed "International Commission" ought to have the "authority to make scientific investigations to determine what regulations are necessary." He appears to have been aware of the Discovery work by this time (1928). On Howell's concern about making sure that the whale conservation issue did not "topple toward the B.S.," see Howell to Kellogg, 13 January (circa 1930), Smithsonian Archives, RU 7170, box 4, file "Howell, A. Brazier, 1930–1931 (2/3)."

assembling a who's who of American Progressive Era conservationists—to protect the whales of the world.

One of the striking things to emerge from this review is the relative unimportance of fisheries sciences in the origin of concerns about whale conservation in the United States. Though W. F. Thompson and Oscar E. Sette had, by the 1920s, more than a decade of experience applying European stock management techniques to heavily exploited marine populations of sardines, tuna, halibut, and salmon, neither their scientific approaches nor their more general awareness of the problems of overfishing appear to have played any role in the emerging US campaign to protect whales.[200] In 1929 Kellogg corresponded briefly with Sette, then the director at the Bureau of Commercial Fisheries Biological Laboratory at Woods Hole, about efforts to secure additional small odontocetes for experimental purposes, but the idea of some collaboration appears to have petered out, and no marine scientists or fisheries biologists feature prominently in the Council for the Conservation of Whales. Woods Hole, for all its importance in other aspects of the history of biology and oceanography, does not feature in the story of the ccw.[201] Rather, its program emerged out of the network that

200. See Arthur F. McEvoy, *The Fisherman's Problem: Ecology and Law in the California Fisheries, 1850–1980* (Cambridge: Cambridge University Press, 1986), particularly pp. 158ff; Tim D. Smith, *Scaling Fisheries: The Science of Measuring the Effects of Fishing, 1855–1955* (Cambridge: Cambridge University Press, 1994); and Joseph E. Taylor, *Making Salmon: An Environmental History of the Northwest Fisheries Crisis* (Seattle: University of Washington Press, 1999). This is in notable contrast to the European case, where Hjort, Maurice, and others with experience of the North Seas herring and plaice fisheries would play an important role.

201. For Sette correspondence, see Kellogg to Oscar E. Sette, 15 July 1929, Smithsonian Archives, RU 7170, box 7, file "Se–Sh." See also Kellogg's exchange with George M. Gray, curator of the Marine Biological Laboratory supply department (which did briefly keep a porpoise alive in its large saltwater pool): George M. Gray to Kellogg, 15 June 1929, Smithsonian Archives, RU 7170, box 3, file "Go–Gra." The institutional history of the Woods Hole scientific establishments is tricky. Consider Dean C. Allard, "The Fish Commission Laboratory and Its Influence on the Founding of the Marine Biological Laboratory," *Journal of the History of Biology* 23, no. 2 (Summer 1990): 251–70. On the general importance of the Marine Biological Laboratory in the history of biology, see Philip J. Pauly, *Biologists and the Promise of American Life: From Meriwether Lewis to Alfred Kinsey* (Princeton, NJ: Princeton University Press, 2000); see also Pauly's chapter "Summer Resort and Scientific Discipline: Woods Hole and the Structure of American Biology, 1882–1925," in *The American Development of Biology*, ed. Ronald Rainger and Jane Maienschein (Philadelphia: University of Pennsylvania Press, 1988), 121–50.

I have begun to reconstruct here: disaffected Bureau of Biological Survey biologists, museum-based vertebrate zoologists with experience as field naturalists, medical physiologists conscripted to the cause, and old-guard conservation grandees. The larger concerns of this diverse group lay, for the most part, with game management and the preservation of birds and mammals. The first masthead of the CCW was not a very salty collection of names.[202]

And yet it was perhaps the greatest coup of the nascent CCW to catch hold of the very best known figure in the American conservation movement just as he was setting sail on the open sea. In 1928 Gifford Pinchot, the photogenic and politically adroit forester who had made conservation a standard of the Republican Party in the age of Roosevelt, was passed over for a presidential nomination he had courted with some energy. Recently trounced in a senatorial election, out of his gubernatorial office in Harrisburg, "America's Forester" began to conceive a theatrical exit strategy that would serve simultaneously as recreation and re-creation: a long (and semi-scientific) voyage to the South Seas.[203] In the autumn of that year—as

202. The first advisory board consisted of Glover M. Allen (president of the ASM), Carlos Avery (president of the American Game Protective and Propagation Association); Barton W. Evermann (chairman of the NRC Commission on Conservation of Marine Life in the Pacific; though he was an ichthyologist, his work dealt almost exclusively with freshwater species; in addition, he was a member of the Cooper Ornithological Club and an avid ornithologist); Frank J. Goodnow (president, Johns Hopkins University); Madison Grant (president, New York Zoological Society); George Bird Grinnell (cofounder and honorary president, Boone and Crockett Club); Gilbert H. Grosvenor (president, National Geographic Society); Chauncey J. Hamlin (conservationist and president of the Buffalo Society of Natural Sciences); William T. Hornaday (founder of the Permanent Wildlife Protection Fund); David Starr Jordan (president emeritus, Stanford University); Vernon L. Kellogg (National Research Council); Clarence C. Little (president, University of Michigan); John C. Merriam (president, Carnegie Institution of Washington); Henry Fairfield Osborn (president, American Museum of Natural History); John C. Phillips (president, American Wild Fowlers); T. Gilbert Pearson (president, National Association of Audubon Societies); Gifford Pinchot (former chief of the Bureau of Forestry and former governor of Pennsylvania); I. T. Quinn (chairman, National Game Conference); George Shiras III (US congressman, amateur naturalist, and popular wildlife photographer); T. Wayland Vaughan (director of the Scripps Institution of Oceanography, a geologist); Henry B. Ward (president, Izaak Walton League of America); Lewis H. Weed (dean of the Johns Hopkins medical school); and Ray Layman Wilbur (president, Stanford University).

203. The story is told in Char Miller, *Gifford Pinchot and the Making of Modern Environmentalism* (Washington, DC: Island Press, 2001), chap. 12; but see also Gifford

Kellogg and Howell, emboldened by their success at the ASM, were working up sharply worded letters to recruit an advisory board for their whale
conservation program—Gifford Pinchot and his wife Cornelia fell in love
with a 148-foot, three-masted topsail schooner, the *Cutty Sark*, in a Maine
shipyard. Rechristened the *Mary Pinchot* and refurbished to accommodate
the Pinchot family, the vessel would become the mobile stage on which
Gifford Pinchot would play out a *Beagle*-esque fantasy of natural history
voyaging while simultaneously courting and spurning the prying eyes of
the newspapermen. The "Pinchot South Seas Expedition—1929" embodied
celebrity nature grandstanding in the best tradition of Teddy Roosevelt and
William Beebe. As Pinchot himself put it, "adventure seasoned with science
is the best kind."[204] He made sure to secure a photographer to memorialize
every facet of the voyage (which put the Galápagos Islands at the top of
the list of ports of call) and to provide illustrations for the projected com-
memorative volume.

In this early stage of work toward the formation of the CCW, when Kellogg and Howell were plotting their campaign, Pinchot started to look like
the ideal figurehead for the movement. The question was how to reach him.
In this, as in other aspects of their publicity, Howell and Kellogg benefited
from their association with Lewis Hill Weed, the neuroanatomist who had
helped Kellogg orchestrate the Hatteras expedition.[205] Weed's role as dean
of the Johns Hopkins University School of Medicine had given him considerable experience in institution building; his skill in administration would
elevate him to the directorship of the medical school in 1929. Working to
promote the growth and financial strength of Hopkins, Weed had made
use of the services of the John Price Jones Corporation of New York City,

Pinchot, *To the South Seas: The Cruise of the Schooner* Mary Pinchot *to the Galapagos,
the Marquesas, and the Tuamotu Islands, and Tahiti* (Philadelphia: John C. Winston,
1930).

204. Cited in Char Miller, *Gifford Pinchot and the Making of Modern Environmentalism* (Washington, DC: Island Press, 2001), 297. For Kellogg and Howell's letters, see
Smithsonian Archives, RU 7170, box 4, file "Howell, A. Brazier, 1920–1929 and undated
(3/3)," and A. Brazier Howell to "Sir," undated, Smithsonian Institution, James Mead
Files, file "Conservation—Whales."

205. For Weed's early engagement, see Howell to Kellogg, undated, Smithsonian
Archives, RU 7170, box 4, file "Howell, A. Brazier, 1920–1929 and undated (3/3)"; for
the Jones Corporation's help with Pinchot, see Howell to Kellogg, 23 December 1928,
in ibid.

"one of the largest publicity and money-raising firms in the country."[206] Weed, still taken with the scientific significance of whales, enticed an associate at the firm, Harold M. Weeks, to take on the whale project for Kellogg and Howell. Weeks would assist them, Weed explained, in directing the publicity campaign and in placing articles in the nation's newspapers. The Jones Corporation had a good sense of how to handle these matters and would help make sure that they got the attention they needed without letting things "get out of hand, for it might well degenerate into a tabloid horror if not properly directed."[207] Weed also emphasized the importance of building up a consortium of respected supporters before making any sudden moves. While Howell and Kellogg were keen to do most of the actual work themselves (Howell complained that a big active board would just mean a lot of "bickering and would probably not get anywhere for a long time"), it was key to have a powerful roster of names.[208]

To this end, Kellogg, Weed, and Howell began working their connections, circulating a carefully edited letter that laid out the initiative of the American Society of Mammalogists and detailed the steps that were needed. Kellogg wrote the first draft of the letter, pointing out that nearly 30,000 whales were being killed each year, that the rate of replacement for all large mammals was known to be slow, and that several whale species were clearly on the verge of extinction. Invoking the collapse of the nineteenth-century New England industry and alluding to the fact that modern whalers had now penetrated into the Antarctic ("the cetacean's [*sic*] last stronghold)," Kellogg recommended a set of specific regulatory actions to curb "malpractices in whale hunting," including

- "absolute protection of such species as are in serious danger of extermination";
- "protection of whales on certain rather circumscribed areas which they now use as breeding grounds, where they are now killed at every opportunity";

206. Benjamin Stolberg, "Vigilantism, 1937—Part II," *Nation* 145, no. 8 (21 August 1937): 191.

207. Harold M. Weeks to Lewis Weed, 10 November 1928, Smithsonian Archives, RU 7170, box 8, file "Wa–Wh."

208. Howell to Kellogg, 13 November 1928, Smithsonian Archives RU 7170, box 4, file "Howell, A. Brazier, 1920–1929 and undated (3/3)."

- "protection of females nursing young";
- "prohibition of whaling in certain tropical areas along migration routes";
- "prohibition of the use of airplanes in whale hunting."

Predictably, many of these recommendations—from breeding preserves to limitations on airplane hunting—recapitulated the platform of successful terrestrial conservation programs in the United States. Calls for licensing, regulatory supervision, and further scientific research rounded out Kellogg's proposal.[209] Howell took this document in hand and reworked it slightly, adding a preamble that firmly framed the whole undertaking as an exercise in use-oriented, rational, "gospel-of-efficiency"-style conservation; soppy sentimentalism would not be part of the program. The aim, as Howell spelled it out, was "adequate conservation of the breeding stock of whales," and he assured potential supporters that "the work will be carried forward in a thoroughly dignified manner and the members of the Advisory Board are assured that their names will not be connected with any sort of propaganda which to them might be objectionable."[210] From the beginning, Howell—whose personal tendencies leaned toward preservation, and who would go on to pen affecting diatribes like "The Plight of the Whale" for the *National Humane Review*—was savvy about translating practical efforts at whale conservation into the language of "perpetual supply" and "complete utilization."[211] For instance, as early as 1926, in his first article touching on modern whaling, Howell went out of his way to emphasize that "our whale supply is a matter of economic importance and not merely of aesthetic concern to those with emotional tendencies."[212]

Armed with this sturdy revised document, Kellogg and Howell made a first round of solicitations to about two dozen of their notable friends and associates, including Merriam, Vaughan, and Jordan. But all of this was

209. This is a summary of Harold M. Weeks to Lewis Weed, 10 November 1928, Smithsonian Archives, RU 7170, box 8, file "Wa–Wh."

210. See two letters, both A. Brazier Howell to "Sir," undated, Smithsonian Institution, James Mead Files, file "Conservation—Whales."

211. See A. Brazier Howell, "The Plight of the Whale," *National Humane Review* (July 1935): 7, though it appears from archival material that Kellogg did originally do some drafting on this piece.

212. See A. Brazier Howell, "Visit to a California Whaling Station," *Smithsonian Miscellaneous Collections* 78, no. 7 (1926): 71–79.

merely the prelude to making a pitch to Pinchot. Kellogg had been charged with that task, and he made his approach through the venerable Albert Kenrick Fisher, a mainstay of bird conservation and government biological service who had been the founder of the forerunner programs in food habits research that had eventually taken shape as part of the Bureau of Biological Survey.[213] Fisher was a personal friend of Pinchot's and was involved in the planning of the vaunted South Seas Expedition. He made it clear that if Kellogg and Howell were successful in lining up a strong list of supporters first, he would take the whole package to Pinchot. As Kellogg put it to Howell, "Fisher is quite optomistic [*sic*] and I feel certain that it will go through if you get acceptances from a sufficient number of those prominent guys."[214] The bigwigs came through, and in December 1928 Fisher laid the whole business before Pinchot, who quickly saw the fortuitous tie-in with his new maritime ambitions. Soon Kellogg was having direct conversations with the great man himself, and the subject was, perhaps surprisingly, how to kill porpoises. Pinchot not only wanted to serve on the Advisory Board of the Council for the Conservation of Whales, he also wanted to put the *Mary Pinchot* into the service of cetology: he and his son Giff would be happy to heave harpoons from the bowsprit in the name of science.

By the time the Pinchot South Seas Expedition set sail, at the end of March 1929 (a departure attended by two dozen movie cameras), the vessel had been outfitted with scores of porpoise irons and whale harpoons designed to Kellogg's exact specifications, along with all the necessary line, floats, and kegs—everything arranged just as Kellogg had learned from his correspondence with old New England whalemen in 1927 and 1928.[215] Kellogg even sent along instructions for preparing cetacean specimens, literature for help with identification of the catches, and a hundred metal and a hundred wooden numbered tags (already tricked out with twist ties) to ensure good record keeping—and to cement his relationship with the CCW's

213. For a biography of Fisher, see Francis M. Uhler, "In Memoriam: Albert Kenrick Fisher," *Auk* 68 (April 1951): 210–13.

214. See Howell to Kellogg, 13 November 1928, and Kellogg to Howell, 16 November 1928, both in Smithsonian Archives, RU 7170, box 4, file "Howell, A. Brazier, 1920–1929 and undated (3/3)."

215. On the departure, see Char Miller, *Gifford Pinchot and the Making of Modern Environmentalism* (Washington, DC: Island Press, 2001), 297; on the outfitting of the vessel, see Kellogg to Gifford Pinchot, 14 January 1929, Smithsonian Archives, RU 7170, box 7, file "Pe–Pi," and two letters from Gifford Pinchot to Kellogg, both dated 17 January 1929, Smithsonian Archives, RU 7170, box 7, file "Pe–Pi."

newest and most important patron.[216] Pinchot expressed great enthusiasm at "the prospect of being of real use in the matter of porpoises," and he promised the National Museum "first call" on any cetacean specimens.[217]

In fact, it did not take long for the *Mary Pinchot* to strike gold. No sooner had the vessel's keel split the cerulean seas of the Gulf Stream than the Pinchot family played out the classic scene from *Down to the Sea in Ships*, spearing from the bowsprit a good-sized *Prodelphinis plagiodon*, which succumbed to two shots to the head from Giff's .38 revolver after an exhilarating fight (figure 3.10).[218] Its liver was delicious, Pinchot noted. Before the "scientific fishing trip" was over, the Pinchots would sock away in the increasingly fetid hold of the *Mary Pinchot* a fair amount of cetacean material for Kellogg and his colleagues (figures 3.11 and 3.12).[219]

The high drama of this kind of scientific collecting suggestively links American whale conservation efforts between the wars to the larger culture of sportsmen-conservationists in the period. Even a cursory review of the leading figures who lent their names to the Council for the Conservation of Whales reveals a strikingly high proportion of men like Henry B. Ward, the president of the Izaak Walton League of America (a club of trout and salmon fishermen) and I. T. Quinn, a game commissioner who had been responsible for inland fishing in Virginia's lakes and streams before rising to national prominence as a defender of sportsmen, fish, and ducks. What drew such characters to an interest in preserving whales? To be sure, there was the general link between outdoorsmen and conservation efforts in the 1920s.[220] But there was more than that. Gary Kroll has written persuasively

216. For tags, see Kellogg to Gifford Pinchot, 29 January 1929, Smithsonian Archives, RU 7170, box 7, file "Pe–Pi"; for instructions, etc., see Kellogg to Gifford Pinchot, 14 January 1929, in ibid.

217. Two letters from Gifford Pinchot to Kellogg, both dated 17 January 1929, in ibid.

218. Char Miller, *Gifford Pinchot and the Making of Modern Environmentalism* (Washington, DC: Island Press, 2001), 300. Bill Perrin tells me he used this specimen (now *Stenella frontalis*) in his revision of the taxonomy and systematics of the spotted dolphins.

219. Gifford Pinchot, *To the South Seas: The Cruise of the Schooner* Mary Pinchot *to the Galapagos, the Marquesas, and the Tuamotu Islands, and Tahiti* (Philadelphia: John C. Winston, 1930), 18.

220. See John F. Reiger, *American Sportsmen and the Origins of Conservation* (New York: Winchester Press, 1975), and James B. Trefethen and Peter Corbin, *An American Crusade for Wildlife* (New York: Winchester Press, 1975).

FIGURE 3.10 Scientific sport: The first porpoise harpooned on the *Mary Pinchot*.
(From Pinchot, *To the South Seas,* p. 20.)

FIGURE 3.11 Weigh-in: More spoils from the *Mary Pinchot*. (From Pinchot, *To the South Seas*, p. 67.)

FIGURE 3.12
Salty boffins: Victory on the
Mary Pinchot. (From Pinchot,
To the South Seas, p. 73.)

that the early twentieth century saw a "gaming" of the sea in the United States, as big-game sportfishing for marlin, tuna, and tarpon rose in prominence and new coastal clubs and tournaments catered to freshwater anglers eager to test their tackle against the larger quarry of the salt.[221] Andrews's *Whale Hunting with Gun and Camera* strikingly demonstrates how this "new frontier" of the oceans affected perceptions of whales and dolphins in the period. While Andrews's text was a cry for increased protection of

221. See Gary Kroll, *America's Ocean Wilderness: A Cultural History of Twentieth-Century Exploration* (Lawrence: University Press of Kansas, 2008). See also John F. Reiger, *American Sportsmen and the Origins of Conservation* (New York: Winchester Press, 1975), 248, on the founding of the Catalina Tuna Club, and Charles Frederick Holder, *The Game Fishes of the World* (London: Hodder and Stoughton, 1913). It is worth recalling that it was in these very years, the later 1920s, that Zane Grey published a series of three popular books on saltwater sport fishing: *Tales of Fishing Virgin Seas* (New York: Grosset and Dunlap, 1925); *Tales of the Angler's El Dorado, New Zealand* (New York: Harper and Brothers, 1926); and *Tales of Swordfish and Tuna* (New York: Harper and Brothers, 1927).

Pacific cetaceans, its author—brandishing his cowboy credentials—itched to get behind the trigger of one of the massive Svend Foyn–style whale cannons, and he described the drama of the chase with the enthusiasm of a hunter.[222] The skill of plugging belugas and other odontocetes with a shoulder weapon from the bridge of a small tossing vessel was a worthy test of any marksman.[223] In the broader context of the new sea sportsmanship, cetaceans—wily, big, and strong—could readily take shape as the ultimate sea trophy.[224] There were turn-of-the-century precedents for this notion. After all, Prince Albert of Monaco, an ardent huntsman and debonair embodiment of the enlightened yachting nobleman, outfitted his first oceanographic research vessel, the regal *Princess Alice*, with a harpoon cannon, and this royal patron of the sciences set out after the mighty sperm whale

222. These scenes appear in Roy Chapman Andrews, *Ends of the Earth* (New York: Knickerbocker Press, G. P. Putnam's Sons, 1929), which includes the following remarkable passage: "To shoot a whale in a rough sea with the tiny vessel doing a Charleston on the crest of the waves and sliding down into the troughs as though she were headed for the bottom, is quite another matter. It is not pot-hunting. In fact it is more like shooting a bird on the wing" (p. 47). He also complained about dolphins that "never got near enough to shoot" (p. 91). Interestingly, Andrews is here replying, in a sense, to an attack on whaling as a "slaughter industry" like commercial bird hunting for meat. See William Temple Hornaday, *Thirty Years War for Wild Life* (New York: Arno, 1970 [1931]), 107: "During the last 50 years I have seen and heard much of guns for the wholesale killing of game, and other forms of wildlife than the commercial food fishes. We know the cannon gun of the market duck-hunter, the many-barreled shotgun for waterfowl that gave birth to the French *mitrailleuse* of 1870, and the automatic shotgun of today that for one cocking gives five shots. All of those, however, sink into utter insignificance beside the powerful whale killing cannon, that is fixed in the bow of each whaleship, or gasoline whaleboat, firing a bomb lance weighing 100 pounds, and effective at *a mile* in every direction!" Emphasis in original (the actual range was closer to 100 yards).

223. See chap. 22 of Roy Chapman Andrews, *Whale Hunting with Gun and Camera: A Naturalist's Account of the Modern Shore-Whaling Industry, of Whales and Their Habits, and of Hunting Experiences in Various Parts of the World* (New York: D. Appleton, 1916).

224. The following quote from Pinchot gives a glimpse of how whales were associated with game fishing in this period: "About that same time a Whale blew within half a mile of us, showing his sharp dorsal fin and 15 or 20 feet of back at the surface, and there were Sea Bats also in these populous straits, which offer the finest stretch of smooth water we saw anywhere in the Galápagos, and bid fair, I think, to become one of the famous game fish grounds of all the world." Gifford Pinchot, *To the South Seas: The Cruise of the Schooner* Mary Pinchot *to the Galapagos, the Marquesas, and the Tuamotu Islands, and Tahiti* (Philadelphia: John C. Winston, 1930), 287.

in the 1880s, eventually taking a forty-footer in a dramatic encounter off the Azores.[225] And not to be outdone, Kaiser Wilhelm II also set out a-whaling in the North Sea at the end of the nineteenth century.[226] With such stories in wide circulation, it is not difficult to understand how commentators writing before World War II could describe the large whales as "the most splendid game in existence."[227] This was more than rhetoric in the United States in this period, as evidenced by the remarkable proliferation of so-called "dude cruises," guided charter boat operations (mostly on the West Coast) that took deep-pocketed sportsmen out on cutters equipped with harpoon guns to give them a shot at the largest game the planet had to offer. These operators disdained commercial use of their catches, and the carcasses were merely set adrift after the thrill of the chase, the pop of the grenades, and the spectacle of blood had come to a silent close.[228]

While Kellogg and Howell worked to eliminate this sort of activity, branding it as wasteful and unsportsmanlike, they were not blind to the sporting pleasures of hunting the smaller whales. Kellogg even tried to encourage potential specimen collectors by pointing out that "the prospect of capturing porpoises should afford a pleasant divertisement for the crew and staff and relieve the monotony of the regular routine on board the ship."[229] And Kellogg was explicit about trying to tie whale conservation to issues of sportsmanship, closing one of his important public lectures on the topic with the following anecdote:

225. Susan Schlee recounts this story in *The Edge of an Unfamiliar World: A History of Oceanography* (New York: Dutton, 1973), 134–35. The prince was particularly interested in using sperm whales to collect deep-sea squid. Thus, he shot the whales ostensibly in order to recover their regurgitation, treating them as proxies for the exploration of the ocean floor. On the "discovery" of the deep sea, see Helen Rozwadowski, *Fathoming the Ocean: The Discovery and Exploration of the Deep Sea* (Cambridge, MA: Harvard University Press, 2005).

226. See "There He Blows," *Punch, or the London Charivari* 103 (23 July 1892): 25.

227. Birger Bergersen, "The International Whaling Situation," *Le Nord* 1 (1938): 112, cited in L. Larry Leonard, "Recent Negotiations toward the International Regulation of Whaling," *American Journal of International Law* 35, no. 1 (January 1941): 90–113, at p. 112.

228. For information on dude cruises, see Kellogg to Howell, 25 January 1935, Smithsonian Archives, RU 7170, box 4, file "Howell, A. Brazier, 1932–1943 (1/3)."

229. Kellogg to unknown, undated, Smithsonian Archives, RU 7170, box 2, file "F–Fo." Kellogg also corresponded with East Coast hunting and fishing guides about securing specimens. See Aycock Brown to Kellogg, 12 May 1931, and Kellogg to Aycock Brown, 19 May 1931, both in Smithsonian Archives, RU 7170, box 1, file "Bra–Bry."

[The killing of females accompanied by young] is not only a matter of economics, but also to a lesser degree one of sportsmanship, which brings to mind the story about a small boy who, after reading an article on the need for conservation of whales remarked to his father, "isn't it all right to catch whales, if you're careful to throw the little ones back?"[230]

In an era that saw American sportsmen taking to the sea in pursuit of new and superlative game (Pinchot was particularly enthusiastic about taking what he thought was the first marlin ever to fall for a feather jig), the movement for whale conservation could draw on the support of an existing infrastructure of sportfishing enthusiasts for whom no quarry, however exotic, was beyond the pale, and for whom game conservation (particularly opposition to excessive "meat hunting" by commercial operators) was already an established collective enterprise.[231]

With Pinchot embarked on his highly visible disappearance, American conservation energies can be understood to have taken to the sea. It was thus an ideal moment to make a major push for attention to the world's whaling situation. Placing Pinchot's name prominently on their new letterhead, Kellogg and Howell launched the CCW's full-scale publicity campaign in the spring of 1929 (figure 3.13). A detailed strategy sheet offers a unique window onto how Kellogg, Howell, and their coconspirators (including powerful figures like Vernon Kellogg of the National Research Council), aided by a professional publicist, plotted a consciousness-raising barrage of news items, feature stories, and magazine pieces[232]

230. See Kellogg's contribution to the record, U.S. Congress, Hearing Before a Special Committee on Wild Life Resources, United States Senate, Seventy-Second Congress, First Session, 20 March 1931, p. 29.

231. To get a feel for this kind of support, see the correspondence between Kellogg and Arthur H. Henn at the Carnegie Museum in Pittsburgh. Henn was the secretary of the Izaak Walton League, Chapter 13, and he offered the "moral support" of his membership to the cause of the CCW. His grasp of the niceties of the issue was questionable, however: he suggested that the CCW might add the "whale shark" to its conservation campaign. The whale shark, of course, is not a whale, but a fish.

232. "Schedule for Whaling Publicity," Smithsonian Archives, RU 7170, box 10, file "Information—Whale Press Releases." It appears that Weeks drafted this document after a meeting with Howell in New York. See Kellogg to H. M. Weeks, 28 November 1928, Smithsonian Archives, RU 7170, box 8, file "Wa–Wh."

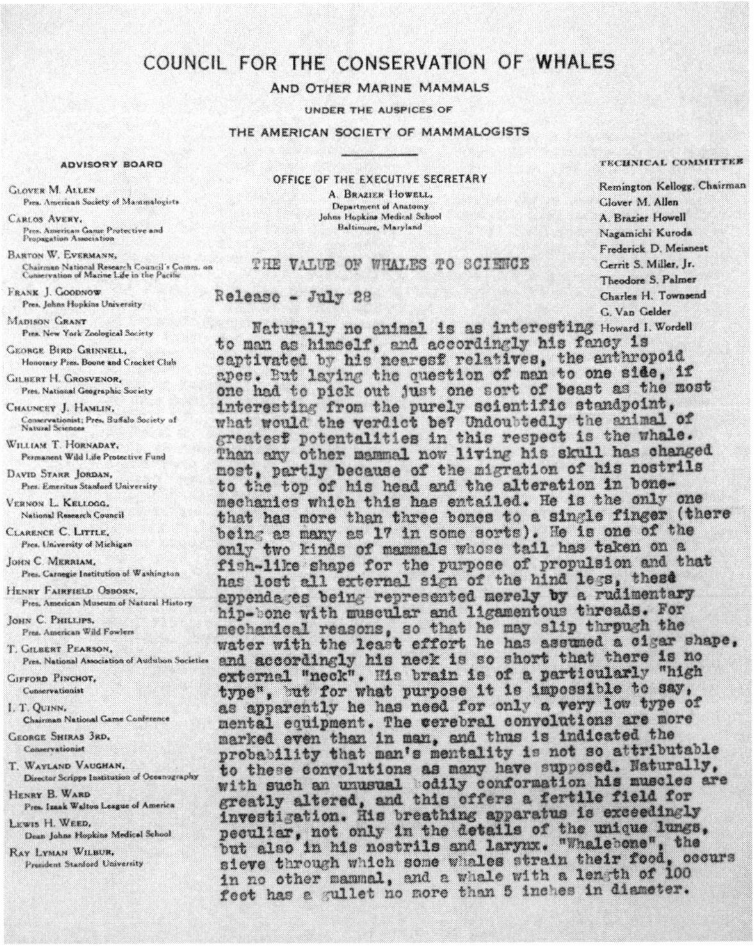

COUNCIL FOR THE CONSERVATION OF WHALES

AND OTHER MARINE MAMMALS

UNDER THE AUSPICES OF

THE AMERICAN SOCIETY OF MAMMALOGISTS

ADVISORY BOARD

GLOVER M. ALLEN
Pres. American Society of Mammalogists

CARLOS AVERY,
Pres. American Game Protective and
Propagation Association

BARTON W. EVERMANN,
Chairman National Research Council's Comm. on
Conservation of Marine Life in the Pacific

FRANK J. GOODNOW
Pres. Johns Hopkins University

MADISON GRANT
Pres. New York Zoological Society

GEORGE BIRD GRINNELL,
Honorary Pres. Boone and Crocket Club

GILBERT H. GROSVENOR,
Pres. National Geographic Society

CHAUNCEY J. HAMLIN,
Conservationist; Pres. Buffalo Society of
Natural Sciences

WILLIAM T. HORNADAY,
Permanent Wild Life Protective Fund

DAVID STARR JORDAN,
Pres. Emeritus Stanford University

VERNON L. KELLOGG,
National Research Council

CLARENCE C. LITTLE,
Pres. University of Michigan

JOHN C. MERRIAM,
Pres. Carnegie Institution of Washington

HENRY FAIRFIELD OSBORN,
Pres. American Museum of Natural History

JOHN C. PHILLIPS,
Pres. American Wild Fowlers

T. GILBERT PEARSON,
Pres. National Association of Audubon Societies

GIFFORD PINCHOT,
Conservationist

I. T. QUINN,
Chairman National Game Conference

GEORGE SHIRAS 3RD,
Conservationist

T. WAYLAND VAUGHAN,
Director Scripps Institution of Oceanography

HENRY B. WARD
Pres. Izaak Walton League of America

LEWIS H. WEED,
Dean Johns Hopkins Medical School

RAY LYMAN WILBUR,
President Stanford University

OFFICE OF THE EXECUTIVE SECRETARY
A. BRAZIER HOWELL,
Department of Anatomy
Johns Hopkins Medical School
Baltimore, Maryland

TECHNICAL COMMITTEE

Remington Kellogg, Chairman
Glover M. Allen
A. Brazier Howell
Nagamichi Kuroda
Frederick D. Meinest
Gerrit S. Miller, Jr.
Theodore S. Palmer
Charles H. Townsend
G. Van Gelder
Howard I. Wordell

THE VALUE OF WHALES TO SCIENCE

Release - July 28

Naturally no animal is as interesting
to man as himself, and accordingly his fancy is
captivated by his nearest relatives, the anthropoid
apes. But laying the question of man to one side, if
one had to pick out just one sort of beast as the most
interesting from the purely scientific standpoint,
what would the verdict be? Undoubtedly the animal of
greatest potentialities in this respect is the whale.
Than any other mammal now living his skull has changed
most, partly because of the migration of his nostrils
to the top of his head and the alteration in bone-
mechanics which this has entailed. He is the only one
that has more than three bones to a single finger (there
being as many as 17 in some sorts). He is one of the
only two kinds of mammals whose tail has taken on a
fish-like shape for the purpose of propulsion and that
has lost all external sign of the hind legs, these
appendages being represented merely by a rudimentary
hip-bone with muscular and ligamentous threads. For
mechanical reasons, so that he may slip through the
water with the least effort he has assumed a cigar shape,
and accordingly his neck is so short that there is no
external "neck". His brain is of a particularly "high
type", but for what purpose it is impossible to say,
as apparently he has need for only a very low type of
mental equipment. The cerebral convolutions are more
marked even than in man, and thus is indicated the
probability that man's mentality is not so attributable
to these convolutions as many have supposed. Naturally,
with such an unusual bodily conformation his muscles are
greatly altered, and this offers a fertile field for
investigation. His breathing apparatus is exceedingly
peculiar, not only in the details of the unique lungs,
but also in his nostrils and larynx. "Whalebone", the
sieve through which some whales strain their food, occurs
in no other mammal, and a whale with a length of 100
feet has a gullet no more than 5 inches in diameter.

FIGURE 3.13 A CCW press release: Reframing whales in the public consciousness. The founding members are listed on the letterhead. (Courtesy of the
Smithsonian Institution.)

(figure 3.14). A letter circulated in March laid out the rationale for this effort
explicitly:

At the present time steps are being taken to begin . . . a campaign to
arouse the interest of the public in the question of whale conservation
and to inform it regarding the urgent need for the adequate protection of these mammals. It is thought that only in such manner can we

SCHEDULE FOR WHALING PUBLICITY

News releases-		Thru	To	Date
1. First general announcement (prepared by Weeks)		H. F. Osborn	All N.Y.	Apr. 22
2. Whaling in Antarctic: thru N.Y. Times and Russell Owen, with Byrd		Weeks	Times	" 24
3. Announcement of plans for whale conservation by other countries (leave to Kellogg)		Meisnest	A.P.	" 29
4. Announcement that movement has become international (send out letters to foreign biol. Socs. asking coop. Apr. 20 (V.L.Kellogg)		V. L. Kellogg	A.P.	May 13
5. Interview with Lucius Eastman on what the league of Nations is doing and might do.		Weeks		May 20
6. Past and potential economic value of whales (R. Kellogg)		Meisnest	A.P.	" 27
7. Story on the scientific value of whale		Howell	A.P.	
8. Tariff Comm announces economic activities in gathering data in other countries (R.Kellogg)		Meisnest	A.P.	June 10

Feature articles - any time after May 10		Attended to by	For
1. Article on New Bedford Whaling Museum, by Watson		Weeks	N.Y.Time
2. On Col. Green's Whaler "The Morgan". (Information thru Watson or Weeks)		C. Jones, A. P.	
3. Article on 1 or more members of Board, record etc.		C. Jones, A. P.	
4. Evolution of the whales, R. Kellogg		C. Jones, A. P.	
5. Short articles on ambergris		Weeks	
6. " " " whalebone		Weeks	
7. Fishes that are not fish (aquat mams.)		Townsend, Sci.Ser.	
8. Can whale be saved from overspecialization ?		Weeks	
9. Scrimshander (whaling figures on ivory)		Watson	
10. Airplanes bring final menace to whales (thru Weeks)		Mingos	
11. Whale a kindly parent		Weeks	

FIGURE 3.14 CCW's publicity strategy: Nurturing public interest in whale welfare. (Courtesy of the Smithsonian Institution.)

create the demand for the action that will be necessary before we can hope to secure desirable legislation.[233]

All those on the Advisory Board were encouraged to prepare "interviews with the press, radio talks, articles or short notes for publication, and

233. A. Brazier Howell to "Gentlemen," 15 March 1929, Smithsonian Institution, James Mead Files, file "Conservation—Whales."

resolutions."[234] Pieces were projected on a host of topics, from emotional appeals like "Whale a Kindly Parent" to dystopian items on how the airplane would prove the "final menace" to the whales of the world. Kellogg would contribute a piece on whales and evolution, and broader pieces framing the imminent loss of whales in the context of a tragic litany of "American life extinct or nearing extinction" promised to link the whale crisis with the fate of the bison and the passenger pigeon.[235] A digest of relevant statistics and facts about the industry would be circulated to all concerned as soon as Kellogg had finished pulling it together.[236] It would put the relevant talking points in everyone's hands. The upshot? "Talk whales all you can, stressing their plight."[237]

The opening salvo was fired from Henry Fairfield Osborn's office at the American Museum of Natural History. On 20 May 1929 Osborn released a five-page brief declaring that "whales are now in danger of extinction" and "can only be saved by a concerted international effort."[238] Announcing the formation of the CCW, Osborn vowed that this "prominent group of scientists and conservationists" would not only work to preserve whales, but would also aim for "the preservation of the romantic industry of whaling." Asserting that "whales once swarmed in every ocean of the world," the press release went on to foreground the scientific and biomedical importance of the "secrets of the whale." "The whale is scientifically one of the most interesting of mammals," Osborn wrote (in a text quite probably

234. Ibid.

235. Smithsonian Institution, James Mead Files, file "Conservation—Whales." The last of those pieces was to be written by Anthony, who had collaborated with Osborn on "Close of the Age of Mammals."

236. For Kellogg's work on the statistics, see Kellogg to Frederick D. Meisnest, undated, Smithsonian Archives, RU 7434, box 6, file "Correspondence, 1928–1940." He used the library of the Fish Commission, as well as a network of commercial informants. See Kellogg to G. Van Gelder, 14 February 1929, in ibid.

237. A. Brazier Howell to "Gentlemen," 15 March 1929, Smithsonian Institution, James Mead Files, file "Conservation—Whales."

238. It is interesting to note that Kellogg and Howell appear to have made use of the Science Service on several occasions for distributing their press releases. For details on this organization, see the valuable study by David J. Rhees, "A New Voice for Science: Science Service under Edwin E. Slosson, 1921–1929," master's thesis, University of North Carolina, 1979. There is also a brief discussion of its workings in Philip J. Pauly's *Biologists and the Promise of American Life: From Meriwether Lewis to Alfred Kinsey* (Princeton, NJ: Princeton University Press, 2000), 4–6.

ghosted by Kellogg and Howell). How were these "warm-blooded" creatures able to "stay underwater for more than an hour, traveling to depths of perhaps a mile, where a pressure of more than one ton per square inch is exerted upon their surfaces"? No one knew, but if scientific investigation ever revealed "how the whale is able to accomplish such feats, and to neutralize the effects of the poisonous gases that at such times are generated in his lungs, it might lead to discoveries that would save many men from death in mine and submarine disasters."[239]

The CCW's campaign repeatedly stressed this theme of the whale as a scientific mystery. Howell himself seems to have drafted a press release entirely dedicated to this topic. Entitled "The Value of Whales to Science," this memo rehearsed the peculiar anatomical changes whales had undergone during their evolutionary development—the loss of hind legs, the refined systems for hearing and sight, the strangely large and convoluted brain (seemingly of "a particularly 'high type'")—before hypothesizing that these creatures might contain rare and powerful chemicals that could be used in medicine:

> It is not unlikely that they produce a special substance, of the nature of an enzyme, that helps overcome the effects of high blood pressure, experienced when they dive deeply, similarly they must have some special provision for overcoming the deleterious effects of high carbon dioxide content in the blood. When these and similar questions have been investigated, it is very probable that the knowledge gained can be applied in correcting a number of serious and even fatal derangements of bodily functions to which man is often subject.[240]

239. Osborn press release, Smithsonian Institution, James Mead Files, file "Conservation—Whales."

240. "The Value of Whales to Science," Smithsonian Archives, RU 7170, box 10, file "Information—Whale Press Releases" (see figure 3.13). Kellogg and others discussed the possibility that some porpoise extract might someday confer unique powers of respiration on human beings. For an entertaining exchange on this matter, see H. I. Wordell to Kellogg, 21 June 1929 forward, Smithsonian Archives, RU 7170, box 8, file "Wi–Wy," involving an "old sea dog" in New Bedford engaged to pursue "blackfish" (pilot whales) as specimens, who volunteered "that he would live on porpoise meat and black fish means for about a week," after which, it was surmised, he might "swim like a fish."

This bulletin, issued on behalf of the CCW, concluded that "it would be a scientific as well as an economic catastrophe if whales should be exterminated—a situation that is now speedily approaching." International action was desperately needed to save "the greatest beasts that ever lived."[241]

These—and an array of other position papers and news releases—poured out to an extensive list of press contacts as 1929 wore on.[242] Kellogg himself composed his researches on the history of the industry into an essay entitled "The Last Phase in the History of Whaling," which he and Weeks worked to place in the *Saturday Evening Post*.[243] Weeks, using materials supplied by Kellogg and Howell, drafted a piece for the *American Magazine* and gave some CCW authors advice on how to tune their submissions for a general audience.[244] Kellogg kept after others, like Witmer Stone at the Academy of Natural Sciences in Philadelphia, jockeying for contributions. Noting rumors of the stranding of a 70-foot whale on the New Jersey coast, Kellogg prodded Stone: "We expect to call upon you for some sort of a sob story lamenting the passing of the whale, which can be released to these hard-working newspaper guys."[245] In the end, the major releases of the CCW were, with the help of the John Price Jones Corporation, circulated to no fewer than fifteen hundred newspapers in the United States.[246] Even with this firehose distribution at his command, Kellogg personally worked his own connections right down to the most local scale, writing a note to his Uncle George back in Davenport, Iowa, encouraging him to try to get the

241. "The Value of Whales to Science," Smithsonian Archives, RU 7170, box 10, file "Information—Whale Press Releases."

242. For another example, see "An Industry Worth Millions Threatened by Present Methods," in ibid. For the full press list, see Smithsonian Archives, RU 7170, box 10, file "Information—Whale Press Releases."

243. "The Last Phase in the History of Whaling," which drew heavily on Harmer's work, can be found in ibid. On the effort to get it into the *Saturday Evening Post*, see Kellogg to Henry Fairfield Osborn, 3 December 1927, Smithsonian Archives, RU 7170, box 6, file "Ol–Os." The effort was apparently not successful.

244. H. M. Weeks to Kellogg, 13 February 1929, Smithsonian Archives, RU 7434, box 6, file "Correspondence, 1928–1940"; Kellogg to H. M. Weeks, 6 February 1929, in ibid.; and H. M. Weeks to Howell, undated, Smithsonian Archives, RU 7170, box 10, file "Information—Whale Press Releases."

245. Kellogg to Witmer Stone, 5 May 1929, Smithsonian Archives, RU 7170, box 8, file "Stockt–Stre."

246. Kellogg to "Uncle George," 21 June 1929, Smithsonian Archives, RU 7170, box 5, file "Man–Max."

releases into the hands of the "editor of the paper with the largest circulation" in that area.[247]

Orchestrating a publicity campaign at this scale gave Kellogg an extended tutorial in the workings of the press and tuned his strategies for success in the arena of public opinion. It was the job of newspapermen, he learned, to dish up "applesauce" for their readers, and getting things on the wire demanded something called "news value."[248] Just what this meant was never entirely clear, but, as one helpful adviser put it, "straight propaganda for a whale protection is apt to fall flat," since no one really cared about "causes" or "uplift movements."[249] Interestingly, it appears that the CCW emphasized the scientific value of whales in their press releases in large part in an effort to capitalize on the prevalent conception of scientific "discovery" as inherently newsworthy: if you needed to make whales into "news" (and this was exactly what the CCW aimed to do), then science was a powerful tool.[250] Pictures helped too, as Kellogg rapidly discovered, and he set himself up as a clearinghouse for choice whale images suitable for weekend editions.[251] Nevertheless, as Kellogg complained privately, it was by no means easy to get attention: "It is a hard racket to break into," he noted, facing the prospect of a rewrite on one of his submissions. The endless cycle of "publicity stunts" (as Howell called them) was making them both "gray headed and stoop shouldered" by the summer.[252]

247. Ibid. Kellogg and his collaborators also made use of the social networks of their wives—as, for instance, when R. M. Anderson's wife linked them to the women's press club of Canada. See R. M. Anderson to Kellogg, 21 May 1929, Smithsonian Archives, RU 7170, box 1, file "And–Ant."

248. For "applesauce," see Kellogg to "Uncle George," 21 June 1929, Smithsonian Archives, RU 7170, box 5, file "Man–Max"; for discussion of news value, see R. M. Anderson to Kellogg, 21 May 1929, Smithsonian Archives, RU 7170, box 1, file "And–Ant."

249. R. M. Anderson to Kellogg, 21 May 1929, Smithsonian Archives, RU 7170, box 1, file "And–Ant."

250. There is a healthy and expanding literature on the relationship between scientific discovery and "news"; Bernard Lightman, James Secord, Tom Broman, and others have done much to draw attention to the issue. For an introduction, consider Geoffrey Cantor and Sally Shuttleworth, eds., *Science Serialized: Representations of the Sciences in Nineteenth-Century Periodicals* (Cambridge, MA: MIT Press, 2004).

251. R. M. Anderson to Kellogg, 21 May 1929, Smithsonian Archives, RU 7170, box 1, file "And–Ant."

252. Kellogg to Chas. C. Adams, 29 July 1929, Smithsonian Archives, RU 7170, box 1, file "Charles C. Adams 1916–1944"; Kellogg to Howell, 12 January 1929, Smithsonian Archives, RU 7170, box 4, file "Howell, A. Brazier, 1920–1929 and undated (3 / 3)." Kel-

Yet their efforts did begin to pay off. By May 1929 editorials with titles like "Pity the Whale!" started to turn up in regional newspapers, complete with respectful mentions of the CCW and scaremongering statistics from Remington Kellogg, "an authority in the United States National Museum," who gave out that 370,000 whales had been killed since 1900, in comparison with fewer than a million in the previous three hundred years combined.[253] The *Philadelphia Ledger* ran a splashy Sunday supplement later that summer,[254] and no piece better embodied the CCW's mission than the foldout picture essay "Science Seeking the Secrets of the Fast Vanishing Whales," produced for the *American Weekly* (figure 3.15).[255] Replete with invocations of scientific mystery and pressing the imminent peril of extinction, this piece liberally quoted Kellogg and Howell on the need for international regulation and even discussed the cetacean research program under way at the Johns Hopkins University School of Medicine. It was exactly this kind of press attention that the CCW sought: binding scientific discovery and conservation sentiments into a newsworthy confection, it pushed whales into public consciousness as special and threatened creatures. In view of the fact that Kellogg and Howell were by this time working with insiders in the US Tariff Commission to draft proposals for US legislation protecting cetaceans, this kind of broad popular interest would—it was hoped—eventually translate into the necessary democratic support for practical regulatory efforts.[256] Pinchot's return in November 1929 and his successful reentry into politics strengthened their hand in legislative circles.[257]

---

logg joked about needing "to go into seclusion for several months" when the publicity campaign got going.

253. *The Morning Mercury*, 9 May 1929, Smithsonian Institution, James Mead Files, file "Conservation—Whales."

254. On the *Philadelphia Ledger* piece, see Kellogg to Witmer Stone, 5 May 1929, Smithsonian Archives, RU 7170, box 8, file "Stockt–Stre."

255. "Science Seeking the Secrets of the Fast Vanishing Whales," *American Weekly*, undated, Smithsonian Archives, RU 7170, box 9, file "Information—Newspaper Clippings."

256. On collaborations with Meisnest (who served on the US Tariff Commission) to draft regulatory language, see Howell to Kellogg, 5 February 1929, Smithsonian Archives, RU 7170, box 4, file "Howell, A. Brazier, 1920–1929 and undated (3/3)."

257. For Pinchot's later involvement, see Howell to Kellogg, undated, Smithsonian Archives, RU 7170, box 4, file "Howell, A. Brazier, 1920–1929 and undated (3/3)"; Kellogg to Gifford Pinchot, 6 December 1929; Gifford Pinchot to Kellogg, 7 December 1929; Kellogg to Gifford Pinchot, 9 January 1930; Gifford Pinchot to Kellogg, 10 Janu-

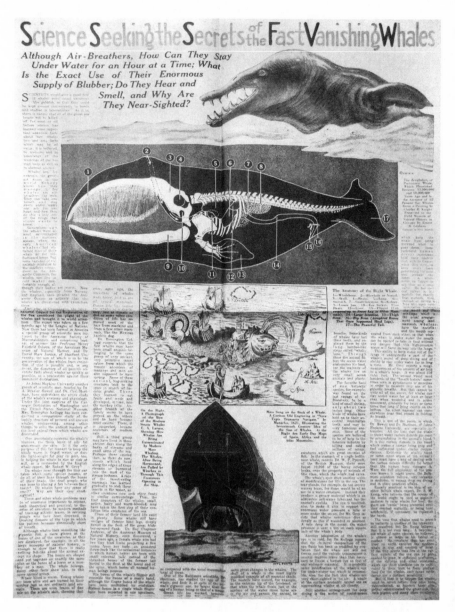

FIGURE 3.15 Science as public relations: An article in *American Weekly* emphasizes the mysteries of whales. (Courtesy of the Smithsonian Institution.)

The ccw's campaign did a great deal to heighten general awareness in the United States of the new character and intensity of modern whaling, but probably the greatest long-term significance of the ccw's work in the late 1920s and early 1930s was the platform it created for Remington Kellogg himself. He used its name and letterhead and its network of supporters to gather information and build new alliances at a critical stage of his career. By 1930, with Howell's help, the ccw had positioned Kellogg as the leading spokesman, adviser, and expert on marine mammal conservation in the United States (displacing other plausible candidates for this role).[258] This new standing opened the way to Kellogg's participation in a flurry of national and international regulatory efforts concerned with whaling in the period before World War II. Since it was in this capacity—as a delegate, as a negotiator, and as a scientific diplomat—that Kellogg would leave his most significant mark on whale politics and whale science in the twentieth century, these aspects of his work will be the primary concern of the pages that follow. Tracing Kellogg's rapid accession to positions of influence in the legislative and regulatory arenas will be the aim of this final section.

BECOMING A SCIENTIFIC STATESMAN

Close ties to developments in Europe were essential to Kellogg's rising trajectory. The United States, after all, had an essentially negligible stake in the world whaling industry at this time (something on the order of one-fiftieth of the world catch in 1929). The action was emphatically elsewhere, with Great Britain and Norway dwarfing all the other entrants in the period.[259]

---

ary 1930; and Kellogg to Julius Pin, 12 September 1938, all in Smithsonian Archives, RU 7170, box 7, file "Pe–Pi."

258. This idea—that the ccw succeeded in placing Kellogg in a position to serve as an expert adviser on policy and legislation—broadly confirms the larger argument of Samuel P. Hays in *Conservation and the Gospel of Efficiency: The Progressive Conservation Movement, 1890–1920* (Cambridge, MA: Harvard University Press, 1959), in which he maintains that Progressive Era conservation movements were (despite protestations and historiography to the contrary) fundamentally oriented toward the concentration of power in the hands of trained elites who used science to solve social problems by administrative means. "Conservation," as he put it, "above all, was a scientific movement" (p. 2).

259. On the statistics for the US industry in this period, see Kellogg to Howell, 1 January 1931, Smithsonian Archives, RU 7170, box 4, file "Howell, A. Brazier, 1930–1931 (2/3)."

The absence of major monied whaling interests in the United States facilitated the work of the ccw in many ways and gave Kellogg a relatively clear
field on which to gather support. But it also meant that any actual movement for the conservation of whales was going to demand contacts and
actions beyond American shores.

In the 1920s, as I have shown in chapter 2, British and Norwegian efforts to gather scientific data for use in guiding the regulation of the industry were already well under way. By the late 1920s these programs had
taken on an increasingly collaborative character. In 1927 the International
Council for the Exploration of the Sea—that unique body dedicated to the
coordination of oceanographic and fisheries research—convened the first
meeting of its "Whaling Committee" under the leadership of Johan Hjort.
Its goal was not only the standardization of the methods of the different
expeditions probing the Antarctic waters, but also (increasingly) the stimulation of a multinational regulatory initiative.[260] The push toward fully pelagic whaling from stern-slipway factory vessels—which freed the industry
from dependence on Antarctic shore stations, and in so doing increasingly
took Antarctic whaling outside the jurisdiction of any state—heightened
the urgency of the matter. The League of Nations, pressured particularly by
Great Britain (which, as the dominant land power in the Southern Ocean,
stood to lose the most from the shift to open-water whaling), took up the
problem in earnest after 1925, conveying the whole thorny matter to its
Committee for the Progressive Codification of International Law. By 1928,
this committee, under the chairmanship of the visionary Argentine jurist
José León Suarez, had issued a number of reports making extraordinary
recommendations, including the institution of a concept of "*res communis*"
in marine resources. Suggesting that the time had come to discard the idea
of the absolute freedom of the seas as well as antiquated conceptions of
"*res nullius*," Suarez asserted that "the riches of the sea, and especially the
immense wealth of the Antarctic region, are the patrimony of the whole
human race."[261] Rejoinders and position papers from a host of international

260. On ICES generally, see Helen M. Rozwadowski, *The Sea Knows No Boundaries: A Century of Marine Science under ICES* (Copenhagen: International Council for
the Exploration of the Sea in association with University of Washington Press, 2002);
on the Whaling Committee in particular, see pp. 95–96. ICES's whaling material can
be found in the ICES Records, Rigsarkivet (Danish National Archives), Copenhagen,
Denmark, box 85, folder 6Q.

261. Cited in Patricia W. Birnie, *International Regulation of Whaling: From Conservation of Whaling to Conservation of Whales and Regulation of Whale-Watching* (New

legal experts quickly followed as the issue of what to do about runaway Antarctic whaling came before the League's Economic Committee for action in the late 1920s.

Kellogg was tipped off about these developments in his correspondence with the coterie of English and continental cetacean scientists with whom he had developed epistolary relationships as a young student of marine mammals. Indeed, no sooner did Kellogg have the resolution of the American Society of Mammalogists in hand than he fired off copies to Sir Sidney Harmer (just retired from the British Museum) as well as to Stanley Kemp (Harmer's hand-picked director of the Discovery Investigations).[262] It was Kemp who apprised Kellogg of the rapid developments at the League of Nations and the work of the ICES Whaling Committee, recommending that "your society in America" ought to put itself in touch with Hjort. Howell, interestingly, expressed misgivings about getting too close to the ICES committee, which he suspected would be biased by its strong links to the Norwegian industry.[263] Nevertheless, the Council for the Conservation of Whales had soon forwarded some of its materials—particularly statistics on the history and value of the US catches—to the Economic Committee of the League of Nations, and the CCW worked relentlessly to strengthen its ties to the relevant European scientists and scientific organizations.[264]

In forging links with those outside the United States who were strongly interested in whaling, Kellogg's industry friend Howard Wordell of the Nye corporation (who had, recall, orchestrated the Hatteras expedition) was invaluable. His position as a buyer and seller of whale products and proprietor of the only active whaling firm on the East Coast of the United

---

York: Oceana Publications, 1985), vol. 1, 112. All of chapter 3 provides useful background on this early stage of international legal negotiations.

262. Kellogg to Harmer, 8 September 1928, Smithsonian Archives, RU 7170, box 3, file "Harmer, Sidney F.," and Kellogg to Stanley Kemp, 9 July 1928, Smithsonian Archives, RU 7434, box 1, folder 11, "Correspondence, 1930s." For references to these communications, see Kellogg to G. Van Gelder, 3 October 1928, Smithsonian Archives, RU 7434, box 6, file "Correspondence, 1928–1940."

263. Stanley Kemp to Kellogg, 2 August 1928, Smithsonian Archives, RU 7434, box 6, file "Correspondence, 1928–1940."

264. E. M. K. Geiling to Kellogg, 14 July 1939 [*sic*, for 1929], Smithsonian Archives, RU 7170, box 3, file "Ga–Ge"; Kellogg to T. S. Palmer, 7 February 1929, Smithsonian Archives, RU 7434, box 6, file "Correspondence, 1928–1940" (seeking cooperation with "foreign biological societies"); see also Kellogg's letter on CCW letterhead to Ernst Stromer, 5 May 1929, Smithsonian Archives, RU 7170, box 8, file "Stromer, Ernst von."

States meant that he had an extended network of friends and partners in the industry, and his enthusiasm for Kellogg's work prompted him to do everything in his power to broker the CCW's contacts with major whaling figures abroad. In fact, Wordell even brought one of his international partners, the proprietor of a Scottish whale-oil refining company, along to attend the whale symposium at the American Society of Mammalogists in April 1928.[265] Other introductions to commercial agents and principals soon followed.[266] To ensure that the CCW did not look too "soft" or preservationist, and to seal their good relations with those in the know (and in the money) on whaling matters, Howell and Kellogg decided to supplement their Advisory Board with a "Technical Committee," which would draw heavily on industry figures. Wordell was recruited to serve the CCW in this capacity, and he helped fill out the committee's ranks. By 1929 the CCW numbered among its technical advisers Nagamichi Kuroda, a member of the Japanese royal family closely associated with the expanding Japanese industry; G. Van Gelder, the proprietor of a medium-sized Dutch-German whale-oil trading house (who had himself been present at the dawn of the Antarctic industry, having shipped on a whaling expedition in 1912); and Frederick H. Meisnest, an officer of the Atlantic Coast Fisheries Company, who had experience with the Pacific industry.[267] Of the new technical group, Howell wrote wryly to Kellogg, "What do you want to call this pack of hyenas—consultants, cetologists, heavy thinkers, or chair warmers?"[268]

In fact, some members turned out to be quite active. Van Gelder proved

265. H. I. Wordell to Kellogg, 30 March 1928, Smithsonian Archives, RU 7170, box 8, file "Wi–Wy." Wordell claimed that this man was the largest refiner of whale oil in Europe. He is not named, but was presumably associated with Salvesen. See Wray Vamplew, *Salvesen of Leith* (Edinburgh: Scottish Academic Press, 1975).

266. Frederick H. Meisnest to Kellogg, 1 July 1928, Smithsonian Archives, RU 7170, box 5, file "Mc–Merriam, C. W." (Meisnest introduced Kellogg to Van Gelder.) For a discussion of Kuroda, see Kellogg to Howell, 12 January 1929, Smithsonian Archives, RU 7170, box 4, file "Howell, A. Brazier, 1920–1929 and undated (2/3)."

267. On Van Gelder's experience, see Frederick H. Meisnest to Kellogg, 1 July 1928, Smithsonian Archives, RU 7170, box 5, file "Mc–Merriam, C. W."; on Meisnest's position, see unknown to Kellogg, 24 April 1939, Smithsonian Archives, RU 7170, box 5, file "Man–Max."

268. Howell to Kellogg ("Remmy"), 14 January 1929, Smithsonian Archives, RU 7170, box 4, file "Howell, A. Brazier, 1920–1929 and undated (2/3)."

to be a particularly tireless promoter of the cause: he worked to get word of the CCW's mission to the ears of Harold Salvesen (Lord Salvesen, the veritable king of the Antarctic whaling industry), who was a personal friend; introduced Howell and Kellogg to Sigurd Risting, the venerable secretary of the Norwegian Whaling Association (who had done early biological research on cetacean reproduction); and labored with considerable enthusiasm to collect and correlate catch statistics to supplement the CCW's briefs.[269]

Where regulation was concerned, Van Gelder's sense of urgency is evident in his correspondence not only with Kellogg and the CCW's other "mammalogicians" (as he called them), but also with Lewis Radcliffe, the acting commissioner of the Bureau of Fisheries in the Department of Commerce, to whom Van Gelder wrote a number of letters in the late 1920s, imploring him to take a leadership role in the drive for international restrictions on the industry.[270] As he put it to Radcliffe shortly after meeting with Kellogg in Washington in the summer of 1928:

> If all governments interested should appoint commissions for investigation, it will probably be the year 2028 that we might see any results, and those who fear that the present whale catching will eliminate the number of whales to practically nihil, will see that by the time the commissions have finished their work, no more whales will be left in the sea.[271]

Coordinated international action was, in his view, the only hope, since the fundamental problem, as he identified it, was that "the ocean is free outside the 12 mile limits." Reprising Suarez's view, Van Gelder lamented that "since Hugo Grotius wrote his 'mare liberum' in the 17th century, the questions of the free sea are not to be discussed again."[272] For this reason, the League of Nations appeared to be the most promising avenue for change, as Van Gelder insisted to Radcliffe and Kellogg both: "Since the League of Nations

269. G. Van Gelder to Howell, 1 March 1929, Smithsonian Archives, RU 7434, box 6, folder 10, "Correspondence 1928–1940." See also G. Van Gelder to Kellogg, 1 March 1929, in ibid. There is a whole folder of Van Gelder's statistical material in the Kellogg papers.

270. For "mammalogicians," see G. Van Gelder to Kellogg, 25 August 1928, in ibid.

271. G. Van Gelder to Lewis Radcliffe, 31 July 1928, in ibid.

272. Ibid.

probably will be unable to prevent the extermination of men through war, she might be interested in a more easy enterprise."[273]

If Van Gelder's earnest desire to see restrictions on Antarctic whaling is beyond doubt, his motivations were almost certainly quite different, at bottom, from those of Howell and Kellogg. In Van Gelder, the CCW had latched onto the drive for regulations sweeping through the established elements of the whaling industry in the late 1920s, a period that saw a massive expansion in productive capacity and matériel (between 1928 and 1929 no fewer than fifteen massive floating factories and seventy-five powerful new whale catchers were added to the Antarctic fleet) and a corresponding slide in prices.[274] It was a boom that terminated in the grim "black season" of 1931–1932 when overproduction, market glut, and the reverberations of a global financial crisis led to a collective agreement on the part of the Norwegian whaling companies to suspend the hunt, temporarily cashiering 10,000 workers and forfeiting what had been a steady source of nearly nine million pounds per annum.[275] In this context, Van Gelder's strong language—he called it "a most disgusting thing that so many companies are formed now"—must be understood to proceed primarily from dismay about the future of the increasingly overcapitalized industry.[276] For many who came to conservation questions from a background of dealing in whale products, the greatest promise of strict new whaling regulations lay in the glimmer of hope that such measures might help bring back the good old days of whale oil at £70 per ton and double-digit dividends.

Kellogg and Howell were no fools where these matters of realpolitik were

273. G. Van Gelder to Lewis Radcliffe, 25 August 1928, in ibid. Van Gelder discouraged them from making this point in Geneva.

274. Prices averaged £32 per ton for number one whale oil between 1924 and 1929 (See J. N. Tønnessen and Arne Odd Johnsen, *The History of Modern Whaling* [Berkeley: University of California Press, 1982], 370), down from all-time highs above £80; the crash and production crisis of 1929 to 1931 would drop the price to the low teens. See Karl Brandt, *Whale Oil: An Economic Analysis* (Stanford, CA: Food Research Institute, 1940), 224, and compare with Tønnessen and Johnsen, *History of Modern Whaling*, 753.

275. The Antarctic whaling season ran through the southern summer—corresponding to the northern winter—and therefore generally ran from roughly November to March. Hence seasons are usually indicated by two calendar years. For details on the buildup and collapse in this period, see J. N. Tønnessen and Arne Odd Johnsen, *The History of Modern Whaling* (Berkeley: University of California Press, 1982), chaps. 21, 22; the figures cited here are from pp. 386, 391.

276. G. Van Gelder to Lewis Radcliffe, 31 July 1928, Smithsonian Archives, RU 7434, box 6, folder 10, "Correspondence 1928–1940."

concerned. I have already shown how they carefully measured the tone of their whale conservation propaganda so as to defend against the charge that they were merely sentimental preservationists. Schooled in the infighting and crossed alliances of American wildlife campaigns after World War I (in which ammunition companies co-opted conservation organizations for their own purposes and accusations about selling out to special interests flew thick and fast),[277] Kellogg and Howell played their cards very carefully as they built the CCW consortium. For instance, when, in the summer of 1929, Norway unilaterally passed legislation prohibiting vessels flying the Norwegian flag from killing right whales, calves, or nursing females, the CCW circulated a favorable brief on the initiative, praising the measure as "a cause for no little elation among those interested in maintaining a commercial supply of whales."[278] Similarly, Kellogg was careful in his correspondence with Wordell and others in the industry to emphasize that the CCW was fundamentally interested merely in what he called "remedial legislation . . . necessary to perpetuate the industry."[279] Playing both ends against the middle, however, Kellogg was also more than willing to keep up extensive correspondence with a director at the American Humane Association, guiding him in raising his membership's ire about the grisly fate of whales (blown up alive, electrocuted, turned into livestock feed) and ensuring that these animal crusaders knew to whom in the State Department or in Congress they ought to write their angry letters.[280] Howell, for

277. For a taste of the bickering over these matters where terrestrial wildlife was concerned, see William Temple Hornaday's score-settling manifesto, *Thirty Years War for Wild Life* (New York: Arno, 1970 [1931]).

278. G. Van Gelder to Kellogg, 25 August 1928, Smithsonian Archives, RU 7434, box 6, folder 10, "Correspondence 1928–1940." (The law also set forth other requirements about the utilization of carcasses and the payment of gunner bonuses.)

279. Kellogg to H. I. Wordell, 20 February 1929, in ibid.

280. This was W. E. Sanderson: his correspondence with Kellogg can be found in the Smithsonian Archives, RU 7170, box 7, file "S–San." The reference to the use of whale products in cattle food raises an interesting feature of twentieth-century whaling that has not been explored: the whole episode can perhaps be best understood as the elaboration of an extensive "meat economy" associated with intensified agricultural practices after World War I. The scope of the undertaking can be grasped when we consider that the major use of baleen whales in the United States in the twentieth century was as protein and fat supplements in chicken feed; in Canada, the Soviet Union, and the United States, dehydrated whale meat was used to feed mink and silver foxes raised on large fur farms. Strikingly, the demand for these products was created in part by the rise of commercial automobiles and the end of the use of horses for transportation

his part, was never afraid to throw a bomb if he thought it could do some good (as in the case of his increasingly scorched-earth campaign against the Bureau of Biological Survey), but he could also be immensely savvy when it came to legislative advocacy. He knew how to tiptoe around the deep pockets and how to pick them whenever possible. For instance, he recommended a tweak to one of Kellogg's briefs, suggesting that the way Kellogg had worded the CCW's position "smacks of government regulation of private business and someone may get off on the wrong foot." Howell insisted that they stick to a wise-use "theme song" that ran like this: "protective measures needed for perpetuation of a North American breeding stock of whales."[281]

Still, despite these careful efforts to balance conservation and commerce, the CCW was eventually rattled by conflicting ideas about how best to stand with respect to the massive and powerful whaling industry. This crucial episode, which unfolded at the end of 1929 (and resulted in the withdrawal of two members of the council), is instructive for what it reveals about the tensions Kellogg and Howell faced in holding together their diverse consortium. At issue in the conflict, interestingly, was scientists' access to whales and whalers. In the summer of 1929, as the CCW's publicity campaign was in full swing, Charles Haskins Townsend, director of the New York Aquarium and member of the CCW's Technical Committee (he had been, recall, an adviser on the Hatteras expedition), arranged a trip to the West Coast to visit a shore whaling station, collect specimens, and investigate the situation of the expanding industry in California waters. Townsend was no stranger to the study of Pacific fauna, having contributed to the fur seal investigations in the late nineteenth century. The timing of his 1929 reconnaissance was fortunate, since the California Sea Products Company had recently arranged a lease agreement by which the Norwegian factory

---

and farm labor in the first decades of the twentieth century; up to that time, horsemeat could be used for these purposes, and there was enough of it to meet the demand. On whale meat in the United States (mostly before 1900), see Nancy Shoemaker, "Whale Meat in American History," *Environmental History* 10, no. 2 (April 2005): 269–94. For a taste of the meat economy, see Alexander Cockburn, "A Short Meat-Oriented History of the World," *New Left Review* 1, no. 215 (January–February 1996): 16–42. On the fate of horses in the United States, see Ann Norton Greene, "Harnessing Power: Industrializing the Horse in Nineteenth Century America" (PhD diss., University of Pennsylvania, 2004).

281. Kellogg to Howell, 16 December 1929, Smithsonian Archives, RU 7170, box 4, file "Howell, A. Brazier, 1920–1929 and undated (3/3)."

vessel *Lansing* (recently returned from gray whaling in the lagoons of Baja California) gained access to the waters off San Clemente Island.[282] Intrigued by the prospect of being the first American naturalist to take passage on a Norwegian factory vessel and to witness Antarctic-style whaling upclose, Townsend succeeded in working out a visiting arrangement with the help of officers in the California State Fisheries Laboratory at Terminal Island, who ferried him out to the *Lansing* in the department's patrol boat.

During the visit Townsend got to witness the taking of an 80-foot female blue whale from the catcher boat *Saunders,* and he enjoyed the run of the factory vessel, collecting a set of unique photographs of the butchering operations and talking informally with the captain and members of the crew. He came back to New York loaded with material for an exceptional article on the industry (it would run, with color illustrations, in the first 1930 issue of the *Bulletin of the New York Zoological Society* under the title "Twentieth-Century Whaling" [figure 3.16]).[283] But he was also acutely sensitive about keeping doors open with his new network of Norwegians. Among other things, he had negotiated an arrangement by which statistics would be forwarded by the Globus Company of Sandefjord, Norway, directly to him at the New York Zoological Society. It appears that Townsend was beginning to imagine that the New York Aquarium might secure for itself a role something like that of the Bureau of International Whaling Statistics, an international clearinghouse for cetacean catch data based in Norway and coordinated by Norwegians. Anxious to preserve his relationships, and apparently persuaded by the unanimously laissez-faire view among the whalemen with whom he spoke (who argued that modern whaling was so expensive that reduced catches would necessarily end the industry long before the world's whale stocks were in danger of extinction), Townsend decided to sever his connections with the CCW, distancing himself from any potentially embarrassing or overzealous moves by Howell or Kellogg. It also seems quite likely that Townsend—considerably older than the core group at the CCW,[284] with decades of experience working on the biology and conservation of marine mammals—harbored some degree of resentment toward the rapid success of their organization, and thus aimed to

282. This vessel is not to be confused with the *Lancing*, the first practicable stern slipway vessel; the *Lansing* was not so equipped.

283. Charles H. Townsend, "Twentieth-Century Whaling," *Bulletin of the New York Zoological Society* 33, no. 1 (January–February 1930): 3–31.

284. Townsend was 33 years older than Kellogg.

January-February, 1930

BULLETIN
NEW YORK ZOOLOGICAL
SOCIETY

FIGURE 3.16 Thar she blows: Cover illustration for Charles H. Townsend's article, "Twentieth-Century Whaling." (Courtesy of Firestone Library, Princeton University.)

re-center US whale conservation work on New York City, and on his own terms. Such a plan would explain his move to pull the society lawyer and platinum philanthropist Madison Grant (a member of the CCW Advisory Board and president of the New York Zoological Society) away with him.

In a confidential letter to Howell tendering his resignation, Townsend explained that he felt he "must continue on good terms" with the whaling companies, and he expressed pessimism about the immediate prospects for regulatory legislation.[285] It was better to wait, in his view, and forge alliances on the inside of the industry, instead of going off half-cocked before there was a real battle to be won. Howell was furious, writing Kellogg that Townsend was "all wet" and that he had "Grant buffaloed." Howell encouraged Kellogg to make no secret of Townsend's indefensible position: "If people felt that they must at all cost maintain friendly relations with those who are responsible for the extermination of wildlife, what a nice situation would obtain," he fumed.[286] At core, it was a generational conflict, though it manifested itself in a contest over who was more "in the know" about the world whaling situation and who had more links to British and Scandinavian informants. Kellogg wrote a blistering critique of Townsend to Grant, accusing the senior sea naturalist of being "far from . . . familiar with present developments" and pointing out that others in the old guard of the seal conservation crowd (like Stejneger) were much closer to the CCW than to Townsend. "It will not speak well for any efforts toward a rational conservation of our resources of pelagic mammals," Kellogg pointed out, if it "becomes known that the New York Zoological Society" was courting those "exterminating what few whales remain along our coasts."[287] Grant stuck by Townsend nevertheless, and the CCW had to get new letterhead made up, minus two prominent names.[288]

Townsend's bid for leadership on the whale issue, however, came to

285. Howell to Kellogg, undated, Smithsonian Archives, RU 7170, box 4, file "Howell, A. Brazier, 1920–1929 and undated (3/3)."

286. Ibid.

287. Kellogg to Madison Grant, undated, Smithsonian Archives, RU 7170, box 3, file "Ge–Gre."

288. A. Brazier Howell to Kellogg, 14 December (year unknown), Smithsonian Archives, RU 7170, box 4, file "Howell, A. Brazier, 1920–1929 and undated (3/3)." Howell to Kellogg: "Hereafter when using the whaling stationery be sure to draw a bold stroke through Townsend's name, else we may have a libel suit on our hands." See A. Brazier Howell to Members of the Council, 3 September 1930, Smithsonian Archives, RU 7170, box 10, file "Information—Whale Press Releases" for an example of this strikeout;

naught.[289] He was defeated, at least in part, by geography. Kellogg and Howell had the advantage being located in Washington, DC, with its proximity to the institutions of federal power. They husbanded that resource carefully, cultivating their relationships with the director of the Bureau of Fisheries and, as international action took shape, with the State Department as well.[290] In fact, it is at least plausible that Townsend's rearguard action in late 1929 was part of a last-ditch effort to squeeze himself into the international advisory role that Kellogg would shortly win.

By the time Townsend and Grant withdrew from the CCW, it had become clear that the League of Nations was going to convene a special session of experts in Berlin to draft preliminary treaty language for a whaling convention. Around January 1930 the American representative on the Economic Committee of the League, Lucius R. Eastman, received authorization to put forward the name of a suitable American whale expert. At this point the CCW's two years of agitation paid off, since Kellogg and Howell had been in touch with Eastman going back to 1928, when his request for statistics on the US whaling industry found its way to Kellogg's desk after bouncing around the Department of Commerce.[291] And Eastman was not the CCW's

---

A. Brazier Howell to Kellogg, 1 March (year unknown), Smithsonian Archives, RU 7170, box 4, file "Howell, A. Brazier, 1930–1931 (2 / 3)" for the revised stationery.

289. Although, interestingly, one (historically) significant legacy of his effort is the "Townsend charts," which appear as a part of his "The Distribution of Certain Whales as Shown by Logbook Records of American Whaleships," *Zoologica* 19, no. 1 (3 April 1935): 1–50. These were Townsend's elaborate plottings of nineteenth-century New England whaling catches by season and by species—maps that have continued to intrigue whale scientists as a source of historical data. What is not well known about these charts is that they represent Townsend's "takeover" of a project called for by Kellogg at the American Society of Mammalogists meeting in 1928, where he gave a talk calling for a review and collation of all old logbooks in American collections. To be fair, Townsend had done plots of catch data for the fur seal investigation (see volume 3 of *Zoologica*; see also Kellogg to Charles H. Townsend, 12 November 1927, Smithsonian Archives, RU 7170, box 8, file "To–Ty"), and the final format of the whale charts must be understood to derive from this earlier work.

290. Kellogg wrote to Howell, concerning Radcliffe, "It seems to me a matter of prime importance that the Council for the Conservation of Whales should not work at cross purposes with the aims of the Bureau of Fisheries." Kellogg to Howell, 1 January 1931, Smithsonian Archives, RU 7170, box 4, file "Howell, A. Brazier, 1930–1931 (2/3)."

291. See Arthur Salter to Lucius R. Eastman, 17 April 1928; Julius Klein to Kellogg, 14 May 1928; and Kellogg to Julius Klein, 16 May 1928, all in Smithsonian Archives, RU 7170, box 2, file "Ea–El."

only contact in international legal arenas by this time. The distinguished American jurist Philip C. Jessup, professor of law at Columbia University, had become the leading legal figure in the United States on the law of the sea in this period, and by 1929 he was preparing a set of lectures to deliver at The Hague on the exploitation of the riches of the sea. Jessup, too, via Howell and the ccw, found his way to Kellogg, initiating an exchange that would be invaluable to Kellogg as he stepped onto the unfamiliar terrain of international treaty making.[292] Jessup was in contact with Eastman as well, and by February 1930 Kellogg had an unofficial letter in hand, letting him know that he would be invited to participate in the Berlin meeting.[293]

As a government employee, Kellogg required State Department approval before he could participate in any official business of the League of Nations, but Stejneger and Wetmore helped arrange things on that side, and Howell, euphoric, passed the hat among the ccw's supporters to collect pocket change for their envoy.[294] Moreover, he pressed Kellogg to write to Merriam immediately to see if the Carnegie Institution might "shower down to have you visit all the cetacean fossils" while in Europe.[295] Kellogg did just that, listing the "major fossil cetacean collections" that he hoped to visit and respectfully requesting any available funds.[296] Merriam ponied

292. See Philip C. Jessup to Kellogg, 25 May 1929, and Kellogg to Philip C. Jessup, 22 June 1929, both in Smithsonian Archives, RU 7170, box 5, file "In–Jon." See also the remarkably detailed letter from Jessup to Kellogg giving him all of the precedents he could find for treaties that obliged individual nations to exercise their enforcement powers. Philip C. Jessup to Kellogg, 13 October 1930, Smithsonian Archives, RU 7165, International Whaling Commission Papers, etc., 1930–1968, box 7, file "Department of State, 1930–1945 Correspondence."

293. Unknown to Lucius R. Eastman, undated, Smithsonian Archives, RU 7170, box 2, file "Ea–El."

294. Kellogg to Howell, 17 February 1930, Smithsonian Archives, RU 7170, box 4, file "Howell, A. Brazier, 1930–1931 (2/3)." Kellogg was in fact sent in a "purely personal capacity" and "not in any way representing his government"; this simplified matters greatly. Kellogg to Lucius R. Eastman, 15 March 1920, Smithsonian Archives, RU 7170, box 2, file "Ea–El." For Howell's efforts to raise additional funds, see Howell to Kellogg, 30 January 1930, and Howell to Kellogg, 30 March (year unknown, presumably 1930), both in Smithsonian Archives, RU 7170, box 4, file "Howell, A. Brazier, 1930–1931 (2/3)."

295. Howell to Kellogg, 7 January 1930, and Howell to Kellogg, 22 January 1930, both in Smithsonian Archives, RU 7170, box 4, file "Howell, A. Brazier, 1930–1931 (2/3)."

296. Kellogg to Merriam, 17 February 1930, Smithsonian Archives, RU 7170, box 2, file "Ea–El." The League of Nations would cover his steamer fare, and he received a per diem for the period of the conference itself.

up several hundred dollars, along with a number of letters of introduction, including a valuable one to Johan Hjort, his old friend and classmate from student days in Munich.[297]

On 24 March Kellogg, armed with a trunk of details on "what ails whales" (as he put it to Howell), and carrying a position paper on behalf of the CCW (to be circulated among the Berlin delegates), set sail on the Holland-America liner *Statendam* for Rotterdam, whence he preceded to Berlin by train.[298] Just before his departure, Lucius Eastman gave him some advice about the proceedings. While momentum was surely gathering, he should not be overly optimistic:

If you have any experience in these matters, you recognize that things move very, very slowly and it takes a good deal of preparation before you can accomplish much by united action. We have been at this matter for two or three years before being able to get this conference of experts together.[299]

Howell was less diplomatic:

Don't let that bunch put anything over on you. You can pound the table and yell as loud as any of them, and I doubt if any of them has had better diplomatic training than almost ten years in the B.S.[300]

297. For Merriam's contribution, see Merriam to Howell, 27 March 1930, Smithsonian Archives, RU 7170, box 6, file "Merriam, John C. (2/3)"; for details on letters of introduction, see Kellogg to Howell, 17 February 1930, Smithsonian Archives, RU 7170, box 4, file "Howell, A. Brazier, 1930–1931 (2/3)."

298. For his comments about what ails whales, see Kellogg to Howell, 17 February 1930, Smithsonian Archives, RU 7170, box 4, file "Howell, A. Brazier, 1930–1931 (2/3)"; for the CCW's position paper, see Howell to Council for the Conservation of Whales and Other Marine Mammals, undated, Smithsonian Archives, RU 7165, box 3, folder 4, "Berlin—Whaling Conference under auspices of League of Nations, 1930–1931"; for his vessel, see Kellogg to Merriam, 19 March 1930, Smithsonian Archives, RU 7170, box 6, file "Merriam, John C. (2/3)." Marguerite, it should be noted, accompanied him on this trip (so we are without his letters to her about the journey); see Kellogg to Harmer, 30 March 1931, Smithsonian Archives, RU 7170, box 3, file "Harmer, Sidney F."

299. Eastman to Kellogg, 18 March 1930, Smithsonian Archives, RU 7170, box 2, file "Ea–El."

300. Howell to Kellogg, 30 March 1930, Smithsonian Archives, RU 7170, box 4, file "Howell, A. Brazier, 1930–1931 (2/3)."

.   .   .   .

There can be little doubt that the 1930 European trip represented a formative experience for Kellogg, who, at 37, found himself with an all-expenses-paid whirlwind tour of the great zoological and paleontological collections of Britain and the Continent, along with a unique opportunity to play at being a scientific plenipotentiary in international relations.[301] Over a month and a half, he made a circuit of fourteen cities, visiting more than a dozen major collections and meeting personally with nearly every foreign scientist to whom he had ever written a letter.[302] And this grand tour began with a bang: three days at the stately Berlin offices of the League of Nations on the Hedmannstrasse, working with a select handful of senior fisheries officials, biologists, and technocrats as they hammered out a draft convention for the regulation of whaling (modeled in large part on the recent Norwegian law).[303] While he was by no means starstruck by the company (he later wrote somewhat dismissively to an old friend about the "so-called experts" who haggled over the final language), he was nevertheless sharing a place at the table with men like E. R. Darnley and J. O. Borley, poised regulatory diplomats who had created and funded the Discovery Investigations (and provided hundreds of thousands of pounds for cetacean research), not to mention Johan Hjort himself, already a legendary figure in fisheries biology.[304] In the end, the "Preliminary Draft for the Regulation of Whaling" covered many of the restrictions Kellogg and Howell had pressed for, such as a full ban on the killing of right whales and nursing females. In addition, the proposed convention required signatory nations to ensure that their whalers collected biological data on any whale they killed anywhere in the world.[305] To be sure, plenty of difficult issues—the definition of an

301. With the emphasis here on "play"—since Kellogg was not, of course, a formal US delegate.

302. His itinerary can be found in Kellogg to Wetmore, 17 March 1930, Smithsonian Archives, RU 7170, box 8, file "Wa–Wh."

303. His annotated drafts on this document can be found at the Smithsonian Archives, RU 7165, box 3, folder 4, "Berlin—Whaling Conference under auspices of League of Nations, 1930–1931."

304. The other members of the committee were M. A. Hodt (Germany), Ed. Le Danois (France), M. Ohta (Japan), and A. M. Ramalho (Portugal).

305. That is to say, whalers on vessels under the flags of the signatory nations. The draft can be found as League of Nations, Economic Committee, *Report to the Council on the Work of the Thirty-Second Session, Held at Geneva from June 2nd to 6th, 1930*

"immature" whale, protections in tropical waters—were left open in Berlin. Kellogg wasn't overly impressed, and he mugged for his friends back home, complaining to A. Brazier Howell of the pointless pomp and bureaucracy: "How about organizing a gas attack on the League of Nations?" wrote Remmy from the Continent.[306] It could be a project for the CCW. Might be better for the whales, anyway, in the end. But he was showing off—a little American bravado.

In fact, the experience had been remarkable. Kellogg had had his first taste of international negotiations, and he had actually participated in the laborious process of achieving consensus on the conservation of natural resources. After that, he packed his trunks, and it was Stuttgart, Munich, Vienna, Venice, Padua, Bologna, Turin, Paris, Brussels, Haarlem, and finally London, where, before shipping for home, Kellogg spent almost two weeks perusing the cetacean collections of the British Museum in the company of Harmer, Hinton, Tate Regan, and a number of other scientists associated with the Discovery work.[307] Everywhere he went, he was received by his fellow "Cetaceenforscher" (cetacean researchers) with courtesy and enthusiasm.[308] As he wrote to a friend, "Meeting the men who are working in this field was a pleasure, and I'm sure will be helpful in the years to come." There were a few disappointments (Professor Abel was away on vacation when Kellogg passed through Vienna), but the opportunity to study the actual type specimens for so many familiar cetacean fossils was invaluable.[309] He devoured collections, filling four hundred notebook pages with tight-handed commentaries on thousands of bones.[310] These notebooks reveal

---

(Publication II, Economic and Financial, 1930 II 24 C.353.M.146). It is interesting to note that CCW correspondence remains in the ICES Records, Rigsarkivet, box 85, folder 6Q, as well as in Oslo, in the Johan Hjort Papers, Nasjonalbiblioteket MS 4° 2911: XIX A, file "Howell, A. Brazier."

306. Kellogg to Howell, 2 June 1930, Smithsonian Archives, RU 7170, box 4, file "Howell, A. Brazier, 1930–1931 (2/3)."

307. Kellogg to James Ritchie, 4 June 1930, Smithsonian Archives, RU 7170, box 7, file "Re–Ri," and Smithsonian Archives, Accession 99–012, Remington Kellogg Papers, 1926–1950 and undated, box 1, "European Museums, research notes, volume 2."

308. The German appears in a letter from Ernst Stromer to Kellogg, circa 1946, Smithsonian Archives, RU 7170, box 8, file "Stromer, Ernst von."

309. Kellogg to Baron Fr. Nopesa, 5 June 1930, Smithsonian Archives, RU 7170, box 6, file "Ne–Nu."

310. For his actual notes, see Smithsonian Archives, Accession 99–012, box 1, "European Museums, research notes, volume 2." See Kellogg to Harmer, 30 March 1931,

the hard work that Kellogg did to make good use of his time in these exotic holdings. Vocabulary lists penciled on the flyleaves gave foreign terms for such tools of the trade as calipers and recorded useful polite phrases for the visiting whale scientist: *"Avez vous le memoire de Portis?"* Mais oui: *"a la table de laboratoire."*[311] In Belgium, at the Musée Royal d'Histoire Naturelle (which held the type specimens of Van Benden), he was bowled over by ten large mounted cetacean skeletons, a display unlike anything he had ever seen before.[312] He declared it "most impressive" in his notebook and lamented only the paucity of periotics in the private collection.[313] Stuttgart was perhaps even more memorable.[314]

At the same time, this midwestern American, accustomed to plain, sturdy displays and clean, prairie Gothic buildings, found these actual Gothic buildings, vaulted and venerable as they were, a little cluttered sometimes, even fusty, particularly in the cabinets and public areas: "I never saw such quantities of junk on exhibition in all my life," he commented to a friend back home. "The museums over there remind one of Dickens' Old Curiosity Shop."[315] Other aspects of European life were equally dismaying: Kellogg fretted over the costs of porters and hotels and puzzled over being served "fish poached in milk" for breakfast.[316] Nevertheless, everywhere he stopped, he left offprints of his *Quarterly Review* article, and by the time he made it back to Washington and began sorting his papers and clearing his accounts, he knew he had the makings of a major monograph on the prehistoric whales in his thick sheaf of notebooks. Those that survive reveal how minutely he annotated his European collection notes in the years to come, clearly returning to them again and again as he worked up the manuscript that would become the major scientific publication of his research life, *A Review of the Archaeoceti*, published in 1936. When this

Smithsonian Archives, RU 7170, box 3, file "Harmer, Sidney F.," for his boasting of the number of pages.

311. *Sic*, without diacritical marks, etc.

312. Smithsonian Archives, Accession 99–012, box 1, "European Museums, research notes, volume 2."

313. Ibid.

314. Kellogg to Roy L. Moodie, 27 June 1930, Smithsonian Archives, RU 7170, box 6, file "Mi–Moo."

315. Ibid.

316. On expenses, see Kellogg to Howell, 2 June 1930, Smithsonian Archives, RU 7170, box 4, file "Howell, A. Brazier, 1930–1931 (2/3)." On the strange breakfasts, see Kellogg to Dr. Fraser, 30 March 1940, Smithsonian Archives, RU 7170, box 3, file "Fra–Fuj."

massive "landmark in cetology" came out, with its detailed survey of all major cetacean fossils and its synthetic essays on the morphological and phylogenetic history of the order, John C. Merriam, now 70, wrote him with (undeniably self-regarding) pride:

Perhaps I flatter myself when I say that this is carrying out an idea which I once had in mind and it can therefore be considered as part of my thinking and research interest at this time of my life. Incidentally, I am quite sure that the whole project as you are carrying it forward is doing more toward development of acquaintance of this great problem that I would have been able to do. Being committed to the idea that the world should grow better and better, I am glad to think of myself as only a stepping stone in the sequence.[317]

As if in confirmation of this torch-passing sentiment, Kellogg included a homage to Merriam in "Adaptation of Structure to Function in Whales," his summary of the physiological researches of the late 1920s and early 1930s. Published in 1938, the piece was effectively Kellogg's last technical publication on whales and dolphins.[318] Increasingly, administrative responsibilities would shape his professional life from that point forward.[319]

Even as those final scientific labors absorbed him in the 1930s, Kellogg remained active in the Council for the Conservation of Whales, which continued to press for ratification of the international treaty Kellogg had helped to draft (and to support related US legislation). Increasingly, however, Kellogg preferred to work behind the scenes on these matters. By using Howell and others as his mouthpieces, Kellogg managed to stay just

317. Merriam to Kellogg, 6 November 1939, Smithsonian Archives, RU 7170, box 6, file "Merriam, John C. (1/3)." "Landmark in cetology" comes from Frank C. Whitmore Jr., "Remington Kellogg (1892–1969)," *Year Book of the American Philosophical Society* (1972): 205–10, at p. 207.

318. Remington Kellogg, "Adaptation of Structure to Function in Whales," *Cooperation in Research, Carnegie Institution of Washington Publication no. 501* (1938): 649–82.

319. This is not to say that Kellogg did not have plenty of additional publications: the typescript full bibliography prepared by Elwyn J. Simons in April 1962 (copy in Smithsonian Institution, Remington Kellogg Library) lists 56 items after 1938 (out of a total of 167); nevertheless, the vast majority of these are reviews, briefs, or edited works. His administrative responsibilities included heading the Division of Mammals and, later, serving as director of the United States National Museum, and eventually as assistant secretary of the Smithsonian as a whole.

beyond the muddy fray of legislative politics, thereby protecting his status as a staid and reliable civil servant and avoiding any hint of rabble-rousing. Berlin showed Kellogg what it took to become an insider on these questions, and in the years to come he directed his steps to that end. In 1937 and 1938, when the major whaling nations of Europe agreed to revisit the question of international whaling regulation (prompted in large part by the push of Germany and Japan into the industry), Kellogg received a formal appointment from the State Department to serve in the US delegation to meetings in London and Oslo.

By that time Kellogg really was an expert, having for years played a subtle and important role in giving legislative teeth to the wish list of whale protections he and Howell first conjured up back in the late 1920s. A legislative history of these competing bills would be a book in itself, and a serious diplomatic history of progress toward international conventions in this period would require yet another volume, but a précis of these intersecting stories is necessary in order to show how Kellogg, Howell, and the ccw helped steer regulatory initiatives through complex political processes.

The draft convention finalized by Kellogg, Hjort, and the other experts in April 1930 at Berlin was approved by the Economic Committee that summer and transmitted with various countries' comments to the League of Nations Assembly in Geneva on 9 September. Shortly thereafter, with minor revisions, this draft convention was signed by eleven national representatives, setting in motion the individual nations' ratification processes.[320] These processes unrolled slowly, and with considerable bickering among the stakeholders, who were simultaneously engaged in private quota negotiations.[321] Thanks in part to the work of the ccw, the United States led the way, completing its ratification by the summer of 1932.[322] Great Britain's Parliament took until January 1935, and it was only

320. Kellogg to Glover M. Allen, 9 April 1932, Smithsonian Archives, RU 7170, box 1, file "Allen, Glover M., 1921–1941," for a summary of this process.

321. See J. N. Tønnessen and Arne Odd Johnsen, *The History of Modern Whaling* (Berkeley: University of California Press, 1982), 400ff.

322. For a summary of progress on ratification in the early 1930s, see Carl D. Shoemaker to Howell, 12 April 1934, Smithsonian Archives, RU 7170, box 4, file "Howell, A. Brazier, 1932–1943 (1/3)." For Howell and Kellogg's work toward US ratification, see Kellogg to Howell, 18 December 1931, Smithsonian Archives, RU 7170, box 4, file "Howell, A. Brazier, 1930–1931 (2/3)"; Howell to Kellogg, 17 February (year unknown, presumably 1932), Smithsonian Archives, RU 7170, box 4, file "Howell, A. Brazier, 1932–1943

with their formal approval of the document that the Geneva Convention for the Regulation of Whaling became effective, binding all the signatory nations.[323]

During these years Kellogg and Howell used the CCW to push for passage of the right kind of enabling legislation under US law to ensure that the convention, when it became effective, would be adequately buttressed by regulations grounded in US federal legislation.[324] To this end, Kellogg (as a curator of mammals in the United States National Museum) and Howell (as "executive secretary" of the CCW) gave testimony to the Senate's "Special Committee on Wild Life Resources," chaired by Frederick C. Walcott of Connecticut. Together, Howell and Kellogg drafted other CCW members into this hearing, where they were assured of a favorable reception, since Walcott was Philip C. Jessup's uncle, and sympathetic to their cause.[325] These audiences gave Kellogg an opportunity to put his researches on the history and statistics of the industry into the public record, and to do so in a way that could directly affect political action. One of his friends wrote him enthusiastically after the performance. "You certainly are the vibrissae of Felis cattus [i.e., the "cat's whiskers"] when it comes to whale hearings. When the public finds out that Camay soap is made out of nasty whale blubber, perhaps whales will not be pursued so

---

(1/3)"; and Howell to Kellogg, 6 April (year unknown, presumably 1932), Smithsonian Archives, RU 7170, box 4, file "Howell, A. Brazier, 1932–1943 (1/3)." It should be noted that Norway and the United States were together the earliest ratifiers.

323. The convention was constructed in such a way as to require eight nations to ratify to become effective, but it was stipulated that both Norway and Britain had to be among those eight.

324. For the active lobbying of individual legislators by the CCW, see Kellogg to Howell, 1 January 1931, Smithsonian Archives, RU 7170, box 4, file "Howell, A. Brazier, 1930–1931 (2/3)"; unknown to F. M. Barnes, 21 January 1931, and Kellogg to Howell, 18 December 1931, in ibid. Howell also worked to bring US corporations into the process, particularly Procter & Gamble, a major US buyer of whale oil for soap and cosmetics. See unknown to F. C. Walcott, 21 January 1939, and Kellogg to Howell, 18 December 1931, both in ibid.

325. For a transcript of this testimony, see U.S. Congress, Hearing Before a Special Committee on Wild Life Resources, United States Senate, Seventy-Second Congress, First Session, 20 March 1931. I found this, and related materials, in the Smithsonian Institution, Remington Kellogg Library, Kellogg binder books 3/14. T. S. Palmer, one of the CCW's technical advisers, also gave testimony. For a discussion of the work of this committee, see William Temple Hornaday, *Thirty Years War for Wild Life* (New York: Arno, 1970 [1931]), 253ff.

energetically."[326] Kellogg would later offer similar testimony to the House Foreign Affairs Committee.[327]

These public performances were just a small part of Kellogg's effort, however, since backdoor approaches proved increasingly important. For instance, Kellogg did *not* sign a letter sent by the American Society of Mammalogists to the president of the United States, expressing their "grave concern" about the whaling situation (although he almost certainly wrote it).[328] Dissembling still further, Kellogg encouraged Howell to structure the ccw as a kind of "shell" by which he (Kellogg) could anonymously feed information to legislators, an arrangement that would build the ccw's standing and keep Kellogg from getting tangled up in restrictions on his activities as a government employee. As Kellogg put it, "Now in regard to the matter of furnishing data to whoever introduces the bill . . . I can furnish you I believe with whatever data may be desired and you can furnish it in the name of the Council."[329]

This was the arrangement they used as the whaling regulation bill (S. 3413) made its perilous journey through the Senate and House in the mid-1930s, struggling out of several committee traps where shadowy opponents (perhaps linked to Alaskan fisheries?) hoped to have it killed.[330] Increasingly, Kellogg stuck close to the wishes of the senators he had come to know and trust (like Peter Norbeck of South Dakota, who was sympathetic to wildlife conservation), and he cultivated his connections in the Coast Guard, the Department of Commerce, and the Treaty Division of the State Department, all of which had to speak with one voice if the bill was to be saved. More and more, he moved with caution around his older and more vociferous allies, writing, for instance, to his main contact at the American Humane Association in February 1936 (as the whole of the legislation hung

326. Marcus W. Lyon to Kellogg, 19 February 1932, Smithsonian Archives, RU 7170, box 5, file "Lyn–Lyc."

327. See U.S. Congress, The Whaling Treaty Act, Hearings, Seventy-Fourth Congress, First Session (S. 3413), February 1936.

328. M. W. Lyon to The President of the United States, 19 January 1932, Smithsonian Archives, RU 7165, box 1, folders 5–7, "Laws, treaties, etc. relating to marine mammals, circa 1929–1941."

329. Kellogg to Howell, 26 September 1930, Smithsonian Archives, RU 7170, box 4, file "Howell, A. Brazier, 1930–1931 (2/3)."

330. For Kellogg on the trajectory of the bill, see Kellogg to Howell, 21 February 1936, Smithsonian Archives, RU 7170, box 4, file "Howell, A. Brazier, 1932–1943 (1/3)." See also Kellogg to W. E. Sanderson, 21 February 1936, Smithsonian Archives, RU 7170, box 7, file "S–San."

in the balance) that "in matters like this, newspaper publicity and the like is not very helpful." Above all, it was key that "the nature of the support" not be "misinterpreted" by those involved—code for "pipe down."[331] Attuned to these delicacies, an older and wiser Kellogg now preferred to work from within the establishment wherever possible: closing that letter, he wrote pointedly, "please keep this letter *confidential*, since I am prohibited from expressing any opinion in regard to current legislation." He therefore needed to be kept out of any public connection to the proceedings, except where he might be called on to provide information or answer questions.[332]

Meanwhile, Howell continued to push the CCW's "straight ticket" of cetacean protections, to field correspondence from supporters on CCW letterhead, and to help coordinate alliances with the International Wildlife Protection Organization and other well-connected conservation groups.[333] Hewing closely to Kellogg's directives, however, Howell held his fire when it looked as if less would be more. The CCW even kept out of an ugly spat over bounty hunting for sea lions on the Pacific Coast (against Howell's better judgment) because Kellogg wanted to hold the CCW in reserve as an ally in the whale regulation battle and was concerned that the organization not look too scrappy.[334]

By December 1936 all the hard work paid off: the whale act passed, and the new Department of Commerce Circular No. 300, "Laws and Regulations for the Protection of Whales," spelled out in detail the new federal protections for cetaceans in US waters. As Kellogg wrote to Wordell with some pride, "so that committee on which you have been a member for so many years has at last finished its job."[335]

---

331. Kellogg to W. E. Sanderson, 21 February 1936, Smithsonian Archives, RU 7170, box 6, file "S–San."

332. Ibid.

333. For the language of the "straight ticket," see Howell to Kellogg, 23 January 1935, Smithsonian Archives, RU 7170, box 4, file "Howell, A. Brazier, 1932–1943 (1/3)"; for Howell's work rallying supporters, see Kellogg to Howell, 30 March 1935, in ibid.; on the International Wildlife Protection Organization, see Kellogg to Merriam, 18 March 1922 forward, Smithsonian Archives, RU 7170, box 6, file "Merriam, John C. (3/3)" (the chairman of this group, John C. Phillips, had a brother with a high-ranking appointment in the State Department).

334. Undated sheet, "Palmer has now outlined . . . ," in Smithsonian Archives, RU 7170, box 4, file "Howell, A. Brazier, 1932–1943 (1/3)."

335. Kellogg to H. I. Wordell, 17 December 1936, Smithsonian Archives, RU 7434, box 6, folder 10, "Correspondence, 1928–1940."

. . . .

With this success, and with Kellogg's increasingly formal participation in international regulatory efforts in the late 1930s, the Council for the Conservation of Whales gradually ebbed away. Now and again Howell brought it out of retirement for a spasm of protest or a yap of agitation, but by 1938 the organization—which had always lived out of its address book—needed fresh blood and new addresses. In that year Howell, infuriated by the discovery that commercial pet food manufacturers had begun using significant quantities of marine mammal meat in their products, began rejiggering the Advisory Board as part of a push for new action: a consumer boycott. He wanted to get "the old ladies not to buy Ross products for their cats."[336] But Kellogg, "mixed up with State, Treasury, and Commerce" and about to leave for London (for an international conference on the regulation of whaling), begged off: "Be sure that you remove my name . . ." he wrote Howell, since "it will make sort of a mess of things here, if my name appears on letters you send out which in part will eventually be forwarded to interested government departments for comment. . . . I am called over usually to discuss these questions [and] . . . letterheads with my name on it do not help matters much."[337] Howell complained about having to omit him from the "illustrious roster of the whaling council," but he acquiesced, only because it was clear that, as he put it, "you can do more off than on."[338]

As indeed proved the case. Even as Kellogg was drawn more deeply into regulatory initiatives, he continued to work to publicize the whale issue. Most significantly, at the same time that Kellogg was easing out of the CCW, he was approached by Gilbert Grosvenor, editor of *National Geographic*, about authoring a major piece for the magazine on whales.[339] The celebrated marine natural history artist Else Bostelmann—who had worked with William Beebe for a decade and had pioneered the use of

336. Howell to Kellogg, 12 February 1938, Smithsonian Archives, RU 7170, box 4, file "Howell, A. Brazier, 1932–1943 (1/3)."
337. Kellogg to Howell, 3 February 1938, in ibid.
338. Howell to Kellogg, 12 February 1938, in ibid.
339. I have not been able to find the earliest correspondence on this project, but it seems likely that the connection was made either by Wetmore or by George Shiras, both of whom were on the Board of Trustees of the National Geographic Society, and were also members of the CCW. Correspondence on the editorial work and the preparation of the illustrations can be found in Smithsonian Archives, RU 7170, box 1, file "Bor–Boy," and in RU 7434, box 6, file "Correspondence 1928–1940."

a black background to depict abyssal creatures in searchlight hues—was engaged to produce, under Kellogg's supervision, a full-color gallery of every known cetacean. Kellogg, in turn, received a commission to write a 40,000-word article: 1,000 words on the "occurrence, characters, habits, life history, food habits, and actions in water" of each of some forty different species. The editors hoped it would all be, as Kellogg mused, "jazzed up and in story form." This was not an easy task, as he pointed out to his fellow cetologist at the British Museum, Francis C. Fraser.[340] Given how little was known, it would be, Kellogg complained, a little like "writing the same thing 45 different ways."[341]

Nevertheless, Kellogg took to the task with gusto, and for more than a year, as he hammered out no fewer than 65,000 words for *National Geographic* on the history of whaling and the natural history of whales, he corresponded regularly with Bostelmann in increasingly detailed exchanges on cetacean behavior, form, and coloration. This correspondence is a remarkable archive of what Lorraine Daston has called "four-eyed seeing": Kellogg annotated Bostelmann's sketches on tracing paper and wrote long memos on numbers of teeth, the shape of spouts, and the position of blowholes and flippers.[342] He even went so far as to send along color wheels and specimens to assist her in her preparation of what were, in the end, three dozen final watercolors (figures 3.17–3.21).

The collaboration appears to have been one of considerable enthusiasm and mutual respect. Kellogg wrote Bostelmann in February 1939, as they were closing in on camera-ready copy:

> You have put life into your sketches and they are not merely inanimate colored figures such as have appeared in the references that I gave you. You have caught the spirit or should I say the natural actions of the cetaceans and that is far more difficult than getting correct outlines. I'm convinced that this is going to be for many years the finest set of cetacean illustrations available. On that account if for no other reason I should like to see each cetacean drawn as accurately as present data permits.

340. Kellogg to Francis C. Fraser, 25 January 1939, Smithsonian Archives, RU 7170, box 3, file "Fra–Fuj."

341. Kellogg to Else Bostelmann, 7 April 1939, Smithsonian Archives, RU 7170, box 1, file "Bor–Boy."

342. Four-eyed seeing is discussed in chap. 2 of Lorraine Daston and Peter Galison, *Objectivity* (Cambridge, MA: MIT Press, 2007).

FIGURE 3.17
(*above*, and see PLATE 9)
Else Bostelmann sees
under the surface: A right
whale family. (From Kel-
logg, "Whales, Giants of
the Sea," P. 43. Courtesy of
the National Geographic
Society.)

FIGURE 3.18
(*left*, and see PLATE 10)
Bostelmann's vision: Gam-
boling dolphins. (From
Kellogg, "Whales, Giants
of the Sea," P. 75. Courtesy
of the National Geographic
Society.)

FIGURE 3.19
(*right,* and see PLATE 11)
Bostelmann pictures the
happy giants: Playful
humpbacks. (From Kellogg,
"Whales, Giants of the Sea,"
P. 46. Courtesy of the National Geographic Society.)

FIGURE 3.20
(*below,* and see PLATE 12)
Bostelmann depicts the
threat: A whaler takes aim
at a blue whale surrounded
by its smaller kin. (From
Kellogg, "Whales, Giants
of the Sea," P. 42. Courtesy
of the National Geographic
Society.)

FIGURE 3.21
(and see PLATE 13)
Bostelmann's para-
disiacal grotto: Gray
whales frolic. (From
Kellogg, "Whales,
Giants of the Sea,"
P. 47. Courtesy of the
National Geographic
Society.)

As if to prove the point, Kellogg then sketched an outline of the caudal flukes of True's beaked whale on the corner of the letter, since he had been "troubled" by the question of the exact shape of the notch between the two lobes.[343] In the final scramble for publication, Bostelmann even abandoned her studio in New York City, setting up and painting in a hotel in Washington to facilitate their collaboration, and she took to leaving sections of her paintings blank if she was unsure about an anatomical detail—they filled these in together.[344]

The success of the piece, which finally appeared in January 1940, outstripped all expectations. While Kellogg was irritated by the editing process (they had pared his workmanlike prose considerably), he had managed to

343. Kellogg to Else Bostelmann, 12 April 1939, Smithsonian Archives, RU 7170, box 1, file "Bor–Boy."
344. Else Bostelmann to Kellogg, 6 April 1939, in ibid.

get a full paragraph into the introduction on how whales appeared to be "heading toward the same fate that pursued the once-vast herds of American Buffalo."[345] He was thus content to have gotten his message across. In this task, Bostelmann helped, using her visual images: she depicted, for instance, a blue whale, in all its glory, coming nose to nose with a Norwegian gunner just slipping out from behind an iceberg (figure 3.20). Her contribution was essential because in many ways the prose served a kind of supporting role: the images were the real tour de force. The illustrations editor thought that the whale plates were among "the most satisfactory that the magazine has ever handled," and many thought the whole sequence "the finest the Geographic has ever published."[346] Kellogg himself was certain that Bostelmann had "set a standard for cetacean illustration."[347] Instead of blurry photographs showing a bit of black skin breaking rough water, or tongue-lolling portraits of sickly beached leviathans already in decay, Bostelmann had created, with Kellogg's help, a gallery of colorful, vital, and even *playful* whales and dolphins, nursing their young, gamboling sociably among their kin, and frolicking in grottolike shallows.

Public response was overwhelming. Kellogg was "swamped" by a veritable "deluge" of correspondence, including a remarkable number of "enquiries [that] indicate an intelligent interest."[348] From Portland, Maine, to Portland, Oregon, Americans wrote to Kellogg with whale stories, whale wagers, and whale questions, until he despaired of the daily mail.[349] This was the effect of reaching over a million readers.[350] Yet Kellogg was relentless: rather than throwing up his hands at the influx of queries and tips, he

345. Concerning the editing process, Kellogg wrote, "selecting sentences and paragraphs at random produces a rather novel result." Kellogg to Else Bostelmann, 30 November 1939, in ibid. For the quote, see "Whales, Giants of the Sea," *National Geographic* 77, no. 1 (January 1940): 35.

346. Kellogg to Else Bostelmann, 3 February 1940, and Else Bostelmann to Kellogg, 22 January 1940, both in Smithsonian Archives, RU 7170, box 1, file "Bor–Boy."

347. Kellogg to Else Bostelmann, 27 January 1940, in ibid.

348. Ibid.

349. For examples of these letters, see Joseph M. Winterhalter to unknown, 1 January 1940, Smithsonian Archives, RU 7434, box 6, folder 10, "Correspondence, 1928–1940," and additional material at this location.

350. This is the number that Kellogg believed read every issue of *National Geographic*, and I believe he got it from the editorial staff. For a general introduction to the cultural resonance of the publication in this period, consider Susan Schulten, *The Geographical Imagination in America, 1880–1950* (Chicago: University of Chicago Press, 2001), esp. chaps. 7–9.

ordered and distributed several hundred offprints of the article, which he used "to good advantage in stimulating the collection of porpoise skeletons in localities where additional material is urgently needed."[351] How many Massachusetts yawls, Gulf shrimp boats, and Pacific tuna cruisers carried a copy of the 1940 *National Geographic* tucked away somewhere as a marine field guide cannot be known, but the odds are there were a good number of such vessels in American waters. This, anyway, was surely what Kellogg intended. Moreover, the issue did a different, but certainly no less important, duty on inland coffee tables across the nation.

The success of the article reveals the degree to which Remington Kellogg succeeded in shaping how Americans came to "see" whales and dolphins at midcentury. And a close look at the preparation of the piece reveals just how hard he worked to control in minute detail the way these animals came into the American imagination. Left and right, he fought off editors' desires to sensationalize cetacean "monstrosity." As he put it to a colleague, "The publisher has the idea that all the pictures should be exciting, such as a whale running its head through a steamer and then winking its eye at the astonished crew."[352] But Kellogg worked with Bostelmann to "get these notions toned down," and at the eleventh hour he even succeeded in getting the title of the whole piece changed. The editors wanted to call it "Whales: Lions of the Sea." Kellogg strenuously objected: "Whales are very distantly, if at all, related to the cat tribe," he complained, and more importantly, "except when mortally wounded," they "are inoffensive and noted for their timidity."[353] In this and in other ways, Kellogg, using Bostelmann's brush, took advantage of a unique publishing opportunity, using it to paint whales and dolphins for the American reader: he picked their colors, edited their contours, and even depicted their characters—all in defiance of the traditional image of whales as deep sea "monsters."

There is a last point to make about this process, and it is perhaps the most surprising of all: reviewing Kellogg's meticulous work shaping the whale images of the 1940 *National Geographic* article, it would be easy to assume that he had by this time a great deal of direct personal experience

351. Kellogg to J. R. Hildebrand, 13 November 1939, Smithsonian Archives, RU 7434, box 6, folder 10, "Correspondence, 1928–1940."

352. Kellogg to Francis C. Fraser, 25 January 1939, Smithsonian Archives, RU 7170, box 3, file "Fra–Fuj."

353. Kellogg to J. R. Hildebrand, 13 November 1939, Smithsonian Archives, RU 7434, box 6, folder 10, "Correspondence, 1928–1940."

with cetaceans in the wild. Nothing, however, could be further from the truth. Kellogg's encyclopedic whale knowledge was exactly that: encyclopedic—the product of a massive and ever-expanding library of offprints, books, note cards, indexed correspondence, and cross-referenced photographs. He mobilized this archive for Else Bostelmann's benefit, but his own direct experience of living whales and dolphins was practically nil. Stranded odontocetes he had seen—several days old, as a rule, and shipped to Washington by truck or train. Remarkably, the Hatteras dolphins appear to have remained his only direct, close contact with live cetaceans. In reviewing thousands of pages of Kellogg's correspondence, manuscript notes, and published writings, the only reference I have found to his having been up close to a large whale was in 1931, when he visited a somewhat oozy embalmed fin whale that a showman named Griffith took around the United States on a flatbed railcar as part of a carnival-like traveling exhibition hawking perfume.[354] Kellogg knew what whales looked like, to be sure, but his collaboration with Bostelmann was less "four-eyed seeing" than "forty-eyed seeing"—a curious process of collatory and centripetal visualization, by which an image of the whole came to be composed over years of fragmentary views: the fortuitous snapshot sent by a boat captain, a piece of pickled skin sent by a colleague, the vivid description of a close encounter culled from a nineteenth-century travelogue.

There is thus something quite affecting about several of Kellogg's efforts to persuade Bostelmann to adjust her images. Concerning the submarine coloration of several of her preliminary watercolors, for instance, he wrote, "Are you quite sure that one cannot see the colors of submerged whales," an unusual hint of doubt creeping into his correspondence: "I have watched schools of common dolphin (*Delphinus delphis*) and white sided dolphin (*Lagenorhynchus acutus*) in the water alongside the 'Aquitania' and one could see the color pattern distinctly even when the Dolphins were submerged."[355] The *Aquitania*, of course, was a Cunard ocean liner, with a deck some eight stories above the surface of the water. What was Kellogg doing on the *Aquitania*? He was headed to Europe for an international whaling conference. From a Hatteras launch, pursuing splashing, netted

354. Kellogg to Glover M. Allen, 25 February 1931, Smithsonian Archives, RU 7170, box 1, file "Allen, Glover M., 1921–1941."

355. Kellogg to Else Bostelmann, 12 February 1939, Smithsonian Archives, RU 7170, box 1, file "Bor–Boy."

Tursiops at some risk to life and limb, to the deck of a Cunard liner surveying the animals in the bow wake far below: it is a striking trajectory, and
one that offers a ready-made epitome of what had happened to Kellogg's
whale work over a decade. An invidious commentator, one intent on showing how the IWC of the 1950s (Kellogg's brainchild) ended up enamored of
protocol and indifferent to whales, might use the juxtaposition to accuse
Kellogg of swapping active engagement with his subject creatures for the
satisfactions of diplomatic status. Even putting such a tendentious reading
aside, the image of Kellogg—one of the leading cetologists in the world by
the late 1930s—collecting natural history observations from the prow of a
luxury liner will not quite leave the mind.

To be fair, Kellogg did grow increasingly vociferous at those European
gatherings. While he made limited contributions to the International Whaling Conference of 1937 (deferring to senior scientists like Birger Bergersen
of Norway), he was rapidly assigned to the technical committees that considered biological issues like sanctuaries and the impact of shore whaling
stations on breeding grounds. He played an important role in clarifying
cetological nomenclature across several languages, and he also helped push
for one of the major successes of the 1937 conference, a ban on the commercial killing of gray whales.[356] In 1938, at a preparatory meeting in Oslo
and again at the full conference in London, Kellogg's contributions were
sharper as he pressed the delegates for action on an Antarctic sanctuary.
He went so far as to take the floor at the fifth session of the latter meeting
to deliver a strongly worded statement:

> Conservation of whales is the objective of the American delegation, and
> we will support measures which promise an effective accomplishment.
> While practically all the principal whaling countries are represented
> here and in general they have indicated a desire to reach an effective
> solution of this problem, the commercial aspects seem to have out-
> weighed the biological in the actual consideration of the proposals.[357]

356. The file that contains the proceedings of this gathering can be found in the
Smithsonian Archives, RU 7165, box 3, folders 5–6, "London—International Whaling
Conference, 1937."

357. "International Whaling Conference, London—June, 1938. Fifth Session
(ICW / 1938 / 29)," pp. 1–2, Smithsonian Archives, RU 7165, box 5, folders 3–4, "London—
International Whaling Conference, 1938."

It was a statement he would repeat, in one form or another (with, it should be noted, little effect) for the next twenty-five years as he gave shape to the IWC at the end of World War II, attending with loving care on its administrative maturation while observing with growing dismay its tenacious inability to implement coherent, effective measures for the protection of whales.

That story will be the subject of the next two chapters, but to close the present episode, it is significant to note that the US role in founding the IWC—and Kellogg's work coordinating the massive interagency initiative that yielded a draft convention in the 1940s—was all foreshadowed in the literature of the Council for the Conservation of Whales. As early as 1935, Kellogg and Howell were circulating a memorandum to US federal agencies and CCW supporters calling for "the appointment of a departmental committee of the federal government to arrange for the calling of an international conference, or to determine the advisability of this government assuming the leadership in the calling of such a conference, to provide for international uniformity, in so far as is feasible, in the regulation of pelagic whaling."[358] By 1946 that wish would become a reality in the form of a watershed international whaling conference in Washington, a meeting Kellogg made happen, chaired, and brought to a successful conclusion with the signing of the modern International Convention for the Regulation of Whaling (ICRW), the charter of the International Whaling Commission.

Nineteen forty-six was a year that put whales much on American minds. In August, several months before that IWC-founding gathering, Walt Disney released *The Whale Who Wanted to Sing at the Met*, an animated short film drawing on a number of the animators, directors, and adapters who had collaborated on *Bambi*, the wartime blockbuster that had such lasting repercussions for American ideas about terrestrial wildlife.[359] *The Whale Who Wanted to Sing at the Met* played a parallel role for marine mammals. Reviewed in the *New York Times* as the saga of a "fabulous ce-

---

358. A. Brazier Howell to unknown, undated, Smithsonian Archives, RU 7170, box 4, file "Howell, A. Brazier, 1932–1943 (1 / 3)."

359. Overlapping personnel on the two projects included Richard Kelsey, Al Zinnen, Ray Huffine, Art Riley, Josh Meador, and George Rowley. On *Bambi*, see Ralph H. Lutts, "The Trouble with Bambi: Walt Disney's Bambi and the American Vision of Nature," *Forest and Conservation History* 36 (October 1992): 160–71, and the discussion in Gregg Mitman, *Reel Nature: America's Romance with Wildlife on Film* (Cambridge, MA: Harvard University Press, 1999).

tacean" with a "triple gaited voice," the film used cutting-edge acoustic technologies to permit Nelson Eddy to ventriloquize a whale singing in multiple-part self-harmony, emulating to comic effect the recent discoveries (made by other new acoustic technologies) that some whales were capable of complex phonations, replete with haunting overtones.[360] Willie, the whale gifted with unique powers of song, whose only ambition is to please fellow sea creatures and beguile sailors, is eventually done in by a modern harpoon gun. Assumed into heaven in an angelic chorus, Willie became an extraordinary poster child for the anti-whaling forces in the United States, and the tear-jerking film—in fact, a high-toned parody of *Citizen Kane*—seduced critics even as it won the hearts of the children who would grow up to call for an end to commercial whaling in the 1960s and 1970s.

The CCW was, by this time, no more, but its project of winning public sympathy for the "plight of the whale" had emphatically taken on a momentum of its own.

CONCLUSION

This chapter set out to examine how cetaceans became the objects of scientific scrutiny in the United States before World War II. To that end, I have considered in some detail the intellectual and professional development of Arthur Remington Kellogg, whose expanding program of biological research and spreading network of collaborators (scientific, industrial, political) gave rise in the interwar period to a unique initiative to study and protect whales. That initiative eventually took concrete shape in the Council for the Conservation of Whales, which after 1928 drew a diverse range of biologists, crusaders, and Progressive Era conservationists into a formal campaign to heighten public awareness of marine mammals in general and the large whales in particular. Emphasizing the scientific importance and perilous condition of these special animals, the CCW recruited champions to its cause and agitated for protective regulations in the United States and beyond. Using this organization to gain a foothold in legislative and diplomatic arenas in the early 1930s, Kellogg succeeded in becoming a

360. The quotes are from the review by Bosley Crowther in the *New York Times*, "The Screen in Review: 'Make Mine Music!' Animated Cartoon by Walt Disney," 22 April 1946, 32.

well-connected scientific expert, uniquely positioned to advise and lobby national governments and international bodies alike on the conservation of cetaceans.

Recovering this story has brought several larger issues to light. It is noteworthy—particularly in comparison with contemporary developments in Britain and Scandinavia—how little the American push for whale conservation in this period had to do with fisheries biology or the ocean sciences. Kellogg found his way to whales, as I have shown, out of a general interest in zoology and vertebrate paleontology, and those he attracted to cetaceans in the early years of his work came out of the Bureau of Biological Survey, medical physiology, ornithology, and even forestry. Interest in Kellogg's research was driven by distinctly American concerns in the interwar years, including anxiety about the demise of large mammals in the West, and concern about how scientists could best serve governments in the management of wildlife. If the whale conservation campaign could (and did) tap new ideas about sea sportsmanship that were taking hold in early twentieth century, this was only because the CCW could mobilize an established tradition of American sportsmen taking a leading role in preserving large and noteworthy fauna. While polar exploration, imperial ambitions, and North Sea fisheries problems shaped Anglo-Scandinavian interest in whales in the 1920s and 1930s, Kellogg and his associates in the United States mobilized anticruelty sentiments prevalent in the American humane society movement by emphasizing that whales were mammalian kin, not mere fish. At the same time, American whale researchers were able to take advantage of controversies over evolution in the United States in this period to draw public attention to whales as exemplary instances of adaptive change.

One of the more interesting points that emerges in the second half of this chapter, I believe, is the discovery that a highly coordinated publicity campaign aimed at raising whale awareness among American citizens actually appears to have played a prominent role in shaping cetaceans as significant scientific objects.[361] That is to say, it is less that the distinctive biology of whales came to public attention in these years as a result of major cetological discoveries, than that Kellogg and his collaborators—as part of a general effort to make whales newsworthy—generated and disseminated the image

361. It is interesting to consider this episode in the context of the analysis offered by Marcel C. LaFollette in his *Making Science Our Own: Public Images of Science, 1910–1955* (Chicago: University of Chicago Press, 1990).

of the whale as an exceptional scientific mystery. If this characterization of the process is correct, it certainly merits further attention, since it promises to help explain how whales became *the* paradigmatic organisms for popular understanding of ecology and environmental science in the late twentieth century. Rather than the by-product of particular scientific research (or the outcome of agitation by fringe activists), this development may well be the result, at least in part, of a positive program by scientists intent on packaging these organisms for popular consumption and trading on tacit modern conceptions of science as "news" to that end.

The first part of this chapter also raises, I think, some broader issues for further consideration. The development of the life sciences in the United States between 1880 and 1930 has been the subject of vigorous historical debate for more than two decades. Did the period see a "revolt from morphology" in favor of a "new biology"? Did it perhaps see a still sharper rejection of the whole "naturalist's tradition" in the form of a decisive turn toward experimental methods borrowed from the physical sciences? At stake in this historiography are questions of continuity and discontinuity, evolution and revolution, and the very nature of scientific change. No sketch of whale science, or of Remington Kellogg's professional development in the 1910s and 1920s, is going to resolve these knotty problems. However, it is striking how well much of the story I have told in this chapter rehearses themes that have exercised historians of biology looking at this period. For instance, it would be easy to argue that Kellogg exemplifies the kind of continuity and durability of the morphological tradition that some commentators have emphasized.[362] As I have shown, he clearly sought to reinvigorate anatomical and comparative investigations with borrowings from physiology and the newest embryology, and he was rewarded for doing so. Like others in the struggling zoological / paleontological world of the museums in this period, Kellogg found himself drawn increasingly into conservation concerns, and this was a way that a number of the "old" morphologists succeeded in mobilizing their expertise on a changing professional landscape.[363]

Kellogg negotiated these developments very successfully: by the late

362. See n. 16 above.

363. See particularly chap. 5 in Ronald Rainger, *An Agenda for Antiquity: Henry Fairfield Osborn and Vertebrate Paleontology at the American Museum of Natural History, 1890–1935* (Tuscaloosa: University of Alabama Press, 1991); and also the introduction to *The Expansion of American Biology*, edited by Keith Rodney Benson, Jane Maienschein, and Ronald Rainger (New Brunswick: Rutgers University Press, 1991).

1930s he was well on his way to claiming the mantle of a scientific states-man. And in the wake of World War II, he would play a leading role in the foundation and development of what is arguably the most important institution in the twentieth-century history of marine conservation, the International Whaling Commission. It is with that organization, and its tormented efforts to create a mechanism for the "rational" management of a pelagic ocean resource, that the next two chapters will be concerned.

FOUR

A CETACEOUS PARLIAMENT

WHALING, SCIENCE, AND
INTERNATIONAL REGULATION

1945 – 1956

"And by that time"—the chemist took up the fantasy and followed it along
his own obsessional lines—"every whaling expedition will be commanded
by a scientist, as the *Discovery II* is today, and twice as many whales will
be taken every year, without harming the species, and the seamen will
be made to conduct the whaling at long last in a rational way—"

R. B. ROBERTSON, *Of Whales and Men*, 1954

INTRODUCTION

Shortly after lunch on Friday, 20 July 1956, on the last afternoon of what
had been four difficult days of wrangling at the eighth annual gathering of
the International Whaling Commission, the lead delegate from the United
Kingdom, R. G. R. Wall, took the floor to air his views. Despite mounting evidence that the stocks of large, commercially valuable whales in the
Antarctic were in a perilous condition—blue whales, the largest and most
valuable of all, had dropped from more than 80 percent of the annual catch
(before World War II) to less than 6 percent in the last season; humpbacks,
once the mainstay of the industry, were no longer found across much of
the deep southern Atlantic; and the 25,000 or so fin whales taken annually
by the $ 200 million industry (on which it had come to depend) were, it
seemed, smaller, younger, and harder to find than ever—the representatives
of seventeen nations who had gathered a few steps from Whitehall in the
conference rooms of 10 Carlton House Terrace had thus far been unable to
reach consensus on reducing the international catch quota for the season

ahead. Dissent among the scientific advisers had not helped matters, and it was for the squabbling scientists that Wall had reserved these frustrated remarks:

> The scientists in effect have told us that in the Scientific Committee there is some argument between two schools of thought. The one school, which is in the majority, thinks that the fin whale stocks are in fact decreasing, and showing signs of serious strain. The other school thinks that the fin whale stocks may possibly be increasing, and the situation is not serious.

Turning to the Dutch delegate, whose scientific experts had spearheaded the rejection of the idea that biological and statistical data pointed conclusively to the depletion of the fin whale stocks, Wall continued:

> Well, Sir, the fin whale stocks may be increasing, we cannot of course be sure. We cannot go down there and count them one by one year by year. The moon may be made of green cheese, Sir, we cannot be sure of that either, we cannot go there and find out whether it is or not. Many things are possible, and some are less likely than others. We are skeptical as to whether the moon is made of green cheese, and we believe it is not. Some of us are equally skeptical as to whether the fin whale stock can possibly be increasing.

Pressing the delegates to reconsider the urgency of the matter, and to accept at least a token reduction in the annual quota, Wall gestured wryly at the IWC's larger obligations—not only to regulate the industry in a "rational" manner consistent with "optimum" yields in perpetuity, but in doing so to guarantee a perennial supply of whale products (mainly oil, but also meat) to feed the people of the world: "We should not risk going hungry on this earth," he concluded, "because we think that if we do we may be able to get a cut of the green cheese on the moon."[1]

In the closing hours of this pivotal meeting, then, courtesy and diplomatic bonhomie took a backseat on the floor of the IWC as a final round of proposals and counterproposals pushed the patience of some delegates to the breaking point. The Netherlands stuck to its guns, arguing that there

---

1. This and the quotes above are from transcripts of the 1956 IWC meeting: IWC / 8 / 13, pp. 108–9.

was "not sufficient evidence" to reduce the quota at all, not even by a pid-dling 500 "blue whale units" (or BWU)—the metric used to regulate the Antarctic catch.[2] Such a reduction would bring the quota down to 14,500 BWU, which was still considerably higher than the 10,000 or 11,000 recommended by some members of the Scientific Committee. As the Norwegian delegate put it, pressing for support for the small cut: "Maybe I am wrong, but it is my conviction that some years will come when we will think that 14,500 is an impossible number of whales to catch in the Antarctic in a reasonable length of time."[3] His concern was prescient, as the next decade would starkly reveal.

When, in the end, the small reduction did pass in a vote of the plenary session, the chairman of the Scientific Committee, Neil A. Mackintosh, rather than expressing enthusiasm, went so far as to request that a note appear in the IWC's annual report, a note that would make clear that the majority of the advisory scientists felt a still greater reduction was needed. He believed his scientific reputation was at stake: "After all," he announced, "there are many people who know there are scientists taking part in this commission's work, and they may be wondering what we are doing."[4]

When the dust had settled and the participants were about to disperse, the chairman of the meeting called on Dr. D. B. Finn, the director of fisheries at the Food and Agriculture Organization of the United Nations (or FAO), who had been a quiet observer of the long days of horse trading. Would Dr. Finn care to make any closing comments before the plenary session adjourned? He would. Rising, he expressed thanks for the invitation to attend the gathering, and he then offered a brief peroration on the vital importance of managing the world's food resources wisely, even in the face of mounting pressures of exploitation. How could the specific reproductive capabilities of a particular organic resource be reconciled with the intense competition of international enterprises pursuing economic gain? That was the problem:

I think that anyone who has sat through these meetings would real-ize that this problem is going to become more and more difficult as competition increases, and it is going to become somewhat *like a*

---

2. The Dutch quote is from IWC / 8 / 13, p. 11. One BWU = one blue whale or two fin whales or two and a half humpbacks or six sei whales. The origin of the BWU lay in the calculus of a cartel of Antarctic whaling companies in the 1930s (see chap. 2 above).

3. Ibid., p. 110.

4. Ibid., p. 121.

race between the acquisition of the biological fact on which alone wise regulation can be framed, and the deprecation which competition might bring on that particular stock. This is not only a biological problem—as matter of fact, the only justification for biological interest is the use that man can make of the product.[5]

D. B. Finn had put his finger on a very serious problem—one that would not go away in the years to come.

.　.　.　.　.　.　.　.　.　.　.

That race, between those seeking biological facts and those seeking the last of the great whales of the Antarctic, eventually came down to the wire, and more than biology was indeed at stake. Ten years later the Antarctic industry would be in a terminal condition, and a backlash against the International Whaling Commission would throw into doubt not only its own future as an institution, but also the future of the treasured ideal of rational international management of ocean resources, an ideal the IWC had embodied from its creation. In 1946, when representatives of the world's whaling nations gathered in Washington, DC, to sign the International Convention for the Regulation of Whaling (or ICRW, the legal framework that called the IWC into being), the IWC represented all the promise of postwar, multilateral, collaborative, global conservation, guided and optimized by scientific expertise. By 1965, however, the organization had become a byword for failure and irresponsibility, and its brief, broken up and reconfigured, was being redistributed among competing cohorts of concerned citizens, national governments, UN agencies, and emerging activist groups. Questions that might have been answered in the committee rooms of 10 Carlton House Terrace in 1956 were, by 1965, being transferred to a new and sprawling venue: the cacophonous and unpredictable court known as "world opinion." There, a gathering effort to "save the whales"—arguably the first global campaign of the modern environmental movement—would rally around the premise that it was from the IWC and its scientists that the earth's largest mammals must above all be saved.

This chapter and the one that follows aim to make sense of these complex developments and their enduring legacy. As the cantankerous exchanges in London in 1956 suggest, the story of the IWC between 1945 and 1965 is a story about biological research and international politics, scientific con-

5. IWC/8/13, p. 123. Emphasis added.

sensus and economic exigencies, the power of numbers and the power of individual (and national) interests. A deeper understanding of these dynamics will shed light on what amounts to a remarkable (and undoubtedly strange) feature of postwar science and social life: namely, that by the late 1960s, cetaceans—an anomalous order of elusive, air-breathing marine mammals—had begun to serve as nothing less than "a way of thinking about our planet," having become, for many, powerful symbols of "the fight to stop the continuing destruction of man's own ecosystem."[6]

At the same time, a close study of science and the IWC in the two decades following World War II offers an opportunity for investigations that go beyond cetology, "cetapolitics," and whale huggers, since the whole episode amounts to an early, intricate, and in many ways fateful case study of the fortunes of scientific expertise in a regulatory setting.[7] A number of scholars have taken up social-scientific analyses of this important subject, examining what has come to be called "the fifth branch" of government in the modern administrative state: scientists as policy makers and advisers. Yet there remains a need for historical work on the development of this critical feature of modern scientific and political life in the twentieth century.[8]

With these ends in view, this chapter and the next survey the history of the first two decades of the International Whaling Commission, with an eye on the workings of its system of scientific advisers. I begin here below with a brief sketch of the early history of the International Whaling Commission, and then turn to a discussion of the secondary literature

6. "Way of thinking" is from the president of the National Geographic Society, Gilbert Grosvenor, in an editorial in *National Geographic* 150 (December 1976), 721. The other quote is from Robert Boardman, *International Organization and the Conservation of Nature* (London: Macmillan, 1981), 135.

7. "Cetapolitics" is from Peter J. Stoett, *The International Politics of Whaling* (Vancouver: UBC Press, 1997).

8. Points of departure in this literature would include Sheila Jasanoff, *The Fifth Branch: Science Advisers as Policymakers* (Cambridge, MA: Harvard University Press, 1990); Liora Salter, Edwin Levy, and William Leiss, *Mandated Science: Science and Scientists in the Making of Standards* (Dordrecht, Holland: Kluwer Academic Publishers, 1998); Alvin M. Weinberg, "Science and Trans-Science," *Minerva: A Review of Science, Learning and Policy* 10, no. 2 (1972): 207–22; Chandra Mukerji, *A Fragile Power: Scientists and the State* (Princeton, NJ: Princeton University Press, 1989). Also relevant is Naomi Oreskes's work, particularly her recent coauthored book: Naomi Oreskes and Erik Conway, *Merchants of Doubt* (New York: Bloomsbury, 2010).

that has been concerned with this institution—and particularly with its use of science before 1965. Reviewing this material will permit me to develop my preferred approach to this important topic while showing how contentious these issues remain for a variety of former participants and recent commentators. Having established the need for an investigation of this period grounded in the methods of a sociologically sensitive history of science (and particularly alert to the "boundary work" by which scientists and others delimited the sphere of scientific expertise), I move in for a close look at the first decade, 1945–1956, and work to unfold the changing relationships between science and policy making at the IWC in those years. Chapter 5 picks up the story on the cusp of the crisis of the mid-1950s and carries an analysis of these problems though to the mid-1960s. By attending both to the actual production of whale knowledge and to the finely geared machinations of whaling politics over this whole period, a striking story of science at work in the world will emerge.

THE IWC AND ITS CRITICS

Addressing a gathering of farmers and cottonseed oil manufacturers in Myrtle Beach, South Carolina, in June 1940, C. E. Lund, director of the fats and oils section of the US Department of Commerce, sounded a somber note. Asked what the effects of war would be on the world trade in edible fats, Lund could only gesture at the "kaleidoscopic events transpiring in Europe" and announce ominously that "the future is in the laps of the Gods."[9] A few years later, the gods, with considerable help from complicit men, had visited unprecedented destruction on continental Europe, carved a path of ruin across the Pacific, and left the prewar networks of trade, finance, and production in blasted fragments. Facing a smoldering and ruined peace, the US leaders charged to convert wartime resources into the instruments of effective occupation and reconstruction set to work identifying the pressing problems that promised to shape the immediate postwar world. Chief among these was the prospect of massive food shortages in Europe and Asia, the looming and direct result of the violent collapse of commercial and agricultural systems across much of what had been the developed world.

Postwar internationalism met the logistical reach of the Allies' global quartermasters exactly six years after Lund's speech, when representatives

9. Smithsonian Institution, James Mead Files, file "Whaling—Technology."

of the new Food and Agriculture Organization gathered in Washington, DC, for a "Special Meeting on Urgent Food Problems" in May and June of 1946. Surveying the latest evidence, this body of experts concluded that continued "scarcity of feeding stuffs" promised serious shortages of fats for human consumption for the foreseeable future. Among other recommendations, the committee called for a new and concerted effort to increase the haul of "marine oil" by the world's whaling industry, which had been decimated by naval warfare: "more abundant catches" were needed, and the United States, in conjunction with the major whaling nations (Norway and the United Kingdom), ought to press for "a rapid increase in the number of factory ships" and "cooperate in the provision of technical personnel and other equipment in order to obtain the maximum increase in all waters where whaling is permitted."[10]

The thrust of that message was intended for those diplomats, whaling industry magnates, and agency functionaries who were scheduled to convene later that year, also in Washington, in order to try to hammer out a new and comprehensive international regime for the regulation of whaling. The FAO report's passing mention of "water where whaling is *permitted*" is a reminder that a sequence of prewar (and indeed, even mid-war) meetings and agreements had already aimed to coordinate and regulate the volatile (if lucrative) industry that had boomed in the 1920s, collapsed disastrously during the Depression, and climbed again to considerable economic importance in the late 1930s.[11] By that time more than 40 percent of the fats used in the production of margarine in Britain came from whales, and similar rates were prevalent across northern Europe.[12] In the manufacture of household and industrial soaps, whale oil was even more important, despite the expansion by Unilever and other companies into

10. This and the quotes above are from the Smithsonian Archives, RU 7165, box 8, folder 3, file "Washington—Informal Inter-agency Committee on the Regulation of Whaling (IICRW)." On the loss of whaling vessels during the war, see the confidential "Explanation of Protocol 23 November 1945," 3, Smithsonian Archives, RU 7165, box 7, folder 7, "London—International Whaling Conference, 1945."

11. For a full treatment of this era, see J. N. Tønnessen and Arne Odd Johnsen, *The History of Modern Whaling* (Berkeley: University of California Press, 1982), chaps. 19–27. For greater detail on the economics, see Karl Brandt, *Whale Oil: An Economic Analysis* (Stanford, CA: Food Research Institute, 1940).

12. Leonard Carmichael to Robert Murphy, undated, Smithsonian Archives, RU 7165, box 23, folder 6, "Correspondence, Fifth International Whaling Commission Meeting (London)."

the plantation-based cultivation of vegetable oils in Africa during these years.[13]

As I have discussed in chapter 2, conservation of whale stocks emerged early as a problem for the modern whaling industry as well as for those governments that derived revenue from taxes on whaling expeditions. Concerns about the rapid decline in humpback catches in particular stimulated the 1931 Geneva Convention for the Regulation of Whaling, which was itself based on an earlier Norwegian national law that came into effect in 1929.[14] While these initiatives called for certain protective prohibitions (for instance, forbidding the killing of calves and females nursing young), the core of the regulations consisted of requirements for licensing and the maintenance of catch records, along with stipulations concerning complete utilization of whale carcasses—reflecting consensus on the desirability of reducing at least wasteful killing, if not killing per se. Minimum lengths, the total protection of nearly exterminated species (like right whales), closed seasons in certain regions, and even limitation on the number of catcher boats to be used in connection with factory vessels were all introduced in the 1930s, either through private and voluntary agreement among companies trying to divide up the spoils or by supplementary conventions signed by participating national governments. The effectiveness of these measures was never great, despite some efforts at mutual inspection and collective enforcement. By the mid-1930s, however, the forceful entrance of both Japan and Germany into the Antarctic industry increased the sense of urgency among the cozy Anglo-Norwegian cliques who ran the dominant concerns (and who maintained close relationships with their national governments and fisheries departments).[15]

13. For a discussion of the difficulties of these operations, see Smithsonian Archives, RU 7165, box 4, folders 5–6, "Oslo—Preliminary Whaling Conference, 1938."

14. For a useful review of pre–World War II regulatory efforts, see Wray Vamplew, "The Evolution of International Whaling Controls," *Maritime History* 2, no. 2 (September 1972): 123–39. Considerably more detailed is L. Larry Leonard, "Recent Negotiations toward the International Regulation of Whaling," *American Journal of International Law* 35, no. 1 (January 1941): 90–113.

15. Germany and Japan both entered the Antarctic pelagic hunt in 1936–1937 (though shipbuilding and commercial plans had made their intentions clear several years earlier). By 1937–1938 four Japanese floating factories took more than 12 percent of the total Antarctic catch, and by 1938–1939 the German fleet of seven floating factories (and fifty-three catcher boats) was taking a similarly large cut of the total catch. See Karl Brandt, *Whale Oil, an Economic Analysis* (Stanford, CA: Food Research Institute, 1940),

Goaded by anxiety that new and un-cozy entrants would suck the profits out of a capital-intensive business (and make it even harder to protect the stock from overexploitation), the primary stakeholders came together in 1937 in the hopes of devising a stricter regulatory regime. Conservation of the whales themselves (as opposed to the conservation of the whaling industry) was much discussed at this international conference, and whale scientists like Remington Kellogg and Neil Alison Mackintosh were brought forward to make recommendations for a sanctuary region in the Antarctic and to expound on the need for additional scientific research. Nevertheless, the place of such considerations in the larger calculus of competing interests is nicely captured by an early exchange at this meeting. Toying with the language of the preamble to the new agreement, one of the delegates suggested they might "add something about the preservation of whales," to which the buoyant British chairman replied, "If you want to get something in about the preservation of the stock I think one might say in the preamble that 'desiring to secure the prosperity of the whaling industry, and for that purpose to maintain the stocks of whales . . .'" This prompted another delegate to suggest flipping the order of the phrases to read, "desiring to maintain the stocks of whales and to secure the prosperity of the whaling industry . . ." But this proposal met with a cold reception. Didn't the emphasis on the industry "rather leave you open to criticism that the only purpose of your protecting [whales] is to make money out of them?" asked the Canadian delegate of his reticent peers. "I am afraid it is brutally true," shrugged the chairman.[16]

In the end, the 1937 conference,—riven by competing factions (pelagic whaling nations versus those limited to using land stations; ambitious newcomers versus established players) and overshadowed by the prospect of insurgent whaling initiatives by nonsignatory nations,—failed to implement restrictions that could actually reduce the number of whales killed in the years that followed. The conference also brought to the fore nagging structural problems: Even if the delegates could agree on restrictions, what legal form should their agreement take? Would domestic legislation

---

and Lars U. Scholl, "Whale Oil and Fat Supply: The Issue of German Whaling in the Twentieth Century," *International Journal of Maritime History* 3, no. 2 (December 1991): 39–62. Also useful for detail on the German situation is Karl Brandt, *The German Fat Plan and Its Economic Setting* (Stanford, CA: Food Research Institute, 1938).

16. Proceedings, fifth session, p. 17, Smithsonian Archives, RU 7165, box 3, folder 6, "London—International Whaling Conference, 1937."

in participating countries always be necessary to buttress such an inter-governmental convention? How would enforcement work? At one point, despairing of such problems, the chairman announced to the delegates his "fear" that it might appear to outsiders as if they were "passing a sort of international legislation."[17] The prospects of success for such an initiative were, in his view, very poor. The 1931 Convention for the Regulation of Whaling had, after all, taken years to come into force due to delays occasioned by the complexities of domestic politics in the ratification processes of the signatory nations. And that had been in an era of considerably greater optimism about the institutions and legal structures of internationalism. Behind that original convention lay years of adroit lobbying by the International Council for the Exploration of the Sea (or ICES, a distinctive consortium of maritime powers with a thirty-year tradition of negotiating over the resources of the Baltic and North Seas) as well as energetic committee work in the League of Nations (both the Committee for the Progressive Codification of International Law and the Economic Committee pressed for action on the whaling problem in the late 1920s).

While there was little hope in 1937 of again stimulating such a push, those earlier agitations had, before they expired on the rocks of continental realpolitik, succeeded in realizing an institution of international scope that would serve an enduring role in efforts to control the industry: the Bureau of International Whaling Statistics, brainchild of the distinguished Norwegian fisheries biologist Johan Hjort, who skillfully built a national Norwegian effort to codify catch data into a global clearinghouse for the world's whaling data.[18] By collecting seasonal returns and collating them into annual reports on the industry, the BIWs had become, by the late thirties, the main way interested parties—both those worried about overexploitation and those planning new exploitative expeditions (and there was certainly some overlap)—kept track of the changing dynamics of Antarctic whaling. It was to the potential of this organization that Remington Kellogg referred when, in 1938, at yet another international gathering on the whaling problem, he proposed a radically new scheme for international regulation. Because his proposal is very close to what later took shape under the IWC,

17. Transcripts, Smithsonian Archives, RU 7165, box 3, folder 5, "London—International Whaling Conference, 1937."

18. The founding documentation is in the ICES Records, Rigsarkivet, box 85, folder 6Q.

it will be worth taking a moment to describe it, despite the cool reception it received in Oslo and London before the war.

What Kellogg put forward was a global quota system, preferably segmented geographically and monitored over the course of the season by a central statistical registry. As he put it,

> Under such an arrangement the whaling inspector on each factory ship would report each week by radio the number of barrels of whale oil processed. In as much as these returns are received week by week and tabulated at the Central Statistical Office, that office could predict when the quota for the whole Antarctic whaling area would be reached, and notify the respective governments and managers of the floating factories that all operations should cease on a specified date.[19]

The explicit model for this arrangement was the bilateral convention between the United States and Canada for the preservation of the halibut fishery of the North Pacific Ocean and Bering Sea, which provided for what came to be called an "International Fisheries Commission," staffed by both nations and empowered (among other things) to call for a seasonal halt to fishing on the basis of its own stock assessments.

The whaling delegates at the 1938 meeting were quick to point out the difficulties that such an arrangement would present in the more complicated setting of a multinational industry working not immediately offshore, but quite literally at the very ends of the earth. While the regulatory meetings of 1938 and 1939 reliably genuflected in the direction of increased conservation measures, in the end the majority of the discussion dealt with quibbling efforts to block rivals and reserve commitments; scholastic parsing of definitions (was a ship that had its keel touching the sand a "land station"?) attenuated any impetus toward real action. In essence, it was becoming increasingly clear that no one was going to let his ox get gored at a moment when the rumbling of troop movements in central Europe suggested that everyone would soon need all the oxen they could find. As Kellogg put it, expressing some frustration, "The commercial aspects seem to have

19. Meeting 4, p. 95, Smithsonian Archives, RU 7165, box 4. For the proposal as presented in London later that year, see Smithsonian Archives, RU 7165, box 5, folder 2, "London—International Whaling Conference, 1939—U.S. Delegation Correspondence."

outweighed the biological in the actual consideration of the proposals."[20] While delegates expressed plenty of enthusiasm for additional research on whale migrations, reproductive habits, and general biology, serious reductions in whaling would not be contemplated by the participating countries, all of which had the prospect of imminent hostilities in view by the time the last of the prewar whaling conferences broke up in late July of 1939.

Kellogg, who by the 1940s had come to take a dim view of annual jaw-jaw by delegates mostly intent on using the language of conservation to angle for a leg up on rivals, alerted the US State Department long before the guns of World War II stopped sounding that calls for unregulated whaling would be forthcoming from America's allies as soon as victory seemed assured. In his view, the radically changed circumstances occasioned by the war provided an opportunity for expanded US leadership in whaling affairs, both to protect its own interests (the United States had by this time a negligible whaling industry and made little use of whale oil as an edible fat, but spermaceti oil—a lubricant with several unusual properties—was a modestly important industrial commodity allocated exclusively to the military in the period) and, above all, "to further the cause of conservation." A postwar whaling conference in Washington, he advised, would have the advantage of concentrating the delegates where they "will not be subject to direct pressure from the British Board of Trade, the Norwegian Whaling Association, and similarly interested shipping interests [sic]." Moreover, the United States might be in a position to help install the kind of "scientific management" that was the only hope for what he called the "last remaining reservoir of whales."[21]

---

20. ICW / 1938 / 19, fifth session, p. 2, Smithsonian Archives, RU 7165, box 5, folder 5, "London—International Whaling Conference, 1938."

21. All quotes in this paragraph are from Kellogg to Leo Sturgeon, 20 November 1943, Smithsonian Archives, RU 7165, box 6, folder 2, "London—International Whaling Conference, 1939—Includes International Whaling Agreement." It is interesting to note that a US-hosted meeting of whaling nations for the purposes of producing a conservation-oriented treaty had been in the platform of the Council for the Conservation of Whales, the organization founded by Kellogg and Howell in 1928 (see chapter 3). As Howell and the CCW wrote to F. T. Bell (US Bureau of Fisheries) on 28 January 1935, "There should be provision made for the appointment of a departmental committee of the Federal government to arrange for the calling of an international conference, or to determine the advisability of this government assuming the leadership in the calling of such a conference, to provide for international uniformity, in so far as is feasible, in the regulation of pelagic whaling. (As whales are seasonally migratory no single govern-

Packed off to London again in January 1944 for an Allied conference on the prospects for postwar whaling, Kellogg carried with him instructions to push for the kind of arrangement he had originally proposed in 1938: a total catch quota, regulated through a reporting system centered on the Bureau of International Whaling Statistics.[22] He was also briefed by the State Department on confidential negotiations with the British and the Norwegians that suggested that—with catching and production capabilities at very low levels (because of war damage)—commercial resistance to conservation measures would be weak, and therefore that "sound principles of regulation of international whaling in the future could be introduced at once."[23] He was also authorized to formalize the US invitation to host a postwar conference aiming at the installation of "a long range conservation program that would give effective protection to existing stocks of whales."[24]

A commitment to a total quota was indeed secured at the 1944 International Whaling Conference in London. In briefing Senator Guy Gillette (Iowa), who chaired the subcommittee responsible for putting the treaty protocol forward for ratification by the US Congress, Kellogg emphasized the significance of this new precedent, calling attention to "one important provision never heretofore incorporated in these international agreements for the regulation of whaling, and that is the annual limitation on the number of whales that can be killed during commercial operation in Antarctic waters." Explaining that in his view the limit had been set too high actually to restrict the 1944–1945 season (16,000 BWU had been settled on, which amounted to two-thirds of the annual prewar catches, but reduced personnel and equipment made it seem unlikely that this level could be achieved for several years), Kellogg nevertheless emphasized the importance of this new "principle of limitation," which "in the future will permit the regulation of the annual catch in such a manner that the stocks of whales will not be destroyed by ruthless operations."[25]

---

ment can adequately control the situation)." Smithsonian Archives, RU 7170, box 4, file "Howell, A. Brazier, 1932–1943 (1/3)."

22. See Kellogg's instructions of 28 December 1943, signed for the Secretary of State by Breckinridge Long, Smithsonian Archives, RU 7165, box 7, folder 6, "London—International Whaling Conference, 1944."

23. The quote is a paraphrase of Birger Bergersen cited in a State Department memorandum to Kellogg, in ibid.

24. Kellogg at 1944 meeting, 13 January 1944, in ibid.

25. Kellogg to Guy M. Gillette, 31 May 1944, in ibid. Private correspondence makes it clear that a number of other participants also considered the initial quota much too

Kellogg's latest trip to London had been, if anything, a still sharper reminder of this danger than his previous sojourns. Sniffing around the British delegation, Kellogg got the sense that two whaling company executives were effectively "managing the fats and oils division of the British Ministry of Foods," and that one of them "stood to make a couple of million profit, if relaxation of the provisions of these international whaling agreements could be agreed on."[26] There was money in the air, together with the clamor of a world shattered by war and desperately in need of calories. It was an awkward time to peddle restrictions. Hence the devil's bargain: a seasonal catch limit that was no limit at all (since it stood beyond the resurgent industry's actual killing capacity), in return for the installation of the precedent for an overall annual catch quota (and a chance to demonstrate that a system of monitored catching could be implemented).[27]

---

high. See, for instance, the view of Birger Bergersen: "It will presumably become increasingly clear to all biologists that this figure is an outside one, and one is inclined to dread the catch statistics which will be forthcoming after this season." Bergersen to Kellogg, 4 December 1948, Smithsonian Archives, RU 7165, box 11, folder 6, "Birger Bergersen Correspondence, 1947–1953 (Ambassador from Norway to Sweden)." Bergersen was also committed to the importance of establishing and solidifying the precedent for a total quota: "He realised that there was no possibility of this figure being reached in the first season but considered it very important to establish the principle from the outset of the resumption of whaling." Summary of verbal proceedings, p. 3, Smithsonian Archives, RU 7165, box 7, folder 6, "London—International Whaling Conference, 1944." Kurkpatrick Dorsey discusses the way postwar oil needs trumped the impetus toward conservation in "Compromising on Conservation: World War II and American Leadership in Whaling Diplomacy," in *Natural Enemy, Natural Ally: Toward an Environmental History of Warfare*, edited by Richard P. Tucker and Edmund Russell (Corvallis: Oregon State University Press, 2004), 252–96.

26. Kellogg to Coolidge, 25 May 1945, summarizing discussion with Henry Maurice, Smithsonian Archives, RU 7170, box 2, file "Cob–Cool."

27. See Kellogg's report to the secretary of state summarizing negotiations, 17 October 1949, Smithsonian Archives, RU 7165, box 28, folder 4, "Reports to the Secretary of State by the U.S. Commissioner to the IWC, Meetings #1–11, 1949–1959." This sort of close work with period sources evidences the weakness of casual assertions like that of Gare Smith, who suggests blithely that "in 1946 'conservation' was viewed only as a means to achieve the larger end of industry development," and that therefore, "it is possible that treaty drafters did not realize that there was an inconsistency and . . . saw no reason to provide for the resolution of conflicts between the goals of conservation and industry growth." Gare Smith, "The International Whaling Commission: An Analysis of the Past and Reflections on the Future," *Natural Resources Lawyer* 16, no. 4 (1984):

At the London meetings of 1945 and 1946 as well, hardship pleadings and diplomatic foot-dragging won various concessions and special exemptions.[28] These unsatisfactory outcomes reinforced for American officials (and their sympathetic British and Norwegian counterparts) the importance of restructuring the whole regulatory regime. Doing so at a gathering geographically removed from the industry's traditional strongholds and convened explicitly with a view to longer-term arrangements (rather than the annual ad hoc scrambles to configure regulations for the forthcoming season, even as investors were madly readying their multi-vessel expeditions in the neighboring harbor) had become essential. It was also evidently desirable to attack the problem before a large number of new ventures were afloat, and to do so while wartime emergency powers remained in effect, since they could facilitate the speedy execution of the necessary legislation.[29] By 1946, then, plans began to take shape for a major new international whaling conference in Washington, DC, the aim of which would be nothing less than a complete overhaul of the international legal framework guiding the regulatory diplomacy in a massive and enormously valuable open-water industry.

With the prospect of a major new international treaty on the horizon, the US State Department undertook a confidential interagency initiative to gather and collate all the relevant documentation (former treaties, their provisions, and their statuses), tap technical expertise, and ponder larger diplomatic considerations. This new "Informal Inter-agency Committee on the Regulation of Whaling" (IICRW), made up of eight civil servants drawn from the Departments of Treasury, Interior, and State, worked intensively for four months to digest thousands of pages of transcripts, data, and correspondence, and then to draft a preliminary version of a new whaling

---

543–67, at p. 550. It is remarkable how much of this sort of hypothetical non-history there is out there about the IWC.

28. For some of these tweaks, see Smithsonian Archives, RU 7165, box 28, folder 4, "Reports to the Secretary of State by the U.S. Commissioner to the IWC, Meetings #1–11, 1949–1959." Note that the 16,000 BWU total limit was kept in force (though it did not, in this period, actually stop the hunt, since catches did not reach this level).

29. For discussion of emergency powers, see the report for the 1945 conference, pp. 32–34, Smithsonian Archives, RU 7165, box 8, folder 5, "Washington—Informal Inter-agency Committee on the Regulation of Whaling (IICRW), circa July 1946–Oct. 1946—Minutes."

convention.[30] It was this document that served as the point of departure for the 44 delegates from 19 nations who came together in November 1946 to debate, fine-tune, and finally sign the International Convention for the Regulation of Whaling—the treaty that has guided most of the word's whaling activities ever since.[31]

At the heart of the new arrangement lay an administrative innovation originally outlined by the British fisheries director, A. T. A. Dobson, in a secret memo conveyed to the US secretary of state back in May 1945.[32] Dobson had envisioned "setting up a steering committee composed of representatives of the U.S., U.K., and Norway." The idea being that such a body might "be authorized to make minor changes and adjustments in the international whaling regulation as may be necessary from time to time," an arrangement that would have the enormous advantage of "eliminating, whenever such minor changes were needed, the necessity of calling an international conference."[33] It was precisely such an "empowered" committee that the ICRW installed, since the documents signed in Washington on 2 December 1946 consisted of two essential parts: first, the "Convention" itself (the ICRW), which provided for the creation of an "International Whaling Commission" (to be composed of representatives of the signatory nations); and second, a "Schedule" of whaling regulations (stipulating catch quota, size limits, prohibited species and regions, length of season, etc.), which could be modified by the commission at its annual meeting. Such modifications, or "amendments of the schedule," were to be guided by several principles, including, most importantly, the need to provide for

30. The invaluable minutes and briefs of this committee can be found in ibid.

31. As of noon on 2 December 1946, the State Department calculated 44 delegates, 13 advisers, 2 additional officials (for a total of 59 officers attended by a staff of 3), plus 3 FAO observers, 11 other observers and consultants, and a total of 64 members of the "international secretariat" (presumably the term used for staff dedicated to support, translation, and administration of the event; the designation was used widely in connection with the 1945 conference in San Francisco that generated the UN Charter, and there may have been some personnel overlap), for a grand total conference participation of 140. See "Recapitulation—Restricted," Smithsonian Archives, RU 7165, box 9, folder 4, "Washington—International Whaling Conference, 1946—Conference Proceedings."

32. See Winant to secretary of state, 19 May 1945, Smithsonian Archives, RU 7165, box 7, folder 4, "Department of State, 1930–1945."

33. Ibid.

"optimum utilization of the whale resources" and an obligation to make decisions "based on scientific findings."[34]

The ICRW as a whole, and the provisions for amendment of the schedule in particular, have been the subject of tens of thousands of pages of commentary, critique, and animadversion for sixty years. A detailed account of the myriad diplomatic negotiations, serial revisions, and strategic misunderstandings that attended its making and deployment would occupy an army of multinational historians for a decade; such an exercise is well beyond the scope of this book. But before I turn to a discussion of the most important recent scholarship on the workings of the IWC in its formative years, I want to draw attention to what I take to be the most important feature of the ICRW—the feature that will preoccupy me in the pages that follow. To wit: The ICRW represented *a novel legal-cum-administrative effort to implement scientific management of natural resources on an international basis.*[35]

34. These stipulations are drawn from paragraph 2 of Article V of the ICRW, which reads in full:

These amendments of the Schedule (a) shall be such as are necessary to carry out the objectives and purposes of this Convention and to provide for the conservation, development, and optimum utilization of the whale resources; (b) shall be based on scientific findings; (c) shall not involve restrictions on the number or nationality of factory ships or land stations, nor allocate specific quotas to any factory or ship or land stations or to any group of factory ships or land stations; and (d) shall take into consideration the interests of the consumers of whale products and the whaling industry.

I will return to the conflicts engendered by these different provisions below.

35. For a discussion of some of the legal precedents, see Patricia W. Birnie, *International Regulation of Whaling: From Conservation of Whaling to Conservation of Whales and Regulation of Whale-Watching* (New York: Oceana Publications, 1985), vol. 1, chaps. 3, 4 (she points out on p. 143 that the IWC was the first global-scale commission to regulate migratory resources). On the precedents for the conservation aspects, see W. M. Adams, *Against Extinction: The Story of Conservation* (London: Earthscan, 2004), and Kurkpatrick Dorsey, *The Dawn of Conservation Diplomacy: U.S.-Canadian Wildlife Protection Treaties in the Progressive Era* (Seattle: University of Washington Press, 1998). For a wider framing of US efforts to implement international legal mechanisms in this period, consider Yves Dezalay and Bryant G. Garth, eds., *Global Prescriptions: The Production, Exportation, and Importation of a New Legal Orthodoxy* (Ann Arbor: University of Michigan Press, 2002). To get a sense of what was understood to be the essential scientific basis for the new regulatory system, it is worth reviewing IICRW

Even a casual review of the relevant archives dramatizes the emphasis that participants and planners placed on "scientific management" in framing the new whaling regime. Remington Kellogg was, as I have suggested, a leading exponent of this effort to place world whaling on a scientific basis. As he wrote to his friend and scientific coconspirator, the Norwegian statesman Birger Bergersen (trained as an anatomist), in 1946, "I consider it vital that we have adequate information on which to base such decisions as may be agreed upon. Guess work is always unsafe and this is particularly true of a natural resource that may be unduly depleted by unwise exploitation."[36] Kellogg was by no means alone, however, in emphasizing the role of science in regulation. The head of the British delegation, A. T. A. Dobson, was explicit about the importance of "biological considerations" in the new agreement, and there remained a small community of whale scientists from Britain and Norway who had been active in stock problems before the war and who saw the ICRW as a unique opportunity to place the industry on a newly scientific footing—a dream that reached back to the turn of the century.[37]

Moreover, the whole notion of formally rationalized schemes, aimed at the sustainable regulation of natural resources and based on the advice of scientific experts, was a major feature of US policy in the period.[38] Even

---

Paper no. 23, "Source Material for Guidance of Contracting Governments and Drafting National Whaling Regulations: Scientific Work As It Has Been Applied to Regulations and As It Has Been Proposed for Investigation," which summarizes the status of scientific research oriented toward regulatory initiatives from 1937 through 1944 for the use of the Informal Inter-agency Committee on Regulation of Whaling. See Smithsonian Archives, RU 7165, box 8, folder 3, "Washington—Informal Inter-agency Committee of the Regulation of Whaling (IICRW), circa July 1946–Oct. 1946—Papers."

36. Kellogg to Bergersen, 8 April 1946, Smithsonian Archives, RU 7165, box 7, folder 7, "London—International Whaling Conference, 1945—Includes International Overfishing Conference, 1946." Bergersen, who had published a monograph on pinnipeds in the 1930s (*Beiträge zur Kenntnis der Haut einiger Pinnipedien unter besonderer Berücksichtigung der Haut der Phoca groenlandica* [Oslo: I Kommisjon Hos J. Dybwad, 1931]), held a professorship in the Anatomy Department of the Norwegian School Of Dentistry before World War II.

37. On Dobson's view, see Kellogg to Bergersen, 27 December 1948, Smithsonian Archives, RU 7165, box 11, folder 6, "Birger Bergersen Correspondence, 1947–1953 (Ambassador from Norway to Sweden)." The scientists referred to included Mackintosh and Johan T. Ruud. For the early history of the "dream," see chapter 2 above.

38. For discussions of this broader context, see Arthur F. McEvoy, *The Fisherman's Problem: Ecology and Law in the California Fisheries, 1850–1980* (Cambridge: Cambridge

while the war was still on, the issue of how to reframe legal and administrative structures in such a way as to guarantee "the perpetuation of ocean fisheries" had already become a major preoccupation of American international lawyers, West Coast fishery executives, and bureaucrats in the Fish and Wildlife Service, all of whom had looked on nervously as competition for salmon, sardines, and halibut created serious international tensions in the North Pacific in the 1930s.[39] Directly addressing this recent history, the United States circulated to the delegates at the 1946 conference a detailed memo on US conservation programs, with particular reference to ocean resources. In this document the United States made explicit the principle of "maximum sustainable yield" (or MSY), which would guide US fisheries policy for decades.[40] In this sense the founding of the IWC represents an important example of what Edward W. Allen, US commissioner of the International North Pacific Fisheries Commission, would call, in a 1954 brief, "The New International Fishery Commission concept," in which "adequate scientific research is a prerequisite to regulation."[41] It was in fact the biologist Wilbert McLeod Chapman, the architect of US Pacific fisheries

University Press, 1986), particularly part III. Also of value are Harry N. Scheiber, "Pacific Ocean Resources, Science, and Law of the Sea: Wilbert Chapman and the Pacific Fisheries, 1935–1970," *Ecology Law Quarterly* 13 (1986): 381–534; Mary Carmel Finley, "The Tragedy of Enclosure: Fish, Fisheries Science, and U.S. Foreign Policy, 1920–1960" (PhD diss., University of California, 2007); and Joseph E. Taylor, *Making Salmon: An Environmental History of the Northwest Fisheries Crisis* (Seattle: University of Washington Press, 1999).

39. Particularly with Canada and Japan, but also with the USSR. See Edward Allen to Willard Cowles, 10 July 1942, Smithsonian Institution, James Mead Files, file "Conservation—Whales": "It is timely now to develop new principles to deal with international fisheries control."

40. Significantly, Kellogg explicitly invoked the concept of "sustained yield" in his brief opening remarks to the first IWC meeting in 1949 (see his note cards for this presentation in the Smithsonian Archives, RU 7165, box 22, folder 7, "First International Whaling Commission Meeting [London], 1949. Material, circa 1945–1955"). For an invaluable history of MSY, see Tim D. Smith, *Scaling Fisheries: The Science of Measuring the Effects of Fishing, 1855–1955* (Cambridge: Cambridge University Press, 1994), and Mary Carmel Finley, "The Tragedy of Enclosure: Fish, Fisheries Science, and U.S. Foreign Policy, 1920–1960" (PhD diss., University of California, 2007). I will discuss MSY in greater detail below. See also "Sanctuaries as Conservation Measure," 21 November 1945, Smithsonian Archives, RU 7165, box 7, folder 6, "London—International Whaling Conference, 1944."

41. Smithsonian Institution, James Mead Files, file "Conservation—Whales."

policy and the patron saint of MSY science, who (as a special assistant in the Department of State) engineered the US law that put the provisions of the ICRW into effect within the US legal system. He steered the draft bill into the statute books with help from Kellogg himself.[42]

In view of the enormous failure of the IWC in the decades that followed, it is important to emphasize the optimism and enthusiasm that attended its creation. In the early years of the organization—the IWC itself met for the first time in 1949, and at least annually thenceforth—it was held up as an esteemed example of the kind of international, scientific conservation program that the world needed, so much so that a variety of other interested groups that might have taken up the protection of cetaceans after the war formally ceded this issue to the new dedicated body, confident that all would be well. For instance, shortly after the 1949 UN "Scientific Conference on the Conservation and Utilization of Resources" at Lake Success, New York, the International Union for the Protection of Nature (IUPN, later IUCN) resolved, concerning whales, that "protection being assured—and successfully—by a specialist international organization, it is decided that the Union will not undertake any separate action in favour of this mammal."[43] Similarly, the conservation energies of Julian Huxley's UNESCO and those of the emergent fisheries program of FAO were both directed elsewhere, on the understanding that the world's whales were in good hands.[44] A number of commentators took great pride in the fact that the ICRW put science at the center of a conservation protocol. For instance, in

42. W. M. Chapman to Kellogg, 28 December 1948, with annotated drafts, in ibid. For helpful context, consider [Mary] Carmel Finley, "The Social Construction of Fishing, 1949," *Ecology and Society* 14, no. 1 (2009): 6.

43. Minutes of IUPN Executive Board, eighth session, fourth sitting, twenty-second meeting, 9 September 1952, Caracas. Cited in Martin W. Holdgate, *The Green Web: A Union for World Conservation* (Cambridge: IUCN, the World Conservation Union/ Earthscun, 1999), 49.

44. There was considerable early debate about whether the IWC ought to be formally integrated into FAO; the United Kingdom initially supported such an arrangement, but the United States resisted. See "Proposed Relationship of International Whaling Commission to Food and Agriculture Organization," 20 November 1946 (IWC / 13), Smithsonian Archives, RU 7165, box 9, folder 5, "Washington—International Whaling Conference, 1946—Conference Documents #1–45." It seems likely (although I have not seen documentary support for this assertion) that the UK delegation, deeply influenced by industry desiderata, preferred the idea of folding whale management into the brief of the FAO so as to ensure that the United States could not turn the new IWC into some sort of preservation organization. Kurk Dorsey appears to have turned up additional

characterizing the "principal accomplishments" of the young IWC in 1953,
Leonard Carmichael, then director of the Smithsonian Institution, repeatedly returned to the centrality of science to its workings, concluding that
"the Commission has inaugurated and continued scientific studies relating
to the biology of the whale and its migratory habits in order to provide a
basis for more effective conservation measures." The result of the IWC's
work, he asserted, was "continual protection to the whale stock."[45]

In view of these early expressions of confidence, it is all the more striking
to realize that twenty years later, the ICRW would be held up by conservation activists as a study in how *not* to structure a treaty for the rational utilization of a natural resource. Kellogg himself would submit his final IWC
report to the secretary of state in 1964 (he was, by then, in his seventies,
and increasingly frail), a swan song that ended by calling into question the
"future usefulness of the Commission" and suggesting that perhaps another
international organization could step in where the IWC had failed.[46] By this
point the scientific advisory system of the commission was in shambles—
the chairman of the Scientific Committee having resigned in disgust after a
last-ditch set of proposed conservation measures met with disregard from
the voting delegates; in solidarity, all of his scientific colleagues refused
to replace him, so whale science at the IWC was effectively acephalous by
1965. The organization's conservation record was dismal. Under the IWC's
watch, the number of catcher boats working in the Antarctic had quadru-
pled, the number of floating factories had more than doubled, and both
kinds of vessels had gotten larger and more powerful. Despite these significant increases in reach and effort, the overall catches had fallen sharply (by
an order of magnitude at least). By the mid-1960s something like a million

---

material on this issue in Foreign Office archives, so my analysis here should be treated
as provisional.

45. Carmichael to Murphy, circa November / December 1953 (on the basis of internal
evidence), Smithsonian Archives, RU 7165, box 23, folder 6, "Fifth International Whaling
Commission Meeting (London), 1953." NB: It is possible that this letter was drafted by
Kellogg for Carmichael's signature.

46. For comments citing the ICRW's failure as a treaty, see the IUCN *Bulletin* 12, New
Series (July / September 1964): 6–7. (I have consulted the copy of this document found
in the Smithsonian Archives, RU 7165, box 27, folder 2, "Correspondence, Sixteenth
International Whaling Commission Meeting [Sandefjord], 1964.") For Kellogg's final
pessimistic comments, see his report for 1964, p. 46, Smithsonian Archives, RU 7165,
box 28, folder 6, "Reports to the Secretary of State by the U.S. Commissioner to the
IWC, Meetings #15–20, 1963–1968."

large whales had been removed from the Antarctic, and an original stock that might have yielded tens of thousands of animals per year in perpetuity had been largely liquidated. Tellingly, by 1965, most of the industry had its sights on new regions, like the North Pacific.

. . . .

Such public failure, and on such a scale, has not gone unconsidered. Since a detailed study of the interaction of scientific research and whaling policy over this period is my intention in this chapter and the one that follows, I would like to take a moment here to review recent scholarship on this problem. My emphasis will be on the need for a new treatment and a revised approach.

For starters, the IWC has not been the subject of any work at all by historians or sociologists of science, who have never been especially attentive to applied science and have only recently begun to look seriously at biological oceanography and fisheries research.[47] Similarly, while environmental historians have touched on the IWC and the history of whaling, they too, surprisingly, have not yet produced a sustained analysis of the subject.[48] This can perhaps be explained in part by the relatively slow de-

47. But see Eric L. Mills, *Biological Oceanography: An Early History, 1870–1960* (Ithaca, NY: Cornell University Press, 1989), and Helen M. Rozwadowski, *The Sea Knows No Boundaries: A Century of Marine Science under ICES* (Copenhagen: International Council for the Exploration of the Sea in association with the University of Washington Press, 2002).

48. Indeed, the handling of the topic in several of the more ambitious surveys is perfunctory and erroneous. John McCormick does not treat the IWC in *The Global Environmental Movement: Reclaiming Paradise* (London: Belhaven, 1989), and his discussions of whaling are misinformed (e.g., p. 7). Robert Boardman states, without explanation or justification, that whaling and its regulation fall "outside the scope" of his book (Boardman, *International Organization and the Conservation of Nature* [London: Macmillan, 1981], 36). Samuel P. Hays and Barbara D. Hays, in *Beauty, Health and Permanence: Environmental Politics in the United States, 1955–1985* (Cambridge: Cambridge University Press, 1987), manage only a paragraph (on p. 113) on the subject in 630 pages, and that paragraph is none too accurate. Richard L. Andrews, in *Managing the Environment, Managing Ourselves: A History of American Environmental Policy* (New Haven, CT: Yale University Press, 1999), entails the whole question to a short discussion of "dolphin-safe tuna" (pp. 342–43). W. M. Adams, in *Against Extinction: The Story of Conservation* (London: Earthscan, 2004), does better, and though the treatment is brief (pp. 190–94), he does describe whaling as "a classic conservation issue."

velopment of an environmental history dealing with marine habitats and populations.[49] Instead, work on the history and development of the IWC has largely emerged from three different (if at times overlapping) communities: scholars of political science and international relations (who have been particularly interested in the legal and political features of the organization), commentators with scientific training and experience on the inside of IWC advising practices (who have written a number of participant narratives and mobilized earlier history as part of various reform efforts), and finally, "activists" both for and against whaling (who have composed a variety of more and less journalistic treatments for the purpose of proselytizing and propaganda).[50] Throughout these diverse writings there is

49. Though this seems to be changing rapidly. See, for instance, the work of HMAP (History of Marine Animal Populations) and the recent conferences and sessions on marine environmental history at Rutgers, Harvard, and elsewhere. For the leading statements on the problem at present, see W. Jeffrey Bolster, "Opportunities in Marine Environmental History," *Environmental History* 11, no. 3 (July 2006): 567–97, and Bolster, "Putting the Ocean in Atlantic History: Maritime Communities and Marine Ecology in the Northwest Atlantic, 1500–1800," *American Historical Review* 113 (February 2008): 19–47.

50. I will discuss some of these treatments in some detail below, but the following are relevant (I list them in an order that corresponds roughly to progress through the three groups I list above): Peter Gidon Bock, "A Study in International Regulation: The Case of Whaling" (PhD diss., New York University, 1966); Patricia W. Birnie, *International Regulation of Whaling: From Conservation of Whaling to Conservation of Whales and Regulation of Whale-Watching*, 2 vols. (New York: Oceana Publications, 1985), Peter J. Stoett, *The International Politics of Whaling* (Vancouver: UBC Press, 1997); William Aron, "Science and the IWC," in *Toward a Sustainable Whaling Regime*, ed. Robert L. Friedheim (Seattle: University of Washington Press, 2001), 105–22; J. A. Gulland, "The Management of Antarctic Whaling Resources," *Journal du Conseil, Conseil International pour l'Exploration de la Mer* 32, no. 3 (1968): 330–41; Ray Gambell, "International Management of Whales and Whaling: An Historical Review of the Regulation of Commercial and Aboriginal Subsistence Whaling," *Arctic* 46, no. 2 (June 1993): 97–107; John Laurence McHugh, "The Role and History of the International Whaling Commission," in *The Whale Problem: A Status Report*, ed. William Edward Schevill, G. Carleton Ray, and Kenneth S. Norris (Cambridge, MA: Harvard University Press, 1974), 305–35; George L. Small, *The Blue Whale* (New York: Columbia University Press, 1971); and David Day, *The Whale War* (San Francisco: Sierra Club Books, 1987). In view of the overwhelming significance of the IWC in the larger history of whaling, the standard histories also have valuable treatments of the organization, particularly, of course, J. N. Tønnessen and Arne Odd Johnsen, *The History of Modern Whaling* (Berkeley: University of California Press, 1982).

considerable shared interest in the problem of where to place the blame for the failure of 1945–1965.

In the most systematic effort to ask—though not, in the end, to answer—this question, the biologist and historian Tim D. Smith, former chair of the US scientific delegation to the IWC, lists no fewer than seven possible "hypotheses" to explain the failure of the kind of optimistic management principles on which the IWC was originally based, including in the tally everything from the folly and greed of human actors to the complexity of the biological environment itself. Data uncertainty and the failure of early population models also make his list of possible culprits.[51] Assessing the precise admixture of explanatory factors, he suggests, would require the kind of fine-grained historical study of the whole episode that, he notes, has never been done—though he is perhaps quixotic in the hope that such a study, properly conducted, might itself constitute a new evidentiary basis for rethinking "Marine Exploitation System Science." He does, however, acknowledge puckishly that the acronym he imagines for this new field (MESS) sums up a great deal about the whole topic.

It is Patricia Birnie's two-volume *International Regulation of Whaling: From Conservation of Whaling to Conservation of Whales and Regulation of Whale-Watching* that offers the most extensive discussion of the IWC's rise and fall in the twentieth century.[52] Like Smith, Birnie wants to pinpoint culprits for its collapse (and to suggest revisions that could lead to what she hoped, writing in the heady days of the push for a moratorium on commercial whaling, would become a reformation and resurgence along new lines); also like Smith, she sees a host of factors at work. Trained in international law (she went on to work in intergovernmental organizations), Birnie is particularly attentive to legal and institutional factors, and when she digs in on a critique of the ICRW, she lists no fewer than fifteen structural features of the treaty that she believes were "serious deficiencies" of the IWC arrangement and precipitated its many difficulties.[53] While she

51. Tim D. Smith, "'Simultaneous and complementary advances': Mid-Century Expectations of the Interaction of Fisheries Science and Management," *Reviews in Fish Biology and Fisheries* 8, no. 3 (1998): 335–48.

52. Patricia Birnie, *International Regulation of Whaling: From Conservation of Whaling to Conservation of Whales and Regulation of Whale-Watching*, 2 vols. (New York: Oceana Publications, 1985). This is largely a reprint of her 1979 dissertation at the University of Edinburgh.

53. These deficiencies were (1) no inspection system; (2) inadequate sanctions system; (3) prolonged objections procedure; (4) noncomprehensive membership; (5) incompat-

is inclined to understand the whole debacle as the product of unhappy (if entrenched) principles of international law—particularly the freedom of the seas on the one hand and the idea that its inhabitants were *res nullius*, there for the taking, on the other—she repeatedly returns to the claim that "a lack of scientific knowledge" was responsible for a litany of bad decisions.[54] Here she adopts a posture quite orthodox among commentators who have reviewed the early history of the IWC's regulatory system: the solution, it is not uncommonly suggested, lay in *more science*.[55]

---

ible objectives (as in being charged to serve the industry while also being charged to use science to achieve an "optimum level" for stocks); (6) no mechanism for restricting effort or blocking transfer of vessels; (7) no settlement or arbitration procedures; (8) the "scientific whaling" loophole; (9) no provisions to enact the appealing generalities of the preamble concerning safeguarding the whales for future generations; (10) prohibition on IWC limits on numbers of vessels used in the hunt; (11) isolation from the UN; (12) BWU quota system; (13) small budget; (14) withdrawal clauses; and (15) limited support for international scientific research. See Patricia W. Birnie, *International Regulation of Whaling: From Conservation of Whaling to Conservation of Whales and Regulation of Whale-Watching* (New York: Oceana Publications, 1985), vol. 1, 202–3. When she boils them down (pp. 260–61), she emphasizes the absence of a system for allocating national quotas and the absence of any provision for inspection. On these points she is in accord with the view of a number of industry executives, particularly Gerald H. Elliot (then managing director of Christian Salvesen Ltd.; later Sir Gerald), who hammers away on the lack of enforcement and the futility of any arrangement unable to allot national quotas in his 1979 article "The Failure of the IWC, 1946–1966," *Marine Policy* 3 (April 1979): 149–55. It is perhaps no surprise that Birnie and Elliot reached the same conclusions: she was using Salvesen's private archive to write her dissertation. For a fuller account of Elliot's views, see his *A Whaling Enterprise: Salvesen in the Antarctic* (Norwich: Michael Russell, 1998). The United States was resistant to the idea that the ICRW might be configured to limit effort or award quotas to particular nations (on the grounds that "such allocation is not in the interest of free and competitive enterprise"). See US "secret" memorandum, Smithsonian Archives, RU 7165, box 8, folder 3, "Washington—Informal Inter-agency Committee on the Regulation of Whaling (IICRW), circa July 1946–Oct. 1946—Minutes."

54. Patricia W. Birnie, *International Regulation of Whaling: From Conservation of Whaling to Conservation of Whales and Regulation of Whale-Watching* (New York: Oceana Publications, 1985), vol. 1, 128, 292, 364.

55. William Aron adopts a version of this approach in "Science and the IWC," in *Toward a Sustainable Whaling Regime*, ed. Robert L. Friedheim (Seattle: University of Washington Press, 2001), 105–22, though he is more attentive to the need for a particular kind of science: population dynamics, rather than more traditional biological work. I will return to this important suggestion below.

Though how much science there was, and what it looked like, are seldom considered.[56]

In contrast to Smith's and Birnie's willingness to weigh multiple factors, Michael Heazle's *Scientific Uncertainty and the Politics of Whaling*—the most recent and ambitious contribution to the topic—makes a bid to reduce the whole issue to essentially a single factor (albeit not a small one): the nature of science itself.[57] Moreover, in defiance of most of the rest of the commentators, Heazle entirely rejects the idea that "more science" would have helped the IWC out of its numerous regulatory binds of the 1950s and 1960s. A scholar of political science, Heazle is interested in speaking to regime theorists in international relations, particularly those who study the development and significance of epistemic communities and their impact on policy making. Broadly speaking, his book takes aim at the coherence of the "precautionary principle" as a guideline for environmental policy. While he claims to be historically oriented—and the book does deal with past practices of scientific advising in the IWC (as well as the current arrangements)—Heazle's central argument is not historical at all. Rather, it is (though one hesitates to use the term) philosophical. In an introductory chapter Heazle does a rapid circuit through post-positivist philosophy of science in order to return with the finding that "we cannot establish that science can directly describe or correspond with the 'real world.'"[58]

After thus dispensing with the notion that science is some kind of "noble quest for the truth," he can move on to assert that the real issue is not science, but rather *who wants what*:[59]

56. Birnie, for all her other strengths, is herself very weak on these issues: see, for instance, her misunderstanding of the fate of marking in the 1961 Scientific Committee report (Patricia W. Birnie, *International Regulation of Whaling: From Conservation of Whaling to Conservation of Whales and Regulation of Whale-Watching* [New York: Oceana Publications, 1985], vol. 1, 305).

57. Michael Heazle, *Scientific Uncertainty and the Politics of Whaling* (Seattle: University of Washington Press, 2006). This book is a revision of his 2003 dissertation. See my review in *Isis* 98, no. 2 (2007): 425–26. The core of Heazle's argument is laid out in a pair of articles in *Marine Policy*: M. Heazle, "Scientific Uncertainty and the International Whaling Commission: An Alternative Perspective on the Use of Science in Policy Making," *Marine Policy* 28, no. 5 (2004): 361–74, and M. Heazle, "Lessons in Precaution: The International Whaling Commission Experience with Precautionary Management," *Marine Policy* 30, no. 5 (2006): 496–509.

58. Michael Heazle, *Scientific Uncertainty and the Politics of Whaling* (Seattle: University of Washington Press, 2006), 32.

59. Ibid., p. 33.

My argument contends that our choices are fundamentally determined by their perceived utility. And since the notion of utility that I am using is entirely a subjective one, different people, groups, societies, and cultures will perceive the utility of a given choice differently in relation to the values and cultural preferences they subscribe to.[60]

In essence, then, Heazle's point of departure amounts to this: people will pretty much do what they want, and science, which is just a particular kind of rhetoric, can't stop them (much less oblige them). When scientific findings seem to make trouble, it is always possible to escape through the large and unclosable loophole known as "scientific uncertainty." Having established the impotence of science to his satisfaction, Heazle turns to the story of the IWC's use of science less to tell a history than to make an example, since the reader is in the presence of a foregone conclusion: the science will be ignored, and people will do what they want. And so it goes.

For all the shortcomings of this approach (and I will get to some of them in a moment), it does have the virtue of permitting Heazle to tell a satisfying tale of biter bitten, since he can show that the same kinds of claims about scientific uncertainty that were once used to scuttle protective action and delay conservation measures (in the 1950s and early 1960s) have more recently been used to head off a return to commercial whaling (in the 1990s). As far as policy recommendations are concerned, he comes away from his case study equipped to counsel both that naïve faith (on the part of scientists, on the part of politicians) in the ability of science to compel collective action is sure to be disappointed when interests collide, and that the modern IWC would do well to make a more concerted effort to manage scientific uncertainty.

It should be noted that Heazle is not the first person to zero in on scientific uncertainty as a key dimension of conflict in the early years of the IWC. Back in 1992 the Norwegian economist and statistician Tore Schweder precipitated an acrimonious debate (and threats of legal reprisal) when he presented to the IWC's Scientific Committee a paper entitled "Intransigence, Incompetence, or Political Expediency? Dutch Scientists in the International Whaling Commission in the 1950s: Injection of Uncertainty."[61]

---

60. Heazle, *Scientific Uncertainty and the Politics of Whaling*, 33.

61. This paper is IWC SC / 44 / 013, which, as a result of Sidney J. Holt's notice of possible action for libel (as I understand it), is now released from IWC archives only accompanied by an extract from IWC / 45 / 4 (Holt's rejoinder). A later version of this

The furor that attended this dip into the IWC archives gives a flavor of how contentious the history of twentieth-century whaling and whale science remains. Taking up what is in many ways the bleakest episode in the history of the IWC—the debate over the status of Antarctic fin whales in the 1950s—Schweder (himself a recent member of the Scientific Committee) reached a version of what would later become Heazle's conclusion about the function of scientific uncertainty in the period: namely, that uncertainty was invoked by interested parties in order to quash movement toward much-needed conservation measures. In fact, Schweder also saw in this story the opportunity for a tale of the biter bitten—though, unlike Heazle, Schweder himself wanted to do some biting. At issue were the then-current debates (to be fair, they have not actually died) around the "Revised Management Plan" (or RMP), a scheme that emerged out of the IWC's post-moratorium "Comprehensive Assessment" and which, if adopted, would have allowed a highly regulated return to commercial whaling.[62] In Schweder's view, the barrage of sophistical critique and Chicken Little fretting that had stalled the RMP (it remains stalled) looked exactly like a reprise of the 1950s, only now a new wave of "anti-whaling scientists" (as opposed to the old pro-whaling scientists) were deploying scientific uncertainty speciously to delay action they objected to on fundamentally nonscientific grounds. If anything, the new situation was, in Schweder's opinion, worse than what took place in the fifties, since E. J. Slijper—the notorious Dutch scientist and spoiler of consensus in the fin whale debates—had been, in Schweder's view, a pretty lousy scientist (at least as far as stock assessment techniques were concerned).[63] By contrast, the spoilers of consensus around the RMP—men like Sidney Holt, Justin Cooke, and William De la Mare—had given shape to the mathematics and the models at the very heart of the RMP. This made them much more formidable adversaries, to be sure, but it also, Schweder intimated, enhanced their culpability as perverters of good science.

---

tête-à-tête was subsequently published as Tore Schweder, "Protecting Whales by Distorting Uncertainty," *Fisheries Research* 52, no. 3 (2001): 217–25, with a commentary (pp. 227–30) by Holt. See also Tore Schweder's article "Distortion of Uncertainty in Science: Antarctic Fin Whales in the 1950s," *Journal of International Wildlife Law and Policy* 3, no. 1 (2000): 73–92.

62. The moratorium was voted in in 1982 on a temporary basis, with the understanding that such a reevaluation of management practices would be undertaken.

63. Here Schweder ends up laying much of the blame squarely where it was placed by an early generation of "Save the Whales" activists like Scott McVay and George Small.

And here we come to the essential difference between Schweder and Heazle. Although their accounts have much the same shape, Schweder views the obfuscatory and obstructionist use of scientific uncertainty as an ethical, not an epistemological, problem. In his view, when science fails to bring hard-working experts to consensus, it is probably the fault of bad men, and they merit censure for scientific misconduct. Heazle, by contrast, views such situations as inherent to the nature of science itself, and the only relevant censure might be that properly falling on anyone who imagined it could be otherwise. Heazle's position presumably attracts greater sympathy from whatever partisans of strong-program SSK (sociology of scientific knowledge) are left out there; Schweder's position apparently attracts more lawsuits.

It is my view that the character of Schweder's analysis of the fin whale controversy of the 1950s flattens out a great deal of that story (as I hope to show below), but in the end it is Heazle's more theoretically ostentatious approach that stands in greatest need of revision if we are to gain deeper insights into the science, politics, and history of whaling in the last half century.

To understand why, begin by bracketing the nettlesome epistemological problems from which Heazle hangs his treatment—put aside the nature of reality and how we know (or don't know) it; put aside how certainty limps in the sublunary and the difficulty of distinguishing truth from belief. Grant, for the sake of argument, that down here in the messy world of competing interests all science is merely a peculiarly elaborate form of rhetoric. It follows that anyone ostensibly interested in the relationship between "science" and "politics" would at the very least want to understand how science works *as rhetoric*. After all, if modern science really amounts to nothing more than an elaborate technique for "making the weaker speech the stronger," it must by any measure be accounted one of the great triumphs of this genre. And what is the great "rhetorical" power of a scientist on entering the domain of social conflict? Above all, it is his ability to define certain features of that conflict as "scientific" problems (which the scientist is uniquely positioned to solve), in contradistinction to the ordinary muddle of "political" problems (which, happily or unhappily, must be worked out with whatever blunt tools are at hand). If we acknowledge this, it rapidly becomes clear that an analytic posture overly at ease with dismissing scientific claims as "merely political" has given away the game. The issue is *not* whether, as a philosophical matter, it is possible to distinguish the domain of fact from that of value, the domain of true knowledge from that of successful deception. Rather, the issue is this: How have certain questions at certain times come to be understood as "scientific," and therefore solvable outside the domain of ordinary politics? In essence, this is the story of the kind of "boundary work" that has been of such interest to historians and sociologists of science in the last twenty years, since the history of success in this exercise is nothing less than the history of the thing that gets called "science" in any given situation—the history of *science itself* as an autonomous body of fact, theories, practices, and practitioners.[64]

In considering science in regulatory and policy settings, as Sheila Jasanoff has pointed out, it is particularly important to attend to the dynamics of this kind of boundary work, since by definition, regulatory activities straddle domains acknowledged by the participants (though seldom with perfect consensus concerning scope and limits) as representing "science" on the one hand and "politics" on the other.[65] How the boundary between these two domains is placed and buttressed at any given time constitutes the central strategic consideration in such situations: when this task is properly handled, scientific opponents can be accused of having succumbed to "capture" by industry; similarly, adroit figures can earn respect as "scientific statesmen" whose Hermes-like ability to move back and forth across the disputed territory enhances their power to set the herms and terms of the boundary.[66]

64. See Thomas F. Gieryn, *Cultural Boundaries of Science: Credibility on the Line* (Chicago: University of Chicago Press, 1999). By contrast, it is a major liability of Heazle's approach that, having already decided that all science is politics, he is (somewhat perversely) effectively liberated from the need to consider the science—or, for that matter, the politics—in any detail.

65. I am paraphrasing Sheila Jasanoff, *The Fifth Branch: Science Advisers as Policymakers* (Cambridge, MA: Harvard University Press, 1990), 14. It is worth noting that authors as diverse as Jasanoff and Hays both emphasize the degree to which "environmental" disputes in the postwar period took a toll on customary ideas about the role of a "disinterested" and "objective" science in public affairs; see, e.g., Samuel P. Hays and Barbara D. Hays, *Beauty, Health and Permanence: Environmental Politics in the United States, 1955–1985* (Cambridge: Cambridge University Press, 1987), chap. 10. I think this is right, and I see the whaling disputes as a significant episode in this process, as I hope to demonstrate in what follows.

66. Sheila Jasanoff, in *The Fifth Branch: Science Advisers as Policymakers* (Cambridge, MA: Harvard University Press, 1990), discusses strategic uses of boundary work on pp. 122 and 178; for the classic "capture" scenario, see p. 15 passim. On the "scientific statesman," see pp. 93–94. Jasanoff has also taken up the boundary question in greater

There can be little doubt that this critical posture—agnostic on metaphysics, attentive to the social dynamics by which facts get to be facts for the purpose of collective action—offers a much more promising way to approach a complex story of regulatory science than does Heazle's certainty about uncertainty and his overweening confidence in the omnipotence of a garden-variety pragmatism. After all, while it seems to be true that the truth never quite shows up on the half shell, commanding instantaneous allegiance wherever she beaches, it is no less true that the emergence of consensus concerning the "facts of the matter" is an important (and by no means rare) feature of regulatory debates. And scientists play a privileged role in establishing such facts, which alter the topography of policy conflict.[67] It will be my aim in the pages that follow to show this process at work in the whaling debates of the 1950s and 1960s, where I will demonstrate that careful work by politically savvy and technically skilled scientists emphatically made facts that changed the shape of the regulatory arena. Indeed, it is hard to imagine a more palpable instance of victory in the agon of the actual: in the early 1960s a group of biologists and statisticians quite literally silenced their casuistical opponents, and in so doing flushed several whaling nations out from behind the rationale of science and onto the naked plain of commercial imperatives.[68] Marshalling new techniques and technologies, this group of scientists succeeded in branding a set of positions on the whaling problem as "mere politics," though those positions had previously enjoyed scientific support. A better example of the strategic value of boundary work would be difficult to invent.

Before I turn to this story and others, I would like to say one more word about historiography. Having, I hope, persuaded the reader of the advantages of adopting an approach like Jasanoff's, I feel compelled to point to several ways in which I depart from her practice. For starters, it is one cost of her "case study" method that she is prevented from offering much more than a schematic and synchronic account of any of her selected episodes.

detail in "Contested Boundaries in Policy-Relevant Science," *Social Studies of Science* 17, no. 2 (May 1987): 195–230.

67. For a recent sociological model of exactly this process, consider Thomas Gehring and Eva Ruffing, "When Arguments Prevail over Power: The CITES Procedure for the Listing of Endangered Species," *Global Environmental Politics* 8, no. 2 (May 2008): 123–48.

68. In thinking about the significance of "silencing" in scientific debate, I have been helped by the "ideal dialogues" in Bruce Ackerman, *Social Justice in the Liberal State* (New Haven, CT: Yale University Press, 1980).

Her accounts are, of necessity, narratively lean, and largely without human characters whose professional development, scientific ideas or practices, or larger biographical trajectories can be grasped or understood to impinge on the dynamics of the regulatory situations she describes. This makes it difficult for her to convey much of a sense, for instance, of the actual phenomenon of scientific "capture": What does it look like? How does it happen? Answers to these questions are not forthcoming. Similarly, while the idea of a "scientific statesman" is an alluring one, we have no occasion to see such a scientific persona take shape and few opportunities to see what difference such a character can make. And yet the importance of these figures to science and social life in the twentieth century (not to mention the twenty-first) is beyond question.

By contrast, as elsewhere in this book, I will make an effort to pay attention to the intellectual biographies and professional trajectories of several of the scientists whose careers were inseparable from the problems of whale science and whaling regulation at midcentury. This perspective will do more than shed light on what some participants were thinking when they made decisions that remain easy to greet by asking, "What were they *thinking*!?" (Though it should certainly help with that.) More importantly, some effort to evoke scientific actors as characters will help us avoid a dangerous pitfall into which commentators on regulatory science in general (and whaling controversies in particular) too easily fall—namely, a naïve belief in the naïve belief of their subjects.[69] To put this a different way, it is necessary to pay close attention to the ways in which scientific actors are themselves subtle, adroit, and self-conscious participants in the fine strategic maneuvering that is always involved in any effort to set compelling and useful facts in particular positions in the landscape of a regulatory skirmish.[70] To choose just one example, archival work in the correspon-

69. I am borrowing this framing of the problem from Bruno Latour, who uses it in "A Few Steps toward an Anthropology of the Iconoclastic Gesture," *Science in Context* 10, no. 1 (1998): 62–83. See below, chapter 6, n. 228.

70. A paper that steps to the edge of this issue in the context of fisheries science is Robert M. Pringle's "The Origins of the Nile Perch in Lake Victoria," *BioScience* 55, no. 9 (2005): 780–87, which looks at the failure of fisheries scientists affiliated with the East African Fisheries Research Organization, who opposed the introduction of what soon became a highly invasive nonindigenous species, but were not able to arrest the decisions to go ahead with the plan. While Pringle is, in my view, interested in the right question (how do / should scientists engage in complicated political maneuvering?), he is not equipped with a theoretical approach that lets him do more than conclude, basically,

dence of whale scientists in the 1950s reveals that decisions superficial commentators have adjudged hopelessly foolish (such as the early reluctance on the part of scientific advisers to push for a much-reduced Antarctic quota) were in fact motivated by impeccable strategic considerations (in this case the desire to set a precedent, within a new and fluid institutional structure, for having scientific recommendations *accepted* deferentially by the plenary session of the IWC).

Understanding these kinds of moves demands that we look closely not at "science" or "politics" in the abstract, but rather at science and politics in the very concrete, and that we do so by following the fate of statistical analyses as they make their way into the exacting arenas of international diplomacy and parliamentary procedure.[71] It is one of the ambitions of this chapter (and the next) to advance the claim that we can learn more about how science works in regulatory settings by lingering for a while with the niceties of "Robert's Rules of Order" than we can from musing over logical fallacies in the rules of induction. The story of whaling and whale science between 1945 and 1965 is a story of science mobilized within what was, at the time, a novel space: the committee rooms of postwar international diplomacy.[72] The results were not pretty, but the learning curves were steep—

that science must be apolitical, although scientists can / must sometimes function in a political arena (paraphrasing from p. 786). I hope to show in what follows that this sort of (well-intentioned) advice takes the political naïveté of its subjects for granted, and wrongly. I have not been in Pringle's sources, but I feel confident that a very different story could be told from them, one in which fisheries scientists in Africa, operating in the complex context of colonial development programs and anticolonial nationalism, would not have needed Pringle as an energetic and youthful tutor in the exigencies of negotiating a political landscape. I could be wrong, but I doubt it.

71. It is worth noting that Jasanoff's work in this area tends to deal with US domestic policy issues. We have much less on science and regulation in an intergovernmental setting. In this regard, consider *Osiris* 21: John Krige and Kai-Henrik Barth, eds., *Historical Perspectives on Science, Technology and International Affairs* (Chicago: University of Chicago Press, 2006).

72. I have been surprised by how little secondary literature there seems to be on science in the setting of the conference room. In view of the increasing importance of this space as a site for scientists in the twentieth century, one can begin to imagine a sustained effort to make sense of what this development has entailed. There may well be a sociological-anthropological literature on conference spaces and the production of knowledge that I have overlooked, but I have consulted Heidrun Friese's "Thresholds in the Ambit of Discourse: On the Establishment of Authority at Academic Conferences," trans. William Clark, in Peter Becker and William Clark, eds., *Little Tools of Knowledge:*

almost as precipitous as the simultaneous reduction in the world's whales. A "race for the biological fact" indeed.

"THE FINE SPIRIT OF CO-OPERATION": 1945–1956

Wrapping things up at the end of the fifteenth session of the 1946 Washington meeting—after the pageantry of the signing of the ICRW, while the spangles of the magnesium flashes of the official photographer's camera were still in the air and in the eyes of the delegates—the Soviet representative, Alexander Bogdanov, saluted his peers (through his interpreter) with a sendoff in the warm spirit of comity and promise. Tipping his hat to Remington Kellogg, who had by this point chaired no fewer than a dozen days of continuous negotiation, Bogdanov captured the mood of the moment when he praised his assembled fellows "for their friendly cooperation" and declared proudly, "We feel sure that further activities in the field of International Regulation of whaling will be conducted in the atmosphere of *friendly understanding* which has been reached at the present conference."[73] Several hours later the room was empty, the plenipotentiaries having taken away with them a showy piece of paper and a vague sense of the possible. Over the next ten years, a tremendous amount of work would be done to translate that piece of paper and those vaporous good intentions into an actual set of practices, protocols, and precedents for successful, scientifically informed regulation of the world's whaling.

In the remainder of this chapter I will examine that first decade, with particular attention to the central issue laid out in the pages above—namely, the boundary work by which participants vied to define scientific problems

---

Historical Essays on Academic and Bureaucratic Practices (Ann Arbor: University of Michigan Press, 2001), 285–312, which seems a step in this direction. Also interesting in this regard are comments made by Mark Weatherall in his book on academic medicine in the nineteenth and early twentieth centuries: "If we take Latour's advice, and simply follow our actors through society, then we find that they spend a lot of time at meetings. . . . This is the point at which priorities are established, outside interests considered, promises made, visions affirmed or quietly shelved." Weatherall, *Gentlemen, Scientists, and Doctors: Medicine at Cambridge, 1800–1940* (Cambridge: Boydell Press, Cambridge University Library, 2000), 6. My thanks to Alistair Sponsel for the latter reference.

73. Transcripts, minutes of fifteenth session, p. 15, Smithsonian Archives, RU 7165, box 9, folder 5, "Washington—International Whaling Conference, 1946—Conference Documents #1–45." Emphasis added.

and to determine the domains within which those problems would be answered. It will be my contention that these early years saw a concerted effort on the part of scientifically oriented actors to facilitate and nurture a collaborative enterprise within which "reasonable" scientific claims, generated by suitable agents, would be taken up and used as the basis for good decision making in the IWC itself. That such claims were imperfectly supported, and therefore more or less provisional (and that they were liable to look that way for a long time to come), was beyond question—the scientists themselves acknowledged as much, both privately and in their "public" presentations and reports to the IWC as a whole. For this reason, those scientists concerned about controlling exploitation (the majority) worked carefully to strengthen their case, but they also tried to show that they were sympathetic to the needs of the industry and that they were willing to work with its representatives to meet "practical" aims. At issue was the need to build trust and establish precedents for IWC action on the basis of modest scientific recommendations.

The growing realization on the part of key scientists that this "collaborative" model for decision making was not going to work marks the end of this initial honeymoon period. As I will show in the next chapter, which deals with the critical decade 1955–1965, a number of the scientific advisers and delegates gradually came to understand that the bar for regulatory decision making "based on scientific findings" was going to be much higher than they had hoped (and expected). Instead of contributing to a collaborative process, they were rather being asked to provide scientific evidence that "proved" the need for protective actions. This realization set the stage for a series of closely watched and highly choreographed scientific showdowns in the early 1960s. In those years an increasingly savvy and desperate alliance of scientists and diplomats, all intent on curtailing exploitation (if for different reasons), worked together to create the conditions whereby their version of whale science could soundly trounce all comers. In this they succeeded, forcing their opponents to "face the facts" about plunging stocks (the effect, however, was not quite what they had hoped).

In reviewing these phases of the evolving relationship between the IWC and whale science, I will aim throughout to show not only how scientific questions got defined and answered, but also how carefully and strategically scientists and nonscientists alike attended to the boundary between scientific and "practical" (often meaning "pragmatic," sometimes meaning "nonscientific," always meaning "about money") matters. Clever moves were made on both sides, though in the end, as I will demonstrate, a number of those moves had unintended consequences adverse to those who initiated them.

. . . .

There is perhaps no better place to witness the emergence of these issues than the very first substantive session of the 1946 meeting to negotiate the ICRW itself. As they reviewed the US draft convention article by article, the delegates were given an opportunity to comment on its provisions and voice their concerns. No sooner had things gotten going than they alighted on proposed Article III, which read in part:

> The commission shall plan and recommend studies and investiga-
> tions relating to whales and whaling to be undertaken by appropriate
> agencies of the contracting Governments, or by other public or private
> agencies, establishments, or organizations; it shall, in collaboration
> with appropriate agencies, or independently, collect and analyze
> statistical information to reveal the current condition and trend of
> the whale stocks and the effects of whaling activities thereon.[74]

Several hands went up. Kellogg, from the chair, solicited commentary. A number of the delegates made clear that this issue—of scientific research and the control thereof—was going to be of central importance to the ne-gotiations. The United Kingdom, for instance, was concerned to protect the existing institutions through which scientific research and catch data were filtered to national governments—the "Whaling Committee" of ICES, and the BIWS above all. For the Dutch, consideration of Article III threw open the whole concept of a commission: why not keep doing things as they had always been done, with an annual conference to debate the regulations? What was the purpose of this proposed new thing with these potentially expanded powers?

The US delegates, as ringmasters and hosts, were quick to step in with soothing clarifications. They assured the UK delegates that the new com-mission would be equipped to "keep in touch with research developments throughout the world and make observations concerning the areas that may not be given sufficient attention," but that it was emphatically not

74. Washington IICRW confidential draft, p. 5, Smithsonian Archives, RU 7165, box 8, folder 3, "Washington—Informal Inter-agency Committee on the Regulation of Whaling (IICRW), circa July 1946–Oct. 1946—Minutes."

intended to replace existing research institutions or duplicate their functions.[75] Referencing US-Canadian fisheries treaties (now for the benefit of the skeptical Dutch delegation), the US representatives went on to point out the proven advantages of a commission empowered to draft changes in regulations "as are deemed necessary from a biological basis" without the need for a full, formal, "Diplomatic conference."[76] But the proposed whaling commission would be different from the International Fisheries Commission, the American spokesmen continued (with an eye back on the British representatives), since the International Fisheries Commission did in fact possess its own independent and dedicated investigative staff, whereas what was proposed here was merely a kind of forum for the consideration of scientific findings. As Ira Gabrielson put it on behalf of the US drafting committee, "In drafting this proposal we tried to safeguard it [the commission] against the setting up of another organization for which we would have to request appropriations for research work. We tried to draft this proposal in such a way as to leave the carrying on of research to the existing organizations, or to any new research group that any government cared to sponsor." It might be difficult now to imagine how it was all going to come together, he admitted, but concluded with confidence, "I can assure you it works very well in practice."[77]

In fact, however, the debate around Article III reflected real and deep concerns on the part of many of the delegates over precisely how much control scientists were to have under the envisioned IWC arrangement. And those concerns took concrete form in the close attention paid to the precise administrative mechanisms by which biological and statistical information would be generated and conveyed into the deliberative space of the IWC. For adroit diplomats and seasoned agency functionaries, everything hinged on the niceties of reporting structures and the architecture of committees. Since Britain had its Discovery Investigations and Norway had its Whaling Institute (and was the home of the pseudo-international BIWS)—and the scientists affiliated with these bodies were integrated into advisory roles already defined by budgetary amendments and decades of

75. Transcripts, minutes of second session, p. 25, Smithsonian Archives, RU 7165, box 9, folder 5, "Washington—International Whaling Conference, 1946—Conference Documents #1–45."

76. Ibid. These quotes are drawn from comments by Gabrielson (p. 28) and Flory (p. 27).

77. Ibid., p. 28.

habit (as well as familiar intergovernmental bodies like ICES)—there was considerable anxiety about what might lie in store should Kellogg and the Americans restructure the channels of advice and redirect the flow of research support.[78] The threat to entrenched interests was clear enough, but still more troubling was the potential for the new IWC to take shape as some kind of scientific supergovernment, empowered to dictate terms to the industry. That such a body might become a tool of the Americans—who bestrode the postwar world in an unprecedented way, who were known to be preoccupied with conservation, and who, worst of all, had no real whaling industry themselves and thus, in the view of many in the Anglo-Norwegian bloc, little feel for the way these things worked—made for some nervous visiting dignitaries in Washington.

And it was to address those fears that the US proposal, in the end, backed away from the proven scientific advising structures of the International Fisheries Commission and did not press to equip the new IWC with its own research body.[79] In the final convention, a modified version of the

78. As far as the BIWS was concerned, there was some discussion in the late 1940s of formally internationalizing the organization (it remained a *Norwegian* entity that handled *international* statistics, hence the tussle over whether "Internasjonal Hvalfangst-Statistikk" was properly Englished as the "Bureau of International Whaling Statistics" or "International Bureau of Whaling Statistics"—the former eventually winning out as more technically correct), but there was ultimately consensus that the organization worked effectively and produced accurate tabulations. Its archives, including a fascinating correspondence collection of more than 30 boxes, are held in Hvalfangstmuseet (Whaling Museum), Sandefjord, Norway. For a sense of the debate over the name and character of the organization, see the exchange between Bogdanov and Dunn in 1949 in Smithsonian Archives, RU 7165, box 12, folder 4, "London—First International Whaling Commission Meeting, 1949—Proceedings of Scientific and Technical Committee," in which Dunn deflected the Soviet request for representation at BIWS this way: "No one has questioned the accuracy of the statistics; some have questioned the interpretations given to the meaning of some of the statistics." In the late 1940s the BIWS was still a relatively small organization: Gunnar Jahn was chairman, and Birger Bergersen and Harold B. Paulsen (Director of the Association of Whaling Companies, Sandefjord) constituted the board.

79. Kellogg's notes on the IICRW papers reveal that this issue was debated. See the annotated draft convention, p. 6, Smithsonian Archives, RU 7165, box 8, folder 6, "Washington—International Whaling Conference, 1946—Kellogg's information statistics and notes," where Kellogg has written in the margin beside Article III: "Major question of substance: (1) shall commission have authority to plan scientific work (2) shall commission have authority to undertake collection of scientific data . . ." The result of internal discussions on these matters is reflected in the "secret" memorandum prepared in Oc-

proposed Article III did survive (as Article IV), fettered with a number of small adjustments that aimed to weaken its mandate.[80] But even this did not satisfy some participants in the negotiations. Spelling out their fundamental concern in a telling marginal annotation on his copy of the draft of Article III, the supreme king of the whaling industry, Captain Harold Salvesen (a member of the British delegation), wrote pointedly, "I am afraid the Commission will get into the hands of scientists and *not* of practical people."[81] Over the next several years (and particularly from 1949 forward, as the IWC actually took shape and began meeting), this issue—of whose hands would hold what levers in the decision-making process—was worked out in concrete form in a sequence of tweaks and revisions to the committee structure of the organization.

Because it is one of my contentions in chapters 4 and 5 that the highly formalized social architecture and procedural mechanisms of parliamentary life afford a uniquely patent dramatization of the sociology of scientific expertise (i.e., there is, in effect, no need for an elaborate sociological *reconstruction* of the paths and plays for power in such a setting because subcommittee structures amount to the *reification* of such dynamics), I will take a moment here to trace the evolution of the IWC scientific advising system by means of a series of what appear to be, on the surface, little more than administrative shuffles. In doing so I will demonstrate how Salvesen (and his ilk) essentially succeeded in keeping the IWC out of the hands of the scientists in its early years, though it is striking to watch how the scientists—concerned above all to create the conditions for collaborative and nonconfrontational exchanges—colluded in their own marginalization.

The initial exercise in erecting a formal distinction between an arena for scientific questions and an arena for political ones emerged, predictably, at the very first real trouble spot in the 1946 negotiations. Shortly after testing the waters on the 16,000 BWU limit, the delegates turned to

---

tober 1946 for the Commodity Problems Committee (Smithsonian Archives, RU 7165, box 8, folder 3, "Washington—Informal Inter-agency Committee on the Regulation of Whaling [IICRW], circa July 1946–Oct. 1946—Minutes"), which abstracted the drafting intentions of the IICRW. On Article III, it emphasizes that there is no intention to "duplicate" any research functions already handled by existing bodies, and that the IWC will *not* have responsibility for "undertaking field or laboratory work."

80. As, for example, the alteration of "shall" to "may" in the initial clause.

81. Cited by Patricia W. Birnie, *International Regulation of Whaling: From Conservation of Whaling to Conservation of Whales and Regulation of Whale-Watching* (New York: Oceana Publications, 1985), vol. 1, p. 178 n. 83.

the fraught topic of minimum catch sizes. Should size limits—such as the 40-foot minimum on sperm whales—be globally uniform or differentiated by region and type of hunting? Norway asked that the delegates consider reducing the limit to 35 feet for land stations, which had smaller ambits of operation, slimmer pickings, and less impact on populations overall. Sensing the potential for a widening conflict on this matter, Kellogg, from the chair, proposed shunting the whole issue into committee:

> Since this is a matter of substance, I propose that we have one or more committees for substance, as distinct from the Drafting Committee. There will be a number of items of this sort where biological or other material will have to be analyzed. If the committees on substance can report on matters of substance, it will simplify the work of the Drafting Committee. The Drafting Committee can then confine its efforts to the niceties of language which express our desires in this matter, and leave to the committees that handle substance the problem of bringing in the best opinion available in such matters.[82]

It was an elegant formulation: there would be, in essence, a "form" committee and one or more "content" committees. The latter would effectively report to the former.

No sooner was this motion on the floor than ears pricked up—everyone wanted to know just what was going to count as "substance." And who would be appointed to this committee? Would it do only "biological matters" or perhaps a somewhat more expansive array of "substantive" issues? The initial scope of the ad hoc "substance committee" was defined (to include particular regulations on sperm and sei whales) and a broad slate of delegates was nominated. The thorny issue of sperm size limits thus sidelined, negotiations in the plenary session could continue.

But very soon other contentious matters threatened to derail progress in the main hall. Would factory operators be required to process *every* part of the whale? What about the parts with very little oil? How much oil was "very little"? What about whaling in tropical regions, which seemed to take animals from calving grounds? At each of these potential impasses, Kellogg reached for the same solution: declare the problem a "matter of substance"

82. Minutes of fourth session, p. 11, Smithsonian Archives, RU 7165, box 9, folder 5, "Washington—International Whaling Conference, 1946—Conference Documents #1–45."

and push it out into the "substance committee."[83] Eventually several such committees were formed, including a "Committee on Biological Data" and another on cetacean nomenclature. Given the critical responsibilities of the Committee on Biological Data (which included reviewing recommendations for total quota and catch sizes), Harold Salvesen and other industry representatives made sure they had a place at the table. Final decisions in these committees were reached in closed-door sessions for which transcripts do not exist.

As successful as Kellogg's strategy was for protecting the collaborative and congenial atmosphere in the conference room (by avoiding awkward open debates), it set a dangerous precedent for the way in which the category of "biological question" would function in the arena of IWC negotiations. It is no exaggeration to say that here, in these earliest moments of the 1946 meeting, the boundary between scientific and political matters was manipulated in such a way as to *protect the domain of politics from itself*. Any fractious question—one that had the potential to disrupt the flow toward diplomatic consensus—got defined as a scientific question and thrown out of the room. The result was that disruptive conversations were conveniently pushed away from the center of the deliberative process, and negotiations were streamlined. In other words, those commentators who have harped sadly on the way scientific questions came to be "politicized" in the early IWC have it exactly wrong: it is not that some set of questions that were already understood to be "scientific" were gradually sullied with the mud of competing interests, but rather that the most difficult political questions in that muddy arena got called scientific in order to tidy and defend a fundamentally diplomatic initiative. In the process the committees dedicated to working out reports on these substantive matters became key battlegrounds where the ugly business of hashing out deals that everyone could live with was done. Limiting membership to "scientists" was impossible under these initial conditions—and no one even made an attempt to do so. The first reports of these committees came back to the plenary session wearing the scars of bruising dissent.[84]

The legacy of these original strategic decisions about how science and politics were to be parsed was durable, and can be seen clearly in the "Rules of Procedure" for the 1949 IWC meeting, which stipulated that a single "Sci-

---

83. Ibid., para. 154–59, 168–72, 262, etc.

84. See, for instance, "Report of Committee on Biological Data," 25 November 1946, IWC / 36, particularly the last paragraph, featuring Soviet reservations.

entific and Technical Committee" should be formed to "keep under review the statistical, biological and other technical information . . . and make recommendations thereon for the consideration of the Commission."[85] Because "technical" was construed widely, and meant something like Salvesen's "practical," this committee rapidly took shape as a mixture of industry executives (along with national fisheries representatives closely tied to the industry) and "scientific types" like Mackintosh and Kellogg. It would be, in essence, a collaborative "substance" committee. However, since this committee was once again used as a dumping ground for all the most difficult questions facing the IWC at its first full gathering, it may not come as a surprise to discover that the need was soon felt within the Scientific and Technical Committee for another committee—a "special scientific committee"—that might go off and apply itself to several of the more vexatious problems confronting the committee as a whole. It was 1946 all over again, but one step further from the center. Tellingly, this time it was Captain Salvesen himself who made the suggestion that a new scientific subcommittee might be in order.

The issue that prompted this new fission was the configuration and extent of sanctuaries where whaling would not be allowed. Salvesen and several other representatives of the whaling industry were keen on limiting protected areas to tropical waters (where, in fact, a tiny fraction of world whaling was being done, so protection would cost the major players in the industry nothing), rather than closing sectors of the Antarctic. Kellogg, by contrast, was anxious to see a flexible and responsive program of sector closings in the heavily whaled waters of the deep south, preferably focusing on regions known to have good concentrations of krill and plentiful stocks of whales. Mackintosh, wavering, acknowledged that such an arrangement "would be desirable," but then noted, in deference to the composite nature of the Scientific and Technical Committee, that "we should balance what is most desirable against whaling interests."[86]

85. IWC 18 (Final Copy), rule XVIII, p. 3, Smithsonian Archives, RU 7165, box 12, folder 1, "London—First International Whaling Commission Meeting, 1949—Conference documents; Kellogg's preparatory notes."

86. Scientific and Technical Committee transcripts, morning session, 2 June 1949, p. 33, Smithsonian Archives, RU 7165, box 12, folder 4, "London—First International Whaling Commission Meeting, 1949—Proceedings of Scientific and Technical Committee."

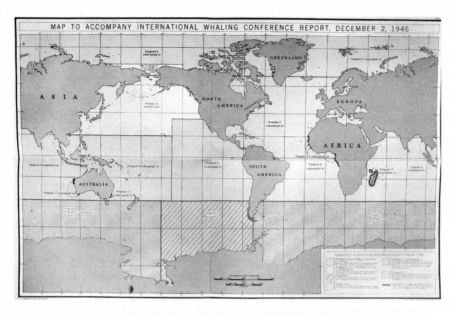

FIGURE 4.1 (and see PLATE 14) The sanctuary: Map to accompany IWC report, 2 December 1946. (Courtesy of the Smithsonian Institution.)

Kellogg pushed, however, announcing, "I question the desirability of setting up a sanctuary in one area and forgetting it." It was a position buttressed by his awareness that the only existing sanctuary—a sector of the deep southern Pacific originally set aside in 1938 and formalized in 1946 (figure 4.1)—had been settled on because no significant whaling had ever been done in the region, which was believed by the industry to be too poor to support paying expeditions.[87] But Salvesen, who from the start of the discussion in the Scientific and Technical Committee had been focused on deferring any serious reconsideration of sanctuary arrangements (he had immediately suggested referring the whole issue to some sort of standing committee), finally closed the door on the topic, announcing firmly: "I do not feel this committee is a suitable body to

---

87. "The South Pacific was set aside not on the basis entirely of what it would do for whales but rather because it was the only area in which no whaling was going on and which could be agreed upon at the time." Ira N. Gabrielson to Mackintosh, 12 April 1950, Smithsonian Archives, RU 7165, box 23, folder 1, "Second International Whaling Commission Meeting (London), 1949."

consider the question."[88] He wanted to see a "suitable scientific committee" composed to ruminate on the matter for the next year and draft a report for the following year's meeting. He got his way, since it was difficult for the scientists to argue against further research and reflection by scientists.

It was the first instance of a tactical maneuver that would characterize IWC deliberations for the next decade: the use of scientific committees not only to push difficult problems to the margins of decision making, but also (more or less nakedly—and it became more naked in the years to come) to delay movement toward decisions seen to be undesirable by certain participants. By the time the 1949 IWC meeting closed, a new, vaguely denominated "Committee of Science" (aka the "Subcommittee of Scientists") had been delivered a fifteen-item agenda consisting of every single controversial issue faced by the new organization, from the 16,000 BWU quota to the issue of sanctuaries, from the issue of penalties to the issue of limiting factory vessels, catchers, and land stations.[89] On the one hand it seemed as if the scientists were being given the reins. On the other hand, as Kellogg and Birger Bergersen began to suspect, they might well have been set up—tightly corralled with and holding the reins of fifteen wild horses. Bergersen wrote Kellogg right after the meeting, "What is worrying me is how we shall get all of this work started."[90] A few months later he was even more desperate:

I am beginning to be seriously troubled by all the work allocated to the so-called Scientific Committee, and I suppose you remember that you are to be a central and highly esteemed member of this commit-

88. Scientific and Technical Committee transcripts, morning session, 2 June 1949, p. 33, Smithsonian Archives, RU 7165, box 12, folder 4, "London—First International Whaling Commission Meeting, 1949—Proceedings of Scientific and Technical Committee."

89. Mackintosh calls it the "committee of science" on p. 40 of the transcripts of the Scientific and Technical Committee for the afternoon session on 3 June 1949. The full agenda can be found on two loose sheets of blue paper (with US annotations) in Smithsonian Archives, RU 7165, box 12, folder 4, at the end of the file "London—First International Whaling Commission Meeting, 1949—Proceedings of Scientific and Technical Committee."

90. Bergersen to Kellogg, 29 August 1949, Smithsonian Archives, RU 7165, box 11, folder 6, "Birger Bergersen Correspondence, 1947–1953 (Ambassador from Norway to Sweden)."

tee. Can you give me some tips on how we shall actually come to grips with our heavy tasks?[91]

Kellogg replied that he had been pestering Mackintosh, the committee chairman, "regarding the work of the scientific committee, or rather lack of activity," but there had been a stony epistolary silence.[92]

I will return to the reasons for this silence at the end of this chapter, since a closer look at Mackintosh's own situation and priorities in these years is crucial to a broader understanding of how science functioned in the early IWC. But before I turn to that issue, let me play out the story of committee structures through the early 1950s, by which time the pattern of scientific advisory procedures had settled down into a regular cycle.

By the IWC's third meeting, at Cape Town in 1951, the ad hoc 1949 move to split off some sort of dedicated scientific committee from the more formal Scientific and Technical Committee was itself formalized in a revision of the rules of procedure. Henceforth there would be two separate entities: a "Scientific Committee" and a "Technical Committee," configured (at least in principle) symmetrically with respect to the Commission as a whole.[93] Their briefs were, however, significantly different. The Scientific Committee, it was decided,

shall review the current scientific and statistical information with respect to whales and whaling, shall review current scientific research programs of Governments, other international organizations or of private organizations, shall consider such additional matters as may be referred to it by the Commission, and shall submit reports and recommendations to the Commission, or by the Chairman of the Commission [*sic*].

Whereas the Technical Committee

shall review and consider the laws and regulations of various governments, the annual reports on infractions submitted by Governments,

91. Bergersen to Kellogg, 21 December 1949, in ibid.
92. Kellogg to Bergersen, 18 January 1950, in ibid.
93. Confusingly, the two committees did continue to meet together as well for several years, but this tradition died in the early 1950s (for a joint meeting in 1952, see Smithsonian Archives, RU 7165, box 14, folder 1, "London—Fourth International Whaling Commission Meeting, 1952—Instructions, proceedings, correspondence, notes, etc."). There was also some overlapping membership.

questions involving the time, manner and intensity of whaling operation, and such additional matters as may be referred to it by the Commission, or by the Chairman of the Commission, and shall submit reports and recommendations to the Commission.[94]

Because the Technical Committee had explicit authorization to handle "questions involving the time, manner and intensity of whaling operation," it soon became the dominant forum out of which recommendations for quotas, seasons, and lengths came to the full commission. In practice, by the early 1950s, the Scientific Committee's reports were "consulted" by the Technical Committee as it prepared its annual recommendations to the commission as a whole. This chain of committee relations ensured that, as Salvesen had originally hoped, the IWC remained firmly in the hands of the "practical men," who had, in effect, veto power over the suggestions of the scientific community.[95]

This formal alteration in the structure of the organization was handled early in the 1951 meeting, and it did not attract much debate. The United Kingdom and the United States supported the amendment, which passed unanimously (after a few tweaks to clarify that membership on both of these committees—which would "deal with the major matters of substance"— was open to a delegate from each participating nation).[96] With Kellogg backing the amendment, and with Birger Bergersen in the chairman's seat at the plenary session, it was clear that the scientists had no objection to the formalization of a committee of their own. A number of them clearly felt like the new director of the (US-supported) Japanese "Whales Research

94. Emphasis added. The language here is taken from Kellogg's 1951 report to the secretary of state, p. 10, Smithsonian Archives, RU 7165, box 28, folder 4, "Reports to the Secretary of State by the U.S. Commissioner to the IWC, Meetings #1–11, 1949–1959."

95. It is important to note that, formally, both committees reported directly to the full commission; however, the pattern of the actual meetings after 1952 always saw the Scientific Committee pull together a report on the previous year's catch data, a report that preceded the Technical Committee's report on proposed amendments to the schedule. Hence, in practice, the Technical Committee functioned both as the real forum for decisions about regulations and as a kind of filter for the work of the Scientific Committee.

96. This membership distinguished these committees from the much smaller finance and administration committee, which consisted of three delegates appointed by the chairman.

Institute," who wrote to Mackintosh that the best hope for good research on whales was "if a subcommittee which can keep *the purely scientific attitude is organized in the present committee*." This way the scientists might create a space outside the hurly-burly of the general IWC meetings, where, as the Japanese researcher noted ruefully, "the political trend is superior."[97]

In this sense the emerging architecture of scientific committees in the IWC—with science increasingly removed from the central forum—must be understood as a product of both push and pull. In these early years, when the mechanics and authority of the organization were still very much up for grabs, the category of "scientific problems" could be used, as we have seen, to push hard decisions out of the thick of delicate international negotiations. At the same time, the desire on the part of the scientists to defend a space "apart" from politics—and their sense that they could, there, finally, get to their business in peace and produce findings that would win respect back in the negotiating arena of the full commission—facilitated the codification of separate (and ever more peripheral) spaces for the consideration of science. Nor was the process complete: yet another iteration of committee budding was required before the advisory system settled into a stable pattern.

The impetus for this final centrifugal move—the creation of a "Scientific Sub-Committee" that met annually between IWC meetings and reported to the Scientific Committee (which met during the annual IWC gathering)—lay in Birger Bergersen's and Remington Kellogg's growing concern in the early 1950s about the total Antarctic quota. In Cape Town in 1951 Bergersen took advantage of the chair to make a brief opening speech in which he praised the promising new spirit of "fruitful international cooperation" and extolled the "common understanding that it is necessary to render effective protection to the biological foundation of a large and important industry."

---

97. T. Maruyama to "Secretary of NIO," 24 May 1952, Smithsonian Archives, RU 7165, box 14, folder 1, "London—Fourth International Whaling Commission Meeting, 1952—Instruction, proceedings, correspondence, notes, etc." Emphasis added. It is perhaps worth noting here that the United States played an important role in rebuilding the Japanese whaling industry after World War II, in large part in the hopes that cetacean protein and fat could be used to address serious food shortages during reconstruction. For an introduction to these issues, consider the correspondence in the Smithsonian Archives, RU 7165, box 11, folder 3, "Reports from Central Liaison Office (Tokyo) to General Headquarters of Supreme Commander for the Allied Powers—Whaling in the Antarctic, 1946–1947."

But he then spelled out the increasingly urgent need to put the cooperation into actual practice as far as the quota was concerned:

> I feel rather certain that every biologist who has examined catch statistics, as they stand today, sees very well that the total catch of 16,000 blue whale units is too high. These 16,000 blue whale units must be looked upon as a kind of interest derived from capital—and I admit that in 1944 we were a little too optimistic about the size of that capital.[98]

Here, it should be noted, Bergersen was being both falsely modest and a little cagey. It is one of the telling ironies of the founding of the IWC that several of the key scientific players in the early negotiations—signally Kellogg and Bergersen—were already quite confident in the mid-1940s that a quota of 16,000 BWU was biologically unsound, but they pressed for it anyway. Why? *In order preemptively to defend themselves against the charge of using biology to advance "political" goals.* As Bergersen put it in a striking note to Kellogg in 1946,

> My personal view—after having studied for weeks the statistics from the last season—is that 16,000 blue whale units is on the top side; but I am afraid lest, if we suggest a reduction in this quantity, we shall be accused of using biological considerations as a pretext for keeping other nations outside of whaling.[99]

A letter like this is a salutary reminder that the scientists were exquisitely sensitive to the political work that would be necessary to erect and protect what they hoped would eventually take shape as an apolitical domain—an apolitical domain to which disputants could recur for equitable solutions to the perennially contentious problems of stock management and assessment. Clearly Bergersen and Kellogg had no illusions about the kind of diplomatic exercises that were required to lay the groundwork for such a space within a complex and polarized political arena. First and foremost, it was

98. Verbatim transcript 3, p. 4, Smithsonian Archives, RU 7165, box 13, folder 5, "Cape Town—Third International Whaling Commission Meeting, 1951—Numbered Conference Documents."

99. Bergersen to Kellogg, 28 October 1946, Smithsonian Archives, RU 7434, box 1, folder 10, "Assorted News Clippings."

necessary to convince everyone that they were impartial—that they were not biased toward preservation of the animals or opposed to the industry. Winning this confidence might even require initially putting forward scientifically unwise recommendations (like the 16,000 BWU initial quota), since earning for science the proper reputation in the whaling community would, they expected, pay much higher dividends later. With their prewar experience of regulatory problems, Kellogg and Bergersen harbored no naïve faith that biological or statistical arguments could do their Archimedean labors unaided. Such seasoned whale scientists knew that they were not going to move the earth (or lower the quota) until they had a number of friends prepared to help them get where they needed to go, set the fulcrum in the regulatory arena, and lend a hand with the lever. No fragile claim about the "truth," they recognized, dropped into the scrum of a multinational scramble for millions of dollars, was likely to shut down the boilers on dozens of factory vessels. At the same time, these men, and others in their cohort, nevertheless retained considerable faith that, if the right sorts of precedents, institutions, and relationships could be built, scientifically informed regulation was not only possible, but actually offered the only chance of arresting a relatively rapid extermination of whale stocks. Relationships, institutions, and precedents were critical to making the system work.

Further evidence that they therefore elected to invest heavily in what we might call the "practical sociology" of their science can be found in the records of their early conversations about *where* conservation-oriented whale science should be done. For instance, in 1948, just before the first full IWC meeting, Mackintosh, Bergersen, Kellogg, and Johan Ruud all discussed the desirability of mobilizing a concerted effort to "write objective articles in scientific periodicals, indicating the present situation of whaling." Indeed, perhaps a broader campaign of "educational activity" in learned periodicals could help "prepare world opinion for the necessity for a reduction of the total limit."[100] Interestingly, however, this plan for an extensive scientific and popular campaign never developed. Instead, this group of potential scientific rabble-rousers all opted, over the next several years, to concentrate their work *within* the emerging structures of the IWC: they chaired the committees and subcommittees, prepared in-

---

100. Bergersen to Kellogg, 4 December 1948, Smithsonian Archives, RU 7165, box 11, folder 6, "Birger Bergersen Correspondence, 1947–1953 (Ambassador from Norway to Sweden)."

ternal reports, and lobbied at the cocktail parties and industry-sponsored dinners.[101] Rather than stirring up "world opinion" (as they had briefly projected), they carefully avoided making noise in the press about the problems, and they shied away from extensive publication in general scientific journals.

On the one hand, it is tempting to see in these developments the "capture" of scientists by industry within a regulatory setting. But tacking this label on what happened actually obscures more than it clarifies, since in these early years, as the documentary evidence reveals, it is more correct to understand these moves as part of a larger strategy for accomplishing their goals—one that proceeded from a general confidence that the cultivation of the new institution of the IWC, and of relationships within it, held greater promise of long-term success than external leverage. Over the next fifteen years, this optimism sustained a series of rude shocks. By the end of that period, some of the participants had indeed been "captured" by the industry in a relatively direct way (they were paid, and they said what the payers wanted to hear), but others had been "captured," I will argue, by the promise of the regulatory institution itself, and still others by the promise of a particular kind of regulatory science deeply enmeshed in that institution and the network of relationships they had built there. Teasing apart these different experiences is illuminating, and I will return to these difficult issues in the latter part of chapter 5.

Historically speaking, however, it is essential that the decisions of the late 1940s and early 1950s not be evaluated in light of those later developments. In these early years it was not only possible to be sanguine about the prospects of scientific regulation, it was easy to be enthusiastic about what felt like the dawn of a new era of possibilities. Flexibility and deference were called for in the hopes of nurturing a new kind of collaboration between science, international diplomacy, and wildlife management. As far as the IWC's early inertia on the total quota, why not try to make the best of it, from the scientific perspective? As Bergersen put it to Kellogg shortly before the first IWC meeting, "We ought to risk maintaining the high total

---

101. These dinners became a regular feature of the annual IWC meetings. See Kellogg's notes to himself for his remarks at a 1951 dinner sponsored by the Hector Whaling Company and Gearing and Jameson in the Smithsonian Archives, RU 7165, box 15, folder 1, "London—Fifth International Whaling Commission Meeting, 1953—Instructions, preparatory material, notes, correspondence, clippings."

limit for a few years, in order to obtain the most reliable statistical data possible to work on."[102]

This posture could be construed as making scientific lemonade out of industry lemons. No one could deny that the data sets would improve over time, and the statistical argument for reducing catches would surely be strengthened by a longer sequence of declining returns. Moreover, the comparability of the annual returns would be greatly enhanced by having a run of hunts conducted under the same catch parameters. Whereas, if the total quota were to be changed (or, indeed, if minimum sizes were to be altered or new sanctuaries opened), it would be necessary to apply various arithmetical fixes to the data in order to normalize the numbers. Such weighting games were always going to be subject to dispute. Thus the net effect of a small quota reduction might be the spoiling of the very data that promised to deliver (eventually) the most compelling evidence in favor of protective measures.

Despite these rationales (or rationalizations), there were some serious downsides to thinking of the problem this way. After all, this kind of argument risked offering an all-purpose "scientific" reason for never making any amendments to the schedule at all, thereby perpetually deferring any conservation-orientated steps.[103] Kellogg himself saw the potential danger early: in 1949, during discussions about possible adjustments to the quota arrangement, Mackintosh noted the value of not making too many changes, since "we could use the comparison with 16,000 units from year to year." Kellogg quickly parried, setting up the problem as one of "protecting the whales" versus "protecting the data":

There is no question but that the blue whale stock is declining and has been overtaxed. It seems as biologists we should give some consideration to protecting any species that seems to be in the same danger. While I'm not a statistician, *the issue is whether or not we are going to*

102. Bergersen to Kellogg, 4 December 1948, Smithsonian Archives, RU 7165, box 11, folder 6, "Birger Bergersen Correspondence, 1947–1953 (Ambassador from Norway to Sweden)."

103. For instance, this argument was used to put off an increase in the minimum size for fin whales in 1953, and by 1955 E. J. Slijper was using a version of it to resist protective measures for blue whales (see the Scientific Committee report for that year, IWC / 7 / 20). It was also repeatedly deployed to resist a move from the BWU to species quotas.

try to give additional protection to the blue whale or maintain a catch to provide statistics.[104]

It was a very short step from the slippery argument that a catch ought to be maintained in order to preserve the data sets to the still more troubling argument that a new catch ought to be initiated in order to begin to gather data on an unexploited region or population about which nothing was known.[105] In fact, in the thick of these same 1949 debates in the Scientific and Technical Committee, the emerging argument about "preserving the data" was seized on by the French delegate and extended in exactly this way. He pointed out that no sooner had the French delegates reflected on the new scientific spirit of the convention and the expressed need for a proper scientific basis for the regulatory regime than they "thought of establishing land stations in tropical Africa *with the aim of getting as soon as possible*

104. Morning session, 1 June 1949, pp. 18–19, Smithsonian Archives, RU 7165, box 12, folder 4, "London—First International Whaling Commission Meeting, 1949—Proceedings of Scientific and Technical Committee." Emphasis added. It is important to note that Kellogg is using this logic to support a rather perverse "conservation" measure: relaxing a ban on pelagic hunting of humpbacks in the Antarctic. Norway had put forward this proposal, defending it in part on the grounds that more access to humpbacks would mean less pressure on blue whales (humpbacks were easier to catch, and thus it was assumed they would be taken more readily than the increasingly scarce blues). It was at best a stopgap measure as far as the blues were concerned, and it soon proved very bad for the humpbacks. The US position in this case must be understood as yet another instance of the kind of political / scientific compromise that distinguished this early phase in the history of the IWC. Kellogg excused the measure (which was voted in) in his report to the secretary of state using tellingly prospective language: "Although some of the results of the meeting, particularly with respect to amendments to the schedule, do not provide more effective control of whaling operations, it is believed that the work of the meeting *paves the way* for continued and more effective conservation of whale stocks." Meeting 1, p. 35, Smithsonian Archives, RU 7165, box 28, folder 4, "Reports to the Secretary of State by the U.S. Commissioner to the IWC, Meetings #1–11, 1949–1959." Emphasis added.

105. This argument became the dominant public defense for the decision to open the Antarctic sanctuary to hunting (on a "limited" three-year basis) in 1955. See Paul Budker, *Whales and Whaling* (New York: Macmillan, 1959), 170, explaining that this was done "in order that the state of the stocks which had been left undisturbed for seventeen years might be ascertained." There was, in fact, as I discuss below, a power-politics rationale for the decision: a game of brinksmanship with the Latin American nations of the Santiago Declaration, which remained outside the IWC.

the biological data which are mentioned in said Article."[106] The bad news was that it had taken a little longer than they had hoped to get the new stations up and running. The good news was that they would start having data from healthy catches of tropical humpbacks to report the following year. With luck these data would help with the difficult question of whether the hunt for these animals was more detrimental in warm (breeding) or cold (feeding) waters. Later Kellogg confided to Bergersen that this expanded tropical industry was likely to be a disaster for the stocks,[107] but it was hard to argue that additional data would not strengthen this very case.

As far as the total Antarctic quota was concerned, by 1952 Bergersen was not only making speeches at the plenary session, he was also working behind the scenes, through correspondence with key diplomats, to push the issue of a reduction onto the IWC's formal agenda.[108] For instance, he wrote privately to Dobson in the UK in January 1951, "This very year there are signs that the catches of certain expeditions will be so poor that we can even start to speak about very great losses." He went on to conclude that "one thing we can be sure about, and that is the necessity very soon of reducing the present total catch of 16,000 blue whale units—if the stock is to be rescued."[109] And yet, in the 1951 and 1952 Scientific Committee meetings, Bergersen was unable to attract supporters to his view, which could easily be construed as a reflection of Norwegian anxiety about the rapid expansion of other nations into what had long been a Norwegian-dominated industry. Full minutes for these meetings were not recorded, so it is impossible to reconstruct in detail the discussions that yielded the decision in both these years not to recommend to the full commission any

106. He was referring to Article III, cited above. Morning session, 1 June 1949, p. 17, Smithsonian Archives, RU 7165, box 12, folder 4, "London—First International Whaling Commission Meeting, 1949—Proceedings of Scientific and Technical Committee." Emphasis added.

107. "The inroads made in humpback stocks by these operations may soon be reflected in the Antarctic catch. The possibility that we may discover that the humpback stock has suffered a marked diminution too late disturbs me." Kellogg to Bergersen, 16 January 1951, Smithsonian Archives, RU 7165, box 11, folder 6, "Birger Bergersen Correspondence, 1947–1953 (Ambassador from Norway to Sweden)."

108. Learning to work effectively in the setting of international organizations meant learning how to control the agenda: see, for instance, Kellogg's instructions to Bergersen about how to arrange to get the quota on the agenda for the next meeting, since "otherwise we cannot review the situation." Kellogg to Bergersen, 27 March 1951, in ibid.

109. Bergersen to A. T. A. Dobson, 11 January 1952, in ibid.

change to the overall quota. However, an "informal note" on the Scientific Committee meeting of 1952 (prepared exclusively for the use of committee members) offers a window into how these delicate negotiations proceeded. With fifteen individuals from nine countries around the table and Mackintosh in the chair, Bergersen went ahead and "suggested that the Committee might care to consider reducing the limit to 15,500 blue whale units," since it "seemed likely that populations in two areas were near danger point."[110] His evidence lay in the recent report on the 1951–1952 season prepared by the BIWS, which had configured the catch data by sector and species in a sequence of tables. Kellogg followed up on Bergersen's suggestion and wanted to know if similar tables were available for earlier seasons, since that "would probably give a clear indication of the need for reducing the present limit." At this point the Soviet representative on the committee moved that more data were exactly what was needed, arguing that "it would be premature to make a decision now" since "the question needed further study." Recommendations for additional collation of data by the BIWS were put forward and unanimously approved by the committee.

Back in the plenary session of the IWC, on the final day, the question arose as to who would undertake this additional research, when, and how. The other most contentious matter raised in the 1952 meeting—whether to continue allowing the pelagic hunt for humpbacks in the Antarctic—had also proved impossible to resolve (nations like Australia and France, which took humpbacks from land stations in warmer waters and had access to few other species, hated the idea that their main quarry was being picked off during their southern migration by factory vessels in the Antarctic). That discussion had similarly ended in a recommendation for further study. Soon the delegates were debating the composition of some sort of scientific subcommittee, or perhaps two, that would meet before the next IWC gathering in order to ready reports on these difficult matters. But how should such a subcommittee work? When should it meet? Who should pay? The Dutch delegate now hedged about the appropriateness of having such crucial issues farmed out to an ad hoc group that would meet privately out of cycle. As he put it, "We are quite willing to give full support to the study, but I would prefer to have inside talks at home with my people who have to handle this stuff before getting in such a committee." In his view, these

110. This document appears in Kellogg's papers in the Smithsonian Archives, RU 7165, box 14, folder 1, "London—Fourth International Whaling Commission Meeting, 1952—Instructions, proceedings, correspondence, notes, etc."

matters ought to go straight to the actual Scientific Committee. Bergersen (who wanted to chair the quota subcommittee in the hopes of preparing a compelling brief on the matter) countered that, since the quota was likely to come up on next year's agenda, it would be better to have a chance to reflect and deliberate before the IWC again found the matter in its lap. Debate followed, with some arguing for a large committee with a broad brief and others contesting the suitability of such an arrangement.

As it all fell out, the IWC authorized a new subcommittee (of the Scientific Committee) to meet outside the confines of the annual IWC gathering, to be convened by Bergersen, and to be funded by those countries wishing to participate (so as to avoid undue financial burden on the IWC itself).[111] Bergersen, with the support of Kellogg, had won. But the whole issue had by this point occasioned some irritation among various delegates—some of whom had wanted the proposed subcommittee to have a narrow purpose (so that they did not have to bother with it), while others hoped it might have a wide agenda (so that it could consider all the trade-offs between lowered quotas, new sanctuaries, and the protection of particular species); still others adopted various permutations on these extremes, according to their regional preoccupations and national interests. In the end, addressing an audience more or less exasperated by the proceedings, the delegate from New Zealand took a moment to spell out exactly where the new "sub" committee fit into the greater scheme of things. His analysis was a consolation to those who had begun to fret about the potential power of this new entity

111. In fact, it was a bit more complicated than that. The 1952 IWC meeting actually authorized *two* subcommittees of the Scientific Committee, one to deal with the question of whether the humpback catch should be regulated via geographically segmented limits (in keeping with the mounting evidence of geographically discrete populations), the other to deal with the issue of an expanded "sanctuary" area in the Antarctic (i.e., a possible no kill zone in Area II, the South Atlantic sector) and, loosely, the general issue of the BWU quota as a whole. Both subcommittees were to meet off-cycle (which is to say, between the 1952 and 1953 IWC meetings); Bergersen was appointed to chair each. Despite extensive and tedious fretting in the plenary session about the separate briefs and compositions of these two subcommittees, they actually met together in early 1953, and issued a joint report. In 1954 that body (an elision of the two subcommittees authorized in 1952) was renamed the "Special Scientific Sub-Committee" with terms of reference essentially the same as those of 1952. By 1955 the "Special" was dropped, and this midyear, off-cycle meeting (now called the "Scientific Sub-Committee" and reporting to the Scientific Committee) became a fixture of the IWC calendar.

and a quick reminder that, in view of the general architecture of IWC decision making, the whole thing was not really all that significant:

I understand that this will be a sub-Committee of the Scientific Committee and that, perhaps, does affect membership on it. There are a number of countries, presumably, . . . who want to ensure, by some form of membership, that their interests are safeguarded; but since the sub-Committee will be reporting to the Scientific Committee—and then of course there is still a further stage where the Scientific Committee would have to report to the full Commission—some countries who are directly interested might feel it is not completely necessary to be on the sub-Committee.

It was an astute and clear-sighted summary of what was, over the next two years, codified (if not very officially) into the basic structure for the scientific advisory system.[112] Each year a small group of scientists would meet (usually in London, almost always with Mackintosh in the chair, generally several months before the summer IWC gathering) to discuss scientific matters related to whaling. This Scientific Sub-Committee would generally draft a report, which would be submitted to the Scientific Committee when the full IWC convened. The Scientific Committee (again chaired by Mackintosh, and generally featuring all the same individuals who served on the Sub-Committee plus a number of additional representatives from nations that did not bother to send a participant to the Sub-Committee) invariably "accepted" this report (which is to say they added it to the stack of paper they forwarded to the full commission). Then, in the scramble occasioned by the arrival of the new statistics on the recent season's catch, the Scientific Committee drafted a new report of its own, making whatever recommendations about catch limitations were able to win the support of the Scientific Committee as a whole. This report then received the attention of the Technical Committee as it hammered out what would and would not be acceptable to the whaling nations in the season to come. Yet *another* report summarizing the outcome of *those* negotiations then went into the plenary session, where it was generally voted in without a great deal of debate, since

112. I say "not very officially" because by 1959 there would be serious questions raised about the status, under the rules of procedure, of what had come to be standardized (in practice) as the annual midyear meeting of the Scientific Sub-Committee. I briefly discuss the resolution of this conflict in the next chapter.

by this point—generally the last of four days of meetings—the positions of the major players were clear and there was little left to discuss.

I belabor the development of this structure and its precise workings because understanding these issues is essential to making sense of what can at times appear to be a dismaying reticence on the part of several of the leading scientists to push harder for the protective measures that they acknowledged, in private, were increasingly urgent. But a close review of the evolution of this scientific advisory system reveals that its origin in a joint "Scientific and Technical Committee" (where stock assessments and quota adjustments were to be made collaboratively) left an enduring mark on the procedures of the IWC and the habits of its participating scientists. As a sequence of more specifically and narrowly "scientific" committees took shape (at progressively greater remove from the central forum of the IWC itself), they became the locus of increasingly frank expressions of concern, but as the resulting reports made their way back up the ladder of the IWC's committee structure, it was understood that consideration of "practical matters" (read: "industry desiderata") could not be disregarded. For compelling evidence of this dynamic, one need look no further than the draft report of the Scientific Committee at the sixth IWC meeting in 1954, in Tokyo. With subcommittee documentation in hand that called for a cessation of blue whale and humpback hunting in the Northern Hemisphere, along with a reduction in the total quota for the Antarctic and various protective amendments to minimum lengths, the Scientific Committee (under Mackintosh) was unsure how to proceed, and it actually spelled out a desire for clearer signals from its neighbors on the Technical Committee:

> With certain qualifications . . . [the members of the Scientific Committee] agreed that from the biological point of view they could generally endorse the recommendations of the Subcommittee; but *recognizing that some of these proposals may involve practical difficulties or problems, they would welcome further discussion with the Technical Committee before definite recommendations are made to the Commission.*[113]

For one unacquainted with the contingencies and microevolution of the scientific advising system of the IWC in these early years, such a comment is utterly flummoxing and tempting to read as a dereliction of duty. But

---

113. "Draft Report of the Scientific Committee," IWC / 6 / 15, para. 4. Emphasis added.

read in the broader institutional context I have just reviewed, it makes perfect sense. Nevertheless, as far as the effort to cultivate a "fine spirit of cooperation"[114] is concerned, we might say that we here discover that spirit being left waiting in rather pathetic friendly anticipation.

While such deference, in the wisdom of hindsight, may seem perverse and self-defeating, I think it is best understood as evidence for the durability of a cooperative model for how scientific information and experts would contribute to the new regulatory system. It was a dream that died hard. By the mid-1950s, however, as I will show in the next chapter, it had become clear to a number of the participants (though perhaps not to Mackintosh) that a different model was called for if the overexploitation of the world's whales was to be stopped.

And yet even as late as 1954, there remained a defensible strategic case for caution. In the early 1950s, as the passage quoted above suggests, the IWC's scientific advisers took great care not to put forward recommendations for restrictive measures without first "testing the waters" in the Technical Committee and sounding the delegates to the plenary session.[115] This kind of thing had gone on for years, and it reflected a deep preoccupation with establishing some kind of precedent for getting the scientific recommendations accepted.[116] Better to be cautious and lay the ground-

114. The quote is from Bergersen to Kellogg, 7 September 1950, Smithsonian Archives, RU 7165, box 11, folder 6, "Birger Bergersen Correspondence, 1947–1953 (Ambassador from Norway to Sweden)," concerning the work of the Scientific Sub-Committee.

115. The Scientific Committee report for 1953 (IWC/5/17, para. 6) even permits an explicit reference to this sort of back-and-forth to creep into the document: "The above are the recommendations of the Scientific Committee which arise from the report of the special Subcommittee. The committee have since been informed that the Commissioners are prepared to agree unanimously that there should be no taking of blue whales before January 16th, that the total limit be reduced to 15,500 units but that the protection of blue whales in Area II should be postponed. The Committee agreed that these proposals, although they do not go quite so far as the Committee's recommendations, will at least go some way toward conservation."

116. See, for instance, Kellogg to Bergersen, 27 November 1950, Smithsonian Archives, RU 7165, box 11, folder 6, "Birger Bergersen Correspondence, 1947–1953 (Ambassador from Norway to Sweden)," in which Kellogg recommends that Bergersen check in with Salvesen before proposing a complete ban on blue whaling in the North Atlantic, since it was essential to know "if he intends to oppose such a proposal." Similar strategic considerations dominated the first meetings of the old "Subcommittee of Scientists" in 1950 (i.e., what became the Scientific Committee itself), in which floated proposals for additional protection of sperm whales foundered on concerns about the "great

work for being heard than to end up sidelined. It was a reasonable view. In fact, when, in 1955, the Scientific Committee went ahead and finally pressed formally for a small quota reduction, it was simply disregarded by the plenary session—a moment that definitively confirmed the failure of the collaborative model and initiated an increasingly urgent scramble on the part of Kellogg and several others to reconfigure the way scientific advising worked in the organization.

Perhaps the clearest mark of the strains occasioned by these early efforts to carve out a workable place for science in the IWC lies in the increasingly nervous emphasis participants placed on signaling those moments when they were speaking "as biologists." This emphatic phrase and a host of proxies ("on biological grounds," "from the biological point of view," "on purely biological grounds") came to litter the reports produced by the various scientific committees in the period and to pepper the speech of those scientific advisers who addressed the plenary sessions of the IWC. The 1951 report of the Scientific Committee offers a choice example. In considering the 16,000 BWU quota,

> the Committee considered the existing limit. From a biological point of view, the Committee believe that present limit is too high, *but in all the circumstances, they do not recommend any change.*[117]

This curious phrasing (if the limit is too high, why not recommend a change?) speaks volumes about the effort to stake out and maintain an "apolitical" identity that could serve as a resource ("a biological point of view") while gingerly working to bring that identity into the political arena ("in all the circumstances") in such a way as actually to make something happen. No one took up the offer.

---

opposition" a call for such measures would surely provoke. Smithsonian Archives, RU 7165, box 12, folder 6, "Oslo—Second International Whaling Commission Meeting, 1950—Committee reports, notes, correspondence, clippings, conference documents, etc." To get a sense of Kellogg's and Bergersen's efforts to feel out (behind the scenes) Salvesen's positions, see Kellogg to Bergersen, 1 May 1951, Smithsonian Archives, RU 7165, box 11, folder 6, "Birger Bergersen Correspondence, 1947–1953 (Ambassador from Norway to Sweden)."

117. Block quote from IWC/3/6. Emphasis in original. For the quote above, see Scientific Committee report for 1952, p. 2; Scientific Committee report for 1951, p. 1; and Scientific and Technical Committee minutes for 1949, p. 20, Smithsonian Archives, RU 7165, box 12, folder 4, "London—First International Whaling Commission Meeting, 1949—Proceedings of Scientific and Technical Committee."

. . . .

Up to this point I have for the most part written about the scientists involved in the early IWC as if they were a kind of bloc. And while in certain
respects this is fair (shared membership on scientific committees definitely
helped establish a shared identity), it is at best a convenient shorthand,
since "the scientists" were by no means a monolithic group. Moreover, even
as the institutional structures of the emerging IWC consolidated something
like a "community" of cetologists in these years, the challenges of building
a working regulatory regime opened divisions within this expanding community that would deepen by mid-decade. A closer look at some specific
research programs will shed light not only on these emerging conflicts but
also on the broader problems that increasingly confronted whale science
in the 1950s.

And it is here, of course, that we must return in earnest to Neil Alison
Mackintosh, who chaired every Scientific Committee meeting held by the
IWC from 1951 (when it was invented) to 1963, as well as all but two or three
Scientific Sub-Committee meetings.[118] It was he who drafted the cautious
(and at times seemingly tortured) language of the various scientific reports
that failed to leverage action in the plenary session, and it was he who presided (without great success) over the increasingly divided and contentious
Scientific Committee meetings of the mid-1950s. Having spent a good deal
of time in what remains of his papers, I am inclined to believe that he bears
a much more significant portion of the responsibility for the way events
unfolded in this decade than commentators have hitherto appreciated. The
temptation to focus on an easy villain—the vociferous Dutch dissenter E. J.
Slijper—has significantly distorted our understanding of this period.[119]

118. Whether it was two or three depends on how one counts: Birger Bergersen
chaired the two subcommittees in 1953 and another one in 1954, but the first two met
jointly (see n. 111 above). From that point through to the creation of the "Committee
of Three" in 1961, Mackintosh chaired everything. Note that in 1949 he also chaired the
first (joint) Scientific and Technical Committee.

119. Until Slijper's papers come to light and find their historian, it will be difficult to
give a balanced assessment of his role in this period. However, it merits mention that
committed conservation-oriented scientists like Bergersen and Kellogg betray no ill will
toward him in their surviving correspondence. In 1953 Bergersen wrote to Kellogg that
"Professor Slijper was to be sure very positive and cooperative during the discussions,"
and added, "but Lienesch [the head of the Dutch delegation and the government fisheries officer], as you will understand, is very accommodating concerning the demands of

Chapter 2 has dealt in some detail with Mackintosh's early training as a biologist and his rise through the ranks of the Discovery Investigations, the largest and best-endowed whale research organization in the world before World War II. We must now pick up that story, since the fate of the Discovery Investigations in the immediate postwar period is essential to the early history of the IWC's scientific advising structure. A brief sketch of the machinations around Discovery in the late 1940s and early 1950s will provide an occasion to consider Mackintosh's professional circumstances, and that will prove instructive as we look more closely at the actual science that was being done in and around the IWC in those years.

Mackintosh himself drafted (circa 1960) a highly confidential personal memo on his affairs in this period, notes that capture the anxiety arising at the close of the war and the attendant reconfiguration of Britain's scientific institutions. "After the war (1946–48)," he wrote in this document, "there was much uncertainty about the future, and working conditions for the small remaining [Discovery] staff were difficult and discouraging." He went on to explain that "in this period . . . my main preoccupation was in preserving the existing *Discovery* organization and assets."[120] In this endeavor it is fair to say Mackintosh basically failed. Having served as director of research in the Discovery Investigations since 1936, Mackintosh had had the misfortune, as we have seen, to preside over the organization through a period of institutional disaggregation and spiraling financial collapse

---

the industry." Bergersen to Kellogg, 5 June 1953, Smithsonian Archives, RU 7165, box 11, folder 6, "Birger Bergersen Correspondence, 1947–1953 (Ambassador from Norway to Sweden)." Even into the late 1950s, Kellogg remained in friendly correspondence with Slijper (see, for instance, Kellogg's warm acknowledgment of Slijper's first edition of *Walvissen*: Kellogg to Slijper, 2 June 1958, Smithsonian Archives, RU 7434, box 6, folder 9, "Correspondence 1956–1959"). This is difficult to fit together with a broad perception in the period that he was the consummate spoiler.

120. "Personal Statement by Dr Mackintosh," June 1960, NOL GERD.M2/1. This document was prepared in connection with Mackintosh's demotion in this period and his effort to preserve his pension. It is interesting to note that the outcome of this episode was the creation of the Whale Research Unit (WRU) at the British Museum. This was, effectively, furlough for Mackintosh, a way for Deacon to get him out of NIO (and replace him as director of biology); in return, Mackintosh would get continuation at the pay rank of SPSO (Senior Principal Scientific Officer) through his sixtieth birthday (in 1961), permitting him to secure his full pension. Thus the whole WRU had its origin as, in essence, a retirement plan. For details on this, see the "confidential" memorandum in Mackintosh's hand in blue ink, NOL Discovery MS, file "NAM Personal." I discuss these institutional maneuvers below in chapter 5.

(recall that Discovery was funded through a tax levied on whale oil landed in UK possessions—largely South Georgia—but the end of the humpback industry in the South Atlantic and the rapid rise of pelagic whaling meant plummeting income through this channel in the 1930s). During the war Discovery specimens were mothballed, and while Mackintosh did work to mobilize his expertise in the service of the defense effort—putting Discovery maps and photographs of the Antarctic at the service of the Admiralty and drafting a document to help aviators distinguish cetaceans from submarines—zoologists who knew a great deal about the reproductive habits of mysticetes were not in great demand in the escalating conflict.[121]

Physical oceanographers, however, saw their stock rise dramatically in the same period, since the new cat-and-mouse game of submarine warfare had quickly taken shape as a race to understand how the temperature gradients and macrostructure of water masses affected acoustics in the deep sea. Success in three-dimensional oceanic warfare depended on performing, evading, and interpreting sonar surveillance of the underwater world. It is for this reason that, while Mackintosh was being sent on rather pathetic errands (for instance, the Admiralty asked him to find some whale skin on which they might run a battery of acoustic tests[122]), his junior Discovery colleague George Deacon, trained in physics and experienced in efforts to map deep-sea thermoclines, found himself swept into highly classified naval research on submarine detection and evasion. Little wonder, then, that shortly after the war, with the Royal Navy Scientific Service in the driver's seat, the Colonial Office—traditional home of the Discovery Investigations—in shambles, and Mackintosh and his cluster of whale researchers facing a mounting budgetary crisis, momentum gathered behind a large-scale effort to consolidate oceanographic sciences in Britain. Nor is it surprising that the directorship of the new entity that resulted, the National Institute of Oceanography (or NIO), went to Deacon, not to Mackintosh.[123] Being

121. NOL Discovery MS, green filing cabinet, file "Whales, not submarines" and Mackintosh's 1942 "Contact Register" with the Admiralty in the same location.

122. Letter to Mackintosh from Navy, 21 November 1939, NOL Discovery MS, green filing cabinet.

123. On the financial problems at Discovery in these years, see Mackintosh's memorandum of 25 June 1948, NOL GERD.M2/1. The reorganization was part of the development of Britain's system of "Research Councils," as can be seen from Stanley Kemp's letter to Mackintosh, 9 June 1944 (NOL GERD.M2/1), advocating that the proposed NIO take shape as a Marine Research Council (like the Agricultural Research Council), directly under the administrative authority of the Lord President of the Privy Council. This view

leapfrogged galled Mackintosh deeply and lastingly, and there was no love lost between the two men for the rest of their lives.[124]

The realignment also had lasting implications for whale research in Britain and beyond. Deacon soon secured a coveted position among the Fellows of the Royal Society (FRS), joining his Discovery colleague Alister Hardy, a biological oceanographer, in the scientific pantheon of the British Isles (they would both soon be knighted). Mackintosh's nomination to FRS foundered. Hardy himself penned his old superior a painfully friendly note explaining why, in the end, he too had felt compelled to support Deacon's bid for leadership of the new NIO: "I do hope . . . you will not feel passed over," Hardy wrote optimistically, "because from the nature of the new institution and its relationship to the Admiralty I think it likely they will want to have a physical oceanographer as head."[125] A glance at Mackintosh's neat prepara- tory notes for his (unsuccessful) interview for the post leaves little doubt that the ultimate decision was quickly reached. After reminding himself of the need "to consider very carefully" and "balance the various interests" in consolidating the new oceanographic institution, Mackintosh (in many ways a cautious and retiring person) skipped swiftly onto his old familiar ground, his loopy hand recording and underlining the basic platform of

lost out, though it might have done better had Kemp not died a few months later. The advertisement for the NIO director's post ran in *Nature* in April 1949.

124. Richard Laws, interview by the author, 19 August 2005; Rosalind Marsden, in- terview by the author, 21 March 2004. For documentary evidence of the clashes between Mackintosh and Deacon, see Mackintosh's memorandum "Some False points in D's first Draft," NOL GERD.M2/1 (III/729), and, shockingly, Deacon's letter to H. F. Willis in the Admiralty, 11 October 1957, NOL GERD.M2/1 (III/739), which includes a scurrilous insinuation (apropos of nothing) that Mackintosh was queer: "Whenever we went to a party in the ship or ashore he had all the attractive girls round him but he never married one, and he has had all the attractive scientific problems round him and has not been able to devote himself wholeheartedly to any one of them." (This particularly blatant entailment of intellectual to sexual potency speaks volumes about the miasma of public- school hazing and football-pitch competition that hangs over all of these documents, and indeed, over this whole era of British field science.) The rumor that Mackintosh was homosexual never went away, and dogged the larger issues of his effectiveness and leadership. Rosalind Marsden, interview by the author, 12 March 2004.

125. Hardy to Mackintosh, 13 June 1949, NOL Discovery MS, file "NAM—Personal." Hardy did in fact support Mackintosh's (failed) bid for FRS; see Hardy to Mackintosh, 7 September 1945, in ibid. It is worth noting that Mackintosh's fate after the war was not wholly unrelated to the death of his predecessor as Scientific Director at NIO, Stanley Kemp, one of Mackintosh's mentors.

his (stillborn) candidacy to lead NIO: "*need to expand work on whales.*"[126] It was a vision for the future of British ocean science that was soundly rejected by the hydrographer of the navy, executives from the Treasury, and the other stakeholders building the new NIO.[127] Mackintosh's idea of ocean research—big expeditions to mark whales while collecting information about the plankton blooms on which they fed—seemed a *Challenger*-era relic to these men. As, in fact, in many ways, it was. Indeed, such exercises in what amounted to applied fisheries research seemed much better suited to a home in the Ministry of Agriculture and Fisheries (MAF), and doubtless a number of the men organizing NIO were befuddled by Mackintosh's sustained efforts to keep what was left of Discovery away from MAF. But there was a history there: Mackintosh resented the earlier efforts of the head of that ministry to strong-arm the Discovery budget under his own control, and by resisting MAF Mackintosh hoped to preserve and rekindle the spirit of the late 1920s and early 1930s, when Discovery was the flagship enterprise for British science on the word's stage.[128] It was not to be.

The broader significance for IWC whale science of these tedious interministry machinations starts to become clear when we consider that the inaugural secretary of the International Whaling Commission as a whole—occupying its sole standing administrative position—was none other than A. T. A. Dobson, the very MAF director who had tried to shanghai Discovery out from under Mack in the mid-1940s. Having served as the head of the British delegation to the Washington conference in 1946, and holding the IWC secretary's perch until 1958 (with its subtle power to shape the agenda and manage the budget of the organization), Dobson loomed over whaling matters in these early years. He and Mackintosh had an uneasy and asymmetrical relationship, particularly after Mackintosh's failed bid to

126. "Notes for Interview," in ibid. It is interesting to note that Mackintosh here characterizes the research of the Discovery Investigations at the start of the war as two-fifths whales, two-fifths "pure oceanography," and one-fifth "other."

127. For a summary of this decision and the early history of the NIO (emphasizing its foundation as an effort to preserve in peacetime the navy research nexus established during the war), see Margaret Deacon, A. L. Rice, and C. P. Summerhayes, *Understanding the Oceans: A Century of Ocean Exploration* (London: University College of London Press, 2001), 14–15, and A. L. Rice, "Forty Years of Landlocked Oceanography," *Endeavour* 18 (1994): 137–46.

128. Kemp to Mackintosh, 9 June 1944, NOL GERD.M2/1, complaining of Dobson's "power politics tack" trying to pull the "Discovery Committee" away from a weakened Colonial Office.

direct the NIO left him running a rump Discovery group tucked into a dim corner of a sprawling military-industrial research complex run by his arch-nemesis, Deacon.[129] Twenty years earlier the Discovery Investigations filled two research vessels and two land-based laboratories, spanned the Southern Hemisphere, and employed hundreds. By 1952 what was left would fit in a rowboat and had been exiled to a suburb of (landlocked) Milford, in Surrey. With the whole institutional base of the grand old Discovery Investigations dismantled, his own professional status indelibly tarnished, his prewar allies either dead (like Kemp and Harmer) or moving on from whales (like Hardy and L. Harrison Matthews—both anxious to keep some distance from what was looking more and more like a floundering enterprise), Mackintosh found himself in a precarious position throughout the early 1950s, circulating desperate memos like "Threatened abandonment of Whale Research" and musing to himself in private notes about where the money was going to come from to pay his skeleton staff and train new whale researchers.[130] He beat the bushes looking for help, even going back to MAF in 1952, hat in hand, seeking support, and complaining that "because whales are not the most important object of study from the purely oceanographical point of view," the Discovery program—ironically considered a "main pillar" of the new NIO at its inception—was coming under an increasingly cruel axe.[131] Tabulating the relative values of the whaling and fishing industries in the United Kingdom and the world, Mackintosh produced briefs arguing that whale research was proportionately funded at

129. In 1953 the NIO moved to a large Admiralty installation in Surrey (referred to as both Witley and Wormley in the documentation). Discussions about this move were well under way by 1951, and Mackintosh fought Deacon at every step of this process, complaining that the whale research "has the most to lose by removal from central London, and the least to gain by association with physical oceanography." See Mackintosh's comments on Deacon's memorandum of 19 June 1951, NOL GERD.M2/1. In the end a compromise was reached: a stash of Discovery material was left behind at the British Museum, but the core staff were moved to Surrey. The "Discovery" name and the "Discovery Reports" were conserved despite the reduced circumstances. Mackintosh himself got an arrangement that allowed him to continue living in London and a rail pass for the commute (on the order of an hour and forty-five minutes each way).

130. For examples of these memoranda, see NOL GERD.M2/1, particularly the sheaf dated 19 June 1951. On Mackintosh's acute sense of the damage to his reputation, see his "Personal Statement," June 1960, NOL GERD.M2/1: "On Dr. Deacon's appointment as Director of the NIO I suffered some loss of authority."

131. The first quote is from Mackintosh to H. J. Johns at MAF, 7 July 1952; the second quote comes from Johns's reply, 30 July 1952, NOL GERD.M2/1.

shillings on the pound.[132] Britain's role in the new IWC, he suggested, was bound to suffer accordingly.[133]

Only against the backdrop of this personal and professional upheaval does Mackintosh's ambivalent role in shaping the scientific advising system of the IWC begin to make sense. While his stature as the leader of the old Discovery Investigations, and as the author of fundamental publications on the biology of the big Antarctic baleen whales, guaranteed him pride of place within the relatively small global community of scientists with exposure to cetaceans, he knew (as did many of those around him at the annual meetings—with Dobson in the secretary's chair and Salvesen in the UK delegation) that his foothold in Britain's postwar research establishment was tenuous at best. Read in this context, Mackintosh's activities within and around the young IWC—his voluminous reports, his projected programs of study, his recommendations for the means by which data might be gathered and assessed—must be understood as an extended and elaborate effort to create a new institutional home for the kind of whale science he knew and which had fallen on such hard times in the United Kingdom. A brief review of his contributions to IWC scientific advising before 1955 will show clearly how he angled to resurrect the Discovery research in this period. This undertaking had important implications for his position with respect to industry leaders: his dogged pursuit of new research-oriented alliances with whaling companies eventually raised some eyebrows among his scientific colleagues.

To start with, then, it is helpful to read Mackintosh's various internal memos about his institutional struggles in the early 1950s alongside the briefs on whale research he was simultaneously submitting to the IWC. These efforts moved in tandem, and indeed, there was even some overlap, as, for instance, when portions of the memorandum that Mackintosh prepared to summarize the successes of the Discovery Investigations up through 1948 (part of his efforts to build a case for protecting the institution through the looming reorganization of British ocean sciences) ended up serving as the base for his 1951 report to the IWC, "Research on the

132. For these calculations see SGB Papers, folder "Notes on Whale Research." Mackintosh tallied these figures in October 1950 as follows: Fisheries in England and Wales (annual value) £30,750,000; research £260,000 (or about 0.85 percent). UK whaling (annual values) £11,000,000; Empire whaling £14,000,000; world whaling £36,000,000; the equivalent UK research budget ought to thus have been about £93,500, but Mackintosh put the world whale research budget below £30,000 in 1950.

133. Mackintosh to Johns, see n. 131 above.

Natural History of Whales."[134] At the center of this document—and in-deed, nearly every document that Mackintosh submitted to the IWC for the next decade—was a plea for greater investment in the two core activities of Discovery-style whale science: whale marking and the collection of bio-logical material that would contribute to a more complete understanding of life histories (ovaries were the organ that seemed to reveal the most, since they could be used to glean information about both aging and reproduc-tive cycles).[135] Together these two enterprises promised to elaborate a full account of cetacean life cycles—their migrations, breeding, and life span.

Thus, as Mackintosh was working hard back home to get some leverage on Deacon (in the hopes of forcing him to maintain an NIO commitment to whale marking), he was simultaneously drafting scientific reports for the IWC that trumpeted the importance of further marking work.[136] In a con-tinuation of the trend at Discovery in the 1930s, Mackintosh (despairing of institutional support at NIO) pursued industry collaborations in his effort to maintain and expand the marking program in the postwar years.[137] It was not, however, a particularly easy sell: at least one of the major companies,

134. IWC / 3 / 3. For evidence as to how this report took shape, see NOL GERD.M2 / 1, particularly "Summary of the Results of the Discovery Investigations" and the exchange of letters with J. Z. Young at UCL in the autumn of 1950.

135. See chapter 2 for more on these two prongs of Discovery research.

136. See, for instance, the minutes of the meeting of the Development Committee on 26 March 1953 (NOL GERD.M2 / 1), where Mackintosh pushed within the UK research establishment for marking support. Compare Part II, "Whale Marking," of his 1953 report to the IWC (IWC / 5 / 3) as well as para. 8–12 of the Scientific Committee's report for that year (IWC / 5 / 17). See also "Memorandum on Whale marking from the U.K. Delegation" (IWC / 5 / 12) and the related material from the 1952 meeting: i.e., Mackintosh's "Memorandum on whale marking" (IWC / 4 / 5). Mackintosh also published articles stumping for increased support for this program. See, for instance, "The Marking of Whales," *Nature* 169, no. 4298 (15 March 1952): 435–37, which was reprinted in *Norsk Hvalfangst-Tidende* 5 (1952): 236–40.

137. Mackintosh proposed such a collaboration to Willis in a letter of 7 July 1952 (NOL GERD.M2 / 1), writing, "Whale marking can be done if the whaling companies undertake to meet the cost of a marking vessel." And he floated the same idea at the private Development Committee meeting of 26 March 1953, saying, "The best immediate policy should be to get whale marking done at the expense of the industry." The issue had come up as early as 1950. See Mackintosh's report as chairman of the Subcommit-tee of Scientists, 14 July 1950, para. 19, Smithsonian Archives, RU 7165, box 12, folder 6, "Oslo—Second International Whaling Commission Meeting, 1950—Committee reports, notes, correspondence, clippings, conference documents, etc."

approached privately about the possibility of "borrowing" a catcher boat for this purpose, replied that "taxation was heavy" and that monies for such research should come from public funds via NIO.[138] Dobson was equally pessimistic about raising funds for such operations through the IWC, which, he was quick to point out, had a total budget of less than £2,500 per year.

As far as the collection of specimens was concerned, Mackintosh had better luck with the industry strategy. Since, as he put it, "it takes time to give a new biologist experience and training for whale research," and since he no longer had either a field station or a research vessel for this purpose, he promptly resurrected and promoted the scheme of getting his young scientists the exposure they needed as inspectors on factory ships.[139] Into the early 1950s the collection of ovaries aboard whaling vessels was facilitated by such irregular arrangements as the tipping of the bosun to encourage his help.[140] Mackintosh's slavish deference to Harold Salvesen at the annual IWC meetings makes better sense when we consider that these schemes were being organized on Salvesen vessels.

Even a cursory reading of the reports of the various IWC scientific committees over these years affords overwhelming evidence of Mackintosh's dogged efforts to stir up support for specimen collecting and marking work.[141] But still more striking are his efforts to use these documents to

138. See Captain Haines's comments in the minutes of the Development Committee meeting of 26 March 1953, NOL GERD.M2/1.

139. For a sense of Mackintosh's efforts in this direction from 1945 onward, see SMRU MS 38600, box 680, which contains about 50 files on "biologists and inspectors," including instructions for the collection of samples, training cards, kit lists, and correspondence with various zoology professors at UK institutions recruiting possible candidates.

140. For a colorful account of this kind of dealing (in 1950), see Robert Blackwood Robertson, *Of Whales and Men* (New York: Knopf, 1954), 156, which depicts a group of whalemen making jokes at the expense of "Ovary Joe"—"a very distinguished London biologist"—who may well be a thinly veiled Mackintosh. Note that threats of action for libel dogged the publication of this book.

141. See, for instance, his submitted "List of Recommended Researches" of 1950, which appears as an appendix to the "Report of the Subcommittee of Scientists" presented at the second IWC meeting in Oslo. This document appears in the Smithsonian Archives, RU 7165, box 12, folder 6, "Oslo—Second International Whaling Commission Meeting, 1950—Committee reports, notes, correspondence, clippings, conference documents, etc." Mackintosh calls there for the "collection of small embryos" as well as examinations of stomach contents and the traditional range of Discovery anatomical observations. See also the discussion of marking at that year's third meeting of the subcommittee scientists, in ibid.

reinforce, and indeed even to vindicate, the importance of what he came to think of as a "system of research on whales" that he had built. He would later claim that this system had "been widely followed by other biologists," and to the degree that this was true, it was in no small part a result of his success in organizing and promoting his kind of whale research from his pulpit as chair of the IWC's various scientific committees.[142]

As early as 1950 Mackintosh was using the opportunity to draft reports like that of the first "Subcommittee of Scientists" to put in a plug for more centralized coordination of "all current research on whales proceeding in different countries," a line item that positioned Mackintosh himself to begin soliciting and assembling progress reports from scientists in other countries doing the sort of whale work he knew and respected.[143] Furthermore, there is evidence that he pursued this path in the face of some disagreement on the part of his colleagues about the exact kinds of whale research that ought to be encouraged by the IWC. One hint of these objections lies in the final document to emerge from the 1950 meeting of the Subcommittee of Scientists, which not only spelled out in cautionary tones that Mackintosh's own program of Discovery-style work "was not intended to be a programme" (but rather a "gathering of outstanding problems"), but also included a suspiciously supererogatory paragraph stipulating that "research workers should be encouraged to investigate problems with which they are specially qualified to deal, rather than . . . be pressed to undertake work outside their ambit." Such a mysterious irruption within a committee document almost certainly points to dissent, and it is most likely that the division reflected here was between Mackintosh on the one side (keen to coordinate marking and specimen gathering) and Bergersen on the other (who hoped to get the IWC more focused on the statistical reports coming out of Norway's BIWS).

142. Mackintosh wrote, in 1957, in a memo circulated within NIO, that "I can personally claim to have founded a system of research on whales which is largely new . . . which has been widely followed by other biologists, and which has substantially helped (and continues to help) the regulation of whaling and the conservation of the stock." It is only with the last part of this claim that one might properly quibble. See NOL Discovery MS, file "NAM—Personal."

143. For an example of this sort of document, see "Scientific progress reports," IWC / 4 / 10. For Mackintosh's push for this sort of coordination, see Subcommittee of Scientists, report of 14 July 1950, para. 14, Smithsonian Archives, RU 7165, box 12, folder 6, "Oslo—Second International Whaling Commission Meeting, 1950—Committee reports, notes, correspondence, clippings, conference documents, etc."

Yet Mackintosh was relentless about seeding the IWC's burgeoning committee documentation with what amounted to casual ratifications of the field biological and anatomical / physiological research that he had helped pioneer. For instance, a very last line crept into the minutes of the Scientific Committee meeting on 30 May 1952: "Agreed that the work on corpora lutea [recall that these are scars of ovulation in the ovary] was of outstanding importance. The meeting then ended."[144] And a similarly enthusiastic pitch found its way into that year's Scientific Committee report itself:

> The reports [of different countries' research, assembled by Mackintosh] were studied and compared with great interest, and it was noted that good progress is being made in certain directions, among other things, in breeding, age determination, distribution and populations. It was also noted that in most countries the corpora lutea of the ovaries are regarded as an essential object of study.[145]

I do not mean to suggest that Mackintosh merely made up asides like this without regard for what actually transpired at these meetings (these were, after all, committee documents, reviewed by all participants), but it must be recalled that he served as both chairman and recorder at all of these sessions. While he was emphatically not a strong chair, he nevertheless had relatively free rein in the post-meeting composition of committee reports, and there can be little doubt that he used this freedom to underline the research methods and problems to which he was most firmly attached.[146]

But was Mackintosh's "system of research on whales," which he worked so tirelessly to promote in these years, what the IWC really needed? Was it,

144. "Subcommittee of Scientists—Minutes of First Meeting, 10 July 1950," Smithsonian Archives, RU 7165, box 14, folder 1, "London—Fourth International Whaling Commission Meeting, 1952—Instructions, proceedings, correspondence, notes, etc."
145. IWC / 4 / 15, p. 4.
146. Mackintosh's qualities as a chair are manifested in several verbatim transcripts of a number of IWC gatherings. For instance, he is interrupted twice in eight pages of transcripts from the meeting of the Scientific and Technical Committees of 31 May 1949 (no one else is interrupted at all, and he was in the chair). He is elsewhere lost to the stenographers in a mumble (see IWC / 14 / 10). Oral histories confirm that he was infuriatingly methodical as chairman of the Scientific Committee and the Scientific Sub-Committee, stopping discussion continually to record notes as his own rapporteur. Richard Laws, interview by the author, 25 March 2004.

PLATE 1.
(*top*) Into the belly of the beast: A stern-slipway factory vessel engulfs its prey. (From the collection of the Natural History Museum, London.)

....

PLATE 2.
(*below*) A sea of blood: Hardy's watercolor of a South Georgia whaling station. (From Hardy, *Great Waters: A Voyage of Natural History to Study Whales, Plankton, and the Waters of the Southern Ocean*, p. 161.)

PLATE 3.
(*right*) A good wooden ship:
The *Discovery*. (From Great
Britain, Discovery Committee, *Report on the Progress
of the Discovery Committee's
Investigations*, opp. p. 8.)

....

PLATE 4.
(*below*) A research ship to
run down whales: The *William Scoresby*. (From Great
Britain, Discovery Committee, *Report on the Progress
of the Discovery Committee's
Investigations* opp. p. 12.)

PLATE 5.
(*above*) The floating
laboratory: The *Discovery II*. (From Great Britain,
Discovery Committee,
*Report on the Progress of
the Discovery Committee's
Investigations*, frontispiece.)

PLATE 6.
(*left*) The postmortem: A
hip-booted cetologist checking fusion in the vertebral
column. (Courtesy of the
Sea Mammal Research Unit,
University of St. Andrews.)

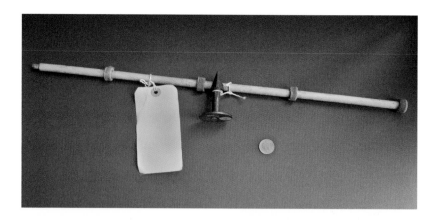

PLATE 7.
(*above*) Rethinking delivery: Photographs of a modified fletch (to be fired from a shotgun). (From the collection of the Scott Polar Research Institute, University of Cambridge.)

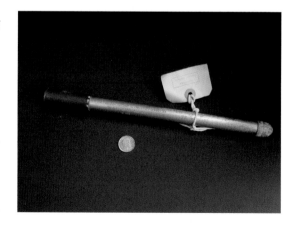

PLATE 8.
(*right*) The "slug": Photo of an early "Discovery Mark." (From the collection of the Scott Polar Research Institute, University of Cambridge.)

PLATE 9.
(*above*) Else Bostelmann sees under the surface: A right whale family. (From Kellogg, "Whales, Giants of the Sea," p. 43. Courtesy of the National Geographic Society.)

PLATE 10.
(*left*) Bostelmann's vision: Gamboling dolphins. (From Kellogg, "Whales, Giants of the Sea," p. 75. Courtesy of the National Geographic Society.)

PLATE 11.
(*right*) Bostelmann pictures the happy giants: Playful humpbacks. (From Kellogg, "Whales, Giants of the Sea," p. 46. Courtesy of the National Geographic Society.)

PLATE 12.
(*below*) Bostelmann depicts the threat: A whaler takes aim at a blue whale surrounded by its smaller kin. (From Kellogg, "Whales, Giants of the Sea," p. 42. Courtesy of the National Geographic Society.)

PLATE 13.
(*left*) Bostelmann's paradi-
siacal grotto: Gray whales
frolic. (From Kellogg,
"Whales, Giants of the Sea,"
p. 47. Courtesy of the National Geographic Society.)

PLATE 14.
(*below*) The sanctuary: Map
to accompany IWC report,
2 December 1946. (Cour-
tesy of the Smithsonian
Institution.)

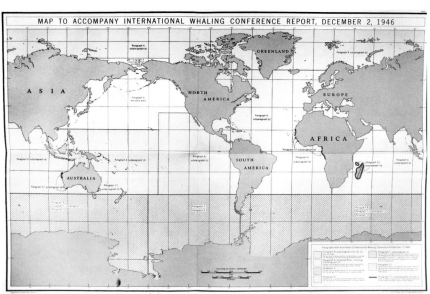

MAP TO ACCOMPANY INTERNATIONAL WHALING CONFERENCE REPORT, DECEMBER 2, 1946

PLATE 15.
(*right*) Soviet whalers meet the
freaks: Greenpeace in mid-hunt.
(Courtesy of Rex Weyler.)

PLATE 16.
(*below*) The miracle of plastics:
The Smithsonian's fiberglass blue
whale is prepared to preside over
Ocean Hall. (Courtesy of the
Smithsonian Institution.)

to push the question further, what the whales of the 1950s needed? Several of his colleagues harbored doubts, though in this sensitive initial period in the development of the IWC's scientific advisory system they made every effort to mute their Fretting in order to present the scientific community as a unified front. Nevertheless, in private, serious concerns began to be raised. Perhaps the most dramatic instance came in 1952, in connection with Mackintosh's expanding plan to coordinate a series of marking expeditions with the whaling industry. While by the early 1950s the annual north–south migrations of Antarctic humpbacks had been largely worked out (on the basis of old Discovery marking and reproductive physiology data), the question of where the southern stocks of blue and fin whales went during the southern winter remained open. It was broadly assumed, by analogy to the humpbacks, that they too moved north into warmer tropical or subtropical waters to breed and calve, but land stations in equatorial regions seldom took these larger animals, and there were no marking or sighting data to confirm this supposition.

It is little surprise, then, that this problem came to the fore in Mackintosh's prospective research program: it had the same shape as the central problem of the prewar Discovery Investigations; it could be resolved by marking campaigns and flensing-platform reproductive physiology; and the question at issue—where are all the big whales in the Southern Hemisphere during the nine months of the year they are not (under the regulatory regime of the early 1950s) being hunted?—could not but pique the curiosity of key figures in the industry. In addition, the research could be done in the off-season, which increased the chances that whaling executives might be persuaded to contribute idle equipment to the initiative. As early as 1950 Mackintosh was carefully crafting scientific reports to highlight this alluring mystery:

> In the course of discussion of various problems, not included in this list, certain points emerged among which may be mentioned the problem of the as yet unknown winter breeding grounds of Blue and Fin Whales, and special importance was attached to the acquisition of information on this matter.[147]

147. Subcommittee of Scientists, report of 14 July 1950, para. 16, Smithsonian Archives, RU 7165, box 12, folder 6, "Oslo—Second International Whaling Commission Meeting, 1950—Committee reports, notes, correspondence, clippings, conference documents, etc."

By 1952 he had succeeded in rousing a fair bit of interest in the problem among both Norwegian and British whaling companies, and there was talk of sending out a factory ship and fleet of catchers to scout warm waters and deliver the trademark "Discovery" darts to any whales they found. Apprised of the momentum behind these plans through an associate in the Norwegian Whaling Association, Birger Bergersen shot Kellogg a "Confidential" letter laying out the situation and raising some tough questions: "After all is said and done—has the time really come when we can freely reveal the whereabouts of the blue and fin whale between the seasons?"[148]

It is in many ways a remarkable moment in the archival residue of this era, both for what it reveals about ambivalence within the scientific community concerning the proper course for scientific research and for how clearly it shows the scientists thinking strategically about the boundary between political and scientific problems. Bergersen was seeking Kellogg's opinion about whether they ought to try to push a particular scientific question—the location of blue and fin breeding grounds—off the research agenda in order to protect not only the whales, but also the new IWC. Bergersen spelled out the real danger when he wrote, "Certain people might learn a bit too much about the whereabouts of the animals during the Antarctic winter" and "get the idea of equipping smash-and-grab expeditions"; unfortunately, "the International Whaling Commission was hardly strong enough, as yet, to do anything to prevent it."[149] Strategic scientific ignorance could thus play a crucial role in protecting the young international regulatory system itself.

Remington Kellogg was immediately sensible of the threat, and he replied that he was "inclined to agree" with Bergersen on the "disturbing question about the desirability of whale marking in areas assumed to be centers of concentration during the winter season."[150] In the end, a crisis

148. Bergersen to Kellogg, 20 February 1952, Smithsonian Archives, RU 7165, box 11, folder 6, "Birger Bergersen Correspondence, 1947–1953 (Ambassador from Norway to Sweden)." These concerns were well founded: Aristotle Onassis was, at this time, well embarked on his project of "pirate whaling" using the notorious *Olympic Challenger*, which had been outfitted in Germany in 1950 for expeditions outside the IWC framework.

149. Ibid.

150. Kellogg to Bergersen, 26 March 1952, Smithsonian Archives, RU 7165, box 11, folder 6, "Birger Bergersen Correspondence, 1947–1953 (Ambassador from Norway to Sweden)."

was averted when the off-season marking program was diverted to "known whaling grounds off West Australia and off the west coast of Africa."[151] As long as the operations were restricted to areas already "actively exploited," it was difficult to object (though Kellogg noted that "a rather large proportion of the marked whales" were likely to be taken in the local operations before much interesting migratory data could be gained).[152]

If this tremor around the marking campaigns of the early 1950s is a particularly telling instance of anxiety about Mackintosh's larger program for regulatory whale science, it was neither the first nor the last such episode. All the way back in 1945, at one of the London gatherings of Allied powers, Bergersen voiced astute concern about Mackintosh's proposed plan to integrate biologists into the first postwar expeditions as factory-ship inspectors. While Bergersen acknowledged that inspectors should have some training in "histology and biometry," he was nevertheless "not in favor of the biologist also assuming the duties of inspector because, in the interests of obtaining the maximum cooperation for his [biological] work, it was essential that he maintain the good-will of the ship's officers, which is often somewhat difficult in the case of an inspector."[153] An exchange like this again highlights early concerns about how the boundary between political problems (like regulatory enforcement) and scientific ones (like specimen gathering) ought best be erected and managed. It is evident that Bergersen was here doubly concerned: proper deference to the vessel's officers might facilitate scientific work at the expense of rigorous oversight; conversely, strict adherence to regulatory responsibilities might undermine a scientific mission. The kinds of "independence" or "impartiality" to be hoped for in scientists, on the one hand, and inspectors, on the other, were not, in the final analysis, the same, and it was essential to the cultivation of both roles, in Bergersen's view, that they be prized apart. Mackintosh's appetite for data—in the form of meaty bits of whales and marks recovered from their boiling blubber—muted his attention to such matters. So much remained

151. Ibid.
152. Ibid.
153. "Report on the International Whaling Conference held in London, November 20 through November 26, 1945," p. 23, included in the briefing packet prepared for the IICRW, Smithsonian Archives, RU 7165, box 8, folder 6, "Washington—Informal Inter-agency Committee on the Regulation of Whaling Conference, 1946—Kellogg's information statistics and notes—Includes IICRW material."

unknown, he regularly opined in committee documentation—more marking was needed, and more specimens.[154]

It was difficult to dispute the point. There were certainly lots of problems (stock segregation, age and size at sexual maturity, debates over specific identity) that remained unresolved, and it was easy to imagine that clarification of these matters would help refine and buttress regulatory recommendations. But by the mid-1950s—when Bergersen had fallen away from active participation in the IWC and Kellogg had been increasingly drawn into administrative matters (he served on the Finance Committee, for instance, and more or less ceased to attend the Scientific Sub-Committee meetings held between annual IWC gatherings)—the dominance of Mackintosh-style whale biology began itself to be a subject of some concern to those eying the falling proportion of blue whales in each year's catch. Not only had E. J. Slijper begun, by 1955, to resist any new protective measures for blue whales (on the old grounds that such measures would distort the comparability of the long-term data sets), but familiar scientific rationales (collecting whale marks and specimens) had been cited by the Scientific Committee itself as part of the controversial decision to recommend opening the southern Pacific sanctuary to hunting for the first time in more than a decade.[155] It was in this context that an increasingly frustrated Remington Kellogg took the floor at a Scientific Committee meeting to remind his colleagues pointedly that "*there was more than one form of research apart from whale marking*" that might attract their collective attention.[156] Mackintosh's notes spell out Kellogg's insurgent suggestion: "He [Kellogg] had in mind espe-

154. For just one example (among countless others), see Mackintosh's 1953 report to the IWC, IWC / 5 / 3, para. 6: "The reports . . . draw attention to certain specific problems relating to conservation which cannot be settled without further knowledge of the biology of whales."

155. For Slijper's position (he was joined by J. M. Marchand, representing South Africa), see the Scientific Committee report for 1955, IWC / 7 / 20, para. 6; for the scientific benefits of opening the sanctuary, see the Scientific Sub-Committee report for 1955, IWC / 7 / 3, para. 7; see also Mackintosh's "Notes on the agenda" for the 1955 meeting of the Scientific Sub-Committee, particularly para. 3 (this document is dated February 1955 and appears in the Smithsonian Archives, RU 7165, box 24, folder 2, "Seventh International Whaling Commission Meeting [Moscow], 1955") and Kellogg's report to the secretary of state for 1955, which reads on page 24, concerning the opening of the sanctuary, "One object is a study of the stock of whales in this area." See Smithsonian Archives, RU 7165, box 28, folder 4, "Reports to the Secretary of State by the U.S. Commissioner to the IWC, Meetings #1–11, 1949–1959."

156. Cited in Scientific Committee report for 1955, IWC / 7 / 20. Emphasis added.

cially the possible employment by the commission of one or two suitable qualified persons to undertake a fuller analysis of all the available statistics of catches."[157]

Touching the point lightly in his report, Mackintosh moved crisply from this vexatious matter of statistics onto more familiar ground, concluding with a push for more complete study of blubber layers: "The committee would like to urge that further measurements of blubber thickness should be taken as a routine" aboard factory vessels and at land stations.[158] Here was a subject of considerable interest to whale biologists and whalers alike. It was the sort of data that Mackintosh might correlate with his ongoing effort to complete a coherent life history table for the large whales of the Southern Hemisphere— the sort of table that he would issue in the early 1960s (figure 4.2), which, by integrating information on migration patterns and reproductive physiology, functioned as a concise visual epitome of Discovery-style whale research.

But as the whole IWC structure teetered in the late 1950s, Kellogg's urging—that there were other kinds of whale science besides blubber measurements, ovaries, and whale marks; that serious statistical analysis of catch data might yield results of real importance to the regulatory process— would not go away.[159] Forcing it to the fore was the work of the next decade.

CONCLUSION

At the beginning of this chapter, I proposed to examine the first decade of scientific advising within the emerging postwar system of international

157. Ibid.

158. Ibid., para. 13.

159. It is interesting to note that Mackintosh had been nudged several times in the postwar period by colleagues who encouraged him to expand his vision of whale science beyond Discovery-style research. A choice instance came as early as 1950, when the distinguished biologist J. Z. Young (at University College London) wrote Mackintosh that more work on sensory physiology and behavioral adaptations might be rewarding. Young, who had reviewed a program for postwar whale research, closed the letter with a prescient warning: "I should be somewhat sceptical whether really important advances would come from these routine investigations themselves and it might be worth while keeping constantly in mind that the purpose of routine is not only to collect masses of data but to provide the opportunities for the oblique insight that leads to really new advances. I suspect that this warning is necessary in any routine research investigation and I should be inclined to push it fairly hard in this case, so as to encourage young men to be using their heads as well as their hands in these laborious investigations." Young to Mackintosh, 29 September 1950, NOL GERD.M2/1.

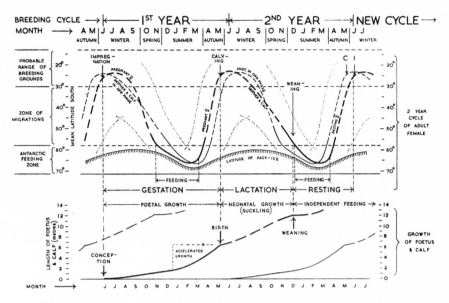

FIGURE 4.2   A summation: Mackintosh's seasonal migration/breeding cycle diagram for the southern Fin Whales. (From Mackintosh, *The Stocks of Whales,* p. 79.)

whaling regulation, and to do so by paying particular attention to the nego-tiation of the boundary between politics and science. It was my contention from the outset that this period was marked by various efforts to cultivate and nurture new collaborative means by which scientific recommendations could serve as the basis for regulatory initiatives within the IWC. The nov-elty of the forum, the promise of the ICRW, the formal commitments of the whaling nations, the active involvement of conservation-oriented scien-tists, and an explicit US interest in promoting multilateral initiatives for the scientific management of ocean resources all contributed to an emphasis on cooperation and consensus building. In this context, as I have shown, scientists worked with diplomats and industry representatives to develop the administrative architecture and procedures of the new organization.

Boundary work took concrete form in the codification of a committee structure that specified where science would get done and how scientific findings would enter the larger forum of the commission. Interestingly, the resulting arrangement—in which a series of scientific committees at progressively greater remove from the deliberative center of the IWC were obliged to pass their recommendations up through increasingly "practi-cal" bodies en route to the commission floor—can be understood to have emerged in response to a pair of convergent imperatives that nevertheless

proved very difficult to integrate happily: on the one hand, diplomatic negotiations within the plenary session were streamlined by shunting contentious matters out to the "substance" committees (staffed, at least in part,
by scientists); on the other hand, the scientists were themselves intent on
defining a space of their own, a space "beyond the politics" where they
might generate the kinds of findings that would serve as the basis of good
regulatory decisions.

In the end, the system that had solidified by 1955 worked better as a
means of holding scientific questions at bay than as a way of protecting
scientific independence or of installing presiding scientific oracles. Above
all, as I have suggested, different (and at times contradictory) efforts to
cultivate a cooperative model for the production and use of whale science made the already difficult business of defining a distinctive sphere
for scientific expertise even tougher. Efforts to define a particular committee as, in effect, a "politics-free zone" could be quite successful, but the
unintended consequence of this success might well be that what happened
in this special space could be written off as irrelevant to pressing political
matters; anyone speaking "as a biologist," in order to get above the fray of
conflicting interests, might discover that "out of the fray" could amount to
"out of the game." In sketching these dynamics I have worked to show that
several whale scientists were acutely sensitive to the gamesmanship that
was necessary to conjure up the promise of a privileged perspective outside
of the horse-trading world of international diplomacy, and they engaged
in careful strategic decision making in the hopes of enhancing respect for,
and the effectiveness of, the scientific community.

Not that all the members of this expanding scientific community agreed
about how cooperation and independence were to be achieved, balanced,
or managed. Closer attention to Neil Alison Mackintosh, the active scientist with the most visible role in the IWC's various scientific committees in
these years, has dramatized the way that specific research priorities could
be tailored—because of financial exigencies, traditional research commitments, or ideas about the proper ways to delimit scientific problems—to fit
industrial interests. This part of the story not only calls for a reappraisal of
responsibility for the early failures of the IWC's scientific advisory system,
it also nuances our sense of what "capture" of scientists in a regulatory
setting might mean. It also points, I think, to two larger conclusions that
merit attention here.

First, much has been made by recent commentators of the importance
of "scientific uncertainty" in regulatory debates in general and in the his-

tory of the IWC in particular. While critics have been quick to seize on ways in which uncertainty has been cultivated by opponents of particular scientific recommendations (in order to undo consensus or derail the implementation of new programs), I have found no discussion at all of the more subtle role that the cultivation of uncertainty plays in the framing of any research program. That is to say, a scientist like Mackintosh—aiming to promote a "system of research" and garner support for the application of his methods—has an enormous stake in scientific uncertainty: it is, in effect, his stock-in-trade. Only where there are unresolved questions of the sort he asserts an ability to address can he advance a claim on collective resources. To the degree he succeeds in resolving those uncertainties, it is incumbent upon him to locate some others that are similarly amenable to his program. The alternative is obsolescence. An appreciation of this feature of what we might call "the promotional life of research schools" sheds considerable light on the construction and deployment of uncertainty in complex regulatory scenarios. It is not that uncertainty is a kind of perpetual echo of some big bang in the universe of epistemology (as Heazle might have it), or that it is the sinister veil behind which scientific malefactors obscure troublesome truths (as Schweder might insist). Rather, the cultivation of uncertainty is a basic feature of scientific practice, and, perhaps even more importantly, an essential tool for justifying particular programs of scientific research. Thus, when Mackintosh larded his 1951 "Report on the Natural History of Whales" to the IWC with sentences like "the existing regulations are partly based on the findings of past research, but to a large extent they rely on probabilities or guesswork, and it is certain that they could be greatly improved if more definite facts could be ascertained," he was palpably trying to write himself into a job.[160] Indeed, when Kellogg boldly redlined one of Mackintosh's early scientific reports in such a way as to downplay uncertainty—Mackintosh had written "many of the regulations are *still to some extent arbitrary*" as part of a call for more research, and Kellogg changed the line to read that "many of the regulations are *by no means finalized*"—Kellogg was thinking less like a scientist trying to build a research program than like the "scientific statesman" he had by this point become.[161]

160. IWC/3/3, p. 1.

161. This remarkable annotation appears on page 3 of the minutes of the first meeting of the Subcommittee of Scientists, Smithsonian Archives, RU 7165, box 12, folder 6,

The second larger issue raised by this chapter—and worthy, in my view, of further consideration—is the strong evidence it provides that the makings of what became, by the second half of the 1950s, a crisis in the regulation of whaling (as well as a crisis in whale populations) can be traced back to the broader revolution in the ocean sciences occasioned by World War II. As we have seen, Mackintosh's orientation to the IWC, and his pursuit of particular avenues of whale research, were substantially the product of his sudden displacement from the pinnacle of research oceanography in the United Kingdom in 1949. With the rise of the postwar funding structures and research priorities of the new NIO, Mackintosh found himself awkwardly dependent on the whaling industry (and the IWC) to support and fund the kind of biological sea research that was really all he knew, and that had once held pride of place in Britain and beyond.

It is to the fate of his salvage efforts, and the accompanying realignment of scientific advising within the IWC—from a collaborative model to what we might call a model of compulsion—that I now turn.

"Oslo—Second International Whaling Commission Meeting, 1950—Committee reports, notes, correspondence, clippings, conference documents, etc." Emphasis added.

FIVE

TRIALS OF FORCE

SCIENTIFIC ADVISING AND THE INTERNATIONAL WHALING COMMISSION

1956 – 1965

> However, we have seen that all the arguments are still based on
> very little evidence, and that whalers can and will restrict their
> activities only on the most incontrovertible of arguments.
>
> E. J. SLIJPER, *Whales*, 1962

INTRODUCTION

At the beginning of the last chapter, I introduced the problem of scientific advising in the IWC by dropping the reader off in media res, in the midst of a set of caustic 1956 exchanges about the status of the Antarctic stocks. The British delegate, R. G. R. Wall, took the IWC's scientific advisers to task with his memorable "green cheese" speech, and the visiting dignitary from FAO summed up what he had seen by warning that the IWC was headed into a "race" for "the biological fact" that was necessary to make sustainable regulation of the industry a reality. Those remarks gave voice to the emerging crisis faced by the IWC in the mid-1950s. By that period it was becoming increasingly clear that the idea of an international regulatory regime "based on scientific findings" was proving enormously difficult to realize. In chapter 4 I retraced the early evolution of this regime, with particular emphasis on the efforts to build administrative structures and procedures by which the whale scientists could make whale science available to the delegates in such a way that they might use it in framing regulatory programs. But as I have shown, the emphasis on collaboration and consensus-driven decision making, along with various more and less successful efforts to define the boundary between political and scientific problems, resulted in the gradual marginalization of the scientists, the attenuation of their recommendations, and a progressive (and arguably pernicious) convergence of their interests with those of the industry. In the first section of this chapter, "Square That Circle,"[1] I follow the intensification of these problems into the mid-1950s by showing how three increasingly contentious subjects—scientific whaling, "ecological" management, and statistical analyses—together sounded the death knell of the collaborative model for scientific regulation. The debates over these three issues dramatized the great difficulty that attended any effort to establish boundaries everyone could agree on, and across which successful exchanges (of regulation-relevant facts) might occur. The last of the three in particular, which went to the heart of the increasingly urgent question facing the IWC (how many whales were left?), set the stage for a push to replace the collaborative model with something new: a model whereby scientific findings would compel the assent of member nations. This change would demand new people, new methods, and new administrative arrangements. By the end of this section, I hope to have shown how American scientists and diplomats succeeded in arranging a finely staged showdown—a kind of formal "battle royale" between those intent on using science to delay regulatory measures and those who proposed to use science to oblige such measures. The victory of the latter, as I will show in the final section of this chapter, was not unqualified, nor is it best described as a victory (such as it was) of "science over politics," as commentators have uniformly suggested.

To get to this showdown and all it implied, let me first evoke the increasingly dire situation of the IWC's scientific advising system at mid-decade by considering its handling of three contentious topics: scientific whaling, ecological management, and the emergence of seemingly irresolvable conflicts over stock statistics.

"SQUARE THAT CIRCLE": 1956–1959

The acrimonious disputes that arose in the mid-1950s over what it meant to "take whales for purposes of scientific research" both foreshadow more

---

1. The quoted section heading is taken from R. G. R. Wall, speaking at the eighth meeting of the IWC in 1956—see proceedings, p. 104, Smithsonian Archives, RU 7165, box 17, folder 5, "London—Eighth International Whaling Commission Meeting, 1956—Numbered conference documents, #1–19."

recent controversies concerning scientific whaling (Japan continues to take whales today via this ɪᴡᴄ loophole, in what is seen by many as a blatant circumvention of the current moratorium) and throw into high relief the increasingly dire difficulties faced by those who had hoped in the early 1950s to define whale science as a helpful politics-free zone whence regulatory initiatives might arise. In essence the episode can be summed up thus: freedom from politics proved to be a very mixed blessing. That is to say, in this instance, success in carving out a politics-free zone for whale science brought unintended (and undesirable) consequences, since such a space could be seized and utilized as what amounted to an "anything-goes" zone. As this began to happen, a number of whale scientists and whaling diplomats looked with dismay on what they had unwittingly wrought.

At issue was Article vɪɪɪ of the ɪᴄʀᴡ, which read in relevant part,

> Notwithstanding anything contained in this Convention any Contracting Government may grant to any of its nationals a special permit authorizing that national to kill, take and treat whales for purposes of scientific research subject to such restrictions as to number and subject to such other conditions as the Contracting Government thinks fit, and the killing, taking, and treating of whales in accordance with the provisions of this Article shall be exempt from the operation of this Convention.[2]

This scientific exemption was not new in 1946. It had been borrowed almost verbatim from the final act of the 1937 international whaling agreement, where it appeared as Article 10.[3] As one of the US drafters of the ɪᴄʀᴡ language (possibly Kellogg himself) commented in an aside on the provision, it was originally introduced in 1937 in order "to legalize and encourage scientific investigations on the order of those carried on by the Discovery

---

2. This wording is basically identical to that laid out in the original draft proposal by the United States—compare Patricia W. Birnie, *International Regulation of Whaling: From Conservation of Whaling to Conservation of Whales and Regulation of Whale-Watching* (New York: Oceana Publications, 1985), vol. 2, 692, and the "confidential" draft, pp. 10–11, Smithsonian Archives, ʀᴜ 7165, box 8, folder 4, "Washington—Informal Inter-agency Committee on the Regulation of Whaling (ɪɪᴄʀᴡ), July 1946–Oct. 1946—Papers."

3. For a summary of the precedents, see "Brief: suspension of regulations for scientific investigation," ɪɪᴄʀᴡ briefing packet, p. 208, Smithsonian Archives, ʀᴜ 7165, box 8, folder 5, "Washington—Informal Inter-agency Committee on the Regulation of Whaling (ɪɪᴄʀᴡ), July 1946–Oct. 1946—Papers."

Committee."[4] In fact, there was even earlier precedent for such a special dispensation, and there is excellent evidence that this earliest suggestion for spelling out the supra-regulatory character of scientific work came from the young Remington Kellogg. Reaching all the way back to the work of the League of Nations Economic Committee in 1930, it becomes clear that, while the 1931 Geneva Convention for the Regulation of Whaling did not mention a scientific exception, the "Preliminary Draft Convention" put together by the Berlin "Committee of Experts" did. That draft contained a final "Observation" that "provision would have to be made to give the necessary exemptions for scientific investigations."[5] Still more interesting is to discover, in Kellogg's holograph notes from the Berlin meeting, a marginal annotation inserting this language into a circulated rough version of the document.[6] This marginalia cannot, of course, establish with certainty that the move to pull science out of the very first international regulatory arrangement was Kellogg's, but, given his attention to securing similar exemptions under the various US enabling laws, there are good grounds for believing this to be the case.[7] Be that as it may, a foretaste of the trouble Article VIII would make for regulators in the years to come was already palpable at the 1946 Washington conference that adopted the ICRW, where no sooner was the draft article introduced than the delegates began squabbling over the indelicate question of just who would secure the market value of any whales taken "for scientific purposes."[8] This question led to an interesting tug of war over whether that issue would best be addressed by the subcommittee working on what was to be done with the value of whales taken illegally or by the more general drafting committee. The tacit issue was, ought the IWC to consider all "whaling outside the convention"

4. Ibid.

5. See Observation VI in League of Nations, Economic Committee, "Report to the Council on the Work of the Thirty-Second Session, Held at Geneva from June 2nd to 6th, 1930" (Publication II, Economic and Financial, 1930, II 24 C.353.M.146).

6. It appears in blue ink as an inserted "Remark IX," in Smithsonian Archives, RU 7165, box 3, folder 4, "Berlin—Whaling Conference under auspices of League of Nations, 1930–1931."

7. See Kellogg's annotations on S. 1045 (9 March 1939), sec. 9, as well as those on the "restricted" 1949 mimeograph of the bill to give effect to the ICRW (two drafts), sec. 9(c), in Smithsonian Institution, James Mead Files, file "U.S. Regulations."

8. Van Dijk (Netherlands) originally raised the question; see minutes of the third session, IWC / 20, p. 10.

at once—lumping science together with rule-breakers—or did science deserve to be "outside the convention" in a different way? Since all the whales
had fat—regardless of who killed them—they all ultimately needed to be
written into the ledger books in pounds, shillings, and pence. In a sense,
then, the initial effort to define a special domain outside the reach of the
IWC was instantly reentangled with the looming questions of monetary
value and mercenary interest that the special scientific domain was meant
to transcend. It was a specter of what would follow.

By the early 1950s the USSR had settled into a rhythm of taking half a
dozen large baleen whales before or after (and sometimes both) the official
opening of the season. Rumors circulated that these whales were being
used to get a jump on the hunt (that they were serving, for instance, as
"fender whales"—animals taken at the start of the catch to serve as giant
bumpers between the factory vessel and the catchers, their carcases literally lashed to the sides of the vessel like huge inner tubes). Other whaling
nations might have been tempted to let this activity slide, or to try it themselves, but by 1955 the issue of enforcement had heated up generally, and
there was considerable suspicion that—in the face of the continuing failure
to implement some sort of international observer scheme—illegal hunting was rampant.[9] Captain Salvesen complained mightily about the Soviet
"scientific whaling" boondoggle, asking what scientific purposes could demand taking six whales in a single day, not once, but *twice*: "One would
imagine that scientists would have preferred their whales one (or ? two) at a

9. This was subsequently discovered to be the case. See R. L. Brownell and
A. V. Yablokov, "Illegal and Pirate Whaling," in *The Encyclopedia of Marine Mammals*,
ed. W. F. Perrin, Bernd G. Würsig, and J. G. M. Thewissen (San Diego: Academic
Press, 2002), 608–12. See also A. V. Yablokov, "Validity of Whaling Data," *Nature* 367,
no. 6459 (13 January 1994): 108. Salvesen suspected extensive Soviet cheating (beyond
that already glimpsed by neighboring vessels in 1950 and again in 1955) on the basis of
what he took to be their impossibly good (in view of their equipment) oil production
statistics (on a per whale basis—essentially an efficiency rating). See his correspondence
with Kellogg, Smithsonian Archives, RU 7165, box 24, folder 1, "Sixth International
Whaling Commission Meeting (Tokyo), 1954." The spectacle of the renegade Onassis
endeavor had done much to sour international goodwill on enforcement generally. This
is a complicated story, and there is a great deal of interesting archival material related
to it in the Smithsonian Institution, James Mead Files. For a general discussion, see
J. N. Tønnessen and Arne Odd Johnsen, *The History of Modern Whaling* (Berkeley:
University of California Press, 1982), 533–61, and Richard Ellis, *Men and Whales* (New
York: Knopf, 1991), 431–33.

time.... How the Russians must laugh!" he wrote to Kellogg in fury.[10] Then, during the 1955–1956 season, a Norwegian catcher crew caught sight of a Soviet factory vessel hauling an obviously undersized whale aboard. When confronted, red-handed (literally), the Soviets "defended the action on the ground that it had been caught for scientific purposes."[11] The event gave rise to a diplomatic fracas, and the Norwegians promptly took the matter to the IWC, drafting a memo calling for more stringent oversight on scientific whaling. The question was taken up in earnest by Mackintosh's Scientific Sub-Committee in 1957. Should the taking of whales for scientific purposes be permitted out of season? Were reports of results necessary? Were whales thus taken (the Soviets had by this point pushed their scientific cull above 65 whales in a single season) to be counted against the Antarctic quota? Was there any possibility that the Scientific Committee might have a hand in defining licit and illicit uses of the scientific exemption?

Noting that a "large number of permits has been issued, but the results have not ... reached the Commission," the Scientific Sub-Committee was inclined to encourage the IWC to nudge participating governments on this matter.[12] Moreover, there seemed to be no reason to think, in the Sub-Committee's view, that it was ever necessary to conduct such catches outside of the stipulated season. And finally, they recommended that such catches (which should always target "as few whales as possible") ought, on balance, to be tallied in the total Antarctic quota.[13]

At the full Scientific Committee meeting three months later, however, things no longer seemed so clear. Further consultation with the delegates and the international lawyers revealed that counting scientific catches in the quota would be impossible without an amendment to the ICRW itself, since Article VIII spelled out very neatly that whaling under this provision "shall be exempt from the operation of this Convention." That careful early effort to hold the domain of regulatory politics away from the activities

10. Salvesen to Kellogg, 12 August 1955, Smithsonian Archives, RU 7165, box 24, folder 1, "Sixth International Whaling Commission Meeting (Tokyo), 1954." Kellogg concurred that "a technical examination of six whales on each of two days would be impossible." Kellogg to Salvesen, 8 September 1955, in ibid.

11. See Foreign Service dispatch no. 1335, 9 December 1955 (sent to Amembassy London), para. 4, Smithsonian Archives, RU 7165, box 24, folder 6, "Eighth International Whaling Commission Meeting (London), 1956."

12. For a list of all permits granted under the Article VIII provision through 1955, see IWC / 7 / 14.

13. See Scientific Sub-Committee report for 1957, IWC / 9 / 2, para. 10.

known as "science" had succeeded all too well: it was ironclad, having been drafted in such a way as to make any whaling called "scientific" untouchable. Could the Scientific Committee perhaps remedy this matter by offering a "practical definition" of "scientific research in this context"? For instance, did it include "such things as technical methods of the industry" (an assumption under which several nations had already begun to operate)? If it did, then it might be possible to take whales for the purpose of a "scientific investigation" such as cooking them at slightly different temperatures, or with the addition of new catalysts. And what about testing new harpoons? Was that whale science?[14]

Private notes prepared by the US delegate to the Technical Committee in 1957 (where this matter was also discussed) open a window onto these testy and finally fruitless disputations.[15] It was evident from the exchanges that "the Norwegian and U.K. companies believe that the Russians and Japanese have taken whales out of season not only for scientific purposes but for processing," and that a "failure to publish their scientific findings" amounted to a de facto invalidation of Article VIII authorization. But there was in fact no requirement within the article for formal publication, and thus this effort to invoke (tacitly) a conception of science grounded in the idea of peer review foundered on the reef of "discretionary power." When it became clear that several industry representatives "believed that scientific purposes should include exploration of industrial techniques," the US representatives were forced to acknowledge that the language of Article VIII admitted such an expansive interpretation, but they nevertheless pressed for a reading more in line with what they characterized as its original intent: "The point was raised by the U.S. that . . . at the time of preparation of the Articles of the Convention it was in the minds of the framers that scientific research is *essentially biological research*."[16]

There was, however, not much the United States (or Norway, or the United Kingdom) could do. Kellogg's own instructions from the State De-

14. The above quotes come from the Scientific Committee report for 1957, IWC / 9 / 13, para. 12. New harpoons (which worked—or rather did not work—by electrocution) were, in fact, tested under Article VIII.

15. The notes are unsigned, but I would guess that they were prepared by Captain D. T. Adams (one of the two US representatives to the Technical Committee). These notes appear in the Smithsonian Archives, RU 7165, box 18, folder 2, "London—Ninth International Whaling Commission Meeting, 1957—Documents, preparatory material, correspondence, reports, notes."

16. Ibid. Emphasis added.

partment as to how he was to handle this issue left little wiggle room, charging him to "oppose any attempt to secure Commission endorsement" of the kinds of limitations on scientific whaling envisioned by the Scientific Sub-Committee, particularly as they related to coastal waters (where territorial jurisdictions would come into play—an issue roiling the State Department in this period, as I discuss below).[17] Moreover, Kellogg was reminded that any real effort to go after this problem would involve reengineering Article VIII by amendment—something no one was eager to attempt.

In the end, the Scientific Committee and the Technical Committee both balked at the idea of trying to define whale science. As the Scientific Committee put it (abandoning the Sub-Committee's recommendations), "it would be very difficult to make any such definition or to draw a line between one branch of science or another."[18] Defending this newfound agnosticism before the IWC as a whole, Mackintosh went so far as to root a maximally expansive notion of research in the very idea of science itself, which was ever ready to push against familiar boundaries. As he put it, "We felt that it would be *anti-scientific* to make any limit on scientific research of any kind," and therefore scientific whaling must remain an unregulated domain, open to free and wide-ranging inquiry.[19] It was a striking use of the rhetoric of science as an ever-expanding enterprise, inimical to stricture and orthodoxy: here such notions were invoked by way of justification for a wholesale refusal to define the boundary between "disinterested" science and "interested" whaling. It was left (ironically) to the *Technical* Committee (i.e., *not* the Scientific Committee) to urge each member government to "interpret the term 'scientific research' just as narrowly as it can," and the IWC as a whole put the messy business aside with a few platitudes concerning the high importance of interpreting the phrase "scientific research" with "the fullest sense of responsibility."[20]

17. J. W. Hanes Jr. to Kellogg, 21 June 1957, "for official use only," para. 12, Smithsonian Archives, RU 7165, box 18, folder 2, "London—Ninth International Whaling Commission Meeting, 1957—Documents, preparatory material, correspondence, reports, notes." It is important to note that Kellogg often had a hand in drafting his own instructions.

18. Scientific Committee report for 1957, IWC / 9 / 13, para. 12.

19. Scientific and Technical Committee recommendation, transcripts, IWC / 9 / 14C, p. 47. Emphasis added.

20. Ibid., p. 51. The latter quote appears on p. 35 of Kellogg's report to the secretary of state for 1957 as a summary of the matter; see Smithsonian Archives, RU 7165, box 28,

Within a few years whole "schools" of whales were being taken under Article VIII, and the Scientific Committee was again struggling unsuccessfully to position itself to control what it decried as the "taking of much larger numbers of whales under this Article than in the past."[21] The issue remained a lingering and prominent reminder that conservation-oriented scientists had proved totally incapable of controlling the boundary of whale science in a way that facilitated their aims: their early enthusiasm for protecting their work from political interference had resulted in a deeply damaging situation, one that made a continuous mockery of the very idea that IWC science could provide a privileged or disinterested arena beyond the ongoing contest for more whales.

In the wake of the scientific whaling controversy and other struggles at the 1957 IWC meeting, Kellogg saw little room for optimism, and it is interesting in this context to note his heightened emphasis that year on what he had begun to call "moral considerations." While a decade earlier it had seemed plausible that a new international regulatory regime grounded on science would gradually bind diverse nations and interests in common standards for action, by 1957 it seemed imperative to yoke the promise of scientific rigor to the pathfinding power of ethics if anything was going to be done to avert a "catastrophe." A moralistic desperation was replacing an earlier technocratic optimism. Rising to address the delegates with a prepared speech on the final day of the meeting, Kellogg invoked his experience, reminding them that he had been "intimately concerned over the past quarter of a century with international efforts to conserve the world's whale stocks." From this perspective, he announced, he "would vigorously and sincerely dispute any contention that we have achieved a successful balance from the standpoint of moral and scientific considerations" (and he would use the words "moral," "duty," and "obligation" half a dozen times in his brief remarks). Specifically, he had been "profoundly disturbed during the course of this session of the Commission" because "commercial considerations have impinged upon the deliberations" to an extent that placed the whole IWC at risk. "We have lost ground at this

folder 4, "Reports to the Secretary of State by the U.S. Commissioner to the IWC, Meetings #1–11, 1949–1959."

21. For discussion of the taking of whole "schools" (and a "full harem of sperm whales"), see IWC / 16 / 3, para. 9 (1964). For quote, see the 1963 "Report of the Scientific Sub-Committee," IWC / 15 / 8, para. 22–23.

session," he concluded, "in our quest for a scientifically defensible balance of biological facts and commercial desires." What lay ahead was not promising.[22]

He was right to be concerned, since by 1957 the wheels looked very likely to come off the IWC as a whole. Increasing resistance to the Antarctic quota (which by this point had been squeaked down to 14,500 BWU) and sharp disagreements about how that quota was to be divided among the whaling nations (in negotiations outside the IWC, since the ICRW prohibited quota allocations to particular countries within the IWC regulations) had led to grumblings in Norway, Japan, and the Netherlands about a possible withdrawal from the organization.[23] Once any of these major whaling states pulled out of the agreement, it would be very difficult to keep the others in, and genuinely unregulated whaling loomed as a real possibility. Even a formal objection to a given year's quota by any one country (permitted under the ICRW, and resulting in a release from the obligation for the objector) would probably blow the top off the whole arrangement, since all the other nations (again, by rule) were guaranteed the opportunity to revisit their commitment to whatever principle another signatory ultimately rejected, and once anyone was outside of the quota system, no one else could afford to be.[24] This was the "catastrophe" that Kellogg and others saw clearly on the

22. This and the quotes above are from "Remarks by Dr. Remington Kellogg at plenary session," Smithsonian Archives, RU 7165, box 18, folder 2, "London—Ninth International Whaling Commission Meeting, 1957—Documents, preparatory material, correspondence, reports, notes." Tønnessen and Johnsen refer to this speech, calling it "the deep felt words of the father of the Convention faced with the tragedy that an important part of his life's work had failed." J. N. Tønnessen and Arne Odd Johnsen, *The History of Modern Whaling* (Berkeley: University of California Press, 1982), 586.

23. All of these maneuvers are recounted in the basic histories of this period, but for a sense of how things looked from within the diplomatic process at the time, it is interesting to consult "Memorandum of a Conversation, Department of State, Official Use Only, 25 October 1956," Morgenstierne (US Ambassador to Norway) to various, Smithsonian Archives, RU 7165, box 25, folder 1, "Ninth International Whaling Commission Meeting (London), 1957." This memo details the vote for a withdrawal in the Whaling Council. Comments on the views of the Norwegian Seamen's Union, etc. give a feel for how things were shaping up in this period.

24. This sort of provision has recently been the subject of sustained investigation; see Howard S. Schiffman, *Marine Conservation Agreements: The Law and Policy of Reservations and Vetoes* (Leiden: Martinus Nijhoff, 2008), and the review of that work by Jaye Ellis (*Global Environmental Politics* 9, no. 2 [May 2009]: 131–33).

horizon in the wake of the 1957 meeting, when objections to the quota by Japan and the Netherlands (who voted against 14,500 B W U but lost) were narrowly averted.[25]

Before I return to 1958 and the endgame of this mounting crisis, I want to circle back and consider another way in which scientific principles had proved unstable ground for regulatory initiatives at mid-decade. A look at the emergence of conflicting ideas about "ecological" management in this period will shed light on the deep issues of territoriality and sovereignty that were percolating through the I W C as it faced 1958.

. . . .

To get a sense of the controversial way in which ecological principles were introduced into the regulatory arena of cetacean management in the 1950s—at further cost to the initial promise of grounding the I W C on "scientific findings"—it will be worth taking a moment to listen to the Chilean foreign minister, Fernando Garcia Oldini, as he addressed a trilateral gathering of representatives from neighboring Peru and Ecuador, the "Conferencia sobre Explotación y Conservación de las Riquezas Marítimas del Pacífico Sur," which opened on 11 August 1952. "We must remember," this gentleman told the assembled Latin American scientists, statesmen, and fisheries executives, "that the unique conditions prevailing along the shores of Chile, Ecuador and Peru have permitted the formation and maintenance of the resource which this conference will study." Expanding on this claim, Garcia went on to sketch the broad geophysical and biodynamic processes that were at least nominally the occasion of this special diplomatic-cum-scientific summit:

> In effect, the seas which border our shores constitute a biological environment formed by a double displacement of water: (1) from the Chilean Antarctic northward to Ecuador and (2) from the great depths of the Pacific upwards toward the surface along the coast. Therefore, conditions in the territorial waters of the three countries are identical and the sea becomes a single breeding ground for cetaceans of the most diverse types.

25. On Dutch-Japanese negotiations for a joint objection to the quota, see Dobson to Kellogg, 12 November 1957, Smithsonian Archives, R U 7165, box 25, folder 2, "Tenth International Whaling Commission Meeting (The Hague), 1958."

Moreover, to this "double displacement" of water it was necessary to add a third flow, which bound this fertile sea reach firmly to the mainland:

> Our continental shores provide untold quantities of food stuffs, enriched by mineral, vegetable, and animal materials which the rivers and streams of Ecuador, Peru and Chile bring to the sea and which serve to generate and maintain our marine fauna and flora.

Carefully summarizing a stirring evocation of land and sea rolling together fruitfully in eternal productive cycles, Garcia looked out to the whales both as the lords of this realm and as the tenant farmers of a distinctively Peruvian-Ecuadorian-Chilean domain:

> From one extremity to the other, the unity inherent in our coastline and our seas contributes jointly to the process of regeneration, development and conservation of the species which will occupy the attention of this conference, as is confirmed by the circumstance that, as all whale fishers have observed, whales reproduce in the warm waters of the north but must immediately return south and search out an environment and food supply which permits the development of their young. It is thus to be expected that to preserve a common resource the interested governments will find it desirable to take coordinated action.[26]

Reasonable enough, no? Indeed, even enlightened. And yet translated copies of this speech, later circulated among the delegates to the International Whaling Commission, aroused considerable anxiety, and the US State Department looked on in worried dismay. Understanding why this should be so requires a brief review of the simmering Northern Hemisphere–Southern Hemisphere resentments that basted the early negotiations around the ICRW. In the process we will begin to see how whales and a more ecologically oriented whale science were together involved in an oceanic land grab with lasting repercussions in international law.

Although both a Chilean and a Peruvian representative attended the 1946 conference in Washington, and both signed the International Conven-

---

26. This and the quotes above are taken from "Amembassy-Santiago dispatch 158, 12 August 1952," Restricted, Smithsonian Institution, James Mead Files. The translation is largely that of the State Department (I have modified punctuation).

tion for the Regulation of Whaling on behalf of their countries, neither nation moved forward with the requisite ratification process, leaving the legal
status of their adherence ambiguous. By 1950 the cause of their reservations
had been spelled out in a Chilean memorandum to the IWC requesting a
series of exemptions from the ICRW's schedule of whaling regulations.[27]
In addition to seeking a release from the strictures of a limited season (in
return for a commitment to stay within a special national BWU quota) and
permission to take sperm whales smaller than the IWC-stipulated minimum (on the grounds that they were used "for the feeding of the country,"
which was partly true—they were in fact used in cattle food), the Chilean government was adamant that the rule prohibiting the installation of
new land stations within 1,000 miles of existing establishments would be a
crushing blow to Chile's small but growing coastal whaling industry.[28] It did
not escape the notice of the South American representatives that this last
regulation openly discriminated in favor of the capital-intensive pelagic
whaling expeditions run by the wealthy industrialized nations (located in
the far north, where comparatively few whales remained) while effectively
blocking the Chileans from developing the means to exploit resources just
off their own coast. Moreover, the structure of IWC regulations permitted factory vessels that had exhausted the Antarctic season in pursuit of
baleen whales to top up their catches with geographically unlimited access to sperm whales in the off-season. Thus, in the late 1940s and early
1950s, their bumper hauls of sperms in the South Pacific further infuriated
Chilean officials, who complained (not unreasonably) that if these animals
were cleaned up in offshore waters, there would be little left for their short-
range catchers. When their continued protests met dismissive retorts from
the IWC subcommittees, the ICRW was basically shelved in Santiago and
Lima alike, and these countries refused to ratify the convention until the
late 1970s.[29]

It is against the backdrop of this mounting frustration that the 1952 remarks of the Chilean foreign minister—and indeed the whole first "Conference on the Exploitation and Conservation of the Marine Wealth of

27. See IWC / 2 / 13, report of the special sub-committee on the Chilean proposals.
See also "Whaling and Fishing in the Southern Pacific," *Norsk Hvalfangst-Tidende* 43,
no. 12 (1954), 690–705. Note that Chile had also raised concerns at the 1949 meeting (see
IWC / 1 / 28 "Report of Panel Set up to Consider the Chilean Memorandum").

28. On the elaboration of the "meat economy," see chapter 3, n. 280 above.

29. Chile and Peru first participated as full members in 1979.

the South Pacific"—must be understood.[30] Since this meeting represented nothing less than a declaration of independence from the IWC: before the conference ended, Ecuador, Peru, and Chile would together announce the formation of a new "Commission," the "Permanent Commission on the Exploitation and Conservation of the Marine Wealth of the South Pacific," a breakaway organization created as a rival regional forum that asserted the right to regulate whaling in the waters adjacent to western South America. The US State Department eyed this development with grave concern and wondered what the implications of what they called the "South Pacific Whaling Commission" would be for the ongoing efforts to nurse the IWC into an effective international organization.[31] The immediate threat to the IWC was clear: even though the actual regulations promulgated by the Permanent Commission were quite close to those of the IWC (on which they were obviously based), the mere existence of a counterconfederation of whaling nations moistened the tissue-paper alliances that held the International Whaling Commission together. Moreover, it was a considerable embarrassment to the IWC's champions that the Permanent Commission harped so incessantly on what they alleged to be the "sad experience of irrational exploitation of the seas" in general and in particular, the failure of the IWC to halt the explosive growth of the Antarctic whaling industry, or to protect the interests of "poor, small countries."[32] Under the IWC, as the chairman of the Santiago conference pointed out indignantly, some twenty factory ships and three hundred catchers were annually taking twenty thousand whales out of what he considered "local waters," and

30. Note that the US State Department (Smithsonian Archives, RU 7165, box 23, folder 6, "Fifth International Whaling Commission Meeting [London], 1953") gives the title of the event as the "Conferencia de la Explotacion y Conservacion de las Especies [sic] Maritimas del Pacifica [sic] Sur" (without diacritics).

31. For the US State Department's concerns (and a close analysis of conflicts between Permanent Commission regulations and those of the IWC), see William C. Herrington to Kellogg, 12 April 1954, with attachments, Smithsonian Archives, RU 7165, box 24, folder 1, "Sixth International Whaling Commission Meeting (Tokyo), 1954." This is also where the Permanent Commission is called the "South Pacific Whaling Commission."

32. The "sad experience" quote comes from a translation of Garcia's speech, included in "Amembassy-Santiago dispatch 158, 12 August 1952," Restricted, Smithsonian Institution, James Mead Files. For the other phrases, and the spirit of the remarks, see the summary of the speech by Ruiz Bourgeois in the annex to "Confidential" correspondence from the British Embassy, Santiago to Foreign Office, 17 October 1952, Smithsonian Archives, RU 7165, box 23, folder 6, "Fifth International Whaling Commission Meeting (London), 1953."

thereby producing 90 percent of the world's whale oil. Meanwhile, all the land stations of the western South American countries combined barely reached 1 percent of this total production.[33]

But if the Permanent Commission's direct threat to the IWC was palpable, there remained an even more grave matter that whispered through the documentation coming out of this new organization, a matter that pushed Latin American whaling policy (and whale biology) right to the forefront of the US foreign policy agenda in this period: the small matter of those novel and expansive *claims to state property interests* in vast swaths of the ocean. At stake there was the fraught issue of territorial seas, a problem that rippled in concentric rings across the geopolitics of the Cold War world in the years to come.

Hints of this issue percolated through Garcia's opening address, and the properly tuned ear heard the oblique references: those invocations of the special conditions of the Humboldt Current, the commingling of three waters off the coasts of the three nations, and the migratory paths of the whales that circulated through this unique ocean region (defining it as an integral whole by means of their peregrinations). All this tended to reinforce the key notion—namely that, as Garcia put it, "conditions in the *territorial waters* of the three countries are identical."[34] And just how far did these "territorial waters" extend? That was the sensitive question. Garcia's remarks gave a clue when he announced that "for the conservation and protection of their natural resources, and for the regulation of their exploitation, in order to obtain greater advantages for the collective welfare, Ecuador, Peru and Chile *assert their sovereignty over the splendid wealth that nature has provided them.*"[35] Where the whales went, Garcia seemed to be saying, the flags of the Andes followed.

But reading between the lines like this was not even necessary. Very close to the top of the agenda for the Santiago conference was the issue of establishing a common policy with respect to claims of jurisdiction over "marginal seas within 200 miles of the coast," and before the conference had closed, snickering remarks were being offered about the old—indeed,

33. These statistics are from "Confidential" correspondence from the British Embassy, Santiago to Foreign Office, 17 October 1952, Smithsonian Archives, RU 7165, box 23, folder 6, "Fifth International Whaling Commission Meeting (London), 1953."

34. "Amembassy-Santiago dispatch 158, 12 August 1952," Restricted, Smithsonian Institution, James Mead Files. Emphasis added.

35. Ibid. Emphasis added.

"archaic"—limit of territorial waters ("derived from the three-sea-mile range of the cannons of the 17th century"), which urgently needed to be "amplified" in accord with modern times.[36] Recent negotiations within the Inter-American Juridical Committee at Rio de Janeiro were cited in support of the propositions that a much-expanded "right to the resources of the sea, soil and sea beds" had become good law across the American continent, and that two hundred miles was taking shape as a proper zone within which to "enforce legal rights of control and vigilance."[37] The traditional principles of freedom of the seas would, naturally, be preserved (in the form of a broad right to unmolested navigation), but it appeared that international "freedom to whale" was going to come off the table. This threat was serious enough, but US State Department officials opined sagely that the broader "freedom to fish" was also (silently) at issue, and that while "no mention has been made so far of tuna fisheries," this "omission may be to spare the feelings of the United States, in view of the fact that the United States is not active in whaling, while it has strong interest in tuna matters."[38] Presumably much else, from submarine listening stations to petroleum resources, was similarly up for grabs.[39]

The origin of 200-mile offshore zones, the various legal rights asserted over these spaces, and the numerous efforts to codify these diverse claims in the formal instruments of international law constitute an enormously complex (and in some respects unresolved) episode in twentieth-century international relations.[40] My purpose here is not to rehearse these issues

36. See remarks of Ruiz Bourgeois, cited in the "Confidential" correspondence from the British Embassy, Santiago to Foreign Office, 17 October 1952, Smithsonian Archives, RU 7165, box 23, folder 6, "Fifth International Whaling Commission Meeting (London), 1953."

37. For these quotes, see conference documents cited in "Whaling and Fishing in the Southern Pacific" (n. 27 above), in Smithsonian Archives, RU 7165, box 24, folder 2, "Seventh International Whaling Commission Meeting (Moscow), 1955."

38. These views are expressed by Philip M. Davenport, First Secretary of the US Embassy in Santiago, in "Amembassy-Santiago dispatch 158, 12 August 1952," Restricted, Smithsonian Institution, James Mead Files.

39. On underwater listening systems such as SOFAR and SOSUS, see chaps. 15–17 of Gary E. Weir, *An Ocean in Common: America Naval Officers, Scientists, and the Ocean Environment* (College Station: Texas A&M University Press, 2001).

40. There is, of course, an extensive literature on this topic. I have made use of the following works in particular: S. N. Nandan, "The Exclusive Economic Zone: A Historical Perspective," in *The Law and the Sea: Essays in Memory of Jean Carroz* (Rome: Food and Agriculture Organization of the United Nations, 1987); Douglas M. Johnston,

or to try to puzzle out how the Truman proclamations of 1945 concerning coastal fisheries and the continental shelf were echoed and reworked by Arab and Latin American states in subsequent years.[41] The full range of diplomatic, strategic, and economic considerations threaded through these developments is comfortably beyond the scope of this book. What I am concerned to show—particularly in light of the considerable importance accorded the Santiago declaration of 1952 in standard accounts of the origin of what became 200-mile "Exclusive Economic Zones"—is the extent to which this increasingly disruptive extension of national rights into the oceans was grounded (perhaps better, "anchored") in scientific claims about marine ecology and the proper management of cetacean resources.

Much to the infuriation of the US officials who had, in the wake of World War II, worked to install their vision of global fisheries regulation orchestrated by international commissions ruled by scientific findings, the insurgent program of the nations of the Permanent Commission was both to declare that project (in the form of the IWC) a colossal failure in practice and to call into question the scientific principles on which it rested. The question of whether their competing scientific program is best understood as merely a rickety prop to "petty nationalism"[42] (as those trying to shore

---

The International Law of Fisheries: A Framework for Policy-Oriented Inquiries (New Haven, CT: Yale University Press, 1965); and Ann Hollick, "The Origin of 200-Mile Offshore Zones," *American Journal of International Law* 71, no. 3 (July 1977): 494–500. More generally, see N. Papadakis, *International Law of the Sea: A Bibliography* (Alphen aan den Rijn, Netherlands: Sijthoff & Noordhoff, 1980) and Andrés A. Aramburú Menchaca, *Historia de las 200 millas de mar territorial (evolución de una doctrina continental)* (Piura, Perú: Universidad de Piura, 1973). Since the "Exclusive Economic Zone" is an important principle of the 1982 Convention on the Law of the Sea (which emerged out of the nearly decade-long process initiated in 1973 with the convening of UNCLOS III), a great deal of literature on the topic can be found in the documentation related to that convention (which has not yet been ratified by the United States and a number of other states).

41. The Truman proclamations are numbers 2667 and 2668 of 28 September 1945. For more on this complicated story, see Harry N. Scheiber, ed., *The Law of the Sea: The Common Heritage and Emerging Challenges* (The Hague: Kluwer Law International Publishers, 2000).

42. For the phrase "petty nationalism" as applied to the Permanent Commission nations in this regard, see Confidential correspondence from the British Embassy, Santiago to Foreign Office, 17 October 1952, Smithsonian Archives, RU 7165, box 23, folder 6, "Fifth International Whaling Commission Meeting (London), 1953."

up the IWC were inclined to believe) or rather as part of an ambitious and early move toward a more holistic and integrated conception of natural resource management is perhaps best left in suspension at present. What is certain, however, is that the scientific vision of the Permanent Commission in the mid-1950s—so widely at variance with that of the IWC (and the US State Department) in fundamental respects—did little to enhance overall confidence that scientific consensus could serve as the point of departure for international regulatory agreements in the period.

For a clearer sense of this rival scientific vision, we can turn to a brief prepared in 1955 by representatives of the Permanent Commission nations as they sought to clarify their distinctive approach to the problems of marine conservation. In contradistinction to the US / IWC view that ocean resources were best thought of as a vast basket of goods, there for the taking by everyone interested and capable (provided they submitted to cooperative regimes of maximum sustainable yield management on a per species basis), the Peruvians, Ecuadorians, and Chileans were offering a claim that a particular superorganismal biological entity—a *bioma*, as they called it— "appertains fundamentally to the coastal country to the ecologic system of which it belongs."[43] The bold move here is the use of an expansive conception of "ecology" to authorize an expanded state jurisdiction. Or, to put it another way, the Permanent Commission nations were drawing on cutting-edge ideas in the biological sciences as part of a defiant and (in their view) counterhegemonic bid for control of ocean resources. As they explained it in a didactic memorandum:

> We should like to note that, for some time now, neither the sea, nor the aggregate of living beings that inhabit it, nor the lands adjoining it, nor the existing climate and hydrological conditions, are any longer the subjects of separate and independent studies. Modern biology and ecology have determined the close interrelation which exists between the above-mentioned elements, and also how they exert their influence to create, in the course of time, a special condition of life which is proper to the region under study.

43. This quote and those that follow come from CEP (Chile, Ecuador, Peru) Document (2), 22 September 1955, in "Santiago Negotiations on Fishery Conservation Problems," Department of State, Smithsonian Archives, RU 7165, box 17, folder 4, "London—Eighth International Whaling Commission Meeting, 1956—Preparatory material, correspondence, documents, notes, etc."

This special kind of scientific object, the memo went on to explain, was known as an "eco-system," and the ecosystem in turn supported a whole "complex of living communities"—the *bioma*.[44] The most recent ecological studies thus having recognized the need for a reflexive and totalizing science (one that included even human beings within a properly expansive study of nature), it was clear to the delegates of the Permanent Commission that "a perfect unity and inter-dependence exists between the communities that live in the sea, which supports their life, and the coastal population which requires both to survive." Summarizing what they took to be the legal implications of this analysis, the Permanent Commission concluded as follows:

> This is, in short, the concept of biological unity from which is derived, in the scientific field, the preferential right of coastal countries. According to this concept, the human population of the coast forms part of the biological chain which originates in the adjoining sea, and which extends from the microscopic vegetable and animal life (fitoplankton [*sic*] and zooplankton) to higher mammals, among which we count man.

In passing a copy of this statement on to Kellogg for his comments, a prominent State Department official expressed a mixture of skepticism and wry admiration for the whole scheme, which looked likely to entangle US diplomatic ambitions in the subtle webs of abstruse ecological theory: "The attached are about as close as I've been able to get to the Chilean theory on the whale stocks off their coast," the under-secretary explained. "This crew is too shrewd to get specific enough that they can be pinned down."[45]

Much as US officials clearly hoped to undermine and discredit what they saw as the Permanent Commission's eco-ploy, it could not be denied that the Andean nations were mobilizing scientific ideas and terminology attracting considerable interest in the United States in this period. In the late 1940s and early 1950s, as Eric Mills has shown, Gordon Riley—G. Evelyn

44. The reference here is to the "biotic community" that Frederic Clements and Victor Shelford called a "biome" in their 1939 text *Bio-Ecology* (New York: John Wiley & Sons, 1939).

45. Fred Taylor to Kellogg, 29 June 1956, Smithsonian Archives, RU 7165, box 17, folder 4, "London—Eighth International Whaling Commission Meeting, 1956—Preparatory material, correspondence, documents, notes, etc."

Hutchinson's first graduate student—was pioneering a new kind of biological oceanography by applying the integrating insights of limnology and the mathematics of predator-prey cycles in population ecology to the problem of modeling productivity in coastal environments.[46] Moreover, these efforts to unravel the physical and biological processes driving plankton dynamics—and with them the larger food cycles of marine environments—represented merely the latest and most successful contributions to a much older program of studying the "ecology of the seas," a project that had long-standing ties to the study of whales. In fact, Mackintosh's junior Discovery colleague Alister Hardy had launched his (rocket-ship) career by developing new technologies for the study of plankton blooms, and he had expended a great deal of effort trying to correlate water movements, ocean chemistry, krill blooms, and whale concentrations on maps of the Scotia Archipelago in the Antarctic—with some success.[47] He went on to become one of very few marine ecologists to play an active role in the British Ecological Society, serving as vice president under the gray eminence of the field, A. G. Tansley, and developing close intellectual ties to Charles Elton.[48] As far back as the 1920s, the baleen whales—the largest animals

46. See chaps. 10 and 11 of Eric L. Mills, *Biological Oceanography: An Early History, 1870–1960* (Ithaca, NY: Cornell University Press, 1989).

47. For a useful (if late) summary of his thinking on these questions, see chaps. 15, 20, and 21 of Alister Clavering Hardy's *Great Waters: A Voyage of Natural History to Study Whales, Plankton, and the Waters of the Southern Ocean* (New York: Harper and Row, 1967), in which citations to the original papers are given. On his "continuous plankton recorder," a sort of mechanical whale for straining plankton on long transits, see chap. 18. (I discuss the device in "Self-Recording Seas," in *Oceanomania: Souvenirs of Mysterious Seas*, ed. Mark Dion and Sarina Basta [London: Michael Mack, 2011].)

48. See John Sheail, *Seventy-Five Years in Ecology: The British Ecological Society* (Oxford: Blackwell Scientific, 1987), particularly chap. 26. For a general sketch of the context, see Donald Worster, *Nature's Economy: The Roots of Ecology* (San Francisco: Sierra Club Books, 1977), and Sharon E. Kingsland, *The Evolution of American Ecology 1890–2000* (Baltimore: Johns Hopkins University Press, 2005). Alister Hardy's work before joining the Discovery Investigations included a paper entitled "The Herring in Relation to Its Animate Environment" (*Fishery Investigations*, series 2, vol. 7, no. 3 [1924]: 53), so he came to whale work with an eye on grounding the life cycles of macrofauna in planktonic studies like those of the Kiel school. For Hardy on Elton, see Alister Clavering Hardy, "Charles Elton's Influence in Ecology," *Journal of Animal Ecology* 37 (1968): 3–8. The following review essays are useful for thinking about the history of ecology and its relationship to environmental history: Kevin Dann and Gregg Mitman, "Exploring the Borders of Environmental History and of the History of Ecol-

on earth, which made their living on some of the smallest (plankton)—presented an exemplary case study with which to dramatize the remarkable chains of interdependence that bound the natural word. Moreover, the very sparseness of life in the Antarctic as a whole made it a particularly suitable site for the study of ecosystem dynamics. As the defenders of *Discovery II* put it in 1950, pumping for a new expedition, "There is a relative simplicity in the water circulation and distribution of life in the Southern Ocean, so that generalizations can often be made from fewer observations than in any other regions, and principles established which apply to all oceans."[49]

In short, then, the ecological view of the Antarctic and its whales traced its ancestry back to the same early twentieth-century scientific investigations that had given rise to Mackintosh's flensing-platform whale biology. The latter held sway in the IWC, however, while the Permanent Commission's efforts to puff new life into a holistic and integrated whale science were perceived by IWC defenders as little more than an oceanic land grab in the name of the "biome."[50] As I have suggested, much more than the IWC

ogy," *Journal of the History of Biology* 30, no. 2 (1997): 291–302; Sharon E. Kingsland, "The History of Ecology," *Journal of the History of Biology* 20, no. 2 (1994): 349–57; and Malcolm Nicolson, "No Longer a Stranger? A Decade in the History of Ecology," *History of Science* 26 (1988): 183–99.

49. Admiralty press release, 5 April 1950, cited in "Amembassy dispatch 1793, 11 April 1950," Smithsonian Archives, RU 7165, box 22, folder 7, "First International Whaling Commission Meeting (London), 1949." See also Mackintosh's comments on the ecology of whales in his memorandum of 19 June 1951, NOL GERD.M2/1.

50. For a choice instance of an actual showdown of sorts between the two approaches, it is interesting to consider the flurry of correspondence in 1957 concerning how best to place an IWC-sympathetic whale biologist in the service of an American company (Archer Daniels Midland) undertaking a joint venture for sperm whales in Peru. When the subject first came up, a number of IWC partisans objected to the idea of helping the Peruvians study cetacean biology at all, since it seemed likely that "the object of the studies would be to fortify claims for the 200-mile offshore conservation limit." See Norwegian opinion cited in "Amembassy dispatch 131, 15 July 1957," Smithsonian Archives, RU 7165, box 25, folder 2, "Tenth International Whaling Commission Meeting (The Hague), 1958." In the end, the State Department realized the advantages of putting advisers in place who would have strong ties to the kind of whale science pioneered in Britain. See memorandum from Fred Taylor to Kellogg, 18 February 1957, Smithsonian Archives, RU 7165, box 18, folder 2, "London—Ninth International Whaling Commission Meeting, 1957—Documents, preparatory material, correspondence, reports, notes": "My horseback reaction, derived from the built-in suspicion of a lawyer, is that: (1) the Latinos are seriously concerned with proper conservation of the sperm whale stocks in question; (2) they want to accomplish this on the scientific basis within their closed

was at stake, since the mid-1950s saw increasing international concern not only about offshore fishing (witness the 1955 Rome Fisheries conference, sponsored by FAO), but also more generally about the whole legal framework regulating ocean activities.[51] By 1955 the United Nations' International Law Commission was deep into a drafting process that would eventually culminate in the 1958 United Nations Conference on the Law of the Sea (UNCLOS I), the first of three such gatherings over a quarter century—all of which would struggle (and ultimately fail) to achieve consensus on the definition of territorial seas.[52] Over these years, Permanent Commission-

---

corporation." The US view is best summed up in State Department Instruction a-89, 2 August 1957, Smithsonian Archives, RU 7165, box 25, folder 2, "Tenth International Whaling Commission Meeting (The Hague), 1958":

> The United States would like to suggest to British authorities that the educational effect of scientific knowledge of sperm whale life history and conservation, which would be contributed by competent and well briefed biologist, should help to develop a sounder and broader Peruvian attitude toward the problem of whale conservation and the value of cooperation with other whaling countries. The present view of Peruvian authorities is that the sperm whale stock off the coast of Chile, Ecuador and Peru are [sic] independent of stocks fished by other countries in the Antarctic. The international whaling commission experts do not agree with them. Depending upon the qualifications of the scientists selected for the work, the department does not, therefore, view the proposal as one which would necessarily, or even likely, fortify the 200-mile thesis which has been largely predicated on the need for independent whale conservation in the area off the western part of South America.

The right kind of science or scientist, in sum, could help bring the Permanent Commission around. In the end, if I am not mistaken, this position went to Robert Clarke, who eventually made Peru his permanent home. Initial funding of Clarke's position was through FAO. For a sense of his initial struggles in-country—trying to get time on vessels, holding off Japanese rivals, etc.—see "Amembassy Santiago, dispatch, 15 October 1958," Smithsonian Archives, RU 7165, box 25, folder 5, "Eleventh International Whaling Commission Meeting (London), 1959." A similar issue arose concerning Ecuador at about the same time. See State Department Instruction a-63 22, October 1957, Smithsonian Archives, RU 7165, box 25, folder 2, "Tenth International Whaling Commission Meeting (The Hague), 1958."

51. See Philip E. Steinberg, *The Social Construction of the Ocean* (Cambridge: Cambridge University Press, 2001), and Harry N. Scheiber, ed., *The Law of the Sea: The Common Heritage and Emerging Challenges* (The Hague: Kluwer Law International Publishers, 2000).

52. For a sense of the drafting process circa 1955–1956, see "Comments of the United States on chap. II and III of the Report of the International Law Commission," seventh session, Smithsonian Archives, RU 7165, box 17, folder 4, "London—Eighth International

style expanded oceanic jurisdictional zones would creep deeper into the law of nations. Even as the United States worked relentlessly (using the promise of MSY science—which seemed to suggest that everyone could win at once) to head off the kind of "biotic territoriality" espoused by a growing number of small countries (increasingly disgruntled by offshore exploitation of ocean resources by large refrigerated trawlers), the Soviet Union nurtured development alliances with these frustrated Third World nations. The rhetoric of resistance to rapacious capitalist exploitation served the collective interest of this emerging bloc.[53] The ecological protestations of the Permanent Commission thus served as the scientific rallying point for an expanding alliance of developing countries who saw the management regimes of international commissions (forever touting the benefits of "maximum sustainable yield") as little more than window dressing on a hegemonic scheme. If these nations' actual use of ecological ideas and methods had about it something of the magpie (picking here and there the brightest bits in an extensive field, without a great deal of quantitative or experimental work), they could protest that they had neither the ships to do the requisite oceanographic work, nor, indeed, the atomic bombs that were proving useful (in Pacific test sites) for revealing the full complexity and extent of trophic cycles in marine environments. Magpie-like their *bioma* theory might be, and yet these magpies were intent on building not a nest, but rather a bulwark against the depredations of northern factory vessels.[54]

In this broader geopolitical-cum-scientific context, it is interesting to look a little more closely at the machinations within the IWC's Scientific Committee in the mid-1950s. Recall that in 1955 the Scientific Committee recommended the opening of the long-established southern Pacific

Whaling Commission Meeting, 1956—Preparatory material, correspondence, documents, notes, etc." Also useful generally is D. H. Cushing, *The Provident Sea* (Cambridge: Cambridge University Press, 1988), particularly chap. 12.

53. Sergei B. Krylov was a key Soviet figure in these debates, as was Padilla Nervo of Mexico. Other nations in the mix included Czechoslovakia, India, Brazil, and Bolivia. See William C. Herrington's memoranda of 15 and 16 June 1956, Smithsonian Archives, RU 7165, box 24, folder 7, "Eighth International Whaling Commission Meeting (London), 1956."

54. As far as the scientific merits of the *bioma* theory are concerned, it is interesting to consider the laborious refutation assembled by Warren Wooster in 1973: "Scientific Aspects of Marine Sovereignty Claims," *Ocean Development and International Law* 1, no. 1 (1973): 13–20. Since the issue remained hot in this period, it is difficult not to read this paper as a party brief.

sanctuary to whaling on a trial basis. The "conservationist" rationales for this decision—reducing some of the pressure on the other sectors of the Antarctic, collecting biological data on a virgin stock—seemed to bear the fingerprints of a perverse imp. But on deeper digging into the archive, this inscrutable action is revealed as nothing less than a wicked little game of biopolitical chicken. The sanctuary lay in the southern Pacific below 40° S and extended westward from Cape Horn, at 70° W.[55] A closer perusal of figure 4.1 reveals that this region not only enveloped southern Chile, but also extended north to encompass nearly half of that nation's latitudinal extent. This geography did not escape the United States representatives to the IWC. In early April of 1955, Kellogg wrote to Mackintosh, who was shortly to chair the Scientific Sub-Committee meeting that would put forward the recommendation to open the sanctuary. In this private letter Kellogg put the issue plainly: "As regards the Sanctuary, it is possible that the removal of the present prohibition might have a beneficial result if such contemplated action would induce Chile and Peru to adhere to the 1946 Convention." In other words, he continued, "it is conceivable that they might wish to bargain if they became aware that such action was contemplated."[56]

I have not found any positive evidence that Kellogg was put up to this exercise in bare-knuckle brinksmanship by his State Department handler-colleagues, but it seems likely. Whatever the original impetus, it is a somewhat shocking proposition, since it betrays the way in which the opening of the sanctuary served as a kind of triple experiment: in the scientific realm, it was a way to gather new data bearing on cetacean migrations; these very data, however, served as part of a political experiment that aimed to prove to the Permanent Commission nations that their *bioma* was deeply entangled with the larger circum-Antarctic environment and thus with the regulatory regime that was authorized to write the rules for hunting whales in the Southern Hemisphere; finally, for the companies (like Salvesen's) that nosed catcher boats into this new area in the 1955–1956 season, the opening of the sanctuary represented an economic experiment of considerable scope. Only the last of these three experiments proved successful, as superbly profitable returns came out of the Pacific sector in the years to come. The sanctuary was gone for good.

55. The westernmost edge of the sanctuary was set in the original negotiations at 160° W.

56. Kellogg to Mackintosh, 7 April 1955, Smithsonian Archives, RU 7165, box 24, folder 2, "Seventh International Whaling Commission Meeting (Moscow), 1955."

In general, then, it is fair to say that the mid-1950s conflict over ecological management of cetaceans further smudged the scientific advising practices of the IWC with the taint of political opportunism. For historians of science, the whole episode provides an exemplary instance of the speed with which new scientific objects—an ecosystem, say, or a "biome"—that have been, at the cost of considerable scientific labor, put forward as natural objects (things not of our making, things that transcend individual beliefs or national interests) are reintegrated into the world of getting and spending, where, in new combinations and novel garb, they become the subject of intense political rivalries. As the 1958 UNCLOS conference (and that of 1960) would reveal, if an IWC-style global regulatory system was going to survive—and not be elbowed out of the way by territorial claims to ocean wealth—it was imperative that the scientific advising system at the heart of such organizations be shown to work. As a State Department memorandum of 1956 put it,

> The U.S. is a member of The International Whaling Commission and, although it has no whaling industry of its own, it has a very real interest in seeing the Commission's whale conservation program function well. This, because to the extent that program is shown to be inadequate for its purpose, the case of the claimants to 200 miles of territorial water is reinforced. Chile, Ecuador, and Peru in particular have assigned the impotence of the International Whaling Commission as one justification for their 200-mile claims.[57]

By 1958, therefore, with a suite of withdrawals from the IWC pending and the survival of the organization in doubt, the crisis involved much more than just the world's whales: the crisis involved the whole framework of US postwar ocean policy. Getting the IWC to work was necessary in order to

57. Taylor and Parsons, "Memorandum of Conversation," 27 January 1956, Smithsonian Archives, RU 7165, box 24, folder 6, "Eighth International Whaling Commission Meeting (London), 1956." To see the same issue of the strength of the IWC linked more generally to postwar fisheries policy, see State Department Instruction a-151, 1 September 1955, Smithsonian Archives, RU 7165, box 24, folder 2, "Seventh International Whaling Commission Meeting (Moscow), 1955": "To gainsay a decision [in the IWC] which is based upon scientific considerations would, to say the least, do nothing to promote the principle that fisheries conservation should be based on scientific research and investigation, as recommended by the International Technical Conference on the Conservation of the Living Resources of the Sea held at Rome, April 18–May 10, 1955."

prove that international scientific regulation of marine resources was possible. By 1958, moreover, getting the IWC to work had come to look more and more like a numbers game. For this reason, we must now turn to the third fraught feature of IWC science in the 1950s: statistical analysis.

.   .   .   .

So how many whales were out there in the oceans? How many of them could be taken without exterminating the industry and / or the whales themselves? These fundamental questions hung over the negotiations described above, but thus far they have not taken center stage in my analysis. Strange as it may seem, this omission is in keeping with the comportment of the actors in this episode: those questions certainly loomed, but loom was mostly all they did. In the early years of the IWC there was plenty of institution-building work to be done—work that everyone recognized was required to create the conditions of possibility for collaborative, conservation-oriented decisions about regulating the hunt. That work, as I have tried to show, was hard enough, and it had to be completed before any fancy new regulations could happen. Merely making some sort of catch limit system function required a great deal of administrative infrastructure and diplomatic string pulling; getting that system to implement the "right" catch limit was a whole different (and secondary) matter. By the mid-1950s, however, this issue could no longer be deferred, since opposing views about what that limit should be (and why) were rattling the IWC apart. In order to understand how, it will be helpful to review some of the early efforts to build statistical arguments for IWC action.

I wish to place some emphasis on that word "arguments" because, as I have already suggested, it was not that the IWC wanted for statistics. In fact, the delegates and scientists alike were awash (even drowning) in statistics throughout this period. Each year, sometime between the spring closing of the Antarctic season and the opening of the summer IWC meeting, the BIWS rushed into the hands of IWC staff and affiliates a whopping sheaf of fifty pages of small-print numbers that tallied the Antarctic season's catches and oil production in several different ways (by expedition, species, region, period of the season, etc.). A second sheaf, concerned with the rest of the world, followed thereafter (the Antarctic still represented about 75 percent of the world catch in these years, so the emphasis at the IWC was always on the first document). There was therefore emphatically no shortage of data. But despite (or indeed, perhaps because of) this twice-yearly torrent of numbers, there were comparatively few efforts to mobilize these figures in statistical *arguments*.[58] I have already touched on what I take to be the best explanations for the absence of such efforts: a culture of Discovery-style whale biology that dominated (via Mackintosh) the scientific advising structure of the new organization; a general concern for institution building; and an emphasis on collaboration that militated against the introduction of forceful arguments (which could only destabilize emerging alliances by forcing representatives to take sides). But this is not to say that there was no earlier tradition of rigorous mathematical investigation of whaling data. There certainly was. In fact, the early 1930s had seen a florescence of such work, both empirical and theoretical, and some of it made lasting contributions to the whole enterprise of modeling natural populations.[59]

Johan Hjort's 1933 coauthored paper, "The Optimum Catch," here deserves pride of place.[60] In addition to applying to the data on the Icelandic whaling industry a version of Amund Helland's technique for estimating a natural population of bears from hunt records, Hjort and his collaborators also undertook a precocious investigation of the theory of surplus production, and they laid out the core insights that the American fisheries biologist Milner Schaefer would elaborate into the dominant fishery management tools of the 1950s.[61] A full treatment of the development of

58. I have found Theodore M. Porter, *Trust in Numbers: The Pursuit of Objectivity in Science and Public Life* (Princeton, NJ: Princeton University Press, 1995), useful for thinking about this aspect of the IWC's history.

59. The standard work on this subject remains Sharon E. Kingsland, *Modeling Nature: Episodes in the History of Population Ecology* (Chicago: University of Chicago Press: 1995). For the application of this work to fisheries, see the invaluable *Scaling Fisheries: The Science of Measuring the Effects of Fishing, 1855–1955* (Cambridge: Cambridge University Press, 1994) by Tim D. Smith. See also D. H. Cushing, *The Provident Sea* (Cambridge: Cambridge University Press, 1988).

60. Johan Hjort, Gunnar Jahn, and Per Ottestad, "The Optimum Catch," *Hvalrådets Skrifter 7* (1933): 92–127 (see above, chap. 2, n. 239). This paper was part of a special issue of *Hvalrådets Skrifter* dedicated to "Essays on Population," several of which merit attention, including Alf Klems's experimental work on the logistic curve, "On the Growth of Populations of Yeast," 55–92.

61. Helland's method, a sequential population analysis that assumes zero population growth and postulates that catches over a discrete interval represent a constant proportion of the overall population, is discussed in Tim D. Smith's *Scaling Fisheries: The Science of Measuring the Effects of Fishing, 1855–1955* (Cambridge: Cambridge University Press, 1994), 221–25. The skeletal surplus production theory outlined in "The Optimum Catch" proceeds from the following analysis of the behavior of natural populations: If a new population grows over time in a manner that can be represented by some form of

these techniques of population analysis is not necessary for our purposes here, but it is interesting to note that the move in "The Optimum Catch" from Helland's type of a posteriori catch data analysis to the more abstract analysis in terms of a priori models of population dynamics was prompted exactly by the difficulty of moving from the whaling industry of the Northern Hemisphere to that of the south.[62] As Hjort et al. put it, "the advantage of Amund Helland's statistical method is that it bases its results, in a purely empirical manner, upon the series of figures furnished by the statistics of the catches."[63] The Iceland whale fishery, conveniently, had caught its

---

sigmoid curve (accelerating at first before decelerating to some stable level now known as "carrying capacity"), the derivative of that function has a maximum somewhere in the middle zone of the growth curve, where the rate of population growth is highest. A population held at this level, therefore, can be said to be in a state of maximum productivity, and the catches that hold the population at this point are "optimum catches." It follows from this analysis that only by reducing a natural population by some significant amount can it be made to yield at its (sustainable) maximum. While the elaboration of such a model (for particular population growth curves, for instance, or particular ages or sizes at which individuals are first subject to exploitation) gets algebraically complicated in a hurry (and did through the 1950s, reaching an apotheosis in the "immortal algebra" of R. J. H. Beverton and S. J. Holt, *On the Dynamics of Exploited Fish Populations* [London: His Majesty's Stationery Office, 1957]), this basic observation about population size and productivity is at the core of the science of MSY.

62. It would be possible to argue that Helland's technique is really no less a "population model" than the surplus production analysis forwarded in the second half of Johan Hjort, Gunnar Jahn, and Per Ottestad, "The Optimum Catch," *Hvalrådets Skrifter* 7 (1933): 92–127, since Helland's analysis trades on the (biological) assumption (justifiable under conditions of such rapid depletion of a population, where the attrition effects overwhelm other factors) that births and deaths can be treated as perfectly offsetting. By these lights, the difference between the two approaches amounts to whether the rate of population growth is treated as a constant (here, zero) or as a function of population size. While this is fair as a description of the mathematical similarity that underlies the two analyses, eliding them this way belies the large differences in their spirit and application. Helland's technique may secrete a population model in its simplifying assumptions, but it was basically invented as a way to milk an existing data set (hence my suggestion that, like various empirical approaches, it is best thought of as an a posteriori technique). By contrast, the surplus production model forwarded in "The Optimum Catch" is much more explicitly concerned with describing population dynamics by means of the interaction of biological parameters formally identified as such.

63. In connection with n. 62 above, it is worth noting that Hjort, Jahn, and Ottestad here make explicit that they think of Helland's method as "empirical," despite the fact that it is really just, as I note above, in some sense a better-parameterized model.

whale stock down to practically nothing (as Norwegian hunters had done with their bears), and thus the run of data afforded a very suitable base for retrospective analysis of what the population once was. But the really big question now lay elsewhere: "It would of course be extremely useful if we could say how large the stock of whales is in the Antarctic Ocean, and in that connection estimate the maximum catch that is obtainable from this stock without decimating it." The problem was that, given the scope of the Antarctic industry and its booming circumpolar expansion, "it would be a hopeless task to endeavour to calculate the number of whales in the stock." It was with this sense of statistical futility in mind that Hjort and his colleagues then turned to theory in the hopes of creating the means to work, in a way, in reverse: if they could figure out what a sustainable catch would be *in principle*, as a percentage of a stock, then the actual catch could be examined with a view toward determining how big a population would need to exist to ensure that extermination was not under way. Efforts could then be made to assess whether this theoretically "necessary" population was realistic.[64] What is striking here is that the move to self-conscious "theory" accompanies the move south, and reflects what was taken to be the fundamental intractability of the data problem presented by the pelagic whaling industry in the Antarctic.[65]

Even as Hjort thus hypothesized imaginary populations (something like 230,000 blue and 75,000 fin whales seemed to be needed to support the catches of the early 1930s ad infinitum, and neither of these numbers seemed implausible), he and a number of his Norwegian collaborators also struggled to get on top of the new flow of actual Antarctic catch statistics, a flow that Hjort himself had done much to create through his lobbying at ICES and the League of Nations for international conventions for reporting and consolidating catch records.[66] These empirical efforts took shape in a series of monographic articles (coauthored by Hjort, J. Lie, Johan Ruud, and others) entitled "Norwegian Pelagic Whaling in the Antarctic," which

64. For the above quotes, and the paraphrased propositions, see Johan Hjort, Gunnar Jahn, and Per Ottestad, "The Optimum Catch," *Hvalrådets Skrifter* 7 (1933): 92–127.

65. A less charitable analysis might note that the move south really saw nothing more than the supplementing of Helland's (tacit) "model" of fixed (zero) population growth with a model of population-dependent growth rate—which, while certainly defensible by means of biological storytelling (i.e., analogy to a yeast culture, invocations of the mystical omnipresence of the logistic, etc.), also had the happy effect of suddenly adding many more whales to the Antarctic analysis, if not to the actual Antarctic.

66. See ICES Records, Rigsarkivet, box 85, folder 6Q.

ran in the *Hvalrådets Skrifter* between 1932 and 1939.⁶⁷ These studies not only tabulated catches by species and region (mapping the results by sector in a changing sequence of polar charts), they also deployed a measure of catch per unit effort (catch per "catcher's day's work," or what came to be abbreviated CDW).⁶⁸ If these papers thereby presented the most synoptic picture available of the situation in the Antarctic in the prewar period—and by doing so definitively laid to rest Sir Sidney Harmer's contention of the 1920s that the whale population was small and limited to tight migratory paths—the authors were above all impressed by the complexity of the data and their inadequacy for the purposes of stock assessment. Discovery researches had established that the population of whales moving into the southern waters displayed some segregation by size and apparently by age: bigger and older animals arrived earlier, with cohorts of juveniles following. At the same time, the whales got demonstratively bigger (fatter) over the course of the feeding season. Gunners' skill and selection effects (together with the contingencies of weather, vessel course, krill blooms, and the like), not to mention regulations that changed from year to year (such as later opening dates for the season), made it difficult to feel confident that what seemed to be trends in the data (apparently smaller numbers of larger whales in certain regions over time; a small decrease in the catch of blue whales per catcher's day's work) actually reflected changes in the stock as a whole. As the first article in the series put it, the "great problem [of the actual impact of the industry on the stock] cannot be solved by means of the material we possess at present"; the aim had to be to contribute each year "a link in a chain of similar investigations," which would eventually make it possible to "ascertain at an early stage whether a whaling ground

67. In citations they are commonly numbered in Roman numerals, I–IX.

68. This measure was introduced in Johan Hjort, J. Lie, and J. T. Ruud, "Norwegian Pelagic Whaling in the Antarctic II," *Hvalrådets Skrifter* 7 (1933): 128–52. Hjort was, through his work on the North Sea fishery, very familiar with the use of this sort of catch per unit effort index, which had been refined and deployed with considerable success by Walter Garstang in analyzing the bottom fishery off the east coast of the British Isles in the late nineteenth century. See Tim D. Smith, *Scaling Fisheries: The Science of Measuring the Effects of Fishing, 1855–1955* (Cambridge: Cambridge University Press, 1994), 102–7. It is worth noting that there is some inconsistency in period usage of the abbreviation CDW. One will sometimes find it meaning "catch per catcher's day's work," which is also sometimes given its own abbreviation, CCDW.

is exhibiting diminished whaling possibilities."[69] Catch per catcher's day's work ought to be particularly telling in this respect, and it seemed to be holding steady for the data Hjort and his colleagues examined.[70]

These prewar analyses never had any real impact on the prewar industry. This is perhaps not so difficult to explain: after all, the main index they offered, CDW (not seriously corrected at this time for the increasing power and efficiency of vessels), looked stable; even had it not, international regulatory diplomacy was just getting started in the 1930s, so there was little institutional framework within which to mobilize recommendations. Moreover, it is worth noting that the sophisticated analysis in "The Optimum Catch"—while it contributed nothing of immediate practical use for regulating the prewar industry (or, indeed, the postwar industry through the 1950s)—offered a general take-home message inimical to any sense of conservation urgency: in essence, Hjort's analysis amounted to a mathematical demonstration that a significant, real, and overall *reduction* in a population could *increase* its long-term productivity, not reduce it. This had to have seemed like good news to those organizing pelagic expeditions.

Thus the hard question is not, I think, why these early mathematical analyses did not have a greater effect on regulation before the war, but rather why a strong return to this kind of analysis was not seen in the postwar period, when it might have done some good.[71] I must confess that this question still puzzles me to a degree. Clearly the war itself has to have been

69. The first two quotes above are from Johan Hjort, J. Lie, and J. T. Ruud, "Norwegian Pelagic Whaling in the Antarctic I: Whaling Grounds in 1929–1930 and 1930–1931," *Hvalrådets Skrifter* 3 (1932): 1–47, at p. 14; the last from Hjort, Lie, and Ruud, "Norwegian Pelagic Whaling in the Antarctic II," *Hvalrådets Skrifter* 7 (1933): 128–52, at p. 139.

70. This was how the BWU tally system obscured changes in stock composition: a shift from (increasingly scarce) blue whales to (less valuable but more plentiful) fin whales could only be "seen" in the data if CDW analyses were broken down by species.

71. Writing of the post–World War II period, John Sheail notes that "thenceforth every population ecologist needed to have within his technical equipment an armchair and some knowledge of mathematics." It is a felicitous notion, the armchair as "technical equipment," and would make a nice conceit for a serious look at the failure of population ecological techniques to penetrate the world of the hip-booters. See John Sheail, *Seventy-Five Years in Ecology: The British Ecological Society* (Oxford: Blackwell Scientific, 1987), 103.

a major factor. Not only did the conflict break the "chain" of analyses just at the point where it was getting very difficult to explain away (by reference to changing opening dates for the season or by the reduced time for gunner selection afforded by shorter seasons) the declining proportion of mature blue whales in the catch; the war itself also had a devastating impact on Norwegian society and scientific institutions. There can be little doubt that cultural division, violence, and massive shortages of basic necessities destabilized the scientific community around Hjort and the Hvalrådet (Whaling Council). The collection of statistics was suspended, and one presumes the analysis of what statistics there were fell to a very low priority, not least because the nearly total suspension of hunting during the war years led many to suppose that the unmolested southern whales must be rebounding at a brisk clip. The senescence and death (in 1948) of Hjort himself—for so long the driving force behind the study of whales in Norway—cannot have helped the postwar recovery of this part of the Norwegian scientific community. Nevertheless, a number of his prewar collaborators (such as Ottestad and Ruud, coauthors of "The Optimum Catch") survived the conflict and reemerged in the 1950s as contributors to the statistical debates over whale stocks. Why their voices began to be heard only in the mid-1950s, and even then not with anything like the force (or analytic rigor) of Hjort's prewar program of stock analysis, remains something of a mystery, one that is likely to be solved only by further work in the archives of Sandefjord and Oslo.[72] It may simply be that the chaos of postwar Europe, and the urgent need for increased access to food fats, militated against overmuch concern with these matters, but other factors may have been in play.

What is certain, however, is that with the postwar rise of the IWC and the Anglo-American dominance in the new organization, leadership on scientific advising fell to Mackintosh and Kellogg (the latter more remote from research, but at the same time more powerful as a presence in the IWC). Mackintosh may have been "the fountainhead for basic data," and Kellogg nothing less than the "most influential person in the Commission" as a whole, but neither of these men had any great facility with, or appetite for, the kinds of mathematical analysis—either number crunching with large

72. Though I remain stumped as to why there is nothing in the Bergersen-Kellogg correspondence about this problem. Kellogg's own correspondence with Hjort appears to terminate in 1933.

data sets or neat manipulations of differential equations—pioneered by Hjort and his group in the 1930s.[73] While a young Kellogg had written Hjort in 1933 expressing admiration for the "high standard" set by the research in "The Optimum Catch," I can find little evidence in Kellogg's scientific notebooks that he had any deep appreciation for the regulatory possibilities of surplus production theory. His contacts at the State Department and the Department of the Interior, several of whom came out of the world of Pacific fisheries regulation (as I have discussed above), were almost certainly more familiar with the way MSY models worked than was Kellogg himself. While we have seen that by the mid-1950s Kellogg was pushing Mackintosh in the Scientific Committee as to whether some serious statistical analysis might not take precedence over still more marking surveys and organ histology, the fact remains: Kellogg was neither a population ecologist nor a statistician and was ill-equipped to undertake, or even advise, such studies. His IWC paperwork is covered with handwritten tallies of catch data, but never do these scribbles feature a mathematical operation more complex than the calculation of percentages.[74] The IWC's original 16,000 BWU quota was set, as we have seen, as a compromise figure, to establish a principle, and the negotiators arrived at that number not through any model of population dynamics or study of trends in catch per unit effort, but rather simply by multiplying the average of a series of prewar catches by two-thirds.

Mackintosh was, if anything, even less inclined to pursue Hjort-style mathematical analyses of the problem of whale stocks, though he certainly read all of the *Hvalrådets Skrifter* articles of the 1930s. In fact, his reading notes on these materials survive, and they afford useful insights into

73. The quote about Mackintosh appears in Fred Taylor to Kellogg, 18 February 1957, Smithsonian Archives, RU 7165, box 18, folder 2, "London—Ninth International Whaling Commission Meeting, 1957—Documents, preparatory material, correspondence, reports, notes." (NB: The folks at the BIWS would surely have chafed at this formulation.) Kellogg is offered his flattering designation in a forwarded message from the prime minister of New Zealand, pressing for US leadership on the impasse of 1959 (Department of State outgoing telegram, 29 June 1959 [sent to Amembassy, London], Smithsonian Archives, RU 7165, box 25, folder 3, "Eleventh International Whaling Commission Meeting [London], 1959").

74. Though, notably, he did deploy a scratch-pad distribution curve to demonstrate that inspectors had to be fudging minimum sizes (there was a suspicious lump in the curve, representing a large number of just-legal whales).

his attitude toward these approaches.[75] Of "The Optimum Catch," Mack-intosh noted that the upshot seemed to be that "a calculation of the stock of Antarctic whales would be hopeless" using the methods outlined in the paper, and thus Mack concluded his comments to himself with a tellingly dismissive remark about theoretical models: "The whole paper's object is not to draw conclusions but to expound a method of studying the stock and its relation to hunting." Mackintosh seemed to have little sense that such a method could itself be a scientific "conclusion." Mackintosh was evidently more interested in the series of statistical articles by Hjort, Lie, and Ruud (he took a total of seven pages of handwritten notes on the five articles), though he was also inclined to quibble, particularly about the quality of the maps. This is significant, since it underscores the degree to which Mackintosh—trained in general zoology in the early twenties and steeped soon thereafter in the chart-oriented naval culture of Discovery—retained a strong commitment to biogeographic analysis.[76] This feature of his thinking is closely tied to his durable preoccupation with migrations and marking (as we have seen) and is further evidenced by the scores of annotated mimeographed maps that remain in his papers. He liked plotting data on charts; with purely statistical operations or "formulas" (as he called the material in "The Optimum Catch") he was never much at ease.[77]

At the same time, it is interesting to note that Mackintosh's preoccupa-tion with spatial distribution ended up playing a surprising and important role in the larger problem of stock assessment in the 1950s. Even as several

75. I found this notebook in the SGB Papers; see "Abstracts of Literature—Whales," sheaf R.

76. As did Hardy. There is a later version of Rozwadowski's argument about the maritime culture of oceanography to be made here. These men lived with charts for a decade, as they would not have done had they, say, gone into museum work after taking their degrees.

77. In one interesting interview with a former colleague of Mackintosh's (John Bannister, a Hardy student who worked on whales at NIO / WRU in the early 1960s), I was told that Mackintosh was "in many ways a very good geographer, skilled at work-ing with maps and histograms." John Bannister, interview by the author, 26 October 2005. As for the paucity of statistical training among the early Discovery recruits, it is valuable to review the 1928 correspondence between Kemp and Fraser (NOL Discovery MS, green filing cabinet, file "Marine Station—Correspondence, etc. 1928"), such as the letter of 16 October in which Fraser noted, "We thought that a little more knowledge in mathematics might be advisable, our example being based on an example shown in a statistics book we have here." Fraser went on to request that the recruits be sent a primer on mathematical methods useful for biological research.

new efforts were beginning to be made to get a handle on the condition
of the Antarctic stocks in this period (deploying CDW indices or simple
models using reproductive and mortality rates derived from biological
studies and / or catch statistics), it was Mackintosh and a collaborator, Sidney G. Brown, who produced the first actual census of the total Antarctic
population of large baleen whales.[78] They did so by means of an approach

78. By 1953 Birger Bergersen had prepared, as part of the work of his Scientific Sub-
Committee, a CDW analysis of pre- and postwar blue whale catches, which strongly
suggested a much decreased stock. See "The Condition of the Antarctic Stocks," appendix
to IWC / 5 / 2, in which blue whales per CDW were shown to have fallen from 1.05 in
1930–1931 to 0.30 in 1951–1952; an overall reduction in the size of the animals taken was
also demonstrated. This study resulted in a later starting date for the blue whale season
in the Antarctic. Not everyone, however, was persuaded by these analyses, since it was
possible to question the import of the decline in blues per CDW on the grounds that
catcher boats were no longer working as close to the deep southern ice edge (where
blues were known to outnumber fins) as they had once done. Of course, the reason
for this was probably—at least in part—that there were fewer blues there to catch than
once upon a time, and thus the companies opted against taking on the additional risk of
working in this region, but this could not be proved exactly because no one hunted there
anymore. The first effort at an IWC population analysis that worked from fundamental
rates was a joint Norwegian-Canadian paper by Hylen et al., "A Preliminary Report on
the Age Composition of Antarctic Fin Whale Catches 1945 / 46 to 1952 / 53, and Some
Reflections on Total Mortality Rates of Fin Whales" (this was a supplementary document
attached to Scientific Committee materials for the 1955 Moscow IWC meeting; I have
consulted the copy in the Smithsonian Archives, RU 7165, box 17, folder 1). That study
used F. I. Baranov's technique for estimating total mortality rates for a fish population
from the slope of the right shoulder of a distribution of age classes. When applied to
fin whale catch data since World War II, this analysis gave somewhat messy results (the
1950–1951 season looked anomalous, and the rates varied from season to season more
than one would have liked—suggesting sampling problems), but a ballpark (and in the
authors' view, low-side) estimate pointed to an average 25 percent annual mortality.
Having arrived at this figure, it was then possible to play with different tables of natural
mortality and reproductive rates to see whether biologically plausible numbers yielded
stable populations under this total drain on the stock. The authors' answer was no
(even a 50 percent fertility rate for mature females necessitated zero natural mortality
for neonates, which was clearly impossible). They therefore summed up their analysis
thus: "The conclusion seems unavoidable that the stocks of fin whales are overfished
by the present rate of whaling" (p. 11). In a separate document filed at the same time
(also available at Smithsonian Archives, RU 7165, box 17, folder 1), Slijper and Drion
disagreed, arguing that the data in question were not representative (in view of various
selection biases introduced by hunting practices), and that therefore the whole analysis could not be trusted. Nor, in their view, were the broadly similar results found by

very different from the statistical and population modeling techniques being used by others: they plotted sighting data. It happened that during its meandering Antarctic voyages of the 1930s, the RRS *Discovery II* had maintained a watch officer charged with the duty of whale spotting. The result of six years' steaming, then, was some 47,000 miles of clear-weather track charts through the waters of the deep Southern Hemisphere, during which time the appointed lookout spotted 2,602 whales, all of which had been identified (from their spouts) as blue, fin, or humpback. This was the kind of data with which Mackintosh excelled, and he and his young assistant Brown set out to interpret these vectorlike distributions by means of supplementary information on cetacean breathing rates (how often were whales on the surface?), visibility gradients (how far away could you see a whale from the bridge of the *Discovery II*?), and migration patterns (how should the data be corrected for seasonal and geographic biases?). By 1956 they had actually succeeded in producing the first overall estimate of the Antarctic population that could be defended as essentially independent of catch data.[79] They placed the probable tally for the average population (1933–1939) between 142,500 and 340,000 animals. Multiplying these total numbers by the fin whale sighting rate (75 percent) left Mackintosh and Brown with the claim that "the fin whale population was not likely to have

---

Richard Laws (also appended in the same group of documents) using a similar kind of analysis (he compared pre- and postwar mortality curves to show the swing toward a younger stock) confirmatory of Hylen et al., since both studies simply used the same unreliable data. Slijper and Drion also questioned whether the youthward shift in age distribution might not reflect a rapidly growing (rather than a declining) population. As I discuss in the text, this interpretation was difficult to defend within the constraints of density dependence and finite carrying capacity. NB: Laws built his age tables using counts of corpora in the ovaries (the preferred Discovery Investigations technique since the 1920s), rather than growth layers in the baleen plates (the aging technique used by Hylen et al. and developed by Ruud); the former method worked only for females, and the latter ceased to be reliable after about age 6 (because wear at the leading edge of the plate begins to erase the earliest layers). See Laws, "Age Determination of Whales by Means of the Corpora Albicantia," *Proceedings of the 15th International Congress of Zoology*, sec. 3 (1958): 303–5.

79. Mackintosh and Brown, "Estimates of the southern population of the larger baleen whales" (IWC / 8 / 5). An expanded version was subsequently published as N. A. Mackintosh and Sidney G. Brown, "Preliminary Estimates of the Southern Populations of the Larger Baleen Whales," *Norsk Hvalfangst-Tidende* 45, no. 9 (1956): 461–80. Their work is briefly mentioned by Tim Smith and Greg Donovan in "The Development of Cetacean Sighting in a Management Context" (unpublished paper).

been much above 255,000" before the war. It was an actual number that
people could fight over.

And fight they did. No sooner had the results been published than
the whole analysis was subjected to a withering barrage of criticism—of
the underlying data (how good had the spotters been at spotting?) as well
as the method (weren't the correction factors pretty iffy?).[80] But even before
the Mackintosh / Brown analysis saw print, while their number was still
circulating as a rumor, arguments over it were already heating up. These
critics attacked not by fussing over their data or technique, but rather by
plugging their figure into different models that, loaded up with various estimated elemental rates (of reproduction, of mortality), purported to track
the fate of the fin whale population—on which the Antarctic industry had
come to depend—across the hiatus of World War II and into the mid-1950s.
It was possible to make these models tell a sad or a happy story about the
Mackintosh / Brown population over these years, depending on how the parameters of the calculations were set. The Dutch team of Slijper and Drion,
assuming a fecundity rate per female of one calf every 2.16 years (derived
from Discovery data on the percentage of pregnant whales in the catch),
an age at sexual maturity for females of four years (a slight departure from
Discovery data, but consistent with Japanese flensing-platform research),
and a series of estimated natural mortalities by age class that ranged from
20 percent in the first year of life to 2.5 percent in the prime years from
ages two to fifteen (a table more or less borrowed outright from a study of
natural mortality in mountain sheep), were able to crank their slide rules
into gear and present an analysis that (they believed) demonstrated that
a Mackintosh / Brown–sized population in 1934 would have *increased* by
1954. Indeed, they went on to show that even under sustained pressure at
the scale of the modern industry—circa 24,000 fins caught per annum—
further population increases could be expected, and they extrapolated glorious scenarios under which the waters of the Antarctic might teem with
nearly a million fins by 1965.

This miracle was significantly aided by making no provision for density-
dependent changes in population growth and by disregarding the question
of a finite carrying capacity. While it might have been possible to defend a
higher carrying capacity for fins in this region (in view of what looked, on
the basis of CDW analysis, like a significantly reduced blue whale popula-

---

80. For some of these critiques, see comments by Einar Vangstein, Symons, Salvesen,
and Ruud in *Norsk Hvalfangst-Tidende* 45, no. 10 (1956): 483, 611, 615, 625.

tion), Drion and Slijper did not gesture in this direction—on the contrary, they claimed not to be persuaded that the blues were much reduced. Thus it is difficult not to conclude that their grasp of population modeling did not, at this time, extend to the notion of carrying capacity.[81]

By contrast, Hjort's former collaborator, the Norwegian Per Ottestad (considerably more skilled at population modeling, having been a coauthor on "The Optimum Catch"), configured the table of natural mortalities in such a way as to make for a stable stock in the prewar period (i.e., a stock at carrying capacity), tweaked these rates to reflect some density dependence over time, and even went so far as to try slightly varying the birth rate to reflect increased reproduction as the population came under exploitation (Mackintosh's group thought they could see evidence for such an uptick in their flensing-platform data). Ottestad's picture of the situation under a range of such assumptions looked uniformly grim (figure 5.1). And this despite the fact that, as he put it, "I am fairly sure that this result is biased towards an optimistic view of the situation."[82]

As a comparison of these two radically divergent assessments would suggest, the 1956 Scientific Committee, which had both papers before it, was unable to achieve unanimity on the question of reducing the Antarctic catch limit. Admittedly, the Dutch team did not win any actual converts to their analysis, but their emphasis on data uncertainty, as well as the very existence of such incommensurable extrapolations, helped scuttle consensus. While Ottestad had made an effort to meet privately with Drion and Slijper to discuss their differences before the meeting, the two Dutch scientist called in sick on this *tête-à-tête* and used the same excuse to take a pass on the Scientific Sub-Committee meeting that year. Nevertheless, by July's IWC gathering in London, both men were back in fighting form, and Slijper again dissented (for what was now the third straight year) from a Scientific Committee resolution calling for "substantial reductions in the annual catch." A softened version, recommending "no increase" in the then-current 15,000 BWU quota, did eventually carry (unanimously) in the Scientific Committee.[83] In the full IWC a stronger push (led by the United

81. This is the same conclusion reached by Schweder and is confirmed by Laws (interview by the author, 25 March 2004).

82. Per Ottestad, "On the Size of the Stock of Antarctic Fin Whales Relative to the Size of the Catch," *Norsk Hvalfangst-Tidende* 45, no. 6 (1956): 298–308.

83. There had been an earlier effort to bring the quota down to 14,500 BWU, but a string of objectors led by the Netherlands resulted in the quota being bumped back up to 15,000 in the 1955–1956 season. See Patricia W. Birnie, *International Regulation of Whal-*

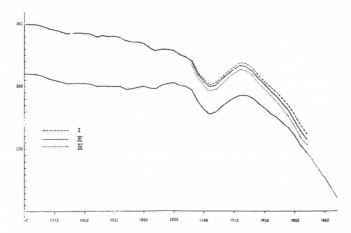

FIGURE 5.1   A grim portrait: Variation in Antarctic stocks of fin whales over time. (From Ottestad, "On the Size of the Stock of Antarctic Fin Whales Relative to the Size of the Catch," p. 304.)

States, but with vocal support from New Zealand and several other nations) to lower the quota to 14,500 BWU succeeded, but only over the dissenting votes of both Japan and the Netherlands, and it required a major diplomatic press behind the scenes to keep these nations from filing lethal objections to the reduced limit.[84]

Meanwhile, Drion and Slijper, undaunted by a rapid tutorial in density dependence, immediately turned their attention to producing a new and exacting brief on all the problems with the available data used in the population assessments by Ottestad and others (catch records were not a random sample, age determination techniques were subject to doubt, natural mortalities were a guessing game, etc.).[85] But even as they did so, US State

ing: From Conservation of Whaling to Conservation of Whales and Regulation of Whale-Watching (New York: Oceana Publications, 1985), vol. 1, p. 229, for a discussion.

84. In fact, the Dutch initially did file such an objection, but were prevailed upon to withdraw it. For US efforts on this front, see Department of State Instruction 1656, A-48, 16 August 1957 (sent to American Embassy at The Hague), Smithsonian Archives, RU 7165, box 25, folder 1, "Sixth International Whaling Commission meeting (Tokyo), 1954," and the draft of F. Taylor's memorandum of 31 July 1956, Smithsonian Archives, RU 7165, box 24, folder 7, "Eighth International Whaling Commission meeting (London), 1956."

85. The Drion and Slijper brief is "Some remarks on our present knowledge of the stock of fin whales in the Antarctic," which is annexed to the Scientific Committee

Department officials were sniffing out what seemed to be the deeper roots of Dutch intransigence on the quota: "The reason ostensibly assigned by the Dutch for their present stand on this matter is a *general* denial of the validity of the scientific premises for it," explained one telling memo on the topic, which then went on to argue that "the real, unadmitted reason is *undoubtedly purely an economic one*."[86] What followed was an elaborate account of the financial arrangements that had apparently entangled the Dutch government in a subsidy arrangement with the builders of the new *Willem Barendsz II*, a recently completed factory vessel under the flag of the Netherlands.[87] It was strongly suspected that under the terms of the deal (which were not public), the government would incur direct costs if catches were not increased.

The suspicious coincidence of Dutch politico-economic exigencies (we need more whales!) and Dutch scientific skepticism about worrisome stock assessments (there's no evidence there aren't a lot of whales!) has not gone unnoticed by commentators on the IWC struggles in this period. Indeed, the idea that Slijper was a "stooge of the whaling industry" who can be held largely responsible for the failure of the Scientific Committee in this period has become a mainstay of the available accounts of the fin whale debacle of the 1950s.[88] And in truth, it is difficult not to raise an eyebrow

---

documents for the ninth IWC meeting (IWC / 9 / 13). By 1957 it was possible to point to inconsistencies between / among not just two, but *three* aging techniques, since after 1955 the Fraser and Purves technique of counting layers in the earplug introduced a whole new aging method (see chap. 6).

86. The quoted memo is that drafted by F. Taylor, 31 July 1956; see n. 84 above. Emphasis added.

87. This was thought to be a guaranteed profit scheme that entitled the vessel's owners to a 6 percent return on their investment—essentially a whaling bond. Such arrangements were by no means unusual, since national governments had long taken an interest in stimulating the industry to provide jobs (in shipbuilding and processing) as well as to secure increased food security, foreign exchange revenue, etc. For more on the economics of the industry and its ties to state policies, see J. N. Tønnessen and Arne Odd Johnsen, *The History of Modern Whaling* (Berkeley: University of California Press, 1982). Kellogg discusses his understanding of the Dutch deal on the *Willem Barendsz II* in a letter to Wall dated 15 September in 1958, Smithsonian Archives, RU 7165, box 25, folder 3, "Eleventh International Whaling Commission Meeting (London), 1959."

88. Schweder and Heazle focus on this aspect of the story, as I discuss above. "Stooge" is from a poem penned by Scott McVay in 1965, "To the Apologist, Defender, and Stooge of the Whaling Industry: EJS," which creepily depicts an imagined world from which human beings disappear continuously and with increasing regularity, tugged away by

when the Slijper and Drion briefs go from raising (legitimate) concerns about data quality (up through 1956–1957) to arguing (circa 1958) for an *increased* Antarctic quota on the grounds (more or less) that no one has *proved* that a higher quota wouldn't be OK.[89] One need not haul modern conceptions of the "precautionary principle" into the witness box to indict the wisdom of such arguments. And while, to the best of my knowledge, no "smoking gun" has yet emerged from Dutch archives that definitively establishes Slijper's conspiratorial complicity with the demands of the Dutch fisheries ministry, his "Whale Research Group, T.N.O." at the University of Amsterdam is known to have been heavily subsidized by the Netherlands Whaling Company.[90]

But if we are to make sense of the larger dynamics of the scientific advising system in the first two decades of the IWC, it is important not to be waylaid by conspiracy theories. After all, the really significant question for the historian of science examining this episode is not "Were the stock assessment contributions of Slijper and Drion financially motivated?" much less "Were Slijper and Drion morally culpable or technically culpable for their modeling errors?" Rather, the question that needs to be addressed is some

---

unknown and alien killers. We are to understand whaling from the whales' perspective, but analogies to the Holocaust in Europe are explicit. The poem is reproduced in Joan McIntyre, comp., *Mind in the Waters: A Book to Celebrate the Consciousness of Whales and Dolphins* (New York: Charles Scribner's Sons and Sierra Book Club, 1974), 93.

89. I am paraphrasing the argument of "Proposal of the Dutch representative concerning the Blue Whale Unit Limit," an appendix to the Scientific Committee report for 1958 (IWC/10/12). In a similar vein, Drion turned his energies in 1957 to demonstrating that *if* things were as bad as Ottestad's analysis suggested, the proposed BWU reductions would be useless—this analysis was advanced, curiously, as an argument for disregarding those proposals. See E. F. Drion, "The Effect of small reduction of the number of Blue Whale Units to be taken in Antarctic Pelagic Whaling," which can be found with Scientific Committee documents for 1957 (IWC/9/20).

90. Slijper himself details these arrangements in "Ten Years of Whale Research," *Norsk Hvalfangst-Tidende* 48, no. 3 (1959): 117–29, at p. 120. The board of directors of the organization included both the president and vice president of the Netherlands Whaling Company, along with several prominent officials in the government fisheries department. I have unfortunately come up dry on detailed biographical material on Slijper himself, having failed to find an actual copy of the only study I could identify: Rogier Lange, Angelique Martens, Kaat Schulte Fischedick, and Floor van der Vliet, "Slijper, Prof. Dr. E. J. Slijper (1907–1968)," in *Lutra* 45, 2002 supplement. There is evidently a study to be done of the Dutch side of this story, if the sources have survived.

version of the following: "Why was the Scientific Committee as a whole unable to generate recommendations capable of seizing the attention of IWC delegates and motivating significant adjustments in the catch quota?" Slijper and Drion are part of this story, to be sure, but, as I hope I have shown in this chapter and the previous one, there was a great deal more going on. Indeed, it should be obvious that the efforts of Slijper and Drion could not have had any leverage at all—indeed, might have been laughed out of the committee room—were it not for the broader culture of research and decision making that Mackintosh had done much to install in the scientific advising system of the IWC. For instance, as I have discussed, Mackintosh retained an overriding preoccupation with organismal biological research on cetaceans. This meant that the kinds of investigation that went into setting the elemental rates in population models—reproductive anatomy and physiology, the fine details of aging techniques—always loomed large in his imagination as well as in his budgetary priorities. With the discovery in 1955 of a promising new method for establishing the ages of baleen whales (by counting the layers in the waxy earplug in the auditory canal), Mackintosh was immediately impressed with the urgent need to collect vast amounts of the relevant tissue, to perfect techniques for sectioning and preparing samples, and to correlate the new approach with the others that had been tried (corpus counts and baleen ridges, both of which had limitations, and which in any case did not perfectly agree). Thus, right in the thick of the mid-decade push for a better statistical grasp of the total stock and its rate of depletion, Mackintosh found himself increasingly preoccupied with a mincing anatomical question that would require extensive labor on the microtome. Slijper too, needless to say, was equally taken with the need for a long-term study of the problem.[91]

This last remark is somewhat snide, since it is of course perfectly true that population models like Ottestad's demand accurate age tables as a prerequisite for mortality analysis.[92] Indeed, all the increasingly refined biological

---

91. Slijper did in fact call for a major initiative aimed at direct verification of the different aging techniques; see Drion and Slijper, "Some remarks on our present knowledge..." (see n. 85 above), and Van Utrecht and Cock, "Provisional report concerning age determination in Fin Whales by means of Baleen plates," app. II to that document.

92. In fact, an important stimulus to the modeling work done in the 1950s was Edward S. Deevey's 1947 article "Life Tables for Natural Populations of Animals," *Quarterly Review of Biology* 22, no. 4 (December 1947): 283–314, which included a discussion of a 1934 monograph from the *Discovery Reports* by J. F. G. Wheeler ("On the Stock of Whales at South Georgia," *Discovery Reports* 9 [1934]: 351–72), which itself endeavored

researches undertaken at NIO and at the other strongholds of "hip-booted cetology" could be justified as contributing directly to stock assessment, insofar as this sort of biological work made possible more refined estimates of the demographic parameters necessary for population modeling. But at the same time, as subsequent developments would reveal, the whole project of meticulously tuning the parameters used in these models ended up functioning more as a distraction from the business of figuring out the basic condition of the stock than as a means to that end. It would soon be demonstrated that, properly handled, the catch data alone (i.e., a corrected catch per unit effort, or CPUE, analysis) could be made to yield compelling evidence that the stock was severely overtaxed. But there was little impetus within the Scientific Committee to undertake the kind of carefully corrected CDW analysis of these data that was needed to mobilize the reams of BIWS numbers into a persuasive argument for reduced catches. In fact, as late as 1958, Slijper could deliver to the committee a totally uncorrected CDW analysis (BWU per CDW, 1947–1958) in defense of the proposition that "the stock of Fin Whales in the Antarctic is more or less stationary," and his claim—absurd in view of the shifting species composition of the total BWU catch, not to mention the vastly increased power and range of catcher boats over this period—went essentially uncontested.[93] The committee's silence probably reflected to some degree a sense, pervasive among the members by this point, that the whole scientific advising system was basically dysfunctional and that the big decisions—about national priorities, about the survival of the IWC itself—were being made elsewhere, in diplomatic exchanges and backroom deals. But the silence also surely reflected the committee's lack of interest in, and lack of sophistication with, CPUE analysis, which, after all, was not "whale science" as the whale scientists understood it. You certainly did not need any whales to do it, and what was a whale scientist without any whales?

---

to generalize about the impact of whaling at South Georgia on the stock of fin whales using age tables. Deevey was able to suggest, on the basis of his general investigation into mammalian life tables, that Wheeler (and the scientist who assisted with his statistical analysis, T. Edser, of MAF) had very likely been overly optimistic about juvenile mortality. If correct, the suggestion had bleak implications for the ability of the stock to withstand what Deevey (who knew next to nothing about whaling, as far as I can tell) called "the alarming exploitation being conducted by the whaling industry in the Antarctic."

93. "Proposal of the Dutch representative concerning the Blue Whale Unit Limit," IWC / 10 / 12.

There can be no better evidence for the durable power of this parochial-ism than the frankly stupefying final paragraph of the 1958 IWC Scientific Sub-Committee report, where, in reply to a circular from FAO on stan-dardizing notation for fisheries statistics, Mackintosh offered the following comment:

> The Sub-Committee agreed that there are advantages in standardizing practice if a new method is convenient to the work, *but this is largely a matter for those who work on population problems of whales.* The Sub-Committee need only draw attention to the existence of the new proposals in case any research worker likes to look further into them.[94]

Come again? Just what did the Sub-Committee think it was doing, if not "working on population problems of whales"? It is as if, for an instant, in the thick of a long discussion about mark recovery and earplug analysis, the whole issue of actual whale stocks in the Southern Hemisphere slipped the collective consciousness. This moment is so odd that it is perhaps best understood as a kind of symptomatic parapraxis-by-committee.[95]

As this bizarre irruption suggests, their preoccupation with refining biological parameters kept even those whale scientists who were generally sympathetic to conservation inordinately focused on the weaknesses (in-deed, possibly even the irrelevance) of BIWS catch data.[96] This point was in fact made forcefully by the US mathematician Douglas G. Chapman, who

94. IWC/10/2, para. 17. Emphasis added.

95. In the Rigsarkivet, Copenhagen, I actually turned up a packet of the FAO materi-als in question. See "Standardization of mathematical notation for population dynam-ics" (ICES Records, Rigsarkivet, box 138, file 9N; indexed as FAO/57/7/4780), which emerged out of a joint resolution by ICES, ICNAF, and FAO in Lisbon in May 1957. See also (in this same file) a pair of interesting cover notes by Holt, as well as a list of papers contributed to the Lisbon meeting (by Gulland, Le Cren, Lundbeck, Parrish, Keir, etc.), which nicely demonstrates the amount of work being done at this time on massaging CPUE data.

96. Even Laws, who was offering the most sophisticated models of population as-sessment by the late 1950s, was perhaps overly impressed by the problems with CPUE analysis, writing in 1960 (in a paper prepared for the Durham Symposium of the British Ecological Society, "The Exploitation of Natural Animal Populations") that "C.D.W. is not a constant unit of effort . . . , nor is there any prospect of correcting for the variation." Laws, "Some effects of whaling on the Southern Stocks of Baleen Whales," IWC/12/8 (1), p. 5. The work of the Committee of Three just a few months later would invalidate this judgment.

was brought in, in the early 1960s, to spearhead a major effort to force consensus on the Antarctic stock assessment (of which more below). As chairman of the newly created "Committee of Three," charged with working up a population analysis that would silence Slijper-style dissent, Chapman met with members of the IWC Scientific Committee in Rome in the spring of 1961, and he offered a private account of their paralysis in a letter that year. The Dutch did not seem to him like the main issue. They might "throw an occasional wrench in the machinery," but the deeper issue lay elsewhere: "The whaling commission scientists have come to no real conclusions," he noted, because "they have been so impressed with the limitations of their data that they have failed to use what they did have."[97] Significantly, Chapman went on to note in an aside that, in general, Mackintosh "handicaps the ad hoc Scientific Committee as a whole,"[98] a damning judgment on Mack's perennially hesitant chairmanship and dogged preoccupation with his version of whale research.

To be fair, however, Mackintosh was by no means alone in displaying a deep-seated ambivalence toward the ultimate value of the available catch statistics. On the contrary, he was in excellent company. The Norwegian economist Gunnar Jahn had not only been a coauthor of "The Optimum Catch" back in 1933, he had also served as director of the Norwegian Central Bureau of Statistics and ran the BIWs through the 1950s. A hero of the anti-Nazi Resistance who had spent time in prison under the Germans, he

97. This quote comes from an invaluable letter from Chapman to William C. Herrington, 9 May 1961, Smithsonian Archives, RU 7165, box 26, folder 5, "Fourteenth International Whaling Commission Meeting (London), 1962."

98. Chapman is here using the name formally given in 1959 to what had up to then generally been called the Scientific Sub-Committee. That year saw a heated debate in the plenary session about the legitimacy of this annual between-IWC-meetings gathering of the scientists. Several countries, notably Australia, objected that this meeting was not provided for in the IWC's bylaws and had therefore been operating "illegally." The anger behind this outburst had much to do with general frustration about the deteriorating situation in the IWC (due to the withdrawal of Norway and the Netherlands). It also reflected the irritation of the geographically remoter nations that had difficulty staying in the thick of decision making without shouldering the additional expense of sending representatives to subcommittee meetings out of cycle. The upshot was really nothing more than a name change for the Scientific Sub-Committee, since the IWC's chairman was empowered by the rules of procedure to convene "ad hoc" committees as he saw fit. The Scientific Sub-Committee was thus rechristened as an ad hoc assembly (this was Kellogg's solution), and the issue evaporated, leaving only a little residue of ill will. For the whole fracas, see the transcripts of the eleventh IWC meeting, IWC / 11 / 14, pp. 89–96.

had the honor of chairing the Norwegian Nobel Prize Committee for more than two decades after the war. In short, he was a serious statistician and a figure of some gravity. And yet in 1956, delivering the annual report on the season to the IWC, Jahn himself spoke somewhat dismissively of the information that could be derived from his own bureau's statistical labors. With a verbal shrug, he put it this way:

> I do not have the impression that many of the Commissioners study international whaling statistics from beginning to end or read through [*sic*]. I think that is not the most important thing. . . . Last year I promised to look into it as to whether we could see from that [i.e., the catch records] the whaling stocks, the fin-whale stock, but in reality the international whaling statistics say very little about it. . . . When the fin-whales are as scarce as the blue whales it will show it, but up to now it has been very difficult to see it from those statistics.[99]

Jahn was here invoking (unknowingly) the tragic truism of research on marine exploitation since the late nineteenth century: the best evidence of stock decline always came when the stock was basically gone.[100] When pressed as to whether he meant to imply that the whole statistics-gathering mechanism was somehow "not very important," Jahn clarified: "What I have said is that you cannot, from international statistics, say anything definite about the stock of fin whales in that district [the Antarctic]; it is impossible so far as the statistics are concerned now."

It is perhaps tempting, with the wisdom of hindsight, to dismiss such a view as the stubborn confidence of a senior figure whose self-assurance and self-importance were not matched by commensurate attention to the development of practices in his field. Both modeling techniques and the mathematics of CPUE had moved on considerably since the early 1930s, but Jahn (who held a directorship at the National Bank of Norway, among other honorific offices) had been plenty busy since then (surviving a war, rebuilding a nation), and had apparently not been keeping up with the literature.[101]

99. Transcripts of the 1956 IWC meeting, IWC/8/13A, p. 50.

100. The point is made by Tim D. Smith in "'Simultaneous and complementary advances': Mid-Century Expectations of the Interaction of Fisheries Science and Management," *Reviews in Fish Biology and Fisheries* 8, no. 3 (1998): 335–48.

101. I am thinking here, among other things, of D. B. De Lury's important paper, "On the Estimation of Biological Populations" (*Biometrics* 3, no. 4 [December 1947]:

There is something to this assessment, and a tone simultaneously defensive and pathetic can be detected in Jahn's finger-wagging reply to an uppity delegate who continued to press him on how the statistics might be better used: "I have worked with this from the beginning, you know," he intoned gravely—which doubtless struck a number of the commissioners as precisely the cause for concern, as opposed to a confidence builder. But it is worth keeping in mind a fact that has been largely forgotten by historians of this period, but which loomed large for old-timers like Jahn and Mackintosh as they scanned the available evidence for clues about the condition of the Antarctic stocks in the 1950s: these men, and others of their generation—a turn-of-the-century generation that largely dominated the proceedings in the IWC in these years—had lived through the late-1920s showdown between Sir Sidney Harmer and the youthful Johan Hjort over the size of the Antarctic stocks. They could remember only too well Harmer's increasingly urgent warnings about the imminent crash of the whole industry, his graphs and charts and statistics that purported to show that the southern stocks of whalebone whales could by no means sustain the level of hunting seen in the 1910s and 1920s. Hjort's insistence that the populations were circumpolar, and much larger than Harmer would allow, had been dismissed (at least by scientists in Britain) as unduly optimistic, if not actually craven. Since that time, of course, those waters had yielded reliable hauls that dwarfed the wildest dreams of those who opened the first land stations on South Georgia. Harmer had not just been wrong, he had been spectacularly wrong, and his posthumous legacy was thus less that of a prescient (if premature) conservationist than that of the neatly mustachioed man who cried wolf.

Nor were the skeptics of the mid-1950s inclined to let his memory rest in peace. Drion and Slijper invoked the enormity of his miscalculation by way of cautionary overture to their 1957 memorandum laying out what they would consider a proper population assessment, schoolmarmishly warning Ottestad, Ruud, and other naysayers that Harmer's error should "stimulate us to search for a sound basis for our statements before giving advices [*sic*] to the whaling industry."[102] As Slijper put it in his book *Walvissen* the following year (where he again invoked Harmer's historic miscue):

145–67), which proved key to the statistical reassessment of the stock in the early 1960s, as I discuss below.

102. Drion and Slijper, "Some remarks on our present knowledge . . ." (see n. 85 above), p. 1.

"An industry which has invested large capital in a whaling fleet and which gives employment to tens of thousands is not so much concerned with whether prohibitive steps are desirable, as with whether they are absolutely *unavoidable*."[103] Kellogg thought this so nicely summed up the "viewpoint of the major operating companies" that he quoted it verbatim in a letter to a colleague in the Fish and Wildlife Service, by way of explanation of how things in the scientific advisory system of the IWC had degenerated.[104]

But what Slijper's observation nicely captured, as Kellogg himself understood very well, was the way the whole framework for regulatory decision making "based on scientific findings" (as called for in Article V of the ICRW) had shifted through the 1950s. Gone was any trace of hope (except perhaps among the most hapless or deluded participants) that the "collaborative model" for scientific advising might eventually yield a working regulatory system.[105] The whole notion that a community of scientists might, through sustained research, strategic relationship building, and careful boundary work, nurture into existence a system of cooperative, rational, international regulation of the whaling industry—that optimistic vision of the forties—had emphatically foundered in the middle passage of the decade

103. Everhard Johannes Slijper, *Whales*, trans. A. J. Pomerans (New York: Basic Books, 1962), 394. Emphasis in original.

104. See Kellogg to J. L. McHugh, 20 December 1962, Smithsonian Archives, RU 7165, box 27, folder 1, "Fifteenth International Whaling Commission Meeting (London), 1963."

105. As far as the hapless or deluded are concerned, one cannot help but be struck by Mackintosh's continued and increasingly sorry efforts to invoke the collaborative spirit, as, for instance, in the 1960 plenary session, where he offered the following on the topic of the quota: "We are always a little doubtful about exactly what we should say about this because we know that there are more factors to take into consideration in fixing the limit than the purely scientific point of view." Transcripts of the 1960 IWC meeting (IWC / 12 / 11), p. 18. See also his conciliatory words in 1962 concerning the idea of a reduction in sperm whale limits: "it is certainly a point in which purely scientific and technical questions are inevitably mixed because the protection of the females is a scientific question and the limitation of the catch and its effect is a technical question." Transcripts of the 1962 IWC meeting (IWC / 14 / 10), p. 30. As these quotes suggest, Mackintosh seems to have retained some hope that a fruitful exchange could be negotiated across the nebulous political / scientific boundary zone he had surveyed for almost thirty years. It appears to me that he was ushered off the IWC stage by the mid-1960s without abandoning this position, which could only have been maintained to this point through self-deception, guile, obliviousness, or some combination thereof.

that followed. The scientific whaling controversy revealed that the apolitical space the IWC's originators had set aside for scientific investigation was a rule-free zone enormously susceptible to misuse. The unresolved clash with Latin American nations over "ecological management" dramatized the speed with which facts about nature could be redeployed as facts about politics—in this case to fissiparous effect. Finally, the crucial numbers game of stock analysis had shaped up in the mid-1950s as a game indeed, in which modestly talented statisticians could veil the vectors of best-effort population models in an opaque broadcloth of confidence intervals.

As a result of these slow and painful lessons, a new question had come to the fore among those who still hoped to bring the whaling industry under control and to establish the merits of international "scientific" regulation of marine resources. By 1958 these men—old-guard types like Kellogg and Ruud, yes, but also a younger group of officials from the United States, Canada, New Zealand, and elsewhere—had ceased to pursue the old mission ("Can working cooperative practices for scientific regulation be developed within the IWC?") and set themselves a new problem: "Can science be used to *compel* consensus, and thereby save not just the whales, but the actual organization itself?" The shift had not been sudden. As early as 1954 Kellogg scribbled a marginal note in a draft copy of one of the Scientific Committee documents, reminding himself of the rationale offered by those resisting a small cut to the quota: "it is not . . . whether or not desire, but whether we *must make* this reduction."[106] In summing up that year's meeting in his report to the State Department, Kellogg alluded to what he perceived as an emerging shift in the burden of proof: "The amendments to the Schedule agreed upon at this meeting, in the opinion of the undersigned, fall short of the minimum protection needed at this time," he wrote, and he went on to explain that "to a great extent, this is due to the absence of sufficient scientific data to convince industry representatives of the need for more stringent measures."[107] In 1954 it was still possible to hope that more re-

106. Kellogg's copy of IWC / 6 / 15, "Draft Report of Scientific Committee," p. 2. Emphasis added. See Smithsonian Archives, RU 7165, box 15, folder 5, "London—Fifth International Whaling Commission Meeting, 1953—Statistical material submitted by the Bureau of International Whaling Statistics." The quote paraphrases the view of Lienesch and Slijper.

107. Report for 1954, p. 12, Smithsonian Archives, RU 7165, box 28, folder 4, "Reports to the Secretary of State by the U.S. Commissioner to the IWC, Meetings #1–11, 1949–1959."

search, channeled through the institutional structures of the still young organization, would eventually tip the balance. But as I have shown, it became increasingly clear through the controversies of the next several years that those structures—the Scientific Sub-Committee and Scientific Committee, the BIWS and the various national research programs—were simply not going to be capable of generating and mobilizing arguments that could force consent. On the contrary, what emerged by 1958 was something very different. Even as the documentation amassed by the scientific advising system deployed increasingly definitive language—urging that "the conclusion seems unavoidable," that "the trend is unmistakable," that a reduction in numbers "can probably not be disputed"—it had become apparent that scientific investigations were every bit as good for *delaying* regulatory action as they were for obliging it.[108] And maybe better.[109]

108. The quoted phrases are from Hylen et al.'s 1955 study (see n. 78 above). By 1956, Ruud was writing, "Scientists have pleaded that the stocks of fin whales . . . are also [along with blue whales] overfished." Johan Ruud, "International Regulation of Whaling: A Critical Survey," *Norsk Hvalfangst-Tidende* 45, no. 7 (1956): 374–87. (NB: This paper was originally published in *Samtiden*, a Norwegian journal of politics and social affairs.) By 1960 Laws, who (fed up) was about to step away from the whole topic, had raised the stakes, declaring in his presentation to the Twenty-Fifth North American Wildlife and Natural Resources Conference in Dallas that "we must conclude that the stocks of fin whales as well as those of blue whales are declining." This statement was based on an uncorrected assessment of catch by species per catcher's day's work. See R. M. Laws, "Problems of Whale Conservation," in *Transactions of the Twenty-Fifth North American Wildlife and Natural Resources Conference* (Washington, DC: Wildlife Management Institute, 1960), 304–19, at p. 316. Some sense of the probable cause of his frustration can be gained by reviewing Mackintosh's 1959 essay "Biological Problems in the Regulation of Whaling," published initially in *New Scientist* (4 June 1959): 1229–31, and republished in *Norsk Hvalfangst-Tidende* 48, no. 8 (1959): 395–404, shortly thereafter. In this piece Mackintosh continued to lean on the complexity of the problem and the ambiguity of the evidence, explaining, for instance, that "the trend, year by year, of the catch per unit effort and of the average sizes of whales caught has looked suspicious, but inferences are confused by changing technical factors," and that "doubts have been expressed . . . on whether some of the data used . . . are sufficiently reliable." The upshot, in his view, was that "there is no final proof that the fin whale stock is now declining." By this point both Ruud and Laws would have disagreed with this claim; only Laws, however, was obliged to report directly to Mackintosh.

109. An excellent foretaste of this dynamic had come as early as 1952, in a blowup over catch regulations in the North Pacific. The United States and Canada supported restrictions on factory vessels in the waters northeast of the Bering Strait, to which

It was this very question that floated in the air when, in the 1958 plenary session—in the thick of what looked like a potentially terminal deadlock over the catch quota—the Dutch commissioner (whose proposal for an increase in the Antarctic limit had just failed) rose to make a modest proposal:

> I wish to inform the International Whaling Commission that the Netherlands Delegation is of the opinion that the scientific evidence required to establish a blue whale unit limit is wholly unsatisfactory. It is clear that the scientists of the various countries differ in their opinion on the interpretation of the data thus far available. . . . *Therefore we propose the appointment of an international scientific commission of specialists in the field of population dynamics to judge whether the data thus far available warrant the conclusion that the whales suffer from over-fishing.*[110]

No one seconded the motion. In fact, everyone pushed back from the desk, and the chairman of the meeting called for an adjournment. It was obvious that this recommendation was not being offered as a means to facilitate resolution, but rather as a blatant—indeed, nearly contemptuous—ploy to delay even further action broadly understood to be necessary. It was as if a gauntlet had been dropped in the regulatory arena: could "scientific findings" ever make anyone do anything?

The stage was thus set for a formal showdown between those who intended to use science to stymie and fetter regulatory efforts and those who intended to use it to tie the hands of opponents of conservation, silence the hecklers, and bind an unruly room into docile accord. It is with that episode—a calling of the Dutch bluff—that the final section of this chapter will be concerned.

---

the Soviets objected strenuously that "there appeared to be no scientific grounds for the amendment." See minutes of Scientific Committee meeting of 30 May 1952, Smithsonian Archives, RU 7165, box 14, folder 1, "London—Fourth International Whaling Commission Meeting, 1952—Instructions, proceedings, correspondence, notes, etc." This dispute eventually led to sharp accusations that the United States and Canada were trying to use the IWC for an oceanic "land grab" in the North Pacific. The Soviet call for more research here cannot but be understood as a delaying tactic.

110. Transcripts, The Hague, Drost speaking, IWC / 10 / 13, p. 85. Emphasis added.

"NO REASONABLE DOUBT": 1959–1965

The basic outline of the showdown—the introduction of the "Committee of Three" and the effort to bring the catch quota into line with their new, independent stock analysis—can be briefly summarized, but to do so, it is necessary to attend for a moment to the broader diplomatic situation among the whaling countries at the end of the 1950s. Despite a flurry of late-decade extra-IWC meetings aimed at divvying up the Antarctic catch quota among the major whaling powers (USSR, Japan, Norway, Netherlands, United Kingdom), it finally proved impossible to achieve a workable consensus on this perennial problem. The tighter BWU limit simply could not provide adequate returns to five overcapitalized national industries. Disputes over enforcement similarly eluded resolution. Thus, shortly after the 1958 IWC meeting in The Hague, the Netherlands (infuriated by their inability to get the quota share they wanted) and Norway (infuriated primarily by what they took to be Dutch unreasonableness) both filed notice of their withdrawal from the organization.[111] Thereby released from the IWC's BWU strictures, each country effectively gave itself the quota it thought it deserved. While "voluntary" arrangements prevented a total free-for-all, the international regulatory system had collapsed.

It was in an effort to rectify this situation that a proposal took shape in the 1960 IWC meeting that would allow for a "temporary" relaxation of the total Antarctic catch limit (a gesture aimed at bringing the Netherlands back into the fold) coupled with a plan to convene a "Committee of Three" experts in population dynamics (aimed at appeasing Norway and putting an end to the annual quota impasse). The new committee would be assembled from nations not involved in Antarctic whaling and provided with the resources necessary to make a new and independent assessment of the Antarctic population. The idea was that the thereby reconvened IWC member nations would agree in advance to bring their total catch quota into line with whatever this new group of scientists determined the stock

111. Japan also filed notice, but was prevailed upon to withdraw the filing shortly thereafter. NB: Kellogg certainly felt the Dutch had overreached. The Soviets, by contrast, had probably been "oversold," receiving a 20 percent share of the Antarctic quota in the five-nation negotiations. This was out of line with their capacity, but was offered as part of a structured deal to get them to stop building additional factory fleets. For Kellogg's view of these wranglings, see his letter to Henry Reiff, 7 April 1959, Smithsonian Institution, James Mead Files, file "Functions, Statistics, Limitations."

could withstand.[112] After various delays (occasioned by, among other things, budgetary problems), the Committee of Three was indeed assembled, consisting of Sidney Holt (FAO), Douglas Chapman (University of Washington) and K. Radway Allen (Fisheries Department, New Zealand). Over the course of two years, this group undertook—to some degree in conjunction with the members of the IWC's existing scientific advising system—a new analysis of the Antarctic stocks.

The Committee of Three's findings and recommendations, delivered into the regulatory arena by early 1963, occasioned considerable gnashing of teeth among the IWC delegates over the two years that followed, but the whaling nations were not, in the end, able to honor their commitment to be ruled by this new assessment, and this despite the fact that the Committee of Three's short-term predictions about declining future catches proved strikingly accurate (which immediately buttressed their credibility).[113] In

112. The degree to which the elements of this "package deal" were structured as interlocking "quid pro quo" arrangements was never made fully explicit. Some countries (i.e., the USSR) objected to suspending the quota for any purpose (ostensibly this objection reflected their concern for conservation, but there were several reasons to question their sincerity, including their continued construction of whaling vessels and the continued trickle of reports—subsequently confirmed to a spectacular extent—of Soviet cheating), and diplomatic deadlocks of the following kind had to be broken (in some cases by nebulous language in the agreements): X won't negotiate on the quota until everyone is in the IWC; Y won't rejoin the IWC until there is a deal on the quota. I do not intend to try to plot every move in these protracted posturings, but will confine myself to what I take to be the relevant scale of analysis for the purposes at hand. It suffices to keep in mind the (large) scale of the efforts and their (nearly absurd) complexity: as Tønnessen and Johnsen point out, between 1958 and 1962 there were no fewer than *eleven* non-IWC international conferences aimed at sorting out the quota allocations (J. N. Tønnessen and Arne Odd Johnsen, *The History of Modern Whaling* [Berkeley: University of California Press, 1982], 602). As Kellogg put it wryly, "Inasmuch as the present controversy involves the commercial interests of five or more nations, one need not anticipate a public airing of differences." Kellogg to Henry Reiff, 7 April 1959, Smithsonian Institution, James Mead Files, file "Functions, Statistics, Limitations." Indeed, none was ever forthcoming.

113. Whether it had been, in fact, a "commitment" came under scrutiny, and depended on how one construed the word "should," which had crept into the language of the original resolution, replacing "will" in the following sentence: "It is the intent of the Commission in setting up this special scientific committee, that the Commission will, within two years after termination of the two-year suspension, bring the Antarctic catch limit into line with the scientific findings." See transcripts of the 1960 IWC meeting

one sense, then, the whole exercise—which was in effect an effort to install the Committee of Three as a version of the kind of "arbitration process" for which the original ICRW made no provision—can be understood to have failed.[114] In a deeper sense, however, the effort must be considered a very real success, since the countries that stepped away from adopting the Committee of Three's quota recommendations ultimately did so *not* on the grounds that they remained unconvinced about the disastrous status of the stocks, but rather with a frank acknowledgment that they could not afford the called-for cuts. As far as the showdown between science-for-delay and science-for-compelled-assent was concerned, the latter camp carried the day. That the assent it compelled did not translate into action had profound implications in the years to come. By 1965 no one could deny that the world's large whales were in serious trouble; those intent on doing something about this new fact left the IWC behind for other venues—FAO, the United Nations, the court of "world opinion"[115]—armed with shocking proof that the foxes were guarding the henhouse.

Given this dramatic upheaval, it is little wonder that the Committee of Three has been treated as a watershed in the history of the IWC and has received a great deal of attention from scholars interested in the relationship between whaling politics and whale science. Because it is my intention to offer a revised account of this episode in the pages that follow, it will be worth reviewing how the available treatments generally handle these events. The following compressed outline seems to me fair: (1) the Committee of Three was a British proposal, aimed above all at saving the whales of the Southern Hemisphere (and thereby the whaling industry); (2) the work of the new committee represented nothing less than a "paradigm shift" in which experts in population dynamics replaced old-style whale biologists in the IWC's scientific advising system; (3) the whole episode amounted to the triumph of a properly scientific use of science over the improperly political use of science, and is thus best understood as a scientific victory (though a political failure).[116] At the risk of appearing gratuitously

---

(IWC / 12 / 11), p. 63. This substitution was a concession to the Soviets. NB: The deadline was also changed before the resolution was finalized; see IWC / 12 / 17.

114. See Patricia W. Birnie, *International Regulation of Whaling: From Conservation of Whaling to Conservation of Whales and Regulation of Whale-Watching*, 2 vols. (New York: Oceana Publications, 1985), on the need for an arbitration process.

115. For a discussion of the use of the term "world opinion," see n. 197 below.

116. For emphasis on the United Kingdom as proposer (and even a suggestion that the US was reluctant), see Patricia W. Birnie, *International Regulation of Whaling: From*

polemical, I will argue that each of these propositions is wrong in its particulars and misleading in spirit.

First, it is by no means the case that the Committee of Three originated
in the United Kingdom. Commentators have been deceived by the fact that
a version of the proposal that would eventually win passage in the IWC
was indeed forwarded by the UK delegation. But even a cursory passage
through the relevant archives for this period reveals that it was in fact the
United States, and specifically US State Department officials, who originally hammered out the idea of a "package deal" linking an independent
scientific investigation to Norway's and the Netherlands' readmission via a
temporary suspension of the catch quota.[117] This would be a minor matter of

---

Conservation of Whaling to Conservation of Whales and Regulation of Whale-Watching
(New York: Oceana Publications, 1985), vol. 1, 257; for "paradigm shift," see Schweder,
"Intransigence, Incompetence, or Political Expediency?" IWC SC / 44 / 013, pp. 13–15. J. N.
Tønnessen and Arne Odd Johnsen, in *History of Modern Whaling* (Berkeley: University
of California Press, 1982), sum up the story as a "parting of the ways—where science
and whaling met and parted" (p. 624).

117. Kellogg originally had a version of the "Committee of Three" proposal in hand at
the 1959 meeting, and he had instructions to float the deal if he felt that it had a chance
of persuading Norway and the Netherlands to retract their pending withdrawal from
the IWC; see his handwritten memorandum in the Smithsonian Archives, RU 7165, box
25, folder 3, "Eleventh International Whaling Commission Meeting (London), 1959,"
which reads in part "U.S. prepared to study how we can help with package deal . . .
provided the question of appropriate catch level, based on maximum sustainable yield
will be ascertained by an impartial ad-hoc committee of experts appointed by chair-
men." The structure of the deal was originally conceived by William C. Herrington in
the State Department; see confidential outgoing telegram, Department of State, 25 June
1959, Herrington to American Embassy, London, for Kellogg, EMBTEL 6720: "Delegation authorized make following compromise two-part integral proposal which must
be package deal: (a) increase in quota for 1959–1960 season only; (b) submit question
of appropriate catch level, based on considerations of maximum sustainable yield to
impartial committee of experts, with quota for following season determined by commission with basis in findings this committee." Smithsonian Archives, RU 7165, box
25, folder 4, "Eleventh International Whaling Commission Meeting (London), 1959."
Herrington was also explicit about the importance of letting one of the other IWC nations, preferably one of the major whaling nations, have credit for the package deal "if
this will help"; see confidential outgoing telegram, Department of State, 27 June 1959,
Herrington to American Embassy (London) for Kellogg, EMBTEL 6791, Smithsonian
Archives, RU 7165, box 25, folder 3, "Eleventh International Whaling Commission Meeting (London), 1959." In the end, Kellogg felt that the 1959 situation was hopeless, and
the US delegation did not put the proposal forward that year; see incoming telegram,

priority and attribution if this misunderstanding did not at the same time obscure a much larger and more significant misconception. The primary aim of the Committee of Three "package deal" was not, in fact, to save the whales of the Southern Hemisphere (or even the industry they sustained), but rather to fend off the increasingly strident moves by countries in the Third World who hoped to use the United Nations Conventions on the Law of the Sea to overthrow the postwar regime of international commission-regulated high-seas fisheries and replace those arrangements (which, as I have discussed above, poorer nations considered fundamentally biased in favor of rich and developed countries) with some version of a collective property model—national, multinational, or perhaps even universal. It was not that US officials were wholly indifferent to the prospect of the extermination of the large whales of the world, but the United States did not really have a horse in that race, being without a pelagic whaling industry. By contrast, as one State Department memorandum put it, "the actions taken by the whaling countries for the coming season may well have an important bearing on the second conference on the law of the sea, to be held in Geneva in 1960." As this memorandum went on to point out, "certain of the coastal countries which have advocated radically extended territorial (or fisheries) jurisdiction over contiguous waters have argued that only through such control by the coastal states can living marine resources be protected against the depredations of the important fishing states with their large and efficient fleets." Insofar as "the United States and other countries interested in maintaining a regime of limited jurisdiction have countered with the argument that proper measures of conservation can be applied and maintained through international agreements," it had become urgent by 1959 to demonstrate that the earliest and most prominent of those agreements, the ICRW (and thereby the IWC), actually *worked*: "In these circumstances, the [US State] Department feels that concrete evidence of lack of restraint, whether on the part of those countries remaining in the Convention or those that have withdrawn from the Convention would be seized upon by the proponents of extended jurisdiction as a justification of their

---

Department of State, Whitney to secretary of state, 1 July 1959, Smithsonian Archives, RU 7165, box 25, folder 3, "Eleventh International Whaling Commission Meeting (London), 1959." The proposal was resurrected the following year and slipped to US allies (both the United Kingdom and Canada) to give it the added legitimacy of being proposed by actual whaling nations.

position." This was a very serious matter: "The breakdown of conservation arrangements for the Antarctic whale stocks surely would encourage and strengthen efforts being made by certain coastal states to reopen the Fisheries Convention signed at the 1958 Geneva Conference in order to extend the jurisdiction of coastal states beyond the territorial sea and contiguous zones in the alleged interest of conservation."[118] It is thus only a slight exaggeration to say that the Committee of Three, at bottom, was mostly about saving US tuna fishermen in the Pacific.[119]

As a result of these concerns, the United States undertook, between 1959 and 1963, a major initiative—diplomatic and scientific—to salvage the regulatory regime of the International Whaling Commission as part of a broader effort to head off moves to nationalize ocean resources. US embassies in a dozen countries around the world were encouraged to make clear to foreign governments the United States' commitment to bailing out the IWC so that "nothing will be done which will create an opportunity . . . to

---

118. This and the quotes above are from Department of State Instruction CA-3816, 3 November 1959 (sent to Tokyo, Oslo, The Hague, London), Smithsonian Archives, RU 7165, box 26, folder 2, "Tenth International Whaling Commission Meeting (The Hague), 1958." William Herrington spelled out these concerns in a letter to Sidney Holt of 7 April 1961, Smithsonian Archives, RU 7165, box 26, folder 5, "Fourteenth International Whaling Commission Meeting (London), 1962" (emphasis added):

> The Antarctic whale stocks provide one of our first and most dramatic examples of the problem we face. . . . Effective action by the Commission to agree upon and apply adequate measures has in the past been blocked by failure to make the most effective use of the statistical techniques now available (partly because of the expensive nature of whale research work) and by the fact that the decisions of some members appear to be considerably influenced by industry pressures. It appears clear that unless we find some means of expeditiously overcoming these limitations the whale stocks will be lost. *This would be a terrific reflection on the scientific approach to conservation of high seas resources and on the cooperative approach by which we are seeking to resolve these problems.*

119. Birnie could therefore hardly have it more wrong when she writes, in connection with the IWC's difficulties in this period, "Developments at the First and Second UN Conferences on the Law of the Sea in 1958 and 1960 respectively had, however, done little to motivate states toward achieving a better balance between exploitation and conservation." Patricia W. Birnie, *International Regulation of Whaling: From Conservation of Whaling to Conservation of Whales and Regulation of Whale-Watching* (New York: Oceana Publications, 1985), vol. 1, p. 261.

point to the situation in the whaling industry as a horrible example of the failure of international conservation agreements and the attitude of fishing states toward conservation."[120]

Turning to the argument that the work of the Committee of Three represented a "paradigm shift" in regulatory science, this contention is not as much an error of fact as of emphasis. It is true, for instance, that Holt, Chapman, and Allen were skilled in the mathematical analysis of population dynamics, particularly as applied to fisheries. Their report did deploy Schaefer's techniques for estimating maximum sustainable yield, a method that had not hitherto been attempted with Antarctic whale data.[121] In addition, with the help of John Gulland (a biologist at Lowestoft Laboratory who was formally added in 1963, making the Committee of Three the "Committee of Four"), they took on a systematic reassessment of catch-per-unit-effort statistics, plausibly correcting this index for changes in the power and efficiency of catching technologies over the first half of the twentieth century. It is not wrong to say, as Schweder does, that followers of these approaches have "ruled the ground in the Scientific Committee for most of the time since 1960."[122]

Nevertheless, this emphasis on a "scientific revolution" both overstates the case (as we have seen, population dynamic models for the Antarctic stocks had circulated in the Scientific Committee since 1955; moreover, a good deal of the Committee of Three's analyses did, in the end, depend on biological data derived from the old-fashioned work of the hip-booted cetologists, including mortality estimates based on mark recovery) and, more problematically, entirely overlooks the extent to which the "revolution" in question was, above all, *administrative*. We here have occasion to return to my introductory suggestion that historians of science interested in the evo-

120. Department of State Instruction CA-3816, 3 November 1959, Smithsonian Archives, RU 7165, box 26, folder 2, "Tenth International Whaling Commission Meeting (The Hague), 1958."

121. On Schaefer's method—which can be thought of as a practical means of deploying the kind of analysis described in "The Optimum Catch" (Johan Hjort, G. Jahn, and P. Ottestad, "The Optimum Catch," *Hvalrådets Skrifter* 7 [1933]: 92–127) by using metrics derived from available fisheries data (catch and catch per unit effort) as proxies for productivity and population size—see Tim D. Smith, *Scaling Fisheries: The Science of Measuring the Effects of Fishing, 1855–1955* (Cambridge: Cambridge University Press, 1994), 225ff.

122. Schweder, "Intransigence, Incompetence, or Political Expediency?" IWC SC/44/O13, p. 15.

lution of modern regulatory systems would do well to pay close attention to "Robert's Rules." As I will show below, several of the scientists involved in the new work of the Committee of Three harbored serious reservations about the applicability of cutting-edge mathematical techniques derived from fisheries biology to the Antarctic whaling situation, and the Committee of Three's final analysis made much less use of these fancier methods than has generally been acknowledged. Moreover, several years later, most of the more involved parts of the committee's analysis would be shown to be incorrect (on account of twofold errors in aging estimates). How, then, was the Committee of Three's report capable of silencing quibblers and compelling universal assent in a way wholly unlike the suite of previous efforts?

The explanation must lie, to a significant degree, in the meticulous strategic work done by US officials not only to configure the Committee of Three itself and its membership, but also to arrange administratively the way it would engage with the preexisting scientific advising structure at the IWC. A great deal of effort went into positioning the new committee so as to ensure its independence and yet at the same time to entail the whale scientists of the whaling nations in its findings. Through a series of joint meetings and exchanges of results, naysayers were forced into collaboration with the process and were thereby neutralized. Their objections were formally incorporated into the final report as questions—questions that were then given answers, a literary convention that accorded the Committee of Three the last word in the most literal way. The result? No minority report, no forum for dissent, no holdouts. By the time the final report came into the hands of the delegates at the IWC, there was no one left to complain.

As this account suggests, characterizing the work of the Committee of Three as the triumph of a properly scientific use of science over the improperly political use of science (a dominant theme in the secondary literature) is more than a little misleading and demands that we adopt a quixotically narrowed conception of the whole episode. Nested in the broader "package deal" of 1960, and situated with respect to the larger geopolitical machinations around the law of the sea in this period, the work of the Committee of Three is better understood as a crafty solution to several sticky political problems by means of a close alliance of scientific and bureaucratic technique. Machiavellian committee work was every bit as important as the IBM punch cards on which the catch data were compiled. This has been overlooked, since if one looks exclusively at the Committee of Three's formal brief, one might easily assume that their job was simply to perform a census of large cetaceans in the Southern Hemisphere and thereby to arrive at an estimate of a sustainable catch. Only when we examine the wider context for their efforts does it become clear that their job was in fact to silence the Dutch scientific delegation, to save the IWC (by securing readmission of erstwhile member nations), and to preserve the very idea of scientific management of ocean resources. To call the result a scientific victory and a political failure is thus to get things exactly backward. By forcing the major whaling nations onto the naked plane of commercial interest, the Committee of Three's (significantly flawed) scientific analysis, carefully channeled through a new administrative arrangement, effected nothing less than a political coup, and created the conditions of possibility for a global backlash against commercial whaling after 1965.

In order to develop these points, I propose to look a little more closely at the development, work, and repercussions of the Committee of Three between 1960 and 1965. It will make sense to divide this discussion into three parts. The first will examine how the committee was structured (emphasizing strategic administrative considerations that motivated its configuration and positioned it for success in forcing consensus). The second will review the actual analysis undertaken by the committee in conjunction with members of the IWC's existing scientific advising community (focusing on how this analysis integrated new techniques with available findings and data). The third will briefly take up the aftermath of the delegates' failure to adopt the Committee of Three's recommendations (dramatizing the political leverage that this failure afforded conservation advocates).

To begin with, then, it is necessary to recover the diplomatic maneuvers that gave rise to the terms of reference for the Committee of Three. The 1960 IWC meeting saw not one but two proposals for some form of scientific arbitration aimed at rescuing the organization from its impasse. One of these (officially forwarded by the United Kingdom) called for "a small independent scientific committee" that would be "appointed in order to carry out an independent assessment" of the Antarctic stocks.[123] This was the proposal that had its origins in the US plan (first proposed in 1959, but ultimately withheld that year) to structure a package deal that would swap a raised quota in return for all nations rejoining the organization and committing to adopt as binding a new scientific analysis prepared by "three independent scientists particularly qualified in the field of population dynamics."[124] By the time this proposal reached the floor of the IWC

123. "Verbatim Report," IWC / 12 / 11, p. 49.
124. Ibid.

in 1960, however, another proposal had been independently floated by Mackintosh's Scientific Committee. This proposal called for the IWC to "establish a 'workshop' "—basically an extended Scientific Committee conference—that would be empowered "to review the data available and make a statement on its discrepancies"; "to study the manipulative and analytic methods employed in whale research and propose ways of improving and standardizing"; "to study the data available on selected problems making the fullest use of developed methods of population dynamics research"; "to review catching methods and other technical aspects of whaling operations to see whether it is practicable to improve the methods of measurement and calibration of fishing effort"; and, finally, "to consider what specific collections and observations could usefully be made as a routine in whaling factories."[125]

This mash up of old-style hip-booter desiderata and cross-our-heart promises to explore much-needed mathematical approaches certainly reflected the Scientific Committee's heightened awareness of the gravity of the regulatory situation. Mackintosh and his colleagues had to have been acutely aware after 1959 that many qualified commentators shared the view of William C. Herrington in the US State Department, who wrote in September 1960 that the "failure of the Commission stems to a considerable extent from failure of the Commission's scientific committee."[126] It is also worth pointing out that the timing and urgency of this proposal for a new Scientific Committee "workshop" had much to do (again) with Mackintosh's increasingly dire personal circumstances at the National Institute of Oceanography. By the end of 1959, Mackintosh's festering relationship with George Deacon had entered its terminal phase. Yet again denied his promotion and increasingly vociferous about the injustices he suffered at Deacon's hand, Mackintosh finally received notice in the spring of 1960 that he would be retired from the NIO. Over the months that followed, Mackintosh threw himself into a campaign of letter writing and protest, which

125. This is an abbreviated list derived from para. 35, sec. B, of the first draft of the Scientific Committee's report for 1960 (IWC/12/14).

126. Herrington to Arnie J. Suomela, 30 September 1960, Smithsonian Archives, RU 7165, box 26, folder 5, "Fourteenth International Whaling Commission Meeting (London), 1962." This was also the view of Harry D. Lillie, who prepared a report for the Norwegian embassy in London in the early 1960s, and who saw primary responsibility for the failure of the IWC to lie with the Scientific Committee. See J. N. Tønnessen and Arne Odd Johnsen, *History of Modern Whaling* (Berkeley: University of California Press, 1982), 635–36.

eventually culminated in Deacon striking a deal to put Mackintosh on furlough at a newly constituted "Whale Research Unit" in London (housed in some swept-out storage space at the British Museum [Natural History]), thereby keeping him quiet and getting him out of Wormley for good (if not off NIO's payroll).[127] Against this backdrop, it is no surprise to find the Scientific Committee's proposal for a big, well-funded cetological "workshop" coupled to a recommendation that "the Commission should urge member governments to ensure that their whale research programs should have field and laboratory staffs sufficient to collect data of the required range, complexity, and accuracy."[128] In short, Mackintosh was yet again endeavoring to use his position within the advising structure of the IWC to angle for an improved position in the professional landscape back home.

The obvious overlap between the proposal for a new scientific assessment by an outside group (a "Committee of Three") and the proposal for a new scientific assessment by the Scientific Committee itself (a "workshop") made for a tricky situation in the plenary session in 1960. The United Kingdom's leading delegate went out of his way to point out that the proposal for an independent assessment "is in no way intended to be any reflection upon the very valuable work which the Scientific Committee has already done." Rather, he put it delicately, it stemmed from the idea that some outside expertise might "confirm from an independent point of view the findings which the Scientific Committee have already put forward."[129] This was, obviously, diplomatese. And yet a serious issue was at stake, not just hurt feelings. As everyone recognized, the members of the Scientific Committee did indeed possess a vast amount of information that would be essential to any so-called "independent" stock analysis. In addition, if members of the Scientific Committee decided to make trouble for such an outside analysis, they would be well positioned to do so. In fact, there was no one positioned better in the whole world (and some of the Scientific Committee's diehard members, for instance Slijper and Drion, had by this point almost a decade's experience cross-examining population analyses).

127. The details of these arrangements can be found in NOL GERD.M2/1, especially in "Personal Statement by Dr. Mackintosh" with cover note of 21 June 1960, and "Whale Research Unit, paper by Dr. Mackintosh."
128. First draft of the Scientific Committee's report for 1960 (IWC/12/14), para. 34, sec. A.
129. "Verbatim Report," IWC/12/11, p. 49.

These concerns were brought sharply to the fore by none other than Sidney Holt in a letter to William Herrington several months later. While Herrington was clearly eager to do an end run around those whale scientists who "deal principally with general life history items that have little bearing on the most important problem of . . . assessment of stocks," and, by doing so, to "obtain objective and reasonable conclusions drawn from the data under conditions that are free from industry pressure," Holt— arguably the world's leading expert in mathematical modeling of fisheries at this point—expressed serious concern about the feasibility of slipping away from the cetologists. As Holt pointed out, to perform a successful stock analysis it was necessary to attack assessment problems with "a group of people . . . familiar both with the methods of analysis and with the general biology of the species and the limitations of the quantitative data." As Holt saw it, the latter kind of information resided with the members of the IWC's Scientific Committee, and no group of outsiders, regardless how skilled in mathematically sophisticated analytic approaches (and that included himself—he was already being solicited for a place in the emerging Committee of Three), could hope to succeed without working closely with the wet biology whale types.[130] Holt was thus emphatic that the analytic approaches he had developed with Raymond Beverton in their 1957 classic *On the Dynamics of Exploited Fish Populations* could not simply be "parachuted in" to address the whale problem. Among other things, the models themselves, as they had been deployed in various North Atlantic fish assessments, had always assumed that the number of recruits entering the population in a given year was independent of the number of their parents.[131] As Holt pointed out, "This cannot, I feel sure, be assumed for

130. Herrington's quotes are from "Confidential Memorandum for the Record," 3 July 1961, and Herrington to Holt, 7 April 1961, both in Smithsonian Archives, RU 7165, box 26, folder 5, "Fourteenth International Whaling Commission Meeting (London), 1962." Holt's comments are from Holt's reply to Herrington, 12 April 1961, in ibid.

131. It was not that spawner-recruit models did not exist at this time. In fact, Herrington himself, as a young student of W. F. Thompson, made major contributions to the development of such approaches, which were subsequently taken up by Ricker, and later by Beverton and Holt themselves. I take Holt here to be saying that such models had not yet been made sufficiently reliable to be used in regulatory settings. On the history of spawner-recruit theory, see Tim D. Smith, *Scaling Fisheries: The Science of Measuring the Effects of Fishing, 1855–1955* (Cambridge: Cambridge University Press, 1994), chap. 8.

whales." At the same time, he continued, "we cannot assume [where whales are concerned] that the number of recruits is proportional to the number of parents, because this leads to population models which have no stable equilibrium, and from which no maximum sustained yield can be calculated." What relationship held between the parent stock and subsequent recruitment was an empirical question, and the "experts in population dynamics" would be wholly dependent on biological data derived from Mackintosh and his colleagues for its resolution.

In short, then, as they considered these competing proposals for new and intensive stock assessment work, US officials and their allies faced a subtle problem: it would be necessary to finesse some sort of collaborative relationship between the existing scientific advisers to the IWC and the Committee of Three—one that nevertheless left the Committee of Three adequately independent to serve as arbitrators in the larger quota impasse.

After many hours of plenary-session maneuvering toward consensus on the larger structure of the package deal—namely, that the quota would be raised,[132] Norway and the Netherlands would rejoin, and a commitment to abide by the recommendations of some sort of new population analysis would be made—the focus of negotiations turned to the precise administrative structure that would generate the new stock assessment. While Mackintosh consented to the principle of "an independent second opinion on the condition of the fin whale stock," he continued to hold out for the position that "in order to make any decision on the sustainable level of yield," it would be necessary to "have the full cooperation of scientists who are familiar with this work"—i.e., he wanted his workshop.[133] In the opening afforded by this modest concession, a new proposal emerged, one that involved the "marriage" of the workshop and the independent assess-

132. It is worth noting that US officials agonized over the ugly trade that the "package deal" necessitated: *raising* the quota to secure the promise to lower it later. Particularly explicit on this trade-off is a confidential 1959 memo to the secretary of state from the US delegation at the London IWC meeting, when Kellogg was considering floating such an offer: "U.S. dilemma appears to lie between actively supporting principles of conservation, with consequent break-up of convention and loss of all conservation mechanisms, and tacitly accepting practical arrangement which though repugnant will at least preserve convention and some hope for future." Department of State, incoming telegram, control number 17525, 24 June 1959, Smithsonian Archives, RU 7165, box 25, folder 3, "Eleventh International Whaling Commission Meeting (London), 1959."

133. "Verbatim Report," IWC / 12 / 11, p. 59.

ment.[134] Under this jury-rigged arrangement, Mackintosh's ad hoc Scientific Committee (formerly the Scientific Sub-Committee, but renamed in 1959) would be formally charged to undertake a "workshop" for the purpose of perfecting their approaches to stock analysis.[135] At the same time, however, a new body, appointed by the chairman of the IWC (in consultation with the vice-chairman and the chairman of the Scientific Committee—Mackintosh himself), would *also* be appointed: "three scientists qualified in population dynamics or other appropriate science and drawn from countries not engaged in pelagic whaling." This "special group of scientists should be asked to work with the ad-hoc Scientific Committee" and to report to the IWC on the condition of the Antarctic whale stocks and the level of sustainable yield that could be supported by those stocks.[136]

After a fair bit of nudge-nudge, wink-wink palaver about whether this "marriage" would be "consummated," a clean copy of the concatenated proposal emerged from the stenographers' pool to be voted in, seven to two. The redrafting and the haste, however, had permitted several tough issues to remain somewhat obscure: Would there be one report or two? Would the analysis of this new "Committee of Three" take the shape of a report to the workshop, which would be charged with assessing the smaller body's findings and reporting in turn to the IWC as a whole? The Japanese commissioner made it clear that this was the way he thought it should work. But such an arrangement would obviously have the effect of yoking what was supposed to be an independent assessment to the very body that had proved incapable of generating compelling recommendations for more than a decade.[137] (Shades here of the by now familiar setup whereby tough problems were pushed out to subordinate subcommittees so as to ensure that any annoying findings could be carefully filtered out en route back to the floor of the IWC as a whole.) The issue was left unresolved as a formal matter, but would be worked out in practice in the year to come as the chairman of the IWC assembled the Committee of Three and initiated its work.

Conveniently, the new chairman of the IWC in 1960 was George R. Clark, the Canadian delegate whose proposal for the "marriage" of the workshop

---

134. This idea was originally floated by Australia (see IWC / 12 / 11, p. 59), but in its final form was drafted by Canada's G. R. Clark (see ibid., pp. 76–78).

135. On the renaming of the Scientific Sub-Committee (and the controversy that gave rise to the change), see n. 98 above.

136. "Verbatim Report," IWC / 12 / 11, p. 77.

137. Ibid., p. 85.

and the US package deal had won acceptance in the IWC. His sympathy for the US position was essential in the year to come, and he stayed in close contact with Kellogg as the membership of the Committee of Three was determined and their work initiated.[138] In fact, a beguiling suite of letters survives in Kellogg's correspondence from the autumn of 1960 detailing how the committee ought to be constituted and how it ought to work with respect to Mackintosh's ad hoc "workshop" committee. What is striking about this series of aerograms is the serial ventriloquism they reflect. Consulted by Clark as to how the committee ought to move forward, Kellogg, in turn, consulted Herrington at the State Department, who replied with a letter suggesting how Kellogg might respond. Not leaving anything to chance, Herrington gave Kellogg two verbatim paragraphs, which he "suggested" Kellogg might use in his letter to Clark. For instance, Herrington urged Kellogg to reply (modestly, but firmly):

> While it is beyond the scope of your inquiry to me, I hope you will not mind my outlining a little of our preliminary thinking regarding the program of work of the special group of three scientists who are to undertake an independent investigation of the condition of the Antarctic whale stocks. It seems to us that it might be tentatively planned that this special three-man group hold three meetings during its existence. The first, of about one month's duration, could appropriately take place at Sandefjord to examine the records of the Commission. This would also include the time required for consultation with those members of the ad-hoc Scientific Committee who have been especially concerned with the problem [i.e., the "workshop" group]. The second meeting might be during May and / or June preceding the annual meeting of the Commission. This meeting, also perhaps of about one month's duration, could be tied in with the meeting of the ad-hoc Scientific Committee, and would include sessions with that body. The third meeting might be tentatively planned for the late summer or autumn of 1961, to provide another month for the three-man group to consult together and prepare its report to the Commission. If the special group wished to carry out analysis of data not possible within these time limits, it might be able to arrange with members of the

138. See Clark to Kellogg, 21 November 1960, and Clark to Kellogg, 6 March 1961, Smithsonian Archives, RU 7165, box 26, folder 5, "Fourteenth International Whaling Commission Meeting (London), 1962."

ad-hoc Scientific Committee for assistance and cooperation in such analysis.[139]

Kellogg, in turn, wrote back to Clark in exactly these words.[140]

While the timing of these meetings was ultimately changed—everything was pushed back by difficulties in securing the funding to support this work and by delays in the readmission of Norway and the Netherlands—this basic outline of the modus operandi of the Committee of Three and its relationship to Mackintosh's ad hoc "workshop" committee proved determinative.[141] The Committee of Three would work closely with Mackintosh's group at a pair of collaborative meetings (one held in Rome in the spring of 1961, the second held in Seattle in December 1962), but the final drafting of the stock assessment report would be done by the Committee of Three alone. While this report was submitted to the IWC directly, Mackintosh's group also prepared their own "report on the report" in the spring of 1963, which was also passed on to the IWC for consideration. By that point, however, largely as a result of the deft management of the whole process, the members of Mackintosh's Scientific Committee were stakeholders in the final product and not positioned to dissent.

It is interesting to consider the degree to which the development of an assessment process that incorporated (and thereby neutralized) dissent-

139. Herrington to Kellogg, 19 October 1960, in ibid.

140. Kellogg to Clark, 31 October 1960, in ibid.

141. The United States succeeded in securing the independence of the Committee of Three—an issue, recall, that had been somewhat finessed in the endgame of the 1960 plenary session negotiations around the package deal—via the strategic position of Clark as chairman of the IWC. Solicited to serve on the Committee of Three, Douglas Chapman inquired early in 1961 about its status: "Do we prepare a report of our own independently of any action of the ad-hoc Scientific Committee, or is it the ad-hoc Scientific Committee that prepares the report with the Special Committee as advisors to and participants with the larger committee[?] Or is it that the final resolution of the question of reports will be left for later decision by the committees concerned?" Chapman to Clark, 8 March 1961, copy, in ibid. To which Clark (still in close contact with Kellogg and Herrington) replied, "With regard to the duties of the Special Committee, it is my view that heir [*sic* for "their"] report is to be an independent one prepared by the three members of the Special Committee. The report of the three scientists does not have to be cleared with or through the Ad-hoc Committee or the Scientific Committee—it is to be an independent report submitted by the three members to the Commission." Clark to Chapman, 13 March 1961, in ibid. It is notable that the opinion of Japan on this matter was never consulted.

ers reflected responsiveness on the part of those committed to securing consent on a workable quota: they had to adjust to the peculiar exigencies of regulatory science in the international arena. For instance, it is clear from their correspondence and other documentation that, initially, Kellogg and Herrington believed that what was most important to secure from the Committee of Three was simply *a number* around which conservation-minded delegates might rally.[142] Indeed, early on, Holt expressed concern that Herrington's enthusiasm for leveraging the IWC by means of a hasty MSY analysis would basically involve a kind of boondoggle: "It . . . has seemed to me that the special three-man committee, even though they make some independent report, would essentially merely make up the numbers." One senses that Herrington and Kellogg just might have been satisfied with such an arrangement (though Holt himself, anxious to protect the credibility and power of the methods he had done so much to develop, was highly ambivalent about throwing them ill-armed into the quagmire of the IWC circa 1960). But over the course of the late 1950s and early 1960s, it became increasingly clear that "the number" that might save the IWC had to be produced in a very special way, not just technically, but also socially and administratively. Perhaps the most forceful lesson concerning the necessary social basis for achieving consensus via science was given to Herrington by the Soviet fisheries negotiator, Aleksandr Akimovich Ishkov, who in a private meeting forwarded a particularly naked power-politics relativism, without apology. In the course of these discussions, as Herrington would report, "it became clear that Ishkov considered that no scientific conclusions, no matter how clear-cut and convincing, would be valid for the Netherlands or apparently for anyone who did not participate in the development of the conclusions." Herrington expressed unconcealed contempt for this conspiracy-theory conception of scientific objectivity: "Presumably in his philosophy there are no basic scientific truths," he noted in dismay.[143] Ishkov's proto-strong-program social-constructivist realpoli-

142. In his 1961 report to the State Department, Kellogg put it this way: "The scientific work so far, although showing a continuing downward trend in the whale stocks, has not been able to establish a quantitative value for the level of exploitation which can be sustained. Thus there is no realistic and objectively established point about which the conservation-minded delegates can rally." See Smithsonian Archives, RU 7165, box 28, folder 5, "Reports to the Secretary of State by the U.S. Commissioner to the IWC, Meetings #12–14, 1960–1962."

143. "Confidential Memorandum of Conversation," Ottawa, 8 February 1962, Smithsonian Archives, RU 7165, box 26, folder 4.

tik proved prescient, however, and while Herrington continued to use the language of objectivity and independence in characterizing the work of the Committee of Three, he nevertheless did everything he could to ensure that their findings were generated and forwarded in such a way as to be diplomatically invulnerable. A technically invulnerable analysis was certainly not sufficient, perhaps not actually necessary, possibly not even feasible.

While, as I have argued above, careful manipulation of committee and reporting structures did much to create the conditions of possibility for this invulnerability, a great deal of the credit for actually managing the process must go to Douglas G. Chapman himself (figure 5.2), a professor of mathematics at the University of Washington who served as the chairman of the Committee of Three and took a leading role in handling its work, both technical and diplomatic. Turning now to the second part of my discussion of the Committee of Three episode—a closer look at its actual activities—it will be appropriate to begin with a brief biographical sketch of its leading figure.[144]

A mathematically precocious Canadian from the backcountry of Saskatchewan, Chapman matriculated at the tender age of fifteen, took a double first degree in economics and mathematics, and went on to receive a master's and a PhD in mathematics from the University of California at Berkeley, writing a dissertation on theoretical approaches to sampling problems.[145] After brief stints in the Department of Mathematics at the

144. While not the chairman, Sidney Holt was in many ways an even more visible presence on the Committee of Three (in view of his pioneering work on dynamic pool models, as well as, more mundanely, his greater geographic proximity to IWC affairs—being in Rome, he was sometimes able to attend meetings Chapman had to miss), and he has remained active in IWC affairs ever since. Holt, with whom I have been in contact, is working on his own history of this period; I have not consulted his private papers. My account here is therefore based to a considerable degree on supplementing IWC archival material and Kellogg's State Department archival material with the thirteen boxes of Douglas G. Chapman papers in the Special Collections of the University of Washington Library (accession number 4581–001). These papers contain Chapman's private notes on the Seattle meeting of 1962 (boxes 9 and 13 in particular), along with some relevant correspondence (although it appears that much of Chapman's early correspondence is not represented in the collection). I have also made use of supplementary material on and by Chapman in the holdings of the National Marine Mammal Laboratory, Seattle, Washington. Particular thanks are here due to Sonja Kromann and Jeff Breiwick.

145. "Program for the Final Examination for the Degree of Doctor of Philosophy of Douglas George Chapman," Friday, 10 June 1949. The dissertation was entitled "Ap-

FIGURE 5.2
The mathematician: Douglas Chapman. (Courtesy of the University of Washington Libraries, Special Collections.)

University of British Columbia and back at Berkeley, he settled into a junior professorship at the University of Washington, where he would remain through an active professional life of more than thirty years, rising to chair the university's biomathematics group and eventually directing the Center for Quantitative Science in Forestry, Fisheries, and Wildlife.[146] Shortly after his arrival in Seattle, Chapman found himself drawn into increasingly applied statistical work through connections to research by US Fish and Wildlife personnel on the Alaska fur seal herds of the Pribilof Islands. During a Guggenheim Fellowship year at Oxford, Chapman received notice of his appointment as a scientific adviser to the US delegation to the North Pacific Fur Seal Treaty Conference, a post that led in turn to a sequence of positions as statistical adviser to the North Pacific Fur Seal Commission and, subsequently, to the International North Pacific Fisheries Commission.[147] It was through this work that Chapman initially came to the atten-

---

plications of the Hypergeometric Distribution to Sample Census Problems; A Class of Lower Bound for the Variance of an Estimate," and was directed by Jerzy Neyman. See University of Washington Library, Special Collections, Douglas G. Chapman Papers, box 1, file "Faculty."

146. He would also serve as dean of the College of Fisheries from 1971 to 1980. See Douglas G. Chapman *curriculum vitae*, University of Washington Library, Chapman Papers, box 1, file "Faculty."

147. For an interesting account of his movement through these advisory roles, see "Testimony of D. G. Chapman—Third Draft," prepared in connection with his service as an expert witness in a trial involving sampling problems and the practices of Pacific

tion of fisheries officials in the Department of the Interior in Washington and eventually found his name forwarded as the United States' top choice to run the Committee of Three. Tall, hardy (he had served as a field meteorologist with the Canadian military during World War II), and a devastating poker player, Chapman was a personable and compelling presence with a keen grasp of quantitative approaches to wildlife management. He would prove an effective leader through the IWC's special assessment process.

For just one example of how Chapman's particular mathematical expertise in sampling enhanced his authority with respect to Mackintosh and the community of IWC scientific advisers, it is worth recalling Mackintosh's durable passion for whale marking. As I have discussed above, Mackintosh continuously directed the energies of the IWC's scientific advising community to the task of delivering and recovering the numbered Discovery-style slugs that had originally been developed to study migration patterns in the 1920s. Over the course of the 1950s, however, Mackintosh increasingly cast about for new and supplementary justifications for this work, and the reports he prepared as chairman of the Scientific Sub-Committee (and the Scientific Committee itself) began to pitch whale marking not simply as a technique for figuring out how whales came and went, or as a means for confirming aging estimates, but rather as an actual tool for stock assessment—increasingly the urgent issue in the IWC. As he put it in the Scientific Committee report of 1954, "The Committee are strongly of the opinion that the most convincing evidence as to the condition of the stock of fin whales is to be sought in whaling marking [*sic*]."[148] While Mackintosh was aware that the scale of the marking campaigns he had succeeded in mounting meant that the sample size was quite small, he continued to plug this new application of his favorite research practice, offering encouraging

---

Telephone Northwest, in the University of Washington Library, Chapman Papers, box 1, file "Speeches and Writings." For a sense of his actual work using mark recovery data and sampling techniques to estimate biological populations in this period, see D. G. Chapman, K. Kenyon and V. Scheffer, "A Population Study of the Alaska Fur Seal Herd," *U.S. Fish and Wildlife Service Special Scientific Report: Wildlife* 12 (1954): 1–77. See also "The Estimate of the Size of a Stratified Animal Population," *Annals of Mathematical Statistics* 27 (1956): 375–89, with C. O. Junge. The National Marine Mammal Library also holds a copy of an unpublished report done by Chapman in 1957: "Population Estimates of Pribilof Fur Seal Pups," National Marine Mammal Laboratory, Douglas Chapman Papers.

148. "Draft Report of Scientific Committee," IWC / 6 / 15, p. 4, para. 18. "*Whaling marking*" certainly feels like a Freudian slip.

words like these in 1957: "Results might come slowly, but it seemed well worthwhile to continue."[149]

With Chapman entering the picture in 1961, however, there was no place to hide. At the Rome joint meeting of the Committee of Three and the "workshop" group of Scientific Committee members, Chapman reviewed Mackintosh's information about the number of whale marks out in the oceans and the number recovered, and he quickly determined that no statistical technique existed by which Mackintosh's marking data could be transmuted into a direct sampling–style population assessment. Chapman had built his professional reputation as a mathematician on his ability to wring the very last drops of information from such data sets, so it was impossible to gainsay his verdict. In the wake of this first joint gathering, then, Mackintosh was no longer able to sell marking under the banner of a sample-based population census, but fell back on suggesting its more limited use in clarifying population structure.[150] The idea of marking for the purpose of sampling had always been something of a post hoc rationale for work born in the Harmer era to answer Harmer's questions about seasonal migration. As I will show below, however, Chapman did spot a clever way to use mark recovery data to estimate mortality rates, and that method would prove useful in the Committee of Three's final report of 1963.

But before I take up that report in some detail, and endeavor to show exactly how the Committee of Three forwarded a compelling case for the depleted condition of the Antarctic stocks, it will be helpful to spend a

149. Scientific Sub-Committee report for 1957 (IWC / 9 / 2), p. 5, para. 9. Admittedly here, Mackintosh had already begun to wobble on how far marking could go toward providing estimates of population sizes.

150. Interestingly, this aspect of the story has been missed by commentators who have accepted Recommendation Number 4 of the First Interim Report of the Special Committee of Three Scientists ("that the scale of marking operations should be substantially increased so that the results will be more useful for stock assessments") as evidence that the Committee of Three endorsed more marking work. However, this is to overlook the second half of that recommendation, which reads, "but if this is not practicable, the marking work should be directed at obtaining data on movements and migrations" (IWC / 13 / 9, p. 2). In that latter half of the sentence, Chapman effectively put paid to Mackintosh's ambition to use marking for population assessment, since the scale of marking that would be needed to ramp up to statistically significant levels for sampling was emphatically *not* practicable. Patricia W. Birnie (in her *International Regulation of Whaling: From Conservation of Whaling to Conservation of Whales and Regulation of Whale-Watching* [New York: Oceana Publications, 1985], vol. 1, 305), misconstrues the episode.

moment reviewing how the committee went about its work during its first two years, paying particular attention to how it interacted with the larger scientific advising community around the IWC and thereby succeeded in organizing the collation of the necessary statistical and biological data.

That Rome meeting of 24 April to 6 May 1961 was the first occasion for Allen, Chapman, and Holt to meet and discuss their assignment. It was also the inaugural convocation of Mackintosh's special "workshop" gathering of his ad hoc Scientific Committee, which saw more than a dozen whale biologists from half a dozen countries assembled under Mackintosh's chairmanship. That this latter meeting convened at FAO headquarters nicely, if informally, made Holt the host, even though, as an administrative matter, neither he nor any of the members of the Committee of Three was formally a member of Mackintosh's group. Feeling out a working relationship, the Committee of Three had the good sense to present themselves as enthusiastic and deferential participant-observers in the workshop, and "asked if they might take full part in the . . . proceedings."[151] This politic (and pragmatic) overture set a collaborative tone and positioned the Committee of Three to direct, to a considerable degree, the efforts of the workshop group.

The bulk of the Rome meeting was dedicated to arranging a series of spreadsheet forms upon which critical data held by different national research programs, as well as data held by the Bureau of International Whaling Statistics, might be tabulated. These included a form on which catches were to be recorded by season, by geographic area (in 10° graticules; see figure 5.3 for the coding system), and by species, along with corresponding biological data on sizes and ages; another form that correlated sexual maturity with age and length; and other forms aimed at collecting the information that would be necessary to develop a more refined index of catch per unit effort (these were forms E1 and E2, which included information on weather, catch matériel, lost hours on specific expeditions due to repairs or a stop-catch condition, etc.) (see figures 5.4–5.6 for examples). It was hoped that these forms, filled out and brought together over the next year, would provide the raw material for a machine-assisted computational analysis: keyed in on IBM data punch cards (see figure 5.7 for the punch card template that corresponds to forms E1 and E2 in figure 5.6), the data could be crunched to derive mortality rates from age-length tables not merely in a general way (as Ottestad, Laws, and others had been doing for years), but

151. "Report of the Special Ad-Hoc Scientific Committee Meeting," IWC / 13 / 8, para. 3.

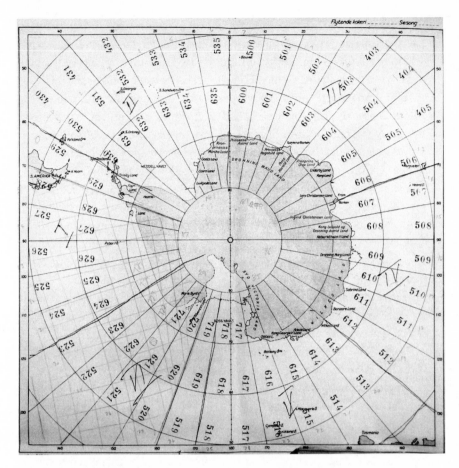

FIGURE 5.3 Spatializing catch statistics: The geographic catch coding system by sector. (Courtesy of the University of Washington Libraries, Special Collections.)

with refined attention to geographic region, season, and even month. It would also be possible to test the different aging techniques against one another in population models and to derive newly robust estimates for recruitment rates (by juxtaposing catches of a defined parent stock and a corresponding offspring class).[152] Conveniently, the basic BIWS catch data

152. See app. 8 of "Meeting of Scientific Committee and Committee of Three Scientists (Seattle, 6th–19th December, 1962), Provisional Report" (IWC / 15 / 6). It is interesting to consider to what degree the use of the electronic computer affected the perception and reception of the Committee of Three's efforts. I have not investigated this question deeply, but there is surprisingly little discussion of the import of the computer-assisted

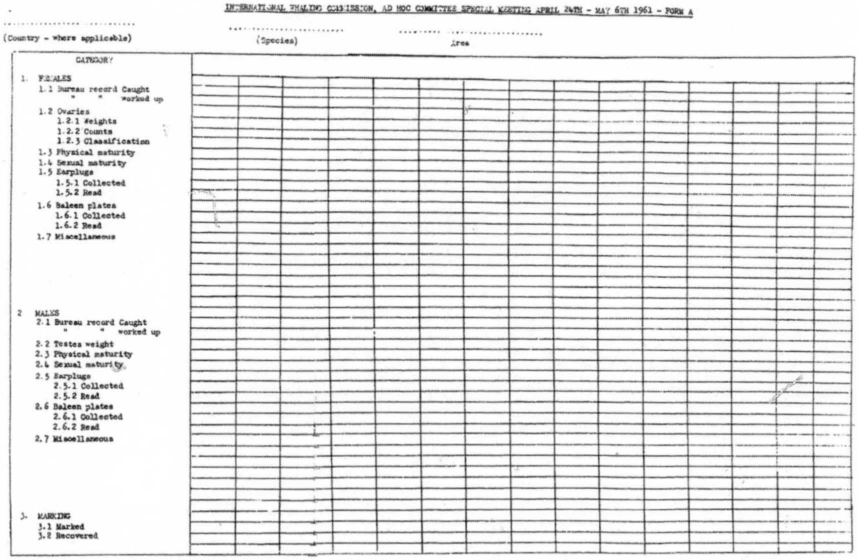

INTERNATIONAL WHALING COMMISSION, AD HOC COMMITTEE SPECIAL MEETING APRIL 24TH - MAY 6TH 1961 - FORM A
(Country - where applicable)
(Species)
Area
CATEGORY
1. FEMALES
1.1 Bureau record Caught
" " worked up
1.2 Ovaries
1.2.1 Weights
1.2.2 Counts
1.2.3 Classification
1.3 Physical maturity
1.4 Sexual maturity
1.5 Earplugs
1.5.1 Collected
1.5.2 Read
1.6 Baleen plates
1.6.1 Collected
1.6.2 Read
1.7 Miscellaneous
2 MALES
2.1 Bureau record Caught
" " worked up
2.2 Testes weight
2.3 Physical maturity
2.4 Sexual maturity
2.5 Earplugs
2.5.1 Collected
2.5.2 Read
2.6 Baleen plates
2.6.1 Collected
2.6.2 Read
2.7 Miscellaneous
3. MARKING
3.1 Marked
3.2 Recovered

FIGURE 5.4 Tabulating the data: Form A. (From "Report of the Special Ad-Hoc Scientific Committee Meeting," IWC/13/8, appendix.)

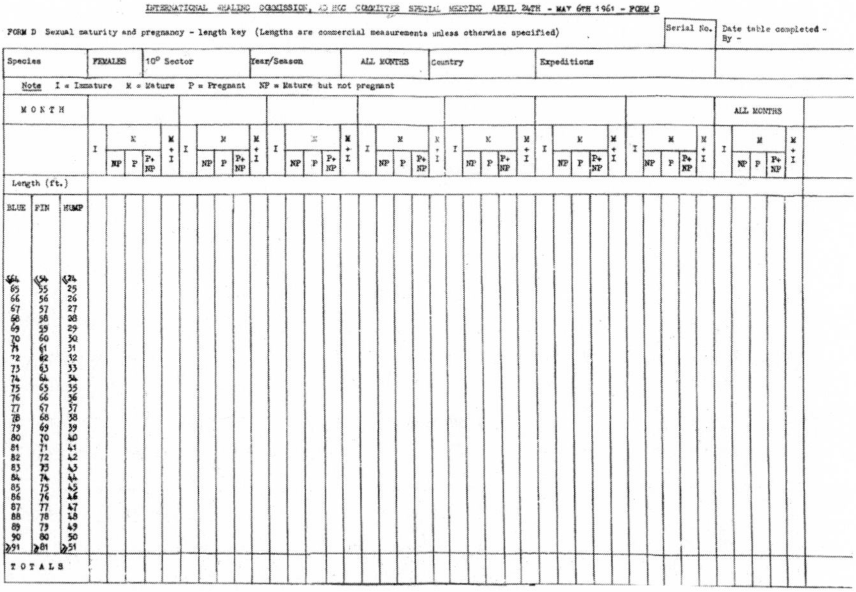

INTERNATIONAL WHALING COMMISSION, AD HOC COMMITTEE SPECIAL MEETING APRIL 24TH - MAY 6TH 1961 - FORM D
FORM D Sexual maturity and pregnancy - length key (Lengths are commercial measurements unless otherwise specified)
Serial No.
Date table completed -
By -
Species
FEMALES
10° Sector
Year/Season
ALL MONTHS
Country
Expeditions
Note I = Immature M = Mature P = Pregnant NP = Mature but not pregnant
MONTH
ALL MONTHS
I
M
NP P P+NP
M+I
Length (ft.)
BLUE FIN HUMP
TOTALS

FIGURE 5.5 Tabulating the data: Form D. (From "Report of the Special Ad-Hoc Scientific Committee Meeting," IWC/13/8, appendix.)

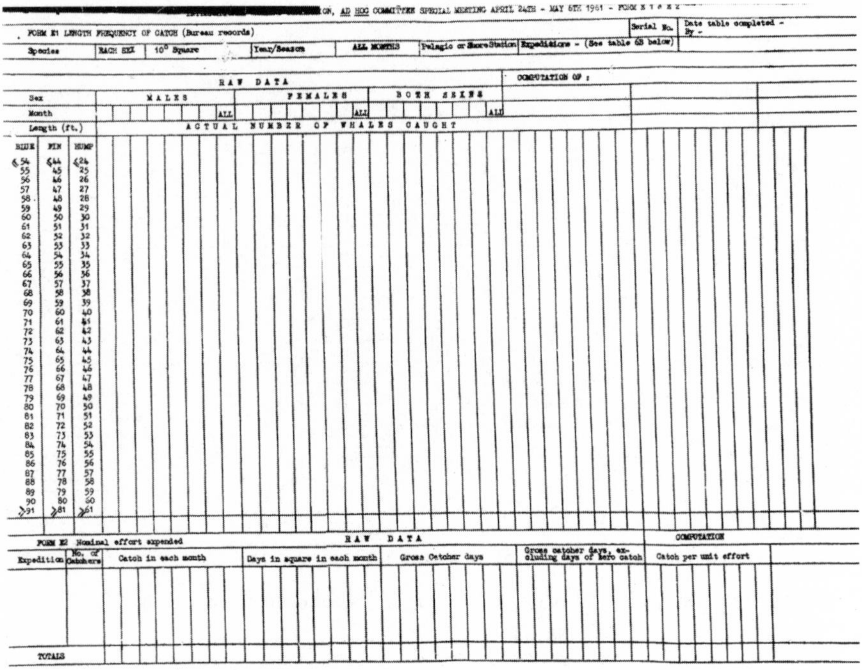

FORM E1 LENGTH FREQUENCY OF CATCH (Bureau records)

| Species | EACH SEX | 10° Square | Year/Season | ALL MONTHS | Pelagic or Shore Station | Expeditions - (See table 6B below) | Serial No. | Date table completed - By - |

RAW DATA / **COMPUTATION OF :**

Sex	MALES		FEMALES		BOTH SEXES	
Month		ALL		ALL		ALL
Length (ft.)	ACTUAL NUMBER OF WHALES CAUGHT					

RIDE	FIN	HUMP
<54	<44	<25
55	45	26
56	46	27
57	47	27
58	48	28
59	49	29
60	50	30
61	51	31
62	52	32
63	53	33
64	54	34
65	55	35
66	56	36
67	57	37
68	58	38
69	59	39
70	60	40
71	61	41
72	62	42
73	63	43
74	64	44
75	65	45
76	66	46
77	67	47
78	68	48
79	69	49
80	70	50
81	71	51
82	72	52
83	73	53
84	74	54
85	75	55
86	76	56
87	77	57
88	78	58
89	79	59
90	80	60
>91	>81	>61

FORM E2   Nominal effort expended

| Expedition | No. of Catchers | Catch in each month | Days in square in each month | Gross October days | Gross October days, excluding days of zero catch | Catch per unit effort |

RAW DATA / **COMPUTATION**

TOTALS

FIGURE 5.6   Tabulating the data: Forms E1 and E2. (From "Report of the Special Ad-Hoc Scientific Committee Meeting," IWC/13/8, appendix.)

were already available on punch cards (figures 5.8 and 5.9), which would simplify the particularly crucial work of adjusting catch-per-unit-effort indices for efficiency.[153]

aspects of the analysis in contemporary documentation. There can be no doubt, however, that the use of computers changed the dynamics of the debate: for instance (as would be clearly revealed in 1963 as the different groups squabbled for control of the data cards), only those with access to the keyed-in data were in a position to second-guess the results. For discussion of the significance of computer modeling in social decision making (largely after 1970), see Donella H. Meadows and J. M. Robinson, *The Electronic Oracle: Computer Models and Social Decisions* (Chichester: John Wiley & Sons, 1985). Also useful is Fernando Elichirigoity's study of the Club of Rome, *Planet Management: Limits to Growth, Computer Simulation, and the Emergence of Global Spaces* (Evanston, IL: Northwestern University Press, 1999).

153. For an excellent idea of the preparatory work that was necessary to key in the BIWS data, see the translated letter from Vangstein (director of the BIWS) to Ruud of 8 June 1961, requesting sixteen clarifications on the required statistical operations (e.g., "If an expedition on the day has caught no baleen whales but

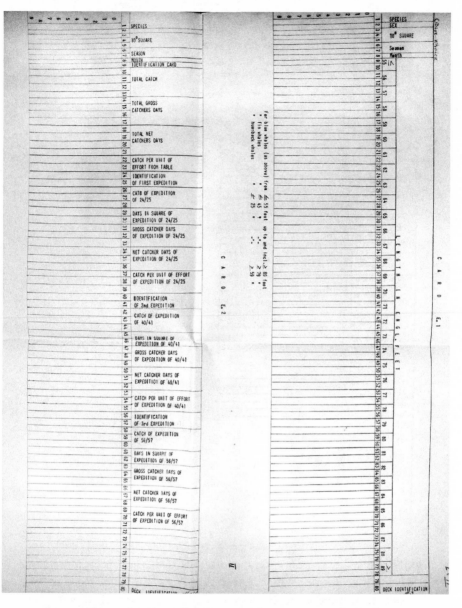

FIGURE 5.7 The punchcard template: Formatting layout corresponding to forms E1 and E2. (Courtesy of the University of Washington Libraries, Special Collections.)

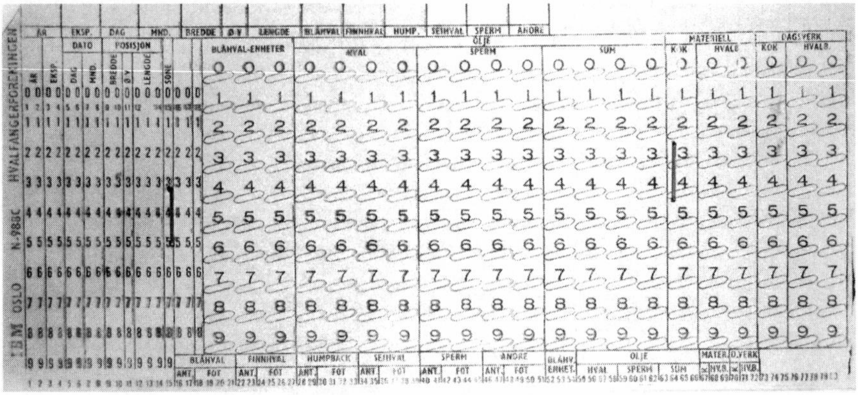

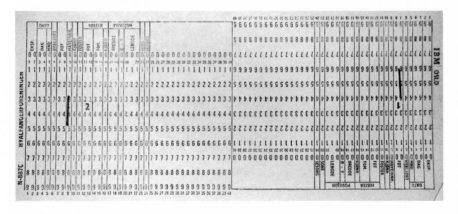

FIGURE 5.8   A new tool: Computing the kills. (Courtesy of the University of Washington Libraries, Special Collections.)

FIGURE 5.9   A new tool: Crunching the numbers. (Courtesy of the University of Washington Libraries, Special Collections.)

While, predictably, the dedicated cetologists were particularly interested in the discrepancies that continued to beset their aging techniques, it is clear from the documentation that emerged from the Rome meeting that already the Committee of Three (which also met separately several times during the two weeks) was urging the workshop group to attend to the question of how catch per unit effort could be adjusted by means of some

one or more sperm whales, is that day to be regarded as 'No Catch Day' or 'Operating day'?"), in Smithsonian Archives, RU 7165, box 20, folder "London, 1961, correspondence."

workable "index of fishing power." Because catch per unit effort served as a proxy for stock size in the basic modeling techniques familiar to Holt, Allen, and Chapman, they were clearly intent on securing better-quality data for this part of their analysis. They also understood that non-catch information—about sea and wind conditions, bunkering, etc.—recorded in expedition logbooks (some of which were available at the BIWS) would be useful for cleaning up CDW data (since it could be used to correct for those perturbations in the CPUE metric that were the product of factors other than the number of whales out there to be caught), and a group was assigned to see what could be turned up for that purpose.

By the time the Rome meeting broke up, then, responsibility for data collection had been distributed to the different members of the workshop, and templates for the collation of those data had been prepared. As Chapman put it in a private letter written shortly after returning to Seattle from the meeting, "Concerning the collection and collation of the biological data, we stressed the urgency of this but our Committee [the Committee of Three] is afraid that there may be some foot-dragging because they [the workshop whale biologists] are afraid the job is too big and can't be undertaken without additional staff, etc." At the same time, Chapman was relatively optimistic that, given the raw catch data already on IBM cards at Sandefjord, "it should take little time to get what is wanted," and (crucially) that "[w]hen a corrected unit of effort has been determined an analysis based on catch and effort alone can be undertaken and will, I am sure, yield satisfactory results."[154]

About a month later, in June 1961, Chapman was back on a plane and headed to London, where, together with Holt, he attended the IWC's Scientific Committee meeting as an "observer" and continued to cajole and pump the members to consolidate their data on the relevant forms. Afterward he was able to make visits both to Mackintosh at NIO (to review what was, after all, the world's largest repository of whale marking and aging information) and to Sandefjord to check in on what would be involved in having one phase of the computational work on CPUE done there.[155] He

154. Chapman to Herrington, 9 May 1961, Smithsonian Archives, RU 7165, box 26, folder 5, "Fourteenth International Whaling Commission Meeting (London), 1962."

155. In the end, Vangstein at Sandefjord worked up the E1 (9,500) and E2 cards (8,000). E1 was a card for length (by species, sex, region, season, and month); E2 was the critical "CPUE" card, which tabulated catch and number of catch days over several intervals (for Chapman's template for these cards, see figure 5.7).

had to have found things at NIO in some disarray, in view of Mack's uncertain professional status and the soon-to-be-announced departure of Dick Laws (the most mathematically sophisticated young scientist in the biology division, who had grown increasingly exasperated with Mackintosh's leadership and the second-class citizenship of biologists under Deacon's directorship). Whatever the American scientist thought of the situation, it is certain that Chapman left Mackintosh with an increasingly sharp sense of the urgency of supplementing the quantitative capabilities of his whale research group, since shortly thereafter Mackintosh got in touch with Raymond Beverton at Lowestoft Laboratory and requested that he be "lent" someone competent in the new fisheries modeling techniques. It was John Gulland who subsequently visited NIO and was briefed on the biological side of cetacean stock assessment by Sidney Brown and the newly hired John Bannister.[156] Gulland subsequently joined the effort and eventually became the fourth member of an expanded Committee of Three (which thereby became, as noted above, the Committee of Four).

While in the spring of 1961 it seemed that the next joint meeting of Mackintosh's group and the Committee of Three might take place in December of that year, financial problems and delays in organizing the data pushed things back by a full year. It was thus not until December 1962 that the whole group again assembled, this time in Seattle, at the University of Washington, where Chapman (not Mackintosh) would chair another two weeks of continuous work—a marathon session that generated the core of the Committee of Three's final report.[157] This meeting saw some twenty-five representatives from nine nations gather to hammer out the final assessments, many of whom came bearing completed forms ready for tabulation. Chapman complained (if quietly) that things would have been facilitated had many of the cards come in sooner, but it was necessary to work with what they had at hand. It is clear from Chapman's files on the meeting, as well as from the notes of another US participant, Dale Rice, that a great deal of the work of the two weeks consisted of actual arithmetic operations—

156. John Bannister, interview by the author, 26 October 2005.

157. Interestingly, Chapman initially adopted the title "Convenor." See University of Washington Library, Chapman Papers, box 13, file "International Whaling Commission." Bannister, who was in attendance at this meeting, recalls that weather delayed the arrival of the UK delegation and put Mackintosh in a very bad mood for much of the meeting. It is not clear to me whether Mackintosh's not being accorded the chair was a product of this delay, but it seems quite possible. John Bannister, interview by the author, 26 October 2005.

tallying, counting up, and keying in the data.[158] Chapman had used his position in the UW Department of Mathematics to secure time from the university's computer center, and the student newspaper, catching wind of the undertaking, published a puckish article on the hundred thousand IBM cards that would be churning through the mainframe before Christmas: a "Whale of a Lot of Data" was hitting campus.[159]

It was not, however, as much data as Chapman and the other members of the Committee of Three would have liked.[160] The question was, could the numbers in hand be made to *talk*? In turning now to a closer look at how the Committee of Three answered this question (in the affirmative), it is important to keep in mind just how opportunistic—how *resourceful* (not to say "unsystematic")—Chapman, Holt, and Allen (and later Gulland too) had to be to get what they needed out of what they had in the time available and with the limited budget they had been provided.[161] The

158. This account is consistent with the memory of Bannister, who recalls sitting at manual calculating machines doing sums for hours during this meeting (interview by the author, 26 October 2005). The C forms (which gave age determinations on animals, again by species, sex, region, season, and month), and the D forms (maturity and pregnancy information, similarly configured) were actually transferred to punch cards in Seattle. Some 46,000 C cards and 11,700 D cards were used in the final calculations. Perhaps the best feel for the work in progress in Seattle is conveyed by Dale Rice's scribbled listing of the status of the different report appendices at the close of the meeting: documents 5, 7, and 8 were "being prepared"; a revision of 6 and a copy of 10 were "being typed"; document 2 was "not yet prep." Together, these appendices constituted more than half of the documentation to come out of Seattle (see National Marine Mammal Laboratory, Seattle, WA, Dale Rice Papers.).

159. University of Washington Library, Chapman Papers, box 13, clipping.

160. For the actual numbers of the different kinds of punch cards produced from spreadsheet forms at the meeting, see University of Washington Library, Chapman Papers, box 13, file "International Whaling Commission."

161. In connection with the question of resources, it is worth noting Chapman's back-of-the-envelope calculations of computational costs for runs of the data cards. The university's computer center could handle 800 cards per minute at a cost of $36 per hour. It appears that the computations were done on two machines, an IBM 709 (a vacuum tube model) and an IBM 1401 (one of the earliest fully transistorized data processing systems, which had come out in October 1959), for a final total of something under £1,500 (circa $30,000 in current US dollars). See accounts in University of Washington Library, Chapman Papers, box 13, file "International Whaling Commission." As far as time constraints were concerned, see William Herrington to Sidney Holt, 7 April 1961, Smithsonian Archives, RU 7165, box 26, folder 5, "Fourteenth International Whaling Commission Meeting (London), 1962."

brief discussions of the technical side of the Committee of Three's report available in secondary sources have tended to emphasize a tidy MSY analysis for fin and blue whales, but the actual document to emerge from the Seattle gathering as the "Provisional Report of the Joint Meeting of the Scientific Committee and the Committee of Three Scientists" was in fact a very baggy collection of some fifteen separate essays that deployed several different and not entirely consistent attacks on the problem of stock assessment. There were quite a few loose ends, some of which remained as the document was hammered into finished form by March 1963.[162] Because this last and most complete version was circulated so late, just a few weeks before the 1963 IWC meeting, it is with the provisional version— the more raw and direct product of the intensive Seattle meeting—that I will primarily concern myself here, since it is that version that had to do the bulk of the work of compelling assent on the seriousness of the stock declines. And compel assent it did: Mackintosh even telephoned Kellogg when his plane stopped over on the East Coast en route back to the United Kingdom from Seattle to inform him that "although the . . . report is not yet in final form, a drastic reduction in the catch must be accepted."[163]

162. To trace the development of the document, it is necessary to compare the first draft, completed in December 1962 (for a copy, see National Marine Mammal Laboratory, Dale Rice papers) with the first circulated form, the first "Provisional Report," IWC / 15 / 6, and both of these in turn with the "Final Report of the Committee of Three Scientists," IWC / 15 / 9. (NB: The two former documents were reports issued "jointly" by the Scientific Committee *and* the Committee of Three; the latter one by the Committee of Three alone. Note further that the latter document circulated in penultimate form as the "Second Interim Report of the Committee of Three Scientists," IWC / 15 / 7.) Significant, for instance, is the fact that a full analysis was done only for female fin whales in the provisional report; male fins, as well as male and female blue whales, were only done later and submitted with the final report.

163. Kellogg to J. L. McHugh, 20 December 1962, reporting Mackintosh's call, Smithsonian Archives, RU 7165, box 27, folder 1, "Fifteenth International Whaling Commission Meeting (London), 1962." This letter also points to what becomes an increasingly serious problem in reconstructing this episode: the growing role of telephone communications among the players. For instance, it is only because William Herrington was again and again unable to get Sidney Holt on the telephone in the spring of 1961 that we have the enormously instructive letter of 7 April, housed in the Smithsonian Archives, RU 7165, box 26, folder 5, "Fourteenth International Whaling Commission Meeting (London), 1962."

So what had been done to make the opposition knuckle under?

In his own reconstruction of the Committee of Three's analysis, written almost a decade after the fact, Chapman neatly summarized the approaches used to achieve population estimates. In his retelling, the success of the undertaking hinged on the converging estimates afforded by several different approaches. A total mortality rate could be derived from the differential rate of recovery of whale marks delivered at two different times.[164] This agreed reasonably well with mortality rates derived from age analysis, a technique that could also provide recruitment rates. A study of changes in catch per unit effort afforded corroborating evidence of stock declines.[165] Not surprisingly, Chapman's version of the story conveys a strong sense of an orderly and systematic approach to, and eventually solution of, the whole problem of southern baleen whale stock assessment. While no part of his story is factually incorrect, it is important to emphasize the degree to which this sort of rational reconstruction belies the improvisational character of the committee's actual attack on the problem. It proved necessary to adjust the approaches used according to what data actually came to hand; there were several false starts, and a good deal of partially complete or abandoned work was left on the cutting room floor.

For instance, after the Rome meeting, the members of the Committee of Three put their heads together to work up a brief outline of a Beverton and Holt–style analysis of population dynamics in baleen whales. The result was a purely theoretical paper that posited several possible formulae for the relationship between female stock size and recruitment and discussed how these formulae might be evaluated using the data the committee hoped eventually to see. While this paper included the fanciest algebraic work of anything in the final report by the Committee of Three, and while it served as a somewhat forbidding initial appendix to the whole analysis, very little of the work represented in this paper entered into the actual calculations

164. That is to say, as Chapman put it, "The ratio of recoveries of whales marked in year 1 to the recoveries of whales marked in year 2 reflects the proportion of marked whales from group 1 that died in the intervening year."

165. For Chapman's account, see "The Plight of the Whales," 1 April 1971, draft paper for the Applications Volume of Joint ASA / NCTM Committee, University of Washington Library, Chapman Papers, box 9. I have also benefited from Kees Lankester's "Thought history of MSY and its consequences in baleen whale management," SC / A86 / CA7, also in the Chapman Papers, box 12, file "Comprehensive Assessment."

made in the course of the committee's work in Seattle, where less slick approaches were found sufficient to make the case.[166]

Similarly, while there had been a great deal of emphasis in Rome on the importance of establishing detailed correction factors that would adjust the CDW index for changing efficiency, by the time the Committee of Three convened with the workshop group in Seattle, none of the more sophisticated efforts to generate such factors had proved successful.[167] It had originally been thought that an efficiency quotient might be applied to each expedition (reflecting its relative effectiveness) and that seasonal quotients that accounted for weather might be applied on discrete geographic bases. But when the time came to get down to work in December 1962, the correction factor actually deployed turned out to be a very gross metric: it seemed, on the basis of a few regression analyses, that the relationship between catching power and catch vessel tonnage was tight enough that, for practical purposes, a CDW index corrected simply for increasing tonnage of whaling matériel on the whaling grounds would have to do.[168] (Later efforts to corroborate the suitability of this rule of thumb with a more comprehensive series of computer computations were in fact not successful.)

What is particularly significant about this decision is the extent to which it reverberated through all of the other analyses undertaken by the committee: Chapman's approach for deriving mortality rates from differential mark recovery made use of a correction factor based on increased catcher efficiency through time; corrected catch per unit effort was essential to the De Lury analysis, which proved so important to the committee's final report; and, finally, it would have been impossible to get going on the Schaefer analysis of sustainable yield without the corrected catch-per-unit-effort indices, which functioned both in making estimates of sustainable yield and as a general index of stock size.[169] Similarly gross tweaking of

166. The basic irrelevance of this initial take on the problem (IWC / 15 / 6, app. 1) was essentially elided in the Committee of Three's final report (see IWC / 15 / 9, p. 4).

167. For the efforts to produce some sort of weather index, see IWC / 16 / 6, app. 3 and 4.

168. This proposal was originally made by Hylen, but it proved very difficult to substantiate later using additional statistical material provided by the Bureau of International Whaling Statistics. For a defense of the approach, see IWC / 15 / 6, app. 5.

169. The basic De Lury analysis involved plotting corrected catch per unit effort against total cumulative catch. Decreasing catch per unit effort against total catch reflects a declining stock, and an extrapolation of the line representing this decreasing catch per unit effort intersects the total catch axis at a point that gives a gross estimate of the total

the data was necessary in several other places (as, for instance, in the application of correction factors to compensate for the distortional effects of gunner preference for blue whales over fin whales).

I do not raise these points to suggest that the Committee of Three's analysis was, somehow, a hack, that it was a sleight of hand, or that it was flat-out wrong (it was, of course, "wrong" in a sense—it was overly optimistic, as would later be revealed—but the report itself was acutely sensitive to its own margins for error and opted, to strengthen its rhetorical hand, for optimistic parameters). My point, rather, is to emphasize that the committee's analysis—despite its robust statistical base, mathematical sophistication, and sheer relentlessness (the final version was nearly eighty pages long)—still left plenty of room for those who wished to quibble about "cer-

---

initial population. In its simplest form, this analysis omits recruitment and mortality, but various tweaks can be applied that permit these rates to be taken into account. See IWC / 15 / 6, app. 10. Part of the appeal of this method in these circumstances was that it did not depend on age classifications and was therefore insensitive to possible errors in aging technique. See D. B. De Lury, "On the Estimation of Biological Populations," *Biometrics* 3, no. 4 (December 1947): 145–67. The Schaefer technique for estimating maximum sustainable yield uses catch data and effort data together with a catchability coefficient (the probability of an animal being caught per unit of catching effort per unit of time, derived from mark recovery rates) to construct arithmetically a table of population sizes from year to year across the period of exploitation. With this table in hand, it is then a relatively straightforward procedure to estimate the net productivity of the population in each year simply by adding the difference between population sizes in successive years to the total catch extracted (if the difference in populations is negative—a declining population—the difference is of course *subtracted* from the catch). Plotted against catch per unit effort (serving here as an index of total population size), this estimated net productivity shaped up as something like the humped curve of sustainable yields that could be derived from taking the derivative of a logistic curve of population growth (as in Johan Hjort, Gunnar Jahn, and Per Ottestad, "The Optimum Catch," *Hvalrådets Skrifter* 7 [1933]: 92–127). By fitting a version of such a curve (by eye) to these data points, some general idea of sustainable yields at different population sizes could be obtained. Once it had been ascertained that a population was significantly depleted (was well below MSYL, or "maximum sustainable yield level," assumed at this time to be 50 percent of an original population on the assumption of a symmetrical sustainable yield curve), it was possible to ascertain instantaneous net recruitment—recruitment minus natural mortality—which set an upper bound on sustainable yield. Recruitment (r) minus mortality (M) could also be estimated in several different ways: a maximum value could be derived simply from the percentage of pregnant females in the catch; elemental rates derived from tag recovery or age table analysis could also be used.

tainty." The correction factor for catch per unit effort, propagated through essentially the whole analysis, could have been attacked as a mechanism for introducing systematic bias. Objections could have been raised to the ad hoc character of several other techniques deployed for handling the data. The ultimate victory of the Committee of Three, however, lay in silencing exactly this sort of quibbling. How did they do it? By subjecting those who tried it to a wonderfully public schooling in their methods. In other words, it was not that such quibbles could not be raised, it was that they could be answered, and much of the force of the committee's analysis lay in answering the quibblers.

To see how, we need only turn to the deliciously dialogic "Supplementary Report by the Committee of Three Scientists," which was appended to their final report as it came before the IWC in 1963. This document contained half a dozen pages devoted to rehearsing and answering the questions and concerns raised by, among others, Professor Slijper. Slijper had not been a major voice at the Seattle meeting (in fact, it is suggestive of his increasing isolation that he stayed in a different hotel than the other scientists attending the meeting),[170] but he continued to nurse his various concerns about the representativeness of the catch data, the uncertainty of age determination methods, and the possibility that significant portions of the Antarctic stock remained just beyond the grasp of the catcher vessels: maybe the wily creatures stayed outside the Antarctic, maybe they were savvy about the vessel routes, who could say for sure? With a certain didactic sadism and more than a little deadpan wit, the Committee of Three trucked out Slijper's preoccupations in turn, only to rotate each slowly over the coals of their final report. Quite right, they noted, that there remained imperfect coordination among the different approaches used in age determination. Excellent, then, that "two of the methods used in the assessments for blue and fin whales do not depend in any way on the age determination data."[171] As for the eternal question of whether "segments of the whale stock remain

170. Dale Rice's papers at the National Marine Mammal Laboratory contain the contact list for the Seattle meeting. Everyone stayed at the Edmond Meany Hotel except Slijper (who stayed at the University Inn, the most expensive hotel in the area) and Ruud, who appears to have stayed with a friend in a private home (417 Harvard Avenue, Apt. 10).

171. "Supplementary Report by the Committee of Three Scientists," IWC / 15 / 14, p. 9. My sense is that this statement is not entirely fair, since the two methods referred to must be Chapman's differential mark recovery technique and the De Lury method, and it is something of a stretch to call the first of these a "method used in the assessment"

outside the Antarctic for protracted periods of time" and thus were being overlooked in the analysis of the catch data, the Committee of Three, armed with its extensive geographically segmented analysis, could retort witheringly, "If such non-Antarctic stocks exist, they are entirely independent of the Antarctic stocks, and . . . not a part of this study."[172] In other words: Go find 'em (and good luck . . .). What about the possibility that the overrepresentation of young animals in the catch reflected not overexploitation, but increased selection for inexperienced whales unfamiliar with the wiles of their human hunters? *Problem*, noted the Committee of Three: such selection would gradually have produced an *underestimate* of mature animals and, therefore, a resulting (artifactual) *increase* in the recruitment rate (calculated, recall, from the ratio of parent stock to young in the catch); since the recruitment rate appeared to dropping sharply, the notion that only dumb young whales were succumbing to the gunners could be dismissed out of hand.

And on it went. Tangling with these characters was clearly not a good idea.

That was very much the sense conveyed by the "report on the report" drafted by Mackintosh's workshop group of the Scientific Committee, who assembled at the beginning of April in 1963 to review the Committee of Three's analysis.[173] Here, gathered in a conference room in Whitehall, not far from the rooms originally occupied by the Discovery Committee, the whale biologists shuffled through the dozens of pages of appendices attached to the Committee of Three's interim report.[174] The workshop group was now on its own: none of the members of the Committee of Three was present. As Mackintosh put it, in framing what he took to be their responsibility, "One of the objects of the present meeting was to consider how far the members of the Sub-Committee felt convinced by the findings of the Committee of Three."[175] To that end, the hip-booters—Slijper, Hideo Omura (from Japan), Viktor A. Arseniev (from the Soviet Union), Ruud, and the whole NIO whale research group (Brown, Bannister, and Gambell, all of whom had recently moved into their newly configured rooms at the

---

of the population as a whole, since it really functioned as a technique for estimating mortality rates, which could then be used in other assessment models.

172. Ibid., pp. 10–11.

173. I am referring here to the 1963 "Report of the Scientific Sub-Committee," IWC/15/8.

174. This was the "Second Interim Report," IWC/15/7.

175. Ibid., p. 1.

British Museum)—considered what they had in hand. The visual presentation of the data made a powerful impression (Kellogg, notably, had been
saying for years that a central problem with the work of the Scientific Committee was that it never presented the stock situation to the IWC delegates
visually). For instance, the Schaefer diagrams that simultaneously depicted
instantaneous recruitment for the current stocks of fins and blues (the lines
$r - M = 0.1$, etc. in figures 5.10 – 5.13; the fin analysis was conducted under
three different initial stock scenarios) while also plotting a sequence of recent catches (the vectors marked with arrows and dates in the same figures)
forcefully conveyed just how far the recent catches were from sustainability.
No amount of leveraging of that $r - M$ diagonal could push the recent
catches down under the population's capacity to reproduce itself. It was
as if a sweep of possible values for recruitment had been used to squeeze
the naysayers into a marginal sector of the diagram. Even the simple De
Lury plots (put together for several of the humpback populations on which
there was little supplementary biological data; see figure 5.14) dramatically
conveyed just how few whales were left for the taking. Moreover, Mackintosh's group had before them the Committee of Three's table laying out
their estimates for what a sustainable catch from the main stocks would be
going forward, along with their estimates of what a maximum sustainable
yield could be were the populations permitted to recover to MSYL (maximum sustainable yield level; see figure 5.15). Here were actual predictions,
which emerged from a mathematically sophisticated and statistically robust analysis, which perhaps the whale biologists did not fully understand,
but which they had, in a way, helped to create.[176]

Battered into submission by the algebra of natural logarithms in the first
appendix, sidelined by the geometry of population diagrams, cowed by the
computational drama of 100,000 cards slipping in and out of the IBM 1401,
the workshop group did not have much to say beyond the rather passive
endorsement they had conferred on the joint report that had emerged from
the Seattle gathering a few months earlier: "Members of the Scientific Committee . . . advised on biological factors and took part in the analysis of the
data, and although they fully agree with the general conclusions, they feel

---

176. It is important to bear in mind the significance of Mackintosh's workshop group
having been made into stakeholders in the Committee of Three's analysis: the whale
biologists had indeed come to feel (at least in the nebulous way that a committee can
"come to feel") that the report was significantly their own.

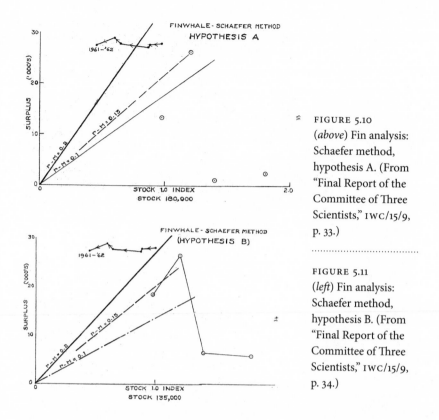

FIGURE 5.10
(*above*) Fin analysis:
Schaefer method,
hypothesis A. (From
"Final Report of the
Committee of Three
Scientists," IWC/15/9,
p. 33.)

FIGURE 5.11
(*left*) Fin analysis:
Schaefer method,
hypothesis B. (From
"Final Report of the
Committee of Three
Scientists," IWC/15/9,
p. 34.)

that the inferences set forth in the appendices are the responsibility of the Committee of Three."[177]

To be fair, the workshop group did spend some time musing on what could be wrong with this monster document: "There are certain possibilities of error affecting the basic data on which all these calculations are built," they noted, "such as the uncertainties in the interpretation of ages, a possible bias in the sample of whale populations as represented by the catches, the possible incomplete representation of the population on the whaling grounds, and the difficulty of a realistic measure of the catching effort"—all the old favorites. However, as they went on to acknowledge, "all these were considered at some length at the Joint Meeting in Rome in 1961, and further at subsequent committee meetings," and therefore, "the

177. IWC / 15 / 6, p. 4.

FIGURE 5.12
(*top*) Fin analysis:
Schaefer method,
hypothesis C. (From
"Final Report of the
Committee of Three
Scientists," IWC/15/9,
p. 35.)

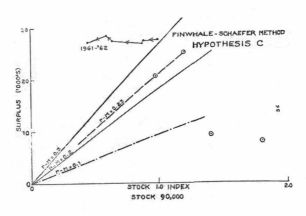

FIGURE 5.13
(*middle*) The Ant-
arctic blue whale:
Schaefer method.
(From "Final Report
of the Committee
of Three Scientists,"
IWC/15/9, p. 44.)

FIGURE 5.14
(*right*) De Lury
plot: Graph of ac-
cumulated catch
for a humpback
population. (From
"Final Report of the
Committee of Three
Scientists," IWC/15/9,
p. 54.)

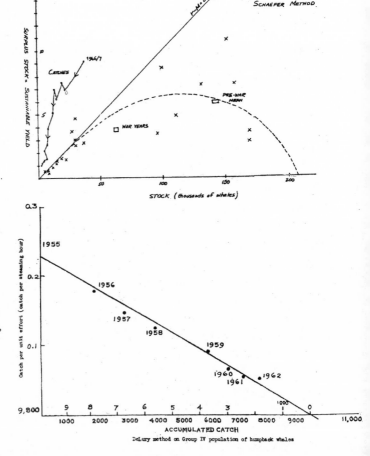

Species	Catch in 1961/62	Sustainable catch in 1962/63, i.e. catch that would not change stock level	Maximum sustainable yield when stocks have been permitted to rebuild	Years to reach the level of maximum sustainable yield if no catch after 1961/62
Blue (other than pygmy blue)	700	less than 200	6,000	50 or more
Humpback Group IV and V*	810	less than 100	less than 1,000	50 or more
Fin	28,000	9,000 or less	20,000	5 or more
Pygmy blue	460	–	probably a few hundred	

*Includes catches in temperate waters in 1961 and refers to total sustainable yields.

10.   On the basis of the tabulated figures sustainable yields of blues and humpbacks of any appreciable magnitude could only be obtained by complete cessation of catching of these species for a considerable number of years. To prevent further decline of the fin whale stock and allow it to build up to levels which will sustain high yields, the catches in the next few years must be drastically reduced to something less than 9,000.

(sd)   D.G. Chapman

on behalf of the Committee.

14th January, 1963.

FIGURE 5.15   The Committee of Three's predictions: Table showing sustainable catch and maximum sustainable yield. (From "Second Interim Report of the Committee of Three Scientists," IWC/15/7, p. 6.)

Sub-Committee believe that they are fully understood by the Committee of Three." Had the data been handled properly? It was hard for the whale biologists to say for sure: "The stock assessments, together with the preparation of data and examination of relevant biological factors, constitute a very large operation, and it is not possible for members of the Sub-Committee to follow and check every step made in the catch analyses and calculation of population parameters." It had to be admitted that "there is latitude for some degree of error, and greater precision could be achieved if some of the uncertain factors could be eliminated."

But—and here was the moment for which many had waited for a good long time—there was no way that anyone in that room at Whitehall was going to try to gainsay the document on the table:

> The Sub-Committee, however, feels satisfied that the findings must be broadly correct, for they are consistent with results from four independent sources, viz., the mortality estimates from age distribution, from marking, from catch per unit of effort, and indirectly from the rough census of pre-war stocks by sightings.

So there it was. There seemed to be nothing left to say.

Actually, though, there were a few more things to deal with. For instance, Mackintosh's group expressed an interest in getting their hands on the data cards themselves, and they set forth the claim that all the material that had been used in the analysis really "should be at the disposal of the Scientific Committee, and anyone having plans for using it should consult that Committee."

And then there was one last thing: the upshot of the Committee of Three's recommendation was that blue whale hunting needed to be stopped for something like half a century if the stock was to recover to MSYL, and that fin whale hunting needed to stop for something approaching a decade if optimum yields were to be had. Did this mean that an actual moratorium on commercial whaling was needed? Consulting with Gulland, the workshop group considered a number of other possible scenarios whereby stocks would be permitted to recover more slowly to their optimum levels while being subjected to considerably reduced (but nonzero) hunting pressure. With such a vision of the near future in mind, Mackintosh appended to the report a brief comment that speaks eloquently to the deep and resilient ties that bound the whale biologists to the whalemen:

> The Sub-Committee is concerned only with the problems of the biology of the stocks and their conservation, but would point out that if the Commission decides that it is desirable to keep the industry in operation, even if on a much smaller scale, *the continuation of catching will be of considerable value to the improvement of knowledge of the*

stocks. They would therefore propose that all possible steps are taken to obtain the fullest possible biological observations on the catches.[178]

These are very much the words of a man just appointed to run the brand-new Whale Research Unit at the British Museum suddenly facing the prospect of not having any whales to research. Hopefully, the working group suggested, some whaling could continue. Otherwise, what was whale science for? And how would the whale scientists get their whales?

. . . .

The workshop group's assessment of the Committee of Three's report could hardly be called a ringing endorsement of the desperateness of the situation, but the whale biologists did acquiesce, if somewhat cautiously (perhaps even grudgingly). Their resolve was stiffened, however, three months later when they reassembled just before the 1963 International Whaling Commission meeting in London. Now again in the company of Chapman, Holt, and Allen, the whale biologists at this gathering—the full Scientific Committee—formalized their approval of the Committee of Three's assessment, passing the following resolution (without dissenters): "The Committee believe that the fullest practical compilation of relevant data has been made for the purposes of this report, and is satisfied that the estimates made by the Committee of Three of the present condition of these stocks are the best possible at this time, and recommend that the Commission act accordingly."[179] Not that their ratification could force the IWC's hand. At the beginning of July 1963, then, the question was, what would the IWC do with the recommendations of the Committee of Three, which now had the stamp of approval of the IWC's own scientific advising community? Would the IWC honor what had been understood by many as a commitment to abide by the Committee of Three's analysis and "not later than 31 July 1964 bring the Antarctic catch limit into line with the scientific findings"?[180]

While the 1960 "package deal" had succeeded in reassembling the full IWC membership (at the cost of several years of unprecedentedly high BWU

178. This and the quotes above are from the 1963 "Report of the Scientific Sub-Committee" (IWC/15/8). Emphasis added.

179. "Report of the Scientific Committee," IWC/15/3, p. 4.

180. This is the phrasing of the Canadian resolution on the package deal that carried without dissent in 1960 (IWC/12/18B).

quotas), the underlying problems—overcapitalized national industries, mutual mistrust about adherence to regulations, continued concern about extra-IWC whaling nations—had not been fixed.[181] The Soviets looked to be increasingly committed to prosecuting the industry on an expanding global scale (1963 was the first year that non-Antarctic returns exceeded those from the Southern Hemisphere), and the Japanese (for whom the economics of whaling were unique in view of the comparatively lucrative trade in whale meat for human consumption) continued to buy up vessels (and their associated quota shares) from the older whaling powers like Britain and Norway, which were less and less competitive. Despite a major US diplomatic push in the lead-up to this fifteenth IWC meeting in 1963, there was a great deal of uncertainty about whether the major pelagic whaling powers would bow under the weight of the Committee of Three's analysis.

In the end, it was not to be. The Technical Committee stayed locked up deep into the night as the meeting wound down, but consensus on a catch quota consistent with the Committee of Three's recommendation eluded them. Taking the condition of the different exploited species into account, the Committee of Three's final recommendation implied that a quota not exceeding 5,000 BWU was the absolute maximum catch consistent with the principles of sustainable yield (they had also strongly encouraged the IWC to set species quotas and to abandon the BWU system altogether). The dominant quota proposals to come out of the logrolling in the technical committee, however, were well over double that.[182]

181. The effective quota in the seasons 1960–1961 and 1961–1962 was 17,780 BWU, which reflected the major pelagic nations each committing to abide by the quota share that it sought in extra-IWC negotiations. Significantly, actual catches fell short of this mark by an increasing margin over the period in question. See J. N. Tønnessen and Arne Odd Johnsen, *The History of Modern Whaling* (Berkeley: University of California Press, 1982), 570.

182. The United Kingdom put forward a 4,000 BWU proposal, which had the support of Norway; the Soviets held out for 12,000, and the Japanese forwarded a compromise proposal for 10,000, which eventually carried in the plenary session. It is difficult to be certain how genuine some countries' commitments to the lowest quotas were, since it rapidly became clear that those numbers had no hope of passing, and therefore protestations of conservation bona fides could be treated as diplomatic freebies. This is the sort of double- and triple-think that makes sorting out the dynamics of the full IWC from a straightforward reading of its documentation (even verbatim transcripts) so difficult.

And yet, significantly, the issue was not the science. The issue, explicitly, was money. As the Dutch delegate put it, as the IWC headed into the final votes:

> We have heard many opinions expressed and we are, I believe, now faced with a very important decision. . . . We have to compare scientific evidence and economic necessity. Each of us can see the situation from one viewpoint, but it is difficult to see it from both sides.[183]

He left little doubt where he stood: economic necessity trumped scientific evidence.

William Herrington himself took the floor to spell out what he had seen over the course of the Technical Committee's negotiations. He stated formally for the record that there were no longer any scientific arguments being put forward by those nations that were now preparing to step away from their general commitment to catch regulations "based on scientific findings" (the language of the original ICRW) as well as from their specific commitment to abide by the findings of the Committee of Three (the core of the package deal of 1960):

> We know of the several proposals made for limits for the coming season. Three of these provide for quotas which are greater than the catch expected during the coming season, using the same fishing effort and size of fleet as operated during the past season. In justification for these quotas we hard [*sic* for "heard"] many explanations, but I believe they all boiled down to what is necessary to provide a profitable operation for the entire fleet which operated last year.[184]

Before the meeting adjourned, a quota of 10,000 blue whale units was adopted, along with, as sop for the hand-wringers, a pious resolution reaffirming the IWC's intent to line up their quota with scientific advice by the summer of the following year (to that end, the Committee of Three, now expanded to the Committee of Four with the formal addition of John Gulland, was commissioned to continue refining and updating its assessment in preparation for next year's meeting). However, in view of the fact that the

183. "Verbatim Report of Plenary Sessions," IWC / 15 / 17, p. 70.
184. Ibid., p. 68.

Soviets, the Japanese, and the Dutch all refused to sign off on this expression of intent, there was little optimism that 1964—the last opportunity to abide by the deadline of the 1960 package deal—would be any different.

As indeed it was not. Despite a beefed-up new report from the Committee of Four, unequivocal support from the Scientific Committee (Mackintosh had finally stepped down as chairman, ending more than a decade of amiable if uncompelling dominance), and a total catch from the 1963–1964 season well below 10,000 BWU, the IWC found itself incapable of doing more than edging the total quota down to 8,000 BWU, a target the scientists again assured them (correctly, as 1965 would reveal) they would be unable to reach. Now it was the turn of the Japanese delegate to spell out explicitly that science was not the issue:

> I quite envy the non-pelagic whaling countries who are not engaged in the Antarctic and can discuss this problem clearly in this Commission purely from the point of view of the conservation of natural resources. However, as you know, Mr. Chairman, Japan is engaged in pelagic whaling in the Antarctic and has many expeditions which go to the Antarctic. I should also like to emphasize that the whaling industry plays a very important role in the national economy in general. Therefore, Japan cannot discuss this problem only in terms of the conservation of whale stocks.[185]

As the transcripts of the 1963 and 1964 IWC meetings make perfectly clear, the work of the scientists did *not*, in the end, compel the delegates to adopt the necessary regulatory measures. But as those same documents also make clear, the work of the Committee of Three *had* compelled the whaling community's assent on the question of fact: the Antarctic whale stocks were in free fall, and no one even bothered to contest the point any longer. As Holt put it forcefully at the end of the 1964 IWC meeting, "There was no reasonable doubt about the correctness in general of the scientific appraisals that had been made."[186]

And this was no small thing, since it placed a new and powerful weapon in the hands of those intent on saving the stocks and protecting the ideal of scientific management of natural resources. There were now actual villains on the stage whose actions could be called out as nakedly mercenary and

185. "Verbatim Report of Plenary Sessions," IWC / 16 / 15, p. 72.
186. Ibid., p. 67.

manifestly contemptuous, not merely of international good fellowship, but of reason itself. Kellogg, Herrington, and others seized immediately on this glinting blade and wielded it to considerable effect in the years to come. As Kellogg explained in his report to the State Department after the 1963 IWC meeting: "The United States Delegation believes that the Fifteenth Meeting of the Commission has produced a record that can be effectively used by those interested in promoting support for survival of the Antarctic whale stocks."[187] And by the following year, he could report the same proposition with even greater conviction: "The verbatim report of the Sixteenth Meeting of the Commission unquestionably shows that conservation principles will not be accepted by three at least of the Antarctic pelagic whaling countries."[188]

By then, Kellogg was working behind the scenes to destroy his baby: only by doing an end run around the IWC itself, he had come to believe, would it be possible to save the whales.[189] Versions of such a plan circulated in 1964 and 1965 and were facilitated by the appointment of a new director in the Fisheries Division of the Food and Agriculture Organization: Roy I. Jackson, an American. The looming prospect of a UN / FAO move into the arena of cetacean conservation promised to be an effective threat to the delegates to the IWC, since their own power would then be at stake: "This, of course, involves the risk of focusing attention on the IWC as a failure, but

187. Report for 1963, p. 44, Smithsonian Archives, RU 7165, box 28, folder 6, "Reports to the Secretary of State by the U.S. Commissioner to the IWC, Meetings #15–20, 1963–1968." It is interesting to note that Kellogg apparently deployed his own copy of the IWC meeting transcript for exactly that purpose: it is the only verbatim transcript missing from Kellogg's collection of IWC documents, which strongly suggests that he sent it to someone to show exactly how bad things were (there seems to be no hard evidence for where this document might have gone, but I am inclined to suspect it went to Hal Coolidge, in view of Kellogg's correspondence with him and his interest in the work of the IUCN). Kellogg also noted in his 1963 report that "at this year's meeting an official record has been made that makes clear: (a) the precarious condition of the whale stocks; (b) which delegations are responsible for failure of the Commission to take effective action; (c) that the reasons for their opposition to effective action are the short-range economic effects on their whaling industries."

188. Report for 1964, p. 46, Smithsonian Archives, RU 7165, box 28, folder 6, "Reports to the Secretary of State by the U.S. Commissioner to the IWC, Meetings #15–20, 1963–1968."

189. As he put it to Hugh Gardner at MAFF in a letter of 3 August 1964, "Personally I an [*sic* for "am"] inclined to believe that the future usefulness of the IWC is questionable."

will spotlight Japan and the U.S.S.R. as responsible for wrecking it." It was a price worth paying: the scientific work had pricked out the bad guys and it was now time to name names in the court of "world opinion." The United States was even prepared to sink the IWC altogether "by threatening to withdraw, or by withdrawing, from the Convention and taking others with us" if the Soviets and the Japanese could not be brought into line.[190]

Other pushes to rally external pressure also gained momentum in the aftermath of the Committee of Three's work. Private letters and telephone calls drew the attention of conservation-minded scientists and nongovernmental organizations to what could increasingly be characterized as nothing less than a crime against humanity. As Holt argued, "it is universally accepted that there is a need to improve nutritional levels in many parts of the world"; therefore an organization intent on liquidating a resource that could provide as much as "two-and-a-half million tons" of food products a year (under the Committee of Three's estimate of MSY) had to be brought to the bar of global justice.[191] By the time the sixteenth meeting of the IWC assembled in Sandefjord in June 1964, the representatives faced a minor deluge of resolutions and correspondence from outsiders expressing dismay about the direction things had gone.[192] As the president of the Wildlife Society put it in a letter to the secretary of the IWC that year, the IWC had an obligation to take "strong action" as "your own facts indicate must be done."[193] A rumbling could be heard in this missive and the others. Indeed, nothing better expressed the direction things would go in the decade to

190. These quotes come from a confidential outline of US strategic considerations drafted by Donald L. McKernan (director of the Bureau of Commercial Fisheries in the Fish and Wildlife Service in the Department of Interior) for Jackson, 8 September 1964. See Smithsonian Archives, RU 7165, box 27, folder 2, "Sixteenth International Whaling Commission Meeting (Sandefjord), 1964."

191. Holt's speech on behalf of FAO, IWC / 16 / 15, pp. 67–70. Interestingly, at this meeting Holt wore two hats, speaking on behalf of FAO separately from his presentation as a member of what was by then the Committee of Four.

192. These communications included a letter from the director-general of the FAO as well as a resolution on the blue whale adopted by the International Union for the Conservation of Nature and Natural Resources, a letter from the vice-chairman of the World Wildlife Fund, and a letter from the president of the US Wildlife Society. For a list of these documents, see IWC / 16 / 3, para. 21. See also Smithsonian Archives, RU 7165, box 27, folder 1, "Fifteenth International Whaling Commission Meeting (London), 1963."

193. Jack H. Berryman to IWC, 4 June 1964, Smithsonian Archives, RU 7165, box 27, folder 1, "Fifteenth International Whaling Commission Meeting (London), 1963."

come than a resolution by the members of the ıucɴ's Survival Service Commission after hearing of the failure of the 1964 ıwc meeting: "It was decided . . . that the ssc would embark upon the most active and militant possible of campaigns against present attitudes with emphasis on the threat of extinction to the world's largest mammal."[194]

The word "militant" spoke volumes.

CONCLUSION

At the start of chapter 4, I proposed to investigate the first two decades of the International Whaling Commission with a view to making sense of the changing relationship between science and policy making across a crucial episode in the history of twentieth-century international regulation of natural resources. This investigation promised to shed light on how cetaceans became a global conservation cause in the postwar period, but from the start I was equally concerned to use the story as the occasion for a detailed consideration of the larger problem of the function of scientific expertise within the regulatory structures of the modern administrative state and the international institutions to which it gave rise. It was my contention at the outset that a sociologically sensitive history of science, diachronically attentive to the strategic contest over the boundary between political and scientific problems (intent on showing, that is, the work that went into drawing and defending, and redrawing and redefending, such a boundary), is uniquely suited to the investigation of this period, and that such a study would improve our grasp of science as a feature of public life in the second half of the twentieth century. In keeping with this approach, my foray into the formation and workings of the ıwc up through the mid-1960s has resisted prosecutorial temptations: less interested in finding out whom to blame for institutional failures than in recovering professional trajectories and divergent commitments, less interested in philosophical pronouncements about the nature of science than in a close study of what got called science when and by whom, I have endeavored in chapters 4 and 5 to recover an intricate story of science at work in the world of regulation. It is now fair to ask: What do we have to show for this exercise?

First, I believe the preceding pages clearly establish the importance of taking administrative and bureaucratic structures seriously when looking

194. Extract from the minutes of the meeting of the ıucɴ, 26 November 1964, Morg, Switzerland. See University of Washington Library, Chapman Papers, box 11.

at the intersection of science and policy. In the last several decades, historians of science have been much concerned to detail the distinctive features of scientific life in particular settings. Certain key sites for the production of natural knowledge (the laboratory, the field, the clinic) have shaped this scholarship, but forays into a variety of other spaces (the courtroom, the city, the salon) have also proved fruitful. As a locus for the resolution of epistemological problems, the committee room has about it little of the drama of, say, the jungle or the asylum, and yet, as far as the postwar evolution of large, socially significant scientific research programs is concerned, there are few spaces more important to understand. As I have shown, the control of parliamentary procedure and the manipulation of subcommittee architecture played an absolutely central role in establishing the "facts of the matter" in the regulatory debates around whaling in the 1950s and 1960s. The boundary work that cordoned off "biological questions" and positioned their answers with respect to "practical matters" was done via a sequence of careful rearrangements of the reporting structures among the different parts of the emerging IWC. The "revolution" of the early 1960s was, as I have demonstrated, as much administrative as it was scientific—though in a deeper sense, I hope my narrative of the Committee of Three's work suggested just how interdependent its technical and diplomatic labors really were. By paying close attention to these mincing features of the administrative workings of the IWC in these critical years, I believe it has been possible to wring additional insights out of the masses of documentation that the organization produced. The vast majority of that documentation—status reports, statistical briefs, position papers—was the work of committees, and it requires a reader sensitized to personalities and administrative dynamics to interpret such writings, which usually have a single primary author but must run a gauntlet of multiple redrafting readers. Here, it has been possible in several instances to meet this hermeneutic challenge by means of valuable archival finds (rough copies, marginal annotations), which have assisted in prizing apart the diversity of views that can squirm together under the veil of an anodyne majority report.

Second, the result of this investigation into the regulatory debates in the IWC between 1945 and 1965 has been a revised account of the formation and evolution of that institution's scientific advising mechanisms. Instead of telling a story about a few naïve scientists waylaid by nefarious plotters, I have emphasized the shifting character of the regulatory arrangement sought by those shaping the new organization. The two decades in question saw the pursuit of a collaborative, cooperative framework for decision making (where scientists would demonstrate their sensitivity to national and industrial exigencies in order to build precedents for having their recommendations accepted and implemented) give way to an increasingly frank contest for dominance in an agonistic forum (where scientific recommendations were required to compel assent and force regulatory action). By showing how expectations and aspirations changed over these years, and why, it has been possible to highlight the politically savvy moves made by a number of the scientists initially intent on nurturing a particular vision of the relationship between whale science and whalers. Decisions that might seem irresponsible or imprudent in hindsight (setting quotas high, declining to adopt forceful resolutions about needed conservation measures) can be shown, when examined in context, to reflect calculated decisions about how to build relationships and model compromise.

This analysis has implications for how scholars of science policy should think about the "capture" of scientists in regulatory settings. Was Neil A. Mackintosh "captured" by British whaling interests? If so, when? In 1956 he sought to distance himself from the I w C's reluctance to make a substantial cut in the Antarctic quota. And yet, in the early 1960s, he was trying to brew up a scheme by which cetologists could have a whole decommissioned factory fleet to themselves so they could rove the oceans and do whale science the good old-fashioned way—by whaling.[195] Was he "captured" by the industry, or trying to "capture" it for his own purposes? How exactly would we know the difference? Research ambitions, means-ends calculations, networks of influence and reciprocity—these are what shaped Mackintosh's relationship with the whaling industry. "Capture" is too blunt a tool to open this box and reveal its workings. A similar point can be made about the scientists' relationship to the I w C itself: institutions are made by human beings, but the human beings are shaped in turn—even "captured"—by their creations. As Herrington put it to Kellogg in 1964, it suddenly seemed as if everybody was "interested in working out a compromise which would maintain the Commission regardless of the whales."[196] They had all, in a sense, been swallowed by a Leviathan of their own making:

195. See his remarks at the 1963 "First International Symposium on Cetacean Research" (which I will take up in more detail in the next chapter), in *Whales, Dolphins, and Porpoises*, ed. Kenneth Stafford Norris (Berkeley: University of California Press, 1966), 144.

196. Herrington to Kellogg, 3 August 1964 (referring particularly to the UK position), Smithsonian Archives, R U 7165, box 27, folder 3, "Sixteenth International Whaling Commission Meeting (Sandefjord), 1964."

the great beast of a state-sponsored administrative agency, which spewed reams of paper before it from which no man could escape. The first moves toward the forum of "world opinion" can be understood as an effort by some to fight back against this strange creature that had gradually trapped them all.[197]

Some readers might protest that the scientists maneuvering around the emerging IWC in the 1950s cannot have been all that "savvy," in view of their ultimate failure to get a cooperative advisory system to work. Fair enough. Some (like Mackintosh) seemed never to catch on. Nevertheless, as a series of (unrequited) overtures came to look increasingly quixotic, and as a series of disputes (over scientific whaling, ecosystem management, and statistical analysis) undermined the optimistic faith that the rubric of "scientific findings" would eventually provide common ground for the cooperative convergence of negotiating positions on regulation, others (like Kellogg) threw their weight behind a new diplomatic, administrative, and scientific initiative. The stock analysis that emerged from this initiative was armed not to win friends, but to defeat adversaries. Defeat them it did, and as a result, by the mid-1960s, those nations unwilling to curtail their exploitation of Antarctic whales were forced to plead economic necessity, not an alternative account of population dynamics.

In fact, a sharp exchange at the 1964 IWC meeting in Sandefjord hammered this point home. When a Soviet delegate that year, Captain G. A. Solyanik, tossed around in the Technical Committee that he had inside information concerning a soon-to-be-released Soviet stock analysis (which he claimed would show that Antarctic fin whales were in better shape than the Committee of Three had warranted), angry formal protests from the Scientific Committee (under the new chairmanship of Johan Ruud) called him to account: Where was the analysis? When it had been presented to the Scientific Committee and had carried the day in that forum, Captain Solyanik would be welcome to marshal it as evidence in the Technical Committee's deliberations; until then, however, the analysis did not exist as far as the IWC was concerned. Under the new institutional order of the mid-1960s, it was put up or shut up, scientifically speaking, and a strongly

197. The expression "world opinion" starts to come up with increasing frequency in the late 1950s. See, e.g., incoming telegram, 30 June 1959, MacArthur (Tokyo) to secretary of state, Smithsonian Archives, RU 7165, box 27, folder 3, "Sixteenth International Whaling Commission Meeting (Sandefjord), 1964."

worded resolution—IWC / 16 / 21—emerged from the Scientific Committee, declaring that

> the Scientific Committee is the scientific forum of the Commission. It regrets that statements about scientific evidence continue to be made in the Technical Committee, which are not made in the Scientific Committee. All such statements should be substantiated by the full scientific data. The Committee therefore recommends that the Commission should not accept, or entertain discussion of, any statements about scientific evidence concerning the condition of the whale stocks unless the evidence on which the statement was based had been made available to the Scientific Committee.[198]

This declaration, which the full commission accepted, marks a fitting place to conclude our investigation of administrative-cum-epistemic boundary work within the IWC at midcentury. By 1965 a reconfigured Scientific Committee had become the obligatory passage point for claims about the facts of the matter.[199] To be sure, the scientists still could not force particular regulatory concessions (1964, for instance, again saw no consensus on the Antarctic quota, and Ruud would step down as chairman in disgust), but they *could* humiliate (and block) unauthorized efforts to speak

198. "Supplementary Report of Scientific Committee," IWC / 16 / 21 (I have slightly modified punctuation, adding a comma to clarify the meaning of the passage).

199. Though it should be said that the Soviets did not just roll over on this issue: in fact, they contested the language of the Scientific Committee's supplementary report and demanded that Ruud remove from that document his account of his conversation with Solyanik. There are a series of fascinating exchanges in the verbatim transcripts of the 1964 meeting (IWC / 16 / 15, see pp. 31ff.) in which the Soviets endeavor to characterize the whole exchange as a difference in the "interpretation" of scientific results and Ruud successfully parries these efforts. In the end, Ruud's victory is a paragon of the strategic use of parliamentary procedure: he carefully reads the disputed paragraphs in the plenary session (putting them into the formal record of the verbatim transcripts), and then agrees to remove them from the supplementary report in question. As this episode demonstrates, Ruud was considerably more forceful as a chairman than Mackintosh, if also less patient, and perhaps as a result, his tenure was shorter. Structurally, it is worth noting that by the early 1960s, the "modern" organization of the Scientific Committee had been formalized: the Scientific Sub-Committee was abolished, and the Scientific Committee meeting was scheduled immediately in advance of each year's IWC gathering.

for what was really real about the natural world. It had taken twenty years of whale science and whale diplomacy to create the conditions of possibility for consensus on a resolution like IWC / 16 / 21. The next twenty years, as I will show in the following chapter, would offer unexpected and exogenous challenges to this consensus.

If these chapters have succeeded in revising our understanding of the early IWC along the lines laid out above, while demonstrating the general historiographic value of this sort of approach, I will be more than satisfied. But I am inclined to hope as well that the excavation of this broad narrative of boundary work and shifting expectations has brought to light several minor key points that may prove of interest to the growing ranks of historians investigating twentieth-century ocean science. For instance, by going beyond the widespread (though false) notion that there was a general "shortage" of whale science before the early 1960s, my study has unfolded the ambitions and objectives of a deeply entrenched research program of hip-boot-style whale biology, a community that played a significant role in shaping what was known about these animals in this period. Moreover, because this type of research had come of age in close association with the industry, it privileged investigations that were concordant with industry imperatives. Because broad shifts in postwar government support for ocean science left the hip-booted cetologists as the physical oceanographers' poor neighbors, good relations with the whaling industry became even more important in the 1950s—at the very moment that the whale scientists were being called on to provide leverage against commercial overreach. Any effort to analyze the policy role of "scientific uncertainty" in these years must take into account these broader dynamics of competing research priorities and altered budgets, since, as I have endeavored to show in some detail, scientific uncertainty plays an important role in the promotional life of scientific research agendas. I am confident that similarly detailed forays into the scientific and bureaucratic workings of other international regulatory bodies—the Inter-American Tropical Tuna Commission (IATTC), for instance, or the Scientific Committee on Antarctic Research (SCAR)— would reveal comparable patterns and problems; comparative work, as the relevant studies emerge, will surely prove revealing.

At the most general level, however, I hope these chapters have demonstrated the value of approaching complex episodes of science and policy making from a genuinely historical perspective. Research libraries brim with policy-oriented monographs intent on tapping historical case studies for real-world lessons about environmental management or the proper organization of regulatory regimes. And yet a core historical insight (that science itself changes through time, and does so in ways that resonate with broader social and cultural patterns) seldom illuminates these analyses, which have an uncanny knack for discovering more or less the same story ("the science got politicized!") wherever they look. Dispensing with an ahistorical and / or programmatic conception of science removes the tacit analytic convention ("the science is what preceded the political conflict") that quietly installs the inevitable conclusion ("after the political conflict, that pre-political science emerged sadly tarnished!") with all the fateful predictability of classical tragedy. Recognizing that the (mobile) boundary between questions understood to be political and those understood to have transcended politics *is* the boundary of science itself in the modern world opens the way to a different kind of story, one that reveals the categories of science and politics aborning, instead of taking them for granted.

. . . .

Nineteen sixty-five did not mark the end of the IWC, but it did mark the end of an era. Remington Kellogg's signature at the bottom of his pessimistic 1964 report to the secretary of state was markedly shaky, and it was the last such document he would sign. In the following year the US delegation would be led by John Laurence McHugh, twenty years Kellogg's junior, a vigorous fisheries biologist who had directed biological research at the Bureau of Commercial Fisheries in the Department of the Interior and would soon become deputy director of the whole bureau.[200] He had served under Kellogg as a deputy delegate to the IWC since 1961, and he would steer US involvement in the organization into the early 1970s.[201] Unburdened by a lifetime of labor on the vexing problems of international whaling, McHugh stepped into the delicate negotiations of the early 1960s all guns blazing. He drafted a spirited internal memo in late 1962 (for Kellogg and Herrington) that laid out a radically different strategy for resolving the regulatory crisis facing the IWC: why not organize a "planned overfish," whereby the industry players could amortize their (over)investment in one clean, brief sweep

200. McHugh, a naturalized Canadian, had taught marine biology at the College of William & Mary before joining government service and had served in 1960–1961 as secretary of the Inter-American Tropical Tuna Commission.
201. He stepped down as commissioner in 1973.

of the Antarctic and then, cashed out, cashier their vessels? As he put it in the memo (which reads like a blast from outer space into the diligent sequence of diplomatic and scientific ruminations of the period):

Do we take the purely aesthetic view that we must save these majestic beasts for posterity? Or do we take the more realistic position that economic objectives must be balanced against aesthetic considerations?

Stumping unhesitatingly for the latter position, McHugh turned to a thoroughly unsentimental and notably maverick economic analysis: the theory of MSY had been developed, he pointed out, to model "a resilient fish stock with high potential replacement capacity." Whales, however, had a very low replacement rate, as the whale biologists had demonstrated with increasing precision. Thus, as he went on to argue, it hardly made sense to manage a cetacean stock with MSY techniques. In fact, it probably made no sense to manage whaling as a "fishery" at all. Better, he argued, "to treat whales as a *mineral resource*, with some modifications." In other words, the regulatory regime ought to be structured around periods of "mining" (as he put it) followed by "periods of recovery." Only "competent economic advice" would permit such a "rational management program" to be framed properly, but there was no reason not to follow the scientific study of the whole problem out "to its logical end point."[202]

In an admonishing and digressive reply that reviewed (yet again) the entire history of the IWC, Kellogg (then in the midst of retiring from the Smithsonian) deflected McHugh's insurgent recommendation and intoned a dirge for the slow demise of right reason that the aging whale biologist felt he had witnessed over his forty years of scientific and diplomatic labor:

There were meetings of the scientific committee when I insisted that as scientists the members should be guided in their deliberations by the available facts and that they would do well to leave policy considerations to their commissioner at plenary sessions. Unfortunately some members of this committee had neither biological training nor appreciation of the rules of evidence.[203]

202. This and the above quotes are from McHugh to "Dr J. [*sic*] Remington Kellogg," 10 December 1962, Smithsonian Archives, RU 7165, box 27, folder 1, "Fifteenth International Whaling Commission Meeting (London), 1963." Emphasis added.
203. Kellogg to McHugh, 20 December 1962, in ibid.

But it is not really clear from this gruffly strained and boundary-preoccupied letter that Kellogg fully grasped the radical reframing of the problem that McHugh was offering. For Kellogg, who had been at the business of "rationalizing" exploitative industries for almost half a century (indeed, in that same year, 1962, he again spelled out that the principal objective of the IWC ought to be "to achieve rational management of the whale resources of the world"[204]), such a departure from the conservation ideals of the Progressive Era—such a dramatic revision to the concept of "rationality" itself—may well have been beyond his reach.

I tip open this misunderstanding not merely to convey a sense of the changing of the guard (though this is palpable in the exchange), but more to gesture at what amounts to an epilogue and epitome of the narrative on offer in these pages: The story of whale science and whaling regulation across the twentieth century is above all the saga of a sequence of bitterly contested and consequential efforts to "be rational" about the largest animals on earth.[205] In the decades that straddled World War I, "rationalizing the industry" meant ensuring, as Johan Hjort put it, "that the entire carcase should be boiled" so as to eliminate waste.[206] But this "technical rationalization," as he called it, this "gospel of efficiency" that passed as a metrical and moral imperative in the first third of the twentieth century, soon betrayed a dismaying tendency toward self-digestion (in the form of spectacular whaling efficiency leading to total industry collapse) that was difficult to reconcile with any garden-variety conception of reason.

Hjort, a sensitive thinker possessed of considerable intellectual ambition, grew inward with this conundrum in the early 1930s, and his "Optimum Catch" paper represented, at its core, an effort to redefine, in mathematical terms, what it meant for human beings—with their particular needs and habits—to be rational about the pursuit of cetaceans. In framing that paper, he posed an unexpectedly vast question that seemed to reach con-

204. This quote appears on p. 48 of Kellogg's 1962 report to the secretary of state, Smithsonian Archives, RU 7165, box 28, folder 5, "Reports to the Secretary of State by the U.S. Commissioner to the IWC, Meetings #12–14, 1960–1962."

205. For a recent book that centers on the idea of rationality in connection with modernization and shifting attitudes toward the natural world, consider Bo Elling, *Rationality and the Environment: Decision-Making in Environmental Politics and Assessment* (London: Earthscan, 2008).

206. Johan Hjort, "Whales and Whaling," *Hvalrådets Skrifter* 28 (1933): 7–29, at p. 24.

siderably beyond the blubber trade: "What . . . is our chief need?" And he answered it in comparably visionary terms: "Security for the future."[207] The rational maximization of returns looked very different, he then set out to show, when charged to reckon with eternity. Summarizing this insight in a philosophical rumination delivered at Harvard in 1937, Hjort offered a précis of the spirit that would guide the leaders in the regulatory arena for the next two decades:

> If we compare on the one side the two curves for a growing whal-
> ing industry and the catch per boat, on the other side the growth of
> a colony of bacteria and its rate of growth, we understand that the
> expansive development, which we sometimes name rationalization,
> sometimes competition, may further the rate of growth or standard of
> life of the individual up to a point from which onwards an increase of
> the active population destroys the industry and lowers the individual
> standard of life. *Rationalization and competition are ideals which*
> *are conceived in a time of expansion, but they become irrational and*
> *destructive to the welfare of the population if the harmony or correlation*
> *between the wealth of nature and the activities of man are forgotten or*
> *lost out of sight.*[208]

His promise of an "optimum catch," a promise enshrined in the ICRW's original commitment to provide for the "optimum utilization of the whale resources,"[209] thus represented an effort to coordinate the wealth of nations with the wealth of nature. It was maximization not over a human lifetime, but over the lifetime of humanity; not maximization within the constraints of human technical ingenuity, but rather within the limited reproductive rates of specific organisms in particular situations.

By the late 1950s, however, when men like John Laurence McHugh were in their prime, maximization within a different set of parameters beckoned. The work of economists like S. V. Ciriacy-Wantrup, Anthony Scott, and Scott Gordon had by then purported to show that "the conservation question, when correctly conceived, becomes simply an aspect of applica-

207. Ibid., p. 28.
208. Johan Hjort, "The Story of Whaling: A Parable of Sociology," *Scientific Monthly* 45, no. 1 (July 1937): 19–34. The quote is from p. 32. Emphasis added.
209. International Convention for the Regulation of Whaling, Article V, sec. 2A.

tion of the traditional theory of capital."[210] Under this analysis, the proper way to be rational about whales was to maximize the *value* to be derived from them, and this might mean, exactly as McHugh pondered, liquidating some or all of the commercially significant species and investing the resulting assets in some more highly remunerative instrument. With such a powerful reconceptualization in the ascendancy (and his name at the top of the reports), McHugh can perhaps be forgiven for thinking that 1965 marked the "turning point" for the IWC: the moment when it went from a litany of well-meaning exercises in acting "too little and too late" to "an achievement of great significance"—namely, securing a commitment to bring the Antarctic quota "below the best estimate of the scientists."[211] Economic rationality had, in his view, finally begun to work its magic. But what did this mean? It seems to have meant that, at long last, the leviathan of capital was moving on, wallowing away on the lookout for new breeding grounds.

Not that McHugh would have put it that way. On the contrary, his enthusiasm for the "turning point" of 1965 must be understood as a plea for the IWC itself, a defense, offered at a fragile moment in the institution's history (1971), by a man who had quickly become one of its own. It was a way of saying: "Now you can trust us. Promise." And, as such, it was a bit of a fudge. After all, it was not until 1967 that the 1965 "commitment" actually came into effect. And anyway the 1965 deal was itself just another version of the (failed) deal of 1960: both amounted to a conservation declaration along the lines of Saint Augustine's notorious prayer ("Lord, make me chaste, but not yet . . ."). On top of that, there was the awkward business that had come to light in 1968: the "appalling" discovery of a flaw in the Committee of Three's models (new biological data on aging overthrew the age table analysis), which meant that the triumphant 1967 "sustainable yield" quota (of a mere 3,200 BWU, a mi-

210. Scott Gordon, "Economics and the Conservation Question," *Journal of Law and Economics* 1 (October 1958): 110–21, at p. 112. See also S. V. Ciriacy-Wantrup, *Resource Conservation: Economics and Policies* (Berkeley: University of California Press, 1952), and Anthony Scott, *Natural Resources: The Economics of Conservation* (Toronto: University of Toronto Press, 1955).

211. John Laurence McHugh, "The Role and History of the International Whaling Commission," in *The Whale Problem: A Status Report*, ed. William Edward Schevill, G. Carleton Ray, and Kenneth S. Norris (Cambridge, MA: Harvard University Press, 1974), 305–35, at p. 331.

nuscule residue of what the Antarctic once produced) was actually *still* too high.

But even so, as he surveyed the work of the IWC through the years of his leadership of the US delegation, McHugh saw reasons for optimism: "In light of these encouraging moves since 1965," he argued, "it is important to support and strengthen the International Whaling Commission, not to destroy it."[212] By 1971 the latter option had a growing number of champions. The danger to the IWC by that point came not from within, as in the early 1960s, but from without. Facing down the swelling clamor of "overzealous and uninformed people" who were calling with increasing truculence for a complete ban on commercial whaling, McHugh expressed concern that public whale enthusiasm had gotten totally out of hand over the last decade—that it had become a rising threat to the very existence of the IWC, and therefore to the whales themselves. How far this had all gone, he ruminated, from the State Department strategy sessions of the mid-1960s:

> I recall events in 1965 at the special meeting of the Whaling Commission and the annual meeting which followed. Ambassador William C. Herrington was then active in whaling matters for the United States government, and he concluded, and I agreed, that the best strategy was to alert public opinion to the slow progress of the Whaling Commission. We did what we could—and our magic incantations, like those of the sorcerer's apprentice, appear to have been successful. The problem now is to halt the forces we have set in motion before they destroy the object of our efforts.[213]

This was, perhaps, to accept both too much responsibility and too little. But by the early 1970s, McHugh had no time for soul-searching. Rather, he forged ahead, appealing to the fond hope "that reason will prevail," since "a total moratorium is not only irrational and unnecessary" but also "impossible to achieve" within the economic and political constraints of the day.[214] It was on this account, as he saw it, that the gathering storm of activism was "threatening the efforts, now beginning to be felt, that several of us have been making to *achieve rational management of world whaling*,"

212. Ibid., p. 332.
213. Ibid., p. 313.
214. Ibid., "Addendum" of 16 October 1972, p. 335.

which would mean, he explained patiently, "full biological productivity for utilization by man."[215]

By 1972, in other words, it was John Laurence McHugh's turn to be befuddled by a sweeping revision of the larger framework in which one might hold forth about the "rational" way to deal with a large air-breathing marine creature. Just a few years earlier, in 1968, the biologist Garrett Hardin had posed what he took to be the most pressing question facing humanity— "What Shall We Maximize?"—and had answered it (*Forget maximizing!*) in the epochal neo-Malthusian think piece "The Tragedy of the Commons." His essay pointed to, among other things, the pending extinction of the overexploited whales in support of his meticulously econometric conclusion that a world full of rational self-maximizers would destroy first the world and then themselves.[216] In fact, the process seemed, in his view, to be well advanced. In this and other alarmist tracts of the late 1960s, the internal contradictions of McHugh's version of "economic rationality" were paraded before an American and European public increasingly preoccupied with environmental pollution and human destructive power.

A closer examination of these developments must be the subject of the next chapter, but it will afford a foretaste of their scope and depth if we take just a moment to consider that, by the early 1970s, Laurence McHugh found himself bearded by a movement that took cold comfort in the calculated pleadings of *human* rationality writ large. For many of these radicals, the rationality relevant to the solution of the problem of hyperintensive whaling—in fact, the rationality that promised to end war, restore the integrity of the ecosystem, and install a new age of peace and love on the blue planet—was *that of the whales themselves.* Here is the neuroanatomist Peter Morgane, writing on the whale brain in 1974:

> Major riddles of nature and relations between species may indeed be answered by these brains, and these opportunities may die with the whales if we do not act now. They could have taught us much if we had only listened. . . . Would that the brains of men could lead them to live

215. Ibid., pp. 331, 312–13. Emphasis added.

216. G. Hardin, "The Tragedy of the Commons: The Population Problem Has No Technical Solution; It Requires a Fundamental Extension in Morality," *Science* 162, no. 859 (13 December 1968): 1243–48. Compare Colin Clark's important later works on whaling: Colin W. Clark, "The Economics of Overexploitation," *Science* 181, no. 4100 (August 1973): 630–34; and C. W. Clark and R. Lamberson, "An Economic History and Analysis of Pelagic Whaling," *Marine Policy* 6, no. 2 (1982): 103–20.

in harmony with Nature instead of ruthlessly plundering the seas that nurtured us.[217]

And the apostate neurophysiologist John C. Lilly—the renegade student of cetacean intelligence and Pied Piper of whale huggers worldwide—was by this time offering actual primers on how the "manipulative" rationality of Earth's most destructive mammals (the ones with hands) could be commensurated with the kinetic, mindful, oceanic intelligence of the creatures he touted as possessing the "largest brains on earth":

> I suspect that whales and dolphins quite naturally go in the directions we call spiritual, in that they get into meditative states quite simply and easily. If you go into the sea yourself, with a snorkel and face mask and warm water, you can find that dimension in yourself quite easily. Free floating is entrancing. . . . Now if you combine snorkeling and scuba with a spiritual trip with the right people, you could make the transition to understanding the dolphins and whales very rapidly.[218]

Thinking with the whales would define, as these quotes suggest, a central ambition of the "Age of Aquarius."

From the rationality of the shop floor to a "mind in the waters," the project of being rational about whales had altered profoundly, irreversibly, over the last half century. Little wonder that McHugh, trained up on the counting-house logic of economic optimality, was irritated and confused. Nor was he alone. The IWC whale scientists who perused Mackintosh's 1957 Scientific Committee report and noticed the brief aside on the final page— "some studies on the brains of whales were also in progress in the U.S.A."— could never in their wildest dreams have imagined that they had just read a harbinger of the research that would undo them in the decades to come.[219] But they had. They definitely had.

It is to this catalytic and complex story that we now must turn.

217. Peter Morgane, "The Whale Brain: The Anatomical Basis of Intelligence," in *Mind in the Waters: A Book to Celebrate the Consciousness of Whales and Dolphins*, comp. Joan McIntyre (New York: Charles Scribner's Sons and Sierra Book Club, 1974), 84–93, at p. 93.

218. Lilly, cited in *Mind in the Waters: A Book to Celebrate the Consciousness of Whales and Dolphins*, comp. Joan McIntyre (New York: Charles Scribner's Sons and Sierra Book Club, 1974), 83.

219. IWC / 9 / 13, p. 5.

SIX

SHOTS ACROSS THE BOW

CETOLOGY AND
THE MIND IN THE WATERS

1960 – 1975

> As the whales go, so go the oceans, and as the oceans go,
> so goes the environment, causing the whales to become the
> symbol of the international environmental movement.
>
> RICHARD D. LAMM, Governor of Colorado, 1976[1]

INTRODUCTION

On 28 May 1963, David Brown, the curator of marine mammals at Marineland of the Pacific, wrote a grumpy note to John C. Lilly, the founder and director of the Communication Research Institute (CRI), a Miami-based center for the study of dolphins. Brown was in a quandary: should he really bother to drag himself all the way across the country in the fat of the summer in order to attend a professional gathering of scientists and specialized animal handlers all dedicated to the great whales and their smaller kin? That meeting, the "First International Symposium on Cetacean Research," would be held in August in Washington, DC, and Brown was scheduled to participate in a session, a roundtable on the practical problems of maintaining small toothed whales in captivity, chaired by the man who had served as his counterpart at Marineland of Florida: Forrest Glenn Wood Jr., known

---

1. Cited in Food and Agriculture Organization of the United Nations, *Working Party on Marine Mammals. Mammals in the Seas: Report of the FAO Advisory Committee on Marine Resources Research, Working Party on Marine Mammals* (Rome: Food and Agriculture Organization of the United Nations, 1978), vol. 1, xxiii.

as "Woody," a trim and punctilious expert on the care and feeding of the bottlenose dolphin, *Tursiops truncatus*.[2] Brown had accepted the invitation to participate about a year earlier, but as the panel took shape, he found himself "singularly unimpressed . . . with Woody's choice of participants" and had begun to toy with the idea of backing out.[3]

Lilly, who was scheduled to chair a session on communication, had been known to hold himself somewhat aloof from the proceedings. A medical doctor by training—a neurophysiologist who had risen rapidly to the rank of section chief at the National Institute of Mental Health (NIMH) before an interest in dolphin brains and behavior drew him to an ambitious project to build an independent research laboratory to study dolphin communication—Lilly had recently become a national figure. In 1961 the publication of his popular-science monograph *Man and Dolphin*, packed with visionary claims about cetacean intelligence and the possibility of interspecies communication, led to a spasm of media attention: a cover feature in *Life*, an appearance on the *Jack Paar Show*, and a host of favorable reviews (figure 6.1). All the publicity was giving private donations to CRI a healthy boost. Moreover, this brush with celebrity had not (yet) undermined Lilly's status with funding agencies and the web of professional colleagues who anonymously vetted the grants that kept CRI's laboratories in Saint Thomas and Miami afloat. Indeed, in addition to support from the Air Force Office of Scientific Research, NASA, NIMH, the Office of Naval Research, and the National Institute of Neurological Diseases and Blindness, Lilly could boast of having just been awarded, in 1962, one of the coveted National Science Foundation "Career" grants, which would, if all went according to plan, fund his work for the next five years.

A well-regarded brain scientist, then, Lilly was not above holding mere cetologists at arm's length, regardless of his own interest in dolphins. A year earlier, still in the first blush of fame from the publication of *Man and Dolphin*, Lilly had received an invitation from the American Institute of Biological Sciences (AIBS) to serve on the planning committee that would organize the Washington symposium on cetacean research. In a letter to a fellow neuroscientist, Frederick G. Worden, at UCLA's Brain Research

2. Marineland of Florida was known in its early years as "Marine Studios" (reflecting its origins in the film industry); in the writings of a number of authors close to this institution in the 1940s through the 1960s, usage is not consistent.

3. Brown to Lilly, 19 April 1963, Stanford University Library, Special Collections, Palo Alto, CA, John C. Lilly Papers, box 3D2, file "Brown, Mr. David."

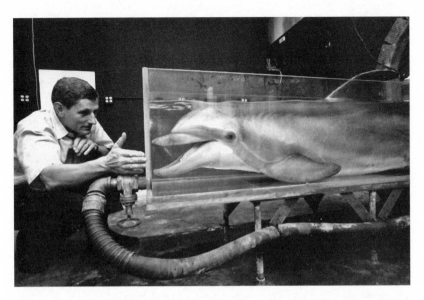

FIGURE 6.1 Man and dolphin: Lilly greets his subject. (Courtesy of the Flip Schulke
Archives. © Flip Schulke.)

Institute, Lilly reported on the honor with a certain insouciance: "You may
be amused to hear," he wrote, "that I have been invited to serve on the planning committee at AIBS for the first International Symposium on Marine
Mammals."[4]

And yet by the spring of 1963, Lilly's enthusiasm for the meeting was
high, and he wrote back to Brown encouraging him to make the trek east
for the big August gathering. With a sly nod to the tail-walking, dog-on-
a-surfboard-pulling, ball-tossing dolphins that were Brown's bread and
butter, Lilly talked up the coming event: "Do come along to the Show and
have a good time and relax in the spirit of the internation [*sic*] occasion.
After all, this *is* another Whale Circus, with Big Names performing to keep

4. Lilly to Worden, copy, 18 April 1962, Stanford University Library, Lilly Papers,
box 3D2, file "Worden, Frederick G., Brain Research Institute, Los Angeles, California."
The planning committee consisted of Lilly and F. C. Fraser, S. R. Galler, K. Norris,
J. R. Olive, I. Rehman, W. E. Schevill, M. C. Tavolga, and H. J. Taylor. See Symposium
Program verso, Scripps Institution of Oceanography Archive, University of California,
San Diego, Carl Hubbs Papers, box 54, folder 56, "West Coast Mammals; Cetacea; Refer-
ences 1952–1977." This file also contains Hubbs's manuscript notes on the first half of the
symposium. For related correspondence, see Scripps Archives, Hubbs Papers, box 4,
MC 5, "American Institute of Biological Sciences, General, 1949–1971."

their big names!" Forget substance, Lilly counseled: sure, there was little to commend Wood's selection of participants in the "practical problems" roundtable, but "real information" wasn't the point. The purpose of the conference had less to do with the sessions and the papers than with the care, feeding, and behavioral analysis of "cetologists" themselves; the point of the event was "to show people working with dolphins and whales one another's faces, and hear how they sound when they have had a few drinks at the bar, and to bury a few hatchets."[5]

In a manner of speaking. What Lilly did not know, apparently, was that several of those headed to the meeting planned to bury their hatchets directly in his person. Nor could anyone back in 1963 have predicted the reverberations that would widen from that conflict in the years to come.

WHALES, DOLPHINS, AND PORPOISES

The First International Symposium on Cetacean Research did indeed take place, as promised, from August 15 to August 18 of 1963.[6] And it did turn out to be an important gathering in several respects. For instance, it resulted in a volume that would constitute a foundational text for modern marine mammal biology, a volume that would dominate the English-language study of cetaceans for more than a decade.[7] But for our purposes, the deeper signifi-

5. Lilly to Brown, 28 May 1963, Stanford University Library, Lilly Papers, box 3D2, file "Brown, Mr. David."

6. See dates on Symposium Program, Scripps Archive, Hubbs Papers, box 54, folder 56, "West Coast Mammals"; see also Hubbs's datebook for 1963, Scripps Archive, Hubbs Papers, box 12, folder 83.

7. This volume was *Whales, Dolphins, and Porpoises*, ed. Kenneth Stafford Norris (Berkeley: University of California Press, 1966). Remarking on the gathering after his retirement more than thirty years later, Norris (by this time the dean of American marine mammal biology) would muse, "I found myself in a cadre of about forty people, not just nationwide, but worldwide, who were concerned with these animals. . . . They came from all over the place, from the British Antarctic Survey, from museums, etc." The mix was eye-opening for the young UCLA PhD, who suddenly found himself, as he put it, "in the scientific swim. . . . It was an exciting time. Sheer physical excitement carried me through there, too." See transcript of interview with Norris by Randall Jarrell and Irene Reti in *Kenneth S. Norris: Naturalist, Cetologist, and Conservationist, 1924–1998*, ed. Randall Jarrell (Santa Cruz: Regional History Project, UC Santa Cruz University Library, 1999), 24. As one of the current leaders in cetacean research in the United States recently recalled, "About the only book around at the time I started my student career was the proceedings of that meeting. That was the only book on cetaceans available

cance of this meeting lies in the fact that it represented an unprecedented encounter between two very different scientific communities. One of these came out of a tradition with which this book has been substantially concerned, the tradition I have called (somewhat irreverently) "hip-booted" whale science. The other, however, emerged from a research enterprise that now comes onstage for the first time in my narrative: military bioscience. In this chapter I will argue that it is the latter tradition (in its various filiations, which were diverse and, finally, as I will show, at odds) that we must examine in some detail if we hope to understand the dramatic transformation of attitudes toward cetaceans in the period 1960–1975.

As a point of departure, it is perhaps worth reminding ourselves just how dramatic that transformation was.[8] Though previous chapters have consistently sifted out a small number of early voices calling for conservation of the world's large whales, these animals remained, in the early 1960s, little more than an industrial commodity of dwindling importance. The Antarctic industry (as we saw in the last chapter) was teetering on the brink of collapse; competitive prices on tropical nut and seed oils (and shifting consumer preferences) had made margarine makers and soap manufacturers less reliant on mysticete fats across the 1950s; and the center of gravity of the whaling industry as a whole was shifting east, to the Soviet Union and Japan, where different market opportunities and economic structures (in Japan, whale meat for human consumption; in the Soviet Union, vast farms of fur-bearing animals in need of low-cost feed—and a command economy not entirely answerable to the demands of profitability) gave the future of whale killing a rosier hue. Sperm whales in temperate waters like the North Pacific, of which there remained a considerable number, were gradually displacing rorquals as industry targets, and with this new focus

---

in the United States, by a U.S. author." See Bill Perrin, interview by Randall Jarrell, in *Kenneth S. Norris*, ed. Jarrell, 116.

8. Two essays deal with this topic in some detail: Paul H. Forestell, "Popular Culture and Literature," in *The Encyclopedia of Marine Mammals*, ed. W. F. Perrin, Bernd G. Würsig, and J. G. M. Thewissen (San Diego: Academic Press, 2002), 957–74; and David Lavigne, Victor B. Scheffer, and Stephen R. Kellert, "The Evolution of North American Attitudes toward Marine Mammals," in *Conservation and Management of Marine Mammals*, ed. John R. Twiss, Randall R. Reeves, and Suzanne Montgomery (Washington, DC: Smithsonian Institution Press, 1999), 10–47. Also of value is Stephen R. Kellert's chapter "Human Values, Ethics, and the Marine Environment," in *Values at Sea: Ethics for the Marine Environment*, edited by Dorinda Dallmeyer (Athens: University of Georgia Press, 2003), 1–18.

for the factory ships came new consumers of whale products: the distinctive liquid wax derived from sperm whales is not good eating, but it does possess useful characteristics as an industrial lubricant (it wound up in transmission fluids, since it held up well at high temperatures and had low viscosity).

Above all, popular interest in the animals cannot be said to have been very great in these years. The annual meetings of the IWC received perfunctory coverage in the national media of the United States, Latin America, and the non-whaling countries of Europe (and there were more and more of these, as, gradually, the whaling companies of England, France, Spain, Germany, and eventually even the Netherlands went under or were bought out by the Japanese). To be sure, the Norwegians continued to take an active interest in the machinations of an industry they had done so much to create and maintain, but that interest was largely confined to questions of employment and finance: the animals themselves could hardly be said to possess significant symbolic power outside their status as the icons of Norway's most remarkable global industry. As for the notion of these creatures possessing "intelligence" or "beauty," one must comb closely through marginal materials to catch even a hint of such preoccupations in the later 1950s. Yes, whalemen and those who wrote about them tended to emphasize that sperm whales could be wily and that rorquals sometimes spooked easily; and yes, the language of superlatives was seldom far away when whales were discussed in popular sources (the blue whale is, after all, the largest beast ever to live on the earth). True, too, that this journalistic preoccupation with the bigness of whales could be translated into a rhetoric of "majesty" on some occasions, as we have seen, but no less frequently the same observation invoked associations of freakishness: "A fairly large monster!" offered R. B. Robertson of a blue whale in his popular book on whaling from the mid-1950s, before going on to note that the carcass of the beast was worth a cool $6,000 and that "its skull is the size and weight of a motor car, but the brain contained therein is not much bigger than the carburetor"—a comment tending to depreciate the mental sophistication of the taxon.[9] Major revisions to this posture would follow in the years to come, with far-reaching implications for conservation efforts.

In popular writings on whales from this period, the technological sublime of the industry itself tended to upstage the natural sublimity of its

9. Robert Blackwood Robertson, *Of Whales and Men* (New York: Knopf, 1954), 139.

quarry. At the close of another successful journalistic treatment of the subject to come out in English in these years, Ivan T. Sanderson submitted to the allure of better living (or is it better killing?) through chemistry:

> I have a friend, a chemical engineer of considerable prominence, who has always been fascinated by sausage machines. . . . His idea is to haul the whale into a structure like a huge coffin and then either cut it up with a power driven super meat slicer, all the blades going simultaneously, straight through the whale, after which the bottom of the box automatically drops all the slices into boilers or alternatively to seal the coffin and just boil the entire whale under pressure. In either case, improved chemical methods would be relied upon to separate whatever grades of whatever kinds of oil are needed from the resultant animal "crude oil."

Sanderson declared the whole scheme visionary and beyond his grasp, but conceded that the "floating cracking plant" would quite possibly replace the "floating factory" as the future of the whaling industry. After all, "if you can get perfume, gasoline, tar, and flavoring for ice cream out of petroleum, I must believe when I am told that chemists should be able to perform a similar miracle with a boiled whale."[10]

Still more fantastic, perhaps, and no less indicative of the technocratic mood on these matters at the start of the period covered by this chapter, is the "modest proposal" dreamed up by the biologist Gifford B. Pinchot (son of the better-known Gifford Pinchot of chapter 3) and some colleagues in the early 1960s. Beguiled by the arithmetic of the trophic pyramid and impressed by experiments suggesting the possibility of dramatically increasing ocean productivity by "fertilizing" marine environments with rate-limiting nutrients, Pinchot laid out a plan to convert a set of Pacific atolls into *whale farms*, where captive rorquals (towed in floating cages into the circular bays as juveniles) would be fattened on artificial plankton blooms until they were large enough to slaughter for meat, oil, and other by-products. By his calculations, a mere $270 worth of phosphate-based fertilizer sprinkled in an atoll of one mile in diameter ought to produce some 57.5 tons of phytoplankton, which would in turn yield roughly 5.75 tons of zooplankton and 0.575 tons of animal flesh. When you did the math, this worked out to

10. This and the quotes above are from Ivan Terence Sanderson, *Follow the Whale* (Boston: Little, Brown, 1956), 344.

about 23 cents a pound for whale meat, which Pinchot conceded was not cheap, but "does not compare unfavorably with beef on the hoof." Moreover, a closed system like an atoll promised nutrient transfer efficiencies higher than traditional open-ocean benchmarks, suggesting the possibility of still lower costs. In a quirky invocation of the logic of having one's cake and eating it too, Pinchot noted that his scheme not only promised to address looming world food shortages, but also stood to "preserve threatened whales from extinction"—domestication being a superb bulwark against extermination. And, best of all, this arrangement would give scientists a unique opportunity to study the largest animals in the world in captivity: "Whales," he suggested, "might prove to be even more interesting than porpoises in behavioral and physiological studies."[11]

I will return to this juxtaposition of whales and dolphins in a moment, since it is a contention of this chapter that thinking about the (ever mysterious) large whales in terms of the (increasingly familiar) small odontocetes—especially *Tursiops truncatus*, the bottlenose dolphin—turns out to be a key element of the larger revolution in thinking about cetaceans that unfolded during the fifteen years in question. But for now, allow me to underline just how far we are (with Pinchot's vision of whale aquaculture, with Robertson's dismissive gesture at whales' disproportionately small brains) from the extraordinary whale-hugging enthusiasms of the 1970s.

We can begin with the formal matters of law and regulation. In 1960 whales were not a matter of legal or legislative concern in the United States.[12] By November 1970, however, largely on the basis of popular clamor (and over the shocked objections of essentially every whale scientist involved in the IWC, and even some outside the organization), the US Secretary of the Interior, a populist Republican from Alaska named Walter J. Hickel, had placed *all* the large whales on the Endangered Species List, single-handedly

11. This and the quotes above are from Gifford B. Pinchot, "Whale Culture—A Proposal," *Perspectives in Biology and Medicine* 10, no. 1 (Autumn 1966): 33–43, particularly pp. 38, 42.

12. As a signatory nation to the ICRW and a promoter of the IWC, the United States did have enabling legislation linked to its international treaty obligations, and this meant that gray whales, like right and bowhead whales, were protected. The unique story of the efforts to protect gray whales off the California-Mexico coast has been admirably handled by Serge Dedina in his *Saving the Gray Whale: People, Politics, and Conservation in Baja California* (Tucson: University of Arizona Press, 2000).

making importation of any whale product into the United States a crime.[13] Promptly bounced from office (though not for that, but rather for his increasingly outspoken criticism of Nixon's Vietnam policy), Hickel would find himself one of the heroes of an increasingly vociferous international crusade to end whaling altogether. A year later, in late 1971, goaded by the

13. The history of this listing is complicated. A stunningly expansive suite of species (sperm whales and "Baleen whales, all species") were initially listed by Interior in an internal document dated 25 May 1970 and circulated over the signature of Fred J. Russell, then recently appointed as undersecretary (for some reason he is listed as "Acting" secretary on the document in question, though Hickel was secretary at this time). See National Archives and Records Administration, College Park, Maryland (NARA) Record Group (RG) 48—Records of the Office of the Secretary of the Interior, "Central Correspondence Files, 1969–1972," box 342, file "Wildlife, Conservation and Control". But procedural objections voided the subsequent publication of this list in the *Federal Register* for June 2 of that year, and the process began anew on 30 July with a revised listing of eight species in total (sperm, finback, sei, bowhead, right, blue, humpback, and gray). Of these, the sperm was by far the most contentious, as no whale biologist then living thought these animals seriously depleted. For a discussion of the procedural error and the relisting, see Glasgow to Horton, 2 September 1970, NARA RG 48, "Central Correspondence Files, 1969–1972," box 59, file "Fisheries, Conservation and Control." After fielding an onslaught of letters from concerned citizens, schoolchildren, and representatives from industry over the summer (and hearing expert testimony from McHugh, Chapman, Scott McVay, and Roger Payne), Interior moved forward with the listing in November. In fact, Hickel finalized the listing and released a press announcement on 24 November, literally within hours of being terminated by the president. See "Department of the Interior News Release: Secretary Hickel Bans Imports of Products from Eight Endangered Species of Whale," NARA RG 48, "Central Correspondence Files, 1969–1972," box 59, file "Fisheries, Conservation and Control." Subsequent congressional and White House hesitation (there were threats to "rescind" the last-minute listing) delayed implementation of the order and led to its enactment by then-acting secretary Russell. Interestingly, in 1969 Hickel had already shown a willingness to meddle in whale matters, and in the teeth of the State Department and the IWC, since in that year his office put an end to gray whaling by Del Monte on the Pacific coast and put in place a quota on fin and sei whales taken by US citizens in the North Pacific. Hickel's appetite for this fight seems to have been largely his own, though his breathless narrative account feels trumped up; see Walter J. Hickel, *Who Owns America?* (Englewood Cliffs, NJ: Prentice-Hall, 1971), esp. chap. 6. On the legislative history of the Endangered Species Act, see Shannon Petersen, "Congress and Charismatic Megafauna," *Environmental Law* 29, no. 2 (Summer 1999): 463–92. Also of interest is Charlotte Epstein, "The Making of Global Environmental Norms: Endangered Species Protection," *Global Environmental Politics* 2, no. 2 (May 2006): 32–54.

newly formed Environmental Defense Fund, the Animal Welfare Institute, and others, Congress would adopt a resolution calling for the secretary of state to negotiate a ten-year global moratorium on commercial whaling (again over the objections of McHugh, Chapman, and other US scientific and diplomatic personnel who had worked on the whaling problem for years). The following June, Hickel himself would take the stage in Stockholm to address a boisterous crowd of young activists in the hopes of getting the landmark UN Conference on the Human Environment to vote in exactly such a ban on the industry. A huge whale puppet would be subsequently paraded through downtown Stockholm as part of an extended Woodstock-style "Celebration for the Whale."[14] The moratorium motion, proposed by the US delegation and structured as a "recommendation" to the IWC, passed by a wide margin at the UN conference.[15] These events left the Japanese and Soviets ever more isolated amid what was fast becoming the most prominent crusade of an increasingly strident environmental movement intricately entangled with antiwar protest and tied to a more general countercultural youth movement.

14. For a flavor of the event, and some context for the youth activism at this gathering, consult Tord Björk, "The Emergence of Popular Participation in World Politics: The United Nations Conference on the Human Environment, 1972" (unpublished seminar paper, Department of Political Science, University of Stockholm).

15. It is interesting to consider the extent to which the emphasis on the whale issue at Stockholm amounted to a kind of PR coup by US officials, who had been enormously concerned about Chinese efforts to focus the event on the environmental dimensions of rapacious US capitalism and militarism in Indochina. Whales, in this sense, were a perfect freebie as an issue, since they promised to refocus attention on the (absent) Soviets. I have not been able to find a documentary "smoking gun" on this issue, but I have no doubt it is out there (*pace* Russell E. Train's contemptuous dismissal of the notion; see his memoir, *Politics, Pandas and Pollution* [Washington, DC: Island Press, 2003], 141). For a synopsis of US views concerning how successfully the US delegation succeeded in parrying Chinese (and Eastern bloc) efforts to make the whole event a "Third World" gripefest, see "Primer on Ocean Jurisdictions: Drawing Lines in the Water," in Report on the United Nations Conference on the Human Environment, Ninety-Second Congress, second session, September 1972. Japanese counter-mobilizations to this emerging movement have been the subject of some recent work, including Anders Blok, "Contesting Global Norms: Politics of Identity in Japanese Pro-Whaling Countermobilization," *Global Environmental Politics* 8, no. 2 (May 2008): 39–66; and Andrew Miller and Nives Dolšak, "Issue Linkages in International Environmental Policy: The International Whaling Commission and Japanese Development Aid," *Global Environmental Politics* 7, no. 1 (February 2007): 69–96.

The IWC, presented with this international mandate, mumbled concerning its unreasonableness (not wholly unreasonably), and in so doing set the stage for one of the most dramatic episodes in the history of environmentalism. On 27 June 1975, a clutch of long-haired pacifist whale lovers who called themselves "Greenpeace," aboard a balky 66-foot halibut seiner out of Vancouver known as the *Phyllis Cormack*, confronted Soviet whalers along the Mendocino Ridge in the North Pacific. Genuine courage, foolhardiness, and good fortune (along with self-conscious attention to the media dictates of fellow Canadian Marshall McLuhan) yielded less than a minute of Super 8 film footage that remains a generational touchstone: it depicts, from water level, the looming Soviet catcher vessel *Vlasny* as it fires a 250-pound Svend Foyn–style grenade harpoon into the back of a small female sperm whale—a harpoon that passes menacingly close to the heads of two young men, George Hunter and George Korotva, perched in a small inflatable powerboat working to disrupt the hunt. Even to a viewer without particular feelings for whales, the film sequence possesses undeniable power, and when it was shown to tens of millions of Americans by a concerned-looking Walter Cronkite on CBS Evening News on 1 July 1975, it galvanized broad and deep public revulsion for modern whaling, and modern whalers (figure 6.2).[16]

It would be a mistake (though one several Greenpeace veterans might not correct) to suggest that this telegenic David-and-Goliath showdown directly secured the IWC moratorium on commercial whaling. In truth, it would take more than a decade of work by many different organizations and individuals in many different places before diplomatic procedures finally squeezed commercial whaling into something like semipermanent suspension (and even then, a variety of exceptions and special arrangements have continued to allow a number of nations and "aboriginal communities"—a hotly disputed designation—to continue to whale). But the "mindbomb" of the Greenpeace encounters—repeated in new forms, and in other oceans, and with increasing truculence into the 1980s—epitomized a new kind

16. There are several valuable sources on this incident and, more generally, on the history of Greenpeace. I have found Rex Weyler, *Greenpeace: How a Group of Ecologists, Journalists, and Visionaries Changed the World* (New York: Rodale, 2004) to be the most useful. But also very instructive is Frank Zelko's "Make it a Green Peace: The History of an International Environmental Organization," (PhD diss., University of Kansas, 2003). To be savored as a primary source, but handled with reservation as an analysis, is Robert Hunter's *Warriors of the Rainbow: A Chronicle of the Greenpeace Movement* (New York: Holt, Rinehart & Winston, 1979).

FIGURE 6.2
(and see PLATE 15)
Soviet whalers
meet the freaks:
Greenpeace
in mid-hunt.
(Courtesy of Rex
Weyler.)

of resistance to whaling specifically and to environmental irresponsibility more generally. The semilegal actions of this peculiar community of grandstanding saboteurs and dope-addled visionaries can be understood both to have signaled and to have leveraged significant political change.[17] By the early 1980s commercial whaling was, in an international legal sense, an endangered species of marine exploitation. It remains so at the time of my writing.

17. For a concise and readable sketch of the campaign to save the whales in these years, authored by a participant in the events, consider the final chapters of Richard Ellis, *Men and Whales* (New York: Knopf, 1991). A drier account, Peter J. Stoett's *The International Politics of Whaling* (Vancouver: UBC Press, 1997), reviews some of this recent history. There are many journalistic treatments, including the very uneven book by David Day, *The Whale War* (San Francisco: Sierra Club Books, 1987).

What can be said about shifting "ideas about whales" across the period of activism and legal-cum-regulatory change that I have just reviewed? And what role do shifts in attitudes toward these animals—if indeed they can be established—play in the political processes that led to something very much like the end, at least for now, of whaling? In one sense these are basic questions, but academic historians, political scientists, and other professionally fretful thinkers will quickly point to a host of problems with this kind of inquiry: *Whose* ideas about whales? Manifested how? And what would it mean to disaggregate the tangle of causal factors that operated at such vastly different scales to precipitate multiparty "action" at the level of international law and national practice? On the one hand, having attended many a seminar in which such methodological matters were threshed at some length, I have a certain respect for the seriousness of these issues, which under certain constructions can be made to touch fundamental philosophical problems concerning human agency and the nature of explanation. In another way, however, much of that seminar time has gone toward persuading me that I became a historian precisely because I am consistently more interested in resolving these Gordian-knotty challenges with the knife of narrative than with the awl of analysis. So I will forgo a preemptive defense of my account on theoretical grounds and see if the story can, by the end, defend itself from excessive methodological heckling.

Back, then, to those basic questions for a moment: Did "ideas" about whales change over the long decade in question? Obviously they did, and with clear implications for the regulatory arena. In important ways the large whales became in this period, for many, "symbolic" organisms: icons of mankind's rapacious carelessness, bellwethers of ecological collapse, rallying devices for social and economic and cultural change. In August 1970 Paul Spong, who would become a key figure in Greenpeace, floated a 48-foot soundstage dubbed the *D'Sonoqua* into the cold waters near Vancouver, carrying a cargo of folk musicians intent on performing a live concert for free-swimming small cetaceans in the bay; it was the first of many similar festival efforts at countercultural interspecies "contact" that would unfold in the run-up to the Mendocino Ridge incident.[18] And by 1976, as NASA prepared to send humpback whale phonation into outer space on a

18. See Joan McIntyre, comp., *Mind in the Waters: A Book to Celebrate the Consciousness of Whales and Dolphins* (New York: Charles Scribner's Sons and Sierra Book Club, 1974), 176.

phonograph disk of gold aboard *Voyager 1* and *Voyager 2*, Gilbert Grosvenor, the president of the (very mainstream) National Geographic Society, could declare that "the whale has become a way of thinking about our planet and its creatures."[19] He was echoing the words of Victor B. Scheffer, the chairman of the Marine Mammal Commission (a whole independent federal "whale" agency created by Congress as part of the 1972 Marine Mammal Protection Act), who had written the year before that whales "have become symbolic of life itself."[20] Clearly these animals were no longer merely aquatic mammals containing large quantities of interesting oils; they had become nothing less than familiar spirits, capable of mobilizing sentiment and action, capable of representing complex and culturally significant assemblages of aspiration and belief.

How did this happen? And why? Genuflecting toward the perils of causal explanation in history, I propose to hazard answers to these questions in the pages that follow, so allow me to outline my thinking here, by way of a précis of the chapter as a whole. In brief, I believe that the most important change in the way a significant number of politically engaged people in Europe and North America thought of the large whales between 1960 and 1975 involved a shift to a view of these creatures as possessed of "intelligence"—defined loosely as cognitive and affective abilities recognizable to human beings as sufficiently like our own (or unlike our own, but in an interesting and important manner) to disqualify them as a prey species. In my view, the single largest factor in this development was the controversial work of John C. Lilly, who made a set of extravagant—and finally unsubstantiated—claims for the mental and linguistic abilities of dolphins starting in the late 1950s. This work was, as I will show, inextricable from (and dependent on) Cold War military bioscience. I demonstrate that this was so in two respects. First, Lilly's general interest in "other minds" was an interest nurtured in the hothouse atmosphere of the *Manchurian*

19. See the editorial by Gilbert Grosvenor in *National Geographic* 150, no. 6 (December 1976), 721. I do not know of a secondary source that deals with the compilation of the golden record, but Carl Sagan's *Murmurs of Earth: The Voyager Interstellar Record* (New York: Random House, 1978) is a participant account.

20. In a speech delivered at the California Academy of Sciences on 1 June 1975, as quoted in Victor B. Scheffer, "The Status of Whales," *Pacific Discovery* 29, no. 1 (January–February 1976): 2–8, at p. 8. To be fair, the Marine Mammal Commission's mandate includes walruses, seals, manatees, and so forth, and therefore calling it a "whale agency" is something of a fib. But it's a fair fib: cetaceans upstaged all these other creatures in the commission's public profile and political activities.

Candidate–era sciences of mind and behavior (he spent much of his professional life before turning to dolphins working on "gaining access" to the inner lives of potentially recalcitrant enemy subjects, and he eventually used the whole spectrum of techniques he learned and refined in this environment in his work on *Tursiops truncatus*). Second, the technical means he used to "hear" what whales and dolphins were "saying" (or "singing")—a notion central to an expanded sense of the cognitive capabilities of these animals—were wholly drawn from the field of naval bioacoustics. Though Lilly would publically and prominently break from these military sources of funding (or, to put it a different way, be broken from them), the US Navy itself would continue to work with small whales, and it would play a significant role in buttressing claims for their intelligence and unique capabilities—doing so, perhaps surprisingly, by advancing programs Lilly himself initially advocated and formally assisted. Lilly's increasingly countercultural persona and his widening feud with the navy would help imbue the disputes over whaling with an unmistakable scent of antiwar zeal. Moreover, as I will argue in conclusion, the filiations of Lilly's influence in the "whale wars" would be quite direct, even as he himself receded from the scene (after all, there was no book better thumbed aboard the Greenpeace vessels of the mid-1970s than Lilly's *Mind of the Dolphin*, for reasons that will be interesting to explore).

By affording Lilly such centrality in this account, I am aware that I am courting the ire of a considerable number of those alive today who care about and work with marine mammals. It is not that they have entirely overlooked Lilly's work. He is, for instance, afforded several judicious pages in a superb jointly authored historical essay on the origins of behavioral research on cetaceans written by two leading scientists in this area, and he remains, posthumously, a kind of guru-sage of a certain subculture of out-there whale lovers who live in the deep reaches of the Internet or entirely off the grid.[21] But most professional practitioners in the field of marine

21. I am thinking of Amy Samuels and Peter Tyack's informative "Flukeprints: A History of Studying Cetacean Societies," the lead essay in *Cetacean Societies: Field Studies of Dolphins and Whales*, edited by Janet Mann, Richard C. Connor, Peter Tyack, and Hal Whitehead (Chicago: University of Chicago Press, 2000), 9–44. Paul Forestell, too, gives Lilly several paragraphs in his helpful essay "Popular Culture and Literature" in *The Encyclopedia of Marine Mammals* (ed. W. F. Perrin, Bernd G. Würsig, and J. G. M. Thewissen [San Diego: Academic Press, 2002], 957–74)—a volume from which Lilly is otherwise largely absent; he does not appear in the brief "Biographies" of those who "made an important and lasting contribution to the study of marine mammals,"

mammal biology have long been at pains to distance serious work on cetacean communication, cognition, and social behavior from the hyperbolic and somewhat unsavory legacy of a man who was, in effect, ousted from the field before it even got started. For much of the last thirty years, any trace of Lillyism (signally, any explicit exuberance about whales and dolphins as "alternative intelligences") has been the kiss of death for a mainstream research proposal. And among conservation-oriented activists and communities committed to the welfare of marine mammals, Lilly again remains an ambivalent, and even divisive, figure. Yes, he was genuinely the first significant voice to invoke, and then to endeavor to substantiate empirically, the claim that dolphins—and by extension and analogy, the large whales—are moral agents, possessed of qualities of awareness (of self and other) and social intelligence (including capacities for relatively complex use of "language") that entail human beings in ethical obligations. But at the same time, he notoriously, vigorously, and largely unsentimentally engaged in forms of experimentation with these very animals—including neurological vivisection and later very loosely controlled psychopharmacological manipulation—that would cause a modern animal rights protester to blanch. His later years (he lived until 2001) as a West Coast B-list celebrity given to wearing jumpsuits and, apparently, binging on mind-altering drugs (and extolling virtues of same) have further vexed his reputation and have by no means facilitated a fair assessment of his place in the history of marine mammal science. In effect, he has been mostly untouchable by serious people for a long time.

It is my intention in this chapter to use Lilly's extensive but as yet largely untapped archive (acquired by Stanford University shortly after his death) to recover the John C. Lilly who preceded the figure who would be the in-

---

and is not referenced in the articles on "Brain" or "Communication." Finally, the whole of his work is dismissed as a withering "total failure" by one of the volume editors, Bernd Würsig, in the essay on "Intelligence and Cognition," which goes so far as to say that "the unscientific nature of Lilly's assertions deterred many others from studying dolphin and whale communication, and . . . addressing intelligence and cognition in an obviously behaviorally flexible taxonomic order of mammals." Lilly gets a single sentence in the comprehensive essay by David Lavigne, Victor B. Scheffer, and Stephen R. Kellert, "The Evolution of North American Attitudes toward Marine Mammals," in *Conservation and Management of Marine Mammals*, ed. John R. Twiss Jr. and Randall R. Reeves (Washington, DC: Smithsonian Institution Press, 1999), 10–47. For a flavor of new-agey Lilly on the Web, consider "Dedicated Dolphin Sites," The Oceania Project, accessed 27 February 2011, www.oceania.org.au/wwwlinks/dolphin.html.

spiration for the film *Altered States*. By reconstructing and contextualizing his work, by examining its reception, and then by following forward the tendrils of his influence into the early 1970s, I hope to clarify his significance while shedding light on some peculiar intersections between whale science and military research during the Cold War.

To that end, and by way of providing context for a critical moment in Lilly's career, let us turn back to the First International Symposium on Cetacean Research of 1963 and examine who was there and what they were studying. Doing so will set up the changing world of marine mammal science in the early 1960s and prepare us to understand how Lilly could be invited into this community and dismissed from it, all in the blink of an eye—with, as I have suggested, significant consequences.

CETOLOGY IN 1963

The first whale scientist to step to the microphone in the John Glenn Suite of the Marriott Motor Lodge at Key Bridge in Washington, DC, shortly after breakfast on Thursday, 15 August 1963, was L. Harrison Matthews, a venerable Discovery hip-booter of the first water.[22] Now risen to scientific director of the Zoological Society of London, Matthews had earned his stripes as a "cetologist" the old-fashioned way, by cutting his way into the innards of hundreds of whales while standing in the icy slurry of an Antarctic whaling station. Invited by the American convener-hosts to kick off this distinctive gathering of international scientists and specially selected "observers," Matthews looked out at a room containing about a hundred men (and a handful of women) who represented the core of what he took to be an emerging scientific field. Admittedly, there was no Soviet or Eastern bloc representation, but with the memory of the Cuban missile crisis of the previous autumn still all too fresh, it was to be understood that scientific and cultural liaisons across the Iron Curtain were at a low ebb. Even so, from the rostrum, Matthews felt he was surveying the leaders of whale science: "Practically everyone taking part in this symposium," he announced, "has carried out original investigations into some aspect of the biology of the Cetacea." And the field was growing apace: "The Cetacea differ so

22. The first session dealt with systematics and distribution; F. C. Fraser was the primary organizer, and he asked that Matthews serve as chair. See Olive to Hubbs, 22 January 1963, Scripps Archive, Hubbs Papers, box 4, MC 5, "American Institute of Biological Sciences, General, 1949–1971."

widely in all aspects of their biology from the other mammals, that cetology has evolved into a specialized branch of zoology," Matthews declared, not without pride.[23] And by plotting a curve of the field's swelling literature, the chairman could illustrate rapid and accelerating attention to cetacean science: in 1961 alone no fewer than 73 full-scale research papers had been contributed to the area, and he and his fellow cetologists could now point to a bibliography amounting to nearly 4,000 works.

So who were these cetologists, anyway? Let's pass an eye over that room for a moment and briskly survey the gathered whale scientists, ostensibly members of this quasi-discipline that Matthews was both conjuring and addressing in his opening remarks. Who, exactly, was out there in the John Glenn Suite on that August morning, rearranging themselves on the Marriot's banquet chairs?

Well, Matthews would have seen plenty of familiar faces in the crowd. For starters, there was a healthy contingent of his old Discovery colleagues: both F. C. "Jimmy" Fraser (who had followed in Sidney Harmer's footsteps and was now the keeper of zoology in the British Museum [Natural History]) and, of course, good old Neil Alison Mackintosh (who had attended the rancorous 1963 London IWC meeting earlier that summer) were in attendance. Together, they had the distinction of being the oldest of the old-school whale biologists, the founders of what Mackintosh, recall, liked to refer to as a "system of research on whales"—one that had been pursued by their hip-booted disciple-colleagues from a host of different nations. And, in fact, a good number of those fellow hip-booted anatomist-physiologists of the large cetaceans were filling the adjacent seats. For instance, there was the familiar spoiler of IWC consensus, E. J. Slijper (professor of zoology at the University of Amsterdam and leader of the industry-funded Dutch Whale Research Group), who had cut his cetological teeth aboard the *Willem Barendsz* in the Antarctic and who had a hand in ensuring that the guests at the 1956 meeting of the Dutch Zoological Association enjoyed juicy fin whale steaks. And there was Åge Jonsgård, too, a Norwegian professor of marine biology who served as deputy director of the State Institute for Whale Research and would soon publish a valuable study of North Atlantic finbacks (based on close collaboration with whaling captains and Norwegian shore station inspectors). And a little farther on,

23. L. Harrison Matthews, "Chairman's Introduction to First Session of International Symposium on Cetacean Research," in *Whales, Dolphins, and Porpoises*, ed. Kenneth Stafford Norris (Berkeley: University of California Press, 1966), 3–6, at p. 5.

there was the Japanese delegation—Hideo Omura, Masaharu Nishiwaki, and Tadayoshi Ichihara—all attached to the Whales Research Institute of Tokyo, an industry-affiliated and state-funded enterprise focused on whale biology and biogeography in the service of the fishery (indeed, Nishiwaki had turned his training as a World War II submarine spotter pilot in the Pacific to profitable postwar use, founding a company of his fellow pilots who served as air scouts for North Pacific whaling fleets).[24]

So the hip-booters were out in force in DC, as one would expect at a whale conference. No surprise there. And no surprise that seated among them, as observers, were a smattering of the scientist-statesmen who had made whale management their business through participation in the IWC. Remington Kellogg himself, for instance, the granddaddy of whale diplomacy, had taken the short drive from his Smithsonian office, and the man who would soon replace him as the US representative to the IWC, John L. McHugh, had also registered to sit in on the meeting. But in reviewing the roster of participants at the event, one promptly notices several contingents of folks who might seem to have little to do with "cetology," at least as Mackintosh might have characterized it. For instance, there was healthy representation from the world of the commercial ocean theme parks that had grown to prominence as entertainment venues in the United States in the 1950s.[25] These were people like Ralph Penner, who was the director of exhibits at one of the new West Coast oceanariums, and Taylor A. ("Tap") Pryor, who directed the Sea Life corporation in Oahu. Perhaps half a dozen trainers and veterinarians were in attendance at the symposium, and while their training was a far cry from the rituals of the flensing platform, it is not all that difficult to explain what they were doing at a symposium on

24. On Slijper, see Everhard Johannes Slijper, *Whales*, trans. A. J. Pomerans (New York: Basic Books, 1962), 42, 58; for examples of the others' work, see Åge Jonsgård, "Biology of the North Atlantic Fin Whale, *Balaenoptera physalus* (L.): Taxonomy, Distribution, Migration, and Food," *Hvalrådets Skrifter* 49 (1966): 1–62; M. Nishiwaki, "Aerial Photographs Show Sperm Whales' Interesting Habits," *Norsk Hvalfangst-Tidende* 51, no. 10 (1962): 395–98; and Peter E. Purves in *Whales, Dolphins, and Porpoises*, ed. Kenneth Stafford Norris (Berkeley: University of California Press, 1966), 123.

25. For some of the history of these establishments, see Susan G. Davis, *Spectacular Nature: Corporate Culture and the Sea World Experience* (Berkeley: University of California Press, 1997), as well as Gregg Mitman, *Reel Nature: America's Romance with Wildlife on Film* (Cambridge, MA: Harvard University Press, 1999). Also useful is Vernon N. Kisling Jr., ed., *Zoo and Aquarium History: Ancient Animal Collections to Zoological Gardens* (Boca Raton, FL: CRC Press, 2001).

cetacean research. In the United States, in the second half of the 1950s, these commercial seaquariums had become important sites for thinking about the biology of "small whales," since these venues provided scientists with unique opportunities to study this difficult taxon in captivity. It was a formal objective of the First International Symposium on Cetacean Research to bring the big-whale people and the little-whale people into the same room and see what they could learn from each other. L. Harrison Matthews, in calling the meeting to order, worked explicitly to characterize the discipline of cetology in terms sufficiently broad to encompass these new spaces for cetacean research and the new kinds of scientists who worked in them. For instance, he invoked the work of the intrepid cetologists who were prepared to "don aqualungs in order to join the objects of their study in a foreign medium," men whose "laboratories are huge oceanariums." Yet the hip-booter in Matthews came out in his remarks nevertheless, in that he felt it was important to remind the audience that what cetologists *really* needed was "an enormous amount of material which is inaccessible to them because it is of no interest to commerce," and that therefore (again invoking the perennial dream of the hip-booters) what all good whale scientists most hoped for was that someone would give them *their own whaling vessel.* But that said, he again tipped his hat to the new community of researchers who had precipitated what he called "the greatest revolution in the study of the Cetacea in recent times" by creating "those remarkable aquatic menageries developed by commercial interests to make money from the amusement of the multitude," the unintended consequence of which had been a wealth of new information about the functional anatomy, physiology, pathology, reproductive activity, and behavior of the living animal. He even went so far as to express a special enthusiasm for the numerous new papers to be presented at the conference that took advantage of what he called "this new and exciting approach."[26]

But the audience in the John Glenn Suite was more than just a mix of American dolphin-park types and foreign hip-booters. It was a much stranger cocktail than that. After all, what was John J. Dreher, a mathematical linguist interested in code breaking and employed by the Lockheed Corporation, the West Coast military contractor, doing in the room? Or

26. Quotes are from L. Harrison Matthews, "Chairman's Introduction to First Session of International Symposium on Cetacean Research," in *Whales, Dolphins, and Porpoises,* ed. Kenneth Stafford Norris (Berkeley: University of California Press, 1966), 4.

Robert S. Gales, the head of the Listening Division of the US Navy Electronics Laboratory at Point Loma, California? Why was William B. McLean, the legendary developer of the Sidewinder missile and the technical director of the US Naval Ordnance Test Station at China Lake, California, attending a *whale conference*? His China Lake facility was located squarely in the *Mojave Desert*—an environment extremely hostile to marine mammals. All the more reason to be curious as to why he had brought along no fewer than half a dozen of his colleagues: engineers, physicists, and a variety of other defense researchers, one of whom gave his title as director of the "Cetacean Research Project" at China Lake. To make sense of this healthy representation of military scientists at the First International Symposium on Cetacean Research, it is necessary to go back for a moment and sketch the development of a shadowy kind of "whale science" that had quietly grown in importance since World War II: naval bioacoustics.

The dramatic significance of submarine warfare during World War II changed the shape of the naval world. Undersea combat added a vast new expanse to the world of sea warfare, and, as a number of scholars have shown, this unsettling "third dimension" prompted major US Navy commitments to oceanographic research in the 1940s and 1950s, much of it linked to antisubmarine warfare (ASW).[27] At the core of these investigations lay problems in underwater acoustics, since the development of sonar, SOFAR, and SOSUS systems for underwater listening, ranging, and target detection meant that whoever achieved greater mastery of the sonic properties of the ocean held the key to victory in the cat-and-mouse game of undersea combat. Physical oceanography benefited first from this new preoccupation with the underwater soundscape. A well-funded genera-

27. See, for instance, Gary E. Weir, *Forged in War: The Naval-Industrial Complex and American Submarine Construction, 1940–1961* (Washington, DC: Naval Historical Center, 1993), and Weir, *An Ocean in Common: American Naval Officers, Scientists, and the Ocean Environment* (College Station: Texas A&M University Press, 2001), particularly chap. 16, "Listening." See also Chandra Mukerji, *A Fragile Power: Scientists and the State* (Princeton, NJ: Princeton University Press, 1989), particularly chap. 3, "War and State Funding in the Twentieth Century," and Susan Schlee, *The Edge of an Unfamiliar World: A History of Oceanography* (New York: Dutton, 1973), particularly chap. 8, "Oceanography and World War II." See also Naomi Oreskes, "Science and Public Policy: What's Proof Got to Do with It?" *Environmental Science and Policy* 7 (2004): 369–83. Doogab Yi composed a useful unpublished review essay, "Oceanography and the Cold War."

tion of geophysicists and engineers (Columbus Islin, Carl-Gustav Rossby, Athelstan Spillhaus, Maurice Ewing, Allyn Vine, and others) went to work on the thermocline and the technologies needed to map the vertical temperature structure of the ocean—essential for sonar operators, since changing temperature zones deflect sound waves and can create sonar "blind spots." But where subaquatic acoustics was concerned, biological oceanography was not far behind the physicists. Already by the 1940s several memorable episodes put noisy marine critters high on the navy's agenda.

Perhaps the best known of these was the notorious problem with the ubiquitous tropical crustaceans known collectively as "snapping shrimp," whose millions of tiny clicks created a deafening din in shallow water, drowning out other sounds. But they were hardly the totality of underwater noisemakers. As Susan Schlee points out, "Within the first six months of the war, American submariners anxiously reported hearing staccato taps, clanks, rumblings, hammerings, croakings, whistlings, and, most eerie of all, the rhythmic beat of sonarlike beeps and pings."[28] The National Defense Research Committee responded by seeking out everything that was known about underwater sound producers and by commissioning new surveys and studies like those done by Martin Johnson, D. F. Loye, D. A. Proudfoot, and others.[29] Immediate postwar work in this area revealed the role of small marine organisms in the phenomenon known as the "deep scattering layer," a kind of meandering "false bottom" observed by depth-probing sonars; this artifact was shown to result from the migrations of small fish and invertebrates, moving in daily and seasonal cycles.

During and after the war, the Woods Hole Oceanographic Institution (WHOI) and Scripps were the major centers of this work. Dozens of researchers with navy-loaned hydrophones applied themselves to the col-

28. Susan Schlee, *The Edge of an Unfamiliar World: A History of Oceanography* (New York: Dutton, 1973), 299.

29. See Martin Johnson, "Underwater Sounds of Biological Origin," UCDWR Rept. U28, PB 48635 (1943): 1–26, and D. P. Loye and D. A. Proudfoot, "Underwater Noise Due to Marine Life," *Journal of the Acoustical Society of America* 18 (1946): 446–49. A significant amount of the most sensitive material in this area was published in the *U.S. Navy Journal of Underwater Acoustics*, which was (and is) a classified journal. I have not been able, thus far, to secure any of these sources (some of which are cited, without titles, in other publications in the period, written by those who had security clearance).

lective task of creating an exhaustive catalogue of the sounds of the sea. Unclassified publications in this area appeared throughout the 1950s, including Winthrop N. Kellogg's "Bibliography of the Noises Made by Marine Organisms," which was completed in 1953 and was followed in 1955 by Kellogg's phonograph record *Sounds of Sea Animals*, released by Moses Asch's Folkways Records.[30] Records like this—and even more significantly for cetology, *Whale and Porpoise Voices*, produced by William Schevill and William Watkins at whoi and released in 1962 (figure 6.3)—represented the commercially accessible tip of an archival iceberg assembled in this period (much of it classified), an audio archive of phonograph records and reel-to-reel tapes providing an encyclopedic repository of every whistle, fizz, grunt, and moan that echoed through the world's oceans. This collatory, systematic enterprise and its attendant analytic activities were much facilitated by the development of off-the-shelf sound spectrographs like the Kay Electronics "Vibralyzer," which made it possible to produce visual representations of sound, often called "sonograms" or "spectrograms" (figure 6.4).[31]

These activities did not constitute the whole of the field of marine bioacoustics, however. Winthrop Kellogg himself made his scientific reputation not by collating the thumps of red drum and croaker, but by a set of experi-

30. Winthrop Kellogg was, as best I have been able to make out, no relation to Remington Kellogg. Asch and Folkways had an interest in this sort of work, and the first "Sounds of the Sea" record (FX6121) appeared in 1943. I know of at least one situation in which Asch loaned a portable tape recorder and reel tapes to a researcher in animal acoustics making a field trip: in 1953 he lent this equipment to Charles M. Bogert, who was studying the sound production of South American frogs. See Charles M. Bogert, "The Influence of Sound on the Behavior of Amphibians and Reptiles," in *Animal Sounds and Communication*, ed. by W. E. Lanyon and W. N. Tavolga (Washington, DC: American Institute of Biological Sciences, 1960), 137–320. Kellogg's bibliography was published as N. W. Kellogg, "Bibliography of the Noises Made by Marine Organisms," *American Museum Novitates* 1611 (20 March 1953): 1–5.

31. For the use of the Vibralyzer in bioacoustics, see Donald J. Borror, "The Analysis of Animal Sounds," in *Animal Sounds and Communication*, ed. by W. E. Lanyon and W. N. Tavolga (Washington, DC: American Institute of Biological Sciences, 1960), 26–37; and Borror and C. R. Reese, "The analysis of bird songs by means of a Vibralyzer," *Wilson Bulletin* 65 (1953): 271–303. Figures 6.3 and 6.4 are from the manual included with William E. Schevill and William A. Watkins, *Whale and Porpoise Voices: A Phonograph Record* (Woods Hole, MA: Woods Hole Oceanographic Institution, 1962). The sonogram (sometimes spelled "sonagram" in this period) is from a Vibralyzer.

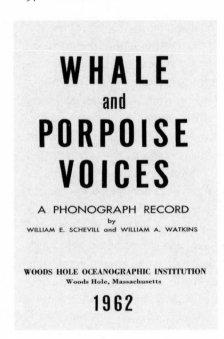

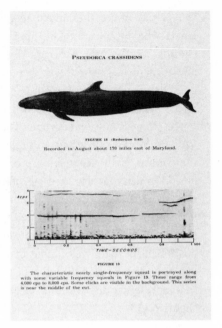

FIGURE 6.3 (*left*) A SW research on the phonograph: Schevill and Watkins's *Whale and Porpoise Voices* was a sampling from the archive of cetacean sounds. (Collection of the author.)

FIGURE 6.4 (*right*) The sonograms: A page from the pamphlet that accompanied *Whale and Porpoise Voices*. (Collection of the author.)

ments done in the mid-1950s demonstrating that the bottlenose dolphin didn't just click and squeal; it was using those sounds to *navigate*, just like the sonar-equipped ships of the world's navies.[32] This finding represented

32. See his summary account of this work: W. N. Kellogg, *Porpoises and Sonar* (Chicago: University of Chicago Press, 1961). For a valuable review of Kellogg's career, see Ludy T. Benjamin and Darryl Bruce, "From Bottle-Fed Chimp to Bottlenose Dolphin: A Contemporary Appraisal of Winthrop Kellogg," *Psychological Record* 32 (1982): 461–82. Priority for the discovery of cetacean echolocation was contested through the 1950s and 1960s, but the sequence of events would seem to be this: In 1947 the first curator at Marine Studios, Arthur McBride, mused in his notebooks on the possibility that wild dolphins were using sound to "see" the netting that fishermen were using to secure specimens for the park. In 1952 Kellogg and Kohler speculated in an article in *Science* on the possibility that dolphins, like bats, might use high-frequency sounds to orient themselves; see W. N. Kellogg and Robert Kohler, "Reactions of the Porpoise

a submarine extension of one of the most dramatic scientific "rediscoveries" of the twentieth century: the use by bats of high-frequency sound in echolocation.[33] The leading figure in this compelling episode, the Cornell biologist Donald Griffin, though not originally involved in marine work, was in fact one of the many researchers to be drawn into the problems of bioacoustics in the postwar period, and he accepted a US Navy commission to draft a report on underwater sound in 1950.[34]

The gradual realization of the power and range of the echolocatory capabilities of the odontocetes attracted considerable interest on the part of the navy, since they suggested an important avenue of "bionic" research, a term that was used for what was seen to be an exciting new intersection of biology and engineering: the study of organic systems for the purpose of improving technology.[35] Echolocation by marine mammals thus consti-

---

to Ultrasonic Frequencies," *Science* 116, no. 3010 (5 September 1952): 250–52. (Florida State University holds an oral history interview with Kohler in its archive, but for whatever reason they were unwilling to grant access to it in 2003 when I contacted them.) Meanwhile, the husband-wife team of William Schevill and Barbara Lawrence at WHOI completed experimental work begun in the late 1940s and published the first experimental evidence supporting the idea of dolphin echolocation. Kellogg published a similar paper in the same year, and the quality of the publications was debated into the early 1960s (Schevill and Lawrence seem to have been the preferred "discoverers" in the communities of correspondents I have reviewed, though this seems to be at least in part a matter of friendships and loyalties rather than anything that one can pinpoint in the source material). In 1960 Norris and his colleagues at Marineland of the Pacific succeeded in configuring dolphin "blindfolds" and published the most compelling "proof" of dolphins' "active sonar." Note that in 1956, seven years after McBride's death, Schevill, who had come into possession of his notebooks, published McBride's manuscript notes from the 1940s in "Evidence for Echolocation by Cetaceans," *Deep-Sea Research* 3 (1956): 153–54, thereby establishing him as in some sense the "first" person to put the idea of dolphin echolocation into writing.

33. For the story of the solution to "Spallanzani's bat problem," see Donald R. Griffin's classic *Listening in the Dark: The Acoustic Orientation of Bats and Men* (New Haven, CT: Yale University Press, 1958; reprint, New York: Dover, 1974). The original finding was published in *Science* in 1944 (see Donald R. Griffin, "Echolocation by Blind Men, Bats, and Radar," *Science* 100, no. 2609 [December 1944]: 590–91).

34. See Donald R. Griffin, *Underwater Sounds and the Orientation of Marine Animals: A Preliminary Survey*, Tech. Rept. 3, Proj. MR 162–429 ONR and Cornell University (Ithaca, NY: Cornell University, Dept. of Zoology, 1950).

35. It is difficult to capture the hold that bionics appears to have had on a significant number of biologists and engineers in this period, particularly those linked to mili-

tuted an active subspecialty for those interested in studying sound under water.

This, then, was the discipline of marine bioacoustics. For a particularly clear snapshot of the nascent field around the time of the First Annual Symposium on Cetacean Research, we can turn to an unclassified naval technical report—NAVTRADEVCEN 1212-1, "Review of Marine Bio-Acoustics: State of the Art: 1964"—prepared for sailors and civilian researchers at the US Naval Training Device Center in Port Washington, New York.[36] The report was commissioned as part of the center's larger program to "compile a library of marine animal characteristics consisting of graphs, audio, and any other recordings" for the use of those preparing to man the navy's sonar and undersea listening stations: as its introduction states, "This knowledge will greatly enhance the training of sonar operators in the ability to detect false targets in the event of military necessity." What follows are a set of short, reference-heavy essays on how marine animals— invertebrates, fishes, cetaceans—make and hear sounds, why they do so, and what the sonograms of such sounds look like; a bibliography of about 200 articles, books, and technical reports completes the volume. The fore- word to this report, signed by the project engineer and the head of the ASW systems trainers branch, offers a glimpse of the speed with which this field was developing: it explains the pressing need for report 1212-1, which up- dated an earlier version done in 1962, by pointing out that "since that time, research and investigation has [sic] uncovered a vast amount of new mate- rial." Things were changing in this area very quickly in the early 1960s.

In fact, just a few months before the cetologists gathered in Washington, most of the bioacoustics types had gathered for a major conference of their own: the Symposium on Marine Bio-Acoustics, held in April 1963 at the Lerner Marine Laboratory, Bimini, in the Bahamas. A glance at the pro- ceedings and participants of this meeting reveals significant overlap with the "Whale Circus."[37] The similarity of the invitation lists had much to do with the fact that both meetings had the same funding source, the Biology

tary funding. Visionary proposals included artificial gills and various other prosthetic extensions of human capabilities. There is a wonderful book to be done on the history of bionics.

36. William N. Tavolga, ed., "Technical Report: NAVTRADEVCEN 1212-1, Review of Marine Bio-Acoustics, State of the Art: 1964" (Port Washington, NY: U.S. Naval Training Device Center, February 1965).

37. More than a dozen of those in attendance at Lerner would subsequently attend the First International Symposium on Cetacean Research.

Branch of the Office of Naval Research (ONR), and both had been husbanded by the same figure, Sidney Galler, the director of that division.[38] He
was in attendance at both meetings, and he had been the force behind the
grants driving much of the US research being presented at both meetings.
It will be worth taking a moment to review some of the cetacean-related
work presented at Lerner in April 1963, since this will help illustrate what
US Navy–linked whale science looked like in the 1960s. Because so many
of these bioacoustic scientists were packing up at the end of the summer to
go to Washington to meet the hip-booters, it will be useful to have a sense
of the world of research from which the navy types came.

As it happens, much of the talk at Lerner had indeed been about whale
sounds, though progress in correlating particular underwater sounds with
particular marine mammal species was still spotty. Communication between researchers with different kinds of security clearance (and different
kinds of access to classified data) made for difficulties on several occasions. It may seem odd to think that whale sounds could be "classified," but
since the data-gathering hydrophone webs were a sensitive dimension of
Cold War monitoring technology, and since it seemed quite possible that
knowledge of the acoustic profiles of cetacean phonation could be used to
mask or disguise military communication systems, there was great concern
about leaks of valuable information. But vetted presentations by military
personnel could still be informative. For instance, G. M. Wentz, from the
US Navy Electronics Laboratory in San Diego, gave an unclassified survey
paper entitled "Curious Noises and the Sonic Environment of the Ocean."
His paper pondered the identity of a Pacific noisemaker that navy listeners
had dubbed "the carpenter," whose hammerings (or "smashings") were a
broadband "nuisance," and which had a sonogram profile not unlike what
were thought to be certain cetacean echolocation signals (figure 6.5). Sperm
whales had been seen in the vicinity of the noise at least once, but a positive
link had not yet been made.

At least four of the other papers presented at Lerner, and a good deal
of the conversation, dealt with another mysterious sound thought to be
linked to whales: a very low-frequency thumping at 20 cycles per second

38. Galler and the military work of the Biology Branch are discussed by Roy Mac-
Leod in his " 'Strictly for the Birds': Science, the Military, and the Smithsonian's Pacific
Ocean Biological Survey Program," in *Science, History, and Social Activism: A Tribute
to Everett Mendelsohn*, ed. Garland E. Allen and Roy MacLeod (Dordrecht, Holland:
Kluwer Academic Publishers, 2001), 307–38.

(cps) that had been picked up around the world. Two researchers from Columbia University, who had been granted time on a Bermuda SOFAR (Sound Fixing And Ranging) hydrophone array, presented a paper on these sounds, which they called BLIPS. In addition to compiling seasonal records of the coming and going of the sounds, these two investigators had actually used the ranging capabilities of the SOFAR array to localize the mysterious sound producers and track them through the listening zone (figure 6.6). Scanning the region with telescopes produced no clear sightings, though a spout here and there had been noted at various times, offering support for one of the going hypotheses: that these sounds had something to do with cetaceans. At the same time, these were certainly not the "usual 'whale' sounds"—the "continuous series of audible whistles, grunts and groans"

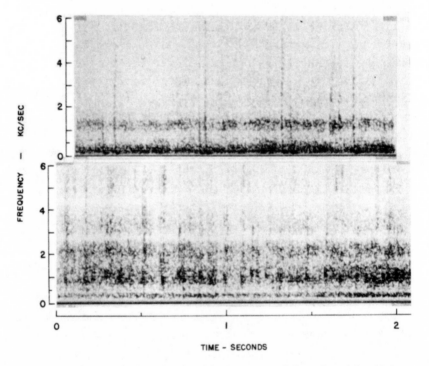

Spectrum-time plot showing "single carpenter" (upper) and "multiple carpenter" (lower).

FIGURE 6.5 Making a racket: Sonogram of the vocalizations of the "carpenter." (From Wentz, "Curious Noises and the Sonic Environment of the Ocean," p. 115. Courtesy of William Tavolga.)

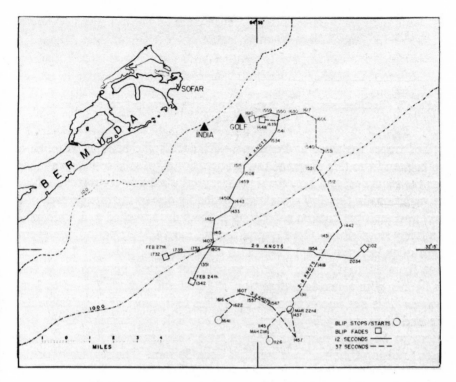

FIGURE 6.6 Unidentified underwater object: A map of BLIP zones in Bermuda. (From Patterson and Hamilton, "Repetitive 20 Cycle per Second Biological Hydroacoustic Signals at Bermuda," p. 134. Courtesy of William Tavolga.)

that "are associated with the northward migration of the humpback whale, which is frequently and easily seen off Bermuda in April and May." Nevertheless, the authors were intrigued by the idea that the thumping might be "caused by some natural continuous life function" of some other species— perhaps, as another researcher had suggested, the heartbeat of fin whales, resonating in their lungs.[39] This hypothesis, however, overlooked the ques-

---

39. The proposal was that of R. A. Walker of Bell Labs, who had monitored the sounds off New England. He published the idea later that year: "Some Intense, Low-Frequency, Underwater Sounds of Wide Geographic Distribution, Apparently of Biological Origin," *Journal of the Acoustical Society of America* 35, no. 11 (November 1963): 1816–24. Walker had also done some air reconnaissance of BLIP sources, without clear results.

tion of how the frequency could remain constant, for if the whales were descending and ascending, the changing pressure would greatly change the size of the resonator (not to mention the fact that different-sized whales ought to have produced different frequencies). Moreover, even if the timing of the pulses could be correlated with what was known of whale hearts, one might well ask about the evolutionary efficiency of a heart that radiated sound energy at such rates: these giants might have wattage to burn, but this seemed extreme. Perhaps, mused the investigators, steeped in the tactics and countertactics of ASW acoustics, "this signal could be used by a whale to stun acoustically or confuse a school of fish or a squid" or to "overload the acoustic defensive or offensive mechanisms of another organism." They continued: "For instance, it is known that the shark homes on the low frequency sound produced by struggling fish. Could it be that the large amplitude 20 cps BLIP signal is used to overload the predator's low frequency receptor system and prevent attacks in such a manner?"[40]

It was the former navy ASW investigator William Schevill, based at WHOI, who eventually produced the BLIP paper that laid the identification problem to rest (he had to get special permission to declassify his findings, which he did in response to learning of the work by the nonmilitary Columbia researchers).[41] As a result of a collaboration with the Canadian Naval Research Establishment, Schevill and his co-investigators—the acoustic technician William Watkins and the WHOI marine biologist and ichthyologist Richard Backus—were able to secure unrestricted access not only to a Nova Scotia listening station, but also to an observation airplane. The result was a set of persuasive observations of *Balaenoptera physalus*, the fin whale, which seemed to be in the area whenever the sounds were heard.

40. B. Patterson and G. R. Hamilton, "Repetitive 20 Cycle per Second Biological Hydroacoustic Signals at Bermuda," in *Marine Bio-Acoustics: Proceedings of a Symposium held at the Lerner Marine Laboratory, Bimini, Bahamas, 11–13 April 1963*, ed. William N. Tavolga (Oxford: Pergamon Press, 1964), 125–45.

41. On Schevill's ASW work, some of which was classified, see Gary E. Weir, *An Ocean in Common: America Naval Officers, Scientists, and the Ocean Environment* (College Station: Texas A&M University Press, 2001), 250, 298, 367. See also the transcript of an interview with William Watkins by Weir and Frank Taylor at WHOI, 29 March 2000, in the Woods Hole Archive. (NB: In my conversations with Watkins before he died, he objected to the public availability of this material.) There are also several mentions of Schevill in a similar interview done the same day with Richard Backus (see pp. 33, 42, of that transcript). Note that, unfortunately for historians, Schevill had Watkins and Lawrence destroy all of his papers after his death. William Watkins, interview by the author, 9 August 2003.

Cosmopolitan, but difficult to spot and identify, the fin whale appeared to account for BLIPs from Hawaii to Bermuda, Newfoundland to San Diego.[42]

While much of the work at the Lerner marine bioacoustics symposium thus dealt with strange underwater sounds and with systems for hearing and identifying them (including an experimental hydrophone array linked to an underwater television camera, which was in turn equipped to release scent attractors to draw marine life), there was, in fact, a whole separate session that dealt exclusively with "special problems in cetacean research." Here too it is remarkable to see the ways in which the preoccupations of Cold War naval engineering shaped investigations of, and ideas about, marine mammals. Nowhere is this more evident than in a remarkable review essay by Kenneth Norris, himself a navy veteran,[43] entitled "Some Problems of Echolocation in Cetaceans." After briefly surveying the literature of the last decade in this area, Norris stepped back for a set of "perspectives and speculations." Warning that scientists had, "in the past, usually underestimated the refinements of animal adaptation," Norris advocated a "working hypothesis for cetacean echolocation and its anatomical correlates"— namely, that it "is probably better to expect sophistication and attribute function to structure, than it is to be forced eventually to such a viewpoint from the more sparse interpretation that structures, even complex ones,

42. W. E. Schevill, W. A. Watkins, and R. H. Backus, "The 20-Cycle Signal and *Balaenoptera* (Fin Whales)," in *Marine Bio-Acoustics: Proceedings of a Symposium Held at the Lerner Marine Laboratory, Bimini, Bahamas, 11–13 April 1963*, ed. William N. Tavolga (Oxford: Pergamon Press, 1964), 147–52. The 20 cps BLIP episode is an important reminder of the degree to which the unclassified Lerner symposium on marine bioacoustics was, on many matters, behind the navy's (secret) curve of discovery. The editor of the proceedings noted of the Schevill, Watkins, and Backus contribution: "This paper represents an extension of remarks made by W. E. Schevill during the discussion period" following two other papers presented at the symposium. But this note does not exactly capture what happened. In fact, Schevill had not been able to say clearly that the WHOI group had already figured out that the BLIPs were finback whales because the finding remained classified. When the conference ended, Galler intervened with the US Navy Bureau of Ships to get permission for Schevill et al. to report their finding in the proceedings volume. William Watkins, interview by the author, 9 August 2003. Roger Payne offers a brief discussion of BLIP research in his *Among Whales* (New York: Scribner, 1995), 178. He does not mention that Richard Backus was a coauthor on the paper that resolved the identification question.

43. Recall that Norris would go on to edit the proceedings of the First International Symposium on Cetacean Research in Washington. It is important not to overstate Norris's naval experience: he was an ensign in World War II and did duty on a troop transport.

need not have a function." Departing from this proposition—which was strongly supported, he argued, by recent work in biophysics—Norris embarked on a speculative functional anatomy of the head structure of the odontocetes in which the whole of the skull and the surrounding soft tissue was interpreted by analogy to a sophisticated directional transducer and sonar detector. "During work on *Phocoena sinus* in 1957," Norris explained, "I became impressed that the scoop-shaped bones of the forehead (also especially prominent in *Hyperoodon* and *Ziphius*) look for all the world like parabolic surfaces whose focal points lie in the general area of the soft anatomy of the forehead, and hence might act as sound reflecting and focussing devices." If this was the case, then perhaps the sounds used in echolocation were "beamed from the forehead of porpoises, rather than being produced in an omnidirectional field from the larynx."[44] And if the echolocatory sounds were generated somewhere in the set of air-filled cavities in the head—the paired tubular sacs and other sinuses found in the complex airways of the odontocetes—then there was circumstantial support for a speculation made by F. G. Wood in a personal exchange with Norris at Marine Studios several years earlier; namely, that the "fatty melon of porpoises and the spermaceti case of sperm whales"—the large lump of liquid wax that so lured nineteenth-century whalers—"might serve as a sound transducer between tissue and seawater." Wood and others had since pressed this point, arguing that the density characteristics of this material might mean that the melon could serve as an "acoustic lens, possibly deformed by musculature," and that this lens might "shape the sound beam coming from the animal's head" (figure 6.7).[45]

In later discussion at the Lerner symposium, this tempting idea attracted further comment. Could the melon really serve as a lens for sound? Was it sufficiently elastic for the kind of deformations that were implied? Had enough attention been paid to the histology of this part of the cetacean anatomy? Was it sufficiently homogeneous to work this way? A scientist from the Ordnance Research Laboratory (at Penn State) drew attention to work being done at General Dynamics on acoustic lensing, where "it was

44. Kenneth S. Norris, "Some Problems of Echolocation in Cetaceans," in *Marine Bio-Acoustics: Proceedings of a Symposium held at the Lerner Marine Laboratory, Bimini, Bahamas, 11–13 April 1963*, ed. William N. Tavolga (Oxford: Pergamon Press, 1964), 317–36, at p. 323.

45. Ibid., p. 329.

shown that you can make a good acoustic lens . . . by using a heterogeneous arrangement of materials"—specifically, "a free flooded sphere with flattened tubes filled with a material of different index of refraction from the water." Changing the distribution of the materials could alter the focal length, and even the axis, of this lens. To any readers of *Moby-Dick*, in which the sperm whale "case" was described as an "immense honeycomb of oil" crisscrossed by "tough elastic white fibers," the image of the acoustic lens must have seemed compelling indeed.

It was in connection with such speculation that William Evans, an acoustics engineer working for Lockheed on a new ASW aircraft (the P3V), undertook an elaborate experiment in collaboration with a group of US Navy engineers, staff from Marineland of the Pacific, and the curator of ichthyology and marine mammals at the Los Angeles County Museum of Natural History. This work—conducted in 1961–1962 at the Naval Missile Center, Point Mugu, California, and also presented at the Lerner symposium—aimed to clarify whether dolphins had the capacity to deploy directional sonar. The setup for this experiment is depicted in figures 6.8 and 6.9. Evans and his collaborators erected a rotary platform cantilevered out from a bayside dock adjacent to the grounds of the Naval Missile Center. To this platform they attached the cadaver of a freshly dead baby *Stenella microps* (now *S. longirostris*, the common spinner dolphin), into whose blowhole they inserted an ARC mode BC-10 transducer, or "miniature sound projector." Thus outfitted, this "echolocating" dolphin was used to make a set of polar ranging diagrams using a fixed receiving hydrophone

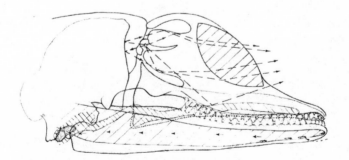

FIGURE 6.7 The bionic whale: Diagram of possible acoustic lensing in an odontocete skull. (From Norris, "Some Problems of Echolocation in Cetaceans," p. 330. Courtesy of William Tavolga.)

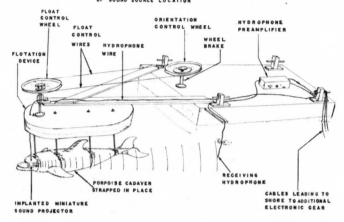

FIGURE 6.8 How to drive a dead dolphin: Schematic of Evans's experimental setup at Point Mugu. (From Evans, Sutherland, and Beil, "The Directional Characteristics of Delphinid Sounds," p. 360. Courtesy of William Tavolga.)

FIGURE 6.9 Cetacean cyborg: Steering the cadaver of a baby dolphin implanted with a sonar system. (From Evans, Sutherland, and Beil, "The Directional Characteristics of Delphinid Sounds," p. 358. Courtesy of William Tavolga.)

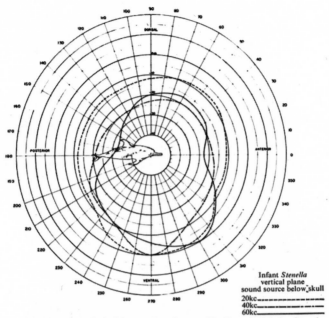

S. microps, vertical plane measurements at 20, 40, and 60 kc/s. Sound source in the area of the larynx.

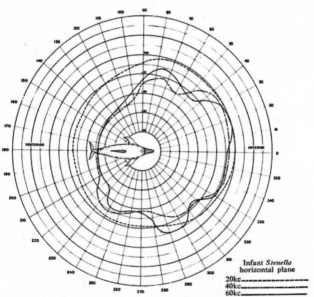

S. microps, horizontal plane measurements at 20, 40, and 60 kcps. Sound source in the area of the nasal sacs.

FIGURE 6.10 Sonar analysis: Polar coordinate plots of an "echolocating" dolphin. (From Evans, Sutherland, and Beil, "The Directional Characteristics of Delphinid Sounds," p. 362. Courtesy of William Tavolga.)

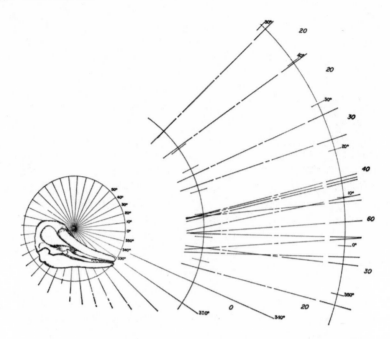

FIGURE 6.11 Sound and skull: Experimental results from the hard parts of an adult bottlenose head. (From Evans, Sutherland, and Beil, "The Directional Characteristics of Delphinid Sounds," p. 365. Courtesy of William Tavolga.)

(figure 6.10). Later, the experiment was repeated using the skull of an adult bottlenose (figure 6.11).

When he presented this work in Bimini, Evans acknowledged that the results were preliminary, but he indicated that there was good evidence of posterior suppression of the signal, suggesting at least the rudiments of sonar "beaming." Both the melon and the parabolic processes on the skull seemed to be involved. More work, though, was clearly needed, with fresher specimens, using a larger acoustic field.

BUT HAVE YOU EVER BEEN INSIDE A WHALE?

Having now reviewed the development of navy-driven, whale-related marine bioacoustics up through the early 1960s, we can return to Washington, to the First International Symposium on Cetacean Research, where the diversity of the participants should now look much less strange. For instance, there is no longer any mystery as to what Robert Gales, the head of the Listening Division of the US Navy Electronics Laboratory, was doing there: he had come to discuss acoustics. He presented a paper entitled "Pickup, Analysis, and Interpretation of Underwater Acoustic Data" as the lead contribution to a session on underwater observation and recording, where a number of others who had attended the Lerner marine bioacoustics symposium also presented work.[46] William Schevill, for instance: the former ASW researcher turned whale expert gave a paper in Washington with WHOI marine biologist Richard Backus presenting new findings on sperm whale "clicks" and their function. Schevill and Backus even played, for the assembled listeners, a tape of the phonations that they had managed to record.

Nor was it exclusively the Americans presenting this sort of work in Washington. The Frenchman René-Guy Busnel, the director of the Laboratoire d'Acoustique Animale, Jouy-en-Josas, and Albin Dziedzic, who worked at a similar institution in Denmark (both of whom had been at the Lerner symposium), also presented work related to cetacean phonation and acoustics. Indeed, the Dutch anatomist-zoologist Dudok Van Heel summarized for the attendees his radical hypothesis on the cause of mass strandings, arguing that these "cetacean suicides" were the product of sonar disorientation that could result from the peculiar seafloor topography in certain coastal regions. He had elaborated this account on the basis of extensive work on a Dutch submarine chaser, investigating the sonar profiles

46. Note that by 1961 the Navy Electronics Laboratory had already put together a sonar training document that introduced navy personnel to the bioacoustic profiles and characteristic movement patterns of a large number of marine mammals. This was one of a number of such guides available in the period. The Office of Naval Intelligence (ONI) had a manual known as "Whale Intelligence," which provided tips on distinguishing large cetaceans from military hardware. In October 1961 Robert L. Eberhardt and several collaborators sought US Navy research funding for a "false-target atlas" that would focus on marine mammals. See Scripps Archive, Hubbs Papers, box 53, folder 54. The other Lerner participants who gave papers in Washington were R. Stuart Mackay, who presented on telemetry of physiological data from living organisms and the use of ultrasound imaging to study structure and performance in vivo; James M. Snodgrass, who presented a paper entitled "Instrument Packaging and Telemetry"; and John C. Steinberg, who presented a paper entitled "Acoustic-Video Techniques for the Observation of Cetaceans." Weir has written an informative essay on Steinberg and his work in conjunction with the navy: "From Surveillance to Global Warming: John Steinberg and Ocean Acoustics," *International Journal of Naval History* 2, no. 1 (1 April 2003).

of the North Sea littoral zone. It was an argument that one of the American researchers working with the navy, Backus, did not like:

> If all the wonderful things we say about odontocetes' sonar are true, it seems to me that the whale with an operative sonar will not really be confused in shallow water. He certainly will be able to differentiate between a bold coast and a shoaling coast unless his sonar is of the crudest sort. . . . The echo ranging in the two situations would sound something like this (I will be a destroyer instead of a porpoise, because I can ping better than I can click): on the bold coast it would be PING-ping, PING-ping; on the shoaling coast it would be P-innng, P-innng, P-innng.[47]

The argument left ever so slightly open the question of what was at issue: the discriminatory capabilities of *Phocoena phocoena* itself or those of the Dutch sonar systems from which the animals' behavior was being inferred.

This exchange points to a larger question we are now ready to ask: What happened when the work of the bioacoustic cetacean researchers (largely American, ASW-trained, US Navy–funded) crossed paths, in Washington, with the hip-booted, whalerish tradition exemplified by the Discovery Investigations and its emulators? Where echolocation was concerned, there were significant intersections. After all, whale sounds had to be made somewhere in the whale head or airways and had to be heard by whale ears (presumably). And, indeed, the anatomy and physiology of cetacean vocalization and auditory systems had become, in the 1950s in Europe, an area of intensive research and rapid discoveries. So much so that when the hip-booted-cetologist chairman of the symposium, L. Harrison Matthews, chose to single out, in his opening remarks, the most radical developments in the study of whales in the last decade, it was this work to which he drew attention:

> Careful anatomical investigations using modern techniques have yielded physiological results of great value; Fraser and Purves, for example, in their work on the complex auditory apparatus of Cetacea have exploded the old myth, based on incomplete observation, that

47. This was Backus, as cited in *Whales, Dolphins, and Porpoises*, ed. Kenneth Stafford Norris (Berkeley: University of California Press, 1966), 603.

whales are hard of hearing, and have shown that hearing in the Cetacea is one of the most efficient and important senses.[48]

Here was a considerable change in human understanding of whales and dolphins that had occurred within the active research life of older the scientists gathered in Washington: cetaceans had gone from being thought of as lumbering deaf-mutes to being understood as chattering creatures with a sense of hearing more refined than the most sophisticated Bell Labs hydrophone.[49] Just how the mammalian auditory apparatus had evolved into a hydrophone, and precisely how it worked—these remained the stuff of anatomical fisticuffs, some of which were on display in Washington in 1963. A moment's attention to this controversy will give a flavor of the sorts of dissent that could be engendered by a meeting of hip-booted anatomists and headphone-wearing ASW researchers.

At issue was the function of the ear canal itself, or the "external auditory meatus." Sixteenth-century anatomists were well aware that whales and dolphins had no "ears" to speak of, and as cetacean auditory capabilities came to be understood, the dominant assumption was that they did not need such an external funnel and sound passageway because their flesh was acoustically transparent, transmitting sound waves in the same manner as the surrounding water. Thus the middle ear could respond directly to sound energy that had passed through the blubber and the flesh of the head. Given that the seemingly vestigial ear canal was plugged, in a number of species, with an "earplug," a long lump of waxy material, it was easy to assume that it did not play a role in cetacean hearing. We have already seen (in chapter 3) that the question of the evolution of the ear bones in cetaceans had helped launch Remington Kellogg on his distinguished career. But he was, as we saw, basically a bone man, and it was only with the development of full-scale hip-booted whale science that the soft tissues of the cetacean hearing apparatus could really come under sustained scrutiny. The received wisdom concerning the irrelevance of the ear canal sustained a serious blow in the mid-1950s, when Francis C. Fraser and Peter Purves,

48. L. Harrison Matthews, "Chairman's Introduction to First Session of International Symposium on Cetacean Research," in *Whales, Dolphins, and Porpoises*, ed. Kenneth Stafford Norris (Berkeley: University of California Press, 1966), 3–6, at p. 5.

49. In fact, Matthews is fudging a bit here, since a number of early anatomists were convinced that whales and dolphins had some capacity to hear, and the squealing of various cetaceans was not unknown. See my discussion of Kellogg's work on this question in the 1920s in chapter 3.

working at the British Museum, demonstrated that the waxy earplug had the capacity to serve as a very efficient underwater sound transmitter (and that muscular control over the tension in the plug seemed to enable the animal to alter the range of received frequencies and, perhaps, to compensate for deep-water pressures). After showing that the internal structure of the ear appeared to be acoustically isolated by means of a mucosal foam filling the sinuses of the auditory system, Fraser and Purves believed that they could rule out general sound transmission to the inner ear via the surrounding tissue.[50] This deeply contentious matter remained unresolved in 1963, leading to the presentation of exotic anatomical experiments at the Whale Circus, experiments that involved passing mechanical vibrations through reconstructions of dead whale ears (figures 6.12 and 6.13).

The ties between this fine anatomical work and the whaling industry were close. Fraser himself, of course, was an old Discovery type, and his experimental enterprise depended entirely on adequate supplies of sufficiently fresh specimens. Furthermore, the whole problem of cetacean auditory thresholds had first emerged in the community of European cetologists in the context of the postwar adaptation of ASDIC (a sonar system developed by the Allied Submarine Detection Investigation Committee, which conferred its acronym on its creation) to the whaling industry. Remarkably, the hostilities of World War II had not yet ended when the idea of hammering the sword of U-boat surveillance technologies into a plowshare for the starving wartime population of the continent came across the negotiating tables. At the International Whaling Conference in London in January 1944, ASDIC systems were mentioned as a potentially revolutionary weapon in the armament of pelagic whaling vessels, and the systems,

---

50. There are two Fraser and Purves articles that detail these findings; they both have the same title, "Hearing in Cetaceans." The first (F. C. Fraser and P. E. Purves, "Hearing in Cetaceans," *Bulletin of the British Museum of Natural History* 2 [1954]: 103–16) is brief; the second (F. C. Fraser and P. E. Purves, "Hearing in Cetaceans: Evolution of the Accessory Air Sacs and the Structure and Function of the Outer and Middle Ear in Recent Cetaceans," *Bulletin of the British Museum of Natural History* 7 [1960]: 1–140) is a book-length monograph. For the competing contemporary accounts, see particularly Reysenbach de Haan, "Hearing in Whales," *Acta Oto-Laryngologica Supplement* 134 (1957): 1–114. See also W. H. Dudok van Heel, "Sound and Cetacea," *Netherlands Journal of Sea Research* 1, no. 4 (1962): 407–507. All of these authors attended the Washington meeting. Also relevant is M. Yamada, "Contribution to the Anatomy of the Organ of Hearing in Whales," *Science Reports of the Whales Research Institute* 8 (1953): 1–79.

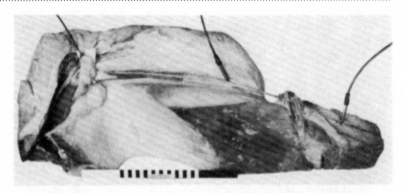

Dissection of ear of fin whale (*Balaenoptera physalus*). Starting from right, note, in succession, open pigmented part of external meatus with blubber adjacent; auricular cartilage with muscles attached; cordlike part of meatus; proximal part of meatus lying in squamomastoid groove with part of corium removed to show earplug and glove finger; tympanic bulla with mesial half removed to show middle-ear cavity and pterygoid sinus. Acoustic probes are shown in three of the positions used for testing sound attenuation.

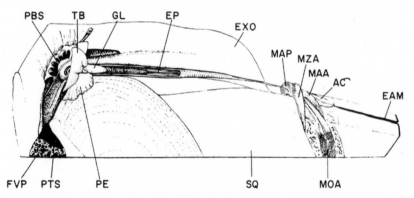

Dissection of ear of *Balaenoptera physalus* showing glove finger and earplug. The earplug has been bisected to show laminations, but in life it would completely envelop the glove finger, of which it is the zona cornea. Extension of middle-ear cavity into pterygoid sinus shows part of the fibrovenous plexus (bottom left) placed on malleus and earplug.

FIGURE 6.12 In one ear . . . : Detailed dissections of whale ears presented at the Whale Circus. (From Peter E. Purves, "Anatomy and Physiology of the Outer and Middle Ear in Cetaceans," p. 331. Courtesy of the University of California Press.)

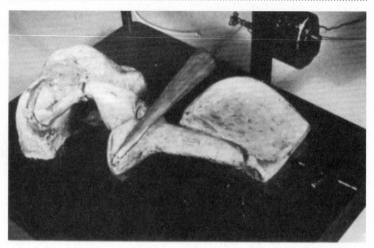

Large working model of ossicular mechanism of ear of fin whale, *Balaenoptera physalus.*

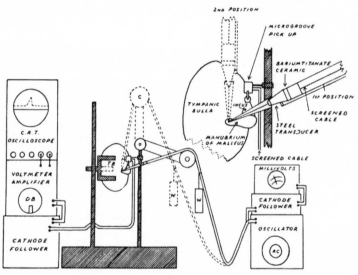

Diagram of apparatus for demonstrating changes in amplitude of torsional oscillations of malleus when actuated by a simulated tympanic ligament vibrating longitudinally at various frequencies and angles of traction.

FIGURE 6.13 ... And out the other: A working model of a fin whale ear. (From Purves, "Anatomy and Physiology of the Outer and Middle Ear in Cetaceans," pp. 355, 357. Courtesy of the University of California Press.)

borrowed from the Admiralty, went south on two English catcher boats in the first postwar season.[51] Other nations soon followed. Trials revealed that the device was of limited use in locating deep-swimming whales, but that it had the most unexpected effect once the pursuit was on: the sound generated by the transducer seemed to cause the whales to surface and flee in a straight line, directly away from the boats. Nothing could be easier for the captain and harpooner than to run them down in this situation. By the early 1950s dedicated commercial "whale-scarers" were being manufactured in England, Germany, and Japan. Whether whales could hear—and in precisely what frequency band they heard—were questions resolved on the high seas of the Antarctic Convergence by Norwegian whalemen long before they were resolved in the anatomical laboratory of the British Museum.

Moreover, the work on whale ears at the British Museum fed back into debates around whaling in a circuitous way. As previous chapters have shown, one of the vexing unresolved problems in cetology (going back to the early twentieth century) was the absence of any reliable technique for establishing the precise age of a given specimen. Because aging information was essential to the efforts to compile statistical portraits of the world's whale stocks, such a technique represented something very like the holy grail of hip-booted cetological investigations (I have already detailed a variety of the approaches pursued by Mackintosh, Ruud, and others). A significant piece was added to the puzzle in 1954, when Purves, working with Fraser at the British Museum on the troublesome ear of the mysticete whales, prepared a histological sample by splitting one of the waxy earplugs lengthwise. What he saw was a set of striations—layers—which looked to be the product of annual or semiannual accretion. A large new area was opened to research on whale ages.[52]

51. J. N. Tønnessen and Arne Odd Johnsen, *The History of Modern Whaling* (Berkeley: University of California Press, 1982), 695. In the Scripps Archive (Hubbs Papers, box 54, folder 58) I found a copy of "Establishment Report #14, E 505–47" authored by M. J. McCarthy under the auspices of HM Underwater Detection Establishment, entitled "The Application of Asdics to Whalecatching," dated November 1946. The document is marked "Restricted," but this classification has been struck out on Hubbs's copy.

52. It took almost two decades before consensus was reached on the mechanism and rate of deposition of these layers. For the original report on the discovery, see P. E. Purves, "The Wax Plug in the External Auditory Meatus of the *Mysticeti*," *Discovery Reports* 27 (1955): 293–302.

If the Anglo-Scandinavian tradition of hip-booted cetology thus had its own reasons for being interested in the auditory apparatus of whales, there remained much room for exchanges of views with the new generation of American acoustics engineers suddenly so concerned with cetaceans, and the gathering in Washington saw some jousting on these matters. Among other things, several of the American researchers raised questions about the conclusions that could be drawn from work on dead specimens. One American medical imaging expert, drawing on hospital experience, pointed out that the gases of decomposition that form in dead tissue radically alter its acoustic characteristics—sometimes so quickly that "investigations became impossible in a pathological specimen by the time we could get it from the operating room into the sonic equipment."[53] This observation suggested the need for further work both on Purves's "artificial ear" and on the claims made by some of his opponents for the acoustic transparency of the tissue of living whales. A sample of fresh tuna hide, noted an American professor of mechanical engineering at Tufts (who would later die on a navy project working with porpoises), had the acoustic opacity of a thin sheet of steel. No one was sure why.

On the subject of phonation, the acousticians and the anatomists ran into still more serious contention. The alluring and elaborate (if admittedly hypothetical) schemes of acoustic lensing and parabolic reflecting dishes embedded in the maxillary bones—the whole image of the "bionic" whale proposed by the acousticians—demanded that the clicks and cracks generated for echolocation originate in the nasal sacs and vestibular passages of the blowhole airway. It was here that Evans and his colleagues had embedded the pinger in their porpoise cadaver, and it was from this point that they experimented with manual ray tracing in an effort to reconstruct the acoustic profiles of different cetacean skulls. Attractive as the resulting diagrams were, a question remained: Just where *did* the sounds come from in the head of an actual whale or porpoise? On this question the ocean marked a deep divide. As the English anatomist Purves pointed out good-naturedly in response to a paper by an American:

I have spoken to a number of people here on this side of the Atlantic who are wedded to the idea that all these noises are made by the

53. *Whales, Dolphins, and Porpoises*, ed. Kenneth Stafford Norris (Berkeley: University of California Press, 1966), 378.

air sac system of the top of the skull, and the larynx appears to be completely neglected. But the larynx is an extremely complex organ, and with careful dissection one can show that there are noisemaking mechanisms there too.[54]

And if this contribution was both offered and received in the spirit of collaborative investigation, things would grow a bit more testy before the symposium was over. After listening to the WHOI researchers Richard Backus and William Schevill play their dramatic tapes of sperm whale "burst pulses" (which they analyzed using oscilloscope photographs), Dr. Purves apparently could not help himself, and he followed up with the first comment on their presentation:

> Well, that was a very interesting recording. For those devotees of the idea that the supramaxillary set of airsacs produces these noises, I will say that in the adult sperm whale there is only one set of supramaxillary sacs. They are positioned mesially on the rostrum and are large enough to take three men lying down. . . . If anyone can tell me how such a system can produce a train of clicks like those we have just heard, I should be very pleased to hear it. On the other hand the larynx is relatively small compared to the size of the body, and I suggest that this is the mechanism that produces the clicks.[55]

As the transcript records the exchange, one of the other European participants then called out: "Dr. Backus, will you answer this question?" To which Backus replied, "That was no question."

Indeed not. It was, rather, a statement of at least one significant response of the Discovery-style hip-booted cetologists confronting the signal-processing whale scientists across the pond—a statement that reflected the different worlds from which students of the cetaceans came in the postwar period. It was a statement that amounted, in the end, to this: "Yes, perhaps, but have you ever been *inside a whale*?"[56]

54. Ibid., p. 508.

55. Ibid., p. 527.

56. For the status of this debate at the end of the millennium, see S. H. Ridgway and D. Carder, "Assessing Hearing and Sound Production in Cetaceans Not Available for Behavioral Audiograms," *Aquatic Mammals* 27, no. 3 (2001): 267–76.

WHISTLING IN THE DARK

How the whale made its clicks would not be resolved in Washington in 1963, but everyone at the conference—hip-booters and bioacousticians alike— was in agreement on two important matters: cetaceans made sounds, and they heard them, too. But all this raised a very real question, a large and looming one, of interest to everyone gathered at the First International Symposium on Cetacean Research: If whales had voices, and whales had ears—if they could talk and they could listen—were they having conversations down there? Nor was this merely a question for specialists in the summer of 1963. After all, it was in the early summer of that year that MGM Studios released *Flipper*, which would become a runaway box office success and spawn a host of imitators, sequels, and an eponymous long-running television series.[57] As Gregg Mitman has argued in his *Reel Nature: America's Romance with Wildlife on Film*, the Hungarian émigré director Ivan Tors's blockbuster about a playful, chattering, "domesticated" marine mammal created a marquee "pet star" who embodied a new American "glamour species."[58] With this film still in theaters, cetacean "language" and cetacean "intelligence" were, without question, the most fraught of all matters cetological in August 1963.

And it was this very matter that had brought the mathematical linguist and acoustics engineer John J. Dreher from Burbank, California, to the Washington conference, where he presented on the ongoing work of the Lockheed group (Dreher had been on Evans's dissertation committee at Ohio State and had drawn him to join the ASW team at Lockheed). Dreher, in collaboration with US Navy personnel and researchers at Marineland of the Pacific, had spent a good portion of the last several years trying to use statistical techniques associated with codebreaking in an effort to make sense of "delphinidese."[59] One version of this approach involved tabulating

57. Richard O'Barry and Keith Coulbourn, *Behind the Dolphin Smile* (Chapel Hill, NC: Algonquin Books, 1988). The television series ran from 1964 to 1968.

58. See particularly chap. 7 in Gregg Mitman, *Reel Nature: America's Romance with Wildlife on Film* (Cambridge, MA: Harvard University Press, 1999).

59. See Dreher and Evans's "Cetacean Communication" in *Marine Bio-Acoustics: Proceedings of a Symposium held at the Lerner Marine Laboratory, Bimini, Bahamas, 11–13 April 1963*, ed. William N. Tavolga (Oxford: Pergamon Press, 1964), 373–93. See also Dreher, "Linguistic Considerations of Porpoise Sounds," *Journal of the Acoustical Society of America* 33, no. 12 (December 1961): 1799–1800, and Dreher's "Cetacean

the occurrences of particular sound patterns and then using information theory approaches to derive an "average information value per symbol" for a range of cetacean phonations. This was, in effect, a noise test, a means of ascertaining the amount of difference that could make a difference in a given signal. It could not prove that there was information there, only that there could be.[60] Tantalizingly, Dreher asserted that he found an information content index (essentially a negative entropy) in porpoise sociability that was roughly equivalent to those found when similar tests were run on human speech. Was he giving *Flipper*-esque talking dolphins the green light? He was cautious, noting that just because the apparent information content of human language and delphinid phonation were similar, "no connection between the two ensembles is implied."[61] Really? Could the same be said for another paper in the same session on communication, the one entitled "Information in the Human Whistled Language and Sea Mammal Whistling"? This study consisted of a review of the three best-studied human "pure-tone" languages: Mazateco (found in Mexico), Silbo-Gomero

---

Communication: Small-Group Experiment," in *Whales, Dolphins, and Porpoises*, ed. Kenneth Stafford Norris (Berkeley: University of California Press, 1966). In his work on word frequency analysis, Dreher was using Fletcher Pratt's *Secret and Urgent: The Story of Codes and Ciphers* (Garden City, NY: Blue Ribbon Books, 1942). In his work on pure-tone languages, Dreher was using Kenneth Pike's book of that title: *Tone Languages: A Technique for Determining the Number and Type of Pitch Contrasts in a Language, with Studies in Tonemic Substitution and Fusion* (Ann Arbor: University of Michigan Press, 1948). What is perhaps most striking is to see Dreher using the idea of "etic" analysis (Tavolga, *Marine Bio-Acoustics*, 376), a term which has here almost certainly been borrowed from Marvin Harris, *The Nature of Cultural Things* (New York: Random House, 1964). According to *Emics and Etics: The Insider/Outsider Debate*, ed. Thomas N. Headland, Kenneth L. Pike, and Marvin Harris (Newbury Park, CA: Sage Publications, 1990), the term appeared in print less than a dozen times from 1961 to 1965. Even if this is an underestimate (as it almost surely is), the discussion suggests that Dreher was interested, as Gregory Bateson would later be, in using anthropological approaches to examining cetacean intelligence.

60. Dreher uses a version of Shannon's information equation in his 1963 paper; for a discussion of the derivation of this formula, see Michael S. Mahoney, "Cybernetics and Information Technology," in *Companion to the History of Modern Science*, ed. R. C. Olby et al. (London: Routledge, Chapman & Hall, 1989), 537–53.

61. John J. Dreher, "Cetacean Communication: Small-Group Experiment," in *Whales, Dolphins, and Porpoises*, ed. Kenneth Stafford Norris (Berkeley: University of California Press, 1966), 529–43.

(from the Canary Islands), and the whistle language of Aas (in the Pyrenees); the paper quite explicitly considered them analogically with respect to odontocete sounds.[62]

We are again very far indeed from what a Discovery-era whale scientist might have understood to be the proper domain of "cetology." Interestingly, in this case, we have lovely evidence of the forehead-slapping dismay with which one of them reacted to the whole business: Mackintosh's manuscript notes from the session on communication at the First International Symposium on Cetacean Research survive, and they save us the trouble of speculating about how a hip-booter felt about all this to-do around porpoise talk. In his loopy hand, Mackintosh carefully recorded his impressions:

> Note: Several speakers on communications, intelligence, and behavior seem not to be clear on what they are really trying to discover, on the solidity of methods, and on what is meant by intelligence.[63]

Mackintosh did not like what he was seeing. And he was not alone.

The chairman of the communication session, of course, was John Cunningham Lilly, the only scientist in the John Glenn Suite who could properly be called a household name, having just emerged from a vortex of publicity around his visionary claims for dolphin intelligence and having received a prominent credit in *Flipper* itself (he consulted on the film, and Tors made contributions to Lilly's research from the proceeds). Lilly was 48 years old, brash and iconoclastic, and he attended the meeting with a small entourage of collaborators and assistants from his Communication Research Institute. As we saw at the outset of this chapter, he had served on the planning committee for the symposium as a whole (having been invited to serve this role by the ONR's Sidney Galler himself, who had funded CRI's work), and he had tight links to many of the US Navy scientists in the room, having been a Caltech classmate of William B. McLean, who had invited him out to lecture to the China Lake group the year before. He was married to an ex-model (it was a second marriage for both of them), lived on an island in the Caribbean, and was known to hang out with Hollywood

---

62. R. G. Busnel, "Information in the Human Whistled Language and Sea Mammal Whistling," in *Whales, Dolphins, and Porpoises*, ed. Kenneth Stafford Norris (Berkeley: University of California Press, 1966), 544–68.

63. SGB Papers, "Cetacean Symposium Notes from Papers." I turned up this four-page document slipped into a file labeled "Buckland Papers."

types. Without question he cut a figure among the cetologists at the Marriott Motor Lodge. Who was he, exactly, and what was he doing there? The answers to these questions will open a window onto just how fast, and how profoundly, the science of whales was changing in these years.

Born in 1915, Lilly, from a well-to-do family in Saint Paul, Minnesota, took a bachelor of science degree from the California Institute of Technology in 1938 and studied at Dartmouth Medical School for two years before moving to the University of Pennsylvania, where he completed his MD in 1942 and remained on the faculty. There, under the influence of Britton Chance and Detlev Bronk, Lilly pursued research in biophysics, including applied investigations into real-time physiological monitoring—work linked to wartime service in military aviation, where techniques for assaying the respiration of airmen were needed.[64] Lilly had contact through his family with the neurosurgeon Wilder Penfield in the later 1940s and developed an interest in neuroanatomy and the electrophysiology of the brain. By 1953 he had been appointed to the neurophysiology laboratory of the National Institute of Mental Health, where he worked under Wade Marshall as part of a joint research program with the National Institute of Neurological Diseases and Blindness.

By the mid-1950s Lilly's lab in Bethesda, Maryland, was performing in vivo electrical stimulation of the brains of macaques—work aimed at cortical mapping by means of correlating point applications of currents at varying thresholds with specific behaviors and reactions in subject animals.[65]

---

64. This work was done through the E. R. Johnson Foundation for Medical Physics, which was run by Bronk, and which had contracts with the army and navy air forces through the Committee on Medical Research of the Office of Scientific Research and Development. Interestingly, in light of Lilly's later work on underwater breathing masks at NIMH, these gas-monitoring technologies were applied, among other things, to detect mask leakage. For a discussion of the nitrogen meter Lilly apparently helped to develop, see "Curriculum Vitae, John Cunningham Lilly, M.D. 1968," p. 2, Stanford University Library, Lilly Papers, box 3C2[D1], file "CRI personnel." See also Lilly and Thomas F. Anderson, "The Nitrogen Meter: An Instrument for Continuously Recording the Concentration of Nitrogen in Gas Mixtures," Report 299, 28 February 1944, National Research Council, Division of Medical Science, Acting for the Committee on Medical Research of the Office of Scientific Research and Development, Committee on Aviation Medicine. This device used photoelectric monitoring.

65. It is important to emphasize the rapid growth of this sort of work in the period. It was in 1954 that Olds and Milner demonstrated that a rat could learn to stimulate its own brain, and later investigations by Delgado and others demonstrated similar behavior

Reporting on some of these investigations at a conference on the reticular formation of the brain, held in Detroit in 1957, Lilly would explain,

> The neurophysiologist has been given a powerful investigative tool: the whole animal can be trained to give behavioral signs of what goes on inside. . . . We are in the position of being able to guess with less margin of error what a man might feel and experience if he were stimulated in these regions.[66]

This was, in many ways, unpleasant business, Lilly acknowledged, pointing out that he had "spent a very large fraction of my working time for the last eight years with unanesthetized monkeys with implanted electrodes." In addressing the nebulous region where neurology, psychology, and animal behavior overlapped, Lilly permitted himself some observations on the affective universe of his scientific subjects:

> When an intact monkey grimaces, shrieks, and obviously tries to escape, one *knows* it is fearful or in pain or both. When one lives day in and day out with one of these monkeys, hurting it and feeding it and caring for it, its experience of pain or fear is so obvious that it is hardly worth mentioning.[67]

It would not be the last time that Lilly would reflect on the inner lives of his experimental animals with considerable confidence. But his experimental animal was about to change. Like a number of American psychology researchers in the mid-1950s—including the echolocation researcher Winthrop Kellogg—Lilly was in the process of leaving monkeys behind for the bottlenose dolphin, *Tursiops truncatus*.

His first brush with cetology came in 1949 when, during a visit to a neurosurgeon friend on Cape Cod, Lilly learned that a recent storm had beached a whale on the coast of southern Maine. A plan took shape for an

---

in cats, as well as the reverse—namely, learning to turn off a current that apparently caused pain/fear/discomfort.

66. John C. Lilly, "Learning Motivated by Subcortical Stimulation: The Start and Stop Patterns of Behavior," in *Reticular Formation of the Brain*, ed. Herbert H. Jasper et al. (Boston: Little Brown, 1958), 705–21, at p. 705.

67. Ibid., p. 719.

impromptu expedition north, with a view toward collecting a novel brain.[68] As it happened, Lilly was acquainted from his days at the University of Pennsylvania with the Swedish-Norwegian physiologist and oceanographer Per F. "Pete" Scholander, who had also worked with Detlev Bronk in aviation physiology during World War II and had then moved to WHOI.[69] Scholander—something of a daredevil, and fascinated by the physiology of extreme environments—had published research on dive physiology and decompression, and while still living in Scandinavia he had conducted a number of pioneering studies on the deep-diving capabilities of marine mammals, particularly whales.[70] Lilly looked up Scholander and recruited him for the trip, and the three men suited up for a drive to Maine. Shortly after reaching the carcass (a large pilot whale), exposing the skull, and beginning to chip away toward the brain, they were joined by two other researchers who had independently made the drive up from Woods Hole: William Schevill and his wife and collaborator Barbara Lawrence. They were, reportedly, somewhat miffed to discover that they had been beaten

68. Lilly recounts this story in *Man and Dolphin* (Garden City, NY: Doubleday, 1961), 40–47.

69. Much can be learned about Scholander's work from his autobiography, *Enjoying a Life in Science: The Autobiography of P. F. Scholander* (Fairbanks: University of Alaska Press, 1990), and a shorter memoir published earlier, "Rhapsody in Science," *Annual Review of Physiology* 40 (1978): 1–17. Scholander attended the First International Symposium on Cetacean Research. His papers are held at Scripps. I have consulted these holdings (5 boxes), which include some interesting material on his work with whales and dolphins, including a set of photographs depicting his visit to Brødrene Saebjørnsen's whaling station in Steinshamn, Norway, in the 1930s. These papers also contain a folder of his notes on the hydrodynamics of dolphin bow riding, work that resulted in a pair of articles in *Science* in 1959: "Wave-Riding Dolphins: How Do They Do It?" *Science* 129, no. 3356 (24 April 1959): 1085–87, and, with Wallace D. Hayes, "Wave-Riding Dolphins," *Science* 130, no. 3389 (11 December 1959): 1657–58.

70. The most substantial early piece of this work was the monograph published in 1940 in *Hvalrådets Skrifter*: P. F. Scholander, "Experimental Investigations on the Respiratory Function in Diving Mammals and Birds," *Hvalrådets Skrifter* 22 (1940): 5–131. I write about this work in "Self-Recording Seas," in *Oceanomania: Souvenirs of Mysterious Seas*, ed. Mark Dion and Sarina Basta (London: Michael Mack, 2011). A valuable discussion of Scholander's research in this area, along with a full bibliography, can be found in John W. Kanwisher and Gunnar Sundnes, eds., *Essays in Marine Physiology, Presented to P. F. Scholander in Honor of His Sixtieth Birthday*, *Hvalrådets Skrifter* 48 (Oslo: Universitetsforlaget, 1965). Some of the early experiments involved the use of pressure gauges affixed to whaling harpoons.

to the punch and particularly concerned that the hacksaw dissection might have damaged the airways of the upper head, which they had come to examine. In the end, however, the cadaver would be theirs, since Lilly and his partners found that the brain had largely been dissolved through autolysis; the smell alone overpowered them.

Though he headed home with little to show for the trip, Lilly had brushed the shores of cetology, and his curiosity did not dissipate. At a meeting of the International Physiological Congress four years later, in 1953, Lilly and Scholander again crossed paths, and Scholander suggested that Lilly get in touch with Forrest G. Wood at Marine Studios. Lilly did, and as a result, he was one of eight investigators to participate in what came to be known informally as the "Johns Hopkins expedition" in the autumn of 1955. It was, in a way, 1928 all over again: a mixed crew of physiologists and medical men gearing up to vivisect some bottlenose, only this time it would be in the carnival environs of a Florida ocean theme park, rather than a remote fishing village on a barrier island.[71]

In preparation for this 1955 trip, Lilly spent the summer in correspondence not only with Wood (securing access to a set of dolphins for experimental work), but also with Schevill at Woods Hole (concerning the anatomy of the airways of the common dolphin)[72] and with Scholander (concerning restraint techniques and the respiratory characteristics of the odontocetes).[73] Using this information, and reaching back to orthello Langworthy's work (discussed in chapter 3), Lilly worked up a dolphin respirator that would, it was hoped, permit the surgeons and neuroscientists of the party to expose the brain of an anesthetized animal in order to begin the work of cortical mapping by neurophysiological techniques.

The Johns Hopkins expedition of 1955, like its counterpart in 1928, was at best a qualified success. Lilly and the other investigators were unsuccessful with their anesthetics and their respirator, and in the end they euthanized, without dexterity, five dolphins, apparently alienating a number of the Ma-

71. The investigators, in addition to Lilly, were J. Rose, V. Mountcastle, and L. Kruger from Johns Hopkins Medical School; C. Woolsey and J. Hind, University of Wisconsin; Karl Pribam, Institute for Living, Hartford, CT; and Leonard Malis, Mount Sinai Hospital. The full records of this work can be found in Stanford University Library, Lilly Papers, box 6A1–B1.

72. Ibid.

73. As early as 1940 Scholander had done respiratory analysis on several restrained and submerged *Phocoena*.

rine Studios personnel in the process.[74] The most significant result of the work was the securing of a set of particularly good (fresh) brain specimens—perfused with preservatives before decomposition could begin—on which Lilly's expedition colleague Lawrence Kruger (who would soon move to UCLA's Brain Research Institute) would later conduct neuroanatomical research, some of which would be presented in Washington in 1963. By that time Lilly was (privately) accusing Kruger of sitting on the specimens and thus inhibiting competing interpretations of the neuron density and other features of the dolphin brain. By 1963, of course, with his name in lights across the idea of dolphin intelligence, Lilly resented Kruger's relentlessly deflationary assessments of the cortical tissue of *Tursiops truncatus* (Kruger himself thought of them as merely dispassionate).[75] Nor was this the only controversy spawned during the Hopkins trip. The visit—and two shorter ones by Lilly that followed—would be a bone of contention between Wood and Lilly for years, finally coming to a head in Washington at the 1963 conference, with, as I will show below, significant repercussions.

If the 1955 investigations were not a triumph, they did deepen Lilly's continuing interest in the cetacean brain.[76] Having heard a set of Wood's recordings of bottlenose at Marine Studios, Lilly was much struck—like a considerable number of others at this time, as we have seen—by the range and apparent complexity of dolphin phonation. In October 1957 and again in 1958—after a visit with Schevill and Lawrence in Massachusetts, where they were conducting work on the auditory range and echolocatory capabilities of a bottlenose dolphin in a facility near Woods Hole—Lilly returned to Marine Studios. This time he was equipped to undertake investigations of the dolphin brain and behavior using techniques like those he had deployed and refined with macaques at NIMH; namely, percutaneous electrodes, driven by stereotaxis, that could probe the brain tissue of an

74. Stanford University Library, Lilly Papers, box 6A1–B1. These records include minute-by-minute logs of each operation and phonograph disks recording the interactions of the scientists during each intervention. Given the broad disagreements that erupted later over this work, closer attention to these materials might prove interesting.

75. For "dispassion," see *Whales, Dolphins, and Porpoises*, ed. Kenneth Stafford Norris (Berkeley: University of California Press, 1966), 237.

76. The 1955 investigations also set in motion the research that would lead, almost a decade later, to the first successful techniques for major surgery on the small whales. See E. L. Nagel, P. J. Morgane, and W. L. McFarland, "Anesthesia for the Bottlenose Dolphin, *Tursiops truncatus*," *Science* 146, no. 3651 (18 December 1964): 1591–93.

unanesthetized, living animal.[77] Over the two visits, three more animals were sacrificed, and Lilly experienced a kind of scientific epiphany that would shape his scientific life, even as its reverberations eventually unmade his scientific reputation.[78]

Compressing a complicated encounter that took place over several days—and that continued to draw Lilly's reflections and reconstructions for years—is not easy, but we can summarize Lilly's sense of his findings this way: First, Lilly persuaded himself that, in comparison to his experience with monkeys, the dolphins appeared to learn very rapidly how to press a switch to stimulate a "positive" region in their brains (and to turn off stimulation to a region causing pain).[79] Second, he claimed to have been much struck by the sense that an injured experimental subject, when returned to the tank with other dolphins, "called" to them and received their ministrations, suggesting an intraspecies "language."[80] Third, on reviewing the tapes made of these investigations, Lilly grew increasingly certain that

77. See John Cunningham Lilly, John R. Hughes, Ellsworth C. Alvord Jr., and Thelma W. Galkin, "Brief, Non-Injurious Electric Waveform for Stimulation of the Brain," *Science* 121, no. 3144 (1 April 1955): 468–69, and Lilly, "Electrode and Cannulae Implantation in the Brain by a Simple Percutaneous Method," *Science* 127, no. 3307 (16 May 1958): 1181–82. Note that Lilly alleged that Schevill and Lawrence were working in a navy facility; Watkins (interview by the author, 9 August 2003) insisted that the work was being done in a private pool on Nonamesset Island, owned by the Forbes family.

78. It was also on these trips that Lilly got interested, through Wood, in the apparent ability of these animals to control the direction of their sound. Using an early AMPEX stereo tape recorder (on loan), Lilly and Wood were able to hear clearly that the click-trains emitted by captive dolphins had directional specificity. Wood discusses this finding in *Marine Mammals and Man: The Navy's Porpoises and Sea Lions* (Washington, DC: Robert B. Luce, 1973). See also Gregg Mitman, *Reel Nature: America's Romance with Wildlife on Film* (Cambridge, MA: Harvard University Press, 1999), 248.

79. This was done by means of a switch, placed within reach of the animal's beak. While I have never seen a reference to this problem, it must be asked whether contact with the switch could have been a product of convulsions and/or efforts by the animal to escape its constraints. Lilly's repeated emphasis on the "purposive" could perhaps be read as special pleading.

80. This issue of the "distress call" was central to later disputes; trainers and animal handlers were well aware of "epimeletic," or caregiving, behavior among these animals. Wood, and before him McBride and Hebb, had raised the subject of the "language" value—"language in the sense that a dog's barking or growling is a language"—of these whistlings. See F. G. Wood, "Underwater Sound Production and Concurrent Behavior of Captive Porpoises, *Tursiops truncatus* and *Stenella plagiodon*," *Bulletin of Marine Science of the Gulf and Caribbean* 3, no. 2 (March 1953): 120–33, at pp. 124–25.

his experimental subjects had been parroting his speech and other human sounds in the laboratory. These three elements—intelligence, an intraspecies language, and (perhaps most significantly) what he took to be fleeting glimpses of an attempt at interspecies communication—left Lilly with a feeling that he was on the cusp of something vast. Reflecting on the work of 1955, 1957, and 1958 in his Lasker Lecture in April of 1962, Lilly tried to explain:

> We began to have feelings which I believe are best described by the word "weirdness." The feeling was that we were up against the edge of a vast uncharted region in which we were about to embark with a good deal of mistrust concerning the appropriateness of our own equipment. The feeling of weirdness came on us as the sounds of this small whale seemed more and more to be forming words in our own language.[81]

After hammering his way into hundreds of mammalian brains, Lilly suddenly heard a voice.

Odd as this breakthrough may seem, Lilly was not alone in his sense of the magnitude of what had happened in the Marine Studios laboratory in the late 1950s. One of Lilly's medical friends who had been in attendance in October 1957, during work on dolphin number six, later mused to him in a letter, "I keep thinking of that first moment when the first, clearly purposeful switch-pressing response occurred. This is one of the extraordinary moments in science."[82] Loren Eiseley, the anthropologist who had become the provost of the University of Pennsylvania, wrote publicly that "the import of these discoveries is tremendous, and may not be adequately known for a long time."[83] And in 1961 Lilly would write of the discoveries in still grander world-historical terms, situating his own research at the cusp of the fourth "great displacement" in the history of science: citing Freud, Lilly explained that although man had, over the last five hundred years, been thrust from the center of the universe, from the center of nature, and finally

81. Joan McIntyre, comp., *Mind in the Waters: A Book to Celebrate the Consciousness of Whales and Dolphins* (New York: Charles Scribner's Sons and Sierra Book Club, 1974), 71.

82. Lawrence S. Kubie to Lilly, 31 July 1961, Stanford University Library, Lilly Papers, box 3A1–C1, file "Kubie, Lawrence S."

83. See Loren Eiseley, "The Long Loneliness: Man and the Porpoise: The Solitary Destinies," *American Scholar* 30, no. 1 (Winter 1960–1961): 57–64.

from the center of his own mind, modern man still thought of himself as the center of all intelligence. This (final?) pillar of human exceptionalism now teetered beside the dolphin tanks. It was his predilection for claims like this—sweeping, visionary, laced with self-aggrandizing enthusiasm— that seriously tried the patience of Lilly's colleagues, who placed increasing pressure on him to deliver some reproducible scientific results.

Those results were slower in coming. Lilly's first published report on his dolphin experiences appeared in December 1958 in the *American Journal of Psychiatry*, in an article entitled "Some Considerations Regarding Basic Mechanisms of Positive and Negative Types of Motivations." It is a deeply strange document, one that opens a window obliquely onto the world of brain research during the Cold War. Because it attracted a spate of articles in newspapers and magazines across the country—and launched the writing project that culminated in *Man and Dolphin* a little over two years later—it is worth examining this initial presentation in some detail.[84]

In view of the paper's reception and impact, it is striking that the discussion of the dolphin work at Marine Studios represents less than one-third of its total length, and that this section is sandwiched in the middle of a wide-ranging discussion of the positive and negative "motivation" regions of the mammalian brain. Lilly's primary concern in this paper was to reflect on the fact that neurophysiological work over the previous five years had established the existence of brain regions that, under stimulation, trigger "negative-painful-stop" responses, whereas other regions trigger "positive-pleasurable-start" responses. At issue, finally, was the balance between the aggregate sizes (and influences) of these two "parts" of the brain. So one reads, for instance, "Of course we like to think that in the total action of the brain, the positive tends to overbalance the negative, and that the intellectual functions might be neutral ones, neither positive nor negative, found in excess of the positive and the negative."[85]

---

84. For a discussion of the flurry of headlines prompted by the original presentation (at the San Francisco meeting of the American Psychiatric Association in May 1958), see Forrest G. Wood, *Marine Mammals and Man: The Navy's Porpoises and Sea Lions* (Washington, DC: Robert B. Luce, 1973), pp. 3, 12 n. 1. William Evans also takes up some of this publicity in *Fifty Years of Flukes and Flippers: A Little History and Personal Adventures with Dolphins, Whales, and Sea Lions, 1958–2007* (Sofia, Bulgaria: Pensoft, 2008).

85. John C. Lilly, "Some Considerations Regarding Basic Mechanisms of Positive and Negative Types of Motivations," *American Journal of Psychiatry* 115 (1958): 498–504, at p. 499. Note that, beginning in the 1940s, Lilly "undertook psychoanalytic training as a student in the Philadelphia Association for Psychoanalysis," where he worked with

Lilly then posed a question: Would point stimulation within the neocortex and the cerebellum—the regions of "higher" brain function—generate neutral, or perhaps mildly positive, effects? It appeared, strangely in his view, that in monkeys it did neither: strong negative response zones could be found in these sophisticated regions. So what? Well, this was troubling, according to Lilly's analysis. Were the centers of *abstract reasoning* laced with "halt" zones? This seemed to raise a striking possibility: the tractability of sophisticated cognitive functioning by means of electrical stimulation. Enter the specter of "mind control."

But perhaps, Lilly went on to reason, these issues—the "positive" versus the "negative" in higher brain function and, crucially, the plasticity of subjects under electrical stimulation—could only really be addressed in a brain larger and more complex than that of the macaque. As Lilly put it, "May not a larger brain be more impervious to such tampering with its innards? May not the trained, sublimating, and sometimes even sublime human mind resist, and even conquer such artificially evoked crassly primitive impulses?" Human experimentation was the only way to get sure answers to these questions. Acknowledging that this remained too risky to contemplate, Lilly then introduced *Tursiops truncatus*: "So far we have found only one animal that has a brain the size of ours who will cooperate and not frighten me to the point where I can't work with him—this animal is the dolphin." A description of Lilly's work at Marine Studios followed, emphasizing the success he and his collaborators had had in finding both positive and negative regions in this large brain. Two paragraphs at the end of this section became the core of the newspaper articles that followed, so I quote them here in their entirety:

> In this abbreviated account, I cannot convey to you all of the evidence for my feeling that if we are to ever communicate with a non-human species of this planet, the dolphin is probably our best present gamble. In a sense, it is a joke when I fantasy that it may be best to hurry and finish our work on their brains before one of them learns to speak our

---

Robert Waelder, a student of Freud's. (See Stanford University Library, Lilly Papers, Lilly typescript 1968, box 3C2 [D1], file "CRI personnel.") Lilly invested eight years in psychoanalysis, a period that overlapped with his neurophysiological and isolation studies. Exactly what role these experiences may have played in the ease with which he moved from electrophysiology of the brain to questions of personality is worth consideration.

language—else he will demand equal rights with men for their brains and lives under our ethical and legal codes!

Before our man in the space program becomes too successful, it may be wise to spend some time, talent, and money on research with the dolphins; not only are they a large-brained species living their lives in a situation with attenuated effects of gravity, but they may be a group with whom we can learn basic techniques of communicating with really alien intelligent life forms. I personally hope we do not encounter any such extraterrestrials before we are better prepared than we are now. Too automatically, too soon, too many of us attribute too much negative systems activity to foreign language aliens of strange and unfamiliar appearance and use this as an excuse for increasing our own negative, punishing, attacking activities on them.[86]

With that, Lilly was finished with dolphins, and he returned to the shadow subject of his presentation: "What does all this mean in terms of us, our species?" Was it possible, without percutaneous electrodes, to investigate the "positive" and "negative" systems of our own brains? "Turning inward, examining our minds, their deep and primitive workings, can we see evidence of the actions and inner workings of the positive, pleasure-like, start, and the negative, pain-fear-like, stop systems?" Lilly's answer was yes, and the technology for doing so preoccupied him for the last third of his talk: isolation tanks—large, temperature-regulated tanks in which neutrally buoyant subjects were confined, in a breathing hood, in total darkness and without sound or sensory input, for as long as they could stand it (figure 6.14).[87]

Here we may seem to be impossibly remote from the study of whales and dolphins, and yet the links are tighter and stranger than one might expect. Lilly was the inventor of the water-immersion technique of sensory deprivation, and he was a significant early contributor to the broader area of sensory deprivation research, opening the third major center for

86. John C. Lilly, "Some Considerations Regarding Basic Mechanisms of Positive and Negative Types of Motivations," *American Journal of Psychiatry* 115 (1958): 501.

87. The best review of this extraordinary research enterprise is the book edited by John P. Zubek, *Sensory Deprivation: Fifteen Years of Research* (New York: Appleton-Century-Crofts, 1969). See also Leo Goldberger, review of *Sensory Deprivation*, edited by John P. Zubek, *Science* 168, no. 3932 (8 May 1970): 709–11.

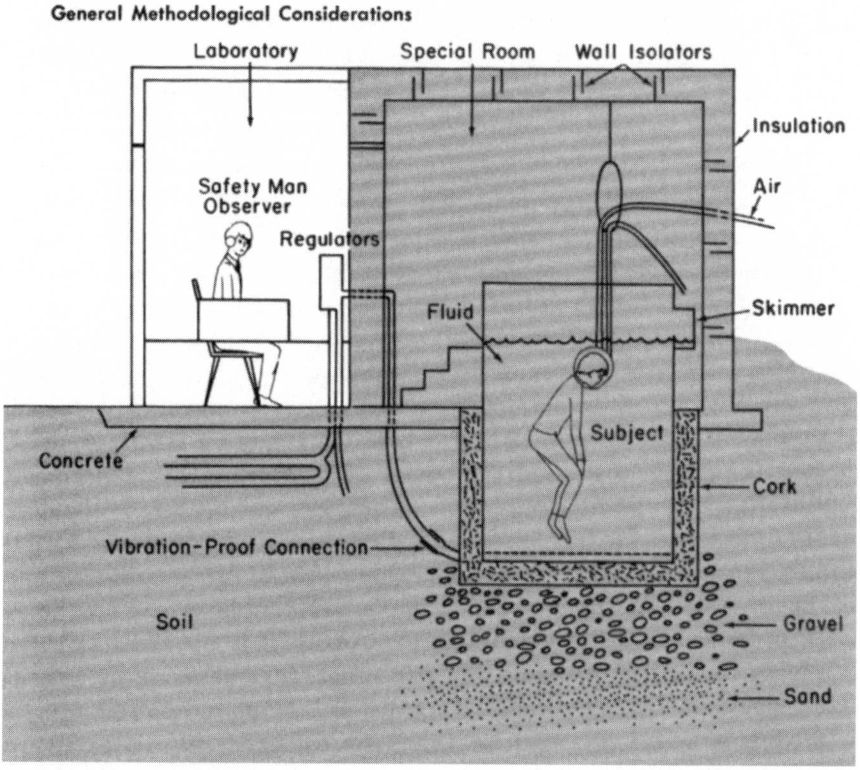

General Methodological Considerations

FIGURE 6.14 The tank: A cross-sectional diagram of a sensory deprivation tank. (From Rossi, "General Methodological Considerations," p. 27.)

such investigations (after McGill and Princeton) in the academic research world and spawning a number of labs (most importantly that of Jay Shurley at the Oklahoma City VA hospital) that built tanks to his specifications. While working on such systems for several years at NIMH (from 1955–1956 forward), Lilly conducted extensive self-experimentation in his tank, developed new equipment for it, advised on the building of similar systems, and lectured around the country on their use.

Such experimental environments seem, in retrospect, macabre, but it is necessary to recall the impetus for this sort of work. The earliest such investigations (at McGill, from 1951 forward) were stimulated by an interest in Russian and Chinese "brainwashing,"[88] and there can be no doubt that

88. See John P. Zubek, ed., *Sensory Deprivation: Fifteen Years of Research* (New York: Appleton-Century-Crofts, 1969), 9, for a discussion.

Lilly's move into this area was linked to such preoccupations, which were firmly established on the national stage in the United States in the mid-1950s. In the Lilly papers, I found a file labeled "Indoctrination, Forced," which contains material on solitude, isolation, and "brainwashing." An adjacent file, labeled "Solitude," contains, most interestingly, Lilly's notes on a conversation with "Dr. Sperling" in the Research and Development office of the Surgeon General's office, dated 23 April 1956. The notes read, "called this date, re: brainwashing, etc, mentioned by Dr. Felix before senate appropriations committee two weeks ago." A set of newspaper clippings make it clear that this exchange took place at the height of *Manchurian Candidate* fears over the practices of Chinese "mind control" scientists.[89] Other correspondence reveals that systems like Lilly's were wanted for two purposes: first, as a tool for screening tests and personality assessments (in order to find individuals particularly resistant to such situations and techniques), and second, as training instruments to improve the resistance of those who might face sensory deprivation conditions—not just soldiers, spies, and diplomats, but others subjected to the rigors of solitary environments, particularly pilots, astronauts, and those manning remote meteorological or monitoring stations.[90]

89. Relevant material in these files includes a typescript by Robert J. Lifton (in the neuropsychiatry division of the Walter Reed Army Institute of Research) entitled "Chinese Communist 'Thought Reform': 'Confession' and 'Re-education' in Penal Institutions," and another essay by Edgar H. Schein (at the Army Medical Service Graduate School) on "Chinese Brainwashing." Published materials represented as clippings include several *New York Times* pieces, including "New Evils Seen in Brain Washing," 4 September 1956, and "Two Challenge Views on Brainwashing," 22 September 1956. See Stanford University Library, Lilly Papers, box 5A1, files "Solitude" and "Indoctrination, Forced."

90. Sensory deprivation screening was used, for instance, in the selection of the astronauts for Project Mercury. In the Lilly papers I discovered that he was for several years in this period a dues-paying member of the Slocum Society (founded in 1955) and received its newsletter. Joshua Slocum (1844–1909), a New England captain, became an international celebrity at the turn of the century after he successfully completed a single-handed circumnavigation in 1895–1898, the first such exploit recorded. He was lost at sea, alone, in 1909. His popular book on his successful voyage, *Sailing Alone Around the World* (New York: Century, 1900), recounts several hallucinatory intervals during long crossings (though it is difficult to assess the tone of these passages, which have a comical quality). The Slocum Society was founded (perhaps paradoxically) to create a community of solitaires, particularly those dedicated to long solo voyages.

Precisely how close Lilly's ties to the world of intelligence operations actually were remains obscure. In his autobiography, he wrote that his decision to leave NIMH in 1958 was motivated by his growing unease with the encroachment of application-oriented, apparently government-linked, investigators seeking information about the work of his Bethesda laboratory.[91] This may be so. Through a Freedom of Information Act request, I secured Lilly's FBI file, and I am therefore able to confirm that Lilly was involved, well into 1959, in advising members of the security establishment about the potential uses of his neurophysiological investigations, though an unseemly flap over his security clearance in May of that year certainly added friction to those relationships.[92] As of 27 August 1959, Lilly had "Secret" clearance, and documentation in Lilly's FBI records indicates both that J. Edgar Hoover personally attended to his file and that he was to be treated as *persona non grata* by the bureau after 1960.[93]

91. See Francis Jeffrey and John Cunningham Lilly, *John Lilly, So Far . . .* (Los Angeles: Jeremy P. Tarcher, 1990), 82–100; see also a brief discussion of Lilly's situation and CIA interest in his work in (the more reliable) John D. Marks, *The Search for the "Manchurian Candidate": The CIA and Mind Control* (New York: Times Books, 1978).

92. "Dr. Lilly's problem concerned a meeting held at the Pentagon in May 1959. This meeting was called in order that ranking officers of the Office of Naval Research, the Air Force, and the Army could hear a briefing by Dr. Lilly on his work on the brain of dolphins, Dr. Lilly explained that the military was interested in this field [TEXT CENSORED] inasmuch as research by himself and other scientists had established that by the use of electrodes placed in the brains of animals and humans the will could be controlled by an outside force. He explained that if an electrode were placed in the brain of a subject. He [*sic*] could make the subject experience great extremes of joy or depression, for example. Dr. Lilly stated that the potential of this technique in 'brain washing' or interrogation or in the field of controlling the actions of humans and animals is almost limitless. He stated that our officials are aware that the Soviets are intensely interested in this field and that they are conducting extensive experiments and that their progress has roughly paralleled that of ours." FBI personal file, "John Cunningham Lilly (Dr.,)" Memorandum, Jones to DeLoach, p. 2. Deletion in category b2, "solely related to the internal personnel rules and practices of an agency."

93. Ibid., p. 4. Lilly apparently found himself caught in a wrangle between the security establishments of the Defense Department and the FBI in late 1959. Having been asked to leave a Pentagon briefing (as noted above) because of a security "problem" with his clearance, Lilly made a set of inquiries and learned (from an unnamed informant) that the problem had originated with the FBI. He followed up, only to be told that this was not the case, and that the FBI wanted to know who had told him this. He refused to divulge his source, despite several visits from agents, both in Miami and in San Juan.

What emerges from a close reading of Lilly's 1958 paper in this context is the remarkable way in which his early dolphin investigations were entangled with this set of seemingly remote preoccupations—"brainwashing," "reprogramming," and "mind control"—that were reverberating through the sciences of brain and behavior in the mid-1950s. For instance, the questions Lilly posed after reporting his work with macaques—Can humans "resist or conquer" such situations? Do different people discover different degrees of "egophilic" or "egophobic" affect in isolation tanks?—were questions wholly tied to the pressing problem of the imperviousness and durability of the Cold War human agent.

These very un-cetological matters might be of merely anecdotal value to our understanding of Lilly's work with dolphins if his 1958 paper were simply a salad of his diverse interests. But this is not the case. Rather, the paper offers palpable clues to the early ties between Cold War psychological-neurological investigations aimed at "accessing" the mind of recalcitrant (and taciturn) enemies and Lilly's research program on interspecies communication. Indeed, by the time Lilly had expanded his brief asides on the promise of dolphin-human communication into the popular book *Man and Dolphin* of 1961, the language of "psy-ops" permeated his porpoise talk. For instance, alluding to the "less well controlled" dolphin training "of the past," which used food rewards, Lilly pointed out that as a result of his work and that of others, "we now have push-button control of the experiences of specific emotions by animals in whose brains we have placed wires in the proper places. . . . Using this 'reward stimulation' technique, we demonstrated quite satisfactorily that a dolphin can vocalize in two different ways." One of these, above the surface ("in the air"), suggested that already the animals were being brought into the sphere of their captors.

Another mind control technique was less invasive, but also promising— "intraspecies solitude":

> If a human being is isolated from other humans for a month or more, and is confined to a small area geographically and a small range of activities, his interest in his surroundings and its minutiae increases radically. . . . Further, if a confined, isolated human is allowed brief

---

The FBI appears to have learned the identity of the source on their own, and marked Lilly as "uncooperative."

contacts with other humans even without a shared language, he begins to find their presence comforting, and a pleasant relief from the "evenness" of his surroundings. If these humans control his only sources of food as well as his sources of intraspecies stimulation, he may adapt to their demands in subtle and not so subtle ways. He may, given time, learn their language, take on their beliefs, etc.

When we catch a dolphin and put him alone in a small tank, we are imposing similar "solitary confinement" strictures on him. Maybe we can thus capture his loyalty, and his initiative.[94]

Later, Lilly would deploy this and other techniques borrowed from the margins of the world of the mind control "spooks"—including a set of extended "chronic-contact" experiments and, finally, experimental psychopharmaceuticals, particularly D-lysergic acid diethylamide (LSD-25), which in the early 1960s was closely linked to the clandestine work of the CIA's "Artichoke" project—in his effort to break though to (or perhaps simply to "break") a dolphin.[95]

This strange imbrication of the techniques of mind control and animal communication in the late 1950s and early 1960s suggests at least one way in which the isolation tank and sensory deprivation research fit with Lilly's program of cetological investigations in this period. But there were others. Among Lilly's papers, in his "Solitude" file, I discovered his copy of the English translation of Jacques-Yves Cousteau's 1953 best seller *The Silent World*, a book (and in 1956, a film) that introduced Cousteau and deep-sea scuba diving to hundreds of thousands of Americans. Lilly's annotations in the margins of this volume suggest that he read the text with care and took particular interest in Cousteau's reflections on the experience of weightlessness and isolation in the silent suspension of the underwater world. The

94. This and the quotes above are from John C. Lilly, *Man and Dolphin* (Garden City, NY: Doubleday, 1961), 190–91. Similar discussions of the importance of "isolation" and "confinement" run through Lilly's early articles on dolphins in *Science*. See, for example, John C. Lilly and Alice M. Miller, "Vocal Exchanges between Dolphins," *Science* 134, no. 3493 (8 December 1961): 1873–76.

95. See Stanford University Library, Lilly Papers, box 3D2, file "Sandoz." On Artichoke, see John D. Marks, *The Search for the "Manchurian Candidate": The CIA and Mind Control* (New York: Times Books, 1978). More generally on the history of LSD in this period, see Jay Stevens, *Storming Heaven: LSD and the American Dream* (New York: Grove Press, 1998).

development of the actual mask and breathing technologies of Lilly's isolation tank must thus be seen in the context of a growing interest in scuba
diving and undersea environments.[96] In fact, in describing his efforts to
communicate with dolphins in this period, Lilly spent a good deal of time
on the need for the experimenter to attempt to commensurate him- or
herself—imaginatively, even physically—with the subject. And here, Lilly
explained, the flotation/isolation tank could be of considerable use, since
it offered a glimpse of the dolphin's perceptual universe.[97] It was for this
reason, apparently, that Lilly saw to it that an isolation tank was built at the
Saint Thomas lab of CRI, adjacent to the indoor dolphin tank. In the early
1960s Lilly's isolation technology appeared poised to give him insight into

96. The edition is the fourth printing, from 1961, so the reading must have occurred in that year or later. In 1961 Lilly and Jay Shurley published their essay "Experiments in Maximum Achievable Physical Isolation with Water Suspension of
Intact Healthy Persons," in *Psychophysiological Aspects of Space Flight*, ed. B. E. Flaherty (New York: Columbia University Press, 1961), 238–74. It was Lilly's last article in
this area. I think it likely that the annotations were made in that year, particularly as
several of them deal with the rubber "Furney goggles" Cousteau describes. Lilly and
Shurley discussed the form of similar latex masks in their correspondence concerning the flotation tank. Other annotations to the text include small marks next to "I
turned over and hung on my back" (p. 5), and "As we submerged, the water liberated
us from weight" (p. 78). For a discussion of the film version of *The Silent World*, as
well as a brief treatment of Cousteau's broader importance in the growing American
fascination with the undersea world in the 1950s, see Gregg Mitman, *Reel Nature:
America's Romance with Wildlife on Film* (Cambridge, MA: Harvard University Press,
1999).

97. "In the course of some experiments I conducted from 1954 through 1956 I was
suspended in water for several hours at a time, and I noticed that my skin gradually
became more and more sensitive to tactile stimuli and an intense sense of pleasure
resulted. However, if the stimulation was carried too far it became intensely irritating.
I reasoned that the dolphin is suspended in water all of his life, twenty four hours a
day, and possibly had developed an intensely sensitive skin." John C. Lilly, *Man and
Dolphin* (Garden City, NY: Doubleday, 1961), 172. The issue of "commensurating" with
the dolphin appears in many places in Lilly's published and unpublished work; see for
instance, ibid., p. 209. Mitman has explored this idea of experimental commensuration
in the study of animal behavior in "Pachyderm Personalities: The Media of Science,
Politics and Conservation," in *Thinking with Animals: New Perspectives on Anthropomorphism*, ed. Gregg Mitman and Lorraine Daston (New York: Columbia University Press,
2005), 175–95. Mitman and Daston together take up the problem in their introduction
to this same volume.

the "mind in the waters" that had become, by 1958, his new experimental subject.[98]

It was with an eye on dedicating himself to this new experimental subject that Lilly, in 1958, departed from NIMH, in the glow of widespread interest in what this hard-driving scientist planned to accomplish with the dolphin. He was jumping several ships at once, in that he simultaneously separated from, and subsequently divorced, his wife of two decades, mother of two of his children. By 1959 Lilly not only had a trade contract for a pair of books on dolphins, he had also begun work opening what would become a dedicated research facility for the study of these animals. From the start, Lilly—keen to reinvent himself at some distance from middle-class respectability (and the smell of monkey cages)—had his eye on the Caribbean, and visits to the Bahamas, Jamaica, and finally, Puerto Rico gave him the lay of the land. He secured a position at the University of Puerto Rico in pharmacology during 1959–1960 as he continued his reconnaissance of the region, and by the end of that year he was remarried, to Elisabeth Bjerg, a former

98. There remains yet another link between sensory deprivation research and dolphin study in this period: as it turns out, the pioneering figure in sensory deprivation work in 1951 was Professor D. O. Hebb, at McGill University. Hebb, who had a simultaneous appointment at the Yerkes Laboratory of Primate Biology in Orange Park, Florida, was the very same Hebb who coauthored the foundational 1948 article (with Arthur McBride), "Behavior of the Captive Bottle-nose Dolphin, *Tursiops truncatus*," *Journal of Comparative Physiological Psychology* 41 (1948): 111–23. This was really the first scientific paper to document behavioral observations on the captive marine mammals of the recently reopened Marine Studios. How Hebb, too, bridged the universes of sensory deprivation and cetology is not absolutely clear. One possibility is that work with captive primates in this period encouraged exploration of the behavioral ramifications of prolonged isolation and boredom, since monkeys respond rapidly and markedly to these conditions; this observation might explain both Lilly's and Hebb's early curiosity. Of the unsavory aspects of some of Hebb's other Cold War work there can be little doubt (see, e.g., Alfred W. McCoy, "Science in Dachau's Shadow: Hebb, Beecher, and the Development of CIA Psychological Torture and Modern Medical Ethics," *Journal of the History of the Behavioral Sciences* 43, no. 4 [Fall 2007]: 401–17). I have found no evidence of a link between Lilly and Hebb, but it seems likely they knew each other through sensory deprivation work, and it is surely possible that Hebb stimulated Lilly's early dolphin interests. Whatever the case may be, the unlikely ties between isolation studies and dolphin studies in this period demand a revised reading of Loren Eiseley's curious and moving essay on Lilly's work, tellingly entitled "The Long Loneliness: Man and the Porpoise: The Solitary Destinies" (*American Scholar* 30, no. 1 [Winter 1960–1961]: 57–64).

fashion model from Saint Croix. He had also settled on Saint Thomas as the location for his new institute, found real estate and agents to help him secure it, and was in a position to entertain visits from two grantmakers from the Office of Naval Research, including, significantly, Sidney Galler. ONR announced that it was willing to top up the funding of his recently incorporated Communication Research Institute (CRI), depending on how the National Science Foundation, National Institutes of Health, and Department of Defense responded to his grant requests. Meetings in Washington, DC, followed, and by 13 June, Lilly could write to his friend Orr Reynolds at NASA that "the Institute has been given a grant by the N.S.F. (with help from O.N.R. and D.O.D.) to build the world's first laboratory devoted to the study of the intellectual capacities of the small, boothed [sic, for "toothed"] whales." The institution's budget was set at about $100,000 a year, though expenses went up when, later that year, Lilly decided to open a second lab in Miami—what he called a "bipedal" arrangement, allowing the work to be "close to our supply of animals."

Shortly thereafter, NASA money was forthcoming through Reynolds, under the Office of Space Sciences' Biosciences Program, which sponsored a "behavioral sciences division" dedicated to, among other things, "investigations on the mechanisms of inter- and intra-species communication of intelligent information, emotional status, and basic drives, in an attempt to discover the mechanisms which nature has evolved, and to supplement these mechanisms by technological devices."[99]

Meanwhile, the building itself was well under way. Navy Underwater Demolition Team frogmen excavated the dolphin pool below the laboratory building with the help of 600 pounds of TNT (figures 6.15 and 6.16). By March 1960 Lilly was ready to bring his first two dolphins—secured through Wood at Marine Studios—down to Saint Thomas. Media coverage of the transfer included a front-page piece in the *Herald Tribune*, but the enthusiasm was short-lived: both dolphins died in a matter of weeks, and Lilly returned to Florida in what was, at least in part, an effort to learn more about how to maintain dolphins in captivity. To this end, he spent time at Theater of the Sea and the Marine Laboratory of the University of Miami. All the while, Lilly was pulling together the manuscript of *Man and Dolphin*, which would be published in August 1961.

99. "Behavioral Biology Program—Biosciences Programs—Office of Space Sciences," sec. 3, Stanford University Library, Lilly Papers, box 3D2, file "Reynolds, Dr. Orr. E., NASA."

FIGURE 6.15 Under construction: CRI's Saint Thomas laboratory in progress. (From Lilly, *Man and Dolphin*, pp. 181, 183.)

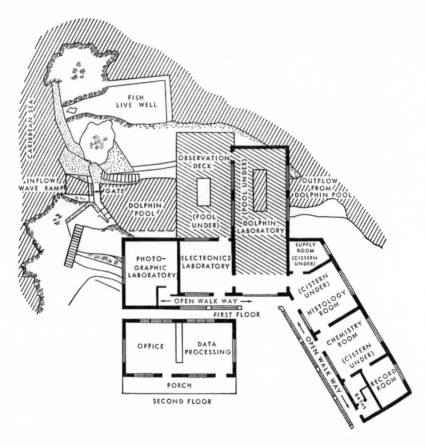

FISH LIVE WELL

OBSERVATION DECK

CARIBBEAN SEA

INFLOW WAVE RAMP

GATE

DOLPHIN POOL

(POOL UNDER)

(POOL UNDER)

DOLPHIN LABORATORY

OUTFLOW FROM DOLPHIN POOL

PHOTO-GRAPHIC LABORATORY

ELECTRONICS LABORATORY

SUPPLY ROOM (CISTERN UNDER)

(CISTERN UNDER)

HISTOLOGY ROOM

OPEN WALK WAY

FIRST FLOOR

CHEMISTRY ROOM

OPEN WALK WAY

(CISTERN UNDER)

OFFICE

DATA PROCESSING

RECORD ROOM

BATHS

PORCH

SECOND FLOOR

FIGURE 6.16
The layout: Plan and ground views of the CRI facility at Saint Thomas. (From Lilly, *The Mind of the Dolphin*, p. 227 [*above*] and after p. 75 [*below*].)

Interestingly, Lilly's talking dolphins beat him into print. In April 1961 the Hungarian physicist Leo Szilard published his biting satire on the arms race, *The Voice of the Dolphins*. This futuristic tale—written in the summer of 1960 (as Szilard fought cancer in Memorial Hospital in New York City) and circulated in various forms among nuclear armament negotiators in 1960—apparently emerged out of Szilard's conversations with Lilly in the late 1950s at NIH.[100] In the story, Lilly's name is explicitly invoked and linked to the founding of a Soviet-American scientific research institute, located in Vienna, which succeeds—it claims—in "breaking through" to dolphins, which prove, according to the scientists, to be brilliant thinkers. Over the next two decades, the Vienna Institute's dolphins serve as (yes) modern "Delphic" oracles who guide the world, via their scientist mouthpieces, through the moves of a chesslike game of nuclear disarmament (using Szilard's controversial scheme for de-escalation). In the end, their work done, the dolphins all die mysteriously, the Vienna Institute dissolves, its records vanish in a fire, and the world has been pulled back from the brink of nuclear annihilation. Szilard winks at the end, alluding to the doubts of some concerning whether the guiding intelligence behind the whole miraculous process had indeed been the dolphins or merely the crafty community of international scientific adepts. The success of this book, which over the next three years sold over 35,000 copies in the United States alone (it was translated into six other languages, becoming a "minor classic of the nuclear age"), makes it clear that the nebulous idea of dolphins as an alternative intelligence—and as potential interlocutors for human scientists— was very much in the air by the time *Man and Dolphin* appeared, and that these ideas were firmly linked to the name John Cunningham Lilly.[101]

And *Man and Dolphin* delivered. Its opening sentence made an extravagant prediction: "Within the next decade or two, the human species will establish communication with another species: nonhuman, alien, possibly extraterrestrial, more probably marine, but definitely highly intelligent, perhaps even intellectual." Lilly's money (quite literally: he had invested his own resources in CRI) was on the dolphin. In a closing chapter he explored

100. A version of the story was circulated among Soviet and American scientists at the Pugwash Conference in 1960.

101. See p. 4 of Barton J. Bernstein's introduction to Leo Szilard, *The Voice of the Dolphins, and Other Stories*, Nuclear Age Series (Stanford, CA: Stanford University Press, 1992). See also William Lanouette and Bela A. Silard, *Genius in the Shadows: A Biography of Leo Szilard, the Man Behind the Bomb* (New York: Charles Scribner's Sons, 1992). Lanouette interviewed Lilly before his death about his interactions with Szilard.

the potential legal, moral, and social problems that would be confronted, going so far as to suggest that in the distant future the world might confront a crisis comparable to that of racialized global human inequality: "For a long time, presumably, they [the educated dolphins] will be in the position of the Negro races who are attempting to become Westernized."[102] At the same time, the encounter with this parallel oceanic intelligence would give human beings "a perspective of which we can be only dimly aware at the present time. Our own communication among ourselves will be enhanced and improved by such contact."[103]

This unstable froth of *Planet of the Apes* futurism and Cold War development theory met with popular enthusiasm (fired in part by an appealing photo-essay in *Life* magazine at the end of July) and a guardedly positive review by the well-known turtle biologist Archie Carr in the *New York Times*. Largely laudatory assessments in a variety of other papers in the United States and the United Kingdom followed. Even the *Quarterly Review of Biology* praised the text for putting "forth in a very readable fashion a study that many people have heard about."[104] Within a year,

102. John C. Lilly, *Man and Dolphin* (Garden City, NY: Doubleday, 1961), 217. Consider, as context, Pierre Boulle's *La Planète des Singes* (Paris: Juilliard, 1963) and its reception in the United States. For a study of the way the race issue played out in the American film world, see Eric Greene, *Planet of the Apes as American Myth: Race and Politics in the Films and Television Series* (Jefferson, NC: McFarland, 1996). By 1965 Lilly was having his CRI staff read and comment on an English translation of Boulle's novel. There is a larger story to be told about the relationship between Lilly's work and the world of science fiction in these years.

103. John C. Lilly, *Man and Dolphin* (Garden City, NY: Doubleday, 1961), 223.

104. Bryan P. Glass, review of *Man and Dolphin*, by John C. Lilly, *Quarterly Review of Biology* 36, no. 4 (December 1961): 311. See also "He barks and buzzes, he ticks and whistles, but can the dolphin learn to talk?" *Life Magazine* 51, no. 4 (28 July 1961): 61–66; Archie Carr, "Have We Been Ignoring a Deep Thinker?," review of *Man and Dolphin*, by John C. Lilly, *New York Times Book Review*, 3 September 1961, 3; Ted Hughes, "Man and Superbeast," review of *The Nerve of Some Animals*, by Robert Froman, and *Man and Dolphin*, by John C. Lilly, *New Statesman* 53, no. 1619 (23 March 1962): 420–21; B. A. Young, "Placid and Self-Contained," review of *Man and Dolphin*, by John C. Lilly, and *The Nerve of Some Animals*, by Robert Froman, *Punch*, 14 March 1962, 443; *New Yorker*, unsigned review of *Man and Dolphin*, by John C. Lilly, 16 September 1961, 178; Robert C. Cowen, "Can We Converse?," review of *Man and Dolphin*, by John C. Lilly, *Dolphins: The Myth and the Mammal*, by Antony Alpers, and *Porpoises and Sonar*, by Winthrop N. Kellogg, *Christian Science Monitor*, 14 December 1961, 11.

the initial run of 10,000 copies had sold out, and a second printing was under way.

During this period Lilly completed work on a handful of scientific papers reporting the results of his dolphin researches (up to the publication of *Man and Dolphin*, Lilly's only published scientific articles mentioning dolphins were the 1958 essay in the *American Journal of Psychiatry* and a brief note in *Science* with Alice Miller, "Sounds Emitted by the Bottlenose Dolphin," which appeared in May 1961). He also expanded his research enterprise at CRI, securing more dolphins, cultivating the board of trustees, and writing grants for new research programs and staff. These years, 1961–1963, represent the waxing of CRI and Lilly's program. Sending out complimentary copies of *Man and Dolphin* to a who's who of old-guard cetology, movie celebrity, and national political power, Lilly soon found himself not only receiving laudatory letters from the respected ethologist-anthropologist Gregory Bateson (who would join CRI shortly thereafter), but also entertaining, in Saint Thomas, President Kennedy's personal physician and a host of other distinguished visitors who wanted to meet the dolphins and the dolphin doctor.[105]

A typescript draft of a program description for CRI, which can be dated to late 1962, captures Lilly's vision for the institute in this heady period, and annotations in his hand indicate his careful efforts to position CRI between psychology, medicine, neurology, and animal behavior:

> The institute is studying intensively one of the unusual creatures of the sea—the bottlenose dolphin. This is a mammal with a brain larger and more complex than the human brain. The unexplored biological territory of the large mammal brain affords an unequaled opportunity for tests in neurophysiology, brain function, communication, and intelligence.[106]

What follows is a list of benefits and objectives of this work, which Lilly carefully renumbered to accord with his sense of their significance, mov-

---

105. For a taste of this, see Lilly's file of correspondence with Dr. Janet Travell, Stanford University Library, Lilly Papers, box 3D2 (where there is also a file of replies to and letters on *Man and Dolphin*).

106. "The Programs of the Communication Research Institute," typescript, Stanford University Library, Lilly Papers, box 3D2, file "Worcester Foundation."

ing "improved techniques of human brain surgery" from first to last and replacing it with, first and second, "improved understanding of human learning and educational processes" and "a more systematic approach to the measurement of interspecies intelligence."

Writing an entry on "Interspecies Communication" in the *McGraw-Hill Yearbook of Science* in 1962, Lilly would briefly mention the work of Cathy and Keith Hayes with chimpanzees before moving on to say that "insofar as the author knows, there have been no systematic and serious attempts to date on the part of the human to learn to speak with another species in its own tongue." This, and the teaching of human language to dolphins, was the work, Lilly reported, of CRI (figure 6.17).

This brief entry captures the combination of dynamic enthusiasm, medico-scientific gravitas, and thinly veiled self- (and CRI) promotion that pervaded Lilly's production in the bustling period that followed the publication of *Man and Dolphin*. During this time Lilly's NSF Career grant came through, he found himself appointed to the planning committee of the "Whale Circus," and his trips to Europe and Washington were frequent. Lilly appeared to be in the vanguard of a vigorous new area of well-supported (and popular) research. At the same time, the *Yearbook of Science* entry reveals how imprecisely Lilly positioned himself with respect to several significant (and even well-established) disciplines—for instance, the fields of animal behavior, ethology, linguistics, and what Thomas Sebeok and Rulon Wells had recently dubbed "zoosemiotics" (the study of signaling behavior in and across animal species). From Darwin's *Expression of the Emotions in Man and Animals* to the bee studies of Karl von Frisch and the investigations of Niko Tinbergen and B. F. Skinner, there existed not just one but several traditions of research within which Lilly's putatively pioneering investigations fit.

There are two possible explanations for this imprecision. First, as part of the posturing that must attend any self-conscious "revolution," Lilly may have chosen willfully to disregard the larger disciplinary constellations within which his institution twinkled. Alternatively, he may simply have been genuinely ignorant of the range and richness of the work that had been done in these areas, and thus allowed himself the celestial arrogance of an elite aeronautical biophysicist and brain scientist coming in for a landing in the provinces of birdsong scientists and those who fretted over the stridulations of mating crickets. Some combination of self-styling strategy and hubristic naïveté seems most probable. The image of Lilly landing his visionary research program on foreign territory—perhaps

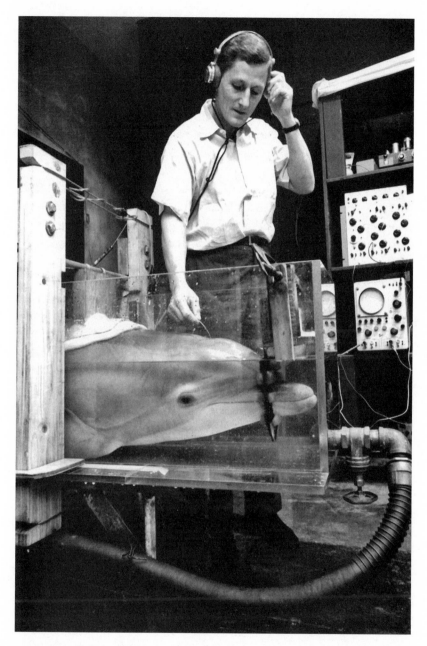

FIGURE 6.17 The interpreter: Lilly at work in the lab. (Courtesy of the Flip Schulke Archives. © Flip Schulke.)

even parachuting in—captures precisely the way that his trajectory into the study of dolphins and animal communication dropped orthogonal to the existing domains where these subjects received attention. He had neither the intimacy with the subject animals and their habits possessed by the animal-handler cetologists affiliated with the ocean theme parks (such as Norris, Wood, or McBride—the last by this time dead), nor did he have any demonstrable expertise in the field of animal communication, a discipline with its own professional meetings and literature. He was not really a bioacoustics person either.

Even as Lilly's network of medical associates, biophysics colleagues from wartime service, and old Caltech friends (many of them in powerful positions within the new institutions of postwar scientific funding) kept him, and CRI, plumped with research contracts and grants in the early 1960s, the established experts in animal communication on the one hand, and captive dolphin handlers on the other, began to undermine his standing and his claims. If they could not deny him the fanfare of his landing in their midst, they could snipe, and they did. Already by the end of 1961, several reviewers had grumbled. James W. Atz (an ichthyologist at the American Museum of Natural History), writing in the bulletin of the New York Zoological Society in December, was at pains to explain—as several hostile reviewers would be—that there was nothing wrong with radical scientific theories, and that Lilly, if correct in his central claims, would rightly take his place beside Darwin as "one of the greatest, most creative innovators in biology." Who could criticize originality and breadth of imagination? However, *Man and Dolphin* presented not a "single observation or interpretation" that could withstand scrutiny. Where this was the case, visionary hypothesis crossed the line into irresponsible deluding of the gullible populace: "Scientists and educators who believe that it is important for members of a democratic society to have a rational view of animal life can only wish that Dr. Lilly had not felt called upon to put himself so prominently in the public eye."[107]

The first shot had been fired, and more blasts were to come, particularly as 1962 came and the scientific papers backing up the "discoveries" de-

107. This and the quotes above are from James W. Atz, review of *Man and Dolphin*, by John C. Lilly, *Animal Kingdom: Bulletin of the New York Zoological Society* 64, no. 6 (December 1961): 190. Atz was an observer at the Washington symposium in 1963. See Symposium Program, Scripps Archive, Hubbs Papers, box 54, folder 46.

scribed in *Man and Dolphin* were not forthcoming. (Lilly alluded there to a number of manuscript articles, but the published papers that eventually emerged did little to buttress his more extravagant claims.) The Cornell-based professor of linguistics and anthropology Charles Hockett, who had written several long essays on the relationship between animal communication systems and the evolution of human language, gave Lilly's book a terse and damning review in *American Anthropologist*, where he acknowledged that dolphins were fascinating, but worried out loud that Lilly might have "dealt a body blow to the important program of research in which he wishes to play a part."[108] The most extensive critique appeared simultaneously, in *Natural History*, jointly authored by William N. Tavolga (who, as it happened, had chaired the Symposium on Marine Bio-Acoustics at Lerner Marine Laboratory) and his wife Margaret, a specialist in animal behavior in the Biology Department at Fairleigh Dickinson University who had also worked with Wood's dolphins at Marine Studios. The Tavolgas declared the book a primer for a young scientist seeking guidance on "how not to do scientific research," and they went after Lilly's "unsound and naïve grasp" of terms like "language," "speech," and "communication." Drawing on the very fields Lilly had disregarded, the two animal behavior researchers wrote,

> Dr. Lilly takes the view that "those who speak are those who are capable of learning language." If "one two three" said with very poor intelligibility by a dolphin is indicative of the giant-brained animal's ability to speak, and therefore to learn language, what is to be said of a parrot's clear-cut, if bird-brained, "Polly wants a cracker"? Furthermore, if the parrot is then given a cracker, have we established communication with an alien species?

In their assessment, the director of the Communication Research Institute needed to do some more thinking about "communication": "There is no doubt in the minds of most of us that dogs, dolphins, and many other animals are able to communicate by sounds or other means (to us or to each

---

108. Charles F. Hockett, review of *Man and Dolphin* by John C. Lilly, *American Anthropologist* 65, no. 1 (February 1963): 176–77. Hockett was the author of "Logical Considerations in the Study of Animal Communication," an invited conclusion (pp. 392–430) to *Animal Sounds and Communication*, edited by W. E. Lanyon and William N. Tavolga (Washington, DC: American Institute of Biological Sciences, 1960).

other) their relatively simple needs and wants, but this remains a far cry from language."[109]

As the fields of animal behavior and linguistics began to close ranks by early 1962, the publication of *Man and Dolphin* also did considerable damage to Lilly's relationship with some of those in the world of the commercial ocean theme parks—a serious problem for a dolphin scientist. Things remained cordial, for the time being, with Kenneth Norris (by this time at UCLA) and with his replacement at Marineland of the Pacific, David Brown. Indeed, Norris even invited Lilly to come out West in 1962 to make an excursion with him to try to net some *Phocoena sinus* in the Gulf of California.[110] But his relations with Wood at Marineland of Florida had seriously soured. Wood figured as a named character in *Man and Dolphin*, and it was he who, as Lilly narrated it, provided CRI with its first two dolphins, Lizzie and Baby, the same dolphins Lilly would be unable to keep alive at the new lab for more than a few weeks. In presenting this anticlimax in his book, Lilly made it clear that he held Marineland and, by implication, Wood himself responsible for these deaths. Shortly after meeting his two dolphins in their tank at Marineland, Lilly noted, he "began to be repelled in earnest by the foul odor coming from their blowholes," and though "some of the people at Marineland had told us this was a normal condition," Lilly's "medical training" told him this was impossible, and that he was smelling "the results of putrefaction by pathogenic kinds of bacteria." In retrospect, he acknowledged, he should have "done something about this" before accepting the animals, but the transfer had its own momentum. After describing the passing of the two dolphins, Lilly closed the chapter by saying that he decided to investigate "this business of the infections" by heading back to Florida.[111]

It is not difficult to imagine how irksome Wood must have found this high-handed tone and thinly veiled blame, appearing as it did in a hugely popular book squarely concerned with his area of expertise and written by a man with no experience whatsoever in the care and maintenance of captive dolphins and porpoises. As it happens, we do not need to imagine, be-

109. This and the quotes above are from Margaret C. Tavolga and William N. Tavolga, review of *Man and Dolphin*, by John C. Lilly, *Natural History* 71, no. 1 (January 1962): 7.

110. Memorandum of 17 October 1961, Stanford University Library, Lilly Papers, box 3D2, file "China Lake—Dr. W. B. McLean."

111. John C. Lilly, *Man and Dolphin* (Garden City, NY: Doubleday, 1961), 157, 155.

cause Wood spelled out some of his frustration in a letter to Atz in October 1961.[112] And while Wood was neither an MD nor a PhD, he was himself well connected among biologists interested in large marine organisms, since he not only presided over a unique research menagerie at Marineland, but had also spent several years on an earlier appointment at Lerner in Bimini working with sharks, doing investigations of interest not only to marine biologists but also to the navy.[113] It was through these links that Wood found himself among the seventy scientists invited to Bimini in April 1963 for the Symposium on Marine Bio-Acoustics, where he had an opportunity to air (publicly and, one assumes, privately) his misgivings about Lilly's work. (Lilly was conspicuously absent from the Lerner symposium.)

In fact, Wood was the commentator at that Lerner session dealing with "special problems in cetacean research," and he used the floor to offer a thinly veiled deprecation of the research going on at an island not far away—at the CRI labs in neighboring Saint Thomas. Speaking of cetacean phonation, he declared, "I have no competence in the matter of linguistics, but my conditioned response to this is that we have to be extremely careful in our interpretation of these sounds. I think this is more likely to be fruitful than certain other work that is currently going on with regard to 'porpoise talk.'"[114] Wood's sly allusion to "conditioned response" played two ways: on the one hand, Wood was suggesting that he, as a lowly animal-care person, was hardly capable of true thought on this matter, and instead was merely parroting; but the reason this was funny was that the "porpoise talk" that so fascinated Lilly, as any good animal trainer knew, was merely a conditioned response—not indicative of intelligence, but perhaps even of the opposite.[115]

112. Gregg Mitman, *Reel Nature: America's Romance with Wildlife on Film* (Cambridge, MA: Harvard University Press, 1999), 176, 248.

113. Wood had a master's degree in biology from Yale University, which he completed after his service in the Army Air Force during World War II. See William E. Evans, "Forrest Glenn Wood 1919–1992," *Marine Mammal Science* 8, no. 3 (July 1992): 324–25.

114. *Marine Bio-Acoustics: Proceedings of a Symposium held at the Lerner Marine Laboratory, Bimini, Bahamas, 11–13 April 1963*, ed. William N. Tavolga (Oxford: Pergamon Press, 1964), 395.

115. One of the central debates joined over the issue of dolphin intelligence was the precise value of mimicry as an indicator of higher intellectual ability. Keller Breland, for instance, insisted that less intelligent animals mimicked more reliably and could be trained to do so faster. See Keller Breland and Marian Breland, *Animal Behavior* (New York: Macmillan, 1966).

When, after the Bimini conference, William Tavolga reviewed the field of cetacean phonation studies for his US Navy report on marine bioacoustics, he reflected the collective misgivings about Lilly's work, and he went out of his way to state that the "humanoid" dolphin sounds described by Lilly were simply "abnormal." As for cetacean communication, Tavolga emphasized the research value of the "quantitative" approaches being used by Dreher and others, but made no mention of CRI.[116]

As the negative reviews of *Man and Dolphin* began to surface, William Schevill penned Lilly a letter that nicely captures the stance of private ambivalence taken by a respected (navy) bioacoustician upon confronting this book—a book that was shaping an international public conversation that had grown so loud that it was getting harder, not easier, for those who studied whales and dolphins for a living to be heard. Schevill's strained collegiality is palpable:

> I've seen a number of hostile reviews of your book (I'm not hostile, just perplexed). It has struck me that many of those complaining are, in part, at least, less bright than you. This may be part of the trouble. My main problem is that I've not found you demonstrating the things that outrage the conventional. Your recent papers in Science, on the other hand, seem easier to understand.[117]

Then, in a closing that thrummed with multiple meanings (outer space, a rocky passage, trippy hipsterism, merely nautical?), Schevill signed off, "Happy Landings . . ."

Despite such quizzical private correspondence and some sniping in the scientific reviews, Lilly remained very much the limelight dolphin scientist in the booming area of dolphin research in August 1963, and he could afford to take the high road at the First International Symposium of Cetacean Research—no need to bicker, no need to slug these matters out, no need to present anything dashing.[118] Who else at the event would draw the spotlight merely by turning up? Moreover, Lilly had a new paper, on the distress

116. William N. Tavolga, ed., "Technical Report: NAVTRADEVCEN 1212-1, Review of Marine Bio-Acoustics, State of the Art: 1964" (Port Washington, NY: U.S. Naval Training Device Center, February 1965), 57.

117. Schevill to Lilly, 12 March 1962, Stanford University Library, Lilly Papers, box 3D2, file of replies to and letters on *Man and Dolphin*.

118. See Lilly to Brown, 13 September 1963, Stanford University Library, Lilly Papers, box 3D2, file "Brown, Mr. David," for Lilly's decision to serve as an uncontroversial in-

call of the bottlenose dolphin, that had just come out in *Science*, and CRI was sufficiently flush that it had just lured Gregory Bateson away from the Mental Research Institute in Palo Alto and placed him in charge of the Saint Thomas laboratory. (Lilly himself was moving back to Miami, in part, it would seem, because of domestic complications—his second wife preferred life in Florida.) To project the strength of his unique and high-profile research institution, Lilly brought what amounted to a delegation to Washington. In addition to Bateson, who, like Lilly, was on the program, CRI was represented by Lilly's longtime collaborator at NIMH, Alice Miller, a biologist with training in sensory physiology who had coauthored nearly all of Lilly's scientific papers; the photographer and filmmaker Frank Essapian, whom Lilly had recently hired away from Marineland of Florida; and Peter Morgane, a PhD in neurology from Northwestern University who had published nearly two dozen scientific articles on the limbic system of rats and done neurophysiological investigations of sleep, hunger, and thirst in rats, cats, and monkeys.[119]

In the session on communication, Lilly presented a short talk, which summarized the work of CRI and stayed firmly on the uncontroversial ground of echolocation clicks, avoiding any of the speculative reaches that attended his discussion of this subject in *Man and Dolphin*—for instance, that the sonar "vision" of dolphins permitted them to see "through" one another and thus to witness the internal anatomical states of their comrades (with attendant affective and nurturing implication—all dolphins were "transparent," for Lilly, in a creepingly moralized sense); or that they

terlocutor and session facilitator at the Washington meeting. It does appear, however, that he did, in the end, present some sort of short paper on the day.

119. Bateson presented a general reflection on research in ethology entitled "Problems in Cetacean and Other Mammalian Communication." This probably differed considerably from his published contribution to Norris's edited volume (Bateson, "Problems in Cetacean and Other Mammalian Communication," in *Whales, Dolphins, and Porpoises*, ed. Kenneth Stafford Norris [Berkeley: University of California Press, 1966], 569–78). The "Index of Observers" in the Symposium Program found in the Hubbs Papers (box 54, folder 46) lists a sixth CRI participant, Charles B. Severn, but he is not mentioned in Lilly's 1963 letter to Brown. I have not been able to trace him and do not believe his link to CRI was long lasting or particularly significant. Scott McVay is nowhere listed in these materials, but in an interview with me (19 September 2003), he reported having attended the symposium and offered several memories of the event (including that Bateson ostentatiously put aside his prepared paper and appeared to speak off the cuff). He was not yet, as I understand it, formally Lilly's employee, though he would join CRI shortly thereafter.

might be capable of blast transmissions of sonar data enabling them to rapidly communicate the configuration of three-dimensional spaces and their contents.

When his former Johns Hopkins expedition colleague Lawrence Kruger presented his paper entitled "Specialized Features of the Cetacean Brain," Lilly did not rise to the mild baiting. As Kruger put it, "Although I am very fond of dolphins too, they do appear unimpressive by these criteria [distribution and number of cell nuclei in the brain tissue], and I feel secure about that statement. My statement is not a unique one, for the same thing has been claimed by virtually all authors who have studied dolphin cortex."[120] Rather than engage on the subjects of brain anatomy or intelligence, Lilly allowed his new colleague Morgane to hold up CRI's position, and Morgane did reply to Kruger at the meeting, promising better-preserved and better-stained specimens of the dolphin brain in the near future, along with an approach to dolphin intelligence in which psychological and ethological tests would supplement a new and thorough neuroanatomy.[121] It was a crossing of swords, but Lilly appears to have kept quiet, preferring to interject statesmanlike comments at other sessions, including supporting (as a "medical man") the idea of collective research on mass strandings by cetaceans.[122] The real showdown would come on the last day of the symposium.

THE DOLPHINS THAT JOINED THE NAVY

In his opening remarks at the First International Symposium of Cetacean Research, L. Harrison Matthews nodded toward Lilly's much-ballyhooed book and some of the controversy it had aroused. In doing so he both adverted to the fact that the hip-booters were not at all oblivious to the brave new world of American dolphin science and offered something of an admonishment concerning what he saw as a potentially unhappy development in the field of cetology:

Now it seems that some people are proposing to prostitute their biological work on Cetacea and involve the animals in human inter-

120. Lawrence Kruger, "Specialized Features of the Cetacean Brain," in *Whales, Dolphins, and Porpoises*, ed. Kenneth Stafford Norris (Berkeley: University of California Press, 1966), 232–54, at p. 252.

121. Ibid., p. 509.

122. Ibid., p. 606.

national strife by training them to act as underwater watchdogs to guard naval installations from frogmen or to act as unmanned submarines. Intelligent as the animals may be, they are, unfortunately, not sufficiently intelligent to refuse cooperation and to treat their trainer to some of those characteristic underwater noises which, if produced in the air, would be regarded as gestures of contempt.[123]

The reference is to perhaps the most superficially outlandish of Lilly's meditations in *Man and Dolphin*, a gesture toward the "implications" of future work on communication with the small odontocetes:

Cetaceans might be helpful in hunting and retrieving nose-cones, satellites, missiles, and similar things that men insist on dropping into the ocean. They might be willing to hunt for mines, torpedoes, submarines, and other artifacts connected with our naval operations. They might also be willing to do scouting and patrol duty for submarines or surface ships, and they might carry their protagonist activities to the point where they can be used around harbors as underwater demolitions team operators.[124]

There was no one in the John Glenn Suite who was unfamiliar with this litany of potential military uses of cetaceans, since Lilly's speculations on these matters had rapidly reverberated through popular channels as well as through communities of specialists.[125]

One suspects that at this moment in the chairman's speech, Forrest G. Wood felt the prickly sense that all eyes were upon him. For in June of that year, just two months earlier, Wood had quit his position as the curator at Marineland of Florida and accepted a new job—as the head of the newly formed "Marine Biosciences Facility" of the Naval Missile Center (NMC), where he would jointly administer (with Thomas G. Lang, from McLean's Naval Ordnance Test Station) the emerging "Cetacean Research Facility" being constructed at Point Mugu. It was this nascent institution that would

123. L. Harrison Matthews, "Chairman's Introduction to First Session of International Symposium on Cetacean Research," in *Whales, Dolphins, and Porpoises*, ed. Kenneth Stafford Norris (Berkeley: University of California Press, 1966), 3–6, at p. 5.

124. John C. Lilly, *Man and Dolphin* (Garden City, NY: Doubleday, 1961), 219.

125. In fact, a newspaper article in the *Staten Island Advance* of 13 March 1962 had already "broken" the story that the navy was working with dolphins at Point Mugu.

grow, through several changes of name and location, into the US Navy's Marine Mammal Program, and it was under the auspices of this program that Wood and his collaborators would succeed, over the next decade, in realizing a not-insignificant number of Lilly's seemingly bizarre ideas about the potential military uses of dolphins and porpoises. As early as 1964 the navy trainers and cetacean researchers (with help from commercial animal-training guru Keller Breland) succeeded in getting a dolphin to work untethered in the open ocean and to return when signaled; in 1965 a navy-trained bottlenose aided the "aquanauts" living in the experimental underwater habitat Sealab II by shuttling matériel and messages to and from the surface; over the next three years several marine mammal deep-water marking and recovery programs were developed and at least two of these were operationalized as navy "systems"; and finally, by 1970–1971, a number of navy dolphins accompanied a specialized team of navy divers to Cam Ranh Bay, in Vietnam, where they were deployed in a program to stop Vietcong sappers.[126] Minesweeping dolphins remain a working component of shallow-water navy ordnance clearance teams, and dolphins still serve in the navy in several other ways.[127]

With this brief enumeration of the "achievements" of this program, I tip open the very most controversial and contested area of scientific research on cetaceans in the period 1960–1975. Because much of the navy's program was classified across these years, and because the program remains active (and substantially classified) at the time of my writing, it is difficult to do historical justice to this work. What can be said for certain is that the existence of the navy dolphin programs did much to heighten the countercultural tensions around cetaceans in the 1970s, and that in different ways the

126. See, for instance, Blair Irvine, "Conditioning Marine Mammals to Work in the Sea," *Marine Technology Journal* 4, no. 3 (1970): 47–52; see also Sam H. Ridgway, *Dolphin Doctor: A Pioneering Veterinarian and Scientist Remembers the Extraordinary Dolphin That Inspired His Career* (San Diego: Dolphin Science Press, 1987), and Forrest G. Wood, *Marine Mammals and Man: The Navy's Porpoises and Sea Lions* (Washington, DC: Robert B. Luce, 1973). The actual deployment in Vietnam (a seven-month tour of duty, apparently not very successful) remains clouded in rumor. The minimal (but presumably reliable) information available appears in a historical essay by the Public Affairs Officer at the Space and Naval Warfare Systems Center in San Diego, Tom LaPuzza, "SSC San Diego Historical Overview," accessed 3 March 2011, http://www.spawar.navy.mil/sandiego/anniversary. See also n. 157 below.

127. D. Graham Burnett, "A Mind in the Water: The Dolphin as Our Beast of Burden," *Orion* 29, no. 3 (May–June 2010): 38–51.

navy's work considerably strengthened—if tacitly—the notion of cetacean intelligence that gathered force across the period in question. The claims of an enthusiast like Lilly might (at least eventually) be dismissed as special pleading, but popular awareness of real, operational, no-nonsense navy mobilizations of trained dolphins tended to silence skeptics. That such awareness was hazy, and that the programs were shrouded in the mystery of armed forces security, enhanced, rather than diminished, the perceived significance of the work.

How did whales and dolphins become the totemic organisms of peaceniks, freaks, and ecoterrorists in the early 1970s? This is, as I have acknowledged, a complicated question, and it is not easy to give a satisfactory reply.[128] But part of the answer, I believe, lies in the crossed trajectories of Forrest Wood and John C. Lilly, who were, in August 1963, still something like colleagues, but whose shifting commitments in the years to come would mark out diametrically opposed positions in an increasingly vociferous set of debates about the relationship between human beings and cetaceans. Before turning to their definitive falling out at the Washington symposium, let me take a moment briefly to review the emergence of the navy's program, and in so doing show how closely entangled Lilly was with early navy work on these animals.

Despite the extent to which cetacean research at Point Mugu systematically delivered on virtually every aspect of Lilly's hypothetical program, Wood himself was at pains to deny, late in his career (in 1972), that "Lilly's ideas provided the motivation for the Navy to enter the field of porpoise research."[129] Rather, Wood insisted that the navy's earliest interest in this kind of work had to do not with communication, or intelligence, or "kamikaze porpoises," but rather with hydrodynamics and a set of biophysical investigations into dolphin propulsion.[130] Here Wood pointed to the fact

128. For a somewhat exasperated effort to answer this question, see Arne Kalland, "Management by Totemization: Whale Symbolism and the Anti-Whaling Campaign," *Arctic* 48, no. 2 (June 1993): 124–33.

129. Forrest G. Wood, *Marine Mammals and Man: The Navy's Porpoises and Sea Lions* (Washington, DC: Robert B. Luce, 1973), 209. However, see William Evans, *Fifty Years of Flukes and Flippers: A Little History and Personal Adventures with Dolphins, Whales, and Sea Lions, 1958–2007* (Sofia, Bulgaria: Pensoft, 2008), and Sam H. Ridgway, *Dolphin Doctor: A Pioneering Veterinarian and Scientist Remembers the Extraordinary Dolphin That Inspired His Career* (San Diego: Dolphin Science Press, 1987).

130. The expression "kamikaze porpoise" emerged in a press flap in 1966–1967 over rumors that dolphins were being trained to attach shell mines to enemy submarines

that in 1960, a full year before the publication of *Man and Dolphin*, the Naval Ordnance Test Station (NOTS) purchased a white-sided dolphin—promptly named "Notty"—from Marineland of the Pacific. Notty was the subject of a set of hydrodynamic and physiological observations through December 1961, when she died. These studies were conducted in several laboratories, including the tanks of Convair Hydrodynamics in San Diego, and the aerospace engineer Thomas Lang was in charge of the program (though a group from General Dynamics was also involved).[131]

Wood was certainly correct that this work was done and that it preceded Lilly's mainstream publications. The hydrodynamics researchers at NOTS (which was responsible for research on the navy's underwater weapons systems, particularly torpedoes) had gotten interested in a biophysical problem known as "Gray's paradox," which highlighted an apparent discrepancy between the observed speed of swimming dolphins and porpoises and the calculated energy output of such an animal.[132] It appeared that porpoises either had a capacity for physiological output wildly surpassing that of any other known animal (unlikely) or that they had achieved a hydrodynamic profile that essentially eliminated drag-producing turbulence. If the latter, which seemed more probable, then it remained to be determined what feature of their anatomy (general shape? skin quality?) or body mechanics (musculature? undulatory movements?) accounted for their remarkable abilities. Navy scientists were hooked.[133] Results of this work through 1963 were, however, inconclusive, in large part because it was difficult to get captive dolphins and porpoises to attain what were thought to be their top

---

and ships. I have a sheaf of newspaper and magazine material on this (mostly quite repetitive), but will omit the references here. It is impossible to say for sure whether this was ever contemplated (what was not contemplated?), though it seems unlikely that it was ever tried.

131. Thomas G. Lang, interview by the author, 18 September 2009. See Thomas G. Lang and Dorothy A. Daybell, "Porpoise Performance Tests in a Sea-Water Tank," Defense Technical Information Center, Tech. Rept. A.D. 0298742, 1 January 1963.

132. On Gray's paradox, see J. Gray, "Studies in Animal Locomotion: VI. The Propulsive Powers of the Dolphin," *Journal of Experimental Biology* 13 (1936): 192–99. For a very general overview, see Sir James Gray, "How Fishes Swim," *Scientific American* 197, no. 2 (August 1957): 48–65.

133. These issues have not been fully resolved. See, for instance, Peter W. Carpenter, "Dolphin Hydrodynamics: Gray's Paradox Revisited," opening keynote lecture, Third International Symposium on Ultrasonic Doppler Methods for Fluid Mechanics and Fluid Engineering, EPFL, Lausanne, Switzerland, September 2002.

speeds in the confinement of testing tanks.[134] Thomas Lang presented a detailed summary report on these investigations, entitled "Hydrodynamic Analysis of Cetacean Performance," at the Washington symposium, concluding that further research was needed.

While there would seem, then, to be grounds for Wood's insistence that navy research had nothing to do with Lilly's work, other evidence suggests otherwise. On the one hand, Lilly himself was involved in these sorts of hydrodynamic investigations from very early on. By 14 February 1960, he had already received a visit in Saint Thomas from a young aeronautical engineer at MIT, Quentin Wald, who wanted to "investigate the hydrodynamics of marine animals, and particularly the dolphin." While that project appears to have been stillborn, Lilly corresponded with both Loyal Goff and Sydney Galler at ONR about this research and its relevance to Gray's paradox, cetacean metabolism, and underwater propulsion. Lilly therefore appears to have been involved, at least peripherally, in early hydrodynamic work like that undertaken at NOTS.[135] Moreover, the navy's work with Notty was not entirely divorced from Lillyesque investigations of phonation and language ability. In fact, NOTS hired an "experimental psychologist" from UCLA, Ronald Turner, to work with Notty and to test different techniques for operant conditioning of her vocalizations.[136] This suggests that from the start, navy investigators at NOTS were interested in more than Notty's hydrodynamic profile—indeed, that they too had been piqued by Lilly's early suggestions about the possibilities of delphinid intelligence and communication.[137]

134. Attempts to get them to reach burst velocity included the playing of prerecorded killer whale phonation through submerged transducers in the testing tank; this was ultimately unsuccessful. In the end it was found that many of the estimated top speeds of porpoises and dolphins had been exaggerated, owing partly to the animal's capacity to bow-ride on ships going faster than they can actually swim.

135. See, for instance, Lilly's correspondence with Loyal Goff in the physiology branch of the Office of Naval Research, Stanford University Library, Lilly Papers, box 3D2, file "Wald, Quentin." Discussions went so far as to prompt Lilly, with ONR's encouragement, to look into the purchase of a boat that would be capable of towing a dolphin cage in open water.

136. See Turner's UCLA dissertation submitted in March 1962, "Operant Control of the Vocal Behavior of a Dolphin." Note that Ken Norris was on Turner's dissertation committee.

137. Lilly explicitly contested, in public, a narrowly "bionic" preoccupation with delphinid phonation in a lecture to an audience at the William Clark Memorial Library in Los Angeles (right in the navy dolphin labs' backyard) on 13 October 1962: "These

In the absence of further declassification of documents and archival finds, it remains impossible to say for certain the degree to which Lilly's early work inspired the first navy interest in porpoises and dolphins, but there can be no doubt that by October 1961, Lilly was in close contact with the most important figure driving the navy's marine mammal research program: Dr. William B. McLean, director of research at NOTS and, as noted above, an old friend of Lilly's from Caltech.[138] A remarkable memo in Lilly's papers dated 17 October 1961 records Lilly's notes on a long telephone conversation with McLean, in which the latter sketched the rapidly shifting profile of West Coast cetacean research. Lilly learned of the work of the Lockheed group and discovered that they were at odds with the China Lake/Tom Lang group, who were investigating laminar flow (Lilly noted to himself, "We have a copy of their report," indicating how closely he was connected to this hydrodynamic research). Moreover, McLean gestured at others who were trying to attract research funding for work on porpoises and dolphins: Ken Norris at UCLA, Pete Scholander at UCSD, and several people at UCLA's Brain Research Institute (all of whom had separate plans afoot for cetacean research programs), not to mention the momentum gathering behind the proposed new cetacean research facility at Point Mugu, about which Lilly was informed when the whole program was still aborning. One senses Lilly's despair at keeping track of who was who in a marginal note: "There are as many plans as there are groups."[139]

A few months later, Lilly would have an even clearer sense of the situation, because in February 1962 McLean invited him out to NOTS for several lectures. It appears that Lilly made a considerable impression there, since he subsequently received a number of letters from NOTS personnel enthusiastically pressing several very curious lines of inquiry. For instance, T. W. Milburn (the head of the psychology division at NOTS and the director of "Project Michelson") followed up in April with some brainstorming:

---

animals are not inanimate cold pieces of sonar apparatus. They use their ultrasounds and their high-pitched sounds interpersonally [sic]." See Ashley Montagu and John C. Lilly, The Dolphin in History (Los Angeles: William Andrews Clark Memorial Library, 1963), 44.

138. McLean was also on the planning committee for the First International Symposium on Cetacean Research, and he would, in 1964, visit Lilly at CRI.

139. Memorandum of 17 October 1961, Stanford University Library, Lilly Papers, box 3D2, file "China Lake—Dr. W. B. McLean."

Reasoning that the machine is such an integral part of man's culture, and that many men find it enormously pleasant to interact with machines, I have wondered whether it might not be possible to develop some mechanical equipment that a dolphin might use. Again, you will see how indegted [*sic*] I am to you for the short conversations we have had thus far, thinking in terms of the short drive, I wondered whether it might be at all feasible (and I realize that the idea may sound a little fantastic) to arm dolphins with some sort of weapon that would enable them more easily to attack shark.

He continued the letter with further thoughts:

Bill McLean and I have been discussing the possibility of developing some dolphin toys, large, complex, mechanical devices, that might be of interest to dolphins even in the open seas, that would involve some kinds of buttons to push that would generate running water, perhaps with one trained dolphin teaching others. I would very much appreciate your own reflections as to whether this sounds too wild to contemplate.[140]

These missives seem so odd that it is difficult to resist the speculation that they are in fact written in some sort of circumlocutory code—a suspicion enhanced by a single NOTS-NMC Off-Base Authorization Form that appears in this file of Lilly's correspondence, indicating that in the fall of 1962, Lilly was scheduled to receive a registered "Secret" visit at CRI from "Louis R. Padberg," a navy electronics engineer working through the NOTS-NMC life sciences department.[141]

140. T. W. Milburn to Lilly, 10 April 1962, in ibid. Note that Milburn, who would later write about the psychological dimensions of deterrence, appears to have attended the First International Symposium on Cetacean Research; in other correspondence he expressed interest in Lilly's isolation tanks. Project Michelson was an integrated research program on strategic deterrence. For a sense of some of the concerns, see Louis D. Higgs and Robert G. Weinland, *Project Michelson Preliminary Report*, Technical Progress Report 309 (China Lake, CA: U.S. Naval Ordnance Test Station, 1963).

141. Stanford University Library, Lilly Papers, box 3D2, file "China Lake— Dr. W. B. McLean." There is, among these materials, a reference to additional documents in a "vault"; I was not able to locate those materials in the Lilly Papers, and it seems

By this time the navy's emerging research program with dolphins and porpoises had been the subject of several newspaper stories and even a short promotional newsreel.[142] The publicity aimed, at least in part, to help raise funds for the work—both within the navy and through dedicated congressional appropriations. Lilly complained to McLean that all this coverage neglected to give proper credit to his own foundational work in the area of communication research, and McLean agreed to see to it that changes were made in the promotional material.

By October 1962 the rudiments of a joint NOTS-NMC research facility for work on dolphins and porpoises had been realized at Point Mugu. After the speedy death of the first two subjects, a full-time veterinarian, Sam Ridgway, was hired. Though at this time he had almost no experience with marine mammals, Ridgway (who had been much taken with Lilly's *Man and Dolphin*) would go on to be a leading figure in the biology of dolphins and sea lions and a mainstay of the navy's program.[143] He would be joined about six months later by Wood, to whom he would report.

According to Ridgway, part of the reason Wood was chosen to run the Naval Missile Center side of the joint NOTS-NMC program was that he was already known to take a dim view of the more visionary elements of Lilly's dolphin science. Apparently McLean (at NOTS) was seen by some at NMC to be overly enthralled by his former classmate and by the future of military dolphins. NMC, pressed by Clinton Maag, the acting head of the life sciences program there, wanted someone "to keep their feet on the ground, and to run a practical dolphin research facility."[144] After an inquiry to Sydney Galler at ONR, Maag and NMC settled on Wood. By these lights, he was

likely they were destroyed. Padberg was among the attendees of the Lerner marine bioacoustics symposium.

142. "Science, A Porpoise with a Purpose," Hearst Metrotone News film reel, black & white, 16 mm., 1962; held by the Cuyahoga Public Library, Ohio. This film is not to be confused with the 1964 production, also US Navy–Supported, "Dolphins That Joined the Navy," narrated by Glenn Ford. Both of these short films are interesting, particularly when viewed in the context of Mitman's work on Cold War nature documentaries. See Gregg Mitman, *Reel Nature: America's Romance with Wildlife on Film* (Cambridge, MA: Harvard University Press, 1999), particularly chaps. 4–6.

143. For Ridgway's early interest in Lilly, see Sam H. Ridgway, *Dolphin Doctor: A Pioneering Veterinarian and Scientist Remembers the Extraordinary Dolphin That Inspired His Career* (San Diego: Dolphin Science Press, 1987), 10.

144. Ibid., p. 41.

hired to be, in a way, the anti-Lilly.[145] For his part, Lilly always insisted that he had a hand in putting Wood's name forward for the position and that he never precisely understood why Wood turned on him.[146]

CETOLOGICAL SHOWDOWN

At this point we are ready to circle back to Washington, to the final day of the First International Symposium on Cetacean Research, and to recover the very public confrontation that closed the festivities. In the end, David Brown—the new curator at Marineland of the Pacific, whom Lilly tried to persuade to attend the meeting—decided to stay in California, pleading the demands of a sick walrus. Because he stayed at home, we have a very detailed record of the closing showdown that took place at the conference between Wood and Lilly, a tête-à-tête that would otherwise be lost, since it was excised from the published proceedings of the symposium. Shortly after returning to Miami, however, Lilly took pen in hand and wrote Brown a long postmortem on the meeting, focusing on a description of how Wood had chosen "to bring his enmity into the open and to try to cast doubt upon me as a scientist."[147] Because Wood appears never to have written about the episode explicitly, we are largely dependent on Lilly's account of the affair, which cannot be accepted at face value, but which certainly conveys the tenor of the confrontation (and his account is both confirmed and extended by oral history).

At issue was Lilly's most recent publication in *Science*, "Distress Call of the Bottlenose Dolphin: Stimuli and Evoked Behavioral Response," where Lilly refined and amplified his characterization of the particular rising and falling tone that, he argued, served as a distress call that brought other *Tursiops truncatus* to the aid of the calling animal. Though the paper summarized observations on twenty-three animals made between 1955 and 1962, Lilly focused the bulk of its narrative on two episodes from the 1955 Johns Hopkins expedition, describing how this call drew attendant captive dol-

145. Sam H. Ridgway, interview by the author, 9 September 2003, Point Loma.

146. Francis Jeffrey and John Cunningham Lilly, *John Lilly, So Far . . .* (Los Angeles: Jeremy P. Tarcher, 1990), 142. Lilly's claim here seems unlikely, however, given that he wrote to David Brown in June 1963 saying that he had "not heard that Woody was joining the Navy at Point Mugu." See Lilly to Brown, 18 June 1963, Stanford University Library, Lilly Papers, box 3D2, file "Brown, Mr. David."

147. Lilly to Brown, 13 September 1963, Stanford University Library, Lilly Papers, box 3D2, file "Brown, Mr. David."

phins to support an injured tankmate, holding it at the surface for several hours to allow it to breathe. The subject of this paper came up after David and Melba Caldwell's presentation on caregiving behavior among bottle-nose dolphins. In the discussion that followed (as Lilly reported it), David Caldwell asked Lilly to review the substance of the *Science* article for those who had not seen it. No sooner had he done so, and sat down, than

> Wood jumped to his feet (obviously quite upset) and said that two of the cases mentioned in my paper in *Science* he knew about at Marine Studios and was present in 1955. Subsequently, he had searched his protocols from the 1955 expedition and had found no reference whatsoever to distress calls; therefore he wished to cast doubt on my observations: those two just didn't occur.

Lilly reported to Brown that at this point, he (Lilly) asked the chairman (Keller Breland, filling in for Margaret Tavolga) for permission to reply immediately, "somewhat angry," but also glad that what he thought of as Wood's simmering resentment over *Man and Dolphin* was now manifest for all to see:

> My reply, in substance, was that:
> 1. The data on which I was basing my observations was on tapes and on motion pictures, taken at the time of the 1955 expedition (and many later recordings).
> 2. Retrospective analysis done months, and in some cases, years later, when we had learned more about the distress call, showed that it was happening unequivocally in those circumstances also.
> 3. The distress call had not yet at that time (1955) been separated from all the other kinds of whistles and clickings that were going on.
> 4. I publicly invited Wood down to Miami to go over the evidence, including motion pictures showing the distressed animal and Wood present!

According to Lilly, Wood neither replied nor accepted.[148] Explaining his position to Brown, Lilly stood on both the veracity and the priority of his

---

148. For Wood's objections to Lilly's writings on his early experiments in Florida, see *Marine Mammals and Man: The Navy's Porpoises and Sea Lions* (Washington, DC: Robert B. Luce, 1973), p. 13 n. 3. Ridgway is the person who most helpfully amplified the

published findings: "I was not talking about distress calls with Wood or with any of the other 1955 participants. He assumes because the 'summary conference' (recorded on my machine) did not include these data, that they didn't exist!"

Lilly felt certain he had deflected the attack, but nothing, he felt, could repair the damage. Now that he understood the extent of Wood's hostility, "further cooperation with him is impossible":

> He is really quite fantastic, and apparently quite unrealistic.
>
> Wood's comments were in extremely poor taste, and obviously he has turned out to be a dedicated opponent.

Moreover, taking the temperature of the gathering in Washington, Lilly found that Wood, however "fantastic," was apparently not alone; rather, he seemed to be gathering allies. "I said later to Ken Norris," Lilly reported, that "each of the four 'cetacean' people have [sic] nominated themselves to the position of my enemy, but I am refusing to elect any of them to that position."[149]

Lilly did not, however, tell Brown about a wisecrack that came from the floor of the symposium at a tense moment in the showdown with Wood about whether *Tursiops* had a distress call. As Lilly insisted that he had unimpeachable evidence that the animals were communicating distress signals, William Schevill piped up from the back of the room, "Well, John certainly has the most distressed dolphins in the world!" The remark, sniping at Lilly's head-hammering experimental observations (as well as his failure to keep his animals alive in Saint Thomas), and perhaps alluding to emerging rumors of Lilly's work with LSD, brought down the house.[150]

Wood-Lilly dustup when I interviewed him on 9 September 2003 at Point Loma; I have also discussed their confrontation at the symposium with Carleton Ray and Thomas G. Lang, both of whom were also in attendance.

149. He is presumably thinking here of Wood and the Tavolgas. The fourth is a little harder to guess, since, in view of later developments, there were plenty to choose from. It may well have been Schevill, who would lose patience with Lilly in the years to come; the same could be said for Norris himself.

150. The incident was recalled by Sam Ridgway in an interview by the author, 9 September 2003, Point Loma. It is not clear exactly when Lilly began his LSD experimentation on the animals, but he was corresponding with C. Henze at Sandoz in January 1963 (about the anesthetic Tricaine methanesulfonate, or MS-222, which Lilly was using in the dolphin work), and it would appear that Henze actually visited CRI that year. The earliest concrete evidence I have of Lilly's dosing the animals with LSD-25 is a letter to

Lilly did write to Brown about another "move" made by Wood at the symposium, one that Lilly saw as a clear bid by Wood to place himself and the new navy research facility at the center of the field, leveraging Lilly and CRI aside. With the "young veterinarian" Ridgway beside him, Wood

> issued a call to serve as a central clearing place for solutions to practical problems on dolphins through the Navy at Point Mugu. I think this would be an inappropriate place (with Wood at the head of it) to do such clearing of information and to issue a bulletin. There is no need in the field for such a place. All of those who are really interested in finding out the true situation in regard to the dolphins and whales ought to steer clear of this unprofessional approach. It looks like an attempt to control the field without really contributing anything himself.

And, for Lilly, this was in keeping with Wood's character. In Lilly's view, Wood was a scientific wallflower: "All through the conference," Lilly explained to Brown, "Wood kept bringing up his 'twelve years' experience at Marine Studios' (none of it published!) . . . he just won't contribute as a scientist."[151]

This was not entirely fair, and indeed, Wood's irritation with Lilly's 1963 *Science* article must have been due, at least in part, to the fact that Lilly nowhere credited Wood in the paper, and nowhere in any of his *Science* publications, from 1961 to 1963, did Lilly cite Wood's 1953 article that dealt with phonation and behavior among captive bottlenose dolphins.[152] Wood thus had a quite legitimate reason for seeing Lilly as something of a scientific interloper, one who had snatched the limelight with borrowed (if reworked) material.

Henze of 26 January 1965, in which Lilly alludes to experimental work already conducted. For all of this, see Stanford University Library, Lilly Papers, box 3D2, file "Sandoz."

151. All citations from this letter are from Lilly to Brown, 13 September 1963, Stanford University Library, Lilly Papers, box 3D2, file "Brown, Mr. David."

152. F. G. Wood, "Underwater Sound Production and Concurrent Behavior of Captive Porpoises, *Tursiops truncatus* and *Stenella plagiodon*," *Bulletin of Marine Science of the Gulf and Caribbean* 3, no. 2 (March 1953): 120–33. This article includes a discussion of a young bottlenose whistling for its mother and cites McBride and Hebb (see n. 98 above) on the potential "language" value of such phonation; if the term distress call is not explicitly used, the concept is absolutely invoked.

Perhaps aware of this, and of the real challenge that Wood and his growing consortium of anti-Lilly colleagues posed, Lilly went to work shortly
after the symposium, trying to undermine Wood's position, not only with
Brown, but with others in still more powerful positions. Several days after
writing his letter to Brown, Lilly sent a long communication to McLean
at NOTS, who had been at the symposium, saying that he was "sorry that
F. G. Wood decided on the last morning to publicly take a shot at me and
at my scientific integrity . . . a show of poor judgment on his part." Not
only was Wood wrong on the facts concerning the distress call, Lilly insisted, but the proposal to make the navy's establishment at Point Mugu a
"central clearing point" for international cetology was a terrible mistake,
and Lilly would never participate.[153] As far as he was concerned, "any bona
fide qualified scientist of good will who wants to work with dolphins or
who is already working with dolphins can come to our laboratory as a
visiting scientist and pursue any aspect of delphinology that he wishes and
learn what we have learned." As for Wood, "apparently he is putting out
a lot of nonsense there [at Point Mugu] about what my research activities have been and are, what my ideas are, and the part that he played in
this in 1955 and 1957."[154] It was not exactly clear what Lilly could do to stop
him, and there is no letter from McLean in the folder replying to Lilly's
complaints.[155]

153. It is unclear exactly what relationship McLean had at this stage to the joint NOTS-
NMC cetacean research facility at Point Mugu. Later, in 1967, McLean would be put in
charge of the whole establishment, which was on its way to San Diego and consolidation
into what would soon be named the Naval Undersea Center (NUC), but it seems likely
that at this point (1963) he saw the emerging institution at Point Mugu as falling into the
sphere of influence of NMC through the appointment of Wood. Without access to navy
correspondence files it is difficult to sort out the various interest groups, pressures, and
conflicting visions that attended the emergence and changing administrative structure
of the US Navy Marine Mammal Program. The best summary I have found is the (not
disinterested) appendix to Forrest G. Wood, *Marine Mammals and Man: The Navy's
Porpoises and Sea Lions* (Washington, DC: Robert B. Luce, 1973).

154. Lilly to McLean, 25 September 1962, Stanford University Library, Lilly Papers,
box 3D2, file "China Lake—Dr. W. B. McLean."

155. McLean would visit Lilly at CRI in early 1964, and he sent along a copy of
Breland's report on training a navy dolphin to wear a harness. Still, it appears that
by 1967, when McLean took over the project, Lilly was untouchable. The McLean
file in the Lilly Papers ends altogether in April of that year. See file "China Lake—
Dr. W. B. McLean" (Stanford University Library, Lilly Papers, box 3D2).

A turf war had emphatically been joined, and Lilly was doing what he could to edge Wood out and to keep CRI firmly at the center of both research and funding contracts. When Ridgway and Wood began distributing "information request forms" with a Point Mugu return address in late October, Brown rallied to Lilly's side, sending his packet back to Point Mugu with a huffy note that called the whole enterprise "an unprofessional concept proposed by people who had little or nothing to contribute to the subject themselves." He copied Lilly on this and sent along an expression of support: "Your brush with Wood must really have been something to behold. I gather from what you and David Caldwell say, he did himself irreparable damage, and this is really quite a pity. Certainly he has not been the most productive man in the field, and has always appeared to me to lack an empathy with dolphins."[156]

In fact, it would work out otherwise. In the end it would be Wood, who died in 1992, who would go on to be remembered as a founding father of the Society for Marine Mammalogy and an active contributor to the Scientific Advisory Committee of the US Marine Mammal Commission. He would publish several dozen articles in scientific journals, and he would successfully steer the navy's Marine Mammal Program through a period of considerable growth and public controversy (including attacks by environmentalist and pacifist vandals who damaged equipment and released trained animals). Putting aside some of the suspect "applications" of navy cetacean research (there have always been unquashable rumors that the navy experimented with "lethal" swimmer interdiction programs—that is, that they trained dolphins to kill enemy swimmers),[157] Wood could always claim that the fundamental science of marine mammals—their behavior, physiology, and therapeutic biochemistry—achieved at Point Mugu and its successor institutions was rivaled by no institution in the United States or elsewhere.

156. Brown to Lilly, 30 September 1963, Stanford University Library, Lilly Papers, box 3D2, file "Brown, Mr. David."

157. For a sense of the controversy around the Vietnam deployment, see the front-page story by William E. Blundell, "Great Divin' Dolphins! Are They in Battle against the Red Peril?" *Wall Street Journal*, 9 May 1972. I handle this episode in some detail in D. Graham Burnett, "A Mind in the Water: The Dolphin as Our Beast of Burden," *Orion* 29, no. 3 (May–June 2010): 38–51. See also David Helvarg, *Blue Frontier: Saving America's Living Seas* (New York: W. H. Freeman, 2001), 59–63.

Lilly, meanwhile, would shortly be anathema to the scientific community, and his passage from leading light to leper came with surprising speed.

THE SHOOTING STAR

The arc of Lilly's fall can be concisely traced on a plane defined by two of the era's significant axes: sex and drugs. Let's do sex first. In 1961, *Man and Dolphin* presented the Saint Thomas laboratory of CRI as a kind of scientific *Swiss Family Robinson*, where Lilly and his beautiful wife and their six children (five from their combined previous marriages, one of their own) shared the labor of scientific life with bottlenose (this sort of thing was already a commonplace in primate work like that of Keith and Catharine Hayes). As the jacket copy on *Man and Dolphin* explained,

> A large fraction of the family home is taken up with the work of the program on the dolphins. The whole family participates: the elder boys have taken motion pictures of the animals in the experimental situations; the younger group have been involved in feeding and in swimming with the animals.
>
> Elisabeth handles a major portion of the administrative load of the institute, and helps with the research and writing.

But this scenario proved more advertising than reportage. Unhappy in their island fastness, Elisabeth and the children had decamped for Miami for good by the end of 1963, and the Saint Thomas lab promptly began its drift from the outer orbit of the military-industrial complex onto the hyperbolic arc of what would come to be called, in a general shorthand, "the sixties." Absent the straitening forms of domesticity, the CRI facility in the Caribbean began to look less and less like an outtake from a wonkish version of *Flipper* and more and more like a bachelor crib for randy scientists seeking sun and surf.[158] The aging roué Aldous Huxley visited in these years, and Bateson (known for his rakish open shirts and nonconformist

---

158. It is interesting to think about this transformation in the context of a growing literature on the question of domesticity and the laboratory. I have been inspired here in part by Deborah Harkness, "Managing an Experimental Household: The Dees of Mortlake and the Practice of Experimental Philosophy," *Isis* 88, no. 2 (1997): 247–62.

sensibility, and having recently embarked on a third marriage to Lois Cam-
mack, twenty-five years his junior) often presided in the dolphin pool, entertaining an irregular flow of the rich, brilliant, and/or curious. A youthful
Carl Sagan made his way down from Harvard, having gotten to know Lilly
through their mutual interest in "breaking through" to nonhuman species
(he and Lilly together created a semisecret society of SETI researchers who
called themselves "The Order of the Dolphins," wore a tie clip–like dolphin insignia, and sent each other coded messages to test their readiness
for extraterrestrial contact).[159] He thoroughly enjoyed himself, and later
claimed that while doing so he made the acquaintance of a young woman
named Margaret Howe, then working at one of the resorts on the island.[160]
Invited into the loose CRI community, the attractive and tomboyish Howe
would eventually be hired by Bateson to "manage" the establishment when
he decamped for Hawaii. She would play the crucial role in Lilly's program of "chronic contact" experiments in 1964 and 1965—experiments (if
this is the right term) that would appear prominently in *The Mind of the
Dolphin*, Lilly's 1967 sequel to *Man and Dolphin*. The published account of
this research—which narrates the way Howe, in a skin-tight leotard, spent

159. This would be wonderful material to pursue, and a point of departure would
be the file entitled "Order of the Dolphins" in Stanford University Library, Lilly Papers,
box 3C2[D1]. This group (born of the first Green Bank conference in November 1961)
eventually included most of the leading figures in what would become the discipline
of exobiology (Frank Drake, Melvin Calvin, J. B. S. Haldane, and others). These gentlemen (and a few women, too) entertained each other by circulating elaborate ciphers
like those "that might be received from another civilization in space" and generally
mused about the possibilities for life on other planets. For some context on all this,
consider James E. Strick, "Creating a Cosmic Discipline: The Crystallization and Consolidation of Exobiology, 1957–1973," *Journal of the History of Biology* 37, no. 1 (2004):
131–80, and Steven J. Dick, *The Biological Universe: The Twentieth Century Extraterrestrial Life Debate and the Limits of Science* (Cambridge: Cambridge University Press,
1999).

160. Lilly himself says that Bateson "discovered" Howe (Francis Jeffrey and John
Cunningham Lilly, *John Lilly, So Far . . .* [Los Angeles: Jeremy P. Tarcher, 1990], 118), but
Sagan's role is outlined in William Poundstone's biography *Carl Sagan: A Life in the Cosmos* (New York: Henry Holt, 1999). Margaret Howe Lovatt denies Poundstone's account
(which is based, according to Poundstone, on Sagan's own writings and, Poundstone
says, confirmed by an interview with Lilly), and says she heard about the CRI work
while working at a hotel on the island and went there on her own initiative, where it
was Bateson who generously folded her into the research program on a volunteer basis.
Margaret Howe Lovatt, interview by the author, 25 August 2009.

several weeks with a male *Tursiops* in a living facility flooded to knee depth (her mouth brightly painted, to help the dolphin read her lips)—wends its way to a "climax" of sexual contact, offered as something like a proxy for increased interspecies mutuality (figure 6.18).[161]

The book, quite possibly because of this frisson of aquatic bestiality, received more public attention than its weaknesses (of organization, clarity, and purpose) perhaps merited, but it was nevertheless not the sort of monograph from which a conventional scientific reputation could easily recover. Not only did the text entertain the notion that dolphins incline toward a "Polynesian" rather than an "American" model of sexuality, it did so in the context of a sweeping indictment of the impoverishingly limited worldview of the human species, particularly in its North American variety. The difficulties of communicating with dolphins are laid firmly at the feet of humanity, who are presented as a sickly lot, driven by fear and violence, inclined to deception and repression.[162] Our only hope, Lilly suggested, was to listen to the dolphins, since (as he put it in an epigraph), "Through dolphins, we may see us as others see us."[163] There is a lingering sense throughout the book that dolphins offer humans a kind of higher psychotherapy.[164] Dredging up everything from Montaigne to William James to the Kinsey reports, Lilly presented sexual liberation and what he called "wet courage" as the preconditions for transcending the species barrier: "The purpose here is to free up one's own mind to see the new possibilities of feeling and

161. See chap. 14 of John C. Lilly, *The Mind of the Dolphin: A Nonhuman Intelligence* (Garden City, NY: Doubleday, 1967). It is interesting to note that Lilly had Howe read *Planet of the Apes* (Pierre Boulle, trans. Xan Fielding [New York: Vanguard Press, 1963]) to prepare for her chronic contact work. This fact, as well as the observations about lipstick and so forth, can be confirmed in the remarkable manuscript files of this work in Stanford University Library, Lilly Papers, box 6A1–B1, file "1965, St. Thomas."

162. John C. Lilly, *The Mind of the Dolphin: A Nonhuman Intelligence* (Garden City, NY: Doubleday, 1967), 128.

163. Ibid., p. xvii.

164. In its original formulation, this idea came from Bateson, who wrote to Lilly shortly after reading *Man and Dolphin* to propose that, if Lilly was right about dolphin intelligence, there was reason to think that these animals had evolved to apply the bulk of their cognitive capacity to the social world, rather than to the material world (roughly speaking, because they had no hands). Bateson suggested this might mean that the dolphins would make, if we could speak to them, ideal psychotherapists for humanity, so obscenely obsessed with things, and so inept in relationships. See Bateson to Lilly, 16 October 1961, Stanford University Library, Lilly Papers, box 3D2, specifically the file of replies to and letters on *Man and Dolphin*.

FIGURE 6.18 Chronic contact: Margaret Howe with her *Tursiops* companion. (From Lilly, *Mind of the Dolphin*, after p. 146.)

thinking without the dry civilized structures."[165] The process amounted to a kind of ecstatic ethnography, one that made use of "hypnosis, drugs," and technology to conjure up the *Umvelt* of creatures that see with sound and can therefore (?) speak in images.[166] It was necessary to develop our own "latent 'acoustical-spatial thinking,'" Lilly asserted, and to this end he sketched the notion of "a dolphin suit with built-in, three-dimensional, sonic and ultrasonic emitters and receivers," and further imagined a fully synesthetic approach to the translation of dolphin phonation:[167]

> The internal picture which the dolphin can then create while sounding slash calls, the internal picture which he creates of his surroundings in terms of beat frequencies coming stereophonically combined from the two ears, must be a very interesting kind of picture. It is as if to us the nearby objects emitted a reddish light and the further objects emitted a bluish light, with the whole spectrum in between. We might see, for example, a red patch very close by and then a dimmer, blue patch in the distance farther away . . . a blue background downward symbolizing the bottom, a red patch up close meaning a fish nearby, and a large green object swimming between us on the bottom meaning another dolphin.[168]

Lilly—suspended in his flotation tank, listening to the whine of the hydrophones—was drifting into the trippy world of stereophonic psychedelia.

And trippy is meant here in the literal sense. By his own account Lilly first took LSD late in 1963, with Constance Tors (wife of Ivan, director of *Flipper*), on a visit to Los Angeles.[169] He wrote subsequently that this initial exposure consisted of a pair of "classic, high-energy" trips "filled

165. John C. Lilly, *The Mind of the Dolphin: A Nonhuman Intelligence* (Garden City, NY: Doubleday, 1967), 170.

166. Ibid., p. 91.

167. Ibid., p. 135.

168. Ibid., p. 152.

169. Lilly had Hollywood links. In 1961 the glamorous actress Celeste Holm, fascinated by news reports of Lilly and his talking dolphins, sought him out while performing a run at the neighboring Coconut Grove. Later her son, Ted Nelson, would spend a year working at CRI, before becoming one of the leading figures at the intersection of information technology and the counterculture. For these links, see Lawrence S. Kubie to Elizabeth and John Lilly, 14 September 1961, and Elisabeth Lilly to L. S. Kubie, 6 December 1962, Stanford University Library, Lilly Papers, box 3A1–C1, file "Kubie, Lawrence S."

with fantastic personal and transpersonal revelations and terrific intellectual breakthroughs."[170] A less fantastic trip followed (complete with an unhappy psychodrama of his failing second marriage), and then, in May 1964, shortly after addressing the annual meeting of the Acoustical Society of America, a near-death experience linked to LSD experimentation.[171] Undaunted—indeed, it would seem, fascinated—Lilly apparently secured additional LSD through his links to NIMH and began a series of some twenty self-dosing experiments at the CRI laboratory in Saint Thomas between 1964 and 1965.[172] These seem to have included at least some injected exposures during which Lilly isolated himself in his flotation tank, Margaret Howe (who refused to try LSD with him) apparently serving as his "safety man" [sic]. Sources suggest that by January 1965 Lilly was also injecting the dolphins with the drug, nominally to test whether it had any effect on their vocalizations, though these experiments do not seem to have been closely controlled.[173] In this period he drafted several versions of research protocols along these lines, which served as applications to Sandoz for relatively large shipments of additional LSD-25.

I think it is important to acknowledge the difficulty of entirely freeing these aspects of Lilly's story from a slightly ludicrous taint: LSD? to dolphins? And a woman in a *tank*, giving them *hand-jobs*?[174] Having pre-

170. Francis Jeffrey and John Cunningham Lilly, *John Lilly, So Far . . .* (Los Angeles: Jeremy P. Tarcher, 1990), 134.

171. In his autobiography (ibid., p. 135) this episode (which involved a coma, hospitalization, and, apparently, some small permanent damage to Lilly's eyesight) is blamed on an improperly washed syringe.

172. The number twenty is Lilly's own (ibid., p. 139); Margaret Howe Lovatt recalls many fewer. Interview by the author, 25 August 2009.

173. Margaret Howe Lovatt recalls one occasion on which Lilly used a jackhammer on the rock wall of the pond in which an LSD-dosed dolphin was swimming, apparently to try to get a rise out of the animal, which was otherwise not behaving in a particularly striking way. Interview by the author, 25 August 2009. Lilly himself cites a "project report" on this work that I have not been able to find in its original form (*The Human Biocomputer: Programming and Metaprogramming [Theory and Experiments with LSD-25]* [Miami: CRI, 1967; Scientific Report #CRI0167]), but it is reasonable to assume that much, if not all, of the content of this document appears in Lilly's later published writings under essentially the same title: *Programming and Metaprogramming in the Human Biocomputer: Theory and Experiments*, available in a second edition (New York: Three Rivers Press, 1987).

174. There is, of course, a fictionalized version of these events: Ted Mooney's novel *Easy Travel to Other Planets* (New York: Farrar, Straus, & Giroux, 1981).

sented lectures on this material on several occasions, I am familiar with the collective smirk by which an audience reflexively relegates the endgame of the Lilly saga to the unproblematic category of period burlesque, and I am myself not wholly immune to this reaction. Nevertheless, it is worth making the effort to see these most extravagant aspects of Lilly's work without the distorting glaze of hindsight. In the early 1960s there were quite a few researchers in biology and psychology departments giving LSD-25 to various animals and observing the effects.[175] There were also, of course, formal experimental investigations with the drug that made use of human subjects, and self-dosing was considered by many a necessary preparation for therapeutic or investigative prescription. Furthermore, as I have suggested above, both Lilly's LSD experiments and the chronic contact experiments with Howe can be understood not as madcap "1960s-style" *antitheses* to the buttoned-down world of military bioscience, but rather as their very apotheosis. LSD was a notable tool in the kit of Cold War scientists of mind and behavior, and they understood it to be an instrument for reducing the inhibitions of those with whom they wanted to talk. Lilly wanted to talk with dolphins; therefore LSD presented itself as a very plausible approach. Similarly, "chronic contact" was a recognized technique for "winning over" recalcitrant or taciturn persons of interest, and careful management of erotic potentials was a nontrivial element of some of these protocols.[176]

These contextualizing observations aside, there can be no doubt that Lilly crossed a set of lines between 1963 and 1966. Were they invisible lines? Perhaps, though a proper answer to this question would thrust us firmly into the middle of some very serious and difficult historical debates about the cultural upheavals of the second half of the 1960s—debates that on the whole have yet to extricate themselves from the political stalemates the era itself did much to define. We do not need to resolve those thorny matters to know that things were changing quickly in these years, and that the transgressive passages of figures like Lilly were doing much to clarify new boundaries. Timothy Leary, of course, is paradigmatic here, and his own

175. Lilly corresponded with some of these scientists. See H. A. Abramson to Lilly, 23 November 1964, Stanford University Library, Lilly Papers, box 3C2[D1], file "Abramson, H. A."

176. It should be noted that Margaret Howe Lovatt claims that she herself was the primary proponent of the chronic contact work. Interview by the author, 25 August 2009.

story parallels (and ultimately intersects) that of Lilly in interesting ways.[177] It was in the spring of 1963, right about the time that Lilly was getting ready for the First International Symposium on Cetacean Research, that Leary and Richard Alpert (later Ram Dass) were dismissed from Harvard under a cloud of opprobrium that had gathered over what appeared to be their overzealous promotion of LSD experiences. Legal restrictions on the drug followed, and by 1966 LSD-25 was so tightly regulated in the United States that there were effectively no open research programs making use of the product. By that time Sandoz had issued a recall for outstanding orders of the drug, and a rising district attorney named G. Gordon Liddy was working overtime to stamp out the Leary-centered experiment in drug-addled alternative living centered at Millbrook, in upstate New York, subject of Tom Wolfe's *The Electric Kool-Aid Acid Test.* Leary himself would soon be jailed. In one sense, the party was over. In another sense, it was just getting going.

And that sums up Lilly's position by 1967 as well. Mounting scandal and critical scientific assessments of CRI's work by increasingly hostile peers (including Kenneth Norris and others in attendance at the First International Symposium on Cetacean Research) backed Lilly into a defensive posture and left him lashing out at his opponents.[178] But to no avail. With the emerging revelations about his unconventional experiments, and no stream of peer-reviewed publications to back up his showboating public claims, his allies could no longer defend him. With his NSF monies soon terminated and other federal agencies requesting the return of loaned equipment, financial matters at CRI reached a critical point. By the end of the year, Lilly had been forced to shutter both the Saint Thomas and Miami laboratories, though not before five of his captive dolphins died (Lilly called it the result of a "hunger strike" on their part; neglect seems more likely); three others were later released into Miami Harbor. By May 1968 Lilly, separated from Elisabeth (they would later divorce), wrote to his literary agent in New York that he was "looking for a job." He headed out to the

177. Lilly followed the Leary story with some care. See Lilly to Worden, 12 December 1967, with clipping, Stanford University Library, Lilly Papers, box 3D2, file "Worden, Frederick G., Brain Research Institute, Los Angeles, California."

178. Consider the suite of very hostile peer reviews he received, filed in Stanford University Library, Lilly Papers, box 3D2–D1, file "National Institutes of Health." Also relevant is Scott McVay's account of overhearing Norris's dismissal of the work of CRI during an on-site evaluation circa 1964, "You see what sort of cockamamie stuff is going on here?" Scott McVay, interview by the author, 7 July 2003.

West Coast, for mind-expanding peripatesis from Berkeley to Esalen and eventually well beyond; he and Leary would meet up on this trip.[179] Interestingly, as he was winging into the California sunset, Lilly stopped by the navy's Point Mugu dolphin laboratory for a quick lunch with an old friend who was working on a vocoder that could "translate" human speech into delphinidese (and vice versa)—Lilly and the navy had both been working on such devices for several years (figure 6.19).[180]

Wood, discovering him on base, had a cow.[181]

As well he might. He and Lilly were emphatically headed in different directions in May 1968. And in this sense, their confrontation back in August 1963 must be understood as the first cracking sound of what rapidly opened into a splitting fissure in cetacean research in the United States in a critical decade. Tectonic plates were moving under that local flare-up of scientific infighting: Lilly, increasingly preoccupied in the mid-1960s with erotic and ecstatic exploration of his own mind and that of his animal subjects, eventually came to feel that the dolphins—sexually liberated, stereophonic, non-manipulative superintelligences—were leading him to a new

179. Lilly to Peter Matson, 2 May 1968, Stanford University Library, Lilly Papers, box 3A1–C1, file "Matson, Mr. Peter H."

180. There is a history to be done on these devices, which were the holy grail of dolphin research in the 1960s. One of them features prominently in the navy film "Dolphins That Joined the Navy" of 1964 (see n. 142 above) where a navy researcher is shown speaking *Hawaiian* into the converter device. It was apparently believed that this language was particularly well suited to dolphin communication. For a technical account of both the (unclassified) navy work and that of Lilly, see the cover story by Richard Einhorn, "Dolphins Challenge the Designer," in *Electronic Design* 15, no. 25 (6 December 1967): 49–64. Also useful is the discussion by Wood in chap. 5 of *Marine Mammals and Man: The Navy's Porpoises and Sea Lions* (Washington, DC: Robert B. Luce, 1973). The great controversy here, at least among conspiracy theorists, involves the untimely death of the navy's main researcher on this vocoder project, Dwight W. Batteau. I bought (from a collector of such things) a copy of the 100-plus-page report filed by Batteau's collaborator Peter R. Markey shortly after his partner was found dead in a lagoon in Hawaii: Batteau and Markey, *Man/Dolphin Communication: Final Report, 15 December 1966–13 December 1967 (Prepared for U.S. Naval Ordnance Test Station, China Lake, California)*, Contract No. N00123-67-C-1103, (Arlington MA: Listening, 1967)." For a somewhat histrionic whirl through the history of the vocoder, consider Dave Tompkins, *How to Wreck a Nice Beach: The Vocoder from World War II to Hip Hop, The Machine Speaks* (New York: Stop Smiling, 2010).

181. This is Lilly's account: Francis Jeffrey and John Cunningham Lilly, *John Lilly, So Far . . .* (Los Angeles: Jeremy P. Tarcher, 1990), 150.

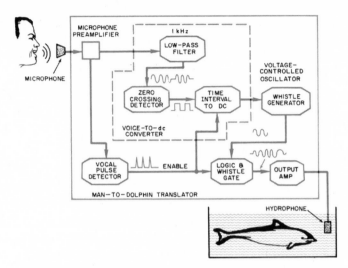

Man-to-dolphin translator converts artificial vocabulary to whistle outputs separated in pitch. The device, developed by Listening Inc., Arlington, Mass., detects speech and measures time delays of vowels in real time. Dolphin's own whistles are converted into synthetic speech by dolphin-to-man translator.

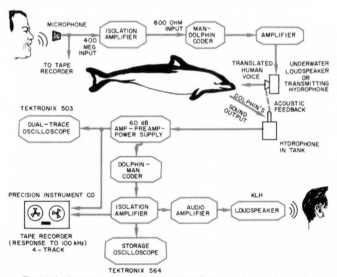

Electronic frequency converter "translates" human band of speech into higher frequency band. Reverse is done with dolphin's output. Four-track recorder tapes unmodified human speech, translated dolphin output, and unmodified dolphin output on separate tracks.

FIGURE 6.19 The power of the vocoder: Human-dolphin translation, circa 1967. (From Einhorn, "Dolphins Challenge the Designer," pp. 61, 62.)

kind of self. Tuning in, turning on, and effectively dropping out, John Lilly left the world that made him—the world of the Cold War biosciences—behind and went on to become a major-minor figure of the pacifist, drug-friendly, eco-sensitized counterculture. Wood, by contrast, still clipped his hair close, and his well-trained dolphins had serial numbers. They, too, possessed remarkable abilities, but about them he and his fellow scientists and trainers could not speak.

It must be said that the parting of their ways has about it the air of a fairy tale for the era—which is in fact what it promptly became. Right after the First International Symposium on Cetacean Research, Lilly entertained four of the European participants down at CRI: the two whale ear specialists Reysenbach de Haan and Peter Purves; the Dutchman Dudok van Heel, who was participating in a new effort to keep captive porpoises in the Netherlands; and the French bioacoustics expert René-Guy Busnel. Did conversation among this rump of the symposium in Miami return to Lilly's showdown with Wood? There is no way to be sure, yet it would appear that Busnel, at least, returned to Europe with an acute sense of what the final scene in Washington represented and a view of the matter quite sympathetic to Lilly. Back in France, Busnel would be in communication with the French novelist and English professor Robert Merle (a recent winner of the Prix Goncourt), and in this period Merle would turn his hand to a dramatic tale that pitted a Lilly-like character (named "Dr. Sevilla") against a navy porpoise-training establishment under the direction of a Wood-like "M. D. Morley"; intelligent dolphins are caught in the middle, and World War III is narrowly averted. Published in 1967 in French as *Un Animal Doué de Raison*, and two years later in English as *Day of the Dolphin*, this thriller would go on to become a major motion picture starring George C. Scott. It would, in a way, immortalize the conflict between Lilly and Wood, fixing it as a confrontation between the forces of peace and war, eros and thanatos—as nothing less than a showdown between the Age of Aquarius and the age of the hydrogen bomb.[182]

Tremors along this strange new fault line would continue for years, and in the smoke and rattle (the miasma of the Vietnam conflict, the mounting

---

182. NB: Robert Merle's *The Day of the Dolphin*, trans. Helen Weaver (New York: Simon and Schuster, 1969) is dedicated to Busnel. For an expanded sense of the way this split in the field looked to a participant, consider the memoir of one of the trainer-divers who became increasingly radicalized across these years: Richard O'Barry and Keith Coulbourn, *Behind the Dolphin Smile* (Chapel Hill, NC: Algonquin Books, 1988).

quakes of antiwar protest, and increasingly truculent environmentalism), dolphins, and their larger cousins, would come to symbolize the aspirations of many who hoped to defy the cultures of death.

A MIND IN THE WATERS: LILLY'S LEGACY

But why their larger cousins? At the beginning of this chapter, I asserted that Lilly's controversial work would have a lasting effect on popular conceptions of the great whales and would influence in important ways the shifting debates around whaling in the decade from 1965 to 1975. It is now time to deliver on those claims. To begin with, then, let me emphasize how quickly Lilly's (as we have seen, very popular) ideas about dolphins were extended, by analogy, to the larger cetaceans.

Lilly himself led the way on this considerable extrapolation of his already tenuous scientific assertions about the smaller odontocetes. The basis of this move was his long-standing neurophysiologist's preoccupation with a simple equation: brain size = cognitive/computational power.[183] While he confronted no small amount of criticism on this point (recall that Lawrence Kruger had sniped at Lilly in 1963 for disregarding careful histology and overlooking differences in neuronal density), Lilly never gave up a somewhat flat-footed cortical reductionism.[184] And by these lights, the four-pound brain of a dolphin was exciting, since it was already a bit bigger, in absolute terms, than a human brain (which averages closer to three pounds), but the twenty-pound brain of a *sperm whale* was genuinely mind-blowing. Lilly piped on the possibilities of such an organ at some length in *The Mind of the Dolphin*:

183. Lilly's work with a loaned LINC computer in the mid-1960s strongly influenced his conflation of cognition and computation. On the device, see Joe November, "LINC: Biology's Revolutionary Little Computer," *Endeavour* 28, no. 3 (September 2004): 125–31. I would also like to thank November for sharing with me some of his archival materials related to the LINC project, including Lilly's final report: Lilly, "An Experimental Investigation of the Dolphin's Communication Abilities by Means of a Dolphin Machine Code," *Proceedings of the Final LINC Evaluation Program Meeting, Washington University, St. Louis (Computer Research Lab), 18–19 March 1965*, sec. 4, pp. 1–22.

184. While he did index cross-species brain size comparisons by reference to body size, he used body *length* rather than body *weight* in this calculation as late as 1967 (John C. Lilly, *The Mind of the Dolphin: A Nonhuman Intelligence* [Garden City, NY: Doubleday, 1967], 18), which suggests something close to contempt for careful analysis of cephalization.

The sperm whales have brains six times the size of ours. . . . Before they are annihilated by man, I would like to exchange ideas with a sperm whale. I am not sure that they would be interested in communicating with me because my brain obviously is much more limited than theirs.[185]

He went on to offer a truly fantastic hypothetical reconstruction of the inner life of an animal thusly equipped. Estimating that the sperm whale would need less than a sixth of its brain for ordinary activities of survival and reproduction, Lilly found himself reeling as he reflected on the possible workings of that much surplus gray matter:

The rest of this huge computer is computing continuous inner experiences beyond our present understanding. If a sperm whale, for example, wants to see-hear-feel any past experience, his huge computer can reprogram it and run it off again. His huge computer gives him a reliving, as if with a three-dimensional sound-color-taste-emotion-re-experiencing motion picture. He can thus review the experience as it originally happened. He can imagine changing it to do a better job next time he encounters such an experience. He can set up the model of the way he would like to run it the next time, reprogram his computer, run it off, and see how well it works.[186]

And this staggering ability effectively to transcend the limits of (human) temporality would in turn lead to capabilities and sensibilities that we could understand only as "religious." With our "relatively small computers," humans are bound to focus mostly on the immediate problems of our bodily conditions, and we thus achieve only with great difficulty the forms of "spiritual" thinking that enable us to consider other human beings within a framework of reciprocity and mutuality. Lilly saw the historical evolution of such "ethical" modes as slow, contingent, and clearly imperfect. By contrast, it seemed entirely possible that "the sperm whale has gone so far into philosophical studies that he sees the Golden Rule as only a special case of a much larger ethic," and that therefore "he probably has abilities here that are truly godlike."[187] Lilly even asserted that he had tested this hypoth-

185. Ibid., p. 55.
186. Ibid., p. 56.
187. Ibid., p. 57.

esis empirically by reviewing the logbooks of nineteenth-century whaling vessels and discovering that sperm whales, almost uniformly, eschewed violence, even when unspeakably tormented by their hunters.

Having posited that humans had something very much like a god in their midst, Lilly went on to propose a national and international research initiative, at or surpassing the scale of the space program, aimed at establishing contact with the superhuman alien that lived right off our shores. The virtue of such an undertaking was that it would mobilize the only domain in which we appeared to be capable of surpassing the whales: social coordination. As Lilly put it, "cooperative activity of the humans is the thing which is impressive, even though no one of us might be impressive to a single sperm whale." And this led him to a further incandescent thought:

> Probably that which would excite the most respect for the human
> species in a sperm whale would be a full symphony orchestra playing a
> symphony. At least this would be an excellent starter to try to convince
> a sperm whale that maybe some of us are better than just in-concert
> murderers of whales. A symphony orchestra playing multiple melodies
> and their complex transformations might keep him interested for at
> least two or three hours. With his huge computer the sperm whale
> could probably store the whole symphony and play it back in his mind
> to himself at his leisure.[188]

The possibilities were indeed endless: such a brain, equipped for echolocation, would of course be able to hear, store, replay, and adjust the "spatial distribution of the sounds"—in effect recreating or even *creating* acoustic environments that would make "stereo" seem laughably primitive. Likening the effect to being able to relive a performance of Mahler's *Messiah* (NB: there is no such piece of music) note by note, emotion by emotion, in space, *all as a pure head trip*, Lilly went on to contemplate that a sperm whale might be able to "modify the music and even further elaborate it beyond any human conception of music."[189]

From these quotes it should be abundantly clear that by the mid-1960s, Lilly was comfortably extending his visionary claims for the "mind" of the large-brained dolphins to the still larger-brained large whales, where the

188. Ibid., p. 115.
189. Ibid., p. 116.

breathless scope of the possible quite literally knew no bounds.[190] And it is essential to appreciate that these ideas, however mad they may seem in retrospect, resonated rapidly and widely, and were incorporated into popular debates closely linked to Lilly's own denunciation of modern whaling, which he called, in the same book, "a black mark on the escutcheon of humanity."[191]

Stripped of some, but not all, of their extravagance, Lilly's claims about the intelligence of these animals appeared again and again in the earliest expressions of popular concern about whaling in what would become the "Save the Whales" campaign of the early 1970s. In 1971, for instance, Karl Erik Fichtelius and Sverre Sjölander published *Människan, Kaskelotvalen och kunskapens träd* in Sweden (just in time for the UN conference in Stockholm). Their well-received book, which would come into English the next year as *Smarter than Man? Intelligence in Whales, Dolphins, and Humans*, was a Lilly-grounded indictment of human beings as dangerous, manipulative, territorial, baboonlike creatures who would do well to get some perspective on the universe from their soulful cetacean neighbors.[192] Discussions of the sizes of whale brains turn up as a component of conservationist arguments in everything from the early and influential exposé on the exterminatory industry published in *Scientific American* in 1966 ("The Last of the Great Whales," authored by Scott McVay, to whom I will return in a moment) to the language of Senate Joint Resolution 115 of 1971 (calling on the State Department to secure a moratorium on global whaling), to the influential 340-page report drafted by Sir Sidney Frost in 1978 as part of a formal inquest into whaling by the Australian government—a report that ended Australian whaling, and which cited Lilly himself a dozen times.[193] When, in the late 1960s, the Canadian outdoorsman, sailor, and nature

190. Lilly tended to emphasize the large-brained odontocetes (like sperm and orca), but he also discussed the rorquals, which also have large brains, if less markedly so when indexed to body size.

191. John C. Lilly, *The Mind of the Dolphin: A Nonhuman Intelligence* (Garden City, NY: Doubleday, 1967), 117.

192. Karl Erik Fichtelius and Sverre Sjölander, *Smarter Than Man? Intelligence in Whales, Dolphins, and Humans*, trans. Thomas Teal (New York, Pantheon Books, 1972).

193. For a reprint of the report, see Sydney Frost, *The Whaling Question: The Inquiry by Sir Sydney Frost of Australia* (San Francisco: Friends of the Earth, 1979). It is interesting to note that Peter Singer also makes a number of appearances in this report. Friends of the Earth worked hard to influence Frost's committee; see the photocopied chapbook/

writer Farley Mowat wrote up his account of a female fin whale trapped in a tidal pond in Newfoundland, and detailed the carnival atmosphere that prevailed in the neighboring village as locals came out to pepper the beast with small arms fire over the course of several days, his outrage at her fate was shaped by his readings of Lilly's books (which he cited) and by Lilly's dramatic claims for the complex inner lives of cetaceans. The resulting work, A Whale for the Killing (1972), would sell several million copies, and the disgust and sympathy it engendered would catalyze the work of Greenpeace, Project Jonah, and other groups that set out to "save the whale."[194]

Surely the high-water mark for conservationist preoccupation with Lilly's vision of the cognitive capabilities of the large whales came in 1974, with the joint publication by Scribners and the Sierra Club of an elegant, large-format, well-illustrated volume celebrating cetaceans, a book that came into the world under the suggestive title Mind in the Waters. The brainchild of Joan McIntyre, a West Coast whale activist employed by Friends of the Earth, this digest of essays, poems, and technical cetology brought together everyone from Carl Sagan to W. S. Merwin, from Gregory Bateson to Pablo Neruda.[195] But in the mix were no fewer than four pieces on cetacean neuroanatomy and/or cognition, including (republished) work by Lilly on the "weirdness" of coming to feel that his dolphins were starting to experiment on him, together with a more empirical contribution from Lilly's old CRI collaborator Peter Morgane on the microstructure of cetacean brain tissue. In this remarkable early mélange of recognizably "new age" spirituality and ecological consciousness, the dominant preoccupation is with reaching out to the whales by means of Lillyesque cerebro-tactile commensuration. As McIntyre put it in a contribution entitled "Mind Play":

---

briefing packet by Friends of the Earth and Project Jonah, Whale Campaign Manual (Kew: Friends of the Earth, ca. 1973).

194. For Mowat's impact on Greenpeace, see Rex Weyler's foreword to the 2005 edition of Farley Mowat, A Whale for the Killing (Mechanicsburg, PA: Stackpole Books, 2005). For the link to Project Jonah, see Mind in the Waters: A Book to Celebrate the Consciousness of Whales and Dolphins, comp. Joan McIntyre (New York: Charles Scribner's Sons and Sierra Book Club, 1974), 13–28, which excerpts Mowat's book.

195. McIntyre was an early hire at the newly founded Friends of the Earth organization, and her initial work focused on a campaign against fur coats; she took up whale work in 1970, having been inspired by listening to Songs of the Humpback Whale (on which, see below).

What then is this other mind, the mind that is in the waters? . . . What is in the mind world of a creature with a brain bigger, and possibly more complex than ours, who cannot act out its will to change the world, if only because it hasn't any hands?

The only way to answer this question, of course, was by means of a meditative, sympathetic, oceanic quietism:

I think the way to enter the mind of the whale, is to enter the water. Whether it's the warm languid floating of the tropical oceans, or the strong cold urging of the surf at higher latitudes, the water is the cradle of Cetacean consciousness.

And this mindful return to the sea offered an escape from ideology, alienation, materialism, and violence:

When a human enters the water, what becomes apparent is the intimate connection between mind and body that the sea forces on her creatures. Without the alienating presence of objects and equipment, with only the naked body encasing the floating mind, the two, split by technological culture, are one again. The mind enters a new modality, where time, weight, and one's self are experienced holistically.[196]

The "mind world" of the whales was thus presented as nothing less than a possible resolution of the mind-body problem itself, since for these transcendent creatures "idea making" and "body experience" appeared to be indistinguishable. It will come as no surprise to the reader to discover that McIntyre, a close reader of both *Man and Dolphin* and *The Mind of the Dolphin*, had actually made a pilgrimage to Saint Thomas in 1971 to see what was left of the CRI laboratory (by then no longer functional) and to seek out Margaret Howe to talk with her about Lilly's ideas and Howe's own experiences with the dolphins.[197] The distinctive

196. Joan McIntyre, "Mind Play," in *Mind in the Waters: A Book to Celebrate the Consciousness of Whales and Dolphins*, comp. Joan McIntyre (New York: Charles Scribner's Sons and Sierra Book Club, 1974), 94–95, at p. 94.

197. Joana Varawa (formerly Joan McIntyre), interview by the author, 28 August 2009.

complex on display in *Mind in the Waters*—peace, mindfulness, aquatic suspension, holism, cetacean consciousness—clearly represents the full florescence of Lilly's isolation-tank dreams of a decade earlier, and it is striking to note the extent to which they had become, by the early 1970s, the rallying cry of a political mobilization to end the killing of the large whales.

All this is, I hope, interesting, and should go some way toward establishing my basic claim: that Lilly's work on dolphins resonated through the whaling debate, and did so by radically transforming attitudes toward the mental lives of these animals.[198] But it is possible to be even more concrete about the filiations of Lilly's influence in the broader arena of the contentious politics around the endgame of commercial whaling. To do so, let me turn to the work of two men directly influenced by Lilly who went on to play decisive roles in the political campaigns to end the industrial exploitation of whales. Both have already been mentioned briefly above: Scott McVay (author of the influential 1966 article in *Scientific American* on the failures of the IWC and the overreach of the world's whalers) and Paul Spong (leader of that Vancouver folk music ensemble that set out to play for the wild orcas in 1970).

For starters, then, who was McVay? Scott McVay was a tall, personable, and well-connected young man who had graduated from Princeton University with a degree in English in 1955 and joined the army, serving for three years in counterintelligence (posted in Germany—his language abilities were superb). He had no scientific training whatsoever but, remarkably, would go on to play a decisive role in what is probably the single most important "discovery" of twentieth-century whale science—if importance is measured in terms of cultural impact. I am referring to the finding formally reported in a coauthored cover story in *Science* on 13 August 1971, under the title "Songs of Humpback Whales" (the discovery itself—that humpback phonation possessed songlike structure—had been made about two years

---

198. For a strikingly clear indication of the extent to which dolphins had become the main tool for thinking about whales by the end of the 1960s, see the comment by J. R. White (the staff veterinarian at the Miami Seaquarium): "The Bottlenose Dolphin is not only the center of all marine attractions throughout the world, *but is also the experimental model from which we have gained what little we know of Cetacea*" (emphasis added not only by me, but also by Carleton Ray, the recipient of this letter). White to Ray, 10 August 1971, Smithsonian Archives, RU 7229, Marine Mammal Program Records, 1967–1973 and undated, box 5, folder 12.

FIGURE 6.20

They are singing to us: The front page of Payne and McVay's landmark article in *Science*. (Payne and McVay, "Songs of Humpback Whales," p. 585. Reprinted with permission from AAAS.)

earlier).[199] It would be difficult to overstate the significance of this work for the campaign to end whaling. The haunting mewls and honks of humpbacks—which, recall, had been well known among military bioacoustics researchers going back to the 1940s, who had thought of them as little more than "noise"—were, in the course of three short years on the cusp of the 1970s, transformed into nothing less than the soundtrack of the "Save the Whales" campaign (figure 6.20).

To give just a little sense of their cultural currency and power, *Songs of the Humpback Whale*, the original 1970 album of these sounds (which were largely recorded on navy equipment), remains the best-selling natural history record of all time. The humpback also holds the title for the single largest one-time pressing in the history of the recording industry: *National Geographic* ordered ten and a half million copies of humpback whale song

199. Roger S. Payne and Scott McVay, "Songs of Humpback Whales," *Science* 173, n.s., no. 3997 (13 August 1971): 585–97. McVay reports that they waited almost 18 months after the acceptance of the original paper to see publication, having decided to hold out for cover placement.

on flexible plastic cards, which were tipped in to every copy of their January 1979 issue. A number of years ago, I set a student researcher on the task of documenting every instance of cetacean phonation incorporated into (or thematized by) popular or classical music since 1970. He compiled an "audiography" of more than 150 items, ranging from Alan Hovhaness's symphonic and exoticizing "And God Created Great Whales" of 1970 to Leonard Roseman's soundtrack to *Star Trek IV* (1986), with entries in between for a veritable who's who of folk and experimental music over more than twenty years: Country Joe McDonald, Judy Collins, Pete Seeger, Paul Winter, George Crumb, Gordon Lightfoot, John Denver, Crosby and Nash, Fred Neil, Jethro Tull, and the Cocteau Twins, to name just a few. By the mid-1970s, as I noted above, these whining aquatic phonations had been selected for inclusion on the gold phonograph record that would wing its way into deep space aboard the *Voyager* spacecraft, a disk intended to serve as an epitome of the vitality of Earth itself. The dissemination of these peculiar sounds into and beyond the sensorium of the 1970s must be understood as one of the very most surprising and significant pop-cultural phenomena of the modern conservation movement.

And the political repercussions were swift and expansive. When Scott McVay and his codiscoverer Roger Payne (a gifted biologist specializing in ornithology, then a junior professor at Rockefeller University) had an opportunity to give testimony at the Department of the Interior in 1970 concerning the suitability of listing the large whales as endangered species, they brought taped songs of humpback whales with them and played them in the conference room. McVay did the same thing the following year at hearings before members of the House Committee on Foreign Relations, which was then debating the idea of a resolution that would force the hand of the State Department in its dealings with the IWC.[200] Shortly thereafter

---

200. This meeting was on 27 July 1971. For a superb window onto how rapidly these tapes were mobilized in an emerging international effort to raise awareness of the need for protective measures for whales, consider the remarkable 46-page internal report McVay authored in August 1970 detailing his three-week trip to Japan in that month (funded by the New York Zoological Society) for the purpose of stimulating and organizing a local whale conservation movement: on the first morning of his first day in Tokyo, McVay "purchased a Sony stereo cassette with stereo headphones" so that he could slip the sounds of humpbacks into the ears of every Japanese scientist, politician, and dignitary he succeeded in meeting over twenty days. McVay was insistent that this musical encounter helped overcome language barriers and met everywhere with "enthusiastic response." See Scott McVay, "Initiation of Whale Campaign in Japan," August 1970, un-

humpback whale phonations would sound from the stage at Stockholm, and those wheezing bleats even got a hearing on the floor of the IWC itself, which by the early 1970s was increasingly besieged by activist-observers, nongovernmental organizations, and assorted rabble-rousers.[201] Suddenly, improbably, it was as if anyone with a radio could hear these most mysterious of animals sounding their own dirge. "They are crying out," declared an emotionally extravagant Walter Hickel shortly after stepping down at the Department of the Interior, and he went on to ask rhetorically, "Is the cry of the whale really the cry of life itself?"[202]

It is more than reasonable to ask just how anyone came to think of these bizarre squeaks and groans as "songs." And it is further worth considering how that notion came to have both such remarkable currency and such explicit eco-political import. Not everyone, needless to say, was so quickly moved. In the Smithsonian Archives I turned up a remarkable set of notes, taken in 1971 by a leading young marine mammal expert named G. Carleton Ray, who had also been called to testify at that same House subcommittee hearing on the moratorium where McVay played his tape. Ray had worked for a decade on walruses, and he had succeeded in the late 1960s in drawing a large number of leading marine mammal scientists in the United States and abroad into an initiative called the US Marine Mammal Program, which was structured as an "integrated research program" under the International Biological Program (IBP, itself an outgrowth of work by the International Council of Scientific Unions in the mid-1960s).[203] He was a friendly colleague of the whole cohort of American

published typescript "intended only for the information of those closely identified with the whale conservation effort." I have this document from McVay himself.

201. For an introduction to the literature on the role of such actors in the workings of the IWC, see Tora Skodvin and Steinar Andresen, "Nonstate Influence in the International Whaling Commission, 1970–1990," *Global Environmental Politics* 3 no. 4 (November 2003): 61–86. For a general account of the shift away from a strictly state-centered approach to management issues, consider Tomas M. Koontz et al., *Collaborative Environmental Management: What Roles for Government?* (Washington, DC: Resources for the Future, 2004).

202. Walter J. Hickel, *Who Owns America?* (Englewood Cliffs, NJ: Prentice-Hall, 1971), 141–42. He also, rather inexplicably, juxtaposed whale song with the cry of black children in American urban ghettos.

203. What seems to be the full archive of this undertaking can be found in the Smithsonian Archives, RU 7229, "United States Marine Mammal Program, Records, circa 1967–1973 and undated." Useful information on the inception and development of the initiative can be found in box 2, folder 3 (Ray to de Carlo, 30 December 1968

whale scientists, especially Kenneth Norris and William Schevill—both of whom he had gotten to know at the First International Symposium on Cetacean Research. By 1971 Ray was an associate professor of biology at Johns Hopkins and was actively involved in pinniped management in the northern Pacific. His administrative and fund-raising work through the IBP had elevated his profile on marine mammal issues, and it was for this reason that he found himself at the table in July 1971 with McVay, McHugh, Chapman, and others, answering questions about whales posed by a set of ranking House Democrats and Republicans. Here are Ray's penciled comments about McVay's testimony and disk-jockeying (after playing a snippet of humpback song for the room, McVay offered to let the politi-

---

[with enclosures]), and box 1, folder 7 (which also includes Ray's CV circa 1970). It is interesting to note that Ray was one of the main organizers of what was known informally as the "Shenandoah Conference" (the International Conference on the Biology of Whales), which was held in June 1971 under the joint sponsorship of the US Department of Interior, the IBP, the New York Zoological Society, and the Smithsonian Institution. This meeting was convened as a semigovernmental affair, growing as it did out of a commitment by Interior to host further biological discussions of the status of whales in connection with their listing as endangered species. Galler, again, was involved in assembling the roster for the event. The proceedings were eventually published as William Edward Schevill, G. Carleton Ray, and Kenneth S. Norris, eds., *The Whale Problem: A Status Report* (Cambridge, MA: Harvard University Press, 1974). An interesting multi-book review that included this volume and Joan McIntyre's *Mind in the Waters* was published in the *New York Review of Books* in 1975: J. Z. Young, "Save the Whales!" *New York Review of Books* 22, no. 12 (17 July 1975), http://www.nybooks. com/articles/article-preview?article_id=9134. What is quite wonderful, in view of the shifting ground on this issue in the early 1970s, is to track, in G. Carleton Ray's notes and correspondence, his efforts (and those of Norris and Schevill as well) to rein in, sanitize, and cabin the contribution by McVay, who was invited to the conference (in Payne's place—he said he could not come, but then subsequently turned up) more or less explicitly as a token conservationist, and who submitted a wide-ranging paper entitled "Reflections on the Management of Whaling" that amounted to a sweeping indictment of whale science in the twentieth century. For instances of all this, see Smithsonian Archives, RU 7229, box 7, folders 15 and 23, together with box 5, folder 4 (featuring a letter from McVay to Norris of 8 October 1971, protesting, "You write that the call for more attention to the brain and nervous system makes you edgy. You are unnecessarily defensive. This is a legitimate and important area for scientific inquiry that deserves first-class attention. That you may be bothered by Lilly's speculations is not germane to the intrinsic significance of research on all mammalian brains, terrestrial and aquatic."). The full records of the conference (including transcripts) are in the Smithsonian Archives, RU 7229, boxes 7 and 8.

cians listen to three whales in stereo over headphones after the close of the hearings):

> Emotional pitch which does not come to grip with the issues. Plays a whale "song"—for emotional impact, I guess.
> [Implication is that this is something which forbids our use—no idea what it means—for "joy"?][204]

It is a wonderful record of the dismay and surprise of someone on the inside of whale science circa 1970 who was at that very moment being wrong-footed on an extremely public stage by an insurgency that he did not fully understand and emphatically could not respect. Ray had published for years on the physiology, phonations, and ecology of marine mammals. Roger Payne (who was credited by McVay as he presented the recordings) knew a great deal about barn owls, but had not a single peer-reviewed publication on whales or seals to his name.[205] And Scott McVay was a university administrator with a BA earned on the basis of an undergraduate thesis on Shaw (as in George Bernard). Ray tried not to sputter his contempt when he had his turn at the mike. Dismissing the idea of a moratorium as unnecessary and damaging to the future of management-oriented research, he went on to disparage sentimental attachment to a set of odd hoots:

> I don't find it very relevant to hear that whales produce music. Cock-a-doodle-doo produces music, too. Whales are smarter than chickens, but it is not relevant to the purpose of this bill.[206]

Here, as the next five years would demonstrate, he was quite wrong. But it is easy to see what he meant. His scare quotes around the word "song" in

204. This document can be found in the Smithsonian Archives, RU 7229, box 3, folder 9.

205. At least not as far as I know. To be fair, however, the landmark *Science* article was in press, as was a coauthored piece on long-range cetacean signaling that would appear later that year: Roger Payne and Douglas Webb, "Orientation by Means of Long Range Signaling in Baleen Whales," *Annals of the New York Academy of Sciences* 188 (December 1971): 110–41.

206. This quote is from p. 35 of U.S. Congress, Hearings Before the Subcommittee on International Organizations and Movements of the Committee on Foreign Affairs, House of Representatives, Ninety-Second Congress, First Session, 1971, which I consulted in the Smithsonian Archives, RU 7229, box 10, folder 7.

his private notes speak clearly to his skepticism about this much-heralded "discovery" concerning the aptitudes of *Megaptera novaeangliae*. Ray, who had coauthored a paper with Schevill on pinniped sounds, was in that cohort of scientists who were having trouble hearing those odd old sounds as "songs," and many of them were a little punchy about the public relations work that this new terminology was doing. By 1973, Ray had been pilloried in the press for his resistance to the moratorium and had seen his Marine Mammal Program, a grand plan for a heavily funded, integrated, international research program on the biology of marine mammals, swamped by the passage of the Marine Mammal Protection Act, which more or less made the creatures untouchable. He found himself forced to begin a grant application by respectfully proposing that Americans needed to "see beyond the 'humanness' of these animals" so that scientists could continue their work.[207] It was a new world.

What does any of this have to with John C. Lilly? Well, we have already seen that Lilly mused at some length, if hypothetically, about what he imagined were the powerful musical aptitudes of the larger whales. Out of this and Lilly's other writings a plausible argument can be advanced that Lilly's claims about the cognitive capacities of the cetaceans set the conditions of possibility for a new hearing of those whale phonations that had been recorded, analyzed, and replayed by navy-affiliated bioacousticians for two decades. But we can make a much stronger link than that. In December 1961 Scott McVay, then working in an administrative capacity at Princeton University, attended a lecture by John Lilly, who had been invited to address the Psychology Department (in connection with his sensory deprivation work).[208] McVay, who had read *Man and Dolphin* earlier that year and been fascinated by it, accompanied Lilly to the Princeton railroad station after his talk and then proceeded to ride the train with him into New York City (where Lilly was spending the night) in order to put to him a list of several dozen questions about his work. The two men struck up a friendship, and a little more than a year later, in the summer of 1963, McVay, together with

207. The quote comes from p. cm3 of Ray's NSF proposal "Marine Mammals: The Biological Basis of Productivity and Conservation," Smithsonian Archives, RU 7229, box 2, folder 9. For Ray's concerns about the "protectionist" lobby and the Marine Mammal Protection Act, see Ray to Senator Ernest F. Hollings, 5 May 1972, Smithsonian Archives, RU 7229, box 4, folder 18. For the failure of major funding efforts for Ray's Marine Mammal Program, together with a flavor of the beatings he took from the "Save the Whales" lobby, see Smithsonian Archives, RU 7229, box 5, folder 1.

208. The invitation came from Jack Vernon.

his wife and children, left Princeton for Miami, where he joined the staff of CRI. He would work with Lilly, and with Lilly's dolphins, until 1965, when McVay returned to Princeton as an assistant to President Robert Goheen.

It was during his employment at CRI that McVay learned to make and analyze sound spectrograms of dolphin phonation using the Kay Vibralyzer. McVay's experimental work involved vocalizing patterned sequences of randomly generated English phonemes for a dolphin in an effort to elicit "parroting" behavior. Traditional techniques of operant conditioning were used to reinforce English-like sound production by the experimental animal, and efforts were made to use vocoder technology to shift the frequencies of human speech into the general range of delphinid phonation. Success was assessed by means of visual comparisons of sound spectrograms of the experimenter's vocalizations and those made responsively by the dolphin. Which is to say, by the mid-1960s, Scott McVay had spent many hundreds of hours looking at spectrograms of cetacean sounds, and he had learned to parse the output of a Vibralyzer into repeating "phonemic" elements and to pore over these graphic outputs looking for patterns of theme and repetition.

It was on the basis of McVay's familiarity with this kind of spectrographic analysis that Roger Payne conveyed to McVay tapes of humpback phonation that Payne himself had been given by Frank Watlington, a navy underwater acoustics researcher stationed at a navy listening station in Bermuda in the 1960s.[209] Payne, who was a serious amateur cellist and who had worked at Cornell on the role of acoustics in owl predation (owls use sound to locate prey in low light conditions), already had a strong interest in animal sounds in general and birdsong in particular, and he had immediately been struck by the idea that these strange cetacean sounds might be thought of as an underwater equivalent to the chirruping of the feathered creatures he knew better. Having studied with Donald R. Griffin, the world-famous (re-)discoverer of bat echolocation, at Harvard while an undergraduate (when Griffin was writing the book on echolocation that would be published in 1958, *Listening in the Dark*), Payne would certainly have been aware of the exciting discoveries being made concerning odontocete echolocatory capabilities in the early 1960s. It seems reasonable

---

209. Watlington conveyed almost a decade's worth of tapes to Payne in 1967, when Payne and his wife visited Bermuda. The link to Watlington was made, it appears, by Henry Clay Frick, who had a vacation home in the area. Katy Payne, interview by the author, 2 September 2009.

to assume that he thought of his emerging interest in whale phonation in 1967 as a direct outgrowth of his work on barn owls. He may also have been interested in conservation issues where whales were concerned, and it is likely that he read McVay's important 1966 article on the overkill in *Scientific American*. Payne was already involved in bird conservation issues and had links to a number of the New York–area lawyers and activists who had worked on the campaign to end the use of DDT as a commercial insecticide. Personable, charismatic, and attractive, he had a reputation as a brilliant experimentalist and a passionate field naturalist. At the same time, he had not done a great deal of work with spectrographs, even though these instruments already had considerable currency among ornithologists. In fact, it was in the lab of another barn owl expert (Mark Konishi, at Princeton) that McVay, on a friendly and informal basis, got access to the equipment he would use to do the spectrographic analysis on Payne's humpback tapes—work he would perform after hours in 1968–1969. And it was in the course of these investigations—printing out, bit by bit, graphic representations of the phonations; parsing them into phoneme-like units; seeking repeating patterns among phonemic elements—that McVay would make the core discovery that lay at the center of the 1971 paper in *Science*: that humpback phonations displayed patterns of repeated themes that met the working definition of "song" then in use among scientists of animal sounds (figure 6.21).[210]

Scott McVay, who is alive as I write, likes to tell the story of receiving, circa 1965, a postcard from Lilly (by this time increasingly erratic) with a melodramatic message along the lines of, "From falling hand I pass the

210. Some discussion in the *Science* paper is given to the proper definition of song, and the work of W. B. Broughton is cited as decisive (W. B. Broughton, "Methods in Bio-acoustic Terminology," in *Acoustic Behavior of Animals*, ed. R. G. Busnel [London: Elsevier, 1963], 3–24). The primary definition used by Payne and McVay is "a series of notes, generally of more than one type, uttered in succession and so related as to form a recognizable sequence or pattern in time." McVay has claimed the central credit for the discovery in several interviews with me dating back to 2003 and in similar interviews he gave to David Rothenberg, who has written an account in his *Thousand Mile Song: Whale Music in a Sea of Sound* (New York: Basic Books, 2008). Though I have talked with Payne informally about some of this, I do not know exactly how he would tell the story. I have heard, however, that he objects to McVay's account of his large role in the finding. It is interesting to note that the transition from cetacean phonation as "noise" to cetacean phonation as "song" was to some extent mediated by the language of "music." See, for instance, William E. Schevill, "Whale Music," *Oceanus* 9, no. 2 (1962): 2–13.

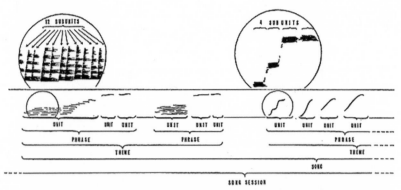

Diagrammatic sample of whale spectrograms (also called sonagrams) indicating terminology used in describing songs. Frequency is given on the vertical axis, time on the horizontal axis. The circled areas are spectrograms that have been enlarged to show the substructure of sounds which, unless slowed down, are not readily detected by the human ear.

FIGURE 6.21 Phonemic analysis: Parsing sonograms of cetacean vocalizations. (From Payne and McVay, "Songs of Humpback Whales," p. 586. Reprinted with permission from AAAS.)

torch . . ."[211] It is by no means impossible to imagine that this really happened, though one might well wonder what, exactly, Lilly imagined the "torch" in question to be. The institution of CRI? Unlikely. The program of dolphin research? Perhaps. The larger vision of breaking through to an alien intelligence in the oceans? Quite possibly. Though it is even more likely, in my view, that by that point in his career Lilly wasn't himself absolutely sure what he meant, other than that he was falling.

Be that as it may, by situating the discovery of "Songs of Humpback Whales" firmly in the context of McVay's work in Lilly's laboratories, I can be understood to have literalized this heroic/pathetic passing of the torch. And something like that is indeed my intention. While parsing priority in cases of codiscovery is a perilous game, I believe that the evidence strongly suggests that McVay's experience working with Lilly and his dolphins—both the technical experience of generating and analyzing sound spectrograms of cetacean phonation, and the broader exposure to a research milieu obsessed with the cognitive, expressive, and musical capabilities of these animals—was essential to the discovery of "Songs of Humpback Whales." Because Roger Payne went on to have a very prominent (if largely nontradi-

211. Scott McVay, interview by the author, 7 July 2003. It is perhaps worth noting that McVay also recalls a meeting with Remington Kellogg, in Washington, DC, circa 1965, in which Kellogg said something similar about the future of the whale problem lying with McVay.

tional) career as a whale scientist, and because he received first authorship on the 1971 paper in *Science*, McVay's role, and his link back to Lilly, have been generally overlooked.[212] And McVay—who was, by the early 1970s, already well on his way to a distinguished career in the nonprofit sector, and who would over the next decade play significant roles in a roster of important conservation organizations, from the Environmental Defense Fund to the World Wildlife Fund—did not tend, as he rose to insider status among the well-heeled of the foundation world, to dwell at any length on his links back to a man who in 1972 could be dismissed witheringly by a former colleague in a front-page article in the *Wall Street Journal*: "I find Dr. Lilly's work very interesting. I like Dr. Seuss, too."[213] But if we are concerned to trace out the tendrils of Lilly's enduring influence in the campaign to end whaling in the 1970s, we would do well to hear echoes of his work in every wail of whale song that sounded through the decade.[214] Indeed, we might hazard a still stronger formulation: through McVay's work, Lilly actually managed to realize his ambition of helping human beings have a resonant, acoustic "encounter" with an alien, aquatic mind. And he gave that experience to much of the listening public of the 1970s.[215]

212. An exception here would be Rothenberg's account. McVay is quoted there as saying that he offered Payne first authorship, a gesture that is presented as a friendly courtesy, given Payne's greater need for the credit as a young biologist.

213. This was the navy-linked bioacoustics expert William Evans, as quoted in William E. Blundell, "Great Divin' Dolphins! Are They in Battle Against the Red Peril?" *Wall Street Journal*, 9 May 1972. See n. 157 above.

214. Interestingly, in the wake of the runaway success of the album *Songs of the Humpback Whale*, Lilly, strapped for revenue, sold a block of CRI recordings of dolphin phonation to Moses Asch's Folkways Records, which issued the album *Sounds and Ultra-Sounds of the Bottlenose Dolphin*" in 1973, with four pages of liner notes by Lilly. For the terms of the deal, see Lilly's letter to his agent: Lilly to Peter Matson, Stanford University Library, Lilly Papers, box 3A1–C1, file "Matson, Mr. Peter H."

215. Payne himself presented humpback song in powerfully Lillyesque terms, writing in *National Geographic* in 1979, "So far, the study of humpback whale songs has provided our best insight into the mental capabilities of whales. Humpbacks are clearly intelligent enough to memorize the order of those sounds, as well as the new modifications they hear going on around them. Moreover, they can store this information for at least six months as a basis for further improvisations. To me, this suggests an impressive mental ability and a possible route in the future to assess the intelligence of whales." Roger Payne, "Humpbacks: Their Mysterious Songs," *National Geographic* 155, no. 1 (January 1979): 18–25.

What about Paul Spong, the other person I described as a more or less direct bridge between the dolphin work of John C. Lilly in the early 1960s and the push to protect the large whales in the early 1970s? Spong's role in the history of Greenpeace has been well documented.[216] Specifically, it is undisputed that it was he who refocused a loose aggregation of pacifistic Vancouver Quakers, Buddhists, mystics, and freaks from a preoccupation with French nuclear weapons testing in the island archipelagoes of the Pacific to a full-time mission to terminate pelagic whaling. It was he who traveled to Japan in 1974 and Norway in 1975 to conduct "espionage" missions aimed at discovering the exact locations of the whaling grounds of the North Pacific in the hopes that Greenpeace might disrupt industry activities in those regions. And in the course of this work he sought out old hip-booted whale scientists like Omura and Jonsgård, presenting himself as a fellow "cetologist" to pump them for information.[217] Without the coordinates for the Mendocino Ridge that he discovered in Sandefjord, Norway (under the somewhat begrudging guidance of Einar Vangstein, the director of the BIWS), the epoch-defining Zodiac showdown of the summer of 1975 would never have occurred.[218]

What is less often noted is that Paul Spong's early career was wholly shaped by Lilly's work.[219] How did Spong "pass" as a cetologist among the hip-booters of Europe and Japan? Well, in a sense, he really was a whale scientist, if of a distinctly renegade variety very much in the Lilly mold. Originally from New Zealand, Spong studied law as an undergraduate before coming to the United States in 1963 as a graduate student in psychology at UCLA, where he joined the circle of the Brain Research Institute

216. See, for instance, Rex Weyler, *Greenpeace: How a Group of Ecologists, Journalists, and Visionaries Changed the World* (New York: Rodale, 2004); Robert Hunter, *Warriors of the Rainbow: A Chronicle of the Greenpeace Movement* (New York: Holt, Rinehart & Winston, 1979); and Frank Zelko, "Make It a Green Peace: The History of an International Environmental Organization" (PhD diss., University of Kansas, 2003).

217. See Rex Weyler, *Greenpeace: How a Group of Ecologists, Journalists, and Visionaries Changed the World* (New York: Rodale, 2004), 242, 262.

218. Ibid., p. 263.

219. Part of the reason this point has not been clearly made is that Spong, who is also alive at my writing, explicitly denies that he was influenced by Lilly, claiming—wholly implausibly—that he "had never heard of Lilly before the early 1970s." Paul Spong, interview by the author, 25 March 2006.

(where, recall, Lilly-linked dolphin brain work was then being done).[220] There, Spong studied physiological psychology, completing major fields in learning and comparative psychology, and served as a research assistant in the BRI's Space Biology Laboratory. In 1966 he defended a dissertation entitled "Cortical Evoked Responses and Attention in Man," a study of the attentiveness and reflex responses of isolated agents subjected to diminished sensory environments and shock stimuli—work funded in part by NASA. John Lilly's close friend and colleague from his days at NIMH, Frederic G. Worden, served on Spong's dissertation committee, and the dissertation itself represented work at the same periphery of the military sciences of mind and behavior where Lilly had been a field leader a decade earlier (evaluation of the mental capabilities and psychic states of subjects in isolation).

In 1967 Spong landed his first job: a position straddling the University of British Columbia's neurology laboratory and the Vancouver Aquarium, which had secured, in February of that year, a 13-foot adolescent female killer whale dubbed "Skana"—a Native American name that apparently means "Supernatural One."[221] Predictably, before leaving Los Angeles, Spong took tuition on the status of brain-and-behavior-oriented research on captive odontocetes from the local expert, Kenneth Norris, who had just brought the proceedings of the First International Symposium on Cetacean Research to press.[222] It is to be supposed that Norris—one of those, recall, who had just helped sink Lilly's government funding—counseled the young researcher to steer clear of Lilly-style work with this new animal, and it may be with this in mind that Norris encouraged Spong to work on establishing the animal's visual thresholds (thus directing him out of the crowded and increasingly divided world of acoustics research). Be that as it may, Spong had certainly familiarized himself with Lilly's dolphin work by this time, and it is impossible that he was unaware of the navy's marine mammal research facility up the coast at Point Mugu, recently brought under classification. By 1968 Spong—whose hair had grown quite long and who

220. Lawrence Kruger, Lilly's nemesis on matters of delphinid neuroanatomy, was a tenured faculty member at the BRI in 1963, and Ken Norris had links to the institute as well. Ronald Turner (discussed above) did a PhD dissertation through the BRI in 1962 entitled "Operant Control of the Vocal Behavior of the Dolphin" and was at that time also writing reports for NOTS at China Lake.

221. Rex Weyler, *Greenpeace: How a Group of Ecologists, Journalists, and Visionaries Changed the World* (New York: Rodale, 2004), 204–5.

222. Paul Spong, interview by the author, 25 March 2006.

at about this time took to wearing a somewhat eccentric wool toque—was getting to the aquarium early in the morning to play his flute for Skana. In the course of his first months of research work he had come to believe—in pure Lilly fashion—that his experimental animal was, in effect, "experimenting" on him (testing his fears, screwing up his results out of boredom or mischievousness, etc.), and by the fall of 1968 he had given a paper at UBC that attracted local media attention, since in it he claimed that orcas were "highly intelligent, social animals" and that aquarium specimens (including his own) should be released into the wild. Interestingly, crossing the Lillyesque wires of Cold War brain science and animal intelligence in yet another way, Spong went so far as to suggest that holding killer whales in captivity amounted, in view of their acoustic orientation, to confining them in a "sensory deprivation environment."[223] He was dismissed from his position shortly thereafter. Briefly committed—apparently consensually, but, it would seem, under pressure from the head of the Psychology Department at UBC—to the university's psychiatric ward, Spong would emerge to acknowledge "taking mescaline during some of his sessions with Skana."[224]

Chanting "Let the whales go" for a reporter covering his story, Spong—now unemployed, but newly impassioned—embarked on his career as a full-time activist on behalf of captive and hunted cetaceans.[225] By 1972, via Farley Mowat, he was connected to Joan McIntyre in San Francisco. Shortly thereafter, Spong, cetacean brain specimens in hand, was lecturing a pair of macho young newspaper reporters named Rex Wyler and Robert Hunter about whale song, neocortical convolutions, and the inherent pacifism of an intelligence evolved without hands. As Wyler—who would, like Hunter, go on to be a foundational figure in Greenpeace—put it, looking through the viewfinder of his camera at an orca brain dropped on his kitchen table by Spong:

> I now looked at the most obvious evidence, the single most astonishing anatomy lesson of my life. Maybe humans aren't the paragon of animal. Maybe evolution has bigger plans and even greater potential than human culture. What if there is an equal or greater intelligence

223. Rex Weyler, *Greenpeace: How a Group of Ecologists, Journalists, and Visionaries Changed the World* (New York: Rodale, 2004), 208.

224. Ibid., p. 209.

225. Ibid.

right here on our earth? I felt as if I had entered a secret lab doing an autopsy of a space invader.[226]

What is really strange is that, in a way, he had. After all, a secret glimpse into the mind of an alien was exactly how Lilly had sold CRI to NASA, the air force, and the navy more than a decade earlier.

So impressed were these men, and eventually others—men and women— by this visual evidence and its penumbra of scientific mystery that they made what has traditionally been understood as the ultimate commitment: they pledged to risk their lives to defend these animals from their human enemies. And risk their lives they did, a few years later, aboard the *Phyllis Cormack*, a vessel outfitted, it is well to recall, with a balky Tcherepnin synthesizer, a quadraphonic underwater sound system, and a full hydrophone array—the better to facilitate the trippy, spatialized, symphonic encounter with these alternative intelligences in the deep sea.[227] The rainbow warriors were at the helm of a kind of Lilly dream vessel.

CONCLUSION

Clearly, by 1975, whales and dolphins were no longer merely aquatic mammals containing large quantities of interesting oils. They had become nothing less than familiar spirits, capable of mobilizing sentiment and action, capable of representing complex and culturally significant assemblages of aspiration and belief. To return to the question from which this chapter departed: What do these totemic figures have to do with science and its history?

I believe the story of John C. Lilly offered in these pages goes some way toward answering this question. It is a peculiar and highly contingent tale. Is it possible to generalize about its shape or content? Perhaps. Part of what is so strange about the story of Lilly and, if you like, the "reenchantment" of the whales (they were, after all, long thought of as mysterious agents of great power) is that we are rightly accustomed to perceiving scientific inquiry as, in significant part, the very opposite process: a process of *demythologization*. In Bruno Latour's terms, it is the hammer of critical

---

226. Ibid., p. 251.

227. On the acoustic technologies of the voyage, see Robert Hunter, *Warriors of the Rainbow: A Chronicle of the Greenpeace Movement* (New York: Holt, Rinehart & Winston, 1979), 167–69.

rationality—the tool of scientific modernity—that strikes those unfortunate fetishes of myth, those totems brought forward by benighted premoderns and their mere belief.[228] Under its iconoclastic blows, the accretions of myth, belief, and legend fall away, revealed to be merely artifacts of human collective fantasy and mythopoesis. The really real, the facts about things, these remain, because they are impervious and durable. What stands where scientists have passed, then, is nature, just nature, without anthropocentric frippery, spiritual corsage, or the frogging and furbelows of human fear and desire. But, as Latour and others in the last decades have been much at pains to show, gestures of iconoclasm have unpredictable effects: not infrequently the hammer rebounds sideways, revealing, for instance, that the iconoclast himself may have even deeper commitments to the fetish he assails than did the putatively deceived in whose midst it stood and in the name of whose liberation the iconoclast has struck (after all, the iconoclast without icons is a lonely figure). Indeed, shattered icons have a strange way of turning up again—lovingly preserved, carefully reworked, and finally elevated on central pedestals in the temples of the iconoclasts.

The story of cetology and the whale in the second half of the twentieth century is fascinating precisely because it represents a lovely instance of this unstable relationship between icons and iconoclasm, between science-doing and mythmaking. Few disciplines can more straightforwardly tell their story as one of gradual demystification: once whales were *the* great unknown—monsters of the deep, Leviathan, known only to God Himself. Gradually the wild legends, the seafarers' yarns, the biblical tales, all gave way—fell away—as men of learning pursued the cetaceans with harpoons, microphones, nets, and neurological probes, leaving a pile of quaint old beliefs in shards (on the one hand) and a proper knowledge of what cetaceans really were (on the other).

And yet somehow, in the process, what emerged was a new creature of extraordinary symbolic power, whose looming significance swelled with each seemingly deflationary blow, whose new iconography was composed, rapidly, mosaic-like, out of the broken bits of the old myths (so navy dol-

228. See the visually compelling compilation of essays on this approach, *Iconoclash*, edited by Bruno Latour and Peter Weibel (Karlsruhe: ZKM, 2002); the earliest presentation is, I believe, Bruno Latour, "A Few Steps toward an Anthropology of the Iconoclastic Gesture," *Science in Context* 10, no. 1 (1997): 63–83. This approach is elaborated in Latour's *Pandora's Hope: Essays on the Reality of Science Studies* (Cambridge, MA: Harvard University Press, 1999).

phins *were* being trained to help sailors, as in the ancient tale of Arion . . .), heightened with bright shards borrowed from the workshops of scientific cetology (so the whale hugger–saboteurs releasing navy dolphins back into the wild could believe that the animals "saw" the fear and goodwill of their liberators by sonar-scanning their racing hearts . . .). It was this puzzling, runaway process that the distinguished professor of linguistics Thomas Sebeok glimpsed in Lilly's *Man and Dolphin* when he reviewed the book in 1963: "Like Blake's Tiger, Lilly's Dolphin is at once something less and something more than man, a visionary creature, symbol as well as thing. With this figure in a double narrative, on the level of science and on the level of myth, he has written a strange, irritating, anecdotal, and provoking book."[229]

Lilly, of course, would be ruled—despite his protests—simply a poor iconoclast and a scientist gone bad. He could (and would) be placed, retrospectively, in the category of the mythmakers, not the cetologists. But as the symbolic significance of whales and dolphins continued to swell, and as the elaborations of that significance became more baroque and versicolor, whale science itself could not but benefit from, and trade on, the ever more glorified status of its beasts. In the process, cetology did indeed rebound in some unpredictable ways: hip-booted cetologists with a hand in the whaling industry might complain about the fuss, but at the same time they were emerging as the keepers of a newly sacred flame. A number of them answered the call, publishing prettified books that fed public fascination, even as the authors postured to retain their credentials as objective men of science, consigning the myths of yore to oblivion as ever (though rehearsing them, custodially, for collective edification).[230] Other figures, like Roger Payne, managed, more successfully than Lilly, to become gurus of whale hugging (and to take international cetacean enthusiasms to the wide screen) while retaining their scientific personae.

All this may seem strange—as Sebeok put it, "strange, irritating, anecdotal, and provoking." But it is strange and provoking—the history of the science of whales in general is, I would argue, strange and provoking—

229. See Thomas A. Sebeok, review of *Communication among Social Bees* by Martin Lindauer, *Porpoises and Sonar* by Winthrop N. Kellogg, and *Man and Dolphin* by John C. Lilly, in *Language* 39, no. 3, pt. 1 (July–September 1963): 448–66, at p. 465.

230. See, for instance, F. D. Ommanney, *Lost Leviathan: Whales and Whaling* (New York: Dodd, Mead, 1971), and *The Whale*, ed. L. Harrison Matthews (London: Allen and Unwin, 1968).

precisely because it brings to the surface a lurking problem that is too easily left submerged when we trace the progress of knowledge making, a problem that is, in the end, central to the history of science: How is it that human investigations of the natural world continuously wipe away the all-too-human fingerprints on the mirror of nature, and in doing so perpetually offer us captivating ways to see ourselves anew?

In the surface of the sea, in Lilly's beloved and addled dolphins, in what Melville called "the awful Chaldee of the Sperm Whale's brow," we catch such beguiling reflections.

SEVEN

CONCLUSION

Dreamt of
Moby Dick the Great White Whale
Cruising about
with a flag flying
with an inscription on it
"I am what is left of Wild Nature"

LAWRENCE FERLINGHETTI, entry in "The Dreambook," 14 October 1977[1]

THE END IS NIGH

In the summer of 1980, my parents, then junior professors of French literature at neighboring universities in the Midwest, rented a cottage on the Outer Banks in North Carolina, not far from Hatteras, where Remington Kellogg and his Hopkins collaborators quietly opened an era of American research on cetaceans back in 1928. I had just turned ten, and the *Tursiops* that (still) moved in arcing pods though the surf in the evening were the first marine mammals I had ever seen. It was the summer of one of the greatest tennis matches of all time: the Wimbledon final that pitted ritually grizzled Björn Borg against a mop-topped and emotionally incontinent John McEnroe in what would become a legendary five-set marathon. We watched it unfold in the peak-ceilinged living room with a visitor who

---

1. Cited in Rex Weyler, *Greenpeace: How a Group of Ecologists, Journalists, and Visionaries Changed the World* (New York: Rodale, 2004), 483.

dropped by, some older friend or colleague of my mother's who pulled up in a convertible looking tanned and comfortable and spoke volubly about politics while drinking a beer. His subject was Ronald Reagan, I recall, and the importance of a strong national defense, but I remember him most clearly because he had a bumper sticker on his car that read "Nuke the Whales." This made an impression on me. Nuke the whales? Why?

I must confess that when I first began thinking about this book, there were moments when I imagined the whole project as a kind of Borgesian effort to trace every thread in the history of the twentieth century that could be understood to be wound up into the tight knot of this sloganeering provocation. In those daydreams I wondered what would happen if one went about treating this paper-thin cultural residue as nothing less than the culmination of all the forces and preoccupations, the technologies and anxieties, the illicit pleasures and the licit violence of the last hundred years. After all, that little irruption of savage Dada hyperbole seemed, as I rotated it in a mind still largely free of whale knowledge, to shiver with the presence (or was it the absence?) of the gravest questions of the age: nuclear war, the end of nature, the politics of nihilism, that peculiar train wreck of hysterical apocalypticism and insouciant ironizing that one could call modernity. And then, too, there was the way those three words (commandment? exhortation? prophecy?) could be read as a kind of three-syllable ultra-haiku on some central and perennial problems in American literature and culture: here was the machine in the garden projected to planetary scale; here was mad Ahab at the helm of a Trident submarine; here was Swiftian satire reborn on our shores as Hollywood excess; here was the hortatory sublime played as a burlesque that was neither quite serious nor quite in jest—and that right there seemed to be a plausible epitome of this wonderful and deranged country.

Well, it was always an absurd notion, and thus to have failed in the execution of such a work may perhaps be accounted a credit. Still, there is a way in which this book has indeed tugged on at least one of the taut cords that gave this slogan its strange resonance in the early 1980s—a resonance I could not feel at the time, but which, in the course of this research, I have come to sense quite keenly. We have seen, in the last chapter, the peculiar way in which, for a growing cohort of activists, reformers, and visionaries in the 1960s and 1970s, cetaceans came to represent—even to embody—an alternative mode of intelligent being, and thus an aspirational ideal for humans, who were in these years in several respects (sociobiology, science fiction, primatology) reimagining themselves as the violent ape-kings of creation. *Homo insapiens* dissembled and escalated; cetaceans saw through each other in a perpetual détente of transparency. We used our hands to make tools and weapons; they used endless play to build relationships and communities. We fought wars and pillaged; they made love and music. We bit the dust that was the residue of death; they stayed in the watery womb that was the cradle of life.[2] We were territorial cave dwellers, born to tribalism; they used the deep sea for long-range communication, reifying the fantasy of a global village.[3]

And there was, moreover, a specifically military dimension to all this, as I have also tried to demonstrate. The story of John Lilly, CRI, Forrest G. Wood, and the navy Marine Mammal Program—and the leakage of that story into the larger cultural preoccupation with thermonuclear Armageddon via best seller and silver screen—must go some way toward accounting for that bumper sticker's bizarre juxtaposition of thermonuclear weaponry and lumbering sea creature. But does it go all the way? Clearly not. The harder one looks, the more perfectly overdetermined that juxtapo-

2. For a choice instance of this imagery, see (leading whale scientist) Roger Payne's ejaculation concerning a stranded whale in *Among Whales*: "As it lies beached and dying, does it not feel for the first time what it is to be like a human infant wailing in rage as it is stranded on dry ground out of its mother's womb? Does the anger that terrestrials show to each other and to the rest of life have its origins in the loss of our marine oneness—a regret for the mistake made by our ancestors who chose to be born out of the sea?" Roger Payne, *Among Whales* (New York: Scribner, 1995), 56. It is tempting here to note, by way of a closing of the circle, that the man who claimed to be the originator of the controversial "aquatic ape hypothesis" of human origins was none other than Alister Hardy—one of the original Discovery scientists. The key to his insight? That human beings have subcutaneous fat deposits structured in a way that Hardy decided was analogous to cetacean blubber!

3. This image is a reference to theories of long-distance cetacean communication, first raised as a possibility by Roger Payne in the early 1970s. For a discussion of this issue (which has recently become a point of controversy because of legal showdowns over ocean noise and cetacean conservation), see David Rothenberg, *Thousand Mile Song: Whale Music in a Sea of Sound* (New York: Basic Books, 2008), esp. chap. 8; Roger Payne, *Among Whales* (New York: Scribner, 1995); Naomi Oreskes, "Science and Public Policy: What's Proof Got to Do with It?" *Environmental Science and Policy* 7 (2004): 369–83; and the forthcoming book by Josh Horwitz, *Whales vs. Navy: The Story of Whales, a Secret Navy Sonar Program, and the Environmental Lawyer Who Exposed It* (New York: Twelve). For what I believe is the earliest published paper to raise such a possibility (though it is more concerned with navigation than communication), see Roger Payne and Douglas Webb, "Orientation by Means of Long Range Acoustic Signaling in Baleen Whales," *Annals of the New York Academy of Sciences* 188 (December 1971): 110–41.

sition begins to feel. Leo Szilard's improbable 1961 Lilly-inspired marriage of cetacean intelligence and nuclear war (in "The Voice of the Dolphins") yoked the mindful whales to the problem of the bomb in a way that proved oddly durable. Perhaps not so strange, then, that Scott McVay headed off to the Pugwash Conference of 1970 to pontificate on whale conservation in a (Melvillean) presentation that would headline in the *Bulletin of the Atomic Scientists*—the community of nuclear intellectuals representing, as they did, the dominant scientific statesmen of the day.[4] And maybe not so odd that Joan McIntyre could be quoted in the *Washington Post* in 1973 likening the moratorium on whaling to the nuclear test ban treaty; her larger claim was that "if we can save the whales we can save ourselves," and doing so would require "a win on the international level."[5] And one cannot really be all that surprised to read in *The Whale Problem* of 1974 a cohort of genuine hip-booted cetologists positing that (assuming adequate samples of whale tissue kept coming in) the cosmopolitan cetaceans, giant bioaccumulating filter systems, could serve as very useful monitors of "radio-nuclides and other fallout products" in the world's oceans.[6] After all, everyone was interested in radioisotopes in those years, why not the whale biologists? Especially the old guard, not quite ready to hang up their flensing knives.

4. Scott McVay, "Does the Whale's Magnitude Diminish?—Will He Perish?" *Bulletin of the Atomic Scientists* 27 (February 1971): 38–41.

5. Nicholas von Hoffman, "Whale Love," *Washington Post*, 16 February 1973, B1.

6. This discussion can be found in the section "Trophic Analysis" in the "Report of the Working Group on Biology and Natural History" from the Shenandoah Conference (see chap. 6, above), which included Ichihara, Jonsgård, and Gambell. William Edward Schevill, G. Carleton Ray, and Kenneth S. Norris, eds., *The Whale Problem: A Status Report* (Cambridge, MA: Harvard University Press, 1974), 7. Was Gambell a hip-booter? There is a rather unflattering portrait of him (is it true?) in *Mind in the Waters*, where Joan McIntyre recalls a conversation between him and Russell Train at the 1972 IWC meeting, in which Train was supposed to have said (in response to the Scientific Committee's rejection of the proposed moratorium on the grounds that research would be impossible if there was no whaling), "Well, I can understand the need for more whales to study, but do you really need thirty-five thousand?" To which Gambell is quoted as responding, with a smile and a stroke of his chin, "Yes, lots more!" The activists in this movement were not, as we know now, above pure fabrication (for instance, the ersatz report by the Greenpeace sailors of having witnessed a sperm whale fighting a giant squid—wholly invented for the purpose of capturing media attention). For the exchange, see *Mind in the Waters: A Book to Celebrate the Consciousness of Whales and Dolphins*, comp. Joan McIntyre (New York: Charles Scribner's Sons and Sierra Book Club, 1974), 106–7.

And there was also the way, of course, that Greenpeace transformed itself, in the early 1970s, from a loose aggregation of antinuke agitators into the shock troops of the "Save the Whales" campaign, leaping to international prominence in the process. But could they have asked for a more perfect convergence of their preoccupations than the emergence of the rumor that one of the privileged uses of sperm whale oil lay in the lubrication of precision mechanisms in the rockets and missiles of the United States and the Soviet Union?

It is not clear to me exactly when this notion first emerged in the whaling debates, but it would appear to have been around 1971 and to have been closely linked to the controversy surrounding the listing of the large whales as endangered species under US law. Walter Hickel gave himself a starring role in a conversation along these lines that purportedly occurred in his office with a pro-whaling visitor from the State Department, who is supposed to have blurted out, as a last-ditch effort to derail the listing, "We have to have whale oil for the space program." To which Hickel quoted himself retorting, "What are you going to use when the whale is extinct?"[7] The listing went forward, with the understanding that stockpiled product would suffice for some time. "Space program" may well have been understood euphemistically, since Rex Weyler reports a conversation in Vancouver in December 1972 between Farley Mowat and Paul Spong in which the former told the latter, "The Russians lubricate their ICBMs with whale oil. So do the Yankees. It's a disgrace."[8] While the vast majority of the sperm-derived lubricants were in fact going into much more innocuous industrial uses (such as automotive transmissions), the link between the transcendent mind in the waters and the transcendent mechanism of destruction was a gift for those promoting whale conservation as one half of a pincer movement against the masters of war.[9] Thus, with "brainwashed" dolphins doing duty in the mangrove swamps of Vietnam and military ordnance being

7. Walter J. Hickel, *Who Owns America?* (Englewood Cliffs, NJ: Prentice-Hall, 1971), 145.

8. See Rex Weyler's Foreword to the 2005 edition of Farley Mowat, *A Whale for the Killing* (Mechanicsburg, PA: Stackpole Books, 2005), viii.

9. The industrial solution to the end of freely available sperm whale products turned out to be the cultivation of the desert plant jojoba, the essential oils of which provide a plausible substitute. This substitution was worked out by a committee of the American Chemical Society under the leadership of Milton Harris. See Jon H. Georg, "Milton Harris: A Biography," which is the front matter to the Milton Harris Papers, held in the Special Collections Department of the Oregon State University Libraries.

used to slay their larger cousins on the high seas, the conditions were set for a community of true believers to raise a rainbow flag over the cetaceans and declare themselves the party of peace and life, standing (in a bobbing Zodiac) against the culture of death.[10] "For a long time now," wrote one of the activists in 1977 in the *New York Review of Books*, "man's awe has been confined to his own capacity for self destruction"; the whales, by contrast, were giving us an opportunity to learn "to revere life."[11] And by 1980, one of the early Greenpeace members could state this position in its most extravagantly dialectical form: "Cetaceans," Michael M'Gonigle would claim boldly, "stand in poetic contrast to human history."[12] That history was the history of war.

Every dialectic, of course, is a contrapuntal conflict waiting to be *aufgehoben*, and that brings us to the Junior Common Room of Kirkland House at Harvard University on 13 April 1979, where a Ramones-inspired punk band called Supreme Pontiff (descendants of The European Liquidators, succeeded by Kid Sonic and the Boom), under the leadership of a Harvard senior who played bass as Töd Venice (he was Robert A. Falk, born in Brooklyn), ground out their new one-minute-and-thirty-nine-second anti-paean "Nuke the Whales," composed by Falk the previous autumn. He and a bandmate (lead singer John Cole, aka "Jean Baptiste"), sensing they had a dance-party sensation on their hands, had gone so far as to buy a silk-screen kit and make up several dozen T-shirts for the gig, emblazoned with the refrain of their kick-over-the-speaker-tower frontal attack on the world of good intentions:

> If grapes aren't union picked don't eat 'em
> Boycott J. P. Stevens, we can't beat 'em
> Water causes cancer, and cancer causes death.
> Jesus Christ, I'm scared to take another breath.

10. This sort of Manichean framing, which is impossible to overlook in the sources from the period, may well owe something to the influence of Norman O. Brown, whose *Life against Death* (Middletown, CT: Wesleyan University Press, 1959) had considerable currency among the eroticizing pacifists of the 1960s.

11. This was Jack Richardson, who had been aboard the *Phyllis Cormack* in 1975 (as a reporter initially rather indifferent to the whales, though he would "convert," as he put it, over the course of the voyage): Jack Richardson, "An Oppressed Group," *New York Review of Books* 24, no. 12 (14 July 1977): 26–27.

12. R. M. M'Gonigle, "The 'Economizing' of Ecology: Why Big Rare Whales Still Die," *Ecology Law Quarterly*, 9, no. 1 (1980): 119–237, at p. 121.

Everybody's got their favorite cause,
Tryin' to pass restrictive laws.
I say can your sad old tales,
And fuck it! NUKE THE WHALES.[13]

Sophomoric? To be sure. But generational? Yes, also. Falk—a major in visual and environmental studies—had hit a nerve, and his catchphrase indictment of the suffocating righteousness of left-liberal orthodoxy, delivered as Frye boot in the ass of the Woodstockers, moved with electrical speed across North America. That very summer it could be found as a bumper sticker in the Deep South as well as on the lips of a quirky non-whale-hugging bluegrass musician named Duck Donald playing the Canadian summer folk circuit (to spasms of dismay).[14] By the following year—when the Pontiffs were no more, but their successor band was fine-tuning another Reaganite non-parody (the underground classic "Nancy Packs a Piece")—there was an actual "Nuke the Whales" country-rock band lighting up the stage in Huntsville, Alabama, and the phrase could be found as a graffito on bathroom walls at UC Berkeley.[15] And that was also the summer I saw it, while standing in the saw grass dunes of a mid-Atlantic barrier island. Which is to say, the viral phrase had quite literally gone coast to coast. It would not be all that crazy to say that Robert A. Falk, sitting

13. My recovery of the origins of the phrase owes a great deal to a letter written by Robert O. Landry to the *New York Times* (published 7 October 1979, p. 12, of the book review section); in it, Landry, who was a recording engineer at WHRB, the Harvard College student radio station, corrected Roy Blount Jr.'s misattribution of the expression and pointed out its earlier use in Falk's song. I followed up the reference into online sources that reviewed the history of the Cambridge punk/new wave scene in the period (see, for instance, the Wikipedia entry on "The Boom [American Band]," last modified 8 December 2009) and eventually found Falk himself, now an entertainment lawyer in Massachusetts, who shared with me ephemera of the moment, including the poster for the Kirkland House gig. Robert A. Falk, interview by the author, 10 September 2009.

14. For the reference to the phrase in the Deep South, see Roy Blount Jr., "Fresh Homebaked Goods," *New York Times*, 12 August 1979, pp. 9, 29, of the book review section; for Duck Donald, see Alistair Brown's write-up of the 1979 Canadian summer folk festival season in his column "Ali's Corner," *Canada Folk Bulletin* 2, no 6 (November–December 1979), 25.

15. On the Alabama band, see *Nuke the Whales*, accessed 7 March 2011, www.nuke25. blogspot.com. And on the graffito, see Daniel W. VanArsdale's fantastic archive, *Eat No Dynamite: A Collection of College Graffiti*, accessed 7 March 2011, www.silcom.com/~barnowl/ graffiti.htm (I am assuming this particular item hails from the original collation).

in his dorm room in Kirkland, wrote the epitaph for the two decades that get called "the sixties," and that he did so by writing a bumper sticker for America in the Age of Reagan.

In 1982 the IWC finally passed a moratorium on commercial whaling,—a mostly unheralded semipermanent "suspension" of the hunt, an agreement hedged around with an intricate and thorny tangle of exceptions, objections, forward-looking delays, and stipulations of future reassessments. Big whaling pretty much came to an end, though the killing of whales did not. Disputes of various kinds—ethical, legal, diplomatic, scientific—continued within and beyond the IWC through that decade and continue into the twenty-first century. But the truth is, the whaling issue passed its sell-by date right about the time of senior house parties, 1979.

A CONDOR'S QUILL

At the start of this book, I promised a study of changing whale knowledge across the twentieth century. Have I delivered on that promise? I'd like to think so. In the five substantive chapters at the heart of this volume, I have traced the development of scientific research on cetaceans from the opening of the modern whaling industry in the late nineteenth century through to its demise in the late twentieth. This has meant recovering the largely forgotten history of several institutions that focused on this sort of work (the Discovery Investigations, the Scientific Committee of the IWC, the Communication Research Institute) as well as sketching the biographies of a number of scientists who made whales central to their life's labor (men like Sidney Harmer, Remington Kellogg, and Neil Alison Mackintosh). Along the way, I have been concerned to give a feel for the changing field practices and research traditions that contributed to changing understanding of this taxon (and they were strikingly diverse, spanning research on reproductive physiology, intricate population dynamic modeling, acoustic technologies associated with antisubmarine warfare, and even Cold War neuroscience). I have also worked to show how the knowledge that came out of these braided investigations entered the world of geopolitical negotiation, economic ambition, and conservation concern. In the end, as I have argued, whale knowledge influenced larger cultural dynamics in surprising and direct ways, particularly in the last third of the twentieth century.

At the same time, as I draw this book to a close, I am acutely conscious of its limitations and failings. Among other things, this study has not made significant use of source material in Japanese, Norwegian, or for that matter, Russian. The precise significance of these omissions is difficult to assess (recall the Rumsfeldian adage about the unknown unknowns . . .). A more complex and ramifying treatment of the diplomatic history around whaling could certainly be composed by an author with access to archival documentation in one or more of these languages.[16] Would a history of twentieth-century whale *science* look different from these perspectives? There would surely be valuable additional things to understand. After all, the Soviet Union banned the killing of small cetaceans (dolphins and porpoises) in 1966, half a decade before comparably comprehensive legislation was forthcoming in the United States, and at the very same time that Soviet whalers were, as we now know, ransacking the world's oceans for larger cetaceans of every sort in secret and sweeping defiance of the ICRW and the IWC.[17] US technical services maintained a translation service covering Russian-language scientific literature through much of this period, and I have seen snippets of this material suggesting that Lilly's work found interested readers behind the Iron Curtain.[18] But a real history of the Soviet attitude toward small cetaceans would surely reveal other factors at play. It is worth noting, of course, that the Russians, too, had a military marine mammal program, and from relatively early on.

Where the Norwegians are concerned, my impression is that a very large percentage of the most important scientific work being done on cetaceans in the twentieth century remained closely connected to English-language

16. For a taste of the possible, consider chap. 5 of Eldrid Ingebjørg Mageli, *Towards Friendship: The Relationship between Norway and Japan, 1905–2005* (Oslo: Oslo Academic Press, 2006).

17. Alexey V. Yablokov, "On the Soviet Whaling Falsification, 1947–1972," *Whales Alive!* 6, no. 4 (October 1997), accessed 11 March 2011, www.csiwhalesalive.org/csi97403.html. See also Yablokov, "Validity of Whaling Data," *Nature* 367, no. 6459 (13 January 1994): 104.

18. See the remarks by Soviet minister of fisheries Alexander Ishkov to the effect that dolphins have brains "strikingly like our own" and that they are the "marine brother of man," cited by Scott McVay in *Mind in the Waters: A Book to Celebrate the Consciousness of Whales and Dolphins*, comp. Joan McIntyre (New York: Charles Scribner's Sons and Sierra Book Club, 1974), 227. But the hypocrisy and bad faith of the declaration has been subsequently denounced by insiders. See the special issue of *Marine Fisheries Review* on Soviet whaling: Yulia Ivashchenko, Phillip J. Clapham, Robert L. Brownell Jr., A. V. Yablokov, and Alfred A. Berzin, "The Truth about Soviet Whaling," *Marine Fisheries Review* 70, no. 2 (2008): 1–59. For an example of the government-sponsored translation work, see "Soviet Studies on Cetaceans," National Technical Information Service Report JPRS 49777, U.S. Department of Commerce, 9 February 1970.

research.[19] This is not to say that there is not a deep primary-source study to be done on Hjort and whaling, since there obviously is, and such a treatment would presumably complement, and quite possibly trouble, my Discovery material. All that would be welcome. By the post–World War II period, US geopolitical dominance (and scientific leadership) had transformed the structures that channeled scientific arguments about cetacean populations. I have no doubt that detailed work on later Norwegian science and scientists (such as Ruud and Jonsgård) would be fascinating, but it is not clear to me that such work would significantly alter the basic story I have told in these pages. Needless to say, I could be wrong.

And then there are the Japanese. Japanese whale science and whale knowledge in the twentieth century? That would surely be a history worth figuring out. But it would not be easy. That, too, I shall happily leave to someone else.

Within the ambit of what was, from the outset, actually achievable in this study, there remain a number of whale books and parts of whale books that I have not written, and their disembodied souls chirp about me at this closing hour. Those readers actively involved with current IWC affairs will surely miss a serious treatment of the contentious labors of the Scientific Committee in the early 1980s, and again in the early 1990s, as the moratorium took shape and saw reassessment. As do I. Environmental historians and American intellectual historians are likely to complain that my handling of the rise of environmental consciousness in the 1960s and 1970s is altogether too oblique (they will be right), and too preoccupied with charismatic megafauna (ditto) to the neglect of other critical dimensions of this period and its problems (significantly, pollution and environmental contamination). A better book would have done better holding these issues together with the efforts to save the whales. And if, for modern environmental activists, I need to acknowledge that species-obsessed conservation efforts have fallen to very low standing among those concerned to save us from ourselves, environmentally speaking (indeed, that the whole business of saving whales and pandas may have been, without our quite knowing, a vast ruse by which collective social action was siphoned off onto gaudy baubles while the world was quietly wrecked by deforestation, ecosystem

19. Mackintosh made the same point in 1965, though it was no less self-serving in his case: Neil Alison Mackintosh, *The Stocks of Whales* (London: Fishing News Books, 1965), 22.

collapse, and the unrestrained burning of fossil fuels), let me do so here, while adding—slightly defensively—that I do not think any of this relegates my study to irrelevance. Historians of modern biology may feel that the development of ecology has been slighted in these pages, and point out that ecological arguments were important to Greenpeace and played an expanding role in the politics of whale conservation after 1970. This too is correct, and a weakness of the present work.[20] On the whole, I would have been delighted to write a better book: the spirit is willing, but the flesh is weak. And truth is, even the spirit can flag a bit at a certain point. I was taught, long ago, that if you fell asleep while praying, the angels would com-

---

20. Though it should be said that Greenpeace's slogan, "Ecology? Look it up—you're involved," had only a glancing relationship to ecology proper. In practice, the attention to ecological rhetoric afforded a powerful way of insisting on the immediate and intimate relevance of seemingly remote issues of conservation or natural resource use. While ecosystem thinking has influenced whale science in the last twenty years, invocations of "ecology" in the whale wars were generally glib. See, for instance, David O. Hill's suggestion that because the whales "regulate the plankton economy" their demise had implications for the production of atmospheric oxygen, leading to the breathless conclusion that "wiping out the whale could jeopardize the supply of oxygen that both marine and human life depend on." These quotes are a paraphrase of Hill's argument by Walter James Miller, in his (wonderful) volume: Walter James Miller, ed., *The Annotated Jules Verne: Twenty Thousand Leagues Under the Sea* (New York: Signet, 1976), 272. Hill was one of the founders of the Rare Animal Relief Effort (RARE), and it was he who focused this nascent organization on whaling in the early 1970s, inspired in part by contact with Joan McIntyre. RARE would become instrumental in organizing boycotts of Japanese products in the United States. Hill was, however, a laid-off airline pilot and skilled amateur birder whose command of technical ecology was nil. David O. Hill, interview by the author, 14 August 2009. Another instance of this sort of loose talk can be found in George L. Small's testimony in U.S. Congress, Hearings Before the Subcommittee on International Organizations and Movements of the Committee on Foreign Affairs, House of Representatives, Ninety-Second Congress, First Session, 1971 (my copy is from the Smithsonian Archives, RU 7229, box 10, folder 7), in which he intimated that if pinnipeds were not protected, killer whales would turn to human bathers for food (see p. 19). Small was a geographer who had written a dissertation on whaling that he turned into the influential book *The Blue Whale* (New York: Columbia University Press, 1971), winner of the National Book Award for 1972. Scott McVay, always clever about mobilizing his BA in English literature, dexterously read Melville's "Monkey-Rope" chapter in *Moby-Dick* into an allegory of ecological interconnectedness: "Can Leviathan Long Endure so Wide a Chase?" *Natural History* 80, no. 1 (January 1971): 36–40, 68–72.

plete (and thereby perfect) your orations. This does not work with whale books, I can assure you.

The foregoing amounts to only a gesture at the shortcomings of this study; reviewers can aid me in fleshing out the list. Let me turn, instead, to a brief recapping of what I take to be salient in the story I have managed to tell.

In addition to elaborating the basic narrative history of an important episode in the history of British science in the first half of the twentieth century, chapter 2 brought to light the forbidding world of the hip-booted biologists. By tracing the evolution of the actual field practices of whale biology in those early years, and by situating those research programs in the context of administrative efforts to get control of the whaling boom in the Southern Hemisphere, I showed how cetologists working on the natural history of whales gradually found themselves sucked into the belly of the beast—growing increasingly dependent on the industry for access to specimens (for work in reproductive physiology) and data (on mark recoveries and migrations). All of this proved to be the unwinding of the fatal compact over which Sidney Harmer agonized in the period before World War I: was it acceptable to trade whales for whale knowledge? Harmer hemmed and hawed, but eventually decided in the affirmative. Indeed, by 1924, as he proudly expostulated on the soon-to-be-launched Discovery expeditions at the Art Workers' Guildhall, he would go so far as to gloat about the savvy of the whole arrangement, boasting that the British taxpayers would not pay a penny for cetacean research "because the whales were paying for it themselves"—a bon mot that met with hearty laughter all around.[21] And pay they did, for the next half century.

In fact, more than fifty years later, in 1985, wags attending the IWC meeting at Bournemouth, UK, issued what amounted to a punchy retort to Harmer's Discovery calculus: a faux press release for the delegates, announcing that

> the International Consortium of Cetaceans—the little-known body that represents all the races of whales and dolphins in the world's oceans—met recently outside Maui. Acting on the recommendation of the homo sapiens sub-committee, the whales agreed to issue a

21. See "Whaling Research: Plans of the 'Discovery' Expedition," *Morning Post*, 3 March 1924. NOL Discovery MS, scrapbook of press cuttings.

FIGURE 7.1 Scientific whaling: We'll just take a few—for research purposes. (Courtesy of the University of Washington Libraries, Special Collections.)

scientific permit for taking up to 65 IWC delegates and observers, "to fill in data gaps and attempt to determine why this species acts as it does."[22]

The accompanying cartoon depicts a harpoon winging up from the deep sea and pegging besuited cetologists; the hidden whales bark that they are taking aim at science itself (figure 7.1). It is an image that bears comparison with the cartoon from which this book departed (figure 2.1), in which science was depicted setting sail, in a very whalerish way, for knowledge of whales.

And in fact it is true that chapter 2 offers something like a prehistory of "scientific whaling," such as it comes down to us in the ongoing debates over current Japanese practices. But it is my hope that my treatment of this early material has a wider reach as well: I believe I have shown, for instance, how the Discovery Investigations got caught in an infelicitous way between the heroic age of ocean exploration (exemplified by the voyages of the *Challenger*) and the dynamics of recognizably "modern" oceanographic research, where military objectives afforded remarkable new opportunities for funding and alliances.[23] Discovery had the misfortune to be conceived

22. "Exclusive to ECO," *ECO* 30, no. 5 (19 July 1985): 3. I have this document from the University of Washington Library, Chapman Papers, box 13, file "Newsletters."

23. In this sense, chapter 2 goes some way toward bridging the periods treated by Helen Rozwadowski (in *Fathoming the Ocean: The Discovery and Exploration of the Deep Sea* [Cambridge, MA: Harvard, 2005]) and Jacob Darwin Hamblin (in *Oceanographers and the Cold War: Disciples of Marine Science* [Seattle: University of Washington Press, 2005]).

in the spirit of the earlier epoch (though this was no longer even then an entirely suitable mode for sophisticated research in the sciences) and to have succeeded in surviving on into the second, by which time the whale biologists—once the avant-garde of the whole undertaking—looked like relics. If the residual glow of nineteenth-century ocean science proved a distraction to Discovery, the era of submarine warfare proved its undoing. In between, Discovery researchers amassed a vast body of data about the large whales of the Antarctic waters, data that, I have suggested, actually went some way toward *obscuring* the seriousness of the problem with the stocks, in that the collection of data came to be (for some) an end in itself—data collection had become a kind of scientific life. It was a strange sort of biology, one where stripes were earned for service in the smoking slaughterhouses of an unprecedented holocaust. But these trials annealed the community of practitioners and heightened their sense of solidarity with others who pursued the whale.

This perverse dynamic—a kind of "saving of the phenomena" (the data, the research trip, the IWC itself) in lieu of the *noumena* (the actual whales, the conservation of which was, as we have seen, supposed to be the point from the very beginning)—recurs in this study. I am reminded of the fate of Remington Kellogg, whose biographical arc constitutes an armature for this book across chapters 3, 4, and 5. Having attended on him so closely, from his youthful exploits at Hatteras to his pioneering efforts to use science to make news and mobilize Americans via the Council for the Conservation of Whales (an important and overlooked episode in the history of Progressive Era concern about nature, and the subject of chapter 3), and finally through to his pained later years at the IWC—shall we see him off?

On the occasion of Kellogg's formal retirement from the directorship of the National Museum of Natural History in 1962, the Smithsonian's feather expert, a spirited Southern woman named Roxie Laybourne, presented her outgoing boss with a cartoon encapsulation of his distinguished career (figure 7.2). The line drawing depicts his life trajectory as a passage from whale (at the start) to whale (at the end), interrupted by a too-lengthy hiatus amidst piles of books and papers. Burning the midnight oil, reading and writing reports at his desk in those lonely middle years, this scientist-statesman has been forced to leave the hand lens of the naturalist behind and to serve out his clerical duties in exile from his beloved subject organism. The whale upon the plinth behind him, obscured by reams of what can only be IWC documentation, sheds a forlorn tear. Kellogg's career thus con-

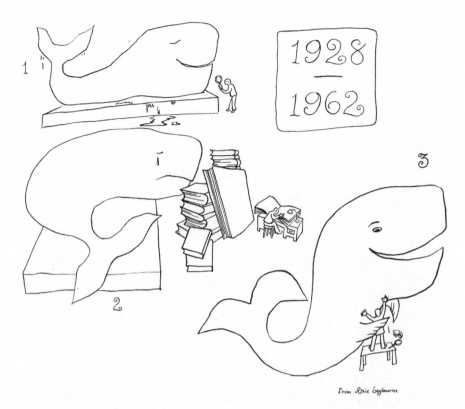

FIGURE 7.2 Cetology and the ages of man: Roxie Laybourne's cartoon depiction of the career of Remington Kellogg. (Courtesy of the Smithsonian Institution.)

figured, Laybourne can cleverly depict his retirement not as a terminus or a launching into outer darkness, but as a welcome homecoming to the love of his younger days: papers aside, hand lens at hand, the senior scientist and a now-cheery beast reunite in an embrace of renewed acquaintance. It is a graphic epigraph that, despite its little fibs (the first panel shows a youthful Kellogg inspecting a large whale still dripping from the sea, but as we have seen, he never really did this with a large whale, spending most of his research life ensconced among museum specimens), possesses some power and more than a little truth.

Perhaps too much truth for its own good, since Laybourne's drawing secretes an irony that cannot have escaped its recipient. To understand how, we must look more closely at her rendition of the whale in the final panel: with its arched body and absence of a slab base, this is not, in 1962, any old whale, but rather is surely intended to refer to the great cetological triumph

FIGURE 7.3 (and see PLATE 16) The miracle of plastics: The Smithsonian's fiberglass blue whale is prepared to preside over Ocean Hall. (Courtesy of the Smithsonian Institution.)

of the Smithsonian Institution in that year, the year Kellogg finally stepped down: the unveiling of the museum's massive, unprecedented, 97-foot-long fiberglass model of a blue whale, which plunged gracefully through the air over the new Ocean Hall (figure 7.3). This "miracle of plastics" (as the *Star Magazine* called it) outstripped in sheer size the museum whales of Brussels, Copenhagen, Stockholm, Leningrad, and even London, whence the templates for its head and tail had originally come.[24] But the true glory of this trophy, everyone agreed, was its unique lifelike shape and distinctive

24. The *Star* quote is from "A Whale of a Job," *Star Magazine*, Washington, DC, 20 May 1962. For the size comparison, see the *Star*, 21 March 1960, B1. For details on the templates, see Kellogg to Francis C. Fraser, 26 September 1956, Smithsonian Archives, RU 7170, box 3, file "Fra–Fui."

hidden suspension: rather than another rigid dirigible of cetacean taxidermy, another giant sausage strapped up to the ceiling or lying on a stage as if waiting for the flenser's blade, the Smithsonian whale, curving toward the depths, appeared to have been frozen in the very act of turning her massive flukes and gliding into a dive. No wires, no base. It had taken an engineer from Andrews Air Force Base to rig up the hidden cantilever that clipped the 9,000-pound sculpture to the wall of the lofty gallery, adjacent to the grand stairway that permitted visitors to descend with the whale into her world.[25]

"Dr. Kellogg used to say with a smile as he looked at this great cetacean that it was his memorial," remembered those who gathered at his funeral in 1969.[26] One is inclined to think it was a wry smile, since no one knew better than Remington Kellogg that the blue whale had been pursued to the very brink of extinction by the mid-1960s—only a few hundred animals probably survived in the whole Southern Hemisphere, and few experts thought they would recover. The whale Kellogg was thus fated to embrace after his long years at the IWC was a simulacrum, a token of loss, an empty shell. The destruction of the blue whales of the Antarctic over the previous fifty years had not just happened on his watch, it had happened while he watched— some 329,946 of them, by his own count.

This was the gloomy and precise number he gave (carefully corrected to include the results of the last season) to the assembled Washington society crowd that gathered for the opening of the new Ocean Hall in the autumn of his last year at the Smithsonian. His notes for his speech, crabbed with cross outs and scribbled over with revisions, betray an orator more than a little uncertain about what should be said on the occasion: a little history of traditional whaling, perhaps; some statistics, always a good idea; a brief paragraph on regulatory efforts, but no mention at all, in the end, of the IWC. There was, he confessed, little reason for optimism: "The Antarctic seas are now the center of a fast fading pelagic whaling industry," he explained. "If history repeats itself, and there is every indication that such is now the case, an important marine resource will have been needlessly

---

25. For a contextual essay that also celebrates Kellogg's whale model, see A. E. Parr, "Concerning Whales and Museums," *Curator* 6, no. 1 (1963): 65–76. Also consider Michael Rossi, "Fabricating Authenticity: Modeling a Whale at the American Museum of Natural History, 1906–1974," *Isis* 101, no. 2 (2010): 338–61.

26. The line appears in the "Testimonial Resolution honoring Dr. A Remington Kellogg," Smithsonian Archives, RU 7434, box 8, folder 1.

dissipated." Then, in closing, he tried several versions of his valediction: "It is quite possible that some of you here today will never see a blue whale," he wrote (in a sentence that must rank high in the annals of understatement), before concluding, "If the present scale of exploitation continues until whaling becomes a losing game, this replica may prove a reminder to the visiting public of the general appearance of a once important marine resource."

Not exactly rousing words. So he tried again, scrawling with a fountain pen:

> Granting that the present scale of killing continues, the visitor here in the not too distant future can get a fairly accurate impression (idea) of the external appearance of the blue whale, the largest mammal known to have existed on this earth.[27]

Better. That would have to do. Not a call to arms, to be sure, but it had a ring to it. A slightly hollow ring, admittedly. But it was a hollow whale that Kellogg, for all his efforts, had left to the world.

Presumably it was one of the minor gods of archive irony who arranged for the same Smithsonian box that contains the manuscript of that speech also to contain a seventeen-page press packet from Joan McIntyre's crusading Project Jonah, complete with mimeographed letters from children around the world begging leaders to save the whales. Though the cover letter is addressed to Kellogg personally, the materials are dated 10 July 1973, by which time Kellogg himself had slipped the mortal coil, like most of the whales he had hoped to protect.[28]

· · · · · · · · · ·

At the close of chapter 5, I showed that Kellogg was one of those brave enough eventually to argue that the problem of excessive commercial whaling would probably need to be solved by some new institution or diplomatic mechanism, the IWC having finally demonstrated to him, by 1964, that it could not be salvaged. It had taken him some time to reach this conclusion, since the record suggests that he harbored serious misgivings for almost a decade. And this points back to that perverse dynamic alluded

---

27. This and other quotes from Smithsonian Archives, RU 7434, box 3, folder 14, "Talk at opening of NMNH Exhibit of Blue Whale." The word "idea" is in superscript in the ms., over the word "impression."

28. Smithsonian Archives, RU 7434, box 3, folder 17, "Project Jonah."

to above, whereby hard work (collecting data, making a complicated institution work) drives out good intentions (say, preserving large cetaceans from excessive exploitation). This pattern seems to me sufficiently pervasive that we would do well to give it a name, since I have come to suspect that it is a phenomenon ubiquitous in the history and practice of collective do-gooding. I propose that we call it the "Nicholson principle" of science and policy making. My moniker hails from the lead character in David Lean's classic 1957 film *Bridge on the River Kwai*, the British Lieutenant Colonel Nicholson, played by Alec Guinness.[29] It is he, recall, who at the film's climax falls victim to a momentary (and very serious) lapse of judgment when he endeavors to save the bridge that he and his men have spent the film laboriously erecting—the problem being, of course, that they are prisoners of war and it is an *enemy* bridge that they have built. Nicholson's eleventh-hour confrontation with his compatriot lying in the bushes with a detonator ("Blow up the bridge? Eeegaads! No!") amounts to a pithy set piece for all those wrong-footed by the virtues of dedication and dogged labor.

It would not be quite fair, I think, to put Kellogg in the Nicholson role with respect to the IWC. After all, when the conservation allies turned up in the bushes, detonator in hand (I am thinking here of figures like Sidney Holt at FAO and the increasingly truculent types at IUCN, who began to argue in the mid-1960s that whales needed to be saved from the IWC itself), Kellogg did not call for backup in an effort to protect the institution he had done so much to build. Rather, he acknowledged that new players were probably needed to resolve the quota impasse and break the resistance of the whaling nations. But we have also seen that he was painfully slow to appreciate the extent to which he was, in the end, basically working for the enemy.

With Neil Alison Mackintosh, the case is, it seems to me, much clearer: his "capture" by the whaling industry amounts to a textbook instance of the Nicholson principle in action. He was, throughout his life, most seriously committed to what he would have called the "rational regulation of the whaling industry," but in the course of building a respected career as a diligent scientist and public servant, he became a reflexive defender of the institutions and research practices to which he had given his life. The prob-

29. The film is based on a novel by the author of *Planet of the Apes*: Pierre Boulle, *Le Pont de la Rivière Kwaï* (Paris: René Julliard, 1952). He had himself been a prisoner of war in Indochina during World War II.

lem was that, by midcentury, an excessive number of whales had also given their lives to those same institutions and practices, and this he had enormous difficulty appreciating. When, in the early 1960s, he spied those who were bona fide allies of rational regulation taking aim at his kind of whale research (and at the Scientific Committee of the IWC), he did not hesitate—they had to be the bad guys, and he yelped for backup from the whalemen. "Blow up the IWC? Eeegaads! No!" It was a moment of truth, and the outcome was more than a little tragic, if also, in a way, understandable. Even someone like G. Carleton Ray, scratching his head while Payne and McVay played whale song at congressional hearings in the early 1970s—even he can be understood to have missed what was going on around him in a distinctively Nicholsonian manner. Were these sorts of whale scientists the "stooges" of the industry? I am inclined to think not. Rather, they were, on the whole, betrayed by the seductions of honest toil.

If the Nicholson principle sums up a theme of chapters 4 and 5, it does not, I think, exhaust what those chapters have to offer. By digging deep into the history of the scientific advising system of the IWC, I have written a detailed account of the ways that whale science did and did not shape the international negotiations around whaling in the two crucial decades following World War II. This is a fundamentally revisionist history, one that argues for nothing less than a reversal of a number of the basic presuppositions and findings of previous scholars to have looked at this episode. So, in my view, it is not that there was "not enough science" to guide IWC decision makers. Rather, there was altogether too much science, if of the wrong sort. And it is not that the work of the Committee of Three represented a victory of scientific wisdom over political treachery. On the contrary, this dramatic refiguring of the debate about the status of the stocks is better thought of as a clever diplomatic victory—one that made brilliant political use of scientific work that was itself nontrivially flawed.

I hope that those who are interested in this history will reckon with my analysis here, but I also hope that the analytic approach mobilized in these chapters will hold the attention of readers not necessarily invested in the vicissitudes of the IWC—readers interested in the broader problem of understanding the complicated and evolving relationship between the production of knowledge and the sphere of collective decision making, readers seeking to tell compelling stories about the intersection of science and politics. Here I do not claim to have elaborated a wholly novel heuristic or to have laid out a radically new theoretical approach. Indeed, I have been deliberately cagey about the virtues of constructivism while at times working to unfold scientific arguments on their own terms (a methodological eclecticism that reflects my primary commitments to historical narrative). Nevertheless, I do believe that I have offered a compelling case study of what it looks like to treat the boundary between political and scientific questions as nothing less than the central historical problem needing to be addressed in such situations. The result, I hope, can serve as an example of how one might approach the workings of science in the complicated arena of intergovernmental environmental regulation. In place of what I take to be largely unfruitful normative disputes about the acceptable measure of "scientific uncertainty," I propose a more nuanced appreciation of the extent to which uncertainty plays a crucial role in the promotional life of research schools and individual scientific agendas. In place of excessive model building or regime theorizing, I propose sustained attention to professional trajectories and the specific content of different intersecting research traditions. In place of a knee-jerk resort to narratives whereby science "gets politicized" in political settings, I propose serious engagement with the ways that scientists themselves have learned to function as political animals when they are in pursuit of tangible goals and scarce resources in complex social situations. For those who already have an appetite for tracking theories of knowledge down into the messy agon we call society, I suggest what I hope is a refreshingly flat-footed technique for reconstructing the intricate sociology of scientific claims: in complicated organizations, these exquisite dynamics are frequently reified in the seemingly tedious dynamics of committee architecture and parliamentary procedure; organizational charts can amount to palpable auto-ethnographies of these strange tribes.

At the end of chapter 5 I offered a kind of résumé of the book to that point, couched as a mini-history of the idea of being "rational" about the largest animals on earth. There I traced the way that this notion wandered across the century from a preoccupation with efficiency to a focus on maximization, with attendant shifts in what was understood to be the relevant scientific discipline: To be "rational" about whales, did you need a biochemist who could harden fat or a mathematician who could do the cascades of differential equations that generated maximum organic matter from a natural population of sea creatures? Or maybe you needed an economist, who could optimize dollar (or kroner) returns over discrete (or maybe indefinite) intervals of time. Was it more rational to treat whales as a liquid asset or as a liquid fat? As a crop or as a mine? As an abstract population that could be modeled by making certain assumptions about stochasticity, or as shifty quarry that learned from experience and required consideration of recursive behavioral characteristics?

In chapter 6, the very idea of human rationality itself came under scrutiny as a result of an increasingly eccentric—but ultimately very influential—effort to reckon with the minds of cetaceans themselves. As I argue in my treatment of John Lilly and the larger reimaginings of whales and dolphins in the 1960s and 1970s, new technologies (scuba, underwater acoustics, in vivo invasive neurophysiology), novel funding opportunities (NASA, the navy, television producers), and unprecedented relationships between science and the media (popular ocean theme parks that were also behavioral laboratories, underwater films that contributed to a reconceiving of the subaquatic world) created the conditions of possibility for a radical reenchantment of the whole taxon. If the bottlenose dolphin was largely thought of in the United States before World War II as the unlovely "herring hog" known to pilfer fishermen's nets, by the late 1960s most Americans were prepared to give this same creature pride of place in the animal kingdom, where it displaced the loyal and familiar dog as the nonhuman animal considered "smartest and friendliest and most highly respected by man."[30] This change was largely, as I have shown, the result of John Lilly's work—both his scientific research and his promotional efforts on behalf of the animals (and himself). Peculiar and reflexive cycles amplified his efforts: the navy's interest in publicly touting its dolphin work was, by the mid-1960s, tangled up with frustration at the extent to which the space race had sucked both money and attention toward the remote cosmos (and the air force), draining the glory from the earth's oceanic "inner space" (and the navy). The Man-in-the-Sea and Sealab programs represented efforts to steal some of the limelight back. Along the way the navy did much to emphasize the parallel between "outer space" and the deep blue beneath the waves: both were vast regions needing intrepid exploration; both harbored secrets that would transform the human future. The analogy did much to amplify the notion that cetaceans might be our alien familiars, intelligent extraterrestrials living right here among us.[31] In view of the navy's efforts

30. This quote is from Forrest G. Wood, remarking in 1973 on the change that had occurred within the previous thirty years. See Forrest G. Wood, *Marine Mammals and Man: The Navy's Porpoises and Sea Lions* (Washington, DC: Robert B. Luce, 1973), 1.

31. To concretize this suggestion, it is worth noting that one of the central figures in the founding of the navy's marine mammal facility at Point Mugu was Lieutenant Commander James H. Berrian, who was serving, circa 1960, in the Life Sciences Division of the Naval Missile Center, where he and his colleagues were working on

to "break though" to these animals (recall that the navy made a film to promote its work on a "Human to Dolphin Translator"), there was reason to think that there might be something to the idea. The idea itself, of course, was first Lilly's (like much else in the navy's program), but it became, if anything, *more* credible as a rumor of classified naval research (in the early 1970s) than as a messianic message from a wiry man in a jumpsuit hyped up on Vitamin K.

How Lilly got from the sober aviation physiologist and pocket-protector-wearing neuroscientist of the 1940s and 1950s to this latter character is part of the story of chapter 6, but even more central to that narrative are the ramifications of his breakout research of the early 1960s. As I hope to have shown, Lilly's use of the toolkit of forensic interrogation in an increasingly extravagant effort to access the inner life of his new subject animals drew his research into a disorienting no-man's-land at the periphery of the Cold War sciences of mind. But there were women there, and drugs, too. Somewhere around 1963 (and in connection, as I have suggested, with the First International Symposium on Cetacean Research), this sort of work hit an inflection point, with serious consequences: Lilly himself went from scientific insider (with strong military ties) to persona non grata in the world of federal funding and scientific bonhomie, and his animals were propelled to unlikely prominence in a tug-of-war between divergent subcultures (hawks vs. doves, frogmen vs. whale huggers). In this swerve, Lilly revised his understanding of both his scientific subject and his scientific subjectivity, reconceiving the problem of cetacean intelligence as less a technical than a psychodynamic challenge, less a problem with the cognitive limits of dolphins and more a problem with the cognitive limits of humans. Pursuing this insight, he embarked on a new course of self-exploration, touting the virtues of liberated eros and the judicious use of psychopharmaceuticals. He had become the whale guru, extrapolating from smart dolphins to quasi-divine sperm whales whose gigantic brains transcended time and space and afforded the properly sympathetic (and lubricated) human an imaginative glimpse of a synesthetic, psychedelic, stereophonic apotheosis

---

"life support systems for closed environments," which they expected would be part of the man-in-space initiative. When the Air Force won the contract for this work, they shifted their efforts to undersea life support systems. For some of this, see Forrest G. Wood, *Marine Mammals and Man: The Navy's Porpoises and Sea Lions* (Washington, DC: Robert B. Luce, 1973), 238. It would be interesting to follow up on Berrian's work, but the archive trail has thus far come up dry for me (correspondence with Jim Berrian, 16 November 2004).

of intelligent life. When, in 1973, the conservation activist Joan McIntyre was quoted in the mainstream press declaring, "I'd like to make love to a sperm whale," she was speaking metaphorically, one presumes, but she was singing John Lilly's tune.[32]

As did others in the period that saw the florescence of "Save the Whales" activism. Chapter 6 ended with an effort to concretize Lilly's influence on this movement by tracing out the filiations of his laboratory techniques (McVay's work parsing sonograms of cetacean phonation into phonemic elements, which issued in the transformative publication/release of "Songs of Humpback Whales") and his visionary preoccupations (Spong's shift from psychological research on captive odontocetes to a radicalized effort to make "contact" with whales using music, a project that gave rise to the decade-defining work of Greenpeace).

In reviewing this material, I am again struck by a larger theme that it brings to light. In a book published in the late 1980s that helped stimulate a minor industry of research on the history of human relationships with animals, Donna Haraway cycled compellingly through a sequence of demanding reflections on the scientific study of primates in the twentieth century. In doing so she aimed again and again to interrogate the ways that such research reflected continuous (if always changing) efforts to naturalize the cultural, and to do so by cabining some of what have been, historically, the most disruptive elements of social and individual life: reproduction, sexuality, race, dominance.[33] She showed convincingly that the scientific study of this particular taxon constituted, among other things, a school for the study of human similarity and difference and a school for the teaching of the same. I will acknowledge that I have written a very different kind of book, but I do think that elements of chapter 6 in particular bear comparison to Haraway's study. In fact, I would go so far as to suggest that cetology, too, served as such a school in the 1960s and 1970s. We have already noted that the early dolphin researchers included a number of former students of primates, and we have seen, too, that certain preoccupations—domesticity, sexuality, intelligence, even race—recur in the behavioral work on cetaceans in these years. But the differences are perhaps more interesting than the parallels or links. I am inclined to suggest that, in the end, primatology served primarily as a domain for thinking about what it is to be human, for

32. Nicholas von Hoffman, "Whale Love," *Washington Post*, 16 February 1973, B1.

33. Donna Haraway, *Primate Visions: Gender, Race, and Nature in the World of Modern Science* (New York: Routledge, 1989).

thinking about *who we are*, whereas cetology in those heady years served most compellingly as a domain for thinking about *who we are not*—it was less a school for the elaboration of similitudes than a laboratory for the exploration of *alterity*. And one hears in that word not just the root term for our notion of radical difference, but also the "alter" of *altered states*: in the whales and the dolphins, we explored both what it would be to be something fundamentally unlike ourselves and how we could *change*.[34] It was, in those years, a timely and compelling sort of investigation.

And this points to what can perhaps serve as a concluding methodological observation. In a number of ways, the approach of this study has been a little odd, at least as far as disciplinary history of science goes. After all, my historical subject has not been circumscribed by the boundaries of any particular scientific field. In pursuing expert whale knowledge in the twentieth century, we have tracked zoologists, biological oceanographers, mammalogists, torpedo designers, paleontologists, bioacousticians, animal trainers, aerospace engineers, and brain scientists, together with a few organic chemists, a physicist of fluid dynamics, a handful of physiologists and medical doctors, and even an anthropologist or two. Oh, and some astronomers. "Cetology" per se barely existed for much of the century I have reviewed, and indeed, no sooner did an actual scientific field dedicated to the study of whales and dolphins really begin to take shape in the 1960s than the terminology was changed, pinnipeds and sirenians were thrown into the mix (along with polar bears, at least by some), and the resulting "field," refashioned as "marine mammal biology," sought to gain greater mainstream scientific prominence by a promiscuous borrowing from ad-

---

34. My thinking here is indebted to Mitman's and Daston's felicitous invocation of the "*morphos* of anthropomorphism." See their introduction to their edited volume *Thinking with Animals* (New York: Columbia University Press, 2004). It is worth thinking, too, about the role that a general rethinking of the oceans themselves played in this process. Sometime between 1950 and 1970, in connection with the work of Jacques-Yves Cousteau and the rise of scuba, color film, and television, the underwater world took on many of the idealized characteristics it retains in the consumer imagination: clarity, warmth, bright colors, and so forth. As any diver can affirm, this is not the way most of the world's oceans look and feel. There is notable parallel and temporal coincidence between this re-visioning of the sea and the reimagining of whales and dolphins. Does one cause the other? Are both stimulated by some third factor? I tend toward the argument that whales and dolphins come in these years to serve in a compelling sort of prosopopoeia: they give the ocean both a face and a voice—enough like ours to seize our attention, different enough to hold it.

jacent and, in many respects, higher-status areas of biology—hence the rapid growth of work in population genetics, with its controversial implications for stock management.[35] One might hazard a formulation along the following lines: "There is no such thing as cetology, and this was a book about it."[36]

It would thus be fair, I think, to charge this book with what amounts to a kind of promiscuity, historiographically speaking, and on this I bow my head. But only for a moment, since I would then like to reply by offering the study as a whole as an example of the value of plotting investigations that are effectively orthogonal to the plane upon which a number of major issues in the history of science have traditionally been contested. In other words, I have tried to show in these pages what happens if, instead of thinking in terms of a scientific subject, we instead take a specific taxon as our historical object and trace out how new technologies, emerging disciplines, and shifting social exigencies interacted to constitute novel and significant scientific knowledge. I do not offer this approach as a substitute for work on the filiations of mentorship, the influence of research schools, the codification of professional identities, and so forth, but I am inclined to suspect that close attention to how a particular group of socially significant organisms came to be understood offers a valuable complementary perspective. What emerged from that process was emphatically not, say, a "new biology," but it was "knowledge of whales": it was what could be said about these creatures from authority, and it was knowledge that would have lasting effects

35. See, e.g., Joe Roman and Stephen Palumbi, "Whales before Whaling in the North Atlantic," *Science* 301, no. 5632 [25 July 2003]: 508–10. On the issue of the "interdisciplinary" nature of marine mammal research, an aside: while slogging my way through eleven boxes of administrative files of the US Marine Mammal Program (Smithsonian Archives, RU 7229), I found myself marveling at the extent to which the whole program of "integrated" biological research in the late 1960s and early 1970s felt, from this perspective, much more like an administrative effort to consolidate and coordinate grants and grantees than a mode of investigation necessitated by new appreciation of the ecological complexity of the organic world. The IBP programs, for instance, feel more like the solution to filing problems at NSF than solutions to scientific problems in nature. This is a cautionary note concerning "interdisciplinary" science as a historical object, and it has shaped the way I have handled whale science in this book: I am, on the whole, less concerned with what the scientists say about discipline-crossing work and more concerned to see who did what with cetaceans.

36. I wish I could take credit for this puckish turn on the beloved opening of Steven Shapin's *The Scientific Revolution* (Chicago: University of Chicago Press, 1996); but the tip of the hat goes to Henry Cowles.

on wide-ranging ideas about the natural world by the end of the twentieth century. I am not alone, of course, in attempting this sort of history, and I have had some valuable models. But I am persuaded that there are others yet to write.[37]

There is reason to believe, too, that this sort of idiosyncratic work can make contributions back to the more traditional disciplinary histories on which it must rely. For instance, in following the shifting character of whale knowledge over the last hundred years, I have been interested to see the ways that earlier scientific findings are frequently "repurposed" by subsequent developments. So, whale-marking investigations that began as a mechanism for tracing migration patterns are later used to calculate mortality rates needed in population models. Similarly, work on aging and reproductive physiology that was initially intended to elaborate a general life history for the rorquals of the Southern Hemisphere was later presented as "for" the calculation of life tables and recruitment statistics. Still later, bioacoustic analyses originally developed to study signal processing and filtering came to be used to investigate the semantics of nonhuman communication. In each of these cases, the methods look the same, but the whole research project in question is totally different. What is perhaps most surprising is the consistency with which such sublated upheavals go more or less unremarked. Indeed, most often there seems to be something like a general amnesia concerning the original intentions and presuppositions of the work. These are slippery shifts within and between disciplines, and they are easy to overlook from the perspective of scientific practice. Seen from the oblique view afforded by a taxon-centric approach, they come into focus more clearly.

AND THE REST IS HISTORY

An astute graduate student who read this manuscript in the endgame asked me, by way of critical encouragement, if I thought the book had an

---

37. Birds, for instance, remain a tempting topic, though Mark Barrow has covered a great deal of this ground in *A Passion for Birds: American Ornithology after Audubon* (Princeton, NJ: Princeton University Press, 2000). It is interesting to note the multiple points at which the story of whale conservation intersects with communities of birders or bird conservationists (recall the bird interests of the early English scientists who signaled the need for Antarctic conservation, and then the links between the CCW and the Progressive Era campaigns to protect birds, and then, still later, Roger Payne's move from ornithology to whale activism).

"exportable analytic." It is perhaps a testimony to the way I think about things that I understood him to mean, in essence, the sort of thing one took away from a book when one had basically forgotten what it was actually about. My answer—to him then, to you now—is two yeses and a no. For the first, minor-key "yes," I would simply point to this conclusion, which has outlined a number of methodological considerations and historical approaches that have been in play in this study and has suggested ways that these could be mobilized in other kinds of inquiry. The major-key "yes," however, is inextricable from the "no."

The fundamental lesson I have taken from the research and writing of this book amounts to nothing less than a kind of sweeping epistemological humiliation. For a number of years, I have read my way through tens of thousands of pages of published and unpublished sources in an effort to make sense of what seemed to me, from the outset, an ambitious, but by no means overwhelming, subject. And I have worked very hard to be faithful to, and responsible to, what happened in the past. But I must here confess that I now harbor grave doubts about this sort of undertaking, doubts that have grown as I prosecuted the task with the fullness of my powers. There is, I think, hardly a sentence in this book that, properly speaking, does not require, in the interests of dutiful nuance and full accuracy, a minor *volume* of reticulated and expanding commentary. Each of the footnotes— and there are more than a few, and they try to be exhaustive—calls out, as I skim the text, for footnotes of its own. In tightening and cleaning the bibliography, I found myself (indeed, I found my assistant) laughing aloud at the unending proliferation of sources and counter-sources, lost cites and untraceable references. When I relinquish this manuscript, I will shelve an entire box of notes for future work: a large box, full to bursting with ideas for additional archives, with further leads requiring attention.

These navel-gazing observations could be heard in the register of postmodern angst, but I do not sing them in that key. All I mean to say is this: it is very hard to say anything that is actually right about anything at all. This feels to me like a premodern point, particularly if it is offered not as a challenge but as a kind of apology; not as a request for first philosophy, but as a point of departure; not as a provocation or an excuse, but rather as something closer to an acknowledgment, or even a request for acknowledgment, of what it is like to be a human being.

All this, if it is anything, amounts, I think, to a kind of exportable antianalysis: it is a generalization not from what I have figured out about my subject, but from what I continue to sense I have not figured out—despite considerable effort. It is a generally applicable proposition (and one you are welcome to recall after you have forgotten what these chapters have to say about the science of whales), but it is not clear that it offers much by way of analytic traction. Indeed, it would seem to be a kind of solvent for such aspirations. The upshot, then, would appear to be this: knowing things is hard. This is perhaps a thin finding for a fat book, but it is also a fitting place to end a book on knowledge of whales. After all, whale ignorance has a distinguished place in the annals of skepticism and the theater of metaphysical humiliation: it was, of course, humanity's inadequate grasp of the ways of Leviathan that constituted the core of Jehovah's testy dismissal of Job's inquiry into first and final things.

To the principle of Job's whale ("whereof one cannot speak, thereof one must remain silent"), we might then add the Jonah corollary: "whereof one *can* speak, one faces a considerable trial."

BIBLIOGRAPHY

UNPUBLISHED SOURCES
(by archive, with relevant collection information)

Federal Bureau of Investigation, Department of Justice, Washington, DC

John Cunningham Lilly Personal File (secured through Freedom of
Information Act request)

Hvalfangstmuseet (Whaling Museum), Sandefjord, Norway

Archive, Internasjonal Hvalfangst-Statistikk (Bureau of International
Whaling Statistics)
Birger Bergersen Papers
Miscellaneous Documents
Records of the United States Whaling Co.

International Whaling Commission (The Red House), Impington, England

Miscellaneous IWC documentation

National Archives and Records Administration, College Park, Maryland (NARA)

Record Group 48—Records of the Office of the Secretary of the Interior
(Central Correspondence Files, 1969–1972)

Natural History Museum, London, England (NHM)

Museum Archives (Central)

Various collections, including Sidney Harmer files, Whale Hall files, Discovery Committee files, etc.

NHM Zoology Collections Archive

Additional Harmer material (NB: located in a separate wing of the museum)

NHM Library

G. E. H. Barrett-Hamilton Journal (a manuscript, shelved with published books)

National Marine Mammal Laboratory, Seattle, WA

Douglas Chapman Papers

Dale Rice Papers

National Oceanographic Library, Southampton Oceanography Centre, Southampton, England (NOL)

George Deacon Papers

Discovery Manuscripts

The National Oceanographic Library at Southampton has an extensive collection of Discovery material and a finding aid of more than 60 pages; however, at the time of my visits (2004), navigating the collection required considerable assistance from the solicitous archive staff. The main finding aid, for instance, was simply reproduced at some point from original Discovery Investigations administrative documentation, but it no longer reflected either the contents or the exact arrangement of a body of material that had undergone several moves and had been subdivided, sometimes hastily, by different individuals and institutions over the better part of 75 years (see relevant footnotes in chapter 2). I was fortunate in my visits to have been afforded extensive access to the shelf area itself, where, in the company of an archivist, I was able to do a kind of sifting impossible to undertake with the finding aid alone. In view of all this, the citation of NOL documents poses something of a challenge. I have adopted throughout the general designation "NOL Discovery MS" to indicate all these holdings, but beyond that I have tended to favor descriptive accounts (including, where possible, location information) of the materials I cite over shelf marks of uncertain value.

Nasjonalbiblioteket (National Library), Oslo, Norway

Manuscripts Collection

Johan Hjort Papers

Johan T. Ruud Papers

Rigsarkivet (Danish National Archives), Copenhagen, Denmark

ICES Records

Riksarkivet og Statsarkivene (Norwegian National and State Archives), Oslo, Norway

Johan Hjort Correspondence and Manuscripts
Fiskeridep., Fangstkontoret (Includes IWC-related materials)

Scott Polar Research Institute, University of Cambridge (SPRI)

Discovery Committee Manuscripts (including minutes and
    associated documentation)
Neil Alison Mackintosh Papers
South Georgia Archives
William L. Allardyce Correspondence
Sidney F. Harmer Correspondence
Discovery Artifacts

Scripps Institution of Oceanography Archive, University of California, San Diego

Carl Hubbs Papers
Per F. Scholander Papers

Sea Mammal Research Unit, University of St. Andrews, Scotland (SMRU)

Whaling Records and Discovery Committee Records (MS 38600 and
WR/DC/series)

At the time of my work in this collection (in 2004), these materials were
effectively uncatalogued (I secured a thick sheaf of papers documenting the
records ostensibly transferred from British Antarctic Survey holdings, but I
was never able to coordinate its listings with what I found). Absent a reliable
indicator of the specific contents of the collection, I reviewed the 20-odd
cardboard packing boxes (including several chart folders, which have different
shelf locations) in their entirety. Citation format again presents a challenge.
I have used the general designator MS 38600 to indicate these items (which
I was given as the call number at the archive), together with a box number,
and, where possible, some description of the file or document (or both, where
relevant).

Sidney G. Brown Papers, England (SGB)

These are papers in the private holdings of Sidney G. Brown, a retired
employee of the Discovery Investigations, now deceased. I am not aware of
what has become of them since he passed away. At the time of my visits to his
home, the collection included a considerable range of unpublished materials

from the second decade of the twentieth century through to the 1970s (I have come to believe that the majority of the collection was Neil Alison Mackintosh's personal culling of materials in Discovery collections that he wished to keep out of public archives); they were, of course, entirely uncatalogued. I have simply specified the documents in question as clearly as possible, since there is no reliable box or file information to give.

Smithsonian Institution, Washington, DC

Smithsonian Archives

Record Unit 7229—United States Marine Mammal Program Records, 1967–1973 and undated
Record Unit 7165—International Whaling Commission Papers, etc., 1930–1968
Record Unit 7434—Remington Kellogg Papers, 1903–1969, and related to 1982
Record Unit 7170—Remington Kellogg Papers, 1871–1965 and undated
Accession 99–012—Remington Kellogg Papers, 1926–1950 and undated

James Mead Files, Marine Mammal Department

These are about four linear feet of miscellaneous papers, some of which relate to Kellogg's work, and all of which pertain to the institution's work on marine mammals. They can be found in the holdings of the Natural History Museum, and at the time of my work in the collection (2005), these materials were overseen by James Mead, curator for marine mammals in the department of vertebrate zoology.

Stanford University Library, Special Collections, Palo Alto, CA

John C. Lilly Papers

At the time of my work in this collection (in 2003), it was both uncatalogued and, in principle, closed to researchers; I am deeply grateful to the archive staff for affording me access. The box designations I have used in citing these sources are those that the library was then using as shelf marks, and the file names are those on Lilly's own files; both of these may have changed if a full-scale cataloguing of these materials has now been completed (as I believe it has).

University of Washington, Special Collections, Seattle, WA

Douglas G. Chapman Papers
William S. Lagen Papers (American Pacific Whaling Co.)

PUBLISHED SOURCES
(including government documents and technical reports)

Ackerman, Bruce A. *Social Justice in the Liberal State.* New Haven, CT: Yale University Press, 1980.

Adams, Jacob. "History and Function of the Alaska Eskimo Whaling Commission." *Proceedings of the First Conference on the Biology of the Bowhead Whale,* Balaena Mysticetus, *Population Assessment* (January 1982): 9–13.

Adams, William M. *Against Extinction: The Story of Conservation.* London: Earthscan, 2004.

Alexander, Louis M. *The Law of the Sea: Offshore Boundaries and Zones.* Columbus: Ohio State University Press, 1967.

Allard, Dean C. "The Fish Commission Laboratory and Its Influence on the Founding of the Marine Biological Laboratory." *Journal of the History of Biology* 23, no. 2 (Summer 1990): 251–70.

Allen, David Elliston. *The Naturalist in Britain: A Social History.* Princeton, NJ: Princeton University Press, 1994.

Allen, Garland E. *Life Science in the Twentieth Century.* New York: Wiley, 1975.

———. "Morphology and Twentieth-Century Biology: A Response." *Journal of the History of Biology* 14, no. 1 (Spring 1981): 159–76.

———. "Naturalists and Experimentalists: The Genotype and the Phenotype." *Studies in the History of Biology* 3 (1979): 179–209.

Allen, Glover M. *Extinct and Vanishing Mammals of the Western Hemisphere, with the Marine Species of All the Oceans.* Cambridge, MA: American Committee for International Wild Life Protection, 1942.

Allen, Kenneth Radway. *Conservation and Management of Whales.* Seattle: University of Washington Press, 1980.

Alpers, Antony. *Dolphins: The Myth and the Mammal.* Boston: Houghton Mifflin, 1961.

Anderson, Katherine. *Predicting the Weather: Victorians and the Science of Meteorology.* Chicago: University of Chicago Press, 2005.

Anderson, Virginia DeJohn. *Creatures of Empire: How Domestic Animals Transformed Early America.* Oxford: Oxford University Press, 2004.

Andrews, Richard N. L. *Managing the Environment, Managing Ourselves: A History of American Environmental Policy.* New Haven, CT: Yale University Press, 1999.

Andrews, Roy Chapman. *Ends of the Earth.* New York: Knickerbocker Press, G. P. Putnam's Sons, 1929.

———. "Monographs of the Pacific Cetacea I—The California Gray Whale." *Memoirs of the American Museum of Natural History,* n.s., vol. 1, pt. 5 (March 1914): 231–87.

———. *Whale Hunting with Gun and Camera: A Naturalist's Account of the Modern Shore-Whaling Industry, of Whales and Their Habits, and of Hunting Experiences in Various Parts of the World.* New York: D. Appleton, 1916.

———. "What Shore Whaling Is Doing for Science." *Nature* 88, no. 2200 (28 December 1911): 280–82.

Anker, Peder. *Imperial Ecology: Environmental Order in the British Empire, 1895–1945.* Cambridge, MA: Harvard University Press, 2001.

Appel, Frederick C. "Deep Thinkers." *Playboy*, August 1968, 101–2, 150–55.

Aramburú Menchaca, Andrés A. *Historia de las 200 millas de mar territorial (evolución de una doctrina continental).* Piura, Perú: Universidad de Piura, 1973.

Aron, William. "Science and the IWC." In *Toward a Sustainable Whaling Regime*, edited by Robert Friedheim, 105–22. Seattle: University of Washington Press, 2001.

Ásgeirsdóttir, Áslaug. "The Environment and International Politics. International Fisheries, Heidegger and Social Method (Review)." *Global Environmental Politics* 7, no. 3 (August 2007): 140–41.

Atz, J. W. Review of *Man and Dolphin*, by John C. Lilly. *Animal Kingdom: Bulletin of the New York Zoological Society* 64, no. 6 (December 1961): 190.

Backus, Richard. Interview by Gary Weir and Frank Taylor at the Woods Hole Oceanographic Institution, Woods Hole, MA, 29 March 2000. Oral History Project of the H. John Heinz III Center for Science, Economics, and the Environment, in conjunction with the Colloquia Series "Oceanography: The Making of a Science, People, Institutions and Discovery."

Bagshawe, Thomas Wyatt. *Two Men in the Antarctic: An Expedition to Graham Land, 1920–1922.* Cambridge: Cambridge University Press, 1939.

Baker, H. H. *Stockholm and Beyond—Report of the Secretary of State's Advisory Committee on the 1972 United Nations Conference on the Human Environment.* U.S. Department of State International Organization and Conference Series, 1972.

Baker, Steve. *Picturing the Beast: Animals, Identity, and Representation.* Urbana: University of Illinois Press, 2001.

Baratay, Eric, and Elisabeth Hardouin-Fugier. *Zoo: A History of Zoological Gardens in the West.* London: Reaktion Books, 2002.

Barnes, Lawrence G. "Search for the First Whale: Retracing the Ancestry of Cetaceans." *Oceans* 17, no. 2 (March 1984): 20–23.

Barrow, Mark V. *A Passion for Birds: American Ornithology after Audubon.* Princeton, NJ: Princeton University Press, 1998.

Barstow, Robbins. "Beyond Whale Species Survival: Peaceful Coexistence and Mutual Enrichment as a Basis for Human/Cetacean Relations." *Sonar, The Magazine of the Whale and Dolphin Conservation Society* no. 2 (Autumn 1989): 10–13.

Barthelmess, Klaus. "An International Campaign against Whaling and Sealing Prior to World War One." In *Whaling and History II: New Perspectives*, edited by Jan Erik Ringstad, 147–67. Sandefjord, Norway: Sandefjordmuseene, 2006.

Bastian, J. "The Transmission of Arbitrary Environmental Information between Bottle-Nose Dolphins." In *Les Systèmes Sonars Animaux: Biologie et Bionique (Cours d'Été O.T.A.N. 26 September–3 October 1966)*, vol. 2, edited by René-Guy Busnel, 803–73. Frascati, Italy: NATO Advanced Study Institute, 1967.

Bateson, Gregory. "Problems in Cetacean and Other Mammalian Communication." In *Whales, Dolphins, and Porpoises*, edited by Kenneth Stafford Norris, 569–78. Berkeley: University of California Press, 1966.

Bateson, Gregory. *A Sacred Unity: Further Steps to an Ecology of Mind.* Edited by Rodney E. Donaldson. San Francisco: Harper Collins, 1991.

Batteau, Dwight W., and Peter R. Markey. *Man/Dolphin Communication: Final Report, 15 December 1966–13 December 1967 (Prepared for U.S. Naval Ordnance Test Station, China Lake, California).* Contract No. N00123-67-C-1103. Arlington, MA: Listening, 1967.

Beck, Peter. *The International Politics of Antarctica.* London: Croom Helm, 1986.

Becker, Peter, and William Clark, eds. *Little Tools of Knowledge: Historical Essays on Academic and Bureaucratic Practices.* Ann Arbor: University of Michigan Press, 2001.

Beddard, F. E. *A Book of Whales.* London: J. Murray; New York: G. P. Putnam's Sons, 1900.

Bell, J. J. *The Whalers.* London: Hodder and Stoughton, 1914.

Benjamin, Ludy T. Jr., and Darryl Bruce. "From Bottle-Fed Chimp to Bottlenose Dolphin: A Contemporary Appraisal of Winthrop Kellogg." *Psychological Record* 32, no. 4 (Fall 1982): 461–82.

Bennett, A. G. *Whaling in the Antarctic.* New York: Henry Holt, 1932.

Benson, Keith R. "American Morphology in the Late Nineteenth Century: The Biology Department at Johns Hopkins University." *Journal of the History of Biology* 18, no. 2 (1985): 163–205.

———. "From Museum Research to Laboratory Research: The Transformation of Natural History into Academic Biology." In *The American Development of Biology,* edited by Roland Rainger and Jane Maienschein, 49–83. Philadelphia: University of Pennsylvania Press, 1988.

———. "Problems of Individual Development: Descriptive Embryological Morphology in America at the Turn of the Century." *Journal of the History of Biology* 14, no. 1 (Spring 1981): 115–28.

Benson, Keith Rodney, Jane Maienschein, and Ronald Rainger, eds. *The Expansion of American Biology.* New Brunswick, NJ: Rutgers University Press, 1991.

Benson, Keith Rodney, and Philip F. Rehbock. *Oceanographic History: The Pacific and Beyond.* Seattle: University of Washington Press, 2002.

Bergersen, Birger. *Beiträge zur Kenntnis der Haut einiger Pinnipedien unter besonderer Berücksichtigung der Haut der Phoca groenlandica.* Oslo: I Kommisjon Hos J. Dybwad, 1931.

———. "The International Whaling Situation." *Le Nord* 1 (1938): 112.

Bernstein, Barton J. Introduction to *The Voice of the Dolphins, and Other Stories,* by Leo Szilard. Stanford, CA: Stanford University Press, 1992.

Berra, Tim M. *A Chronology of the American Society of Ichthyologists and Herpetologists Through 1982.* Lawrence, KS: American Society of Ichthyologists and Herpetologists/Allen Press, 1984.

Bestor, Theodore C. *Tsukiji: The Fish Market at the Center of the World.* Berkeley: University of California Press, 2004.

Beverton, R. J. H., and S. J. Holt. *On the Dynamics of Exploited Fish Populations.* London: His Majesty's Stationery Office, 1957.

Birnie, Patricia W. *International Regulation of Whaling: From Conservation of Whaling to Conservation of Whales and Regulation of Whale-Watching.* 2 vols. New York: Oceana Publications, 1985.

Björk, Tord. "The Emergence of Popular Participation in World Politics: The United Nations Conference on the Human Environment, 1972." Unpublished seminar paper, Department of Political Science, University of Stockholm.

Blake, R. F. "Submarine Signaling: The Protection of Shipping by a Wall of Sound and Other Uses of the Submarine Telegraph." *Transactions of the American Institute of Electrical Engineers* 33, no. 2 (1914): 1549–61.

Blix, Arnoldus Schytte, Lars Walløe, and Øyvind Ulltang, eds. *Whales, Seals, Fish, and Man: Proceedings of the International Symposium on the Biology of Marine Mammals in the North East Atlantic, Tromsø, Norway, 29 November–1 December 1994.* Developments in Marine Biology 4. Amsterdam: Elsevier, 1995.

Blok, Anders. "Contesting Global Norms: Politics of Identity in Japanese Pro-Whaling Countermobilization." *Global Environmental Politics* 8, no. 2 (May 2008): 39–66.

Boardman, Robert. *International Organization and the Conservation of Nature.* London: Macmillan, 1981.

Bock, Peter Gidon. "A Study in International Regulation: The Case of Whaling." PhD diss., New York University, 1966.

Bocking, Stephen. *Nature's Experts: Science, Politics, and the Environment.* New Brunswick, NJ: Rutgers University Press, 2004.

———. "Science and Spaces in the Northern Environment." *Environmental History* 12, no. 4 (October 2007): 867–94.

Boehrer, Bruce Thomas. *Parrot Culture: Our 2500-Year-Long Fascination with the World's Most Talkative Bird.* Philadelphia: University of Pennsylvania Press, 2004.

Boenninghaus, Georg. *Das Ohr des Zahnwales.* Jena: Gustav Fischer, 1903.

Bogert, Charles M. "The Influence of Sound on the Behavior of Amphibians and Reptiles." In *Animal Sounds and Communication,* edited by W. E. Lanyon and W. N. Tavolga, 137–320. Washington, DC: American Institute of Biological Sciences, 1960.

Bolster, W. Jeffrey. "Opportunities in Marine Environmental History." *Environmental History* 11, no. 3 (July 2006): 567–97.

———. "Putting the Ocean in Atlantic History: Maritime Communities and Marine Ecology in the Northwest Atlantic, 1500–1800." *American Historical Review* 113 (February 2008): 19–47.

Bonner, Nigel. *Whales of the World.* New York: Facts on File, 1989.

Booklist. Unsigned review of *The Mind of the Dolphin: A Non-Human Intelligence,* by John C. Lilly. Vol. 64, no. 11 (1 February 1968): 615.

Borror, Donald J. "The Analysis of Animal Sounds." In *Animal Sounds and Communication,* edited by W. E. Lanyon and W. N. Tavolga, 26–37. Washington, DC: American Institute of Biological Sciences, 1960.

Borror, Donald J., and C. R. Reese. "The Analysis of Bird Songs by Means of a Vibralyzer." *Wilson Bulletin* 65, no. 4 (1953): 271–303.

Boulle, Pierre. *La Planète des Singes.* Paris: René Julliard, 1963.

———. *Planet of the Apes*. Translated by Xan Fielding. New York: Vanguard Press, 1963.

———. *Le Pont de la Rivière Kwaï*. Paris: René Julliard, 1952.

Brandt, J. F. "Untersuchungen über die Fossilen und Subfossilen Cetaceen Europa's." *Mémoires de l'Académie Impériale des Sciences de St.-Pétersbourg*, ser. 7, vol. 20, no. 1 (1873): 1–372.

Brandt, Karl. *The German Fat Plan and Its Economic Setting*. Stanford, CA: Food Research Institute, 1938.

———. *Whale Oil: An Economic Analysis*. Stanford, CA: Food Research Institute, 1940.

Breathnach, A. S. "The Cetacean Central Nervous System." *Biological Reviews* 35 (1960): 187–230.

Breland, Keller, and Marian Breland. *Animal Behavior*. New York: Macmillan, 1966.

Brooks, Martin. *Fly: The Unsung Hero of 20th-Century Science*. New York: Ecco, 2002.

Broughton, W. B. "Methods in Bio-acoustic Terminology." In *Acoustic Behavior of Animals*, edited by R. G. Busnel, 3–24. London: Elsevier, 1963.

Brown, Alistair. "Ali's Corner." *Canada Folk Bulletin* 2, no 6 (November–December 1979), 25.

Brown, George G. *The Origins of Natural Science in America: The Essays of George Brown*. Edited by Sally Gregory Kohlstedt. Washington, DC: Smithsonian Institution Press, 1991.

Brown, Norman O. *Life against Death: The Psychoanalytical Meaning of History*. Middletown, CT: Wesleyan University Press, 1959.

Brown, Sidney G. "Dispersal in Blue and Fin Whales." *Discovery Reports* 26 (1954): 355–84.

———. "The Movement of Fin and Blue Whales within the Antarctic Zone." *Discovery Reports* 33 (1964): 1–54.

———. "Whale Marking: A Short Review." In *A Voyage of Discovery*, edited by Martin Angel 569–81. Oxford: Pergamon Press, 1977.

Brownell, R. L., and A. V. Yablokov. "Illegal and Pirate Whaling." In *Encyclopedia of Marine Mammals*, edited by W. F. Perrin, Bernd G. Würsig, and J. G. M. Thewissen, 608–12. San Diego: Academic Press, 2002.

Budiansky, Stephen. *Nature's Keepers: The New Science of Nature Management*. New York: The Free Press, 1995.

Budker, Paul. *Whales and Whaling*. New York: Macmillan, 1959.

Buhs, Joshua Blu. *The Fire Ant Wars: Nature, Science, and Public Policy in Twentieth-Century America*. Chicago: University of Chicago Press, 2004.

Burgess, G. H. O. *The Curious World of Frank Buckland*. London: John Baker, 1967.

Burnett, D. Graham. "A Mind in the Water: The Dolphin as Our Beast of Burden." *Orion* 29, no. 3 (May–June 2010): 38–51.

———. Review of *Scientific Uncertainty and the Politics of Whaling* by Michael Heazle. *Isis* 98, no. 2 (2007): 425–26.

———. "Sea of Fire." In *Tropical Visions in an Age of Empire*, edited by Felix Driver and Luciana Martins, 113–34. Chicago: University of Chicago Press, 2005.

———. "Self-Recording Seas." In *Oceanomania: Souvenirs of Mysterious Seas*, edited by Mark Dion and Sarina Basta. London: Michael Mack, 2011.

———. *Trying Leviathan: The Nineteenth-Century New York Court Case That Put the Whale on Trial and Challenged the Order of Nature*. Princeton, NJ: Princeton University Press, 2007.

Burton, Robert. *The Life and Death of Whales*. 2nd ed. New York: Universe Books, 1980.

Busnel, R. G., ed. *Acoustic Behavior of Animals*. London: Elsevier, 1963.

———. "Information in the Human Whistled Language and Sea Mammal Whistling." In *Whales, Dolphins, and Porpoises*, edited by Kenneth Stafford Norris, 544–68. Berkeley: University of California Press, 1966.

———, ed. *Les Systèmes Sonars Animaux: Biologie et Bionique (Cours d'Été O.T.A.N. 26 September–3 October 1966)*. 2 vols. Frascati, Italy: NATO Advanced Study Institute, 1966.

Butterworth, D. S. "Science and Sentimentality." *Nature* 357, no. 6379 (18 June 1992): 532–34.

Caldwell, Lynton Keith, ed. *International Environmental Policy*. 2nd ed. Durham, NC: Duke University Press, 1990.

Caldwell, M. C., and D. K. Caldwell. "Intraspecific Transfer of Information Via the Pulsed Sound in Captive Odontocete Cetaceans." In *Les Systèmes Sonars Animaux: Biologie et Bionique (Cours d'Été O.T.A.N. 26 September–3 October 1966)*, vol. 2., edited by René-Guy Busnel, 879–936. Frascati, Italy: NATO Advanced Study Institute, 1966.

Calman, W. T. "Sidney Frederic Harmer, 1862–1950." *Obituary Notices of the Royal Society* 7 (1951): 359–71.

Cameron, Jenks. *The Bureau of Biological Survey: Its History, Activities and Organization*. Baltimore: Johns Hopkins University Press, 1929.

Canadian Naturalist. Unsigned review of *The Whale Problem: A Status Report*, edited by William E. Schevill, G. Carleton Ray and Kenneth S. Norris. Vol. 89 (1975): 475–76.

Cantor, Geoffrey, and Sally Shuttleworth, eds. *Science Serialized: Representations of the Sciences in Nineteenth-Century Periodicals*. Cambridge, MA: MIT Press, 2004.

"Captain Ahab in Modern Dress." *Fortune*, October 1932, 52–59, 106–7.

Carpenter, Peter W. "Dolphin Hydrodynamics: Gray's Paradox Revisited." Opening keynote lecture, Third International Symposium on Ultrasonic Doppler Methods for Fluid Mechanics and Fluid Engineering, EPFL, Lausanne, Switzerland, September, 2002.

Carr, Archie. "Have We Been Ignoring a Deep Thinker?" Review of *Man and Dolphin*, by John C. Lilly. *New York Times Book Review*, 3 September 1961, 3.

Cartmill, Matt. *A View to a Death in the Morning: Hunting and Nature Through History*. Cambridge, MA: Harvard University Press, 1993.

Cassidy, Sean Douglas. "Mindbombs and Whalesongs: Greenpeace and the News." PhD diss., University of Oregon, 1992.

Chapman, Douglas G. "The Future of the Great Whales." *Puget Soundings* (March 1976): 18–21.

Chapman, Douglas G., and C. O. Junge. "The Estimate of the Size of a Stratified Animal Population." *Annals of Mathematical Statistics* 27 (1956): 375–89.

Chapman, Douglas G., K. Kenyon, and V. Scheffer. "A Population Study of the Alaska Fur Seal Herd." *U.S. Fish and Wildlife Service Special Scientific Report: Wildlife* 12 (1954): 1–77.

Choice. Unsigned review of *The Mind of the Dolphin: A Non-Human Intelligence*, by John C. Lilly. Vol. 5, no. 8 (October 1968): 986.

Christie, Edward. *Finding Solutions for Environmental Conflicts: Power and Negotiation*. Cheltenham, UK: Edward Elgar, 2008.

Churchill, Frederick B. "In Search of the New Biology: An Epilogue." *Journal of the History of Biology* 14, no. 1 (Spring 1981): 177–91.

Ciriacy-Wantrup, S. V. *Resource Conservation: Economics and Policies*. Berkeley: University of California Press, 1952.

Clapham, Phillip J., and C. Scott Baker. "Whaling, Modern." In *Encyclopedia of Marine Mammals*, edited by W. F. Perrin, Bernd G. Würsig, and J. G. M. Thewissen, 1328–32. San Diego: Academic Press, 2002.

Clark, Colin W. "The Economics of Overexploitation." *Science* 181, no. 4100 (August 1973): 630–34.

Clark, C. W., and R. Lamberson. "An Economic History and Analysis of Pelagic Whaling." *Marine Policy* 6, no. 2 (April 1982): 103–20.

Clarke, Robin, and Lloyd Timberlake. *Stockholm Plus Ten: Promises, Promises?: The Decade Since the 1972 UN Environment Conference*. London: International Institute for Environment and Development/Earthscan, 1982.

Clements, Frederic E., and Victor E. Shelford. *Bio-Ecology*. New York: John Wiley & Sons, 1939.

Clutton-Brock, Juliet. *A Natural History of Domesticated Mammals*. 2nd ed. Cambridge: Cambridge University Press, 1999.

Cockburn, Alexander. "A Short Meat-Oriented History of the World." *New Left Review* 1, no. 215 (January–February 1996): 16–42.

Cocks, A. Heneage. "Additional Notes on the Finwhale Fishery on the North European Coast." *Zoologist*, ser. 3, vol. 9 (1885): 134–43.

———. "The Finwhale Fishery of 1885 of the North European Coast." *Zoologist*, ser. 3, vol. 10 (1886): 121–36.

———. "The Finwhale Fishery of 1886 on the Lapland Coast." *Zoologist*, ser. 3, vol. 11 (1887): 1–16.

———. "The Finwhale Fishery on the Coast of Finmark." *Zoologist*, ser. 3, vol. 8 (1884): 366–70.

Coetzee, J. M. *The Lives of Animals*. Princeton, NJ: Princeton University Press, 1999.

Coleman, Jon T. *Vicious: Wolves and Men in America*. New Haven, CT: Yale University Press, 2004.

Coleman, William. "Morphology between Type Concept and Descent Theory." *Journal of the History of Medicine and Allied Sciences* 31, no. 2 (April 1976): 149–75.

Coleman-Cooke, John. *Discovery II in the Antarctic: The Story of British Research in the Southern Seas*. London: Odhams Press, 1963.

Colonial Office (United Kingdom). *Correspondence [October, 1911–December, 1913] Respecting the Protection of Whales and the Whaling Industry.* Miscellaneous no. 278 (Confidential), March 1914.

———. *Further Correspondence [January, 1914–March, 1915] Related to Whaling and the Protection of Whales (in continuation of Miscellaneous No. 278).* Miscellaneous no. 300 (Confidential), October 1915.

———. *Inter-departmental Committee on Whaling and the Protection of Whales. Minutes of Evidence, &c.* Miscellaneous no. 298 (Confidential), October 1915.

———. *Report of the Interdepartmental Committee on Research and Development in the Dependencies of the Falkland Islands, with Appendices, Maps, &c.* Cmd. No. 657. London: His Majesty's Stationery Office, 1920.

Commoner, Barry. *The Closing Circle: Nature, Man, and Technology.* New York: Knopf, 1972.

Cormack, R. M. "The Statistics of Capture-Recapture Methods." *Oceanography and Marine Biology* 6 (1968): 455–506.

Corner, George Washington. "George Linius Streeter." In *Dictionary of Scientific Biography*, vol. 13, edited by Charles Coulston Gillespie, 96–97. New York: Charles Scribner's Sons, 1970.

Coulombe, Harry N., Sam H. Ridgway, and William E. Evans. "Respiratory Water Exchange in Two Species of Porpoise." *Science* 149, no. 3679 (2 July 1965): 86–88.

Cousteau, Jacques-Yves. *The Silent World.* With Frédéric Dumas. New York: Harper, 1953.

Cousteau, Jacques-Yves, and Philippe Diolé. *The Whale: Mighty Monarch of the Sea.* Translated by J. F. Bernard. Garden City, NY: Doubleday, 1972.

Crawford, Nelson Antrim. *Yearbook of Agriculture, 1927.* Washington, DC: Government Printing Office, 1928.

Creager, Angela N. H., and William Chester Jordan, eds. *The Animal/Human Boundary: Historical Perspectives.* Rochester, NY: University of Rochester Press, 2002.

Credland, Arthur G. "Some Notes on the Development of Cetology, Popular Interest in the Whale Tribe, and a Famous Literary Whale." *Scottish Naturalist* 111 (1999): 93–126.

Crist, Eileen. *Images of Animals: Anthropomorphism and Animal Mind.* Philadelphia: Temple University Press, 1999.

Cushing, D. H. *The Provident Sea.* Cambridge: Cambridge University Press, 1988.

Dallmeyer, Dorinda G., ed. *Values at Sea: Ethics for the Marine Environment.* Athens: University of Georgia Press, 2003.

Dann, Kevin, and Gregg Mitman. "Essay Review: Exploring the Borders of Environmental History and the History of Ecology." *Journal of the History of Biology* 30, no. 2 (1997): 291–302.

Dannenfeldt, Karl H. "Ambergris: The Search for Its Origin." *Isis* 73, no. 268 (1982): 382–97.

Darnley, Rowland. "A New Antarctic Expedition." *Nineteenth Century and After* 93 (May 1923): 718–28.

Darrow, Clarence, and William J. Bryan. *The World's Most Famous Court Trial.* Clark, NJ: The Lawbook Exchange, 2008.

Daston, Lorraine, and Peter Galison. *Objectivity*. Cambridge, MA: MIT Press, 2007.

Daston, Lorraine, and Gregg Mitman, eds. *Thinking with Animals: New Perspectives on Anthropomorphism*. New York: Columbia University Press, 2005.

Davis, Lance E., Robert E. Gallman, and Karin Gleiter. *In Pursuit of Leviathan: Technology, Institutions, Productivity, and Profits in American Whaling, 1816–1906*. Chicago: University of Chicago Press, 1997.

Davis, Susan G. *Spectacular Nature: Corporate Culture and the Sea World Experience*. Berkeley: University of California Press, 1997.

Day, David. *The Whale War*. San Francisco: Sierra Club Books, 1987.

Deacon, Margaret, A. L. Rice, and C. P. Summerhayes, eds. *Understanding the Oceans: A Century of Ocean Exploration*. London: University College of London, 2001.

Dedina, Serge. *Saving the Gray Whale: People, Politics, and Conservation in Baja California*. Tucson: University of Arizona Press, 2000.

Deevey, Edward S. "Life Tables for Natural Populations of Animals." *Quarterly Review of Biology* 22, no. 4 (December 1947): 283–314.

De Haan, Reysenbach. "Hearing in Whales." Supplement, *Acta Oto-Laryngologica* 134 (1957): 1–114.

De Lury, D. B. "On the Estimation of Biological Populations." *Biometrics* 3, no. 4 (December 1947): 145–67.

Devine, Eleanore, and Martha Clark. *The Dolphin Smile: Twenty-Nine Centuries of Dolphin Lore*. New York: Macmillan, 1967.

Dezalay, Yves, and Bryant G. Garth, eds. *Global Prescriptions: The Production, Exportation, and Importation of a New Legal Orthodoxy*. Ann Arbor: University of Michigan Press, 2002.

Dice, Lee R. "The Scientific Value of Predatory Mammals." *Journal of Mammalogy* 6, no. 1 (February 1925): 25–27.

Dick, Steven J. *The Biological Universe: The Twentieth Century Extraterrestrial Life Debate and the Limits of Science*. Cambridge: Cambridge University Press, 1999.

Dimitrov, Radoslav S. *Science and International Policy: Regimes and Nonregimes in Global Governance*. Lanham, MD: Rowman & Littlefield, 2006.

Dion, Mark, and Sarina Basta, eds. *Oceanomania: Souvenirs of Mysterious Seas*. London: Michael Mack, forthcoming.

Donovan, Gregory P. "International Whaling Commission." In *Encyclopedia of Marine Mammals*, edited by W. F. Perrin, Bernd G. Würsig, and J. G. M. Thewissen, 637–41. San Diego: Academic Press, 2002.

———. "The International Whaling Commission and the Revised Management Procedure." Paper presented at the Conference on Responsible Wildlife Resource Management held in the European Parliament, Brussels, 1993.

Dorsey, Kurkpatrick. "Compromising on Conservation: World War II and American Leadership in Whaling Diplomacy." In *Natural Enemy, Natural Ally: Toward an Environmental History of Warfare*, edited by Richard P. Tucker and Edmund Russell, 252–69. Corvallis: Oregon State University Press, 2004.

———. *The Dawn of Conservation Diplomacy: U.S.-Canadian Wildlife Protection Treaties in the Progressive Era*. Seattle: University of Washington Press, 1998.

Doughty, Robin W. *Feather Fashions and Bird Preservation: A Study in Nature Protection.* Berkeley: University of California Press, 1975.

Downs, Anthony. "Up and Down with Ecology: The 'Issue-Attention Cycle.'" *Public Interest* 5 (Fall 1971): 38–50.

Drayton, Richard. *Nature's Government: Science, Imperial Britain, and the 'Improvement' of the World.* New Haven, CT: Yale University Press, 2000.

Dreher, John J. "Linguistic Considerations of Porpoise Sounds." *Journal of the Acoustical Society of America* 33, no. 12 (December 1961): 1799–1800.

Dudok van Heel, W. H. "Sound and Cetacea." *Netherlands Journal of Sea Research* 1, no. 4 (1962): 407–507.

Duffield, Deborah A., Sam H. Ridgway, and Robert S. Sparkes. "Cytogenetic Studies of Two Species of Porpoises." *Nature* 213, no. 5072 (14 January 1967): 189–90.

Dunlap, Thomas R. *Saving America's Wildlife: Ecology and the American Mind, 1850–1990.* Princeton, NJ: Princeton University Press, 1988.

Durant, Robert F., Daniel J. Fiorino, and Rosemary O'Leary, eds. *Environmental Governance Reconsidered: Challenges, Choices, and Opportunities.* Cambridge, MA: MIT Press, 2004.

Eaken, Richard M. "History of Zoology at the University of California, Berkeley." *Bios* 27 (1956): 67–90.

Ehrenfeld, David W. *Conserving Life on Earth.* New York: Oxford University Press, 1972.

Einhorn, Richard. "Dolphins Challenge the Designer." *Electronic Design* 15, no. 25 (6 December 1967): 49–64.

Eiseley, Loren. "The Long Loneliness: Man and the Porpoise: The Solitary Destinies." *American Scholar* 30, no. 1 (Winter 1960–1961): 57–64.

Elichirigoity, Fernando. *Planet Management: Limits to Growth, Computer Simulation, and the Emergence of Global Spaces.* Evanston, IL: Northwestern University Press, 1999.

Elling, Bo. *Rationality and the Environment: Decision-Making in Environmental Politics and Assessment.* London: Earthscan, 2008.

Elliot, G. H. "The Failure of the IWC, 1946–1966." *Marine Policy* 3, no. 2 (April 1979): 149–55.

———. *A Whaling Enterprise: Salvesen in the Antarctic.* Norwich: Michael Russell, 1998.

———. *Whaling 1937–1967: The International Control of Whale Stocks.* Kendall Whaling Museum Monograph Series 10. Sharon, MA: Kendall Whaling Museum, 1997.

Ellis, Jaye. Review of *Marine Conservation Agreements: The Law and Policy of Reservations and Vetoes*, by Howard S. Schiffman. *Global Environmental Politics* 9, no. 2 (May 2009): 131–33.

Ellis, Richard. *The Book of Whales.* New York: Knopf, 2000.

———. "Forrest Glenn Wood." *Journal of Cephalopod Biology* 2, no. 2 (1992): 118.

———. *Men and Whales.* New York: Knopf, 1991.

Elzinga, Aant. "Antarctica: The Construction of a Continent by and for Science." In *Denationalizing Science: The Contexts of International Scientific Practice*, edited by

Elisabeth Crawford, Terry Shinn, and Sverker Sörlin, 73–106. Dordrecht, Holland: Kluwer Academic Press, 1993.

———. "Beyond the Ends of the World: Nationalism and Internationalism in Antarctic Exploration and Imagination, 1895–1914." Paper presented at the History of Science Society Conference, Milwaukee, WI, 7–9 November 2002.

Epstein, Charlotte. "The Making of Global Environmental Norms: Endangered Species Protection." *Global Environmental Politics* 6, no. 2 (May 2006): 32–54.

Eschricht, Daniel Frederik, Johannes Theodor Reinhardt, and Wilhelm Kukkjeborg. *Recent Memoirs on the Cetacea*. Edited by William Henry Flower. London: Robert Hardwicke, 1866.

Evans, William E. *Fifty Years of Flukes and Flippers: A Little History and Personal Adventures with Dolphins, Whales, and Sea Lions, 1958–2007*. Sofia, Bulgaria: Pensoft, 2008.

———. "Forrest Glenn Wood, 1919–1992." *Marine Mammal Science* 8, no. 3 (July 1992): 324–25.

Evans, W. E., and J. J. Dreher. "Observations on Scouting Behavior and Associated Sound Production by the Pacific Bottlenosed Porpoise (*Tursiops gilli* Dall)." *Bulletin of the Southern California Academy of Sciences* 61, no. 4 (1962): 217–26.

Evans, W. E., W. W. Sutherland, and R. G. Beil. "The Directional Characteristics of Delphinid Sounds." In *Marine Bio-Acoustics: Proceedings of a Symposium Held at the Lerner Marine Laboratory, Bimini, Bahamas, 11–13 April 1963*, edited by William N. Tavolga, 353–72. Oxford: Pergamon Press, 1964.

Ferguson, Robert. *Harpooner: A Four-Year Voyage on the Barque* Kathleen, *1880–1884*. Edited by Leslie Dalrymple Stair. Philadelphia: University of Pennsylvania Press, 1936.

Fessler, Aaron L. Review of *The Mind of the Dolphin: A Non-Human Intelligence*, by John C. Lilly. *Library Journal* 92, no. 20 (15 November 1967): 4168.

Fichtelius, Karl Erik, and Sverre Sjölander. *Man's Place: Intelligence in Whales, Dolphins, and Humans*. London: Gollancz, 1973.

———. *Smarter Than Man? Intelligence in Whales, Dolphins, and Humans*. Translated by Thomas Teal. New York: Pantheon Books, 1972.

Finlayson, Alan Christopher. *Fishing for Truth: A Sociological Analysis of Northern Cod Stock Assessments from 1977–1990*. St. John's, Newfoundland, Canada: Institute of Social and Economic Research at the Memorial University of Newfoundland, 1994.

Finley, [Mary] Carmel. "The Social Construction of Fishing, 1949." *Ecology and Society* 14, no. 1 (2009): 6.

Finley, Mary Carmel. "The Tragedy of Enclosure: Fish, Fisheries Science, and U.S. Foreign Policy, 1920–1960." PhD diss., University of California, 2007.

Fitter, Richard. "Whaling: Almost a Victory." *Oryx* 12, no. 2 (1973): 186–87.

Flader, Susan. *Thinking Like a Mountain: Aldo Leopold and the Evolution of an Ecological Attitude toward Deer, Wolves, and Forests*. Columbia: University of Missouri Press, 1974.

Flaherty, Bernard E. *Psychophysiological Aspects of Space Flight*. New York: Columbia University Press, 1961.

Flower, William Henry. "Cetacea." In *Encyclopedia Britannica* 15, 9th ed., edited by T. S. Baynes, 391–400. Edinburgh: A. and C. Black, 1875–1889.

———. *Essays on Museums and Other Subjects Connected with Natural History.* London: Macmillan, 1898.

———. *List of the Specimens of Cetacea in the Zoological Department of the British Museum.* London: Printed by order of the Trustees, 1885.

Fogg, Gordon Elliott. *A History of Antarctic Science.* Cambridge: Cambridge University Press, 1992.

Food and Agriculture Organization of the United Nations. *Mammals in the Seas.* Vol. 1. *Report of the FAO Advisory Committee on Marine Resources Research, Working Party on Marine Mammals.* FAO Fisheries Series no. 5. Rome: Food and Agriculture Organization of the United Nations, 1978.

———. *Mammals in the Seas.* Vol. 2. *Pinniped Species Summaries and Reports on Sirenians. Being the Annex B Appendixes VI and VII of the Report of the FAO Advisory Committee on Marine Resources Research, Working Party on Marine Mammals, with the cooperation of the United Nations Environment Programme.* FAO Fisheries Series no. 5. Rome: Food and Agriculture Organization of the United Nations, 1979.

———. *Mammals in the Seas.* Vol. 3. *General Papers. Large Cetaceans. Selected Papers of the Scientific Consultation on the Conservation and Management of Marine Mammals and Their Environment. Marine Mammals. Marine Ecology. Fishery Management. Potential Yield. Stock Assessment. Cetacea.* FAO Fisheries Series no. 5. Rome: Food and Agriculture Organization of the United Nations, 1981.

———. *Mammals in the Seas.* Vol. 4. *Small Cetaceans, Seals, Sirenians and Otters. Selected Papers of the Scientific Consultation on the Conservation and Management of Marine Mammals and Their Environment.* FAO Fisheries Series no. 5. Rome: Food and Agriculture Organization of the United Nations: 1982.

Forestell, Paul H. "Popular Culture and Literature." In *Encyclopedia of Marine Mammals,* edited by W. F. Perrin, Bernd G. Würsig, and J. G. M. Thewissen, 957–74. San Diego: Academic Press, 2002.

Forman, Charles William. "Science for Empire: Britain's Development of the Empire through Scientific Research 1895–1940." PhD diss., University of Wisconsin, 1941.

Fowler, Charles W., and Tim D. Smith, eds. *Dynamics of Large Mammal Populations.* New York: John Wiley & Sons, 1981.

Francis, Daniel. *A History of World Whaling.* New York: Viking, 1990.

Frangsmyr, Tore, Sten Lindroth, Gunnar Eriksson, and Gunnar Broberg, eds. *Linnaeus: The Man and His Work.* Berkeley: University of California Press, 1983.

Fraser, F. C. "Royal Fishes." In *Functional Anatomy of Marine Mammals,* vol. 3, edited by R. J. Harrison. London: Academic Press, 1977.

Fraser, F. C., and P. E. Purves. "Hearing in Cetaceans." *Bulletin of the British Museum of Natural History* 2 (1954): 103–16.

———. "Hearing in Cetaceans: Evolution of the Accessory Air Sacs and the Structure and Function of the Outer and Middle Ear in Recent Cetaceans." *Bulletin of the British Museum of Natural History* 7 (1960): 1–140.

Freeman, Milton M. R. "Political Issues with Regard to Contemporary Whaling." In

Who's Afraid of Compromise?, edited by Simon Ward, 2–3. Tokyo: Institute of Cetacean Research, 1990.

Friedheim, Robert L., ed. *Toward a Sustainable Whaling Regime*. Seattle: University of Washington Press, 2001.

Friedman, Lawrence A. "Legal Aspects of the International Whaling Controversy: Will Jonah Swallow the Whales?" *N.Y.U. Journal of International Law and Policy* 8 (1975): 210–39.

Friend, Milton. "Conservation Landmarks: Bureau of Biological Survey and National Biological Service." In *Our Living Resources: A Report to the Nation on the Distribution, Abundance, and Health of U.S. Plants, Animals, and Ecosystems*, 7–9. Washington, DC: U.S. Department of the Interior, National Biological Service, 1995.

Friends of the Earth. *The Stockholm Conference—Only One Earth: An Introduction to the Politics of Survival*. London: Earth Island, distributed by Argus and Robertson, 1972.

Friends of the Earth and Project Jonah. *A Report on Whaling and Australia's Role in the Slaughter of Whales*. Melbourne: Friends of the Earth, 1973.

———. *Whale Campaign Manual*. Kew: Friends of the Earth, ca. 1973.

Friese, Heidrun. "Thresholds in the Ambit of Discourse: On the Establishment of Authority at Academic Conferences." Translated by William Clark. In *Little Tools of Knowledge: Historical Essays on Academic and Bureaucratic Practices*, edited by Peter Becker and William Clark, 285–312. Ann Arbor: University of Michigan Press, 2001.

Frohoff, Toni, and Brenda Peterson. *Between Species: Celebrating the Dolphin-Human Bond*. San Francisco: Sierra Club Books, 2003.

Frost, Gary L. "Inventing Schemes and Strategies: The Making and Selling of the Fessenden Oscillator." *Technology and Culture* 42, no. 3 (2001): 462–88.

Frost, Sydney. *The Whaling Question: The Inquiry by Sir Sydney Frost of Australia*. San Francisco: Friends of the Earth, 1979.

Fulton, Thomas Wemyss. *The Sovereignty of the Sea: An Historical Account of the Claims of England to the Dominion of the British Seas, and of the Evolution of the Territorial Waters*. Edinburgh: William Blackwood and Sons, 1911.

Gambell, Ray. "British Whaling and Whale Research." In *British Marine Science and Meteorology: The History of Their Development and Application to Marine Fishing Problems*, edited by the Buckland Foundation, 83–95. Lowestoft: Buckland Foundation, 1996.

———. "International Management of Whales and Whaling: An Historical Review of the Regulation of Commercial and Aboriginal Subsistence Whaling." *Arctic* 46, no. 2 (June 1993): 97–107.

———. "Some Effects of Exploitation on Reproduction in Whales." In *The Environment and Reproduction in Mammals and Birds: Proceedings of the Symposium of the Society for the Study of Fertility, Edinburgh, Scotland, March 1972*, edited by J. S. Perry and I. W. Rowlands, 533–53. Oxford: Blackwell Scientific Publications, 1973.

"Gamboling Dolphin in a New Role: My Pal Flipper." *Life Magazine* 54, no. 23 (7 June 1963): 61–64, 66, 69.

Gaskin, D. E. *Ecology of Whales and Dolphins*. London: Heinemann Educational Books, 1982.

Gehring, Thomas, and Eva Ruffing. "When Arguments Prevail over Power: The CITES Procedure for the Listing of Endangered Species." *Global Environmental Politics* 8, no. 2 (May 2008): 123–48.

Geisler, Jonathan H., and Jessica M. Theodor. "Hippopotamus and Whale Phylogeny." *Nature* 458 (19 March 2009): E1–E5.

Geison, Gerald L. *Physiology in the American Context, 1850–1940*. Bethesda, MD: American Physiological Society, 1987.

Genoways, Hugh H., and Patricia W. Freeman. "Evolution of a Scientific Meeting: Eighty Annual Meetings of the American Society of Mammalogists, 1919–2000." *Journal of Mammalogy* 82, no. 2 (2001): 582–603.

Georg, Jon H. "Milton Harris: A Biography." Milton Harris Papers, Special Collections, Oregon State University Libraries.

Giata, Raymond. *The Philosopher's Dog: Friendships with Animals*. New York: Random House, 2002.

Gieryn, Thomas F. *Cultural Boundaries of Science: Credibility on the Line*. Chicago: University of Chicago Press, 1999.

Gill, Theodore, and J. A. Allen. *Contributions to the Bibliographical Literature of American Mammals*. New York: Arno Press, 1974.

Gilmore, Charles W. "A History of the Division of Vertebrate Paleontology in the United States National Museum." *Proceedings of the United States National Museum* 90, no. 3109 (1942): 305–77.

Gilpin, Robert, and Christopher Wright, eds. *Scientists and National Policy-Making*. New York: Columbia University Press, 1964.

Glass, Bryan P. Review of *Man and Dolphin*, by John C. Lilly. *Quarterly Review of Biology* 36, no. 4 (December 1961): 311.

Goddard, Donald, and Sam Swope, eds. *Saving Wildlife: A Century of Conservation*. New York: Harry N. Abrams in association with The Wildlife Conservation Society, 1995.

Goldberger, Leo. "In the Absence of Stimuli." Review of *Sensory Deprivation*, by John P. Zubek. *Science* 168, no. 3932 (8 May 1970): 709–11.

Golden, Frederic, ed. "Whither the Whales." *Oceanus* 32, no. 1 (Spring 1989).

Goldman, E. A. "The Predatory Mammal Problem and the Balance of Nature." *Journal of Mammalogy* 6, no. 1 (February 1925): 28–33.

Goodwin, George G. "Porpoise—Friend of Man?" *Natural History* 56 (1947): 337, 383.

Gordon, Bernard L., ed. *Man and the Sea: Classic Accounts of Marine Exploration*. Westerly, RI: The Book and Tackle Shop, 1994.

Gordon, Scott. "Economics and the Conservation Question." *Journal of Law and Economics* 1 (October 1958): 110–21.

Gottlieb, Robert. *Forcing the Spring: The Transformation of the American Environmental Movement*. Rev. and updated ed. Washington, DC: Island Press, 2005.

Graham, Michael. "Concepts of Conservation." In *Papers presented at the International*

Technical Conference on the Conservation of the Living Resources of the Sea, Rome, 18 April–10 May, 1955, 1–53. New York: United Nations, 1956.

Gray, Sir James. "How Fishes Swim." *Scientific American* 197, no. 2 (August 1957): 48–65.

———. "Studies in Animal Locomotion: VI. The Propulsive Powers of the Dolphin." *Journal of Experimental Biology* 13 (1936): 192–99.

Gray, John Edward. *Catalogue of Seals and Whales in the British Museum.* 2nd ed. London: Taylor and Francis, 1866.

———. *Synopsis of the Species of Whales and Dolphins in the Collection of the British Museum.* Illustrated with 37 plates, by the late William Wing. London: Bernard Quaritch, 1868.

Great Britain. *"Discovery" Expedition. First Annual Report, 1926.* London: His Majesty's Stationery Office, 1927.

———. *Discovery Investigations: Second Annual Report, January, 1927–May, 1928.* London: His Majesty's Stationery Office, 1929.

Great Britain Discovery Committee. *Report on the Progress of the Discovery Committee's Investigations.* London: The Discovery Committee, Colonial Office, 1937.

Greene, Ann Norton. "Harnessing Power: Industrializing the Horse in Nineteenth Century America." PhD diss., University of Pennsylvania, 2004.

Greene, Eric. *Planet of the Apes as American Myth: Race and Politics in the Films and Television Series.* Jefferson, NC: McFarland, 1996.

Greene, Mott T. "Oceanography's Double Life." *Earth Sciences History* 12, no. 1 (1993): 48–53.

Grey, Zane. *Tales of Fishing Virgin Seas.* New York: Grosset and Dunlap, 1925.

———. *Tales of the Angler's El Dorado, New Zealand.* New York: Harper and Brothers, 1926.

———. *Tales of Swordfish and Tuna.* New York: Harper and Brothers, 1927.

Griffin, Donald R. *Animal Engineering: Readings from* Scientific American. San Francisco: W. H. Freeman, 1974.

———. *Animal Minds: Beyond Cognition to Consciousness.* Chicago: University of Chicago Press, 2001.

———. *Animal Thinking.* Cambridge, MA: Harvard University Press, 1984.

———. "Echolocation by Blind Men, Bats, and Radar." *Science* 100, no. 2609 (29 December 1944): 590–91.

———. *Listening in the Dark: The Acoustic Orientation of Bats and Men.* New Haven, CT: Yale University Press, 1958. Reprint, New York: Dover, 1974.

———. *The Question of Animal Awareness: Evolutionary Continuity of Mental Experience.* New York: Rockefeller University Press, 1981.

———. *Underwater Sounds and the Orientation of Marine Animals: A Preliminary Survey.* Tech. Rept. 3, Proj. MR 162–429 ONR and Cornell University. Ithaca, NY: Cornell University Department of Zoology, 1950.

Grosvenor, Gilbert. Editorial. *National Geographic* 150, no. 6 (December 1976): 721.

Grove, Richard H. *Green Imperialism: Colonial Expansion, Tropical Island Edens and the Origins of Environmentalism, 1600–1860*. Cambridge: Cambridge University Press, 1996.

Guerrini, Anita. *Experimenting with Humans and Animals: From Galen to Animal Rights*. Baltimore: Johns Hopkins University Press, 2003.

Gulbrandsen, Lars H. "The Role of Science in Environmental Governance: Competing Knowledge Producers in Swedish and Norwegian Forestry." *Global Environmental Politics* 8, no. 2 (May 2008): 99–122.

Guldberg, G. A. "Bidrag til Cetaceernes Biologi. Om fortplantningen og draegtigheden hos de Nordatlantiske Bardehvaler." *Christiania-Videnskabs-Selskabet-Forhandlinger* 9 (1886): 56.

Gulland, J. A. "The Effect of Regulation on Antarctic Whale Catches." *Journal du Conseil (Conseil International pour l'Exploration de la Mer)* 30, no. 3 (1966): 308–15.

———. "The Management of Antarctic Whaling Resources." *Journal du Conseil (Conseil International pour l'Exploration de la Mer)* 32, no. 3 (1968): 330–41.

Gunther, Eugene Rolfe. "The Habits of Fin Whales." *Discovery Reports* 25 (1947): 113–42.

———. *Notes and Sketches Made During Two Years on the "Discovery" Expedition, 1925–1927*. Oxford: Holywell Press, 1928.

Hamblin, Jacob Darwin. *Oceanographers and the Cold War: Disciples of Marine Science*. Seattle: University of Washington Press, 2005.

Hamilton, Robert. *The Natural History of the Ordinary Cetacea or Whales*. Edinburgh: W. H. Lizars, 1837.

Hanson, Elizabeth. *Animal Attractions: Nature on Display in American Zoos*. Princeton, NJ: Princeton University Press, 2002.

Haraway, Donna. *The Companion Species Manifesto: Dogs, People, and Significant Otherness*. Chicago: Prickly Paradigm Press, 2003.

———. *Primate Visions: Gender, Race, and Nature in the World of Modern Science*. New York: Routledge, 1989.

Hardin, G. "The Tragedy of the Commons: The Population Problem Has No Technical Solution; It Requires a Fundamental Extension in Morality." *Science* 162, no. 859 (13 December 1968): 1243–48.

Hardy, Alister Clavering. "Charles Elton's Influence in Ecology." *Journal of Animal Ecology* 37 (1968): 3–8.

———. *Great Waters: A Voyage of Natural History to Study Whales, Plankton, and the Waters of the Southern Ocean*. New York: Harper and Row, 1967.

———. "The Herring in Relation to its Animate Environment. Part I: The Food and Feeding Habits of the Herring with Special Reference to the East Coast of England." *Fishery Investigations*, ser. 2, vol. 7, no. 3 (1924): 53.

———. "The Herring in Relation to its Animate Environment. Part II: Report on Trials with the Plankton Indicator." *Fishery Investigations*, ser. 2, vol. 8, no. 7 (1926): 13.

———. "N. A. Mackintosh." *Nature* 249, no. 5459 (21 June 1974): 785.

———. "Whale-Marking in the Southern Ocean." *Geographical Journal* 96, no. 5 (November 1940): 345–50.

Harkness, Deborah. "Managing an Experimental Household: The Dees of Mortlake and the Practice of Experimental Philosophy." *Isis* 88, no. 2 (1997): 247–62.

Harmer, Sidney F. Abstract of "The Present Position of the Whaling Industry." *Nature* 110, no. 2772 (16 December 1922): 827.

———. "The Blue Whale." *Nature* 115, no. 2886 (21 February 1925): 283.

———. "History of Whaling in the South and Statistical Evidence of Breeding Periods Derived from Foetal Records." *The Challenger Society* 1, no. 16 (1919): 6–7.

———. "Is the Whale Being Exterminated?" *Anglo-Norwegian Trade Journal* 9, no. 98 (February 1923).

———. "Modern Whaling." *Reports of the British Association* 89, sec. D (Edinburgh, 1921): 422–23.

———. "Presidential Address, 24 May 1928." *Proceedings of the Linnean Society of London* 140, no. 1 (November 1928): 51–95.

———. "Presidential Address, 24 May 1930." *Proceedings of the Linnean Society of London* 142, no. 1 (January 1930): 85–163.

———. "The Scientific Development of the Falkland Islands Dependencies." *Geographical Journal* 56, no. 1 (July 1920): 61–65.

———. "Scientific Investigation of the Whaling Problem." *Nature* 111, no. 2790 (21 April 1923): 540–41.

———. "The South Georgia Whale Fishery." *Nature* 102, no. 2552 (26 September 1918): 65.

Harris, Marvin. *The Nature of Cultural Things*. New York: Random House, 1964.

Harrison, Richard J., ed. *Functional Anatomy of Marine Mammals*. Vol. 3. London: Academic Press, 1977.

Hart, Ian B. *Pesca: The History of Compañia Argentina de Pesca Sociedad Anónima of Buenos Aires: An Account of the Pioneer Modern Whaling and Sealing Company in the Antarctic*. Salcombe: Aidan Ellis, 2001.

Haskell, P. T., and F. C. Fraser, eds. *Biological Acoustics: Proceedings of a Symposium Held at the Zoological Society of London*. London: The Society, 1962.

Haskell, Ronald J., Jr. "Abandoning Whale Conservation Initiatives in *Japan Whaling Association v. American Cetacean Society*." *Harvard Environmental Law Review* 11 (1987): 551–92.

Hauser, Mark D. *Wild Minds: What Animals Really Think*. New York: Henry Holt, 2000.

Hayden, Sherman Strong. *The International Protection of Wild Life: An Examination of Treaties and Other Agreements for the Preservation of Birds and Mammals*. New York: Columbia University Press, 1942.

Hays, Samuel P. *Conservation and the Gospel of Efficiency: The Progressive Conservation Movement, 1890–1920*. Cambridge, MA: Harvard University Press, 1959.

Hays, Samuel P., and Barbara D. Hays. *Beauty, Health, and Permanence: Environmental Politics in the United States, 1955–1985*. Cambridge: Cambridge University Press, 1987.

Headland, R. K. *The Island of South Georgia*. Cambridge: Cambridge University Press, 1984.

———. "The Natural History of Antarctica: The Early Voyages." *Scottish Naturalist* 111 (1999): 9–36.

Headland, Thomas N., Kenneth L. Pike, and Marvin Harris. *Emics and Etics: The Insider/Outsider Debate.* Newbury Park, CA: Sage Publications, 1990.

Heazle, Michael. "Lessons in Precaution: The International Whaling Commission Experience with Precautionary Management." *Marine Policy* 30, no. 5 (2006): 496–509.

———. "Scientific Uncertainty and the International Whaling Commission: An Alternative Perspective on the Use of Science in Policy Making." *Marine Policy* 28, no. 5 (2004): 361–74.

———. *Scientific Uncertainty and the Politics of Whaling.* Seattle: University of Washington Press, 2006.

"He barks and buzzes, he ticks and whistles, but can the dolphin learn to talk?" *Life Magazine* 51, no. 4 (28 July 1961): 61–66.

Helvarg, David. *Blue Frontier: Saving America's Living Seas.* New York: W. H. Freeman, 2001.

Hendrickson, Robert. *More Cunning Than Man: A Complete History of the Rat and Its Role in Human Civilization.* New York: Kensington Books, 1999.

Henninger-Voss, Mary. *Animals in Human Histories.* Rochester, NY: University of Rochester Press, 2002.

Herman, Louis M. *Cetacean Behavior: Mechanisms and Processes.* New York: John Wiley & Sons, 1980.

Herzen, Bruce C. "Whales Entangled in Deep Sea Cables." *Deep-Sea Research* 4 (1957): 105–15.

Hickel, Walter J. *Who Owns America?* Englewood Cliffs, NJ: Prentice-Hall, 1971.

Higgs, Louis D., and Robert G. Weinland. *Project Michelson Preliminary Report.* Technical Progress Report 309. China Lake, CA: U.S. Naval Ordnance Test Station, 1963.

Hillix, William A., and Duane M. Rumbaugh, eds. *Animal Bodies, Human Minds: Ape, Dolphin, and Parrot Language Skills.* New York: Kluwer Academic/Plenum Publishers, 2004.

Hills, David O. "Vanishing Giants." With paintings by Richard Ellis. *Audubon* (January 1975): 56–90.

Hinton, Martin Alister Campbell. *Reports on Papers Left by the Late Major G. E. H. Barrett-Hamilton Relating to the Whales of South Georgia.* London: The Crown Agents for the Colonies, 1925.

Hjort, Johan. "Fiskeri og hvalfangst i det nordilige Norge." *Aarsberetning Vedkommende Norges Fiskerier* 4 (1902): 1–251.

———. "Human Activities and the Study of Life in the Sea: An Essay on the Methods of Research and Experiment." *Geographical Review* 25, no. 4 (October 1935): 529–64.

———. "The Story of Whaling: A Parable of Sociology." *Scientific Monthly* 45, no. 1 (July 1937): 19–34.

———. "Whales and Whaling." *Hvalrådets Skrifter* 7 (1933): 7–29.

Hjort, Johan, G. Jahn, and P. Ottestad. "The Optimum Catch." *Hvalrådets Skrifter* 7 (1933): 92–127.

Hjort, Johan, J. Lie, and J. T. Ruud. "Norwegian Pelagic Whaling in the Antarctic. I: Whaling Grounds in 1929–1930 and 1930–1931." *Hvalrådets Skrifter* 3 (1932): 1–47.

———. "Norwegian Pelagic Whaling in the Antarctic. II." *Hvalrådets Skrifter* 7 (1933): 128–52.

———. "Norwegian Pelagic Whaling in the Antarctic III." *Hvalrådets Skrifter* 8 (1933): 4–36.

Hockett, Charles. "Logical Considerations in the Study of Animal Communication." In *Animal Sounds and Communication*, edited by W. E. Lanyon and W. N. Tavolga, 392–430. Washington, DC: American Institute of Biological Sciences, 1960.

———. Review of *Man and Dolphin*, by John C. Lilly. *American Anthropologist* 65, no. 1 (February 1963): 176–77.

Holder, Charles Frederick. *The Game Fishes of the World*. London: Hodder and Stoughton, 1913.

Holdgate, Martin W. *The Green Web: A Union for World Conservation*. Cambridge: IUCN, the World Conservation Union/Earthscan, 1999.

Holdgate, Martin W., Mohammed Kassas, and Gilbert F. White. *The World Environment 1972–1982: A Report*. Natural Resources and the Environment Series, vol. 8. Dublin: United Nations Environment Programme/Tycooly International, 1982.

Hollick, Ann L. "The Origins of 200-Mile Offshore Zones." *American Journal of International Law* 71, no. 3 (July 1977): 494–500.

Holm, Poul, Tim D. Smith, and David J. Starkey, eds. *The Exploited Seas: New Directions for Marine Environmental History*. Saint John's, Newfoundland: International Maritime Economic Association/Census of Marine Life, 2001.

Holt, Sidney. "Fifty Years On." *Reviews in Fish Biology and Fisheries* 8, no. 3 (1998): 357–66.

———. "Let's All Go Whaling." *Ecologist* 15, no. 3 (1985): 113–24.

Holt, Sidney, and Edward Mitchell. "Counting Whales in the North Atlantic." *Science* 303, no. 5654 (2 January 2004): 39–40.

Hornaday, William Temple. *Thirty Years War for Wild Life*. New York: Arno Press, 1970 [1931].

Horwitz, Josh. *Whales vs. Navy: The Story of Whales, a Secret Navy Sonar Program, and the Environmental Lawyer Who Exposed It*. New York: Twelve, forthcoming.

Howell, A. Brazier. *Aquatic Mammals: Their Adaptations to Life in the Water*. Springfield, IL: C. C. Thomas, 1930.

———. "Asymmetry in the Skulls of Mammals." *Proceedings of the United States National Museum* 67, no. 2599, art. 27 (1925): 1–18.

———. "The Plight of the Whale." *National Humane Review* (July 1935): 7.

———. "Visit to a California Whaling Station." *Smithsonian Miscellaneous Collections* 78, no. 7 (1926): 71–79.

Howell, A. Brazier, and Lawrence M. Huey. "Food of the Gray and Other Whales." *Journal of Mammalogy* 11, no. 3 (August 1930): 321–22.

Huber, Ernst. "Anatomical Notes on Pinnipedia and Cetacea." *Carnegie Institute of Washington* 447 (1934): 105–36.

Hughes, Ted. "Man and Superbeast." Review of *The Nerve of Some Animals*, by Robert Froman, and *Man and Dolphin*, by John C. Lilly. *New Statesman* 53, no. 1619 (23 March 1962): 420–21.

Hunter, Robert. *Warriors of the Rainbow: A Chronicle of the Greenpeace Movement.* New York: Holt, Rinehart and Winston, 1979.

Hurrell, Andrew, and Benedict Kingsbury, eds. *The International Politics of the Environment: Actors, Interests, and Institutions.* New York: Oxford University Press, 1992.

Ingold, Tim. *The Perception of the Environment: Essays on Livelihood, Dwelling, and Skill.* London: Routledge, 2000.

International Conference on Whaling. International Agreement for the Regulation of Whaling (With Final Act of the Conference), London, June 8, 1937. The agreement has not been ratified by His Majesty's government in the United Kingdom. London: His Majesty's Stationery Office, 1937.

———. Protocol Amending the International Agreement of June 8, 1937, For the Regulation of Whaling, London, June 24, 1938. The ratification of His Majesty's government in the United Kingdom was deposited in London on December 7, 1938. London: His Majesty's Stationery Office, 1939.

———. Treaty Series No. 933. Regulation of Whaling. Protocol between the United States of America and other powers amending the International agreement for the regulation of whaling, signed in London June 8, 1937. Signed at London June 24, 1938. Ratification advised by the Senate of the United States, August 5, 1937. Ratified by the President of the United States, August 13, 1937. Ratification of the United States of America deposited at London, September 3, 1937. Proclaimed by the President of the United States, May 18, 1938. With Certificate of extension and Final act of the conference. Washington, DC: U.S. Government Printing Office, 1939.

———. Treaty Series No. 944. Regulation of Whaling. Protocol between the United States of America and other Powers Amending the International Agreement for the Regulation of Whaling, Signed in London June 8, 1937 (Treaty Series No. 933). Signed at London June 24, 1938. Ratification advised by the Senate of the United States March 8, 1939. Ratified by the President of the United States March 16, 1939. Ratification of the United States of America deposited at London, March 30, 1939. Proclaimed by the President of the United States April 8, 1939. With Certificate of extension and Final act of the conference. Washington, DC: U.S. Government Printing Office, 1939.

International Parliamentary Conference on the Environment. *The parliamentary viewpoint on the Stockholm conference; proceedings. Le point de vue parlementaire sur la Conférence de Stockholm; procès-verbaux. Der parlamentarische Standpunkt zur Stockholmer Konferenz; Bericht.* Beiträge zur Umweltgestaltung, Heft A 9. Berlin: E. Schmidt, 1973.

International Whaling Conference. Whaling. Convention between the United States of America and other governments signed at Washington under date of December 2, 1946. Signed at Washington under date of December 2, 1946. Ratification advised by the Senate of the United States of America July 2, 1947. Ratified by the President of the United States of America July 18, 1947. Ratification of the United States of

America deposited at Washington July 18, 1947. Proclaimed by the President of the United States of America November 19, 1948. Entered into force November 10, 1948. Washington, DC: U.S. Government Printing Office, 1949.

Irvine, Blair. "Conditioning Marine Mammals to Work in the Sea." *Marine Technology Journal* 4, no. 3 (1970): 47–52.

Isenberg, Andrew C. *The Destruction of the Bison: An Environmental History, 1750–1920.* Cambridge: Cambridge University Press, 2000.

Ivashchenko, Yulia, Phillip J. Clapham, Robert L. Brownell Jr., A. V. Yablokov, and Alfred A. Berzin. "The Truth about Soviet Whaling." *Marine Fisheries Review* 70, no. 2 (2008): 1–59.

Jackson, Gordon. *The British Whaling Trade.* London: A. and C. Black, 1978.

Jarrell, Randall, ed. *Kenneth S. Norris, Naturalist, Cetologist and Conservationist, 1924–1998: An Oral History Biography.* Santa Cruz: Regional History Project, UC Santa Cruz University Library, 1999.

Jasanoff, Sheila S. "Contested Boundaries in Policy-Relevant Science." *Social Studies of Science* 17, no. 2 (May 1987): 195–230.

———. *The Fifth Branch: Science Advisers as Policymakers.* Cambridge, MA: Harvard University Press, 1990.

Jasper, Herbert H., Lorne D. Proctor, Robert S. Knighton, William C. Noshay, and Russell T. Costello, eds. *Reticular Formation of the Brain.* Boston: Little Brown, 1958.

Jeffrey, Francis, and John Cunningham Lilly. *John Lilly, So Far . . .* Los Angeles: Jeremy P. Tarcher, 1990.

Jeffries, Sir Charles Joseph, ed. *A Review of Colonial Research, 1940–1960.* London: Her Majesty's Stationery Office, 1964.

Jenkins, James Travis. *Whales and Modern Whaling.* London: H. F. & G. Witherby, 1932.

Jessup, Philip C. "The International Protection of Whales." *American Journal of International Law* 24, no. 4 (October 1930): 751–52.

John, D. D. "The Slaughter of Whales." *National Humane Review* (July 1938): 20.

Johnson, Martin. "Underwater Sounds of Biological Origin," UCDWR Rept. U28, PB 48635 (1943): 1–26.

Johnston, Alexander Keith. *The Physical Atlas of Natural Phenomena: For the Use of Colleges, Academies, and Families.* Philadelphia: Lea and Blanchard, 1850.

Johnston, D. Gordon, and Sam H. Ridgway. "Parasitism in Some Marine Mammals." *Journal of the American Veterinary Medical Association* 155, no. 7 (1 October 1969): 1064–72.

Johnston, Douglas M. *The International Law of Fisheries: A Framework for Policy-Oriented Inquiries.* New Haven, CT: Yale University Press, 1965.

Jones, A. G. E. *Polar Portraits: Collected Papers.* Whitby, UK: Caedmon of Whitby, 1992.

Jones, Susan D. *Valuing Animals: Veterinarians and Their Patients in Modern America.* Baltimore: Johns Hopkins University Press, 2003.

Jonsgård, Åge. "Biology of the North Atlantic Fin Whale, *Balaenoptera physalus* (L.):

Taxonomy, Distribution, Migration, and Food." *Hvalrådets Skrifter* 49 (1966): 1–62.

Kalland, Arne. "Management by Totemization: Whale Symbolism and the Anti-Whaling Campaign." *Arctic* 48, no. 2 (June 1993): 124–33.

Kanwisher, John, and Alfred Senft. "Physiological Measurements on a Live Whale." *Science* 131, no. 3410 (6 May 1960): 1379–80.

Kanwisher, John W. and Gunnar Sundnes, eds. *Essays in Marine Physiology, Presented to P. F. Scholander in Honor of His Sixtieth Birthday. Hvalrådets Skrifter* 48. Oslo: Universitetsforlaget, 1965.

Kellert, Stephen R. "Human Values, Ethics, and the Marine Environment." In *Values at Sea: Ethics for the Marine Environment*, edited by Dorinda G. Dallmeyer, 1–18. Athens: University of Georgia Press, 2003.

Kellert, Stephen R., and Edward O. Wilson, eds. *The Biophilia Hypothesis*. Washington, DC: Island Press, 1993.

Kellogg, Remington. "Adaptation of Structure to Function in Whales." *Cooperation in Research, Carnegie Institution of Washington Publication no. 501* (1938): 649–82.

———. "Ancient Relatives of Living Whales." In *Explorations and Field-Work of the Smithsonian Institution* 3111 (1930): 83–90.

———. "Antarctic Whales." Chap. 14 in *Science in Antarctica, Part 1: The Life Sciences in Antarctica*. Publication 839. Washington, DC: National Academy of Sciences–National Research Council, 1961.

———. "Are Moles Held in Check by Blacksnakes?" *Bulletin of the Green Section of the U.S. Golf Association* 2, no. 5 (17 May 1922): 157–59.

———. "Cetacean Studies in Europe." *Explorations and Field-Work of the Smithsonian Institution* 3525 (1938): 41–46.

———. *Contributions to Paleontology from the Carnegie Institution of Washington: Additions to the Tertiary History of the Pelagic Mammals on the Pacific Coast of North America*. Washington, DC: Carnegie Institution of Washington, 1925.

———. "The Current Whale Oil Industry," review of *Whale Oil: An Economic Analysis*, by Karl Brandt. *Scientific Monthly* 52, no. 2 (February 1941): 178–79.

———. "Description of a New Genus and Species of Whalebone Whale from the Calvert Cliffs, Maryland." *Proceedings of the United States National Museum* 63, no. 2483, art. 15 (5 February 1924): 1–17.

———. "Description of the Skull of *Megaptera miocaena*, a Fossil Humpback Whale from the Miocene Diatomaceous Earth of Lompoc, California." *Proceedings of the United States National Museum* 61, no. 2435, art. 14 (1922): 1–18, pls. 1–4.

———. "Description of Two Squalodonts Recently Discovered in the Calvert Cliffs, Maryland; and Notes on the Shark-toothed Cetaceans." *Proceedings of the United States National Museum* 61, no. 2462, art. 16 (1923): 1–14.

———. "Fossil Cetaceans from the Florida Tertiary." *Bulletin of the Museum of Comparative Zoology at Harvard College* 94, no. 9 (November 1944): 433–71.

———. "A Fossil Physeteroid Cetacean from Santa Barbara County, California." *Proceedings of the United States National Museum* 66, no. 2564, art. 27 (28 February 1925): 1–8.

———. "A Fossil Porpoise from the Calvert Formation of Maryland." *Proceedings of the United States National Museum* 63, no. 2482, art. 14 (1924): 1–36.

———. "The Habits and Economic Importance of Alligators." *Technical Bulletin of the U.S. Department of Agriculture* 147 (December 1929): 1–36.

———. "The History of Whales—Their Adaptation to Life in the Water." PhD diss., University of California, Berkeley, 1928.

———. "The History of Whales—Their Adaptation to Life in the Water." *Quarterly Review of Biology* 3, no. 1 (March 1928): 29–76.

———. "The International Whaling Commission." Paper presented at the International Technical Conference on the Conservation of the Living Resources of the Sea, Rome, 18 April–10 May 1955.

———. "Mexican Tailless Amphibians in the United States National Museum." In *United States National Museum Bulletin* 160. Washington, DC: United States Government Printing Office, 1932.

———. "A New Fossil Toothed Whale from Florida." *American Museum Novitates* 389 (5 December 1929): 495–504.

———. "On the Identity of the Porpoise *Sagmatias amblodon*." *Zoological Series, Field Museum of Natural History* 27 (8 December 1941): 293–311.

———. "On the Occurrence of Remains of Fossil Porpoises of the Genus *Eurhinodelphis* in North America." *Proceedings of the United States National Museum* 66, no. 2563 (23 May 1925): 1–40.

———. "Pelagic Mammals from the Temblor Formation of the Kern River Region, California." *Proceedings of the California Academy of Sciences* 19, no. 12 (30 January 1931): 217–397.

———. "Pinnipeds from Miocene and Pleistocene Deposits of California: Description of a New Genus and Species of Sea Lion from the Temblor Together with Seal Remains from the Santa Margarita and San Pedro Formations and a Resume of Current Theories Regarding Origins of Pinnipedia." *University of California Publications Bulletin of the Department of Geological Sciences* 13, no. 4 (14 April 1922): 23–132.

———. *A Review of the Archaeoceti.* Washington, DC: Carnegie Institution of Washington, 1936.

———. Review of *Whales*, by E. J. Slijper, translated from the Dutch by A. J. Pomerans. *American Scientist* 51, no. 1 (March 1963): 100A–101A.

———. "Studies on the History and Evolution of Whales." *Carnegie Institution of Washington Yearbook* 38 (1939): 311–12.

———. "Studies on the History and Evolution of Whales." *Carnegie Institution of Washington Yearbook* 39 (1940): 294–95.

———. "Studies on the Relation of Sense Organs to the General Problem of Aquatic Adaptation." *Carnegie Institution of Washington Yearbook* 25 (1926): 403–7.

———. "Study of the Skull of a Fossil Sperm-Whale from the Temblor Miocene of Southern California." *Carnegie Institution of Washington Publication no. 346* (1927): 1–23.

———. "Tertiary Pelagic Mammals of Eastern North America." *Bulletin of the Geological Society of America* 35 (30 December 1924): 755–66.

———. "The Toad." U.S. Department of Agriculture, Bureau of Biological Survey Circular Bi-664 (August 1922): 1–7.

———. "Whales, Giants of the Sea," *National Geographic* 77, no. 1 (January 1940): 35–90.

———. "Whaling Statistics for the Pacific Coast of North America." *Journal of Mammalogy* 12, no. 1 (February 1931): 73–77.

———. "What is Known of the Migrations of Some of the Whalebone Whales." In *Annual Report of the Board of Regents of the Smithsonian Institution Showing the Operations, Expenditures, and Condition of the Institution for the Year Ending June 30, 1928*, 467–94. Publication 2981. Washington, DC: U.S. Government Printing Office, 1929.

Kellogg, Remington, and Frank C. Whitmore. "Mammals." *Geological Society of America* 67 (1957): 1021–24.

Kellogg, W. N. "Bibliography of the Noises Made by Marine Organisms." *American Museum Novitates* 1611 (20 March 1953): 1–5.

———. *Porpoises and Sonar*. Chicago: University of Chicago Press, 1961.

Kellogg, W. N., and Robert Kohler. "Reactions of the Porpoise to Ultrasonic Frequencies." *Science* 116, no. 3010 (5 September 1952): 250–52.

Kellogg, W. N., Robert Kohler, and H. N. Morris. "Porpoise Sounds as Sonar Signals." *Science* 117, no. 3036 (6 March 1953): 239–43.

Kemp, Stanley. "The 'Discovery' Expedition." *Natural History Magazine* 6 (1928): 183–97.

Kemp, Stanley, A. C. Hardy, and N. A. Mackintosh. "Discovery Investigations: Objects, Equipment, and Methods." *Discovery Reports* 1 (June 1929): 141–232.

Kete, Kathleen. *The Beast in the Boudoir: Petkeeping in Nineteenth-Century Paris.* Berkeley: University of California Press, 1994.

Kingsland, Sharon E. "Essay Review: The History of Ecology." *Journal of the History of Biology* 27, no. 2 (Summer 1994): 349–57.

———. *The Evolution of American Ecology, 1890–2000*. Baltimore: Johns Hopkins University Press, 2005.

———. *Modeling Nature: Episodes in the History of Population Ecology*. Chicago: University of Chicago Press, 1995.

Kirkwood, G. P. "Background to the Development of Revised Management Procedures." *Report of the International Whaling Commission* 42 (1992): 236–43.

Kisling, Vernon N. Jr., ed. *Zoo and Aquarium History: Ancient Animal Collections to Zoological Gardens*. Boca Raton: CRC Press, 2001.

Klems, Alf. "On the Growth of Populations of Yeast." *Hvalrådets Skrifter* 7 (1933): 55–92.

Klinowska, Margaret. "How Brainy are Cetaceans?" *Oceanus* 32, no. 1 (1989): 19–20.

Koch, Albert. *Description of the* Missorium Theristocaulodon *(Koch) or Missouri Leviathan, (*Leviathan Missouriensis*) Together with Its Supposed Habits and Indian Traditions, Also Comparisons on the Whale, Crocodile, and Missourium, with the Leviathan, As described in the 41st Chapter of the Book of Job*. 5th ed. Dublin: C. Crookes, 1843.

Kohler, Robert E. *Landscapes & Labscapes: Exploring the Lab-Field Border in Biology.* Chicago: University of Chicago Press, 2002.

———. *Lords of the Fly: Drosophila Genetics and the Experimental Life.* Chicago: University of Chicago Press, 1994.

———. "Science, Foundations, and American Universities in the 1920s." *Osiris* 3 (1987): 135–64.

Kohlstedt, Sally Gregory. *The Formation of the American Scientific Community: The American Association for the Advancement of Science, 1848–60.* Urbana: University of Illinois Press, 1976.

———. "Museums on Campus: A Tradition of Inquiry and Teaching." In *The American Development of Biology,* edited by Roland Rainger and Jane Maienschein, 15–47. Philadelphia: University of Pennsylvania Press, 1988.

Koontz, Tomas M., Toddi A. Steelman, JoAnn Carmin, Katrina Smith Korfmacher, Cassandra Mosely, and Craig W. Thomas. *Collaborative Environmental Management: What Roles for Government?* Washington, DC: Resources for the Future, 2004.

Krige, John, and Kai-Henrik Barth, eds. *Historical Perspectives on Science, Technology and International Affairs.* Chicago: University of Chicago Press, 2006.

Kroll, Gary. *America's Ocean Wilderness: A Cultural History of Twentieth Century Exploration.* Lawrence: University Press of Kansas, 2008.

———. "Exploration in the *Mare Incognita*: Natural History and Conservation in Early Twentieth Century America." PhD diss., University of Oklahoma, 2000.

Kruger, Lawrence. "Specialized Features of the Cetacean Brain." In *Whales, Dolphins, and Porpoises,* edited by Kenneth Stafford Norris, 232–54. Berkeley: University of California Press, 1966.

Kuronuma, Yoshihiro, and Clement A. Tisdell. "Institutional Management of an International Mixed Good: The IWC and Socially Optimal Whale Harvests." *Marine Policy* 17, no. 4 (July 1993): 235–50.

LaFollette, Marcel C. *Making Science Our Own: Public Images of Science, 1910–1955.* Chicago: University of Chicago Press, 1990.

Landauer, Lyndall B. "From Scoresby to Scammon: Nineteenth Century Whalers in the Foundations of Cetology." PhD diss., University of California, 1982.

———. *Scammon: Beyond the Lagoon: A Biography of Charles Melville Scammon.* San Francisco: Associates of the J. Porter Shaw Library in Cooperation with the Institute for Marine Information, 1986.

Lang, Thomas G. "Hydrodynamic Analysis of Cetacean Performance." In *Whales, Dolphins, and Porpoises,* edited by Kenneth Stafford Norris, 410–34. Berkeley: University of California Press, 1966.

Lang, Thomas G., and Dorothy A. Daybell. "Porpoise Performance Tests in a Sea-Water Tank." Defense Technical Information Center, Tech. Rept. A.D. 0298742, 1 January 1963.

Lang, T. G., and H. A. P. Smith. "Communication between Dolphins in Separate Tanks by Way of an Electronic Acoustic Link." *Science* 150, no. 3705 (31 December 1965): 1839–44.

Lange, Rogier, Angelique Martens, Kaat Schulte Fishedick, and Floor van der Vliet. "Slijper, Prof. Dr. E. J. Slijper (1907–1968)." Supplement, *Lutra* 45 (2002).

Langworthy, Orthello. "A Description of the Central Nervous System of the Porpoise (*Tursiops truncatus*)." *Journal of Comparative Neurology* 54, no. 2 (1932): 437–38.

———. "Lewis Hill Weed, 1886–1952." *Transactions of the American Neurological Association* 78 (1953): 301–2.

Lanouette, William, and Bela A. Silard. *Genius in the Shadows: A Biography of Leo Szilard, the Man Behind the Bomb.* New York: Charles Scribner's Sons, 1992.

Lansing, Michael J. "'Salvaging the Man Power of America': Conservation, Manhood, and Disabled Veterans During World War I." *Environmental History* 4, no. 1 (January 2009): 32–57.

Lanyon, Wesley E., and William N. Tavolga, eds. *Animal Sounds and Communication.* Washington, DC: American Institute of Biological Sciences, 1960.

Laporte, Léo. "George G. Simpson, Paleontology, and the Expansion of Biology." In *The Expansion of American Biology*, edited by Keith Rodney Benson, Jane Maienschein, and Ronald Rainger, 80–106. New Brunswick, NJ: Rutgers University Press, 1991.

LaPuzza, Tom. "ssc San Diego Historical Overview." Accessed 3 March 2011. http://www.spawar.navy.mil/sandiego/anniversary.

Larson, Edward J. *Summer for the Gods: The Scopes Trial and America's Continuing Debate over Science and Religion.* Cambridge, MA: Harvard University Press, 1998.

Latour, Bruno. "A Few Steps toward an Anthropology of the Iconoclastic Gesture." *Science in Context* 10, no. 1 (Spring 1997): 63–84.

———. *Pandora's Hope: Essays on the Reality of Science Studies.* Cambridge, MA: Harvard University Press, 1999.

Latour, Bruno, and Peter Weibel. *Iconoclash.* Karlsruhe: ZKM, 2002.

Laurie, Alec H. "The Age of Female Blue Whales and the Effect of Whaling on the Stock." *Discovery Reports* 15 (May 1937): 223–84.

Lavigne, David, Victor B. Scheffer, and Stephen R. Kellert. "The Evolution of North American Attitudes toward Marine Mammals." In *Conservation and Management of Marine Mammals*, edited by John R. Twiss, Randall R. Reeves, and Suzanne Montgomery, 10–47. Washington, DC: Smithsonian Institution Press, 1999.

Laws, Richard M. "Age Determination of Whales by Means of the Corpora Albicantia." *Proceedings of the Fifteenth International Congress of Zoology* (1959): 303–5.

———. "The Foetal Growth Rates of Whales with Special Reference to the Fin Whale, *Balaenoptera physalus* Linn." *Discovery Reports* 19 (1959): 281–308.

———. "Problems of Whale Conservation." In *Transactions of the Twenty-Fifth North American Wildlife and Natural Resources Conference*, 304–19. Washington, DC: Wildlife Management Institute, 1960.

———. "'Researches on the Period of Conception, Duration of Gestation and Growth of the Foetus in the Fin Whale': Comments on C. Naaktgeboren's, E. J. Slijper's and W. L. van Utrecht's Article." *Norsk Hvalfangst-Tidende* 49, no. 5 (1960): 216–20.

———. "Reproduction, Growth and Age of Southern Fin Whales." *Discovery Reports* 31 (1961): 327–486.

League of Nations, Economic Committee. *Report to the Council on the Work of the Thirty-Second Session, Held at Geneva from June 2nd to 6th, 1930.* Publication 11, Economic and Financial, 1930 II 24 C.353.M.146. Geneva: League of Nations, 1930.

Leatherwood, Stephen, and Randall R. Reeves, eds. *The Bottlenose Dolphin.* San Diego: Academic Press, 1990.

Legislative Council of the Falkland Islands. "Ordinance No. 3 of 1906: An Ordinance to Regulate the Whale Fishery of the Colony of the Falkland Islands." *Falkland Islands Gazette* 16, no. 11 (5 October 1906): 118–20.

———. "Ordinance No. 5 of 1908: An Ordinance to Regulate the Whale Fishery of the Colony of the Falkland Islands." *Falkland Islands Gazette* 18, no. 9 (1 September 1908): 115–17.

———. "Ordinance No. 6 of 1924: An Ordinance to Provide for the Establishment and Administration of a Research and Development Fund for the Dependencies." *Falkland Islands Gazette* 34, no. 1 (1 January 1925): 15–16.

Leonard, L. Larry. "Recent Negotiations toward the International Regulation of Whaling." *American Journal of International Law* 35, no. 1 (January 1941): 90–113.

Leopold, Aldo. *A Sand County Almanac, and Sketches Here and There.* New York: Oxford University Press, 1987 [1949].

Lillie, D. G. "Observations on the Anatomy and General Biology of Some Members of the Larger *Cetacea.*" *Proceedings of the Zoological Society of London* 80, no. 3 (1910): 769–92.

Lilly, John C. *The Center of the Cyclone: An Autobiography of Inner Space.* New York: Julian Press, 1972.

———. "Critical Brain Size and Language." *Perspectives in Biology and Medicine* 6, no. 2 (Winter 1963): 246–55.

———. "Distress Call of the Bottlenose Dolphin: Stimuli and Evoked Behavioral Responses." *Science* 139, no. 3550 (11 January 1963): 116–18.

———. "Electrode and Cannulae Implantation in the Brain by a Simple Percutaneous Method." *Science* 127, no. 3307 (16 May 1958): 1181–82.

———. "An Experimental Investigation of the Dolphin's Communication Abilities by Means of a Dolphin Machine Code." In *Proceedings of the Final LINC Evaluation Program Meeting, Washington University, St. Louis (Computer Research Lab), 18–19 March 1965,* section 4, 1–22.

———. "Importance of Being in Earnest about Dialogues of Dolphins." *Life Magazine* 51, no. 4 (28 July 1961): 68.

———. "Interspecies Communication." *McGraw Hill Yearbook of Science and Technology* 1962: 279–81.

———. "Learning Motivated by Subcortical Stimulation: The Start and Stop Patterns of Behavior." In *Reticular Formation of the Brain,* edited by Herbert H. Jasper, Lorne D. Proctor, Robert S. Knighton, William C. Noshay, and Russell T. Costello, 705–21. Boston: Little Brown, 1958.

———. *Man and Dolphin.* Garden City, NY: Doubleday, 1961.

———. *The Mind of the Dolphin: A Nonhuman Intelligence.* Garden City, NY: Doubleday, 1967.

———. *Programming and Metaprogramming in the Human Biocomputer: Theory and Experiments.* 2nd ed. New York: Three Rivers Press, 1987.

———. "Some Considerations Regarding Basic Mechanisms of Positive and Negative Types of Motivations." *American Journal of Psychiatry* 115, no. 6 (December 1958): 498–504.

———. "Vocal Behavior of the Bottlenose Dolphin." *Proceedings of the American Philosophical Society* 106, no. 6 (12 December 1962): 520–29.

Lilly, John C., and Thomas F. Anderson. "The Nitrogen Meter: An Instrument for Continuously Recording the Concentration of Nitrogen in Gas Mixtures." Report 299, 28 February 1944, National Research Council, Division of Medical Science, Acting for the Committee on Medical Research of the Office of Scientific Research and Development, Committee on Aviation Medicine.

Lilly, John C, John R. Hughes, Ellsworth C. Alvord, and Thelma W. Galkin. "Brief, Non-injurious Electric Waveform for Stimulation of the Brain." *Science* 121, no. 3144 (1 April 1955): 468–69.

Lilly, John C., and Alice M. Miller. "Sounds Emitted by the Bottlenose Dolphin." *Science* 133, no. 3465 (26 May 1961): 1689–93.

———. "Vocal Exchanges between Dolphins." *Science* 134, no. 3493 (8 December 1961): 1873–76.

Lilly, John C., Alice M. Miller, and Henry M. Truby. "Reprogramming of the Sonic Output of the Dolphin: Sonic Burst Count Matching." *Journal of the Acoustical Society of America* 43, no. 6 (June 1968): 1412–24.

Lilly, John C., and Jay Shurley. "Experiments in Maximum Achievable Physical Isolation with Water Suspension of Intact Healthy Persons." In *Psychophysiological Aspects of Space Flight*, edited by B. E. Flaherty, 238–74. New York: Columbia University Press, 1961.

Lindsay, Debra. *Science in the Subarctic: Trappers, Traders, and the Smithsonian Institution.* Washington, DC: Smithsonian Institution Press, 1993.

Linnaeus, Carl. *Miscellaneous Tracts Relating to Natural History, Husbandry, and Physick.* Translated by Benjamin Stillingfleet. New York: Arno Press, 1977.

Lipps, Jere H. "Success Story: The History and Development of the Museum of Paleontology at the University of California, Berkeley." *Proceedings of the California Academy of Sciences* 55, supplement 1 (October 2004): 209–43.

Lipscomb, Frances. "Whales Alive! In Education and Culture." Paper presented at a Global Conference on the Non-Consumptive Utilization of Cetacean Resources, Boston, 2 June 1983.

Little, Luther. "Alfred Brazier Howell, 1886–1961." *Journal of Mammalogy* 49, no. 4 (1968): 732–42.

Livingstone, David N. *Putting Science in Its Place: Geographies of Scientific Knowledge.* Chicago: University of Chicago Press, 2003.

Lockley, R. M. *Whales, Dolphins, and Porpoises.* New York: W. W. Norton, 1979.

Lönnberg, Einar. "Om Hvalarnes Harstamning." *Svenska Veteskaps-Akademiens Årsbok* (1910): 219–59. Translated by Mark D. Uhen as "On the Origin of Whales." Accessed 11 March 2011. www.paleoglot.org/files/Lönnberg 10.pdf.

Loye, D. P., and D. A. Proudfoot. "Underwater Noise Due to Marine Life." *Journal of the Acoustical Society of America* 18 (1946): 446–49.

Lucas, Frederic A. "The Passing of the Whale." *Forest and Stream* 71, no. 8 (22 August 1908): 291–93.

———. "The Passing of the Whale." Supplement, *Bulletin of the New York Zoological Society* 30 (July 1908): 445–48.

Ludwig, Donald, Ray Hilborn, and Carl Walters. "Uncertainty, Resource Exploitation, and Conservation: Lessons from History." *Science* 260, no. 5104 (2 April 1993): 17, 36.

Lutts, Ralph H. "The Trouble with Bambi: Walt Disney's Bambi and the American Vision of Nature." *Forest and Conservation History* 36 (October 1992): 160–71.

———, ed. *The Wild Animal Story*. Philadelphia: Temple University Press, 1998.

MacDougall, Walter Allan. *The Heavens and the Earth: A Political History of the Space Age*. New York: Basic Books, 1985.

Mack, Arien, ed. *Humans and Other Animals*. Columbus: Ohio State University Press, 1999.

Mackenzie, W. C. "The Whaling Industry: Economic Aspects." *Canadian Geographical Journal* 38 (March 1949): 140–43.

Mackintosh, N. A. "Biological Problems in the Regulation of Whaling." *New Scientist* (4 June 1959): 1229–31.

———. "Biological Problems in the Regulation of Whaling." *Norsk Hvalfangst-Tidende* 48, no. 8 (1959): 395–404.

———. "Distribution of the Macroplankton in the Atlantic Sector of the Antarctic." *Discovery Reports* 9 (1934): 65–160.

———. "The Marking of Whales." *Nature* 169, no. 4298 (15 March 1952): 435–37.

———. "The Marking of Whales." *Norsk Hvalfangst-Tidende* 41, no. 5 (1952): 236–40.

———. "The Natural History of Whales." *Polar Record* 3, no. 24 (1942): 553–63.

———. "Observations on Whales from Ships." *Marine Observer* 22 (1952): 87–90.

———. "The Southern Stocks of Whalebone Whales." *Discovery Reports* 22 (1942): 197–300.

———. *The Stocks of Whales*. London: Fishing News Books, 1965.

———. "The Work of the Discovery Committee." *Proceedings of the Royal Society of London. Series A, Mathematical and Physical Sciences* 202, no. 1068 (22 June 1950): 1–16.

Mackintosh, N. A., and Sidney G. Brown. "Preliminary Estimates of the Southern Populations of the Larger Baleen Whales." *Norsk Hvalfangst-Tidende* 45, no. 9 (1956): 461–80.

Mackintosh, N. A., and J. F. G. Wheeler. "Southern Blue and Fin Whales." *Discovery Reports* 1 (1929): 257–540.

Maehle, Andreas-Holger. "Cruelty and Kindness to the 'Brute Creation': Stability and Chance in the Ethics of the Man-Animal Relationship, 1600–1850." In *Animals and Human Society: Changing Perspectives*, edited by Aubrey Manning and James Serpell, 81–105. London: Routledge, 1994.

Mageli, Eldrid Ingebjørg. *Towards Friendship: The Relationship between Norway and Japan, 1905–2005*. Oslo: Oslo Academic Press, 2006.

Magnolia, L. R. *Whales, Whaling and Whale Research: A Selected Bibliography*. Special literature survey, no. 52. Redondo Beach, CA: TRW Systems Group, 1972.

Mahoney, Michael S. "Cybernetics and Information Technology." In *Companion to the History of Modern Science*, edited by R. C. Olby, G. N. Cantor, J. R. R. Christie, and M. J. S. Hodge, 537–53. London: Routledge, Chapman & Hall, 1989.

Maienschein, Jane. "Shifting Assumptions in American Biology: Embryology, 1890–1910." *Journal of the History of Biology* 14, no. 1 (Spring 1981): 89–113.

Maienschein, Jane, Ronald Rainger, and Keith R. Benson. "Special Section on American Morphology at the Turn of the Century. Introduction: Were American Morphologists in Revolt?" *Journal of the History of Biology* 14, no. 1 (Spring 1981): 83–87.

Malamud, Randy. "How People and Animals Coexist." *Chronicle Review*, 24 January 2003, sec. 2, p. B7.

Mann, Janet, Richard C. Connor, Peter L. Tyack, and Hal Whitehead, eds. *Cetacean Societies: Field Studies of Dolphins and Whales*. Chicago: University of Chicago Press, 2000.

Marks, John D. *The Search for the "Manchurian Candidate": The CIA and Mind Control*. New York: Times Books, 1978.

Marsden, Rosalind. "The Discovery Committee—Motivation, Means, and Achievements." *Scottish Naturalist* 111 (1999): 69–92.

———. "*Discovery* Investigations: An Early Attempt at Ecologically Sustainable Development." *Archives of Natural History* 32, no. 2 (2005): 161–76.

———. "Expedition to Investigation: The Work of the Discovery Committee." In *Understanding the Oceans: A Century of Ocean Exploration*, edited by Margaret Deacon, A. L. Rice, and C. P. Summerhayes, 69–86. London: University College of London Press, 2001.

———. "Investigations of the Humboldt Current Following a Long Series of Misadventures: The Voyage of the William Scoresby, 1931." In *Oceanographic History: The Pacific and Beyond*, edited by Keith Rodney Benson and Philip F. Rehbock, 186–96. Seattle: University of Washington Press, 2002.

Marshall, Francis H. A. *An Introduction to Sexual Physiology for Biological, Medical, and Agricultural Students*. London: Longmans, Green, 1925.

———. *Physiology of Reproduction*. London: Longmans, Green, 1910.

Marshall, N. B. "Francis Charles Fraser, 16 June 1903–21 October 1978." *Biographical Memoirs of Fellows of the Royal Society* 25 (1979): 286–317.

Massachusetts Fish and Game Association and Alfred Otto Gross. *Progress Report of the New England Ruffed Grouse Investigation Committee*. Boston: Massachusetts Fish and Game Association, 1926.

Matthew, William Diller. "The Evolution of the Horse: A Record and Its Interpretation." *Quarterly Review of Biology* 1, no. 2 (April 1926): 139–85.

Matthews, L. Harrison. "Chairman's Introduction to First Session of International Symposium on Cetacean Research." In *Whales, Dolphins, and Porpoises*, edited by Kenneth Stafford Norris, 3–6. Berkeley: University of California Press, 1966.

———. "The Humpback Whale, *Megaptera nodosa*." *Discovery Reports* 17 (1937): 7–92.

———. "Notes on the Southern Right Whale, *Eubalaena australis.*" *Discovery Reports* 17 (1938): 169–82.

———. *Penguins, Whalers, and Sealers: A Voyage of Discovery.* New York: Universe Books, 1978.

———. "Science and Whaling." In *The Whale*, edited by Leonard Harrison Matthews, 160–86. New York: Simon and Schuster, 1968.

———. "The Sei Whale, *Balaenoptera borealis.*" *Discovery Reports* 17 (1938): 183–290.

———. *South Georgia, the British Empire's Subantarctic Outpost: A Synopsis of the History of the Island.* Bristol: John Write & Sons, 1931.

———. "The Sperm Whale, *Physeter catodon.*" *Discovery Reports* 17 (1938): 93–168.

———, ed. *The Whale.* New York: Simon and Schuster, 1968

Matthews, Leonard Harrison and Maxwell Knight. *Senses of Animals.* New York: Philosophical Library, 1963.

Mawer, Granville Allen. *Ahab's Trade: The Saga of South Seas Whaling.* New York: St. Martin's Press, 1999.

Mayr, Ernst, and William B. Provine. *Evolutionary Synthesis: Perspectives on the Unification of Biology.* Cambridge, MA: Harvard University Press, 1980.

McBride, Arthur F., and D. O. Hebb. "Behavior of the Captive Bottle-Nose Dolphin, *Tursiops truncatus.*" *Journal of Comparative and Physiological Psychology* 41 (1948): 111–23.

McCook, Stuart. "'It May Be Truth, But It Is Not Evidence': Paul du Chaillu and the Legitimation of Evidence in the Field Sciences." *Osiris* 11 (1996): 177–97.

McCormick, James G., Ernest Glen Wever, Jerry Palin, and Sam H. Ridgway. "Sound Conduction in the Dolphin Ear." *Journal of the Acoustical Society of America* 48, no. 6B (December 1970): 1418–28.

McCormick, John. *The Global Environmental Movement: Reclaiming Paradise.* London: Belhaven Press, 1989.

McCoy, Alfred W. "Science in Dachau's Shadow: Hebb, Beecher, and the Development of CIA Psychological Torture and Modern Medical Ethics." *Journal of the History of the Behavioral Sciences* 43, no. 4 (Fall 2007): 401–17.

McEvoy, Arthur F. *The Fisherman's Problem: Ecology and Law in the California Fisheries, 1850–1980.* Cambridge: Cambridge University Press, 1986.

McFarland, W. L., and P. J. Morgane. "Dolphin Brain Atlas Project." *Psychiatric Spectator* 3, no. 11 (1967): 17–18.

McHugh, J. L. "The Role and History of the International Whaling Commission." In *The Whale Problem: A Status Report*, edited by William E. Schevill, G. Carleton Ray, and Kenneth S. Norris, 305–35. Cambridge, MA: Harvard University Press, 1974.

———. "The United States and World Whale Resources." *BioScience* 19, no. 12 (December 1969): 1075–78.

McIntyre, Joan, comp. *Mind in the Waters: A Book to Celebrate the Consciousness of Whales and Dolphins.* New York: Charles Scribner's Sons and Sierra Book Club, 1974.

McIntyre, Joan. "Mind Play." In *Mind in the Waters: A Book to Celebrate the Consciousness of Whales and Dolphins*, compiled by Joan McIntyre, 94–95. New York: Charles Scribner's Sons and Sierra Book Club, 1974.

McLeod, Roy. "'Strictly for the Birds': Science, the Military, and the Smithsonian's Pacific Ocean Biological Survey Program, 1963–1970." In *Science, History, and Social Activism: A Tribute to Everett Mendelsohn*, edited by Garland E. Allen and Roy McLeod, 307–38. Dordrecht, Holland: Kluwer Academic Publishers, 2001.

McVay, Scott. "Can Leviathan Long Endure So Wide a Chase?" *Natural History* 80, no. 1 (January 1971): 36–40, 68–72.

———. "Does the Whale's Magnitude Diminish?—Will He Perish?" *Bulletin of the Atomic Scientists* (February 1971): 38–41.

———. "How Hears the Dolphin? A Research Project at the Auditory Laboratories." *Princeton Alumni Weekly* 68, no. 4 (1967): 6–9, 17–18.

———. "Last of the Great Whales." *Scientific American* 215, no. 2 (1966): 13–21.

———. "Reflections on the Management of Whaling." In *The Whale Problem: A Status Report*, edited by William E. Schevill, G. Carleton Ray, and Kenneth S. Norris, 369–82. Cambridge, MA: Harvard University Press, 1974.

———. "To the Apologist, Defender, and Stooge of the Whaling Industry: EJS." In *Mind in the Waters: A Book to Celebrate the Consciousness of Whales and Dolphins*, compiled by Joan McIntyre, 93. New York: Charles Scribner's Sons and Sierra Book Club, 1974.

Meadows, Donella H., and J. M. Robinson. *The Electronic Oracle: Computer Models and Social Decisions*. Chichester, NY: John Wiley & Sons, 1985.

Melville, Charles Scammon. *Marine mammals of the north-western coast of North America; described and illustrated, together with an account of the American whale-fishery*. New York: Dover Publications, 1968.

Melville, Herman. *Moby-Dick, or, The Whale*. Edited by Harrison Hayford, Hershel Parker, and G. Thomas Tanselle. Evanston, IL: Northwestern University Press and the Newberry Library, 1988. First published 1851.

Merle, Robert. *The Day of the Dolphin*. Translated by Helen Weaver. New York: Simon and Schuster, 1969.

Merriam, John C., et al. "Continuation of Paleontological Researches." *Carnegie Institution of Washington Yearbook* 20 (1921): 447–51.

———. "Continuation of Paleontological Researches." *Carnegie Institution of Washington Yearbook* 21 (1922): 398–400.

———. "Continuation of Paleontological Researches." *Carnegie Institution of Washington Yearbook* 22 (1 November 1922–31 October 1923): 351–53.

———. "Continuation of Paleontological Researches." *Carnegie Institution of Washington Yearbook* 23 (1 November 1923–30 June 1924): 293–96.

———. "Continuation of Paleontological Researches." *Carnegie Institution of Washington Yearbook* 24 (1925): 357–59.

———. "Continuation of Paleontological Researches." *Carnegie Institution of Washington Yearbook* 25 (1926): 403–7.

———. "Continuation of Paleontological Researches." *Carnegie Institution of Washington Yearbook* 26 (1927): 363–66.

———. "Continuation of Paleontological Researches." *Carnegie Institution of Washington Yearbook* 27 (1928): 384–89.

———. "Continuation of Paleontological Researches." *Carnegie Institution of Washington Yearbook* 28 (12 December 1929): 388–91.

———. "Continuation of Paleontological Researches." *Carnegie Institution of Washington Yearbook* 29 (11 December 1930): 396–99.

———. "Continuation of Paleontological Researches." *Carnegie Institution of Washington Yearbook* 31 (9 December 1932): 326–30.

———. "Continuation of Paleontological Researches." *Carnegie Institution of Washington Yearbook* 32 (15 December 1933): 323–30.

———. "Continuation of Paleontological Researches." *Carnegie Institution of Washington Yearbook* 33 (14 December 1934): 302–13.

———. "Continuation of Paleontological Researches." *Carnegie Institution of Washington Yearbook* 34 (13 December 1935): 313–29.

———. "Continuation of Paleontological Researches." *Carnegie Institution of Washington Yearbook* 35 (11 December 1936): 316–26.

———. "Continuation of Paleontological Researches." *Carnegie Institution of Washington Yearbook* 36 (10 December 1937): 332–45.

———. "Continuation of Paleontological Researches." *Carnegie Institution of Washington Yearbook* 37 (9 December 1938): 340–64.

———. "Continuation of Paleontological Researches." *Carnegie Institution of Washington Yearbook* 40 (12 December 1941): 316–33.

Mets, David R. "The *Force* in U.S. Air Force." *Aerospace Power Journal* (Fall 2000): 61–62.

M'Gonigle, R. Michael. "The 'Economizing' of Ecology: Why Big, Rare Whales Still Die." *Ecology Law Quarterly* 9, no. 1 (1980): 119–237.

Miller, Andrew R., and Nives Dolšak. "Issue Linkages in International Environmental Policy: The International Whaling Commission and Japanese Development Aid." *Global Environmental Politics* 7, no. 1 (February 2007): 69–96.

Miller, Char. *Gifford Pinchot and the Making of Modern Environmentalism.* Washington, DC: Island Press, 2001.

Miller, Gerrit S. "The Telescoping of the Cetacean Skull (with Eight Plates)." *Smithsonian Miscellaneous Collections* 76, no. 5 [Publication 2822] (1925): 1–55.

Miller, Walter James, ed. *The Annotated Jules Verne: Twenty Thousand Leagues Under the Sea.* New York: Thomas Y. Crowell, 1976.

Mills, Eric L. *Biological Oceanography: An Early History, 1870–1960.* Ithaca, NY: Cornell University Press, 1989.

Milton, Kay. *Loving Nature: Towards an Ecology of Emotion.* London: Routledge, 2002.

Mitman, Gregg. "Evolution as Gospel: William Patton, the Language of Democracy, and the Great War." *Isis* 81, no. 3 (1990): 446–63.

———. "Pachyderm Personalities: The Media of Science, Politics and Conservation."

In *Thinking with Animals: New Perspectives on Anthropomorphism*, edited by Gregg Mitman and Lorraine Daston, 175–95. New York: Columbia University Press, 2005.

———. *Reel Nature: America's Romance with Wildlife on Film*. Cambridge, MA: Harvard University Press, 1999.

Møhl, Bertel, Peter T. Madsen, Magnus Wahlberg, Whitlow W. L. Au, Paul E. Nachtigall, and Sam H. Ridgway. "Sound Transmission in the Spermaceti Complex of a Recently Expired Sperm Whale Calf." *Acoustics Research Letters Online* 4, no. 1 (January 2003): 19–24.

Montagu, Ashley, and John C. Lilly. *The Dolphin in History*. Los Angeles: William Andrews Clark Memorial Library, 1963.

Mooney, Ted. *Easy Travel to Other Planets*. New York: Farrar, Straus, & Giroux, 1981.

Moran, Jeffrey P. *The Scopes Trial: A Brief History with Documents*. New York: Bedford St. Martin's, 2002.

Mørch, Jens Andreas. "Improvements in Whaling Methods." *Scientific American* 99 (1 August 1908): 75.

———. "On the Natural History of Whalebone Whales." *Proceedings of the General Meetings for Scientific Business of the Zoological Society of London* 2 (1911): 661–70.

Mørch, John (Jens Andreas Mørch). "Manufacture of Whale Products." Supplement, *Scientific American*, no. 1772 (2 January 1909): 15–16.

Morgane, Peter. "The Whale Brain: The Anatomical Basis of Intelligence." In *Mind in the Waters: A Book to Celebrate the Consciousness of Whales and Dolphins*, compiled by Joan McIntyre, 84–93. New York: Charles Scribner's Sons and Sierra Book Club, 1974.

Morley, F. V., and J. S. Hodgson. *Whaling North and South*. London: Methuen, 1927.

Morley, John David. *Journey to the End of the Whale*. London: Weidenfeld & Nicolson, 2005.

Moser, Susanne C. "In the Long Shadows of Inaction: The Quiet Building of a Climate Protection Movement in the United States." *Global Environmental Politics* 7, no. 2 (May 2007): 124–44.

Mowat, Farley. *A Whale for the Killing*. With a foreword by Rex Weyler. Mechanicsburg, PA: Stackpole Books, 2005.

Mukerji, Chandra. *A Fragile Power: Scientists and the State*. Princeton, NJ: Princeton University Press, 1989.

Mumford, Louis. *Herman Melville: A Study of His Life and Vision*. Revised ed. New York: Harcourt, Brace and World, 1962.

Munz, Tania. "The Bee Battles: Karl von Frisch, Adrian Wenner and the Honey Bee Dance Language Controversy." *Journal of the History of Biology* 38, no. 3 (November 2005): 535–70.

Murphy, Raymond. *Rationality and Nature: A Sociological Inquiry into a Changing Relationship*. Boulder, CO: Westview Press, 1994.

Murphy, Robert Cushman. "A Desolate Island of the Antarctic." *American Museum Journal* 13, no. 6 (October 1913): 243–60.

———. *Logbook for Grace: Whaling Brig* Daisy, *1912–1913*. New York: Macmillan, 1947.

————. "South Georgia, an Outpost of the Antarctic." *National Geographic Magazine* 41, no. 4 (April 1922): 409–44.

Murray, John, and Johan Hjort. *The Depths of the Ocean, a General Account of the Modern Science of Oceanography Based Largely on the Scientific Researches of the Norwegian Steamer* Michael Sars *in the North Atlantic*. London: Macmillan, 1912.

Myers, Norman. "The Whaling Controversy: A Fresh Look at the Japanese Position Might Lead to a Workable Solution to the Present Impasse between Whalers and Conservationists." *American Scientist* 63, no. 4 (July–August 1975): 448–55.

Naaktgeboren, C., E. J. Slijper, and W. L. van Utrecht. "Researches on the Period of Conception, Duration of Gestation and Growth of the Foetus in the Fin Whale, Based on Data from International Whaling Statistics." *Norsk Hvalfangst-Tidende* 49, no. 3 (March 1960): 113–19.

Nagel, E. L., P. J. Morgane, and W. L. McFarland. "Anesthesia for the Bottlenose Dolphin, *Tursiops truncatus*." *Science* 146, no. 3651 (18 December 1964): 1591–93.

Nandan, S. N. "The Exclusive Economic Zone: A Historical Perspective." In *The Law and the Sea: Essays in Memory of Jean Carroz*. Rome: Food and Agriculture Organization of the United Nations, 1987.

Nash, Roderick Frazier. *The Rights of Nature: A History of Environmental Ethics*. Madison: University of Wisconsin Press, 1989.

National Academy of Sciences. "Abstracts of Papers Presented at the Autumn Meeting, 8–10 November 1956, Washington, D.C." *Science* 124, no. 3228 (9 November 1956): 935–41.

————. "Abstracts of Papers To Be Presented at the Annual Meeting, 22–24 April 1957, Washington, D.C." *Science* 125, no. 3251 (19 April 1957): 746–52.

"The Navy's Program in Hydrobiology." *Naval Research Review* (August 1962): 18.

Nayman, Jacqueline. *Whales, Dolphins, and Man*. London: Hamlyn, 1973.

"Neil Alison Mackintosh." *Geographical Journal* 140, no. 3 (October 1974): 524–25.

Nelkin, Dorothy, ed. *Controversy: Politics of Technical Decision*. 2nd ed. Beverly Hills: Sage Publications, 1984.

New York Zoological Society. *Zoologica: Scientific Contributions of the New York Zoological Society, 1907–1915*, vol. 1. New York: The Society, The Zoological Park.

New Yorker. Unsigned review of *Man and Dolphin*, by John C. Lilly. 16 September 1961, 178.

Nicht, Sanelma. "Plaice and Place." *Cabinet* 32 (Winter 2008–2009): 45–46.

Nicolson, Malcolm. "No Longer a Stranger? A Decade in the History of Ecology." *History of Science* 26 (1988): 183–200.

Nihon Kujirarui Kenkyujo. *Research on Whales*. Tokyo: The Institute of Cetacean Research, 1995.

Nishiwaki, M. "Aerial Photographs Show Sperm Whales' Interesting Habits." *Norsk Hvalfangst-Tidende* 51, no. 10 (1962): 395–98.

Norris, Kenneth S. *The Porpoise Watcher*. New York: W. W. Norton, 1974.

————. "Some Problems of Echolocation in Cetaceans." In *Marine Bio-Acoustics: Proceedings of a Symposium Held at the Lerner Marine Laboratory, Bimini, Bahamas, 11–13 April 1963*, ed. William N. Tavolga, 317–36. Oxford: Pergamon Press, 1964.

———. "Trained Porpoise Released in the Open Sea." *Science* 147, no. 3661 (26 February 1965): 1048–50.

———, ed. *Whales, Dolphins, and Porpoises.* Berkeley: University of California Press, 1966.

"Notes." *Nature* 101, no. 2546 (15 August 1918): 468–72.

November, Joe. "LINC: Biology's Revolutionary Little Computer." *Endeavour* 28, no. 3 (September 2004): 125–31.

Nummela, Sirpa, Tom Reuter, Simo Hemilä, Peter Holmberg, and Pertti Paukku. "The Anatomy of the Killer Whale Middle Ear." *Hearing Research* 133, no. 1–2 (July 1999): 61–70.

Nyhart, Lynn K. *Biology Takes Form: Animal Morphology and the German Universities, 1800–1900.* Chicago: University of Chicago Press, 1995.

Nyhart, Lynn K. "The Disciplinary Breakdown of German Morphology, 1870–1900." *Isis* 78, no. 3 (1987): 365–89.

Oates, David. *Earth Rising: Ecological Belief in an Age of Science.* Corvallis: Oregon State University Press, 1989.

O'Barry, Richard, and Keith Coulbourn. *Behind the Dolphin Smile.* Chapel Hill, NC: Algonquin Books, 1988.

Ohsumi, Seiji. "The Necessity of Employing Lethal Method in the Study of Whale Research." In *Research on Whales,* edited by the Institute for Cetacean Research. Tokyo: Institute for Cetacean Research, 1995.

Ommanney, F. D. *Lost Leviathan: Whales and Whaling.* New York: Dodd, Mead, 1971.

Oreskes, Naomi. "Science and Public Policy: What's Proof Got to Do with It?" *Environmental Science & Policy* 7 (2004): 369–83.

Oreskes, Naomi, and Erik Conway. *Merchants of Doubt.* New York: Bloomsbury, 2010.

Orr, Robert T. "The Tuna-Dolphin Problem." *Pacific Discovery* 29, no. 1 (January–February 1976): 11.

Osborn, Henry Fairfield, and Harold Elmer Anthony. "Can We Save the Mammals?" *Natural History* 22, no. 5 (September–October 1922): 388–405.

———. "Close of the Age of Mammals." *Journal of Mammalogy* 3, no. 4 (November 1922): 219–37.

Oswald, John D. *The Cry of Nature, or, an Appeal to Mercy and to Justice on Behalf of the Persecuted Animals.* Mellen Animal Rights Library, vol. 8. Edited by Jason Hribal. Lewiston, NY: Edwin Mellen Press, 2000. First published 1791.

Ottestad, Per. "A Mathematical Method for the Study of Growth." *Hvalrådets Skrifter* 7 (1933): 30–91.

———. "On the Size of the Stock of Antarctic Fin Whales Relative to the Size of the Catch." *Norsk Hvalfangst-Tidende* 45, no. 6 (1956): 298–308.

Palmer, Stephen Canon. "A. G. Bennett—Naturalist, Whaling Officer, Customs Officer and Postmaster." *Falkland Islands Journal* 6, no. 4 (1995): 91–105.

———. "Far from Moderate." PhD diss., University of Portsmouth, 2004.

Papadakis, Nikos. *International Law of the Sea: A Bibliography.* Alphen aan den Rijn, Netherlands: Sijthoff & Noordhoff, 1980.

Parfit, Michael. "Are Dolphins Trying to Say Something, or Is It All Much Ado About Nothing?" *Smithsonian Magazine* (October 1980): 73–80.

Parr, A. E. "Concerning Whales and Museums." *Curator* 6, no. 1 (1963): 65–76.

Patterson, B., and G. R. Hamilton. "Repetitive 20 Cycle per Second Biological Hydro-acoustic Signals at Bermuda." In *Marine Bio-Acoustics: Proceedings of a Symposium held at the Lerner Marine Laboratory, Bimini, Bahamas, 11–13 April 1963*, edited by William N. Tavolga, 125–45. Oxford: Pergamon Press, 1964.

Pauly, Philip J. "The Appearance of Academic Biology in Late Nineteenth-Century America." *Journal of the History of Biology* 17, no. 3 (Fall 1984): 369–97.

———. *Biologists and the Promise of American Life: From Meriwether Lewis to Alfred Kinsey*. Princeton, NJ: Princeton University Press, 2000.

———. "Summer Resort and Scientific Discipline: Woods Hole and the Structure of American Biology, 1882–1925." In *The American Development of Biology*, edited by Roland Rainger and Jane Maienschein, 121–50. Philadelphia: University of Pennsylvania Press, 1988.

Payne, Roger. *Among Whales*. New York: Charles Scribner's Sons, 1995.

———. "Humpbacks and Their Mysterious Songs." *National Geographic* 155, no. 1 (January 1979): 18–25.

Payne, Roger S., and Scott McVay. "Songs of Humpback Whales." *Science* 173, no. 3997 (13 August 1971): 585–97.

Payne, Roger, and Douglas Webb. "Orientation by Means of Long Range Acoustic Signaling in Baleen Whales." *Annals of the New York Academy of Sciences* 188 (December 1971): 110–41.

Pearce, Fred. *Green Warriors: The People and the Politics behind the Environmental Revolution*. London: Bodley Head, 1991.

Perrin, William F., Bernd G. Würsig, and J. G. M. Thewissen. *Encyclopedia of Marine Mammals*. San Diego: Academic Press, 2002.

Petersen, Shannon. "Congress and Charismatic Megafauna." *Environmental Law* 29, no. 2 (Summer 1999): 463–92.

Peterson, M. J. "Whalers, Cetologists, Environmentalists, and the International Management of Whaling." *International Organization* 46, no. 1 (Winter 1992): 147–86.

Philo, Chris. "Animals, Geography, and the City: Notes on Inclusions and Exclusions." In *Animal Geographies: Place, Politics, and Identity in the Nature-Culture Borderlands*, edited by Jennifer Wolch and Jody Emel, 51–70. London: Verso, 1998.

Pike, Kenneth L. *Tone Languages: A Technique for Determining the Number and Type of Pitch Contrasts in a Language, with Studies in Tonemic Substitution and Fusion*. Ann Arbor: University of Michigan Press, 1948.

Pilleri, Georg, ed. *Investigations on Cetacea*. Vol. 12. Berne: Brain Anatomy Institute, 1981.

Pinchot, Gifford. *To the South Seas: The Cruise of the Schooner* Mary Pinchot *to the Galapagos, the Marquesas, and the Tuamotu Islands, and Tahiti*. Philadelphia: John C. Winston, 1930.

Pinchot, Gifford B. "Whale Culture—A Proposal." *Perspectives in Biology and Medicine* 10, no. 1 (Autumn 1966): 33–43.

Pinsel, Marc I. *150 Years of Service on the Seas: A Pictorial History of the U.S. Naval Oceanographic Office from 1830 to 1980*. Vol. 1 (1830–1946). Washington, DC: Department of the Navy, Oceanographic Office, 1981.

Pirages, Denis. *New Context for International Relations: Global Ecopolitics*. North Scituate, MA: Duxbury Press, 1978.

Porter, Gareth, and Janet Welsh Brown. *Global Environmental Politics*. Boulder, CO: Westview Press, 1991.

Porter, Theodore M. *Trust in Numbers: The Pursuit of Objectivity in Science and Public Life*. Princeton, NJ: Princeton University Press, 1995.

Poundstone, William. *Carl Sagan: A Life in the Cosmos*. New York: Henry Holt, 1999.

Pratt, Fletcher. *Secret and Urgent: The Story of Codes and Ciphers*. Garden City, NY: Blue Ribbon Books, 1942.

Press, Jacques Cattell, comp. *American Men and Women of Science. Cumulative Index, Editions 1–14*. New York: R. R. Bowker, 1983.

Pringle, Robert M. "The Origins of the Nile Perch in Lake Victoria." *BioScience* 55, no. 9 (2005): 780–87.

"Programme of the Final Public Examination for the Degree of Doctor of Philosophy of Arthur Remington Kellogg." 15 October 1928, 10:00 AM, Administration Building, Carnegie Institution of Washington, DC. Berkeley: University of California Press, 1928.

Publishers' Weekly. Unsigned review of *The Mind of the Dolphin: A Non-Human Intelligence*, by John C. Lilly (hardback). Vol. 191, no. 26 (26 June 1967): 66.

———. Unsigned review of *The Mind of the Dolphin: A Non-Human Intelligence*, by John C. Lilly (paperback). Vol. 195, no. 6 (10 February 1969): 78.

Purves, Peter E. "Anatomy and Physiology of the Outer and Middle Ear in Cetaceans." In *Whales, Dolphins, and Porpoises*, edited by Kenneth Stafford Norris, 320–80. Berkeley: University of California Press, 1966.

———. "The Wax Plug in the External Auditory Meatus of the Mysticeti." *Discovery Reports* 27 (1955): 293–302.

Racovitza, Emile G. *Cétacés*. Anvers: J.-E. Buschmann, 1902.

———. *Expedition Antarctique Belge: Résultats du Voyage du S.Y. Belgica en 1897–1898–1899, Rapports Scientifiques, Zoologie, Cétacés*. Anvers: J.-E. Buschmann, 1903.

Rader, Karen. *Making Mice: Standardizing Animals for American Biomedical Research, 1900–1955*. Princeton, NJ: Princeton University Press, 2004.

Radin, Max, and A. M. Kidd, eds. *Legal Essays in Tribute to Orrin Kip McMurray*. Berkeley: University of California Press, 1935.

Ræstad, Arnold. *La Chasse à la Baleine en mer libre: Une question de législation international devant la Socété des Nations*. Paris: Les Éditions Internationales, 1928.

Rainger, Ronald. *An Agenda for Antiquity: Henry Fairfield Osborn and Vertebrate Paleontology at the American Museum of Natural History, 1890–1935*. Tuscaloosa: University of Alabama Press, 1991.

———. "The Continuation of the Morphological Tradition: American Paleontology, 1880–1910." *Journal of the History of Biology* 14, no. 1 (Spring 1981): 129–58.

———. "Vertebrate Paleontology as Biology: Henry Fairfield Osborn and the American

Museum of Natural History." In *The American Development of Biology*, edited by Ronald Rainger and Jane Maienschein, 219–56. Philadelphia: University of Pennsylvania Press, 1988.

Rainger, Ronald, and Jane Maienschein, eds. *The American Development of Biology*. Philadelphia: University of Pennsylvania Press, 1988.

Rayner, George W. "Preliminary Results of the Marking of Whales by the Discovery Committee." *Nature* 144, no. 3659 (16 December 1939): 999–1002.

———. "Whale Marking: Progress and Results to December 1939." *Discovery Reports* 19 (1940): 245–84.

———. "Whale Marking II: Distribution of Blue, Fin and Humpback Whales Marked from 1932 to 1938." *Discovery Reports* 25 (1948): 31–38.

Reddy, Michelle L., J. S. Reif, A. Bachand, and S. H. Ridgway. "Opportunities for Using Navy Marine Mammals to Explore Associations between Organochlorine Contaminants and Unfavorable Effects on Reproduction." *Science of the Total Environment* 274 (2001): 171–82.

Reding, Andrew. "The Parable of the Good Cetacean: Whales, Dolphins, and International Law." *Anima* 4, no. 1 (Fall 1977): 44–54.

Reiger, John F. *American Sportsmen and the Origins of Conservation*. New York: Winchester Press, 1975.

"Resolution Opposing the Poisoning of Predatory Mammals and Rodents." *Journal of Mammalogy* 13, no. 3 (August 1932): 288.

Reynolds, John E., III, and Sentiel A. Rommel, eds. *Biology of Marine Mammals*. Washington, DC: Smithsonian Institution Press, 1999.

Rhees, David J. "A New Voice for Science: Science Service Under Edwin E. Slosson, 1921–1929." Master's thesis, University of North Carolina, 1979.

Rice, A. L. "Forty Years of Landlocked Oceanography." *Endeavour* 18 (1994): 137–46.

Rice, Dale W., and Allen A. Wolman. *The Life History and Ecology of the Gray Whale* (Eschrichtius robustus). Special Publication of the American Society of Mammalogists, no. 3. Stillwater, OK: American Society of Mammalogists, 1971.

Richardson, Jack. "An Oppressed Group." Review of *Dolphins, Whales and Porpoises: An Encyclopedia of Sea Mammals*, by David J. Coffey. *New York Review of Books*, 14 July 1977, 26–27.

Ridgway, Sam H. "The Bottle-Nosed Dolphin in Biomedical Research." In *Methods of Animal Experimentation*, vol. 3, edited by W. I. Gay. New York: Academic Press, 1968.

———. "Buoyancy Regulation in Deep Diving Whales." *Nature* 232, no. 5306 (9 July 1971): 133–34.

———. *Dolphin Doctor: A Pioneering Veterinarian and Scientist Remembers the Extraordinary Dolphin That Inspired His Career*. San Diego: Dolphin Science Press, 1987.

———, ed. *Mammals of the Sea: Biology and Medicine*. Springfield, IL: Charles C. Thomas, 1972.

———. "Medical Care of Marine Mammals." *Journal of the American Veterinary Medical Association* 147, no. 10 (15 November 1965): 1077–85.

———. "Navy Marine Mammals." *Science* 243, no. 4893 (17 February 1989): 875.

Ridgway Sam H., and D. Carder. "Assessing Hearing and Sound Production in Cetaceans Not Available for Behavioral Audiograms." *Aquatic Mammals* 27, no. 3 (2001): 267–76.

Ridgway, Sam H., N. J. Flanigan, and James G. McCormick. "Brain-Spinal Cord Ratio in Porpoises: Possible Correlations with Intelligence and Ecology." *Psychological Science* 6, no. 11 (1966): 491–92.

Ridgway, Sam H., and D. Gordon Johnston. "Blood Oxygen and Ecology of Porpoises of Three Genera." *Science* 151, no. 3709 (28 January 1966): 456–58.

Ridgway, Sam H., and James G. McCormick. "Anesthetization of Porpoises for Major Surgery." *Science* 158, no. 3800 (27 October 1967): 510–12.

Ridgway, Sam H., B. L. Scronce, and John Kanwisher. "Respiration and Deep Diving in the Bottlenose Porpoise." *Science* 166, no. 3913 (26 December 1969): 1651–54.

Ringstad, Jan Erik, ed. *Whaling and History II.* Sandefjord, Norway: Sandefjordmuseene, 2006.

Risting, Sigurd. "Whales and Whale Foetuses: Statistics of Catch and Measurements Collected from the Norwegian Whalers' Association 1922–1925." In *Rapports et Procès-Verbaux des Réunions*, vol. 50. Copenhagen: Andr. Fred. Høst and Sons, 1928.

Ritvo, Harriet. *The Animal Estate: The English and Other Creatures in the Victorian Age.* Cambridge, MA: Harvard University Press, 1987.

———. "Animal Planet." *Environmental History* 9, no. 2 (2004): 204–20.

Robbins, Louise E. *Elephant Slaves and Pampered Parrots: Exotic Animals in Eighteenth Century Paris.* Baltimore: Johns Hopkins University Press, 2002.

Robbins, Paul. "Shrines and Butchers: Animals as Deities, Capital, and Meat in Contemporary North India." In *Animal Geographies: Place, Politics, and Identity in the Nature-Culture Borderlands,* edited by Jennifer Wolch and Jody Emel, 218–40. London: Verso, 1998.

Roberts, Brian. "Neil Alison Mackintosh." *Polar Record* 17 (1975): 422–28.

Robertson, Robert Blackwood. *Of Whales and Men.* New York: Knopf, 1954.

Rodgers, Daniel T. "In Search of Progressivism." *Reviews in American History: Progress and Prospects* 10, no. 4 (December 1982): 113–32.

Rodwell, John. "Human Relationships with the Natural World: An Historical Perspective." *Ecos: A Review of Conservation* 24, pt. 1 (2003): 10–16.

Roe, H. S. J. "Seasonal Formation of Laminae in the Ear Plug of the Fin Whale." *Discovery Reports* 35 (September 1967): 1–30.

Roman, Joe. *Whale.* London: Reaktion Books, 2006.

Roman, J., and S. R. Palumbi. "Whales before Whaling in the North Atlantic." *Science* 301, no. 5632 (25 July 2003): 508–10.

Rome, Adam. "'Give Earth a Chance': The Environmental Movement and the Sixties." *Journal of American History* 90, no. 2 (September 2003): 525–54.

———. "Nature Wars, Culture Wars: Immigration and Environmental Reform in the Progressive Era." *Environmental History* 13, no. 3 (July 2008): 432–53.

Romer, Alfred S. "Vertebrate Paleontology." In *Geology 1888–1938: Fifteenth Anniver-*

sary Volume of the Geological Society of America, 107–35. New York: Geological Society of America, 1941.

Roofe, Paul G. "The Intelligent Dolphin." Review of *The Voice of the Dolphins, and Other Stories*, by Leo Szilard, *Man and Dolphin*, by John C. Lilly, and *Porpoises and Sonar*, by Winthrop N. Kellogg. *Humanist* 22, no. 1 (January–February 1962): 33–34.

Rossi, Michael. "Fabricating Authenticity: Modeling a Whale at the American Museum of Natural History, 1906–1974." *Isis* 101, no. 2 (2010): 338–61.

Rossi, A. Michael. "General Methodological Considerations." In *Sensory Deprivation: Fifteen Years of Research*, edited by John Peter Zubek, 16–43. New York: Appleton-Century-Crofts, 1969.

Rothenberg, David. *Thousand Mile Song: Whale Music in a Sea of Sound*. New York: Basic Books, 2008.

Rothfels, Nigel, ed. *Representing Animals*. Bloomington: Indiana University Press, 2002.

———. *Savages and Beasts: The Birth of the Modern Zoo*. Baltimore: Johns Hopkins University Press, 2002.

Rozwadowski, Helen M. "Fathoming the Ocean: The Discovery and Exploration of the Deep Sea: 1840–1880." PhD diss., University of Pennsylvania, 1996.

———. *Fathoming the Ocean: The Discovery and Exploration of the Deep Sea*. Cambridge, MA: Harvard University Press, 2005.

———. *The Sea Knows No Boundaries: A Century of Marine Science under ICES*. Copenhagen: International Council for the Exploration of the Sea in association with University of Washington Press, 2002.

———. "Small World: Forging a Scientific Maritime Culture for Oceanography." *Isis* 87, no. 3 (1996): 409.

Rudmose Brown, R. N., J. H. Pirie, R. C. Mossman, and W. S. Bruce. *The Voyage of the "Scotia": Being the Record of a Voyage of Exploration in Antarctic Seas*. Edinburgh, London: W. Blackwood and Sons, 1906. Reprint, Edinburgh: Mercat Press, 2002.

Ruffner, James A. "Two Problems in Fuel Technology." *History of Technology* 3 (1978): 123–61.

Russell, Dick. *Eye of the Whale: Epic Passage from Baja to Siberia*. New York: Simon & Schuster, 2001.

Ruud, Johan T. "The Blue Whale." *Scientific American* 195, no. 6 (1956): 46–50.

———. "Further Studies on the Structure of Baleen Plates and Their Application to Age Determinations." *Hvalrådets Skrifter* 29 (1945): 5–69.

———. "International Regulation of Whaling: A Critical Survey." *Norsk Hvalfangst-Tidende* 45, no. 7 (1956): 374–87.

———. "The Surface Structure of the Baleen Plates as a Possible Clue to Age in Whales." *Hvalrådets Skrifter* 23 (1940): 3–24.

Ruud, Johan T., Robert Clarke, and Åge Jonsgård. "Whale Marking Trials in Steinshammn, Norway." *Norsk Hvalfangst-Tidende* 42, no. 8 (August 1953): 293–305.

Ryder, Richard D. *Animal Revolution: Changing Attitudes towards Speciesism*. Oxford: Berg Publishing, 2000.

Sachs, Wolfgang, ed. *The Development Dictionary: A Guide to Knowledge as Power.* London: Zed Books, 1992.

Sagan, Carl. *Murmurs of Earth: The Voyager Interstellar Record.* New York: Random House, 1978.

Salmon, Patrick. *Scandinavia and the Great Powers, 1890–1940.* Cambridge: Cambridge University Press, 1997.

Salter, Liora, Edwin Levy, and William Leiss. *Mandated Science: Science and Scientists in the Making of Standards.* Dordrecht, Holland: Kluwer Academic Publishers, 1988.

Salvesen, Theodore E. "The Whaling Industry of Today." *Journal of the Royal Society of Arts* 60, no. 3097 (29 March 1912): 515–23.

Samuels, Amy, and Peter L. Tyack. "Flukeprints: A History of Studying Cetacean Societies." In *Cetacean Societies: Field Studies of Dolphins and Whales*, edited by Janet Mann, Richard C. Connor, Peter L. Tyack, and Hal Whitehead, 9–44. Chicago: University of Chicago Press, 2000.

Sanderson, Ivan Terence. *Follow the Whale.* Boston: Little, Brown, 1956.

Sandler, Todd. "Intergenerational Public Goods." In *Global Public Goods: International Cooperation in the 21st Century*, edited by Inge Kaul, Isabelle Grunberg, and Marc A. Stern, 20–50. Oxford: Oxford University Press, 1999.

Sapolsky, Harvey M. *Science and the Navy: The History of the Office of Naval Research.* Princeton, NJ: Princeton University Press, 1990.

Savours, Ann. "The Natural History of the *Discovery*'s Antarctic Voyages, 1901–1931." *Scottish Naturalist* 111 (1999): 37–68.

———. *The Voyages of the* Discovery: *The Illustrated History of Scott's Ship.* London: Chatham Publishing, 2001.

Scarff, James E. "Ethical Issues in Whale and Small Cetacean Management." *Environmental Ethics* 3 (Fall 1980): 241–79.

———. "The International Management of Whales, Dolphins, and Porpoises: An Interdisciplinary Assessment. Part One." *Ecology Law Quarterly* 6, no. 323 (1977): 326–426.

———. "The International Management of Whales, Dolphins, and Porpoises: An Interdisciplinary Assessment. Part Two." *Ecology Law Quarterly* 6, no. 571 (1977): 574–638.

Scheffer, Victor B. "The Largest Whale." *Defenders of Wildlife International* 49, no. 4 (August 1974): 272–74.

———. "The Status of Whales." (Adapted from address delivered at "Whales, Dolphins, and Porpoises: What is Their Future," sponsored by California Academy of Sciences; Coastal Marine Laboratory, University of California, Santa Cruz; and Los Angeles County Museum of Natural History, San Francisco, 1 June 1975.) *Pacific Discovery* 29, no. 1 (January–February 1976): 2–8.

———. *The Year of the Whale.* New York: Charles Scribner's Sons, 1969.

Scheiber, Harry N. *The Law of the Sea: The Common Heritage and Emerging Challenges.* The Hague: Kluwer Law International Publishers, 2000.

————. "Pacific Ocean Resources, Science, and Law of the Sea: Wilbert Chapman and the Pacific Fisheries, 1935–1970." *Ecology Law Quarterly* 13 (1986): 381–534.

Schevill, William E. "Whale Music." *Oceanus* 9, no. 2 (1962): 2–13.

Schevill, William E., R. H. Backus, and J. B. Hersey. "Sound Production by Marine Animals." In *The Sea: Ideas and Observations on Progress in the Study of the Seas*, vol. 1: *Physical Oceanography*, edited by Maurice Neville Hill, 540–66. New York: John Wiley & Sons, 1962.

Schevill, William E., and Barbara Lawrence. "High-Frequency Auditory Response of a Bottlenosed Porpoise, *Tursiops truncatus* (Montagu)." *Journal of the Acoustical Society of America* 25, no. 4 (1953): 1016–17.

————. "Underwater Listening to the White Porpoise (*Delphinapterus leucas*)." *Science* 109, no. 2824 (11 February 1949): 143–44.

Schevill, William E., and Arthur F. McBride. "Evidence for Echolocation by Cetaceans." *Deep-Sea Research* 3 (1953): 153–54.

Schevill, William E., G. Carleton Ray, and Kenneth S. Norris, eds. *The Whale Problem: A Status Report*. Cambridge, MA: Harvard University Press, 1974.

Schevill, William E., W. A. Watkins, and R. H. Backus. "The 20-Cycle Signal and *Balaenoptera* (Fin Whales)." In *Marine Bio-Acoustics: Proceedings of a Symposium held at the Lerner Marine Laboratory, Bimini, Bahamas, 11–13 April 1963*, edited by William N. Tavolga, 147–52. Oxford: Pergamon Press, 1964.

Schiebinger, Londa. *Nature's Body: Gender in the Making of Modern Science*. Boston: Beacon Press, 1993.

Schiffman, Howard S. *Marine Conservation Agreements: The Law and Policy of Reservations and Vetoes*. Leiden: Martinus Nijhoff, 2008.

Schlee, Susan. *The Edge of an Unfamiliar World: A History of Oceanography*. New York: Dutton, 1973.

Schmidhauser, John R., and George O. Totten III. *The Whaling Issue in U.S.-Japan Relations*. Boulder, CO: Westview Press, 1978.

Scholander, P. F. *Enjoying a Life in Science: The Autobiography of P. F. Scholander*. Fairbanks: University of Alaska Press, 1990.

————. "Experimental Investigations on the Respiratory Function in Diving Mammals and Birds." *Hvalrådets Skrifter* 22 (1940): 5–131.

————. "Rhapsody in Science." *Annual Review of Physiology* 40 (1978): 1–17.

————. "Wave-Riding Dolphins: How Do They Do It?" *Science* 129, no. 3356 (24 April 1959): 1085–87.

Scholander, P. F., and Wallace D. Hayes. "Wave-Riding Dolphins." *Science* 130, no. 3389 (11 December 1959): 1657–58.

Scholl, Lars U. "Whale Oil and Fat Supply: The Issue of German Whaling in the Twentieth Century." *International Journal of Maritime History* 3, no. 2 (December 1991): 39–62.

Schor, Elizabeth Noble. "Per Fredrik Thorkelsson Scholander." *American National Biography Online*. Accessed on 21 July 2003. wysiwyg://article.141/http://www.anb.org/articles/13/13–02497-article.html.

Schulten, Susan. *The Geographical Imagination in America, 1880–1950*. Chicago: University of Chicago Press, 2001.

Schusterman, Ronald J., T. G. Lang, and H. A. P. Smith. "Communication between Dolphins in Separate Tanks." *Science* 152, no. 3720 (15 April 1966): 387.

Schusterman, Ronald J., Jeanette A. Thomas, and Forrest G. Wood, eds. *Dolphin Cognition and Behavior: A Comparative Approach*. Hillsdale, NJ: Lawrence Erlbaum Associates, 1986.

Schweder, Tore. "Distortion of Uncertainty in Science: Antarctic Fin Whales in the 1950s." *Journal of International Wildlife Law and Policy* 3, no. 1 (2000): 73–92.

———. "Protecting Whales by Distorting Uncertainty: Non-precautionary Mismanagement?" *Fisheries Research* 52, no. 3 (2001): 217–25.

Scientific American. Unsigned review of *The Mind of the Dolphin: A Non-Human Intelligence*, by John C. Lilly. Vol. 218, no. 4 (April 1968): 144.

Scoresby, William. *An Account of the Arctic Regions, with a History and Description of the Northern Whale-Fishery*, vol. 1. Edinburgh: Archibald Constable, 1820.

Scott, Anthony. *Natural Resources: The Economics of Conservation*. Toronto: University of Toronto Press, 1955.

Sebeok, Thomas A. Review of *Communication among Social Bees*, by Martin Lindauer, *Porpoises and Sonar*, by Winthrop N. Kellogg, and *Man and Dolphin*, by John C. Lilly. *Language* 39, no. 3, pt. 1 (July–September 1963): 448–66.

Serpell, James. *In the Company of Animals: A Study of Human-Animal Relationships*. Cambridge: Cambridge University Press, 1996.

Shabecoff, Philip. *Fierce Green Fire: The American Environmental Movement*. New York: Hill & Wang, 1993.

Shapin, Steven. *The Scientific Revolution*. Chicago: University of Chicago Press, 1996.

Sheail, John. *Seventy-Five Years in Ecology: The British Ecological Society*. Oxford: Blackwell Scientific Publications, 1987.

Shoemaker, Nancy. "Whale Meat in American History." *Environmental History* 10, no. 2 (April 2005): 269–94.

Siebenüner, Bernd. "Learning in International Organizations in Global Environmental Governance." *Global Environmental Politics* 8, no. 4 (November 2008): 92–116.

Simon, Noel. "Of Whales and Whaling." *Science* 149, no. 3687 (27 August 1965): 943–46.

Simpson, G. G. Review of *A Review of the Archaeoceti*, by Remington Kellogg. *Journal of Mammalogy* 19, no. 1 (February 1938): 113–14.

Simpson, MacKinnon and Robert B. Goodman. *Whale Song: A Pictorial History of Whaling and Hawai'i*. Honolulu: Beyond Words Publishing, 1986.

Skodvin, Tora, and Steinar Andresen. "Nonstate Influence in the International Whaling Commission, 1970–1990." *Global Environmental Politics* 3, no. 4 (November 2003): 61–86.

Skolnikoff, Eugene B. *The Elusive Transformation: Science, Technology, and the Evolution of International Politics*. Princeton, NJ: Princeton University Press, 1993.

Slijper, E. J. "The Still Unexplained Mystery of the Whales." *Norsk Hvalfangst-Tidende* 50, no. 2 (February 1961): 41–54.

———. "Ten Years of Whale Research." *Norsk Hvalfangst-Tidende* 48, no. 3 (1959): 117–29.

———. *Whales*. Translated by A. J. Pomerans. New York: Basic Books, 1962. Originally published as *Walvissen* (Amsterdam: Centen's, 1958).

Slijper, E. J., W. L. van Utrecht, and C. Naaktgeboren. "Remarks on the Distribution and Migration of Whales, Based on Observations from Netherlands Ships." *Bijdragen tot de Dierkunde* 34 (1964): 4–93.

Slocum, Joshua. *Sailing Alone around the World*. New York: Century, 1900.

Small, George L. *The Blue Whale*. New York: Columbia University Press, 1971.

Smith, Gare. "The International Whaling Commission: An Analysis of the Past and Reflections on the Future." *Natural Resources Lawyer, Journal of the Section of Natural Resources Law, American Bar Association* 16, no. 4 (1984): 543–67.

Smith, Tim D. "Examining Cetacean Ecology Using Historical Fishery Data." *Research in Maritime History* 21 (2001): 207–14.

———. *Scaling Fisheries: The Science of Measuring the Effects of Fishing, 1855–1955*. Cambridge: Cambridge University Press, 1994.

———. "'Simultaneous and complementary advances': Mid-century Expectations of the Interaction of Fisheries Science and Management." *Reviews in Fish Biology and Fisheries* 8, no. 3 (1998): 335–48.

———. "Stock Assessment Methods: The First Fifty Years." In *Fish Population Dynamics*, edited by J. A. Gulland, 1–33. London: John Wiley & Sons, 1988.

Sörlin, Sverker, and Paul Warde. "The Problem of the Problem of Environmental History: A Re-reading of the Field." *Environmental History* 12, no. 1 (January 2007): 107–30.

Soulé, Michael E., and Gary Lease, eds. *Reinventing Nature? A Response to Postmodern Deconstruction*. Washington, DC: Island Press, 1995.

Southwell, Thomas. "The Migration of the Right Whale (*Balaena mysticetus*)." *Natural Science* 12 (June 1898): 397–414.

———. "Notes on the Arctic Whaling Voyage 1905." *Zoologist* 4, no. 776 (February 1906): 41–48.

Southwell, Thomas, and Sidney F. Harmer. "Notes on a Specimen of Sowerby's Whale (*Mesoplodon bidens*) Stranded on the Norfolk Coast." *Annals and Magazine of Natural History*, ser. 6, vol. 11 (1893): 275–84.

"Soviet Studies on Cetaceans." National Technical Information Service Report JPRS 49777. U.S. Department of Commerce, 9 February 1970.

Speert, Harold. "Memorable Medical Mentors." Pt. 9. "George L. Streeter (1873–1948)." *Obstetrical and Gynecological Survey* 60, no. 1 (2005): 3–6.

———. "Memorable Medical Mentors." Pt. 1. "Lewis Hill Weed (1886–1952)." *Obstetrical and Gynecological Survey* 59, no. 2 (2004): 61–64.

Spong, Paul. "Cortical Evoked Responses and Attention in Man." PhD diss., University of California, Los Angeles, 1966.

Sponsel, Alistair. "The Cambridge Natural Sciences Tripos, 1915–1949." M. Sc. diss., Cambridge University, 2001.

Spufford, Francis. *I May Be Some Time: Ice and the English Imagination*. London: Faber and Faber, 1996.

Stearn, William Thomas. *The Natural History at South Kensington: A History of the British Museum, 1753–1980*. London: Natural History Museum, 1998.

Stein, Barbara R. *On Her Own Terms: Annie Montague and the Rise of Science in the American West*. Berkeley: University of California Press, 2001.

Steinberg, John C. *Studies of Underwater Noise*. Final Report to Bell Telephone Laboratories. Miami: The Marine Laboratory, Institute of Marine Science, University of Miami, December 1961.

Steinberg, Philip E. *The Social Construction of the Ocean*. Cambridge: Cambridge University Press, 2001.

Sténuit, Robert. *The Dolphin, Cousin to Man*. Translated by Catherine Osbourne. New York: Sterling, 1968.

Sterling, Keir Brooks. "Builders of the U.S. Biological Survey, 1885–1930." *Journal of Forest History* 33 (October 1989): 180–87.

———, ed. *Contributions to the Bibliographical Literature of American Mammals*. Natural Sciences in America Series. New York: Arno Press, 1974.

———. *Last of the Naturalists: The Career of C. Hart Merriam*. New York: Arno Press, 1974.

Stevens, Jay. *Storming Heaven: LSD and the American Dream*. New York: Grove Press, 1998.

Steward, Frank, ed. *The Presence of Whales: Contemporary Writings on the Whale*. Anchorage: Alaska Northwest Books, 1995.

Stock, Chester. "John Campbell Merriam 1869–1945." *National Academy Biographical Memoirs* 26 (1948): 209–17.

Stoddard, Herbert L. *The Bobwhite Quail: Its Habits, Preservation, and Increase*. New York: Charles Scribner's Sons, 1931.

———. *Report on Cooperative Quail Investigation: 1925–1926 (with Preliminary Recommendations for the Development of Quail Preserves)*. Washington, DC: Published by the Committee Representing the Quail Study Fund for Southern Georgia and Northern Florida, 1926.

Stoett, Peter J. *The International Politics of Whaling*. Vancouver: UBC Press, 1997.

Stolberg, Benjamin. "Vigilantism, 1937." *Nation* 145, no. 7 (14 August 1937): 166–68.

———. "Vigilantism, 1937—Part II." *Nation* 145, no. 8 (21 August 1937): 191.

Straty, Richard R. "Methods of Enumerating Salmon in Alaska." In *Transactions of the Twenty-Fifth North American Wildlife and Natural Resources Conference*, edited by James B. Trefethen, 286–96. Washington, DC: Wildlife Management Institute, 1960.

Strick, James E. "Creating a Cosmic Discipline: The Crystallization and Consolidation of Exobiology, 1957–1973." *Journal of the History of Biology* 37, no. 1 (2004): 131–80.

Sullivan, Robert. *Rats: Observations on the History and Habitat of the City's Most Unwanted Inhabitants*. New York: Bloomsbury, 2004.

Sunday Times Insight Team. *Rainbow Warrior: The French Attempt to Sink Greenpeace.* London: Hutchinson, 1986.

Swithinbank, C. W. M. "Neil Alison Mackintosh, 1900–1974." *Journal of Glaciology* 14, no. 71 (1975): 335–36.

Szilard, Leo. *The Voice of the Dolphins, and Other Stories.* Nuclear Age Series. Stanford, CA: Stanford University Press, 1961.

Takahashi, Junichi. "English Dominance in Whaling Debates: A Critical Analysis of Discourse at the International Whaling Commission." *Japan Review* 10 (1998): 237–53.

Talbot, Lee M. *A Look at Threatened Species: A Report on Some Animals of the Middle East and Southern Asia Which Are Threatened with Extermination.* London: The International Union for Conservation of Nature and Natural Resources, 1960.

Tavolga, Margaret C., and William N. Tavolga. Review of *Man and Dolphin*, by John C. Lilly. *Natural History* 71, no. 1 (January 1962): 5–7.

Tavolga, William N., ed. *Marine Bio-Acoustics: Proceedings of a Symposium held at the Lerner Marine Laboratory, Bimini, Bahamas, 11–3 April 1963.* Oxford: Pergamon Press, 1964.

———. "Technical Report: NAVTRADEVCEN 1212-1, Review of Marine Bio-Acoustics, State of the Art: 1964." Port Washington, NY: U.S. Naval Training Device Center, February 1965.

Taylor, Joseph E., III. "Boundary Terminology." *Environmental History* 13, no. 3 (July 2008): 454–81.

———. *Making Salmon: An Environmental History of the Northwest Fisheries Crisis.* Seattle: University of Washington Press, 1999.

Taylor, Scott. *Souls in the Sea: Dolphins, Whales, and Human Destiny.* Berkeley: Frog, 2003.

Tefft, Carol L. Review of *The Mind of the Dolphin: A Non-Human Intelligence*, by John C. Lilly. *Library Journal* 92, no. 22 (15 December 1967): 4639.

Tesh, Sylvia Noble. *Uncertain Hazards: Environmental Activists and Scientific Proof.* Ithaca, NY: Cornell University Press, 2001.

"There He Blows." *Punch, or the London Charivari* 103 (23 July 1892): 25

Thomas, Keith. *Man and the Natural World: Changing Attitudes in England 1500–1800.* Oxford: Oxford University Press, 1996.

Thompson, William Irwin. "Alternative Realities." Review of *The Mind of the Dolphin: A Non-Human Intelligence*, by John C. Lilly. *New York Times Book Review*, 13 February 1972, 27.

Time. Unsigned review of *The Mind of the Dolphin: A Non-Human Intelligence*, by John C. Lilly. 17 November 1967, 121.

Tomlin, A. J. "On the Behavior and Sound Communication of Cetaceans." *Trudy Instituta okeanologii (Transactions of the Institute of Oceanology)* 18 (1955): 28–47.

Tompkins, Dave. *How to Wreck a Nice Beach: The Vocoder from World War II to Hip Hop, The Machine Speaks.* New York: Stop Smiling, 2010.

Tønnessen, J. N., and Arne Odd Johnsen. *The History of Modern Whaling.* Berkeley: University of California Press, 1982.

Tors, Ivan. *My Life in the Wild*. Boston: Houghton Mifflin, 1979.

Townsend, Charles H. "The Distribution of Certain Whales as Shown by Logbook Records of American Whaleships." *Zoologica* 19, no. 1 (3 April 1935): 1–50.

———. "The Porpoise in Captivity." *Zoologica* 1, no. 16 (1914): 289–99.

———. "Twentieth-Century Whaling." *Bulletin of the New York Zoological Society* 33, no. 1 (January–February 1930): 3–31.

Train, Russell. *Politics, Pandas, and Pollution*. Washington, DC: Island Press, 2003.

Trefethen, James B., and Peter Corbin. *An American Crusade for Wildlife*. New York: Winchester Press, 1975.

True, Frederick W. *The Whalebone Whales of the Western North Atlantic: Compared with Those Occurring in European Waters, with Some Observations on the Species of the North Pacific*. Smithsonian Contributions to Knowledge 33. Washington, DC: Smithsonian Institution, 1904.

Tuan, Yi-Fu. *Dominance and Affection: The Making of Pets*. New Haven, CT: Yale University Press, 1984.

Tucker, Richard P., and Edmund Russell. *Natural Enemy, Natural Ally: Toward an Environmental History of Warfare*. Corvallis: Oregon State University Press, 2004.

Turner, James. *Reckoning with the Beast: Animals, Pain, and Humanity in the Victorian Mind*. Baltimore: Johns Hopkins University Press, 1980.

Turner, Ronald. "Operant Control of the Vocal Behavior of a Dolphin." PhD diss., University of California, Los Angeles, 1962.

Twiss, John R., Randall R. Reeves, and Suzanne Montgomery. *Conservation and Management of Marine Mammals*. Washington, DC: Smithsonian Institution Press, 1999.

Tyack, Peter L. "Communication and Cognition." In *Biology of Marine Mammals*, edited by John E. Reynolds III and Sentiel A. Rommel, 287–323. Washington, DC: Smithsonian Institution Press, 1999.

Ufkes, Frances M. "Building a Better Pig: Fat Profits in Lean Meat." In *Animal Geographies: Place, Politics, and Identity in the Nature-Culture Borderlands*, edited by Jennifer Wolch and Jody Emel, 241–55. London: Verso, 1998.

Uhler, Francis M. "In Memoriam: Albert Kenrick Fisher." *Auk* 68 (April 1951): 210–13.

Umezaki, Yoshito. "Defeat at the IWC." In *Kujira to Inbou*. Tokyo: Seizando Shoten, 2004.

United Nations Conference on the Human Environment. *Development and Environment: Report and Working Papers of a Panel of Experts Convened by the Secretary-General of the United Nations Conference on the Human Environment (Founex, Switzerland, June 4–12, 1971)*. Environment and Social Sciences, no. 1. Paris: Mouton, 1972.

———. United Nations Conference on the Human Environment. *Report of the United Nations Conference on the Human Environment, Stockholm, 5–16 June 1972*. New York: United Nations, 1973.

U.S. Congress. Agreement Amending the International Agreement for the Regulation of Whaling: Report to Accompany Executive D. Seventy-Eighth Congress, Second Session, 1944.

———. Hearing Before a Special Committee on Wild Life Resources. Seventy-Second Congress, First Session, 1931.

———. Hearings Before the Subcommittee on International Organizations and Movements of the Committee on Foreign Affairs, House of Representatives, Ninety-Second Congress, First Session, 1971.

———. International Agreement for Regulation of Whaling: Report to Accompany Executive U. Seventy-Fifth Congress, First Session, 1937.

———. Protocol Amending the International Agreement for the Regulation of Whaling: Report to Accompany Executive C. Seventy-Sixth Congress, First Session, 1939.

———. Protocol Amending the International Agreement for the Regulation of Whaling: Report to Accompany Executive I. Seventy-Ninth Congress, Second Session, 1946.

———. Regulation of Whaling: Hearing Before a Subcommittee of the Committee on Interstate and Foreign Commerce on S. 2080, a Bill to Authorize the Regulation of Whaling and to Give Effect to the International Convention for the Regulation of Whaling Signed at Washington under Date of December 2, 1946, by the United States of America and Certain Other Governments, and for Other Purposes. July 20, 1949. Eighty-First Congress, First Session, 1949.

———. The Whaling Treaty Act. Hearings before the Committee on Foreign Affairs, House of Representatives, on S. 3413, to Give Effect to the Convention between the United States and Certain Other Countries for the Regulation of Whaling, Concluded at Geneva September 24, 1931, Signed on the Part of the United States March 31, 1932, and for Other Purposes. February 11, 18, 25, March 3, 7, and 10, 1936. Seventy-Fourth Congress, First Session, 1936.

———. The Whaling Treaty Act. Hearings before the Committee on Foreign Affairs, House of Representatives, on S. 3413, to Give Effect to the Convention between the United States and Certain Other Countries for the Regulation of Whaling, Concluded at Geneva September 24, 1931, Signed on the Part of the United States March 31, 1932, and for Other Purposes. February 11, 18, 25, March 3, 7, and 10, 1936. Seventy-Fourth Congress, First Session, 1936.

U.S. Congress, Senate Special Committee on Conservation of Wild Life Resources. Economics of the Whaling Industry with Relationship to the Convention for the Regulation of Whaling. Washington, DC: U.S. Government Printing Office, 1933.

U.S. Department of Commerce. "Laws and Regulations for the Protection of Whales." Department Circular no. 300, Bureau of Fisheries. Washington, DC: U.S. Government Printing Office, 1936.

U.S. Department of the Interior. "Scientific Study Team Outlines Plans for National Fisheries Center and Aquarium." News release, Interior 5634, Science Feature, Office of the Secretary, 1 September 1963.

United States Quarterly Book List. Unsigned review of *International Regulation of Fisheries*, by Larry Leonard. Vol. 1, no. 4 (December 1945): 36–40.

Vallance, William Roy. "The International Convention for Regulation of Whaling and the Act of Congress Giving Effect to Its Provisions." *American Journal of International Law* 31, no. 1 (January 1937): 112–19.

Vamplew, Wray. "The Evolution of International Whaling Controls." *Maritime History* 2, no. 2 (September 1972): 123–39.

———. *Salvesen of Leith.* Edinburgh: Scottish Academic Press, 1975.

Van Stuyvenberg, J. H. *Margarine: An Economic, Social and Scientific History, 1869–1969.* Liverpool: Liverpool University Press, 1969.

Van Vaneden, M., and Paul Gervais. *Ostéographie des cétacés vivants et fossiles, comprenant la description et l'iconographie du squelette et du système dentaire de ces animaux ainsi que des documents relatifs à leur histoire naturelle.* Paris: A. Bertrand, 1880 [1868–1879].

Veldman, Meredith. *Fantasy, the Bomb, and the Greening of Britain: Romantic Protest, 1945–1980.* Cambridge: Cambridge University Press, 1994.

Vercors [Jean Marcel Bruller]. *You Shall Know Them.* Translated by Rira Barisse. Boston: Little, Brown, 1953.

Vincent, Howard Paton. *The Trying-Out of Moby-Dick.* Boston: Houghton Mifflin, 1949.

Vogt, William. *Road to Survival.* New York: William Sloane Associates, 1948.

Von Bloeker, Jack C., Jr. "Who Were Harry R. Painton, A. Brazier Howell and Francis F. Roberts?" *Condor* 95 (1993): 1061–63.

Walker, R. A. "Some Intense, Low-Frequency, Underwater Sounds of Wide Geographic Distribution, Apparently of Biological Origin." *Journal of the Acoustical Society of America* 35, no. 11 (November 1963): 1816–24.

Wallace, Richard L., comp. *The Marine Mammal Commission Compendium of Selected Treaties, International Agreements, and Other Relevant Documents on Marine Resources, Wildlife, and the Environment.* Washington, DC: The Commission, 1994.

Wallis, Roy. *On the Margins of Science: The Social Construction of Rejected Knowledge. Sociological Review* Monograph 27. Keele, UK: University of Keele, 1979.

Walton, D. W. H. "The First South Georgia Leases: Compañia Argentina de Pesca and The South Georgia Exploring Company Limited." *Falkland Islands Journal* (1983): 14–25.

Watkins, William. Interview by Gary Weir and Frank Taylor at the Woods Hole Oceanographic Institution, Woods Hole, MA, 29 March 2000. Oral History Project of the H. John Heinz III Center for Science, Economics, and the Environment, in conjunction with the Colloquia Series "Oceanography: The Making of a Science, People, Institutions and Discovery."

Watson, Lyall. *Sea Guide to Whales of the World.* Illustrated by Tom Ritchie. New York: Dutton, 1981.

Watson, Paul, as told to Warren Rogers. *Sea Shepherd: My Fight for Whales and Seals.* Edited by Joseph Newman. New York: W. W. Norton, 1982.

Weatherall, Mark. *Gentlemen, Scientists, and Doctors: Medicine at Cambridge, 1800–1940.* Cambridge: Boydell Press, Cambridge University Library, 2000.

Weber, Max Wilhelm Carl. *Die Säugetiere: Einführung in die Anatomie und Systematik der recenten und fossilen Mammalia.* Jena: Fischer, 1904.

Weinberg, Alvin M. "Science and Trans-Science." *Minerva* 10, no. 2 (1972): 207–22.

Weir, Gary E. *Forged in War: The Naval-Industrial Complex and American Submarine Construction, 1940–1961.* Washington, DC: Naval Historical Center, 1993.

———. "From Surveillance to Global Warming: John Steinberg and Ocean Acoustics." *International Journal of Naval History* 2, no. 1 (1 April 2003): 1–11. Accessed on 14 September 2011. http://www.ijnhonline.org/volume2_number1_Apr03/article_weir_steinberg_apr03.htm.

———. *An Ocean in Common: American Naval Officers, Scientists, and the Ocean Environment.* College Station: Texas A&M University Press, 2001.

Wentz, G. M. "Curious Noises and the Sonic Environment of the Ocean." In *Marine Bio-Acoustics: Proceedings of a Symposium held at the Lerner Marine Laboratory, Bimini, Bahamas, 11–13 April 1963,* ed. William N. Tavolga, 101–23. Oxford: Pergamon Press, 1964.

Wever, Glen Ernest, James G. McCormick, Jerry Palin, and S. H. Ridgway. "Cochlea of the Dolphin, *Tursiops truncatus*: The Basilar Membrane." *Proceedings of the National Academy of Sciences* 68, no. 11 (November 1971): 2708–11.

———. "Cochlea of the Dolphin, *Tursiops truncatus*: General Morphology." *Proceedings of the National Academy of Sciences* 68, no. 10 (October 1971): 2381–85.

———. "Cochlea of the Dolphin, *Tursiops truncatus*: Hair Cells and Ganglion Cells." *Proceedings of the National Academy of Sciences* 68, no. 12 (December 1971): 2908–12.

———. "Cochlea Structure in the Dolphin, *Lagenorhynchus obliquidens*." *Proceedings of the National Academy of Sciences* 69, no. 3 (March 1972): 657–61.

Weyler, Rex. *Greenpeace: How a Group of Ecologists, Journalists, and Visionaries Changed the World.* New York: Rodale, 2004.

"Whaling and Fishing in the Southern Pacific." *Norsk Hvalfangst-Tidende* 43, no. 12 (1954): 690–705.

Wheeler, J. F. G. "The Age of Fin Whales at Physical Maturity." *Discovery Reports* 2 (1930): 403–34.

———. "On the Stock of Whales at South Georgia." *Discovery Reports* 9 (1934): 351–72.

Whitehead, Hal. *Voyage to the Whales.* Post Mills, VT: Chelsea Green, 1990.

Whitmore, Frank C., Jr. "Remington Kellogg (1892–1969)." *Yearbook of the American Philosophical Society* (1972): 205–10.

Wieland, G. R. "The Conservation of the Great Marine Vertebrates: Imminent Destruction of the Wealth of the Seas." *Popular Science Monthly* 72, no. 5 (May 1908): 425–30.

Wilson, Charles Henry. *History of Unilever: A Study in Economic Growth and Social Change.* 2 vols. London: Cassell, 1954.

Winge, Herluf. "Udsigt over Hvalernes indbyrdes Slaegtskab." *Vidensk. Medd. fra Dansk naturh. Foren.* 70 (1918): 59–142.

Winge, Herluf, and Gerrit S. Miller. "A Review of the Interrelationships of the Cetacea." *Smithsonian Miscellaneous Collections* 72, no. 8. Washington, DC: Smithsonian Institution, 1921.

Winner, Langdon. *The Whale and the Reactor: A Search for Limits in an Age of High Technology.* Chicago: University of Chicago Press, 1986.

Wislocki, George. "On the Structure of the Lungs of the Porpoise (*Tursiops truncatus*)." *American Journal of Anatomy* 44, no. 1 (1929): 47–77.

———. "The Placentation of the Harbor Porpoise (*Phocoena phocoena*)." *Biology Bulletin* 65, no. 1 (1933): 80–89.

Wolfe, Cary. *Animal Rites: American Culture, the Discourse of Species, and Posthumanist Theory.* Chicago: University of Chicago Press, 2003.

———. *Zoontologies: The Question of the Animal.* Minneapolis: University of Minnesota Press, 2003.

Wood, Forrest G. *Marine Mammals and Man: The Navy's Porpoises and Sea Lions.* Washington, DC: Robert B. Luce, 1973.

———. "Underwater Sound Production and Concurrent Behavior of Captive Porpoises, *Tursiops truncatus* and *Stenella plagiodon*." *Bulletin of Marine Science of the Gulf and Caribbean* 3, no. 2 (March 1953): 120–33.

Wood, F. G., and Sam H. Ridgway. "Utilization of Porpoises in the Man-in-the-Sea Program." In *An Experimental 45-Day Undersea Saturation Dive at 205 Feet.* ONR Report ACR-124, 407–11. Washington, DC: Office of Naval Research, 1967.

Wooster, Warren. "Scientific Aspects of Marine Sovereignty Claims." *Ocean Development and International Law* 1, no. 1 (1973): 13–20.

Worster, Donald. *The Ends of the Earth: Perspectives on Modern Environmental History.* Cambridge: Cambridge University Press, 1988.

———. *Nature's Economy: The Roots of Ecology.* San Francisco: Sierra Club Books, 1977.

Worthington, Edgar Barton. *The Ecological Century: A Personal Appraisal.* New York: Oxford University Press, 1983.

Würsig, Bernd. "Cetaceans." *Science* 244, no. 4912 (30 June 1989): 1550–57.

———. "Intelligence and Cognition." In *Encyclopedia of Marine Mammals*, edited by W. F. Perrin, Bernd G. Würsig, and J. G. M. Thewissen, 628–37. San Diego: Academic Press, 2002.

Yablokov, Alexey V. "On the Soviet Whaling Falsification, 1947–1972." *Cetacean Society International: Whales Alive!* 6, no. 4 (October 1997). Accessed 18 April 2011. http://csiwhalesalive.org/csi97403.html.

———. "Validity of Whaling Data." *Nature* 367, no. 6459 (13 January 1994): 108.

Yamada, M. "Contribution to the Anatomy of the Organ of Hearing in Whales." *Science Reports of the Whales Research Institute* 8 (1953): 1–79.

Yi, Doogab. "Oceanography and the Cold War." Unpublished manuscript, January 2003.

Young, B. A. "Placid and Self-Contained." Review of *Man and Dolphin*, by John C. Lilly, and *The Nerve of Some Animals*, by Robert Froman. *Punch*, 14 March 1962, 443.

Young, J. Z. "Save the Whales!" *New York Review of Books* 22, no. 12 (17 July 1975). Accessed on 14 September 2011. http://www.nybooks.com/articles/article-preview?article_id=9134.

Zelko, Frank. "Make it a Green Peace: The History of an International Environmental Organization." PhD diss., University of Kansas, 2003.

Zimmerman, Brett. *Herman Melville: Stargazer*. Montreal: McGill-Queen's University Press, 1998.

Zoologica: Scientific Contributions of the New York Zoological Society. Index to Volume I. Numbers 1 to 20 Inclusive, September, 1907–June, 1915. New York: The Society of the Zoological Park, 1915.

Zubek, John Peter, ed. *Sensory Deprivation: Fifteen Years of Research*. New York: Appleton-Century-Crofts, 1969.

Zuk, Marlene. "Why Not Save Jellyfish as Well as Whales?" *Chronicle Review*, 21 March 2003, sec. 2, B13.

FIGURE SOURCES

2.16 From Great Britain, Discovery Committee, *Report on the Progress of the Discovery Committee's Investigations*, frontispiece.

2.17 From Great Britain, *Discovery Investigations: Second Annual Report*, p. 12.

2.18 From Mackintosh, "The Work of the Discovery Committee," p. 7.

2.19 From Mackintosh and Wheeler, "Southern Blue and Fin Whales," p. 423.

2.20 From Mackintosh and Wheeler, "Southern Blue and Fin Whales," p. 425.

2.21 From Mackintosh and Wheeler, "Southern Blue and Fin Whales," p. 442.

2.22 From Mackintosh and Wheeler, "Southern Blue and Fin Whales," p. 443.

2.23 From Great Britain, Discovery Committee, *Report on the Progress of the Discovery Committee's Investigations*, opp. p. 20.

2.24 From Kemp, Hardy, and Mackintosh, "Discovery Investigations," p. 223.

2.25 From SMRU MS 38600, box 643 (containing smaller boxes of negatives and prints).

2.26 From the notebooks filed in SMRU WR/DC/1–4.

2.27 From the notebooks filed in SMRU WR/DC/1–4.

2.28 From the notebooks filed in SMRU WR/DC/1–4.

2.29 From SMRU MS 38600, box 643 (containing smaller boxes of negatives and prints).

2.30 From SMRU MS 38600, box 680, file "Biologists in Factories."

2.31 From SMRU MS 38600, box 642.

2.32 From NHM Zoology Collections, file "Whale-marking, 1919–1924."

2.33 From NHM Zoology Collections, file "Whale-marking, 1919–1924."

2.34 From NHM Zoology Collections, file "Whale-marking, 1919–1924."

2.35 From the artifacts collection of the Scott Polar Research Institute.

2.36 From the artifacts collection of the Scott Polar Research Institute.

2.37 From Rayner, "Whale Marking: Progress and Results to December 1939," pl. LIX.

2.38 From Mackintosh, "The Southern Stocks of Whalebone Whales," p. 247.

2.39 From NOL Discovery MS, green binder, "Scientific Reports: Whale Marking, South Georgia, Hired Catchers."

3.1 From Townsend, "The Porpoise in Captivity," after p. 299.

3.2 From Smithsonian Archives, RU 7434, box 6, folder 14, "Photographs of Remington Kellogg."

3.3 From Smithsonian Archives, RU 7170, box 9, file "Information—Cetacea."

3.4 From Smithsonian Archives, RU 7170, box 9, file "Information—Cetacea."

3.5 From Smithsonian Archives, accession 99–012, box 1, file "Periotics, notes, and photographs, 1927–1950 and undated."

3.6 From Winge, "Udsigt over Hvalernes indbyrdes Slaegtskab," p. 61.

3.7 From attachments to Kellogg to Pinchot, 14 January 1929, Smithsonian Archives, RU 7170, box 7, file "Pe–Pi."

3.8 From attachments to Kellogg to Pinchot, 14 January 1929, Smithsonian Archives, RU 7170, box 7, file "Pe–Pi."

3.9 From Wordell to Kellogg, 11 May 1929, Smithsonian Archives, RU 7434, box 6, file "Correspondence, 1928–1940."

3.10 From Pinchot, *To the South Seas*, p. 20.

3.11 From Pinchot, *To the South Seas*, p. 67.

3.12 From Pinchot, *To the South Seas*, p. 73.

3.13 From Smithsonian Archives, RU 7170, box 10, file "Information—Whale Press Releases."

3.14 From Smithsonian Archives, RU 7170, box 10, file "Information—Whale Press Releases."

3.15 From Smithsonian Archives, RU 7170, box 9, file "Information—Newspaper Clippings."

3.16 From Townsend, "Twentieth-Century Whaling," cover illustration.

3.17 From Kellogg, "Whales, Giants of the Sea," p. 43.

3.18 From Kellogg, "Whales, Giants of the Sea," p. 75.

3.19 From Kellogg, "Whales, Giants of the Sea," p. 46.

3.20 From Kellogg, "Whales, Giants of the Sea," p. 42.

3.21 From Kellogg, "Whales, Giants of the Sea," p. 47.

4.1 From Smithsonian Archives, RU 7165, box 15, folder "London, fifth meeting."

4.2 From Mackintosh, *The Stocks of Whales*, p. 79.

5.1 From Ottestad, "On the Size of the Stock of Antarctic Fin Whales Relative to the Size of the Catch," p. 304.

5.2 From University of Washington Library, Chapman Papers, box 1, file "Bio."

5.3 From University of Washington Library, Chapman Papers, box 13, file "International Whaling Commission."

5.4 From "Report of the Special Ad-Hoc Scientific Committee Meeting," IWC/13/8, appendix.

5.5 From "Report of the Special Ad-Hoc Scientific Committee Meeting," IWC/13/8, appendix.

5.6 From "Report of the Special Ad-Hoc Scientific Committee Meeting," IWC/13/8, appendix.

5.7 From University of Washington Library, Chapman Papers, box 9, file "Whales and Whaling."

5.8 From University of Washington Library, Chapman Papers, box 13, file "International Whaling Commission."

5.9 From University of Washington Library, Chapman Papers, box 13, file "International Whaling Commission."

5.10 From "Final Report of the Committee of Three Scientists," IWC/15/9, p. 33.

5.11   From "Final Report of the Committee of Three Scientists," IWC/15/9, p. 34.

5.12   From "Final Report of the Committee of Three Scientists," IWC/15/9, p. 35.

5.13   From "Final Report of the Committee of Three Scientists," IWC/15/9, p. 44.

5.14   From "Final Report of the Committee of Three Scientists," IWC/15/9, p. 54.

5.15   From "Second Interim Report of the Committee of Three Scientists," IWC/15/7, p. 6.

6.1   From the Flip Schulke Archives; part of a *Life Magazine* photo shoot conducted by Schulke in 1961. Some of the images (including this one) were published in "He Barks and Buzzes, He Ticks and Whistles, but Can the Dolphin Learn to Talk?" *Life Magazine* 51, no. 4 (28 July 1961): 61–66.

6.2   From the private collection of Rex Weyler.

6.3   From the manual included with William E. Schevill and William A. Watkins, *Whale and Porpoise Voices: A Phonograph Record* (Woods Hole, MA: Woods Hole Oceanographic Institution, 1962).

6.4   From the manual included with William E. Schevill and William A. Watkins, *Whale and Porpoise Voices: A Phonograph Record* (Woods Hole, MA: Woods Hole Oceanographic Institution, 1962).

6.5   From Wentz, "Curious Noises and the Sonic Environment of the Ocean," p. 115.

6.6   From Patterson and Hamilton, "Repetitive 20 Cycle per Second Biological Hydroacoustic Signals at Bermuda," p. 134.

6.7   From Norris, "Some Problems of Echolocation in Cetaceans," p. 330.

6.8   From Evans, Sutherland, and Beil, "The Directional Characteristics of Delphinid Sounds," p. 360.

6.9   From Evans, Sutherland, and Beil, "The Directional Characteristics of Delphinid Sounds," p. 358.

6.10   From Evans, Sutherland, and Beil, "The Directional Characteristics of Delphinid Sounds," p. 362.

6.11   From Evans, Sutherland, and Beil, "The Directional Characteristics of Delphinid Sounds," p. 365.

6.12   From Purves, "Anatomy and Physiology of the Outer and Middle Ear in Cetaceans," p. 331.

6.13   From Purves, "Anatomy and Physiology of the Outer and Middle Ear in Cetaceans," pp. 335, 337.

6.14   From Rossi, "General Methodological Considerations," p. 27.

6.15   From Lilly, *Man and Dolphin*, pp. 181, 183.

6.16   From Lilly, *The Mind of the Dolphin*, p. 227 (*above*) and after p. 75 (*below*).

6.17   From the Flip Schulke Archives; part of a *Life Magazine* photo shoot conducted by Schulke in 1961.

6.18   From Lilly, *Mind of the Dolphin*, after p. 146.

6.19 From Einhorn, "Dolphins Challenge the Designer," pp. 61, 62.

6.20 Payne and McVay, "Songs of Humpback Whales," p. 585.

6.21 Payne and McVay, "Songs of Humpback Whales," p. 586.

7.1 From University of Washington Library, Chapman Papers, box 13, file "Newsletters."

7.2 From Smithsonian Archives, RU 7434, box 8, folder 2, "Drawing by Roxie Laybourne executed upon Kellogg's retirement."

7.3 From Smithsonian Archives, RU 7434, box 2, folder 9, "News Clippings."

INDEX

Southern Princess (vessel), 71n, 187n

southern right whales (*Eubalaena australis*). *See* right whales

"Southern Stocks of Whalebone Whales, The" (Mackintosh, 1942), 141, 151, 167, 169n, 170n, 171n

Southern Whaling and Sealing Co., 81

South Georgia, 12, 14n, 18–21, 33, 34, 38, 39, 41, 43, 44, 45, 49, 51, 52, 54, 54n, 56, 57, 58, 62, 64, 68, 69n, 70, 73, 73n, 74, 76, 76n, 79, 81, 86, 87, 88, 91, 92n, 94, 94n, 95n, 97, 101, 103, 104, 107, 114, 117, 120, 123, 123n, 125, 129n, 133, 137, 142, 145, 146, 147n, 163, 164, 168, 169, 175, 176, 178, 185, 185n, 186, 237, 386, 447n, 451, plate 2

South Georgia Company. *See* Christian Salvesen, Ltd.

South Kensington Museum. *See* British Museum (Natural History)

South Orkney Islands, 19, 32, 32n

South Pacific Whaling Commission. *See* Permanent Commission on the Exploitation and Conservation of the Marine Life of the South Pacific

South Pole, 25, 141n, 176

South Shetland Islands, 19, 20, 34, 51, 70, 79, 88, 89, 92n, 101, 103, 123, 123n, 145n, 155n

Southwell, Thomas, 35n, 123n, 124n

Soviet Union, 295n, 343n, 362n, 378, 409, 409n, 410, 411, 427, 455n, 456, 456n, 457n, 458n, 472, 498, 498n, 500, 502, 506, 507n, 521, 526, 526n, 527, 528n, 533, 576, 577n, 651, 655, 662, plate 15

Sowerby's whales (*Mesoplodon bidens*), 35, 35n

Space and Naval Warfare Systems Center (San Diego, CA), 598n

space program, 624, 651, 668

Spain, 522

spawner-recruit models, 467n. *See also* modeling; population, dynamics

Special Committee on Wild Life Resources (US Senate), 280n, 308, 308n

"Specialized Features of the Cetacean Brain" (Kruger, 1966), 596

Special Meeting on Urgent Food Problems (Washington, DC), 331

specific identity. *See various subheadings under* cetacean research

spectrographs, 539, 635, 636, 637

Speert, Harold, 235n, 236n

Sperling (member of 1913 interdepartmental committee on whaling, UK), 54n

Sperling, Dr. (Research and Development office of the Surgeon General's office), 576

sperm whales (*Physeter macrocephalus*), xvi, 10, 11, 12, 41n, 52, 77, 224, 233, 237, 238, 252, 279n, 364, 382n, 413, 417, 425n, 426n, 482n, 521, 522, 525n, 527, 543, 548, 549, 561, 622, 625n, 650n, 651, 651n, 669, 670; as "godlike," 623–24

Spillhaus, Athelstan, 538

spinal fluid, 236

spinner dolphins (*Stenella longirostris*), 549. *See also* dolphins

Spitsbergen (Norway), 10, 41n

Spong, Paul, 529, 626, 639, 639n, 640, 640n, 641, 651, 670

Sponsel, Alistair, 358n

sport, 184, 195n, 273–79, 322. *See also* hunting; management, of wildlife

sportsmen, 255, 256, 274, 277, 278, 279, 280, 322

spotter aircraft, 16, 535

SPRI. *See* Scott Polar Research Institute

Spufford, Frances, 72n

squalodonts, 204n, 207, 209

Sreenivasan, Janani, xix

SSC. *See* Survival Service Commission

SSK. *See* sociology of scientific knowledge

stains, 194